Binnewies · Glaum · Schmidt · Schmidt

Chemical Vapor Transport Reactions

Michael Binnewies · Robert Glaum · Marcus Schmidt · Peer Schmidt

Chemical Vapor Transport Reactions

DE GRUYTER

Authors

Professor Dr. Michael Binnewies
Universität Hannover
Institut für Anorganische Chemie
Callinstr. 9
30167 Hannover
michael.binnewies@aca.uni-hannover.de

Dr. Marcus Schmidt
Max-Planck-Institut
für Chemische Physik fester Stoffe
Nöthnitzer Str. 40
01187 Dresden
mschmidt@cpfs.mpg.de

Professor Dr. Robert Glaum
Institut für Anorganische Chemie
der Universität Bonn
Gerhard-Domagk-Str. 1
53121 Bonn
rglaum@uni-bonn.de

Professor Dr. Peer Schmidt
Hochschule Lausitz (FH)
Fachrichtung Chemie
und Lebensmittelchemie
Großenhainer Str. 57
01968 Senftenberg
peer.schmidt@hs-lausitz.de

ISBN 978-3-11-048349-9
e-ISBN 978-3-11-025465-5

Library of Congress Cataloging-in-Publication Data

A CIP catalog record for this book has been applied for at the Library of Congress

Bibliographic information published by the Deutsche Nationalbibliothek

The Deutsche Nationalbibliothek lists this publication in the Deutsche Nationalbibliografie; detailed bibliographic data are available in the internet at http://dnb.dnb.de.

© 2012 by Walter de Gruyter GmbH & Co. KG, Berlin/Boston

Typesetting: Meta Systems GmbH, Wustermark
Printing and binding: Hubert & Co. GmbH & Co. KG, Göttingen

Printed in Germany

♾ Printed on acid-free paper

www.degruyter.com

Professor Dr. Dr. h.c. *Harald Schäfer*
Dedicated to Professor Dr. Dr. h.c. *Harald Schäfer*
in recognition of his fundamental research to the topic
„Chemical Vapor Transport Reactions".

Preface

The origin of this book has a long history. Back in 1962, *Harald Schäfer* published a monograph with the title *Chemische Transportreaktionen*. It provided a survey of the complete knowledge on this subject back then. The "transport book" was translated shortly after its appearance into English and Russian. For more than half a century the book has served as a highly cited reference for basic principles, understanding, and application of chemical vapor transport reactions.

Research on chemical vapor transport, however, was not at all completed with the appearance of *Schäfer's* book. During the decades since 1962 the results of many new investigations in this field of research have been published. These papers present a plethora of novel examples for the preparative use of chemical vapor transport. One target of these studies was the synthesis of pure, well defined, and well crystallized solids, frequently with the purpose of a detailed characterization of their physical properties. Another aim of such experiments was the preparation of single-crystals for crystal structure analyses. Apart from the application as a synthetic tool, during these years a large body of work was dedicated to the development of improved models for the understanding of various aspects of chemical vapor transport experiments. Building on the foundations laid by *Harald Schäfer*, significant progress in this respect was achieved by *Reginald Gruehn*, *Gernot Krabbes*, and *Heinrich Oppermann* and their research groups.

After becoming professor emeritus in around 1980, *Harald Schäfer* summarized the knowledge accumulated by then, which had drastically grown compared to 1962, into a new manuscript. This manuscript was extended by *Reginald Gruehn*, but never completed. The great interest of many colleagues in chemical vapor transport reactions, from inorganic and solid state chemistry as well as from material science, caused our decision to restart the publishing project. We drew additional motivation from the fact that all of us were students of *Schäfer, Gruehn,* and *Oppermann*.

Eventually nucleation and growth of the project was substantially supported by *Rüdiger Kniep*. His hospitality at the Max-Planck-Institut für Chemische Physik fester Stoffe (Dresden, Germany) allowed frequent meetings in a secluded and very productive atmosphere. This support allowed us to finish the "new transport book" within a period of three years. Here we would like to take the opportunity to express our particular gratitude for this support.

The authors had at their disposal *Gruehns* extended version of *Schaefers* manuscript from 1980. The present book is the result of various re-organizations of the text, substantial extension, and updating, especially for the literature from 1980 to 2010. Thus, for the first time, the models required for an understanding even of very complex chemical vapor transport experiments are presented in a uniform way. Chapter 11 "Gas Specics and their Stability" should improve understanding of the chemical behavior of inorganic compounds in the gaseous phase. Furthermore, we have been particularly concerned to elucidate the *chemical* background and principles of transport reactions, and to document as completely as possible the preparative benefit that the method is able to provide. In this sense, for example, chemical vapor transport of intermetallic compounds and of chalcogenide halides have been for the first time systematically reviewed and described.

We thank Dr. Werner Marx, Zentrale Informationsvermittlungsstelle der Chemisch-Physikalisch-Technischen Sektion der Max-Planck-Gesellschaft, for numerous very careful literature surveys on the subject. Support from Mrs. Claudia Schulze, Hannover, as well as from the library staff at MPI CPfS is gratefully acknowledged for supplying us with copies of more than 2000 references from literature. Mrs. Friederike Steinbach, Dresden, is thanked for her support with the organization and layout of the manuscript. Our particular thanks go to Professor Heinrich Oppermann for his critical review of the manuscript.

We would also like to express our thanks to Dr. Ralf Köppe (KIT, Karlsruhe, Germany) for his contribution concerning the calculation of thermodynamic data with the help of quantum chemical methods.

Last but not least we want to acknowledge the great effort and dedication Sebastian and Manuela Jüstel and Melanie Ahend put into the translation of major parts of the manuscript.

Michael Binnewies, Robert Glaum,
Marcus Schmidt, Peer Schmidt

Hints and Suggestions for the Reader

At the start of this book some hints and suggestions should be given to the reader to allow easier access to the subject. Recommendations are also provided on how to use the extensive lists of references.

Chapter 1 provides the reader with a summary of the subject of chemical vapor transport reactions. Only a basic knowledge of inorganic and physical chemistry will be required to understand this introduction. Reading this chapter might already be sufficient for those interested only in chemical vapor transport as a preparative tool. If the reader wants to acquire an in-depth, state-of-the-art understanding of the thermodynamic background of chemical vapor transport and of the processes determining the mass flow within an ampoule exposed to a temperature gradient, Chapter 2 provides the necessary reading.

In Chapters 3 to 10 the characteristic transport behavior of various classes of compounds is described on the basis of selected examples. At the end of each chapter or section tables are included, which show in alphabetical order a survey of transportable solids of the corresponding compound class. In addition, the tables list, along with the literature references, the transport agents (additives) and temperature gradients that have been used. To allow easier finding of information concerning particular compounds and classes of compounds, as well as to simplify handling of the very large body of references from literature, we deliberately avoided setting up just a single reference list at the end of the book. Instead, at the end of each chapter the corresponding references are given. References that in our opinion are of particular interest in the context of the given chapter are set in bold letters.

For an understanding of the chemical reactions participating in chemical vapor transport some knowledge on gaseous inorganic compounds is required. A survey on this topic can be found in Chapter 11.

Chapter 12 gives some hints on how to deal with thermodynamic data. Questions concerning thermodynamic tables and databases, experimental origin, as well as precision and accuracy of data are addressed. A review on estimation of thermodynamic data completes this chapter.

Computer-aided modeling of chemical vapor transport experiments is described in Chapter 13. The two computer programs dealt with in detail can be obtained free of charge. Sources are provided.

Eventually the final chapters, Chapter 14 "Working Techniques" and Chapter 15 "Selected Experiments for Practical Work on Chemical Vapor Transport Reactions", are meant as an introduction to the practical realization of chemical vapor transport experiments in the laboratory. The examples are carefully chosen and they should provide insight into the various aspects of chemical vapor transport. In addition they should lead, even when conducted by less experienced experimenters, to well developed, aesthetically appealing crystals.

Content

1 Chemical Vapor Transport Reactions – an Introduction

The term chemical vapor transport (CVT) summarizes a variety of reactions that show one common feature: a condensed phase, typically a solid, is volatilized in the presence of a gaseous reactant, the **transport agent**, and deposits elsewhere, usually in the form of crystals.

We understand a solid in this case as a two- or three-dimensional infinite substance, e.g., metals or ionic compounds but not molecular crystals. The deposition will take place if there are different external conditions at the site of crystallization than at the site of volatilization, generally different temperatures. In many cases, CVT is associated with a purification effect. Thus, we are dealing with a formation process of pure and crystalline solids. The formation of single crystals is of particular value because, among other things, it allows the determination of the crystal structure by diffraction methods. Beyond the aspect of basic research, CVT reactions have also gained practical significance: they form the basis of the operating mode of halogen lamps. Furthermore, an industrial process is based on a CVT reaction, the *Mond-Langer* process for the purification of nickel (Hol 2007).

1.1 Historical Development and Principles

Chemical vapor transport reactions are no invention of scientists. During the history of the earth, they often occurred in nature forming stones and minerals without human influence, in particular at high temperatures. Even today, we still find evidence of natural CVT reactions in form of beautifully shaped crystals. *Bunsen* was the first who observed and described this (Bun 1852). He noticed that the formation of crystalline iron(III)-oxide is associated with the presence of volcanic gases that contain gaseous hydrogen chloride. Today, we know that the volatilization and separation of Fe_2O_3 is based on the equilibrium reaction 1.1.1:

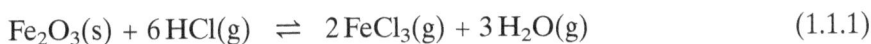

$$Fe_2O_3(s) + 6\,HCl(g) \;\rightleftharpoons\; 2\,FeCl_3(g) + 3\,H_2O(g) \tag{1.1.1}$$

Due to the *forward reaction*, Fe_2O_3 is transformed into the gas phase; this is called **dissolution**. Due to the *back reaction*, it is deposited from the gas phase. The place of the volatilization is called the **source** and the place of deposition the **crystallization zone** or **sink**. The chemical equation describing dissolution and deposition is called the **transport equation**. Reaction enthalpy and entropy of a transport reaction always refers to the dissolution reaction.

Van Arkel and *de Boer* were the first scientists who carried out specific transport reactions in the laboratory from *1925* onward (Ark 1925). They were motivated by the huge interest in finding processes to fabricate pure metals like titanium at that time (Ark 1939). *Van Arkel* and *de Boer* used the so-called *hot-wire method*. In the process, the raw metal M (e.g., a metal of the 4th group) transforms into a gaseous metal iodide (MI_n) in the presence of iodine as transport agent. This exothermic reaction has to be carried out in a sealed vessel. The iodide is formed at the metal surface and then disperses equally in the whole reaction vessel, thus reaching a hot wire, which was heated up electrically to high temperatures. Due to this much higher temperature, the back reaction of the exothermic dissolution reaction is favored by *Le Chatelier*'s principle; thus the metal iodide is decomposed and the metal deposited. Such a vapor transport reaction can be described by the equation 1.1.2.

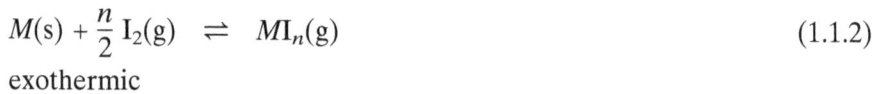

$$M(s) + \frac{n}{2} I_2(g) \;\rightleftharpoons\; MI_n(g) \tag{1.1.2}$$
exothermic

Here, the general principle of transport reactions can be seen clearly again: the source material is transformed reversibly into gaseous products by means of the transport agent. At another temperature, and therefore a changed equilibrium position, the back reaction sets in, which results in the deposition of the solid from the gas phase in the crystallization zone. It is common to characterize the source and deposition temperatures with T_1 and T_2, respectively: T_1 representing the lower temperature. Therefore, exothermic reactions always transport from T_1 to T_2 $(T_1 \rightarrow T_2)$; endothermic reactions from T_2 to T_1 $(T_2 \rightarrow T_1)$.

A CVT reaction is typically expressed as follows:

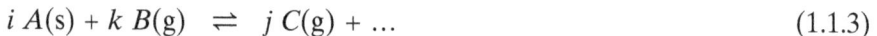

$$i\,A(s) + k\,B(g) \;\rightleftharpoons\; j\,C(g) + ... \tag{1.1.3}$$

There can be further gaseous products in addition to $C(g)$.

No doubt, *van Arkel* and *de Boer* recognized the principle; however, they only used it for the transport of metals. They neither tried to gain a quantitative understanding of the reaction nor the transfer of the principle to chemical compounds. A systematic research and description of CVT reactions was carried out by *Schäfer* in the 1950s and 1960s (Sch 1962). It became apparent that pure and crystalline solids of great variety could be made with the help of CVT reactions: metals, metalloids, intermetallic phases, halides, chalcogen halides, chalcogenides, pnictides, and many others. Today, we know thousands of different examples of CVT reactions. Chemical vapor transport developed to an important and versatile preparative method of solid state chemistry.

Schäfer's effort also showed that CVT reactions follow thermodynamic rules; kinetic effects are rarely observed, which makes a general description easier. Today, the understanding of CVT reactions is well developed; forecasts about alternative transport agents, optimum reaction conditions, and the amount of transported substance are possible and easily accessible via computer programs (Kra 2008, Gru 1997). The proper handling of these programs requires a knowledge of the thermodynamic data (enthalpy, entropy, heat capacity) of all involved condensed and gaseous substances.

A book (Sch 1962), a number of review articles (Nit 1967, Sch 1971, Kal 1974, Mer 1982, Len 1997, Bin 1998, Gru 2000) and an extensive book chapter (Wil 1988) have been published on CVT reactions. These efforts provided an insight into the state of research at that time. *Wilke's* book covers the topic of crystal growth in a more comprehensive way.

1.2 Experimental

In principle, two working methods are applied for the practical realization in the laboratory: transport in open or closed systems. An open system is a tube, made out of glass or ceramic material, which is open on both sides. Inside, a continuous flow of the transport agent is passed over the source material; the solid, which is kept at a certain temperature, deposits at a different site at another temperature under release of the transport agent. In a closed system, typically a sealed ampoule, the released transport agent remains in the system and constantly re-enters the reaction. Thus, in a closed system, a much smaller amount of the transport agent is needed than in an open system. In some cases a few milligrams of the transport agent are sufficient to cause a *transport effect*. These two methods are also observed in nature. Let us remember *Bunsen's* observation: a stream of hydrogen chloride reacts with iron(III)-oxide at hot sites in a volcano. In this process, gaseous iron(III)-chloride and water steam are produced, which react with each other at another place and at another temperature leading to the re-formation of solid iron(III)-oxide and hydrogen chloride. An example of vapor transport in an open system well known to chemists is the preparation of pure chromium(III)-chloride. First, solid chromium(III)-chloride is synthesized from the elements as raw material. Afterwards it is "sublimated" at high temperature in a stream of chlorine (as formulated in *G. Brauer, Handbuch der Präparativen Anorganischen Chemie* (Bra 1975)). In fact, this is not a sublimation but a CVT reaction according to the following equation:

$$CrCl_3(s) + \frac{1}{2} Cl_2(g) \rightleftharpoons CrCl_4(g) \qquad (1.2.1)$$

An example from nature of transport in a closed system is the formation of quartz crystals of different varieties, such as rock crystal, amethyst, or citrine, in closed rock cavities – mineralogists call them druses – caused by the impact of the transport agent water vapor. Here, the vapor transport reaction takes place above the critical temperature and critical pressure of water. That is what is called *hydrothermal synthesis* (Rab 1985). Today, this method is used on a commercial scale for the production of α-quartz.

In the laboratory one predominantly works with closed systems. A simple closed system is a sealed glass tube. Such a *transport ampoule* has a typical length of 100 to 200 mm and a diameter of 10 to 20 mm. It includes about one gram of the solid that is to be transported, and as much transport agent as is needed to raise the pressure in the ampoule to about one bar during the reaction. A rough calculation reveals that the solid exists in surplus.

1.3 Thermodynamic Considerations

It has already been said that the *transport direction* results from the sign of the reaction enthalpy of the transport reaction based on *Le Chatelier*'s principle: in the case of exothermic transport reactions, there will be transport to the zone of higher temperature; in the case of an endothermic reaction, to the cooler zone. However, not every exothermic or endothermic reaction of a solid with a gas is qualified as being a transport reaction. There are two more conditions that have to be fulfilled in order that a reaction between the solid and gas phase can be utilized as a CVT reaction:

- All products formed by the vapor transport reaction have to be gaseous under the reaction conditions.
- The equilibrium position of the transport reaction must not be extreme.

The biggest transport effect is expected when the numerical value of the equilibrium constant K_p at the *mean transport temperature* $[\bar{T} = (T_1 + T_2)/2]$ is about one. We will explain this later with an example.

A vapor transport reaction can be divided into three steps: the *forward reaction* at the source material; the *gas motion*; and the *back reaction* leading to the formation of the solid in the crystallization zone. In most cases, the slowest and therefore the rate-determining step is the gas motion. At a pressure of about 1 bar, the gas motion mainly takes place through diffusion; thus, the diffusion laws determine the velocity of the whole process. A directed mass transfer by diffusion takes place if a spatial gradient of activities da/ds of the involved components exists. As we are observing the diffusion of gases, it is practicable to introduce the partial pressure gradient dp/ds instead of the gradient of the activities. The transported amount of substance per time is proportional to the partial pressure gradient. The differential dp/ds can be approached by the quotient of differences: $dp/ds \approx \Delta p/\Delta s$. Here, Δp is the difference of the partial pressures at the transport temperatures T_1 and T_2 and Δs stands for the length of the diffusion path, i.e., the distance between the source and sink. Furthermore the transported amount of solid is proportional to the diffusion coefficient D. Due to the fact that the diffusion coefficient is inversely proportional to the total pressure Σp in the observed system, the transported mass must be – per time unit at a certain cross section – proportional to the ratio $\Delta p/\Sigma p$.

In order to get an overview of the expected transport effect in a certain system, one often calculates the ratio $p/\Sigma p$ as a function of the temperature and depicts this correlation graphically[1]. This way, one gets illustrative graphs, which in simple cases allow direct conclusions about the transport effect. This will apply if the whole reaction process can be described by only *one* transport equation.

An easy case   Let us consider such an easy case, the vapor transport of zinc sulfide with iodine as transport agent and a temperature gradient 1000 to 900 °C. Here, we make the simplified assumption that iodine is present as $I_2(g)$ and sulfur as $S_2(g)$ at the reaction conditions so that the transport equation is as follows:

---

[1] $p/\Sigma p$ is equal to the molar fraction x

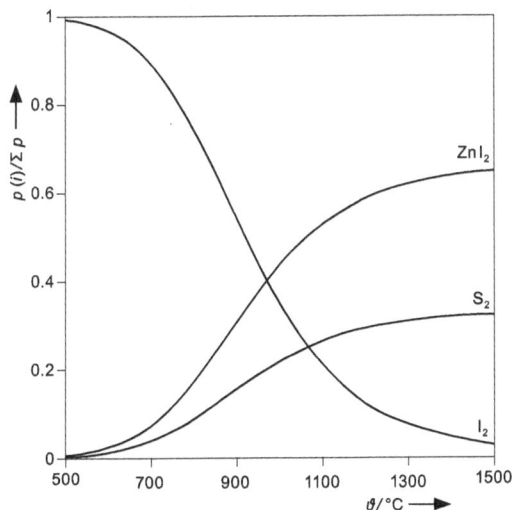

Figure 1.3.1 Temperature dependency of the molar fraction (normalized partial pressure), which is normalized by the total pressure of I_2, ZnI_2, and S_2 in the gas phase in the reaction of zinc sulfide with iodine.

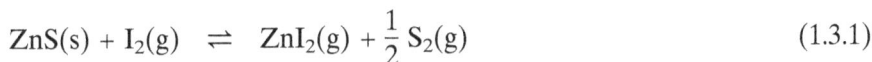

$$ZnS(s) + I_2(g) \; \rightleftharpoons \; ZnI_2(g) + \frac{1}{2} S_2(g) \tag{1.3.1}$$

$$\Delta_r H^0_{298} = 144 \text{ kJ} \cdot \text{mol}^{-1}, \qquad \Delta_r S^0_{298} = 124 \text{ J} \cdot \text{mol}^{-1} \cdot \text{K}^{-1}$$

The reaction is endothermic; which is why the equilibrium position shifts with rising temperature to the side of the reaction products, zinc iodide and sulfur[2].

Figure 1.3.1 shows the graph that emerges if the partial pressures of iodine, zinc iodide, and sulfur (S_2) are calculated from the thermodynamic data of the transport reaction and plotted as a function of the temperature, taking into account the experimental conditions.

As expected, the amount of iodine decreases with rising temperature and the amount of zinc iodide and sulfur increases. Zinc iodide and sulfur (here predominantly S_2) are the so-called **transport-effective species** because they contain the atoms of which the solid zinc sulfide consists. Furthermore, the partial pressure of S_2 is exactly half that of ZnI_2 in the entire temperature range. This corresponds to the ratio of the stoichiometric factors of S_2 and ZnI_2 in the transport equation. The transport takes place from T_2 to T_1. The expected transport effect is largest where the quotient $\Delta p / \Sigma p$ at a defined temperature interval ΔT reaches the highest value for the transport-effective species. In this case, it is close to the inflection point of the illustrated graph for ZnI_2 (or S_2), thus at temperatures around 900 °C. Calculating the equilibrium constant K_p for three chosen temperatures, 700, 900, and 1100 °C, one obtains the following values: 0.03 bar$^{0.5}$, 0.5 bar$^{0.5}$, and 3.8 bar$^{0.5}$. These values show that the expected transport effect is

---

[2] An exact treatment of the vapor transport of zinc sulfide with iodine, including atomic iodine and other sulfur species, leads to a similar result.

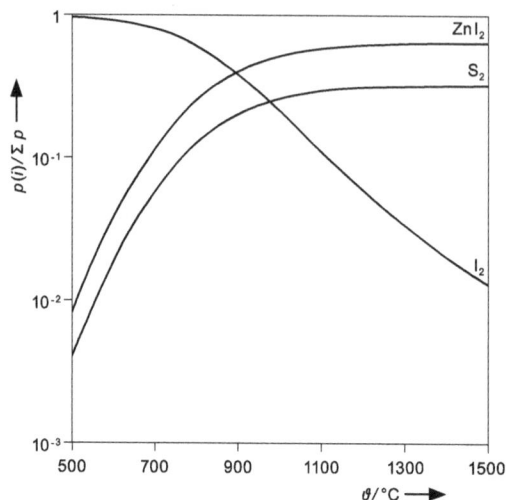

Figure 1.3.2 Temperature dependency of the molar fraction of I_2, ZnI_2, and S_2 in the gas phase for the reaction of zinc sulfide with iodine in logarithmic presentation.

particularly large if the numerical value of the equilibrium constant for the transport reaction is close to unity.

Often, it is advantageous to choose a logarithmic presentation for the graphic illustration of the temperature dependency of the normalized partial pressures. Such an illustration of the discussed case is given in Figure 1.3.2.

A complicated case   According to *van Arkel,* iron can be transported exothermically with iodine from 800 to 1000 °C. At first, the following transport equation comes into consideration:

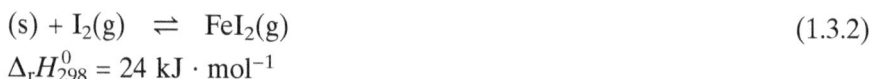

$$(s) + I_2(g) \; \rightleftharpoons \; FeI_2(g) \tag{1.3.2}$$
$$\Delta_r H_{298}^0 = 24 \text{ kJ} \cdot \text{mol}^{-1}$$

This reaction, however, is endothermic. According to *Le Chatelier*'s principle, transport from T_2 to T_1 is expected. Because of the strict validity of this principle, one has to assume that the transport obviously cannot be described, or at least not alone, by the reaction formulated above. A detailed investigation showed that other reactions take place as well. Accordingly, iron(II)-iodide forms monomer and dimer molecules, FeI_2 and Fe_2I_4, in the vapor. The reaction of iron and iodine with the formation of gaseous Fe_2I_4 molecules can be described like this:

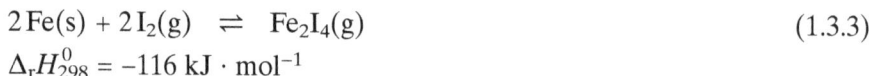

$$2\,Fe(s) + 2\,I_2(g) \; \rightleftharpoons \; Fe_2I_4(g) \tag{1.3.3}$$
$$\Delta_r H_{298}^0 = -116 \text{ kJ} \cdot \text{mol}^{-1}$$

Again, this reaction equation has the character of a transport equation. The reaction is exothermic. Therefore, one expects transport from T_1 to T_2. The situation becomes more complicated because iodine is present partly in atomic form at high temperatures. Iodine atoms can also react with iron to form $FeI_2(g)$ and

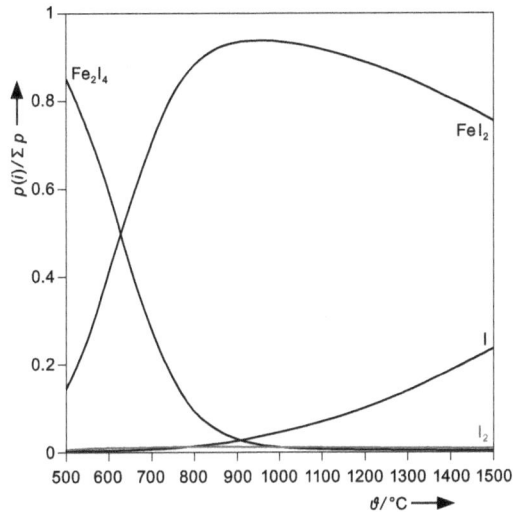

Figure 1.3.3 Temperature dependency of the molar fractions of I, I_2, FeI_2, and Fe_2I_4 in the gas phase formed by the reaction of iron with iodine.

$Fe_2I_4(g)$; for this reaction, one can again formulate a transport equation. The molecules FeI_2 and Fe_2I_4 act as transport-effective species.

Figure 1.3.3 shows the molar fractions of all molecules involved in the transport reaction as a function of the temperature.

Figure 1.3.3 clearly shows the following:

- The fractions of iodine atoms and iodine molecules are small compared to those of iron iodides. The equilibria, which lead to the formation of FeI_2 and Fe_2I_4, are apparently positioned on the product side.
- The fraction of FeI_2 initially increases with rising temperatures. This corresponds to the expectation of the formation of FeI_2 from iron and iodine molecules by an endothermic equilibrium (*Le Chatelier*'s principle). If this equilibrium was the only determinant of the transport of iron, it would migrate from the higher to the lower temperature in the considered interval (800 ... 1000 °C). Above 1000 °C, the fraction of FeI_2 decreases with rising temperature. This is due to the presence of mainly atomic iodine in this temperature range. The reaction of iron with iodine atoms is exothermic, so that the amount of FeI_2 decreases with rising temperature and the amount of iodine atoms increases. The fraction of Fe_2I_4 decreases with rising temperature, too.

Below 1000 °C, we therefore deal with two opposing processes – the increasing formation of FeI_2 and the decreasing formation of Fe_2I_4, both because of rising temperature. The first process lets us expect transportation toward the zone of lower temperature, the second one to the higher temperature zone. Based on the considerations used so far, it is not predictable which process will dominate. A new term – **gas-phase solubility** – is helpful for answering this question.

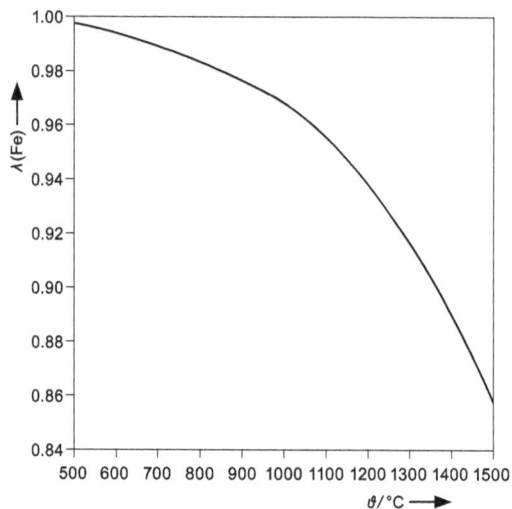

Figure 1.3.4 Temperature dependency of the solubility of iron.

Gas-phase solubility, λ The term gas-phase solubility is used with reference to the solubility of a substance in a liquid. Liquid solutions are used for the purification of the dissolved substance through re-crystallization. One uses the temperature dependency of the solubility, respectively of the solubility equilibrium. From a solution saturated at higher temperature re-crystallization of the solute is achieved by cooling. Generally, this is associated with a purifying effect. A CVT reaction works basically the same way. Here, too, one uses the temperature dependency of the gas-phase solubility in order to crystallize and to purify. In both cases, one deals with heterogeneous equilibria: in the first case between a solid and a liquid, in the second, between a solid and a gas phase. For liquid solutions the term solubility is used independently of the chemistry of the solution process. A solid substance, e.g., a typical molecular compound, dissolves without any recognizable chemical change during the solution process. A gas such as hydrogen chloride, however, dissolves in water under considerable chemical change (dissociation and hydration). Base metals dissolve with hydrogen evolution in acid. For these chemically very different processes one uses terms such as dissolve, solution, and solubility. Hence, it was obvious to introduce the term dissolution for the transfer of a solid through a chemical reaction into the gas phase (Sch 1973).

The example of the transport of iron shows the advantage of the term solubility in the description of complicated transport reactions. According to the transport equations 1.3.2 and 1.3.3, iron is volatilized as FeI_2 and Fe_2I_4, and thus dissolved in the gas phase. The gas phase is the solvent, i.e., all gaseous species together. The description of the solubility of iron in the gas phase in a quantitative way on the basis of this idea has to take into consideration that one molecule of Fe_2I_4 includes two Fe-atoms, whereas the FeI_2 molecule includes only one. This is done in the way that the partial pressure of Fe_2I_4 is multiplied by the factor 2. The same applies to the solvent gas phase. If one primarily assumes iodine atoms, one has to multiply the partial pressures of the other gaseous molecules with the corresponding factors. The solubility of iron in the gas phase can be described as follows:

$$\lambda(\text{Fe}) = \frac{p(\text{FeI}_2) + 2 \cdot p(\text{Fe}_2\text{I}_4)}{p(\text{I}) + 2 \cdot p(\text{I}_2) + 2 \cdot p(\text{FeI}_2) + 4 \cdot p(\text{Fe}_2\text{I}_4)} \qquad (1.3.4)$$

The temperature dependency of the so defined solubility of iron in the gas phase takes both ferrous molecules, FeI_2 and Fe_2I_4, into consideration. Figure 1.3.4 presents the solubility of iron in the gas phase as a function of temperature.

The solubility decreases with increasing temperature. If less iron dissolves at higher temperatures in the gas phase than at lower temperatures, iron must be transported from lower to higher temperatures. This is in line with experimental observations. With the aid of the term *solubility of a solid in the gas phase*, it is possible to describe the transport direction correctly even when several transport reactions occur simultaneously.

1.4 Equilibrium Solids in Source and Sink

It is of prime interest for the preparative chemist whether a certain solid can be obtained using CVT reactions, which transport agents are suitable, and under which conditions transport can be expected. At this point, we want to present some general qualitative considerations.

The easiest case is the CVT of an element, for example the metal nickel, with carbon monoxide (the *Mond-Langer* process). The transport is described by the following transport equation:

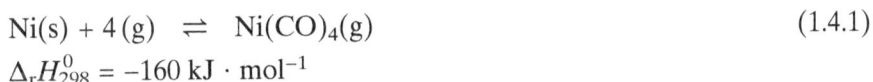

$$\text{Ni(s)} + 4\,(\text{g}) \;\rightleftharpoons\; \text{Ni(CO)}_4(\text{g}) \qquad (1.4.1)$$
$$\Delta_r H^0_{298} = -160\;\text{kJ} \cdot \text{mol}^{-1}$$

The reaction is endothermic; therefore, the transport takes place from T_1 to T_2. The description of the transport of metals with iodine, as mentioned above, is more complicated. Nevertheless, in both cases, the transport of metal with iodine and the transport of nickel with carbon monoxide, the metal is deposited in the sink. If source and sink material have the same composition, it is called **stationary transport**.

Let us now consider the transport of a binary compound A_nB_m. The circumstances are easiest if A is a metal and B a non-metal, for example oxygen or sulfur. Halogens or halogen compounds such as hydrogen halides are often used as transport agent in these cases. Let us have an exemplary view on the following transport equations:

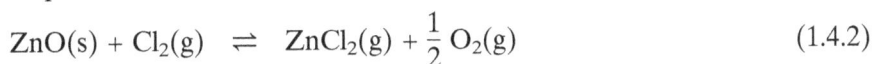

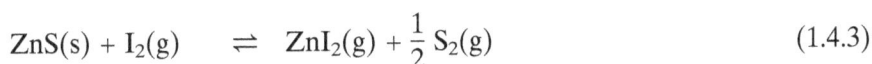

$$\text{ZnO(s)} + \text{Cl}_2(\text{g}) \;\rightleftharpoons\; \text{ZnCl}_2(\text{g}) + \frac{1}{2}\,\text{O}_2(\text{g}) \qquad (1.4.2)$$

$$\text{ZnS(s)} + \text{I}_2(\text{g}) \;\rightleftharpoons\; \text{ZnI}_2(\text{g}) + \frac{1}{2}\,\text{S}_2(\text{g}) \qquad (1.4.3)$$

In these cases, gaseous metal halides emerge by the transport reaction. Under these conditions, the non-metals oxygen and sulfur are present in elemental form in the gas phase. Thus, the transport agent reacts only with A but not with B.

If a binary compound A_nB_m is to be transported in which A and B are both metals or semi-metals with high boiling temperatures, the transport agent will have to react with either components, A and B, to form gaseous reaction products. It is by no means certain that the transport agent reacts to the same extent with A and B. Whether this happens depends on the transport agent, the chemical characteristics of A and B, and the thermodynamic stability of all participating substances.

Under certain conditions the stationary transport of an intermetallic phase A_nB_m is possible. However, it also happens that A or B or even an intermetallic phase with a different composition is transported. Additionally, the simultaneous formation of two different phases is possible. If the solids in the source and the sink have a different composition, it is called **non-stationary transport**. In such cases the composition of the source material will necessarily change during the course of the experiment. This can lead to a change in the transport behavior of the system over time. Thus, sequential deposition of condensed phases at the sink might be observed. With the transport of ternary or multinary intermetallic compounds, the conditions become even more complicated. Similar considerations apply – in a slightly modified form – to the transport of ternary ionic compounds such as spinels or perovskites. Hence, considering the transport of a spinel $A^{II}B_2^{III}O_4$, it cannot be predicted whether one of the binary oxides $A^{II}O$, $B_2^{III}O_3$, the spinel, or a ternary oxide with different composition will be transported.

1.5 Transport Agent

It is a prerequisite for a transport effect that the transport agent reacts with the source material with the formation of one or more gaseous compounds. Particularly often, halogens or halogen compounds are used. The selection of adequate transport agents requires, on the one hand, certain knowledge of the chemistry of inorganic solids and, on the other hand, basic knowledge of thermodynamics. An important aspect has already been mentioned: the equilibrium position must not be extreme. Let us again consider the CVT of zinc sulfide from 1000 to 900 °C in this context. The halogens chlorine, bromine, or iodine are considered as transport agents. The transport equation is formulated as follows:

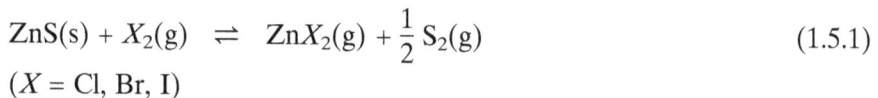

$$\text{ZnS(s)} + X_2\text{(g)} \ \rightleftharpoons \ \text{Zn}X_2\text{(g)} + \frac{1}{2}\,\text{S}_2\text{(g)} \tag{1.5.1}$$

$(X = \text{Cl, Br, I})$

If one calculates the equilibrium constants K_p from the thermodynamic data for the mean transport temperature 950 °C using numerical values of the reaction enthalpy and entropy, one obtains:

$X = \text{Cl: } K_p = 2.9 \cdot 10^5 \text{ bar}^{0.5}$

$X = \text{Br: } K_p = 3.7 \cdot 10^3 \text{ bar}^{0.5}$

$X = \text{I: } \quad K_p = 0.9 \text{ bar}^{0.5}$

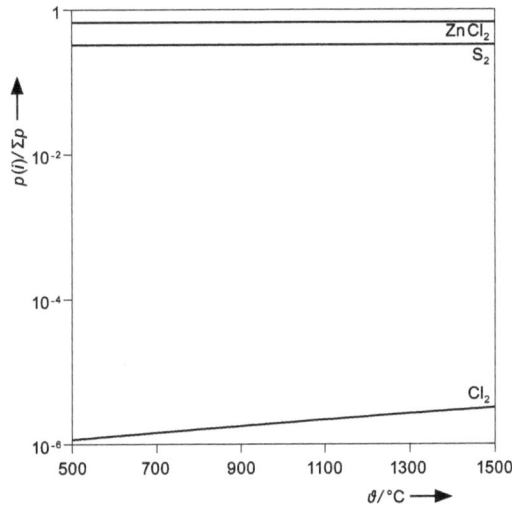

Figure 1.5.1 Temperature dependency of the molar fraction of Cl_2, $ZnCl_2$, and S_2 in the gaseous phase during the reaction of zinc sulfide with chlorine.

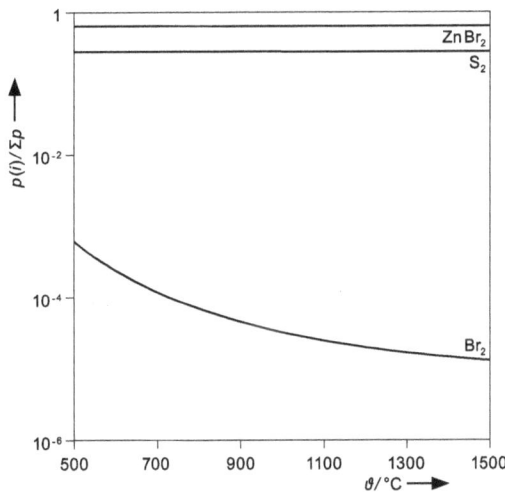

Figure 1.5.2 Temperature dependency of the molar fraction of Br_2, $ZnBr_2$, and S_2 in the gas phase during the reaction of zinc sulfide with bromine.

The reactions of zinc sulfide with chlorine and bromine show a shift of the equilibrium position to the side of zinc chloride or zinc bromide, respectively. For iodine as a transport agent, one calculates the equilibrium constant close to unity. Thus, among the three observed transport agents, iodine is considered the most suitable. Experimental studies confirm this prediction. The graphic illustration of the molar fractions of X_2, ZnX_2, and S_2 as a function of temperature immediately shows that chlorine and bromine are not suitable as transport agents (Figures 1.5.1 and 1.5.2) because their molar fraction is practically the same at $1000\,°C$ and at $900\,°C$. This means that the difference of partial pressure Δp is very low so that a considerable transport effect cannot be expected.

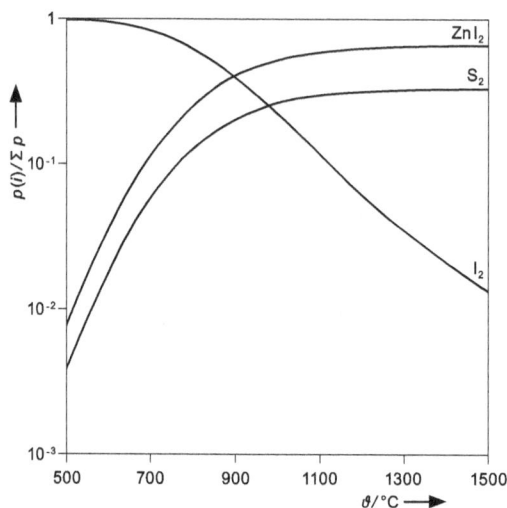

Figure 1.5.3 Temperature dependency of the molar fractions of I_2, ZnI_2, and S_2 in the gas phase during the reaction of zinc sulfide with chlorine.

The circumstances of the reaction of zinc sulfide with iodine are completely different (Figure 1.5.3). The fractions of ZnI_2 and S_2 vary with the temperature in the source and sink. Here, a transport effect is expected.

The optimum transport temperature The temperature at which the numerical value of the equilibrium constant K_p equals 1 is referred to as the optimum transport temperature T_{opt}. The equilibrium constant is easily calculated from the thermodynamic data of the transport reaction. For this purpose it is sufficient to use the reaction enthalpy and reaction entropy at 298 K. *van't Hoff's* equation establishes the link between K and the thermodynamic data of the reaction enthalpy and entropy.

$$\ln K = -\frac{\Delta_r H^0}{R \cdot T} + \frac{\Delta_r S^0}{R} \tag{1.5.2}$$

For $K = 1$ the following expression results:

$$T_{opt} = \frac{\Delta_r H^0}{\Delta_r S^0} \tag{1.5.3}$$

Choosing a suitable transport agent, one does not always have to make preliminary thermodynamic observations. Often it is sufficient to check whether similar solids have already been transported, which transport agent has been used, and at which temperatures the transport has occurred. The numerous tables in Chapter 3 to 10 can be of assistance in that respect.

The **transport rate**, the amount of deposited substance per time in the sink, is highest in a transport gradient at the optimum transport temperature. However, transport reactions are often carried out at higher temperatures than the calculated optimal transport temperature. Usually, such conditions are chosen because at the calculated temperature unwanted reaction products would condense or crystal growth would be kinetically inhibited.

According to *Schäfer* the transport rate $\dot{n}(A)$ can be calculated in simple cases with the help of the following equation, which is based on a diffusion approach.

$$\dot{n}(A) = \frac{n(A)}{t'} = \frac{i}{j} \cdot \frac{\Delta p(C)}{\Sigma p} \cdot \frac{\bar{T}^{0.75} \cdot q}{s} \cdot 0.6 \cdot 10^{-4} \ (\text{mol} \cdot \text{h}^{-1}) \qquad (1.5.4)^3$$

$\dot{n}(A)$ Transport rate $(\text{mol} \cdot \text{h}^{-1})$
i, j stoichometric coefficients in the transport equation
$\qquad i\,A(s) + k\,B(g) \ \rightleftharpoons \ j\,C(g) + \dots$
$\Delta p(C)$ partial pressure difference of the *transport-effective* species C (bar)
Σp total pressure (bar)
$\bar{T}$ mean temperature along the diffusion path (K)
$\qquad$ (practically $\bar{T}$ can be taken as average of T_1 and T_2)
q cross section of the diffusion path (cm^2)
s length of the diffusion path (cm)
t' duration of the transport experiments (h)

1.6 Overview of Vapor Deposition Methods

The vast number of heterogeneous reactions involving the gas phase hardly differ from each other: if a condensed substance encounters a temperature gradient, it moves from the place of dissolution via the gas phase to the place of deposition, from source to sink. One observes the transport of a solid to another place. However, we do not "see" how the substance is led to the gas phase and deposited at another place. In the following, such processes are described and compared.

Sublimation Carbon dioxide, dry ice or CO_2, provides a concise example of sublimation.

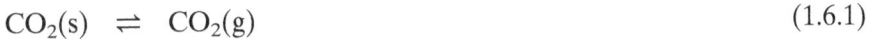

$$CO_2(s) \ \rightleftharpoons \ CO_2(g) \qquad (1.6.1)$$

Other possible gas species in the system carbon/oxygen, such as $CO(g)$, $C_x(g)$, and $O_2(g)$, or $O(g)$, are not important factors due to their significantly lower partial pressures at the given temperature.

Ionic solids can sublime, too. A well known example is aluminum(III)-chloride, which is present in the gas phase in large fraction as the dimeric molecule Al_2Cl_6:

---

[3] In most cases instead of the factor $0.6 \cdot 10^{-4}$ a value of $1.8 \cdot 10^{-4}$ is given, which found entrance to the literature (Sch 1962). According to recent findings the factor $0.6 \cdot 10^{-4}$ results in a smaller numerical value of the diffusion coefficient and corrects a mathematical error.

$$2\,AlCl_3(s) \;\rightleftharpoons\; Al_2Cl_6(g) \tag{1.6.2}$$

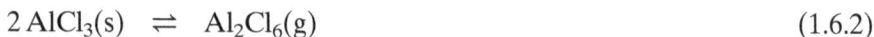

Generally, the sublimation of a compound AB_x is described by the following equilibrium.

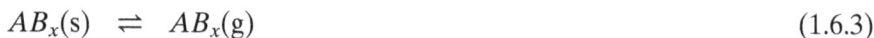

$$AB_x(s) \;\rightleftharpoons\; AB_x(g) \tag{1.6.3}$$

Decomposition sublimation A solid can also decompose into various gaseous products on heating. While on cooling down, the initial solid can be re-crystallized from the gas phase. This is called decomposition sublimation. Ammonium chloride provides an example of this. It decomposes into ammonia and hydrogen chloride in the vapor phase; NH_4Cl molecules do not occur.

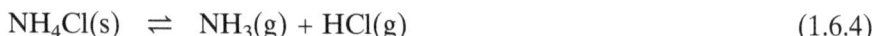

$$NH_4Cl(s) \;\rightleftharpoons\; NH_3(g) + HCl(g) \tag{1.6.4}$$

During cooling, solid ammonium chloride is formed again. Generally, the decomposition sublimation of a compound AB_x can be described by the equilibria 1.6.5 or 1.6.6 respectively.

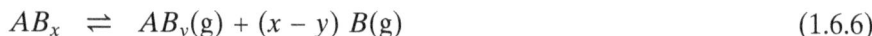

$$AB_x \;\rightleftharpoons\; A(g) + x\,B(g) \tag{1.6.5}$$

$$AB_x \;\rightleftharpoons\; AB_y(g) + (x-y)\,B(g) \tag{1.6.6}$$

A decomposition sublimation can be congruent or incongruent. The solid obtained by a congruent decomposition sublimation (e.g., ammonium chloride) has the same composition as the initial solid. The solid obtained by an incongruent decomposition sublimation has a different composition. A simple example of this is copper(II)-chloride. If heated in a dynamic vacuum to several hundred degrees, the vapor contains the molecules $CuCl$, Cu_3Cl_3, Cu_4Cl_4, and Cl_2. At a cooler site solid copper(I)-chloride is deposited.

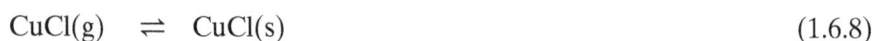

$$(s) \;\rightleftharpoons\; CuCl(g) + \frac{1}{2}\,Cl_2(g) \tag{1.6.7}$$

$$CuCl(g) \;\rightleftharpoons\; CuCl(s) \tag{1.6.8}$$

The phase relations will get more complicated if the decomposition leads to further condensed phases and a reactive gas phase (auto transport).

Auto transport Auto transport resembles the appearance of sublimation or decomposition sublimation. The initial solid is dissolved at a higher temperature without the addition of an external transport agent. The transport agent is formed by thermal decomposition of the solid. This transfers the initial solid into the gas phase. An example of auto transport is provided by $MoBr_3$:

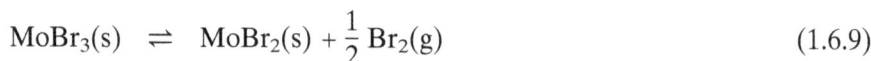

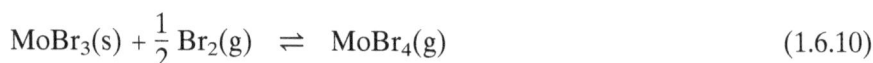

$$MoBr_3(s) \;\rightleftharpoons\; MoBr_2(s) + \frac{1}{2}\,Br_2(g) \tag{1.6.9}$$

$$MoBr_3(s) + \frac{1}{2}\,Br_2(g) \;\rightleftharpoons\; MoBr_4(g) \tag{1.6.10}$$

Based on this example, one can formulate the course of the auto transport in general terms (Opp 2005): a compound $AB_x(s)$ does not generate a transport-effective partial pressure of the gas species $AB_x(g)$ on its own.

Auto transport is based on two co-existing solid phases, $AB_x(s)$ and $AB_{x-n}(s)$, as well as a gas phase, which is generated through the decomposition reaction.

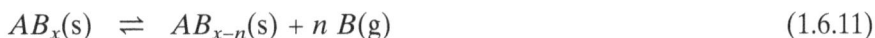

$$AB_x(s) \;\rightleftharpoons\; AB_{x-n}(s) + n\, B(g) \tag{1.6.11}$$

The formed gaseous species B then reacts with the formation of only gaseous products AB_{x+n} in the sense of a CVT reaction. Auto transport reactions are generally endothermic like sublimation and decomposition sublimation ($T_1 \rightarrow T_2$). The transport equilibrium can only be effective if two conditions are met. First, the partial pressure of B must be sufficiently high; and, second, the transporting phase $AB_x(s)$ must remain in equilibrium, thus $AB_x(s)$ must not be decomposed completely. B stands for a gaseous decomposition product in the sense of equation 1.6.11. Thus, B can be an atom (e.g., an iodine atom), a homo-nuclear molecule (Cl_2, S_8...), or a heteronuclear molecule such as $BiCl_3$.

There may be a smooth transition of the described phenomena of sublimation or decomposition sublimation to auto transport. This is the case for the dissolution of $CrCl_3$ in the gas phase. One can find congruent sublimation, the formation of gaseous chromium(III)-chloride, and incongruent decomposition, in solid chromium(II)-chloride and chlorine simultaneously. In a consecutive reaction chlorine can react with the starting solid $CrCl_3$, thus becoming the transport agent. The transport-effective gaseous molecule is $CrCl_4$. In case of the chlorides $MoCl_3$ or VCl_3, the gas molecules MCl_3 are too unstable or unknown and the migration in the temperature gradient takes place only through auto transport according to the equilibria 1.6.12 and 1.6.13 (M = Cr, Mo, V).

$$MCl_3(s) \;\rightleftharpoons\; MCl_2(s) + \frac{1}{2}\, Cl_2(g) \tag{1.6.12}$$

$$MCl_3(s) + \frac{1}{2}\, Cl_2(g) \;\rightleftharpoons\; MCl_4(g) \tag{1.6.13}$$

Auto transport is also known in other substance classes, such as oxides, chalcogenides, and, above all, chalcogenide halides. IrO_2 is an example of the auto transport of an oxide. At a temperature of around 1050 °C this compound decomposes into the metal and oxygen (1.6.14). In equilibrium 1.6.15 oxygen reacts with the starting solid to form the transport-effective gas species IrO_3. At ca. 850 °C the back reaction takes place and IrO_2 deposits crystalline.

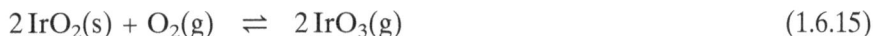

$$IrO_2(s) \;\rightleftharpoons\; Ir(s) + O_2(g) \tag{1.6.14}$$

$$2\, IrO_2(s) + O_2(g) \;\rightleftharpoons\; 2\, IrO_3(g) \tag{1.6.15}$$

Generally, these auto transport reactions are feasible as "regular" CVT reactions. In these cases, the transport is also possible through the addition of the transport agent without the preceding decomposition reaction.

If at least one of the components of the depositing compound AB_x in the gas phase equilibrium (sublimation, decomposition sublimation, auto transport) does not form a gas species with sufficient vapor pressure that is suitable for substance transport, the addition of a transport agent will be necessary.

Chemical vapor transport reaction A CVT reaction, as described in detail above, is characterized by the fact that another substance, the transport agent, is required for the dissolution of a solid in the gas phase. The transport of zinc oxide with chlorine is an example:

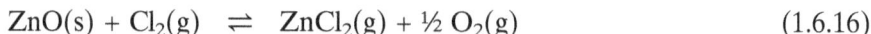

$$ZnO(s) + Cl_2(g) \rightleftharpoons ZnCl_2(g) + \tfrac{1}{2} O_2(g) \tag{1.6.16}$$

Here, the substances that appear in the vapor are different to those in the solid.

Bibliography

Ark 1925   A. E. van Arkel, J. H. de Boer, *Z. Anorg. Allg. Chem.* **1925**, *148*, 345.
Ark 1939   A. E. van Arkel, *Reine Metalle*, Springer, Berlin, **1939.**
Bin 1998   M. Binnewies, *Chemie in uns. Zeit* **1998**, *32*, 15.
Bra 1975   G. Brauer, *Handbuch der Präparativen Anorganischen Chemie*, Enke, Stuttgart, **1975.**
Bun 1852   R. Bunsen, *J. prakt. Chem.* **1852**, *56*, 53.
Gru 1997   R. Gruehn, R. Glaum, O. Trappe, *Computerprogram CVTrans*, University of Giessen, **1997.**
Gru 2000   R. Gruehn, R. Glaum, *Angew. Chemie* **2000**, *112*, 706. *Angew. Chem. Int. Ed.* **2000**, *39*, 692.
Hol 2007   A. A. F. Holleman, N. Wiberg, *Lehrbuch der Anorganischen Chemie*, de Gruyter, Berlin, 102. Aufl. **2007.**
Kal 1974   E. Kaldis, *Principles of the vapor growth of single crystals*. In: C. H. L. Goodman (Ed.)
            Crystal Growth, Theory and Techniques. Vol. 1. **1974**, 49.
Kra 2008   G. Krabbes, W. Bieger, K.-H. Sommer, T. Söhnel, U. Steiner, *Computerprogram TRAGMIN*, Version 5.0, IFW Dresden, University of Dresden, HTW Dresden, **2008.**
Len 1997   M. Lenz, R. Gruehn, *Chem. Rev.* **1997**, *97*, 2967.
Mer 1982   J. Mercier, *J. Cryst. Growth*, **1982**, *56*, 235.
Nit 1967   R. Nitsche, *Fortschr. Miner.* **1967**, *442*, 231.
Opp 2005   H. Oppermann, M. Schmidt, P. Schmidt, *Z. Anorg. Allg. Chem.* **2005**, *631*, 197.
Rab 1985   A. Rabeneau, *Angew. Chem.* **1985**, *97*, 1017.
Sch 1962   H. Schäfer, *Chemische Transportreaktionen*, Verlag Chemie, Weinheim, **1962.**
Sch 1971   H. Schäfer, *J. Cryst. Growth* **1971**, *9*, 17.
Sch 1973   H. Schäfer, *Z. Anorg. Allg. Chem.* **1973**, *400*, 242.
Wil 1988   K.-Th. Wilke, J. Bohm, *Kristallzüchtung*, Harri Deutsch, Frankfurt, **1988**.

2 Chemical Vapor Transport – Models

Transport based on congruent evaporation

Transport via a single equilibrium reaction

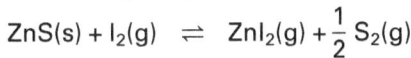

$$ZnS(s) + I_2(g) \ \rightleftharpoons \ ZnI_2(g) + \frac{1}{2} S_2(g)$$

Complex transport behavior based on several equilibrium reactions

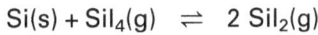

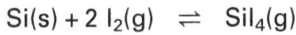

$$Si(s) + SiI_4(g) \ \rightleftharpoons \ 2\,SiI_2(g)$$
$$Si(s) + 2\,I_2(g) \ \rightleftharpoons \ SiI_4(g)$$

Transport based on incongruent evaporation

Transport experiments in a quasi steady-state

Deposition of phases with homogeneity range (FeS_x)

Deposition of a single phase in the sink starting from a two-phase equilibrium mixture in the source (V_nO_{2n-1})

Time-dependent (non steady-state, non-stationary) transport experiments

Sequential migration of a polyphasic solid (CuO/Cu_2O)

In Chapter 2 the theoretical foundations of chemical transport reactions will be laid. Various models that are based on thermodynamic considerations will be explained in detail. It is state of the art to use computer programs for modeling and quantitative description of transport reactions. These programs allow calculations, even without detailed thermodynamic knowledge and understanding. Thus, optimum experimental conditions, the direction of transport in a temperature gradient, and transport rates can be obtained for many transport systems, frequently even in a predictive way. For more complicated cases, however, a detailed treatment of the underlying thermodynamics will be required. Such a treatment is particularly necessary when a condensed phase with homogeneity range or polyphasic equilibrium solids do occur in a transport experiment. In all cases, the simple looking as well as the more complicated ones, prior to an experiment the experimentator has to develop some idea of which condensed equilibrium phases and gaseous species are to be expected for the transport system under consideration. Knowledge of the possibly occuring condensed equilibrium phases and of the dominating gaseous species is an unavoidable prerequisite if modeling of transport experiments is to have an outcome close to reality. Thus, information is required on the equilibrium relations between the various condensed phases of a transport system. Similarly, at least a basic understanding of the expected gaseous compounds and their reactivity is necessary.

In addition to the influence of thermodynamic data and phenomena, the transport behavior can be affected by kinetic effects. While the mass flow via the gas phase is generally assumed to be rate determining, some examples have been observed where the kinetics of one or more elementary reaction steps in the transport process exert a dominating influence. In this context the heterogeneous vaporization reaction, the seed formation, as well as crystal growth during the deposition has to be mentioned.

The most important characteristics for various transport processes are summarized by the following schematics.

Congruent vaporization of a condensed phase:
ratio of the elements in the condensed phase and the gaseous phase of the source are identical.

Congruent dissolution into the gas phase always causes congruent deposition at the sink.
⇒ stationary (steady state), not time-dependent transport behavior

Model for simple transport behavior	*Model for complex transport behavior*
Chemical transport can be fully described by a *single* heterogeneous equilibrium reaction.	The gas-phase composition is determined by *several* independent equilibria.
Calculation of K_p and subsequently of Δp.	Calculation of λ and subsequently of $\Delta\lambda$.
Assessment of the equilibrium state and of the direction of the transport using $\Delta_r H^0$ ($T_2 \rightarrow T_1$ or $T_1 \rightarrow T_2$).	Assessment of the equilibrium state and of the direction of the transport using $\Delta\lambda_{(T_2 - T_1)}$ ($T_2 \rightarrow T_1$ or $T_1 \rightarrow T_2$).

Incongruent dissolution of the source's condensed phase in the gas phase:
ratio of elements in the condensed phase and the gas phase of the source are **not** identical.

The gas-phase composition is determined by *several* independent equilibria.

Calculation of the mass flow of the components A and B, $J(A/B)$, between the equilibrium regions (volumes) by the *flux relation*.

Extended transport model: "quasi-stationary transport"	*Co-operative transport model*: "sequential transport"
Quasi-stationary transport behavior.	Non-stationary (time-dependent) transport behavior.
Determination of the composition of the sink's condensed phase by applying the condition for steady state: ε = const. (stationary relation).	Determination of the composition of the sink's condensed phases and of the deposition sequence by an iteration procedure.

2.1 Thermodynamic Basis for Understanding Chemical Vapor Transport Reactions

In Chapter 1 we described the basic principles of chemical transport reactions. In addition, an extended understanding of the processes involved in chemical transport will allow systematic planning and experimental realization of transport experiments using optimum conditions.

As already mentioned, chemical transport experiments can be carried out in open (flow tube) or closed systems (ampoules). As a consequence of heterogeneous equilibrium reactions one or more condensed phases are transferred into the gas phase; this process can be regarded as *dissolution* (in the gas phase). From the gas phase, condensed phases can be deposited. The deposition can occur spatially separated from the place of dissolution and will be driven by a temperature gradient or, more generally, by a gradient of the chemical potential. In this process several questions are of interest for practical purposes as well as for the chemical understanding of a transport experiment:

- Which reaction(s) determines the transfer of a condensed phase (in most cases of a solid) into the gas phase and how can it (they) be described quantitatively?
- Which gaseous species are formed by the dissolution of a condensed phase in the gas phase?
- What are the rates for dissolution, migration, and deposition of a condensed phase?
- How to estimate the best experimental conditions for a given transport system?
- What (chemical) information can be obtained from transport experiments?

The depth of theoretical considerations that precede an experiment will depend on the system under investigation and on the complexity of the involved heterogeneous (solid–gaseous) and homogenous (gas phase) equilibria. An extensive thermodynamic treatment will not always be necessary. For a more detailed understanding, however, this chapter will provide a survey of models that will allow a quantitative description of the observations made over the course of a transport experiment. These models will provide an understanding of how the gas phase relevant for the transport has been formed, which temperature gradient will be suitable, which condensed phases will be deposited, and if possibly deposition of polyphasic solids in the sink has to be anticipated. The models should always yield the rate of mass flow from source to sink.

In open systems (flow tubes) the transport rate [$\dot{m}$ (g $\cdot$ h^{-1}) or $\dot{n}$ (mol $\cdot$ h^{-1})] for a solid that is in equilibrium with the gas phase will be proportional to the flow rate of the streaming gas (see studies on evaporation equilibria by the *transpiration method* (Kub 1993)). In closed ampoules the diffusion between equilibrium spaces is decisive. Therefore, the partial pressure gradient $\mathrm{d}p/\mathrm{d}T$ – that means thermodynamics – is important. Since diffusion coefficients for gaseous species are all of the same order of magnitude($\approx$ 0.025 cm^2 $\cdot$ sec^{-1}) the influence of individual variations on the transport rate is relatively small. This will be treated in section 2.7.

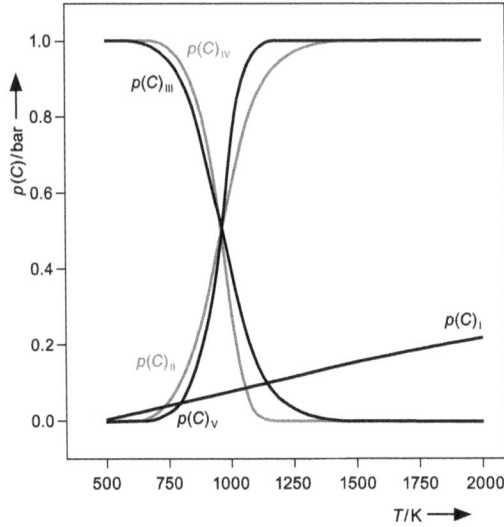

Figure 2.1.1 Influence of $\Delta_r H^0$ and $\Delta_r S^0$ on the magnitude of $p(C)$ as a function of the temperature and depending on the type of reaction (Sch 1956b)[1].

We start by considering single reactions of the following type:

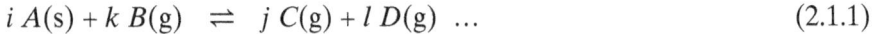

$$i\,A(\text{s}) + k\,B(\text{g}) \ \rightleftharpoons \ j\,C(\text{g}) + l\,D(\text{g}) \ \ldots \tag{2.1.1}$$

The equilibrium partial pressure $p(C)$ over the mono-phasic condensed phase $A(\text{s, l})$ and its temperature dependence are determined by the heat of reaction and the entropy of reaction. For varying numbers of $\Delta_r H^0$ and $\Delta_r S^0$ the resulting pressures are presented in Figure 2.1.1. The sign of the reaction entropy depends decisively on the difference Δn, which follows from the number of gaseous molecules occuring on the product and educt side, respectively. For an approximation, $\Delta_r S^0 = \Delta n \cdot 140$ J $\cdot$ mol^{-1} $\cdot$ K^{-1} can be assumed. Figure 2.1.2 shows the influence of a constant pressure $p(D)$ on $p(C)$ as a function of temperature. From the temperature dependence of the partial pressure of C, $p(C)$, follows for a given temperature gradient, ΔT, a partial pressure difference $\Delta p(C)$.

Figures 2.1.1 to 2.1.7 visualize the relation between $\Delta p(C)$ and $\Delta_r H^0$ for various temperatures. The transport rate $\dot{n}$ (mol $\cdot$ h^{-1}) and $\dot{m}$ (mg $\cdot$ h^{-1}), respectively, is proportional to $\Delta p(\text{C})$ (see section 2.6).

From the considerations that are summarized in Figures 2.1.1 to 2.1.7 various conclusions can be drawn on how the transport process is influenced by thermodynamic parameters.

---

[1] I) $A(\text{s}) + B(\text{g}) \ \rightleftharpoons \ C(\text{g})$ ($\Delta_r H^0 = 20$ kJ $\cdot$ mol^{-1}, $\Delta_r S^0 = 0$ J $\cdot$ mol^{-1} $\cdot$ K^{-1}),
 II) $A(\text{s}) + B(\text{g}) \ \rightleftharpoons \ 2\,C(\text{g})$ or
  $2\,A(\text{s}) + B_2(\text{g}) \ \rightleftharpoons \ 2\,C(\text{g})$ ($\Delta_r H^0 = 140$ kJ $\cdot$ mol^{-1}, $\Delta_r S^0 = 140$ J $\cdot$ mol^{-1} $\cdot$ K^{-1}),
 III) $A(\text{s}) + 2\,B(\text{g}) \ \rightleftharpoons \ C(\text{g})$ ($\Delta_r H^0 = -140$ kJ $\cdot$ mol^{-1}, $\Delta_r S^0 = -140$ J $\cdot$ mol^{-1} $\cdot$ K^{-1}),
 IV) $A(\text{s}) + 4\,B(\text{g}) \ \rightleftharpoons \ C(\text{g})$ ($\Delta_r H^0 = -420$ kJ $\cdot$ mol^{-1}, $\Delta_r S^0 = -420$ J $\cdot$ mol^{-1} $\cdot$ K^{-1}),
 V) $4\text{A}(\text{s}) + B_4(\text{g}) \ \rightleftharpoons \ 4\,C(\text{g})$ ($\Delta_r H^0 = 420$ kJ $\cdot$ mol^{-1}, $\Delta_r S^0 = 420$ J $\cdot$ mol^{-1} $\cdot$ K^{-1}),
  ($\Sigma p = 1$ bar).

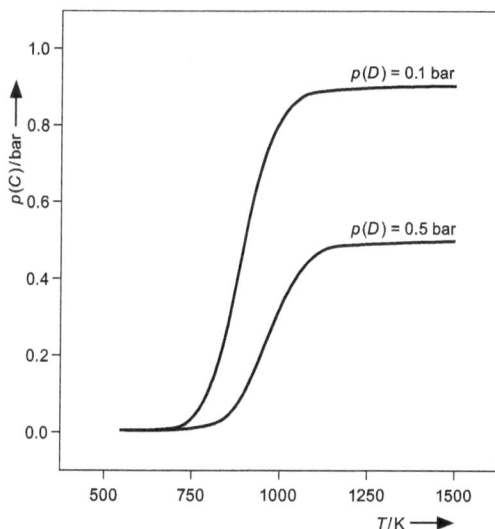

Figure 2.1.2 $p(C)$ as a function of T according to *Schäfer et al.* (Sch 1956b).
Reaction type: $A(s) + B_2(g) \rightleftharpoons C(g) + D(g)$.
($p(D) = 0.1$ and 0.5 bar, respectively; $\Sigma p = 1$ bar; $\Delta_r H^0 = 140$ kJ $\cdot$ mol^{-1}; $\Delta_r S^0 = 140$ J $\cdot$ mol^{-1} $\cdot$ K^{-1})

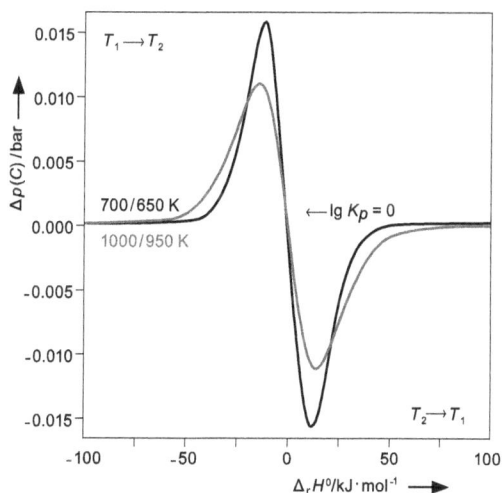

Figure 2.1.3 $\Delta p(C)$ as function of $\Delta_r H^0$ according to *Schäfer et al.* (Sch 1956b).
Reaction type: $A(s) + B(g) \rightleftharpoons C(g)$ (exothermic or endothermic)
($\Delta_r S^0 = 0$ J $\cdot$ mol^{-1} $\cdot$ K^{-1}; $\Sigma p = 1$ bar)

The influence of $\Delta_r H^0$ The sign of $\Delta_r H^0$ determines the sign of $\Delta p(C)$ and thus the direction of the migration of the condensed phase in a temperature gradient. Exothermic reactions lead to transport from T_1 to T_2, endothermic ones from T_2 to T_1. For $\Delta_r H^0 = 0$, $\Delta p(C)$ will also be zero. For this case no transport will occur (see Figures 2.1.3 to 2.1.7).

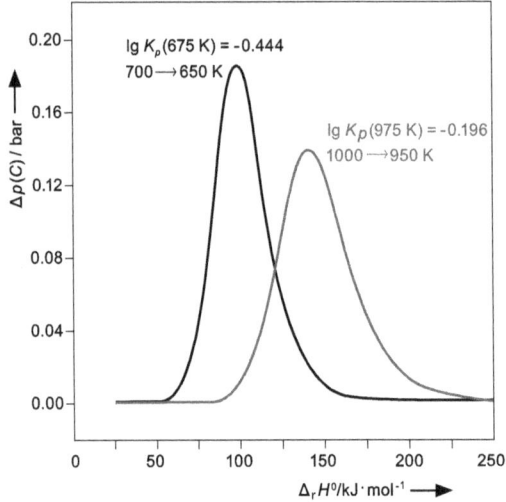

Figure 2.1.4 $p(C)$ as function of $\Delta_r H^0$ according to *Schäfer et al.* (Sch 1956b).
Reaction type: $A(s) + B(g) \rightleftharpoons 2\ C(g)$ (endothermic)
($\Delta_r S^0 = +140$ J $\cdot$ mol^{-1} $\cdot$ K^{-1}; $\Sigma p = 1$ bar) Figure according to *Schäfer et al.* (Sch 1956b).

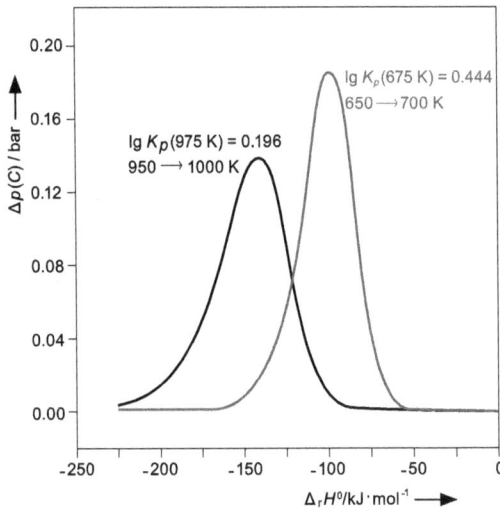

Figure 2.1.5 $\Delta p(C)$ as function of $\Delta_r H^0$ according to *Schäfer et al.* (Sch 1956b).
Reaction type: $A(s) + 2\ B(g) \rightleftharpoons C(g)$ (exothermic)
($\Delta_r S^0 = -140$ J $\cdot$ mol^{-1} $\cdot$ K^{-1}; $\Sigma p = 1$ bar)

The influence of $\Delta_r S^0$ Figures 2.1.3 to 2.1.7 show $\Delta p(C)$ as a function of $\Delta_r H^0$ at various numbers of $\Delta_r S^0$. For small entropies of reaction, transport from T_1 to the higher temperature T_2 or vice versa from T_2 to T_1 can occur depending on the sign of $\Delta_r H^0$. This conclusion is particularly valid if the number of gaseous molecules remains unchanged by the transport reaction (see Figure 2.1.3). The maximum in the transport rate increases with rising $\Delta_r S^0$, regardless of its sign, given that $\Delta_r H^0$ changes correspondingly (Figure 2.1.7).

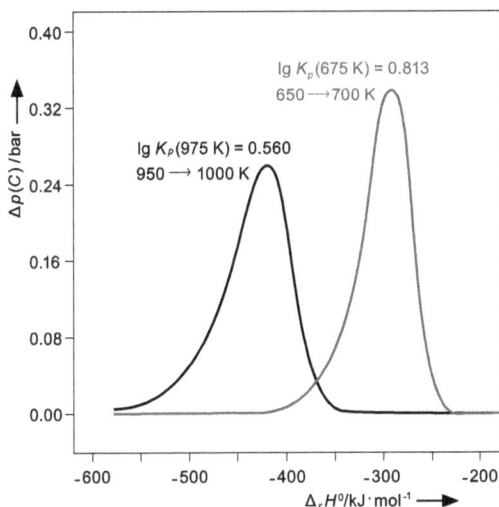

Figure 2.1.6 $\Delta p(C)$ as a function of $\Delta_r H^0$ according to *Schäfer et al.* (Sch 1956b). Reaction type: $A(s) + 4\,B(g) \;\rightleftharpoons\; C(g)$ (exothermic) ($\Delta_r S^0 = -420\ \mathrm{J \cdot mol^{-1} \cdot K^{-1}}$, $\Sigma p = 1$ bar)

For the case of the absolute value of $\Delta_r S^0$ being significantly different from zero ($\geq 40\ \mathrm{J \cdot mol^{-1} \cdot K^{-1}}$), transport is only possible if $\Delta_r H^0$ and $\Delta_r S^0$ show the same sign (Figures 2.1.3 to 2.1.7). For a reaction accompanied by a sufficiently large gain in entropy ($\Delta_r S^0 \geq 40\ \mathrm{J \cdot mol^{-1} \cdot K^{-1}}$) a noticeable transport of the condensed phase is only possible from a higher to a lower temperature ($T_2 \rightarrow T_1$). On the other hand, a reaction having a negative entropy ($\Delta_r S^0 \leq -40\ \mathrm{J \cdot mol^{-1} \cdot K^{-1}}$) can only lead to a significant transport from T_1 to T_2 (Figure 2.1.7).

The influence of the equilibrium position For the special case of $\Delta_r H^0 = 0$ and $\Delta_r S^0 = 0$ the partial pressure difference $\Delta p(C)$ equals zero and consequently the transport rate will vanish ($\dot{n} = 0$). This is true despite $\lg K_p = 0$, which in all other cases is regarded as the prerequisite for a significant transport rate (see Figure 2.1.3). Deviation of $\Delta_r H^0$ from zero leads to $\lg K_p \neq 0$. On increasing the absolute value of $\Delta_r H^0$ (at $\Delta_r S^0 = 0$) the transport rate runs through a maximum and eventually approaches zero. Variation of K_p by choice of other temperatures may cause drastic changes in transport behavior (rate and direction) as is expected from the steep slope of the curves shown in Figures 2.1.3 to 2.1.6.

An increase in Δp can be obtained when the gas-phase composition is selected in such a way that the equilibrium shifts to a less extreme position. For reactions accompanied by a change in the number of gaseous molecules, the same shift can be achieved by variation of the total pressure. In agreement with this conclusion, addition of one of the (gaseous) reaction products might influence the transport rate positively. In the same way variation of the total pressure (e.g., by variation of the amount of transport agent) should have a positive effect on those transport reactions that are accompanied by a change of the number of gaseous molecules.

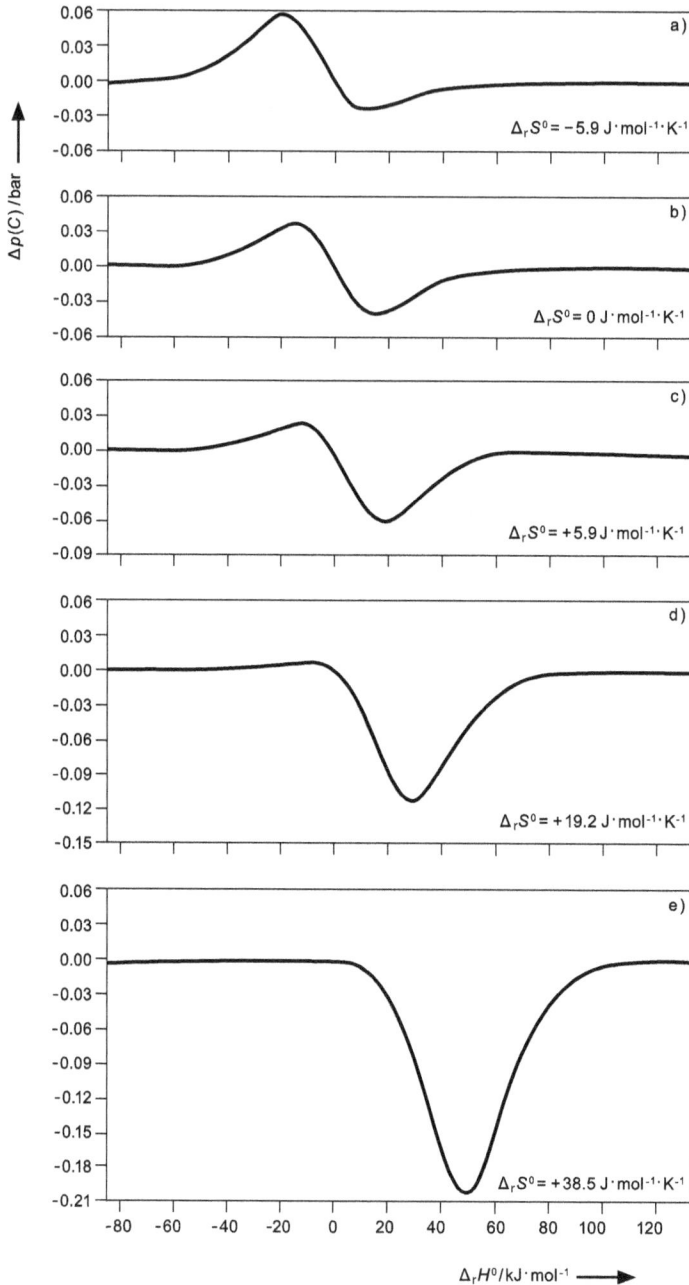

Figure 2.1.7 Transport of a solid $A(s)$ via the reaction $A(s) + B(g) \rightleftharpoons C(g)$ according to *Schäfer et al.* (Sch 1956b).
(temperature gradient $1273 \rightarrow 1073K$; $\Sigma p = 1$ bar)
The curves a) to e) give $\Delta p(C) = p(C)_{T_2} - p(C)_{T_1}$ as a measure for the transport rate in dependence on $\Delta_r H^0$ at constant $\Delta_r S^0$.

Summary The value of Δp is expected to be most important for the calculation of transport behavior (at different experimental conditions) for a certain reaction. Chemical vapor transport will only be observed for appreciable partial pressure differences $\Delta p \geq 10^{-5}$ bar (see section 2.6). The same holds for more complex chemical systems, where advantageously the difference of the solubility in the gas phase $\Delta\lambda$ is used instead of the partial pressure difference Δp.

The aforementioned relations and considerations have proved their value many times in the selection of transport reactions for practical applications. Since the entropy $\Delta_r S^0$ of a reaction under consideration can be estimated in a rather simple way from the number of gaseous molecules in its reaction equation, the values for $\Delta_r H^0$ and T that will allow a significant transport rate can be easily calculated.

Transport systems with a *single* transport agent As an extension to the aforementioned thermodynamic considerations, we compare the results of transport experiments with iron, cobalt, and nickel (Sch 1956b) according to equation 2.1.2 with the conclusions based on Figure 2.1.8.

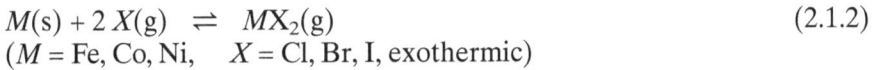

$$M(s) + 2\,X(g) \;\rightleftharpoons\; MX_2(g) \qquad\qquad (2.1.2)$$
$$(M = \text{Fe, Co, Ni}, \quad X = \text{Cl, Br, I, exothermic})$$

The experimental conditions are held constant at temperatures $T_1 = 1073$ K and $T_2 = 1273$ K and the total pressure of $\Sigma p = 0.1$ bar. The entropy for reaction 2.1.2 amounts on average to $\Delta_r S^0 = -85$ J $\cdot$ mol^{-1} $\cdot$ K^{-1}. This value deviates significantly from the abovementioned estimate of ± 140 J $\cdot$ mol^{-1} $\cdot$ K^{-1} since for atomic gaseous species the contributions from rotation and vibration to the partitition function become zero. Using $\Delta_r S^0 = -85$ J $\cdot$ mol^{-1} $\cdot$ K^{-1} the partial pressure differences for varying heats of reaction can be calculated. The transport rates follow then by using equation 1.5.4. The results are shown in Figure 2.1.8. The ob-

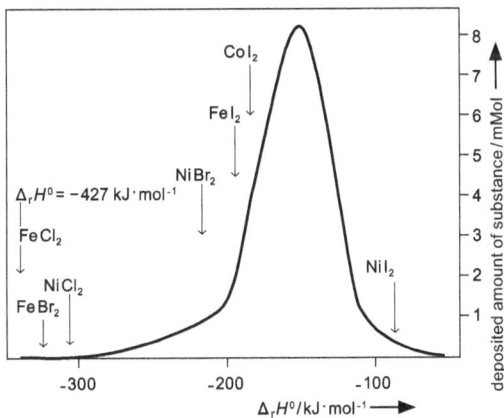

Figure 2.1.8 Metal transport according to reaction $M(s) + 2\,X(g) \rightleftharpoons MX_2(g)$: transported amount of metal as a function of the heats of reaction according to *Schäfer et al.* (Sch 1956b). (Calculations based on transport ampoules according to Figure 2.6.1.1; time duration 10 h; $1073 \rightarrow 1273$ K; $\Sigma p = 0.1$ bar; $\Delta_r S^0 = -85$ J $\cdot$ mol^{-1} $\cdot$ K^{-1})

served metal transport matches nicely the calculated curve. For CVT of the corresponding metals via $FeCl_2$, $FeBr_2$, and $NiCl_2$ experimental evidence has never been found. For the systems Ni/Br, Fe/I, Co/I, and Ni/I, in contrast, migration of the metals is observed. Within this series the transport rate increases from Ni/Br through Fe/I to Co/I and decreases eventually for Ni/I.

Qualitative understanding of these results is straightforward. For $FeCl_2$, $FeBr_2$, and $NiCl_2$ the equilibrium is positioned toward the side of the reaction products: these dihalides are very stable. Following the increased decomposition of the dihalides along the series $NiBr_2$, FeI_2, CoI_2 the transport rate increases, too. For rather unstable halides (e.g., NiI_2) the transport effect decreases again. In agreement with the thermodynamic considerations the maximum transport effect is found close to $\lg K_p = 0$.

Reactions with transport agent mixtures The subsequent examples (equation 2.1.3 to 2.1.6) show, that sometimes two or even more reaction partners jointly act as transport agent.

$$W(s) + 2\,H_2O(g) + 3\,I_2(g) \;\rightleftharpoons\; WO_2I_2(g) + 4\,HI(g) \tag{2.1.3}$$

$$MoS_2(s) + 2\,H_2O(g) + 3\,I_2(g) \;\rightleftharpoons\; MoO_2I_2(g) + 4\,HI(g) + S_2(g) \tag{2.1.4}$$

$$Pt(s) + 2\,CO(g) + Cl_2(g) \;\rightleftharpoons\; Pt(CO)_2Cl_2(g) \tag{2.1.5}$$

$$Cr_2O_3(s) + \frac{1}{2}O_2(g) + 2\,Cl_2(g) \;\rightleftharpoons\; 2\,CrO_2Cl_2(g) \tag{2.1.6}$$

For these cases, a positive effect on the transport rate is observed when the components of the transport agent are added according to the ratio suggested by equations 2.1.3 to 2.1.6. Actually, examples of much more complicated gas-phase compositions are known. Two systems are the transport agent combinations Te/Cl and P/I, which are applied successfully for the transport of vanadium oxides (sections 2.4.2 and 5.2.5) and of phosphates (section 6.2.1), respectively. Frequently, it becomes rather difficult for such systems to write down single transport reactions that are close to reality.

Criteria for the selection of a chemical transport reaction and for its optimum experimental conditions Prior to a chemical transport experiment with a given solid $\Delta_r H^0$ and $\Delta_r S^0$ should be calculated, or at least estimated, for the intended transport reaction (see Chapters 12 and 13). The optimum transport temperature, $T_{opt} = \Delta_r H^0/\Delta_r S^0$, is the temperature for which $\Delta_r G^0 = 0$ and $K_p = 1$. For a temperature gradient around T_{opt} the transport rate is expected to be at its maximum. In the laboratory frequently deviations from the optimum temperature (range) might be necessary due to the following reasons:

- The solid has only a limited thermal stability, above a certain temperature decomposition, phase transformation or melting will occur.
- The solid and/or the transport agent will react above a certain temperature with the ampoule material.

- The temperature has to be sufficiently high, to prevent condensation of transport-active gaseous species.
- The temperature should not be too low, to enable sufficiently high reaction rates and fast setting of the equilibria.

Example: Transport of ZnO

In this case silica ampoules are used, the temperature should not exceed ϑ_{source} $\approx 1000\,°C$, to avoid reaction between the solid and the silica wall. Transport can be achieved by employing various transport agents, as is shown by the survey of various heterogenous equilibria including ZnO as solid phase (Figure 2.1.9). The best transport agent is $HgCl_2$ ($1000 \rightarrow 900\,°C$):

- For the heterogenous transport equilibrium $\lg K$ comes to lie close to zero ($K_p \approx 1$) (based on the reaction with the stoichiometric number of ZnO being unity).
- The temperature dependence of $\lg K$ is comparatively large.
- Condensation of zinc chloride will not occur.
- The temperatures are sufficiently high, to ensure the fast setting of all equilibria at T_2 and T_1.

These predictions have been confirmed by experiments. It is always recommended prior to a CVT experiment to consider the aforementioned points. For the selection of T_1 and T_2 additional aspects, which are related to crystal growth (nucleation, growth rate), might gain importance.

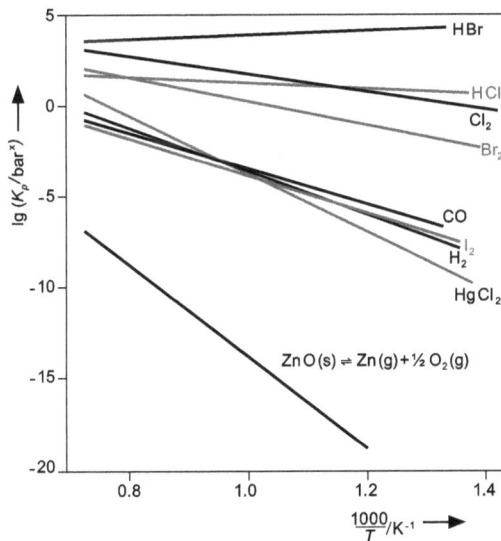

Figure 2.1.9 Heterogenous equilibria including ZnO(s) as solid according to *Schäfer* (Sch 1972).
(K_p is based on the reaction with the stoichiometric number of ZnO being unity.)

2.2 Condensed Phases in a Transport Experiment – the Most Simple Case

The most simple case of a chemical transport experiment is given, if in the source region (region of dissolution) and in the sink region (region of deposition) over the whole period of the experiment single-phase solids with identical composition are present (**source's condensed phase(s): QBK, sink's condensed phase(s): SBK**). The heterogeneous equilibria involving condensed phase and gas phase at the source and the sink region as well as their temperature dependence are fully described by a *single* equilibrium reaction. Under this condition the ratio of the molar numbers of all components (elements) contained in the condensed and the gas phase will be identical. This situation is denominated as **congruent transport**. After setting of the heterogeneous equilibrium in source and sink, the composition of the gas phase remains constant during the whole period of the transport experiment. The gas phase's composition does *not* depend on time. In the same way the transport rate is independent of time, the transport proceeds **stationary** (in a **steady-state**). The aforementioned transport of ZnO by $HgCl_2$ and the transport of metals according equation 2.1.2, as well as a vast number of further transport experiments described in the subsequent chapters of this book proceed in this way.

Knowledge of the equilibrium partial pressures and their temperature dependence is essential for the understanding of a transport process. For the most simple case, calculation of the equilibrium pressures in source and sink can be accomplished on the basis of the law of mass action. Thus, knowledge of the equilibrium constants and of the experimental details, as side conditions, is required. Eventually, calculation of the expected transport rate is possible by applying *Schäfers* diffusion equation, which will be derived later (section 2.6).

For the chemical transport of zinc sulfide by iodine, the course of such calculations will subsequently be presented step by step. The migration of ZnS in a temperature gradient proceeds according to equation 2.2.1 (see section 7.1).

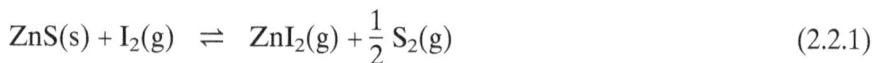

$$ZnS(s) + I_2(g) \rightleftharpoons ZnI_2(g) + \frac{1}{2} S_2(g) \qquad (2.2.1)$$

For equation 2.2.1 the thermodynamic characteristics $\Delta_r H_{298}^0 = 144\ kJ \cdot mol^{-1}$ and $\Delta_r S_{298}^0 = 124\ J \cdot mol^{-1} \cdot K^{-1}$ are obtained by using the thermodynamic data of sphalerite and of all gaseous reaction partners (Bin 2002). The reaction is endothermic, migration of the solid will proceed in a temperature gradient $T_2 \rightarrow T_1$. The calculated heat and entropy of reaction lead to the temperature dependence of $\Delta_r G^0$ and lg K, which are graphically represented in Figure 2.2.1. For their calculation via the *Gibbs–Helmholtz* equation and the *Van't Hoff* equation, the temperature dependence of $\Delta_r H^0$ and $\Delta_r S^0$ have been ignored for the sake of simplicity. Thus, for all temperatures $\Delta_r H_{298}^0$ and $\Delta_r S_{298}^0$ have been used. The temperature dependence can be fully accounted for, if necessary, by using *Kirchhoff*'s equations. For $T = 1160$ K one obtains $\Delta_r G^0 = 0\ kJ \cdot mol^{-1}$ and therefore $K_{p,\ 1160} = 1\ bar^{0.5}$. Following the considerations in section 2.1, transport should be

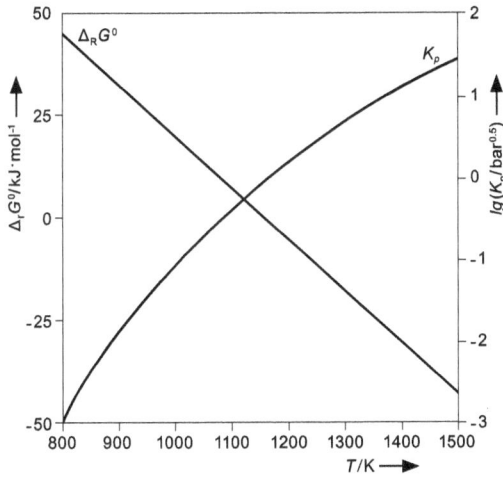

Figure 2.2.1 Temperature dependence of $\Delta_r G^0$ and lg K for reaction 2.2.1.

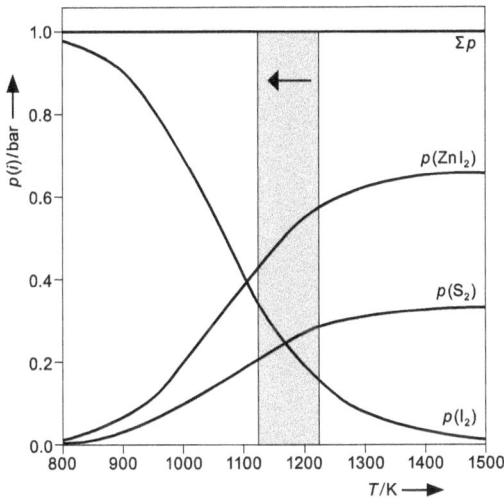

Figure 2.2.2 Equilibrium partial pressures $p(ZnI_2)$, $p(S_2)$, and $p(I_2)$ as a function of temperature. ($\Sigma p = 1$ bar). The direction of the ZnS transport in a temperature gradient is indicated by the arrow.

possible around this temperature. We choose the temperature gradient 950 → 850 °C and the total pressure in the ampoule to be $\Sigma p = 1$ bar.

Using these data the equilibrium pressures (Figure 2.2.2) as well as the partial pressure differences (Figure 2.2.3) are calculated. For calculation of the three partial pressures at a given temperature three unique (mathematical) equations are required. One of these is, as always for such systems, the law of mass action for the transport reaction:

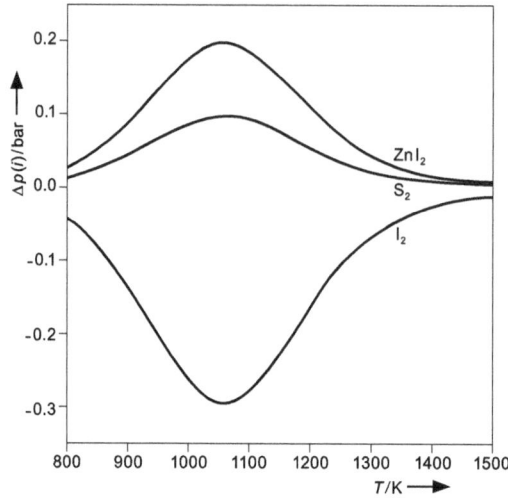

Figure 2.2.3 $\Delta p(\text{ZnI}_2)$, $\Delta p(\text{S}_2)$, and $\Delta p(\text{I}_2)$ as a function of T. ($\Delta T = 100$ K; $T = (T_2 + T_1)/2$)

$$K_p = \frac{p(\text{ZnI}_2) \cdot p^{\frac{1}{2}}(\text{S}_2)}{p(\text{I}_2)} \qquad (2.2.2)$$

The second equation is based on the molar ratio of zinc and sulfur in the gas phase. As a consequence of (2.2.1) this ratio equals unity. Since a S_2 molecule consists of two sulfur atoms, whereas a zinc iodide molecule contains one zinc atom only, the relation between their partial pressures (*stoichiometric relation*) is given by (2.2.3):

$$p(\text{S}_2) = \frac{1}{2} \cdot p(\text{ZnI}_2) \qquad (2.2.3)$$

The third equation follows from the (experimental) side condition that the total pressure in the ampoule should amount to 1 bar.

$$\Sigma p = p(\text{I}_2) + p(\text{ZnI}_2) + p(\text{S}_2) \qquad (2.2.4)$$

Determination of such systems of equations can be achieved iteratively, e.g., by using the built-in macros "GoalSeek" or "Equation Solver" in Microsoft Excel. These procedures, known as the K_p method, for the calculation of chemical equlibria are described fully in the literature (Bin 1996).

The starting amount of iodine to match the desired experimental conditions can now be calculated. Filling the partial pressures $p(\text{I}_2)_{1273}$, $p(\text{I}_2)_{1173}$, $p(\text{ZnI}_2)_{1273}$, $p(\text{ZnI}_2)_{1173}$, and the ampoule volume ($V_{\text{total}} = 20$ cm^3) with the side condition $V(\text{source}) = 2/3\ V_{\text{total}}$ and $V(\text{sink}) = 1/3\ V_{\text{total}}$ into the ideal gas law yields the amount of iodine $m(\text{I}_2) = 37.8$ mg, which is required to fulfill the chosen (experimental) side condition $\Sigma p = 1$ bar. Applying *Schäfer's* diffusion equation (2.2.5 and 2.6.11) eventually yields the expected transport rate:

$$\dot{n}(A)\,(A) = \frac{n(A)}{t'} = \frac{i}{j} \cdot \frac{\Delta p(C)}{\Sigma p} \cdot \frac{D^0 \cdot \bar{T}^{0.75} \cdot q}{s} \cdot 2.4 \cdot 10^{-3} \;(\text{mol} \cdot \text{h}^{-1}) \quad (2.2.5)$$

$\dot{n}(A)$ molar number of the transported solid ZnS
i, j stoichiometric coefficients for zinc in $ZnI_2(g)$ and $ZnS(s)$
$\Delta p(C)$ difference between the equilibrium pressures at T_2 and T_1 of ZnI_2 (bar)
Σp Total pressure in the transport ampoule (bar)
D^0 diffusion coefficient ($0.025 \; \text{cm}^2 \cdot \text{s}^{-1}$)
$\bar{T}$ mean temperature along the diffusion path (K)
q cross section of the diffusion path (cm^2)
t' duration of transport experiment (h)
s length of diffusion path (cm)

$$\dot{n}(ZnS) = \left(\frac{1}{1}\right) \cdot (0.13 \;\text{bar}/1\;\text{bar}) \cdot (0.025 \;\text{cm}^2 \cdot \text{s}^{-1} \cdot 1173^{0.75} \cdot 2.0 \;\text{cm}^2/10\;\text{cm})$$
$$\cdot \, 2.4 \cdot 10^{-3} \;\text{mol} \cdot \text{h}^{-1}$$

$$\dot{n}(ZnS) = 0.3 \cdot 10^{-3} \;\text{mol} \cdot \text{h}^{-1}$$

The transport rate of zinc sulfide $\dot{m}(ZnS) = 30 \;\text{mg} \cdot \text{h}^{-1}$, as it is expected from the aforementioned considerations, lies in the range of the experimentally observed values (Nit 1960, Har 1974).

2.3 Complex, Congruent Transports

There are many examples where chemical transport of a solid can not be completely described by *just one* reaction, since a more complex gas phase is formed. For these cases several unique equilibrium reactions have to be considered. Their number r_u has to be derived by using equation 2.3.1.

$$r_u = s - k + 1 \tag{2.3.1}$$

s is the number of gas species, k the number of components (according to *Gibbs*'s phase rule the number of elements).

   The transport of silicon with iodine might serve as an example for complex congruent transport behavior. The gas species SiI_4, SiI_2, I_2, and I might occur. This leads to $r_u = 4 - 2 + 1 = 3$. By three unique equilibria the partial pressures of all gas species are determined. One set of unique equations is given by 2.3.2 to 2.3.4, others are possible.

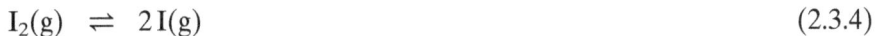

$$Si(s) + 2\,I_2(g) \; \rightleftharpoons \; SiI_4(g) \tag{2.3.2}$$

$$Si(s) + SiI_4(g) \; \rightleftharpoons \; 2\,SiI_2(g) \tag{2.3.3}$$

$$I_2(g) \; \rightleftharpoons \; 2\,I(g) \tag{2.3.4}$$

In principle, both SiI_2 and SiI_4 can participate in the transport of silicon from the source to the sink. We do, however, not know the extent of their contribution.

We have already introduced in Chapter 1 the solubility λ of a solid in the gas phase. This expression summarizes the partial pressures $p(SiI_4)$ and $p(SiI_2)$, and allows in a simple way dissolution of a solid in a gas phase. For $\lambda(Si)$ the equations 2.3.5 and 2.3.6, respectively, are obtained.

$$\lambda(Si) = \frac{p(SiI_2) + p(SiI_4)}{p(I) + 2 \cdot p(I_2) + 2 \cdot p(SiI_2) + 4 \cdot p(SiI_4)} \tag{2.3.5}$$

$$\lambda(Si) = \frac{\Sigma(\nu(Si) \cdot p(Si))}{\Sigma(\nu(I) \cdot p(I))} = \frac{n^*(Si)}{n^*(I)} \tag{2.3.6}$$

The n^* represent the **balance of molar numbers** of the corresponding components.

"Solubility" λ of a solid in the gas phase   Following the reasoning of *Schäfer*, the solubility λ describes the maximum amount of a solid that can be precipitated (recovered) from a given gas phase (Sch 1973). This amount is calculated on a purely stoichiometric basis, regardless whether such a precipitation will be chemically possible. After subtraction of the molar numbers of the components of the solid, the remaining content of the gas phase might be taken as the solvent L.

In a rather simple system, which might, for example, consist of a solid $A(s)$, a halogen $X_2(g)$ as transport agent, and an inert gas I, the gaseous molecules A, AX, A_2X, X_2, X, and I are considered. Thus, for $\lambda(A)$ the following equation is obtained, which is based on the molar numbers n:

$$\lambda(A) = \frac{n^*(A)}{n^*(L)} = \frac{n^*(A)}{n^*(X) + n^*(I)}$$
$$= \frac{n(A) + n(AX) + 2 \cdot n(A_2X)}{n(AX) + n(A_2X) + 2 \cdot n(X_2) + n(X) + n(I)} \tag{2.3.7}$$

The balance of the molar numbers for A, L, X, and I are described by the variables $n^*(A)$, $n^*(L)$, $n^*(X)$, and $n^*(I)$. The definition of the solubility λ presumes that all species are at the same temperature and that all molar numbers are related to the same volume of the gas phase. Therefore, the molar numbers n can be replaced by the pressures p.

$$\lambda(A) = \frac{p^*(A)}{p^*(L)} = \frac{p^*(A)}{p^*(X) + p^*(I)}$$
$$= \frac{p(A) + p(AX) + 2 \cdot p(A_2X)}{p(AX) + p(A_2X) + 2 \cdot p(X_2) + p(X) + p(I)} \tag{2.3.8}$$

A **pressure balance** is described by p^*, similar to the balance of the molar numbers n^*. Whether, as expressed by equations 2.3.7 and 2.3.8, the solubility of A in the gas phase is defined with respect to the total remaining gas phase, to X, or even to the inert gas I, does not matter. One has to keep in mind, however, that only such components (elements) in the gas phase can be considered as solvent, if they do not occur in the solid. Depending on the chosen definition the numeric values of $\lambda(A)$ may vary.

$$\lambda(A) = \frac{p^*(A)}{p^*(X) + p^*(I)} = \frac{n^*(A)}{n^*(X) + n^*(I)} \tag{2.3.9}$$

$$\lambda'(A) = \frac{p^*(A)}{p^*(X)} = \frac{n^*(A)}{n^*(X)} \tag{2.3.10}$$

$$\lambda''(A) = \frac{p^*(A)}{p^*(I)} = \frac{n^*(A)}{n^*(I)} \tag{2.3.11}$$

$$\lambda(A) \neq \lambda'(A) \neq \lambda''(A) \tag{2.3.12}$$

Sometimes difficulties are encountered when the formulation of the solubility is attempted. Therefore, subsequently some examples are elaborated in detail. In general, molar numbers as well as pressures can be used in these calculations.

1. Example: $Si(s) + SiCl_4(g)$

solid: Si; gas phase ($\approx 1000\,°C$): $SiCl_4$, $SiCl_2$

During the transport of silicon with silicon(IV)-chloride (e.g., $1000 \rightarrow 900\,°C$) the following endothermic equilibrium between the solid and gas phase is established.

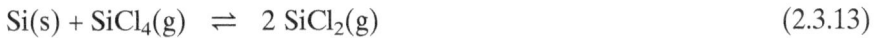

$$Si(s) + SiCl_4(g) \rightleftharpoons 2\,SiCl_2(g) \tag{2.3.13}$$

Given that a solid consisting of silicon is present at T_2 as well as at T_1, the relation $p(SiCl_2)_{T_2} > p(SiCl_2)_{T_1}$ will be valid. At each place along the diffusion path from source to sink within the ampoule equation 2.3.14 will hold:

$$\Sigma p = p(SiCl_4) + p(SiCl_2) = p^*(Si) \tag{2.3.14}$$

Thus, one might be tempted to conclude that there should be no transport of silicon, since with respect to $p^*(Si)$ there exists no gradient. That there is nevertheless a transport can be related to the presence of a partial pressure gradient of $SiCl_2$, which can disproportionate into Si and $SiCl_4$. This situation is expressed by equations 2.3.15 and 2.3.16.

$$\lambda(Si) = \frac{p(SiCl_2) + p(SiCl_4)}{2 \cdot p(SiCl_2) + 4 \cdot p(SiCl_4)} \tag{2.3.15}$$

$$\lambda(Si)_{T_2} > \lambda(Si)_{T_1} \tag{2.3.16}$$

In considering this example the distinction between **reversible** and **irreversible** **solubility** (of a solid in the gas phase) becomes clear. Obviously, only this amount of silicon is reversibly dissolved in the gas phase, which leads to the formation of $SiCl_2$: equation 2.3.13. The amount of $SiCl_4$ that is obtained by shifting equilibrium 2.3.13 quantitatively to the left, e.g., by lowering the temperature, gives the amount of silicon that is irreversibly dissolved in the gas phase. $SiCl_4$ is the solvent and does not, despite its silicon content, contribute to the reversible solubility of Si(s).

2. Example: $Fe(s) + HCl(g)$

solid: Fe; gas phase ($\approx 1000\,°C$): $FeCl_2$, H_2, HCl

$$\lambda(Fe) = \frac{n^*(Fe)}{n^*(Cl) + n^*(H)} = \frac{n(FeCl_2)}{2 \cdot n(FeCl_2) + 2 \cdot n(H_2) + 2 \cdot n(HCl)} \qquad (2.3.17)$$

Equation 2.3.17 describes the solubility of iron. It not only allows the description for Fe + HCl, but also for additional H_2 and $FeCl_2$ that might have been introduced into the reaction volume; $\lambda(Fe)$ follows for each case from the composition of the equilibrium gas phase.

3. Example: $W(s) + Cl_2(g) + Ar$

solid: W; gas phase ($1500 \ldots 3000\,°C$): W, WCl_x ($1 \leq x \leq 6$), Cl_2, Cl, Ar

$$\lambda(W) = \frac{n^*(W)}{n^*(Cl) + n^*(Ar)}$$

$$= \frac{n(W) + n(WCl) + \ldots + n(WCl_6)}{2 \cdot n(Cl_2) + n(Cl) + n(WCl) + \ldots + 6 \cdot n(WCl_6) + n(Ar)} \qquad (2.3.18)$$

All gas species are included in the formulation of λ, without regard to their actual chemical importance. In this sense the inert gas is considered, too.

4. Example: $GaAs(s) + I_2(g) + H_2(g)$

solid: GaAs; gas phase ($\approx 1000\,°C$): GaI, GaI_3, I_2, I, HI, H_2, As_4, As_2

$$\lambda(GaAs) = \frac{n^*(Ga) + n^*(As)}{n^*(I) + n^*(H)} \qquad (2.3.19)$$

Since single-phase GaAs is considered as the equilibrium solid, we get $n^*(Ga) = n^*(As)$. For this case (congruent evaporation) it would suffice to express the molar number of GaAs dissolved in the gas phase by the solubility of just one of its components.

This situation might be modified by introducing additional arsenic into the ampoule, which will occur in the gas phase only and therefore will contribute to the solvent. This molar number of arsenic $n^*(As)_{II}$ has to be distinguished from $n^*(As)_I$, which originates exclusively from the dissolution of GaAs.

$$n*(As)_I = n*(Ga) \tag{2.3.20}$$

$$n*(As)_{II} = 4n(As_4) + 2n(As_2) - n*(As)_I \tag{2.3.21}$$

$$\lambda(GaAs) = \frac{n*(Ga) + n*(As)_I}{n*(I) + n*(H) + n*(As)_{II}} \tag{2.3.22}$$

The distinction of amounts (molar numbers) that have to be regarded as solute and those that act as solvent ($n*(As)_I$ and $n*(As)_{II}$ in our example) appears no longer reasonable for the components of solids that possess a homogeneity range. The individual solubilities of the various components, no longer restricted by stoichiometric relations, may lead to enrichment or depletion of a component in the solid ("shift of composition").

5. *Example:* $ReO_2(s) + Re_2O_7(g) + I_2$

solid: ReO_2; gas phase ($\approx 500\,°C$): Re_2O_7, ReO_3I, I_2, I

For this case two types of oxygen should be distinguished. The total transformation of dissolved rhenium into ReO_2 moves the oxygen with the molar number $n*(O)_I = 2 \cdot n*(Re)$ into the solid, while $n*(O)_{II} = n*(O) - n*(O)_I$ remains in the gas phase. Thus follows equation 2.3.23.

$$\lambda(ReO_2) = \frac{n*(Re) + n*(O)_I}{n*(I) + n*(O)_{II}} \tag{2.3.23}$$

The solubility $\lambda(ReO_2)$ thus defined, consists of an amount of rhenium irreversibly dissolved in the gas phase. This amount of ReO_2 can not be precipitated and remains in the gas phase as Re_2O_7.

6. *Example:* $Fe_2O_3(s) + HCl(g)$

solid: Fe_2O_3; gas phase ($\approx 900\,°C$): $FeCl_3$, Fe_2Cl_6, HCl, H_2O

$$\lambda(Fe_2O_3) = \frac{n*(Fe) + n*(O)}{n*(Cl) + n*(H)} \tag{2.3.24}$$

The solubility of Fe_2O_3 in HCl is described by equation 2.3.24. For cases where additional H_2O is introduced into the reaction volume, equations 2.3.25 to 2.3.27 have to be used.

$$n*(O)_I = 1.5 \cdot n*(Fe) \tag{2.3.25}$$

$$n*(O)_{II} = n*(O) - n*(O)_I \tag{2.3.26}$$

$$\lambda(Fe_2O_3) = \frac{n*(Fe) + n*(O)_I}{n*(Cl) + n*(H) + n*(O)_{II}} \tag{2.3.27}$$

The same expression for $\lambda(Fe_2O_3)$ describes the deposition of Fe_2O_3 from a gas flow consisting of $FeCl_3$, H_2O, and O_2 (*chemical vapor deposition,* CVD). For this case a surplus of H_2O and O_2 might be used to accomplish quantitative deposition of iron from the gas phase. Under such conditions equilibrium is established between the Fe_2O_3 solid and the gas phase containing $FeCl_3$, Fe_2Cl_6, HCl, Cl_2, H_2O, and O_2.

Given that equilibrium has been established the transport rate is proportional to the difference $\Delta\lambda = \lambda(T_{source}) - \lambda(T_{sink})$ of the solubilities of a solid in the gas phase. The latter is regarded as solvent L and comprises all gas species. The solubility λ can be described by the expression $\lambda = n^*(A)/n^*(L)$ or, using the relation between n and p given by the ideal gas law, by $\lambda = \Sigma(\nu(A) \cdot p(A))/\Sigma(\nu(L) \cdot p(L))$. The meaning of L has been explained in the preceding paragraphs. The numbers $\nu(A)$ and $\nu(L)$ denominate the stoichiometric coefficients of A and L in the gas species. The equation for the solubility of a solid in a gas phase holds for systems of any order of complexity in closed as well as in open systems.

2.4 Incongruent Dissolution and Quasi-stationary Transport Behavior

2.4.1 Phase Relations Accompanying Incongruent Dissolution of a Solid

The thermodynamic description/modeling of transport systems get increasingly complicated if the transported compound shows a homogeneity range $AB_{x\pm\delta}$ or the transport occurs in a system with several co-existing condensed phases, e.g., AB_y and AB_z. For these cases congruent dissolution of the condensed phase in the gas phase becomes rather unlikely. Incongruent dissolution should be anticipated. This case is characterized by different ratios of the molar numbers for the components (elements) in the source solid and the corresponding gas phase. As an immediate consequence, the composition of the solid at the sink (deposition) side, expressed by the ratio $(n(A)/n(B)$ in $AB_{x, T(sink)}$ is not longer identical to the ratio of the balance pressures $p^*(A)/p^*(B)$ of the components A and B at the sink. In the same way the ratio of the components of the deposited condensed phase does not need to be identical to that of the dissolved phase. Thus, in the sink a phase might be deposited that is different to the one present at the source. This behavior is comparable to the peritectic melting of a solid and the compositional shift that accompanies the re-formation of a solid from this melt upon cooling. The compositions of melt and solid are different.

In addition to the deposition of co-existing condensed phases (AB_y and AB_z) nearly all transport reactions concerning phases of variable composition within a homogeneity range $AB_{x\pm\delta}$ have to be rationalized as incongruent CVT. The general problem is treated in a vivid way for the transport of $TiS_{2-\delta}$ (Sae 1976). This transport corresponding to the heterogeneous equilibrium 2.4.1.1 is accompanied by the decomposition reaction 2.4.1.2.

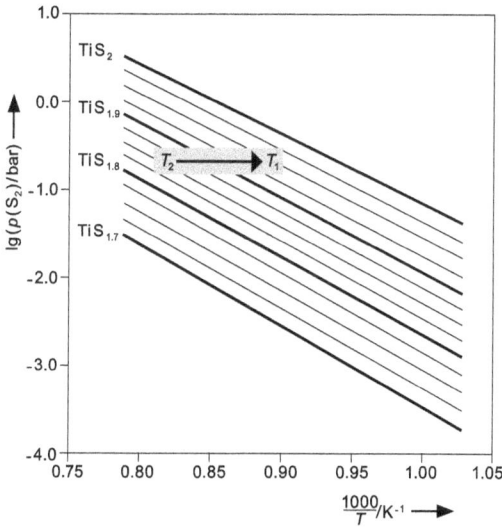

Figure 2.4.1.1 Barogram for the Ti/S system showing the co-existence pressures in the compositional range TiS$_{2-\delta}$. The phase relations observed in chemical vapor transport experiments in a temperature gradient 950 to 850 °C are visualized; graphics according to *Saeki* (Sae 1976).

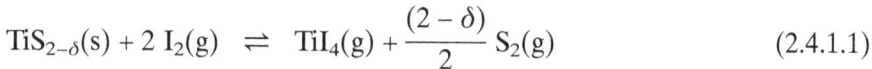

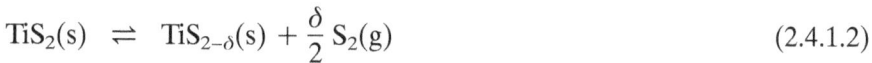

$$TiS_{2-\delta}(s) + 2\,I_2(g) \;\rightleftharpoons\; TiI_4(g) + \frac{(2-\delta)}{2}\,S_2(g) \tag{2.4.1.1}$$

$$TiS_2(s) \;\rightleftharpoons\; TiS_{2-\delta}(s) + \frac{\delta}{2}\,S_2(g) \tag{2.4.1.2}$$

For transport experiments in the temperature gradient 950 to 850 °C, independent of the starting composition TiS$_{2-\delta}$ of the source solid, at T_1 a sulfur-enriched phase will always be deposited. At the same time a sulfur-depleted phase forms at the source. Starting from a source solids composition TiS$_{1.889}$ yields transport of TiS$_{1.933}$; the solid at the dissolution side (source) is depleted of sulfur. For this case the thermodynamic description of the observed phase relations can be based in a rather simple approach on the independent calculation of the equilibrium conditions for source and sink. As both equilibrium regions are linked to each other via the gas phase, solids of corresponding compositions are obtained at T_2 and T_1. These compositions are determined by equation 2.4.1.3:

$$p(S_2)\ (\text{above TiS}_{2-\delta_1}\ \text{at}\ T_1) = p(S_2)\ (\text{above TiS}_{2-\delta_2}\ \text{at}\ T_2)\ (\delta_2 > \delta_1) \tag{2.4.1.3}$$

A valuable means for the analysis of the aforementioned phase relations is provided by the barogram of the corresponding system. The curves $\lg(p/p^0) = f(x, T)$ allow determination of the stoichiometric coefficient x along an isobar for given temperatures T_2 and T_1 (Figure 2.4.1.1).

A farther reaching, more general treatment of the phase relations encountered in transport systems with incongruent dissolution of a solid is based on the fact that the two equilibrium regions (source and sink) are indeed not independent of each other. For example, we consider the case of a system consisting of three components and two phases (solid + gas phase). In such a system the two components A and B may form the solid AB_x; the latter will be transferred by the transport agent X into the gas phase. According to *Gibbs*'s phase rule the system possesses three degrees of freedom for its thermodynamic description: Σp, T_{source}, and $x_{T_{source}}$.

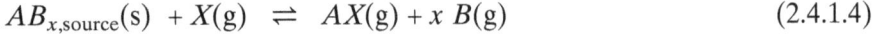

$$AB_{x,source}(s) + X(g) \rightleftharpoons AX(g) + x\, B(g) \tag{2.4.1.4}$$

$$F = K - P + 2$$

$$F = 3 - 2 + 2 = 3$$

From the aforementioned considerations, it follows that the composition of $AB_{x,\,sink}$ at the sink temperature T_{sink} might be variable, but not independent of the equilibrium conditions valid for the dissolution (source) region (T_{source}, x_{source}, Σp). For a congruent CVT, modeling of the transport effect is possible via *independent equilibrium calculations* for the source and sink regions followed by determination of the differences of partial or balance pressures. In contrast to this situation, the equilibrium calculations for source and sink of an incongruent transport have to be linked to each other. Only in doing so, does it become possible to determine the composition $AB_{x,\,sink}$ at T_{sink}. The relation between the two equilibrium regions at T_{source} and T_{sink} can be described in the following way: at the beginning of an incongruent transport, equilibrium between the solid(s) and the gas phase at the source is established. A mass flow via the gas phase from source to sink follows, which yields a ratio of molar numbers of the components in the gas phase identical for source and sink. From the (oversaturated) gas phase at the sink a solid will be deposited, which will show a different ratio of molar numbers $n(A)/n(B)$ as compared to the initial gas phase. As a consequence of the deposition, the ratio of the molar numbers in the sink's gas phase will be changed with respect to the source's gas phase. The resulting differences of the molar numbers in the gas phases of source and sink $n_{source} - n_{sink}$ is referred to as **flux**.

The composition of $AB_{x,\,sink}$ is determined by the ratio of the molar flow for A and B, but not by the the ratio of the balance pressures.

$$\left(\frac{n(B)}{n(A)}\right)_{T_{sink}} = \left(\frac{flux(B)}{flux(A)}\right)_{T_{source} \to T_{sink}} = \frac{J(B)}{J(A)} = x_{sink} \tag{2.4.1.5}$$

The situation might be visualized by a picture of a crowd moving around a city by bus: for sure, the mobility of the people during the day is rather high, the "transport effect" at the end of the day is, however, likely to be rather low since most of the passengers will have returned to their homes. If for the "transport" a small car used, significantly less people will be mobile. The transport effect, however, does not only depend on the mobility of the transported species but is also characterized by a directed motion to a certain destination. Thus, given that the passengers do not return to the location of their departure, even with the small car an effective transport can be achieved. The composition of the crowd at the destination follows from the number of people who have been moved in a directed fashion.

For congruent transport equation 2.4.1.5 is valid, too. Obviously, a transfer with constant molar ratio of the components will occur between the equilibrium regions if the ratio of the balance pressures between source and sink is constant. The validity of the flux relation is assumed for *all* chemical transport reactions.

The steady state is given for an incongruent transport only as long as the equilibrium position at the dissolution region (source) remains constant. For different values of x_{sink} and x_{source}, the compositions of the source's solid and the source's gas phase have to change during the course of the transport experiment. The compositional change of the source solid might proceed in two ways.

- Discontinuous compositional change by formation of two co-existing phases.
- Continuous compositional change within a homogeneity range.

For modeling of the aforementioned, more complicated phase relations in CVT experiments two models with the corresponding software are available:

- The **extended transport model** by *Krabbes, Oppermann,* and *Wolf* allows in particular the closed description of the deposition of solids with homogeneity range (Kra 1975, Kra 1976a, Kra 1976b, Kra 1983). Within the model it is assumed for simplicity that the amount of solid in the source is infinitely large. Thus, deposition at the sink of a solid with deviating composition will leave the steady-state practically unchanged (**quasi-stationary transport**).
- The **model of co-operative transport** by *Schweizer* and *Gruehn* is aiming especially at the description of the time-dependent migration of multi-phasic solids in a temperature gradient (Schw 1983a, Gru 1983). The model allows the complete transfer of the source solid(s) to the sink and thus comprises the description of the compositional change of the solid(s) in source and sink with time (**non-stationary** or **non-steady state transport**).

The two models are implemented in separate computer programs (extended transport model: TRAGMIN (Kra 2008); model of co-operative transport: CVTRANS (Tra 1999)). Both programs are based on equilibrium calculations by the G_{min}-method (Eri 1971). By this method the composition of condensed equilibrium phase and gas phase are derived via a minimization of the Gibbs energy of the system. Chapter 13 provides explanations on the G_{min} method and on both computer programs.

2.4.2 The Extended Transport Model

Considering the fluxes of the individual components for a description of exclusively diffusive gas motion was first introduced by *Lever* for the rather simple system $Ge(s)/GeI_4(g)$ (Lev 1962b, Lev 1966). Accordingly, the flux of germanium $J(Ge)$ is obtained, if one assumes a single diffusion coefficient $\bar{D}$, which is the average of the individual values for all gas species:

$$J(Ge) = \frac{p^*(Ge)}{\Sigma p} J_{ges} - \frac{\bar{D}}{R \cdot T} \cdot \frac{dp^*(Ge)}{ds} \qquad (2.4.2.1)$$

Richardson and *Noläng* made use of this relation when deriving transport rates for the CVT in complex systems (Ric 1977). Their model, which was based on the additional assumption 2.4.2.2, allows reasonable prediction of the transport behavior only for the congruent dissolution of a solid. Relating the flux of component A to the total balance of molar numbers of A (2.4.2.2) turned out to be inappropriate for cases involving incongruent dissolution.

$$n(A(s))_{source} + n(A(g))_{source} = n(A(s))_{sink} + n(A(g))_{sink} \tag{2.4.2.2}$$

According to *Krabbes, Oppermann,* and *Wolf* the steady state of a transport system involving incongruent dissolution of a solid AB_x is determined by linking the balance of molar numbers for A in the sink $[n(A(s))_{sink} + n(A(g))_{sink}]$ to the molar number of A in the gas phase of the source (Kra 1975, Kra 1976a, Kra 1976b, Kra 1983). The molar number of A in the source's solid does obviously not contribute to the flux. This situation is described by the following equation:

$$n(A(g))_{source} = n(A(s))_{sink} + n(A(g))_{sink} \tag{2.4.2.3}$$

Eventually, one arrives at the **stationarity relation** ε by linking the fluxes $J(B)$ and $J(A)$ of the individual components of the systems and assuming that the net flux of the transport agent X will vanish; $J(X) = 0$ (Kra 1975).

$$\left(\frac{p^*(B) - x_{sink} \cdot p^*(A)}{p^*(X)} \right)_{source} = \left(\frac{p^*(B) - x_{sink} \cdot p^*(A)}{p^*(X)} \right)_{sink} = \varepsilon \tag{2.4.2.4}$$

After re-arranging 2.4.2.4, the statement becomes applicable for the description of a chemical vapor transport:

$$\left[\left(\frac{p^*(B)}{p^*(X)} \right)_{source} - \left(\frac{p^*(B)}{p^*(X)} \right)_{sink} \right] =$$

$$x_{sink} \cdot \left[\left(\frac{p^*(A)}{p^*(X)} \right)_{source} - \left(\frac{p^*(A)}{p^*(X)} \right)_{sink} \right] \tag{2.4.2.5}$$

$$\frac{\left[\left(\frac{p^*(B)}{p^*(X)} \right)_{source} - \left(\frac{p^*(B)}{p^*(X)} \right)_{sink} \right]}{\left[\left(\frac{p^*(A)}{p^*(X)} \right)_{source} - \left(\frac{p^*(A)}{p^*(X)} \right)_{sink} \right]} = \frac{\Delta\lambda(B)}{\Delta\lambda(A)} = x_{sink} \tag{2.4.2.6}$$

Comparing equations 2.4.2.6 and 2.4.2.3 shows that the components' fluxes, $J(A)$ and $J(B)$, are proportional to the differences of the corresponding balance pressures in source and sink, normalized by the balance pressures for the solvent. In the same way, the ratio $J(B) : J(A)$ is equal to the ratio of differences of the components' gas phase solubilities. Thus, the composition of the solid deposited at the sink, $AB_{x,\,sink}$, is determined by the difference of its components' solubilities in the gas phase of source and sink (Kra 1975). Based on the extended transport model, this approach to the theoretical treatment of CVT reactions is realized in the software package TRAGMIN (Kra 2008) (see section 13.3).

In addition to calculation of equilibrium partial pressures and condensed phases the extended transport model offers further information on experimental realization and theoretical understanding of transport reactions:

- Calculation of the transport efficiency of gas species and deduction of the prevailing transport reaction(s).
- Calculation of the influence of experimental conditions on the deposition of solids with homogeneity range.
- Calculation of the influence of experimental conditions on the deposition of multi-phasic solids.

Transport efficiency of gas species   While according to the stationarity relation 2.4.2.4 the normalized balance pressures provide information on the transfer from source to sink of the individual components, the use of partial pressures $p(i)$ gives hints on the fluxes of individual gas species. In that way the contribution of the various gas species $i(g)$ to the transport can be elucidated and the formulation of a transport equation (or equations) becomes possible. Frequently, transport is determined by just a single heterogeneous reaction. It will be referred to as the **dominating transport reaction** in the rest of the book. The net flux of one particular gas species is given by the expression $\Delta[p(i)/p^*(X)]$ and is referred to as *transport efficiency* (2.4.2.7). Thus, the contribution – the efficiency – of an individual gas species to the transport is described. Two parts of the transport efficiency have to be distinguished: On one hand the part that originates from the various heterogeneous solid/gas equilibria, which typically are the transport reactions, and on the other hand the part that results from homogeneous gas phase equilibria.

$$w(i) \; = \; \Delta \left(\frac{p(i)}{p^*(X)} \right)_{\text{source} \to \text{sink}} = \left(\frac{p(i)}{p^*(X)} \right)_{\text{source}} - \left(\frac{p(i)}{p^*(X)} \right)_{\text{sink}} \qquad (2.4.2.7)$$

Transport effecting gas species are formed at the source (dissolution region) and consumed at the sink. Their *transport efficiency* $w(i) = \Delta[p(i)/p^*(X)] > 0$ indicates a net flux from source to sink $(p(i)_{\text{source}} > p(i)_{\text{sink}})$. Therefore they facilitate in the temperature gradient the transfer of the individual components of the condensed phase(s). Species having a transport efficiency $\Delta[p(i)/p^*(X)] < 0$ are consumed by heterogeneous equilibrium reactions at the source and released at the sink $(p(i)_{\text{source}} < p(i)_{\text{sink}})$, these species have to be regarded as *transport agents*.

The **absolute transport efficiency** refers to the fraction of the transport efficiency of a gas species, that really contributes to the transport of a solid. The example of the transport of a sulfide using iodine might serve for a better understanding. For this case one can easily calculate the transport efficiency of molecular iodine $I_2(g)$. Of the obtained value, however, only a fraction, the absolute transport efficiency, is related to the heterogeneous transport equilibrium. This fraction results from the calculated transport efficiency minus the fraction of iodine that is lost due to dissociation into iodine atoms. The phrases "transport efficiency" and "absolute transport efficiency" are in their meaning somewhat similar to the earlier introduced differences of reversible and irreversible solubilities in the gas phase (see section 2.3).

The transport efficiency is defined as

$$w(i) = [p(i)/p^*(X)]_{\text{source}} - [p(i)/p^*(X)]_{\text{sink}},$$

where $p^*(X)$ is the balance pressure of an arbitrary gas species X of the system under consideration. Gas species X might be the transport agent (e.g., a halogen), a constituent of the transport additive (e.g., chlorine when using tellurium(IV)-chloride) or even an inert gas (e.g., nitrogen). Depending on the chosen reference species X, its molar number, and thus its balance pressure, different values for the transport efficiency are obtained. Therefore, transport efficiencies calculated for different transport systems, various transport agents, and varying amounts of transport agent, are not directly comparable. Only if an inert gas having always the same balance pressure (e.g., nitrogen as residual gas with, say, $p^*(N_2) = 4 \cdot 10^{-6}$ bar in an evacuated ampoule) is chosen as reference, comparison becomes possible. Were the balance pressure of the transport agent itself (e.g., $p^*(Cl_2)$) or a component of the transport additive (e.g., $p^*(Cl_2)$ when using tellurium(IV)-chloride) chosen for the calculation, relative values for the transport efficiencies of individual gas species are obtained only, never absolute values. As the transport efficiencies calculated are only for the purpose of deriving the dominating transport reaction, the relative values are sufficient. As already stated, direct comparison of transport efficiencies of different transport systems requires calculation of the $w(i)$ always with the same balance pressure of an inert gas. In any case, it should be made clear which balance pressure has been used for normalization, e.g., $w(i) = \Delta[p(i)/p^*(N_2)]$, when providing transport efficiencies.

In this context, we should like to emphasize that equations for transport reactions always have to be formulated in a way that the fluxes of the gas species are in agreement with their direction of motion in a temperature gradient. In other words, the transport agent ($w(i) < 0$) has to occur at the educt side of a transport reaction (together with the transported solid), while the transporting gas species ($w(i) > 0$) have to be formulated as products. Note: the partial pressure ratio of the product gas species will be related to the composition of the transported solid, only for the case of congruent transport. Some rather simple examples might serve for a better understanding of the aforementioned definitions and considerations.

Examples

*ZnO transport by HCl: a **single** heterogeneous reaction*

When using hydrogen chloride for CVT of ZnO, the phase relations are rather simple: the system ZnO/HCl consists of four components Zn, O, H, and Cl. Since HCl practically is not dissociated into the elements under the reaction conditions, it will be regarded as *one* component (pseudo-component). Thus, according to

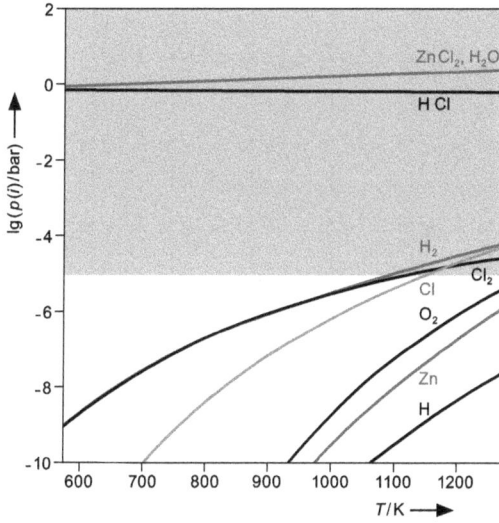

Figure 2.4.2.1 Partial pressures $p(i)$ for the dissolution of ZnO by HCl(g).

2.4.2.8, the system, which contains the three gas species HCl, $ZnCl_2$, and H_2O, is fully described by a single reaction equation, 2.4.2.9.

$$r_u = s - k + 1 = 3 - 3 + 1 = 1 \qquad (2.4.2.8)$$

The only relevant reaction between ZnO(s) and HCl(g) results in a congruent dissolution of ZnO accompanied by formation of $ZnCl_2$(g) and H_2O(g) (2.4.2.9). In the graphical representation of the gas-phase composition (Figure 2.4.2.1) the congruent dissolution of ZnO is reflected by the ratio $p(ZnCl_2) : p(H_2O) = 1$. Further, in principle possible gas species (H, H_2, Cl, Cl_2, O_2, Zn) can be ignored in the temperature range under consideration. They are irrelevant for the transport, since their partial pressures are well below 10^{-5} bar. Calculation of the transport efficiencies yields for w(HCl) a negative value; its normalized partial pressure at the source is lower than at the sink. According to expectation HCl acts as transport agent. This means, it is consumed at the source region and released at the sink. The mass transfer from the source to the sink of the two components of the solid is facilitated via the two gas species $ZnCl_2$ and H_2O both showing $\Delta[p(i)/p^*(X)] > 0$ (Figure 2.4.2.2). Calculation via the temperature dependent equilibrium constant $K_{p,T}$ leads to the same results, since the transport is fully described by a single transport reaction.

$$ZnO(s) + 2\,HCl(g) \;\rightleftharpoons\; ZnCl_2(g) + H_2O(g) \qquad (2.4.2.9)$$

$$\Delta_r G^0_{1000} = -23\;kJ \cdot mol^{-1}; \; K_{p,1000} = 16$$

ZnO transport by Cl_2: **a single** *heterogeneous reaction,* **several** *homogeneous reactions*

On first glance, the transport of zinc oxide by chlorine appears as simple as that using hydrogen chloride. Accounting for the three components of the system

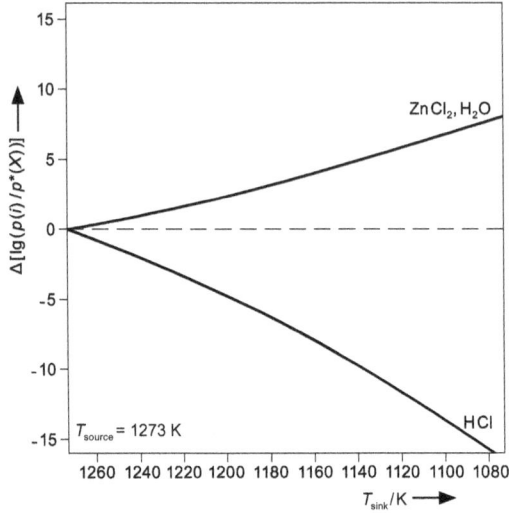

Figure 2.4.2.2 Transport efficiencies $w(i) = \Delta[p(i)/p^*(X)]$ of the gas species that participate in the dissolution and transport of ZnO by HCl(g).

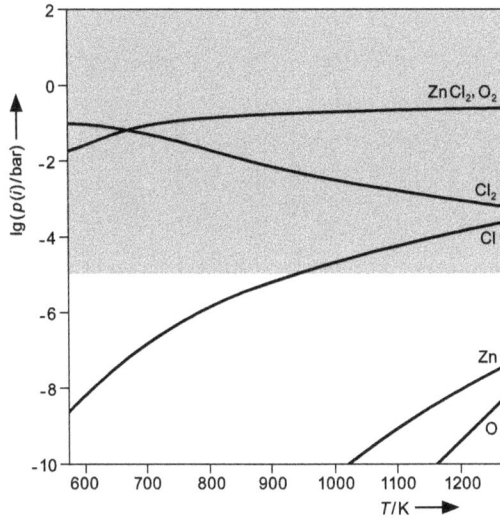

Figure 2.4.2.3 Partial pressures $p(i)$ of the gas species participating in the dissolution of ZnO by Cl_2.

(Zn, O, Cl) as well as for the three gas species Cl_2, $ZnCl_2$, and O_2 one arrives at just a *single* independent reaction, which is required for the complete description (2.4.2.10). However, calculations show that dissociation of Cl_2 into the atoms is not neglectable at temperatures above 1000 °C. The partial pressure of Cl exceeds 10^{-4} bar and therefore might become relevant for the transport (Figure 2.4.2.3). Dissociation of molecular oxygen, however, does not reach an order of magnitude that is relevant for the transport ($p(O) < 10^{-7}$ bar).

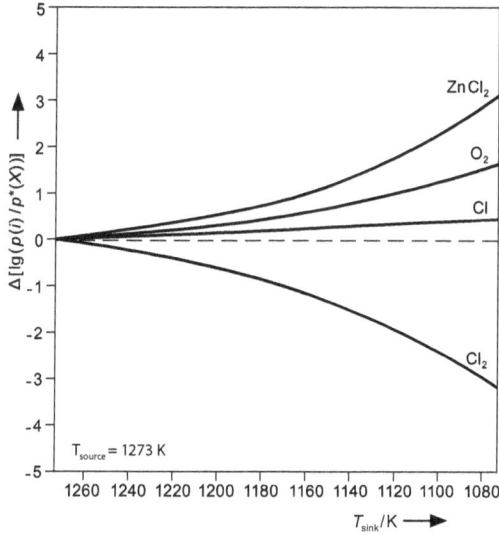

Figure 2.4.2.4 Transport efficiencies $w(i) = \Delta[p(i)/p^*(X)]$ of the gas species that participate in the dissolution and transport of ZnO by Cl_2.

Thus, the number of gaseous species that will have an impact on the transport of ZnO by chlorine rises to four. Consequently, two independent equilibria $(r_u = 4 - 3 + 1 = 2)$ will be required for the description of the system (equations 2.4.2.10 and 2.4.2.11). How these equilibria are formulated in a chemically reasonable way follows from the calculated transport efficiencies (Figure 2.4.2.4).

Calculation of the transport efficiencies $\Delta[p(i)/p^*(X)]$ for the considered gas species yields only for Cl_2 a negative value; Cl_2 has to be regarded as transport agent. The positive values for $w(ZnCl_2)$ and $w(O_2)$ correspond to their role as transporting species. Furthermore, for atomic chlorine a positive transport efficiency $w(Cl)$ is obtained. The two required unique reaction equations have to be set up in a way that gas species with negative value for their transport efficiency occur as educts. Gas species with $w(i) > 0$ have to be formulated as products. Accordingly, for the CVT of zinc oxide by chlorine the following chemical equations 2.4.2.10 and 2.4.2.11 can be given.

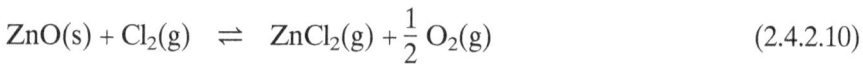

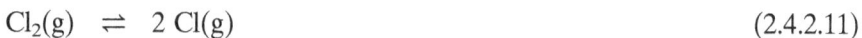

$$ZnO(s) + Cl_2(g) \;\rightleftharpoons\; ZnCl_2(g) + \frac{1}{2} O_2(g) \qquad\qquad (2.4.2.10)$$

$$Cl_2(g) \;\rightleftharpoons\; 2\, Cl(g) \qquad\qquad (2.4.2.11)$$

The calculated transport efficiency of atomic chlorine can be traced back exclusively to the homogeneous equilibrium 2.4.2.11. While indeed a transfer of chlorine from the source (T_2) to the sink (T_1; $T_2 > T_1$) via Cl(g) is facilitated, atomic chlorine does not contribute in any way to the transfer of zinc oxide. The efficiency of Cl_2 as transport agent for ZnO via equilibrium 2.4.2.10 has to be reduced for the fraction of Cl_2 involved in the homogeneous vapor phase equilibrium 2.4.2.11. The absolute transport efficiencies $w(Cl_2) = -3.2$, $w(Cl) = 0.4$, $w(O_2) = 1.5$, and

$w(ZnCl_2) = 3.0$ (Figure 2.4.2.4) allow calculation of the flux of chlorine from source to sink via $ZnCl_2$ (equation 2.4.2.10) and Cl (equation 2.4.2.11), respectively. A ratio of fluxes of $15 : 1$ via $ZnCl_2$ and Cl, respectively, is obtained.

Bi_2Se_3 *transport by* I_2: **two** *heterogeneous reactions,* **one** *homogeneous reaction*

Chemical vapor transport of bismuth selenide is possible by the addition of iodine (Schö 2010). Calculation of the equilibrium gas phase for the system Bi_2Se_3/I yields partial pressures of relevant magnitude for the gas species BiI_3, BiI, Se_n, I_2, and I (Figure 2.4.2.5). To a first approximation we assume that the variety of selenium-containing species in the gas phase is sufficiently represented by the dominating molecule $Se_2(g)$. The remaining gas species relevant for the transport will require three unique equilibria for a complete description of the transport behavior ($r_u = 5 - 3 + 1 = 3$).

Calculation of the transport efficiencies for Bi_2Se_3/I_2 shows $w(I_2)$ to be the only one having a negative value. Thus, in the system, $I_2(g)$ is effective as transport agent (Figure 2.4.2.6). Both BiI_3 and BiI contribute to the transport of bismuth. Selenium is carried as $Se_2(g)$ from the source to the sink.

The contribution of the gaseous bismuth iodides requires the formulation of two unique heterogeneous transport equilibria:

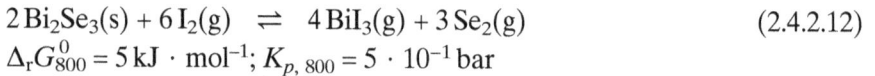

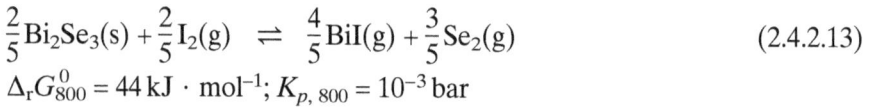

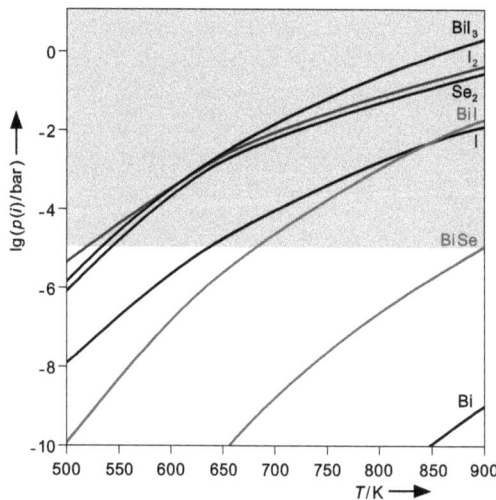

$$2\,Bi_2Se_3(s) + 6\,I_2(g) \;\rightleftharpoons\; 4\,BiI_3(g) + 3\,Se_2(g) \tag{2.4.2.12}$$
$$\Delta_r G^0_{800} = 5\,kJ \cdot mol^{-1}; \; K_{p,\,800} = 5 \cdot 10^{-1}\,bar$$

$$\frac{2}{5}Bi_2Se_3(s) + \frac{2}{5}I_2(g) \;\rightleftharpoons\; \frac{4}{5}BiI(g) + \frac{3}{5}Se_2(g) \tag{2.4.2.13}$$
$$\Delta_r G^0_{800} = 44\,kJ \cdot mol^{-1}; \; K_{p,\,800} = 10^{-3}\,bar$$

Figure 2.4.2.5 Partial pressures $p(i)$ for the gas species formed during the dissolution of Bi_2Se_3 by $I_2(g)$.

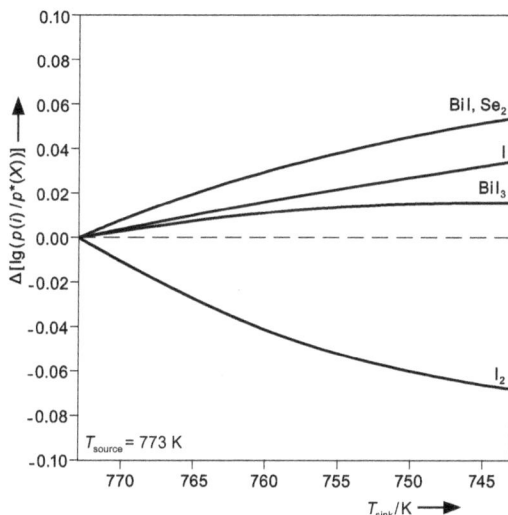

Figure 2.4.2.6 Transport efficiencies $w(i)$ of the gas species that participate in the dissolution and transport of Bi_2Se_3 by I_2.

On first sight, one would expect, due to a more balanced equilibrium position, a higher transport effect via 2.4.2.12 with BiI_3 as bismuth carrier, than for 2.4.2.13 with BiI. The detailed calculation, however, shows a higher transport efficiency for BiI. This result follows from the higher temperature dependence of equilibrium 2.4.2.13 and as a consequence a larger variation of $p(BiI)$ in the chosen temperature gradient (see Figure 2.4.2.5). Still, the transport is not dominated by one heterogeneous equilibrium, and both 2.4.2.12 and 2.4.2.13 contribute to the transport. The fractions of Bi_2Se_3 transported by $BiI_3(g)$ and $BiI(g)$, respectively, can be derived from the corresponding transport efficiencies: $\Delta[p(BiI_3)/p^*(X)]$:

$$\Delta[p(BiI_3)/p^*(X)] : \Delta[p(BiI)/p^*(X)] = 0.34.$$

The homogeneous equilibrium 2.4.2.14 does not participate in the transport of the solid, but reduces the efficiency of I_2 for the transport of Bi_2Se_3 by the amount $1/2 \cdot \Delta[p(I)/p^*(X)]$.

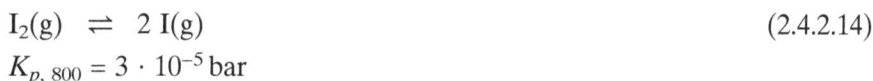

$$I_2(g) \rightleftharpoons 2\,I(g) \tag{2.4.2.14}$$
$$K_{p,\,800} = 3 \cdot 10^{-5}\,\text{bar}$$

Sublimation of Bi_2Se_3

Decomposition of Bi_2Se_3, equation 2.4.2.15, results in the formation of gaseous BiSe. Its partial pressure $p(BiSe) \approx 10^{-5}$ bar at 650 °C just reaches the order of magnitude that is required to be relevant for CVT (Figure 2.4.2.5). We may analyze this equilibrium in a separate calculation: assuming congruent evaporation of the

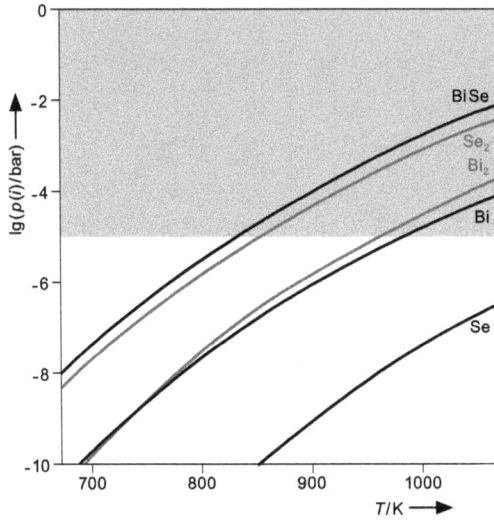

Figure 2.4.2.7 Partial pressures $p(i)$ of the gas species involved in the decomposition sublimation of Bi_2Se_3.

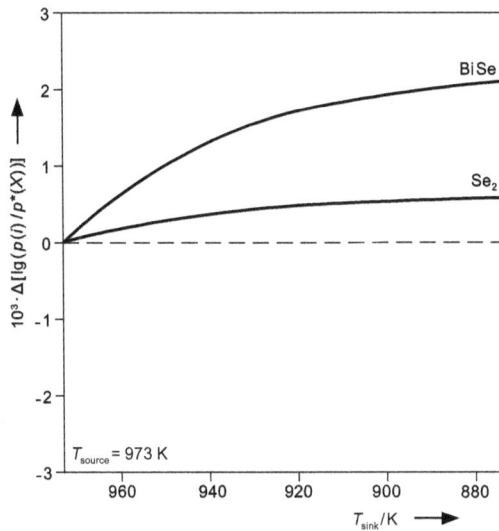

Figure 2.4.2.8 Transport efficiencies $w(i)$ for the gas species that participate in the decomposition sublimation of Bi_2Se_3.

solid by a decompositional sublimation leads to the dominating gas species BiSe and Se_2 the ratio of their partial pressures will be $4:1$.

Partial pressures of BiSe and Se_2 sufficiently high for migration of the solid in a temperature gradient will be reached above 600 °C, which is still below the melting temperature $\vartheta_m(Bi_2Se_3) = 722\,°C$. Besides the transport of Bi_2Se_3 by iodine at lower temperatures around 500 °C, at higher temperatures a decomposition sublimation, without the addition of a transport agent, is anticipated. This behavior is reflected by the transport efficiencies of the gas species: BiSe and Se_2 show $\Delta[p(i)/p^*(X)] > 0$, they act as a carrier. The ratio of the efficiencies of these gas species follows from their partial pressure ratio according to equilibrium 2.4.2.15. As expected for a sublimation process none of the species shows $\Delta[p(i)/p^*(X)] < 0$; there is no transport agent in the system.

The decomposition sublimation can be described by a single chemical equation:

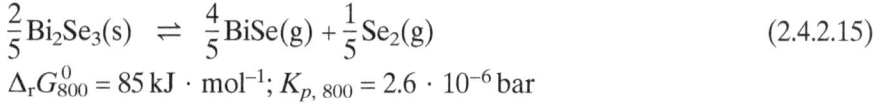

$$\frac{2}{5} Bi_2Se_3(s) \rightleftharpoons \frac{4}{5} BiSe(g) + \frac{1}{5} Se_2(g) \tag{2.4.2.15}$$

$\Delta_r G^0_{800} = 85\ kJ \cdot mol^{-1}$; $K_{p,\,800} = 2.6 \cdot 10^{-6}\ bar$

The theoretical treatment of the aforementioned examples might suffice to show that solid/gas equilibria of any complexity might be described and chemically understood by the extended transport model. For sublimation reactions and simple CVT reactions, the calculation of transport efficiencies provides a direct picture of the transport by a *unique equilibrium reaction*. In complex transport systems with more than one independent reaction, distinction between heterogeneous and homogeneous gas phase equilibria becomes possible; furthermore, the efficiency of the various gas species for the transport can be deduced. In most cases observed so far, based on the calculated transport efficiencies, formulation of one or two dominant transport reactions will fully suffice to describe the *fluxes* of the components between the equilibrium regions.

Application of the extended transport model is without alternative for the theoretical treatment of systems that involve incongruent dissolution (in a gas phase) of a solid with homogeneity range. Only detailed analysis of the molar fluxes (via the gas phas) will allow a prediction of the composition deposited at the sink. Thus, understanding becomes possible, of how the transport of a solid with homogeneity range will proceed, if its composition will be different at the source and sink regions, and if a component will be enriched at one region. The dissolution of multiphasic source solids has to be treated in the same way. By calculation and comparison of the involved molar fluxes it has to be decided whether a *simultaneous transport* or a *sequential migration* (see section 2.5) of several phases will occur.

Transport of solids with homogeneity range   Originally, the extended transport model was introduced for the analysis of the CVT of iron sulfide FeS_x of variable composition by the addition of iodine (Kra 1975), see section 7.1. Iron sulfide FeS_x ($1.0 \leq x \leq 1.15$) exhibits a homogeneity range, which extends to higher sulfur contents. The transport of FeS by iodine should proceed via the following equilibrium:

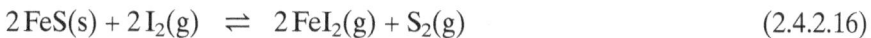

$$2\,FeS(s) + 2\,I_2(g) \rightleftharpoons 2\,FeI_2(g) + S_2(g) \tag{2.4.2.16}$$

The partial pressure $p(S_2)$ expected from (2.4.2.16), however, will be limited due to equilibrium 2.4.2.17 participating in the process:

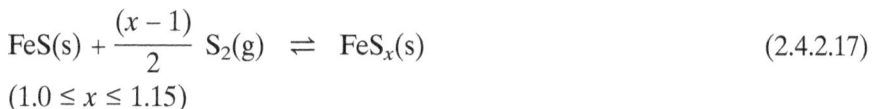

$$FeS(s) + \frac{(x-1)}{2}\,S_2(g) \rightleftharpoons FeS_x(s) \tag{2.4.2.17}$$

$(1.0 \leq x \leq 1.15)$

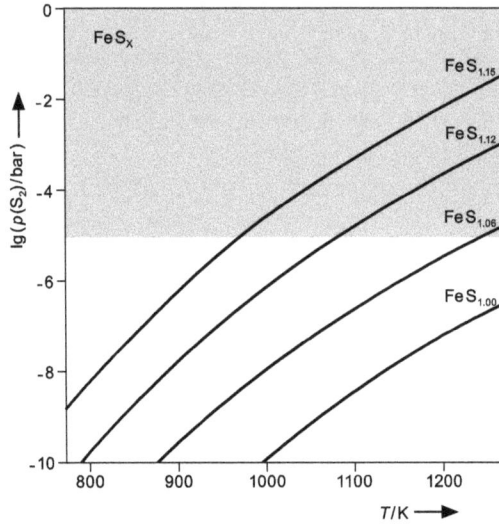

Figure 2.4.2.9 Co-existence pressures $p(S_2)$ above iron sulfide within its homogeneity range FeS$_x$ (1.00 ≤ x ≤ 1.15) according to *Krabbes et al.* (Kra 1975).

For the transport of FeS$_{1.0}$ two problems are encountered. Firstly, the sulfur partial pressure is too low to reach the transport relevant order of magnitude of 10^{-5} bar. Secondly, reaction of FeS$_{1.0}$ with iodine according to 2.4.2.16 in combination with 2.4.2.17 shifts the iron sulfide's composition within the homogeneity range to higher sulfur contents and rather high partial pressures $p(\text{FeI}_2)$ are built up. The equilibrium position of 2.4.2.17 depends on temperature. An even stronger influence on the composition of FeS$_x$ is exerted by the molar ratio of FeS$_{1.0}$ and iodine charged into an ampoule. For a volume of 20 cm^3 and a starting amount of 500 mg FeS$_{1.0}$, addition of more than 60 mg of iodine leads to a significant compositional shift for FeS$_x$ ($x > 1.05$). Charging the same ampoule with 1 g FeS$_{1.0}$ and 10 mg iodine will leave the solid's composition close to FeS$_{1.01}$. Iron sulfide FeS$_x$ at its lower phase boundary ($x = 1.00$) can be obtained by introducing as source solid a two-phase mixture Fe + FeS$_{1.0}$.

Subsequent paragraphs will deal with the questions for which x CVT of FeS$_x$ will be possible and how the iron sulfide's composition in sink and source do correlate with temperature and temperature gradient. To begin with, we will consider the co-existence pressures of sulfur within the homogeneity range of iron sulfide (Figure 2.4.2.9). The "iron-rich" lower phase boundary FeS$_{1.0}$ is characterized by the lowest partial pressure $p(S_2) = f(T)$; its value does not exceed 10^{-6} bar while the temperature is raised up to 1000 °C. Such a partial pressure is not sufficient for transport and, therefore, FeS$_{1.0}$ will not be transportable by equilibrium 2.4.2.16 applying iodine as transport agent. On the other hand, a solid at the upper phase boundary FeS$_{1.15}$ exhibits already above 700 °C a partial pressure of sulfur which should by sufficient for CVT. At higher temperatures even for iron sulfide with lower sulfur content CVT is expected; e.g., at 1000 °C for FeS$_x$ with $x \geq 1.05$ and $p(S_2) > 10^{-5}$ bar (Figure 2.4.2.9). The transport rate observed for FeS$_{1.05}$ is, however, still rather small $\dot{m} < 0.1$ mg · h^{-1} (Kra 1976a).

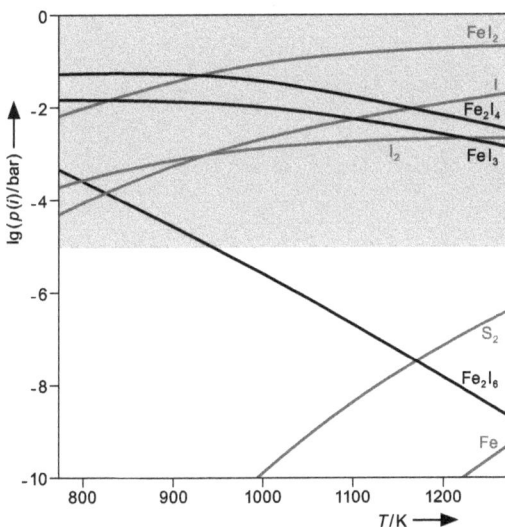

Figure 2.4.2.10 Partial pressures $p(i)$ of the gas species involved in the dissolution of FeS$_{1.0}$ by iodine in the gas phase as a function of temperature; figure according to *Krabbes et al.* (Kra 1975).

Calculation of the equilibrium pressures (Figure 2.4.2.10) shows, that the actual situation is much more complex than expected from consideration of the equilibria 2.4.2.16 and 2.4.2.17. Besides the gas species I$_2$ and I, which originate from the added iodine, the gaseous iron-carriers FeI$_3$, Fe$_2$I$_6$, FeI$_2$, and Fe$_2$I$_4$ occur. S$_2$ is the only sulfur-containing gas species. Above the equilibrium solid FeS$_{1.0}$ addition of iodine (0.5 mg $\cdot$ cm^{-3}) leads to the formation of the gaseous iron-carriers FeI$_3$, FeI$_2$, and Fe$_2$I$_4$ with partial pressures higher than 10^{-5} bar, which are regarded as relevant for transport. In contrast, up to 1000 °C sulfur ($p(S_2) < 10^{-6}$ bar) does not reach a partial pressure that might become transport effective. The pressure $p(S_2)$ in Figure 2.4.2.10 is identical to the sulfur co-existence decomposition pressure of FeS$_{1.0}$ (Figure 2.4.2.9). Conclusions drawn from Figure 2.4.2.10 back the statement that CVT of FeS$_{1.0}$ by iodine will be impossible (Kra 1975).

Starting with a solid of higher sulfur content, e.g., FeS$_{1.1}$, changes the gas-phase composition significantly (Figure 2.4.2.11). Now the partial pressure of sulfur is higher than 10^{-4} bar (Figure 2.4.2.9) and becomes transport effective, like the various gaseous iron-carriers. According to this calculation FeS$_{1.1}$ should be transportable.

The thermodynamic treatment of the system Fe/S/I as complex chemical vapor transport requires four unique (chemical) equations ($r_u = 6 - 3 + 1$). These can be derived from the graphical representation of the transport efficiencies of the various gas species (Figure 2.4.2.12). From $\Delta[p(I_2)/p*(X)] < 0$ follows that I$_2$ has to be regarded as a transport agent, while FeI$_2$ and S$_2$, which show $\Delta[p(i)/p*(X)] > 0$, are determining the flux of iron and sulfur to the sink. The heterogeneous equilibrium 2.4.2.18 results as the dominating transport reaction. The mass flow via this equilibrium is, however, limited by the rather low transport efficiency $\Delta[p(S_2)/p*(X)] = 10^{-2}$ of S$_2$.

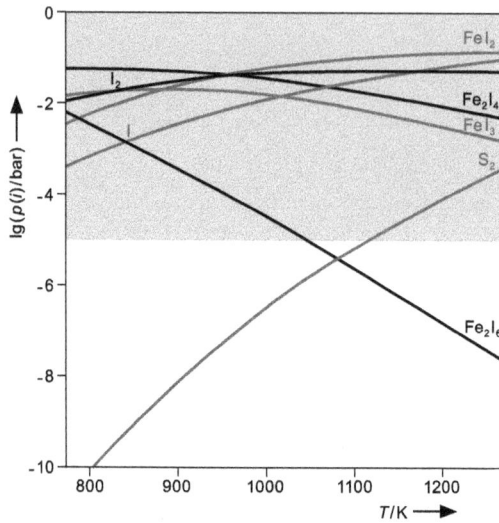

Figure 2.4.2.11 Partial pressures $p(i)$ of the gas species involved in the dissolution of FeS$_{1.1}$ by iodine in the gas phase as a function of temperature; figure according to *Krabbes et al.* (Kra 1975).

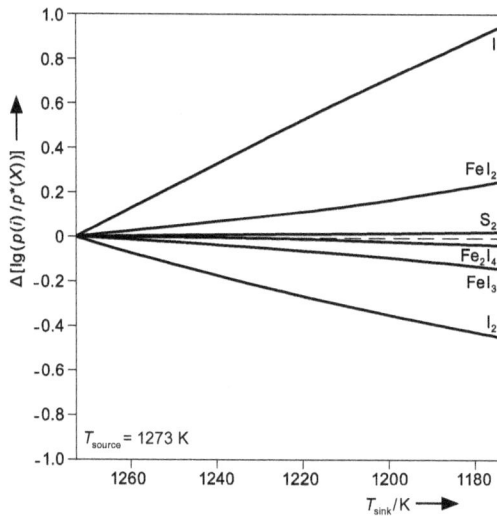

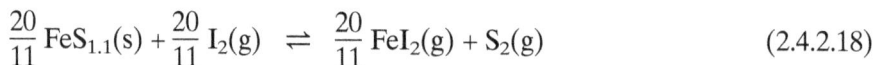

Figure 2.4.2.12 Transport efficiencies $w(i) = \Delta[p(i)/p^*(X)]$ of the gas species involved in the dissolution of FeS$_{1.1}$ by iodine in the gas phase as a function of temperature; figure according to *Krabbes et al.* (Kra 1975).

$$\frac{20}{11}\,FeS_{1.1}(s) + \frac{20}{11}\,I_2(g) \;\rightleftharpoons\; \frac{20}{11}\,FeI_2(g) + S_2(g) \tag{2.4.2.18}$$

According to the transport efficiencies of the gas species (Figure 2.4.2.12) over solid FeS$_{1.1}$ the homogeneous equilibria (2.4.2.19) to (2.4.2.21) are prevailing. Spe-

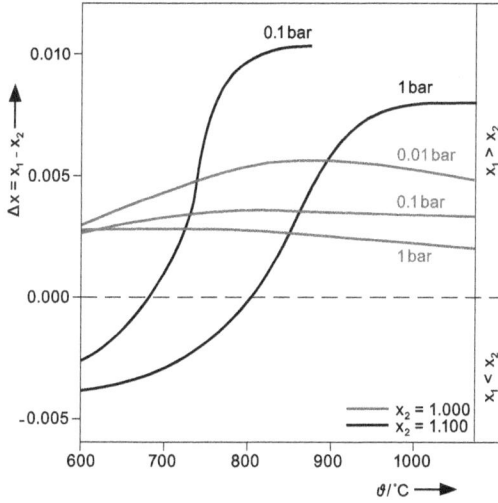

Figure 2.4.2.13 Calculated difference $\Delta x = x_1 - x_2$ of the stoichiometric coefficients for FeS_x in source and sink according to *Krabbes et al.* (Kra 1975).

cies having a negative value $\Delta[p(i)/p^*(X)] < 0$ $(i = I_2, Fe_2I_4, FeI_3)$ are consumed at the source region (at T_2). They have to occur at the educt side of the equilibria.

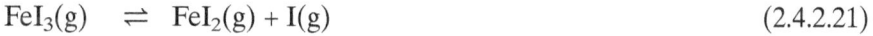

$$I_2(g) \quad \rightleftharpoons \quad 2\,I(g) \tag{2.4.2.19}$$

$$Fe_2I_4(g) \quad \rightleftharpoons \quad 2\,FeI_2(g) \tag{2.4.2.20}$$

$$FeI_3(g) \quad \rightleftharpoons \quad FeI_2(g) + I(g) \tag{2.4.2.21}$$

As already explained (equation 2.4.2.17), the composition FeS_{x1} of the solid deposited at the sink will depend on the source solid's composition FeS_{x2}, the temperature of source and sink, and, eventually, on the amount of added transport agent. The latter influence leads to an indirect correlation of total pressure and composition FeS_{x1}. FeS_x at or close to the *lower phase boundary* furnished at the source will react during equilibration with iodine under formation of solid FeS_x enriched in sulfur. Eventually, solid FeS_x, possibly with even higher sulfur content will be deposited at the sink. Besides the very low transport efficiency of $S_2(g)$ over solid $FeS_{1.0}$ the enrichment of sulfur in the sink's solid provides another reason why FeS_x at the *lower phase boundary* can not be transported using iodine. Transport experiments at temperatures below 700 °C, starting from solid FeS_x with higher sulfur content, however, within the homogeneity range, might lead to a small enrichment of iron in the sulfide deposited at the sink. Above a temperature of around 800 °C an enrichment of sulfur in the sink's solid will always be observed (Figure 2.4.2.13).

While transport of FeS_x at its *lower phase boundary* using iodine is impossible, successful use of HCl might be possible (Kra 1976b). Equilibration of the starting solid with hydrogen chloride will lead to hydrogen sulfide of sufficiently high, trans-

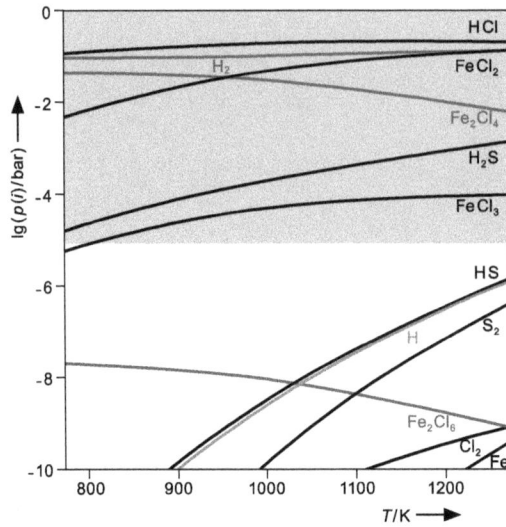

Figure 2.4.2.14 Partial pressures $p(i)$ of the gas species involved in the dissolution of $FeS_{1.0}$ by HCl in the gas phase as a function of temperature; figure according to *Krabbes et al.* (Kra 1975).

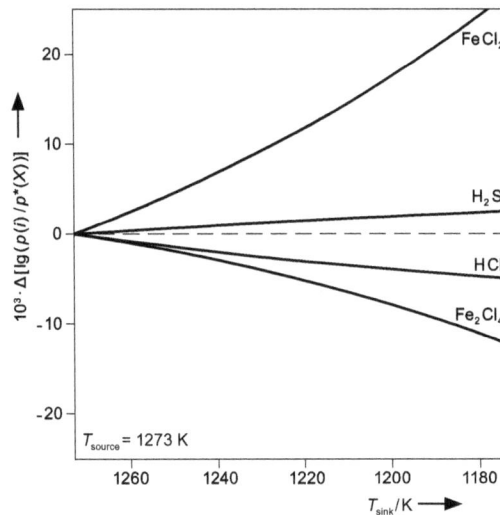

Figure 2.4.2.15 Transport efficiencies $w(i) = \Delta[p(i)/p^*(X)]$ of the gas species involved in the dissolution of $FeS_{1.0}$ by HCl(g) in the gas phase as a function of temperature; figure according to *Krabbes et al.* (Kra 1975).

port-relevant partial pressure ($p(H_2S) \leq 10^{-3}$ bar). Iron will occur in the gas phase predominantly as monomeric and dimeric iron(II)-chloride (Figure 2.4.2.14).

The transport efficiencies of the various gaseous species (Figure 2.4.2.15) reflect their fluxes via the gas phase. Hydrogen chloride, $\Delta[p(HCl)/p^*(X)] < 0$, acts as the transport agent. The mass transport from source to sink of the solid dissolved by HCl in the gas phase is facilitated by $FeCl_2$ and H_2S. Both show $\Delta[p(i)/p^*(X)]$ > 0. The transport efficiency of $FeCl_2$ is reduced by dissociation reaction 2.4.2.23, which causes the formation of $Fe_2Cl_4(g)$ at lower temperatures, without contributing to the transport.

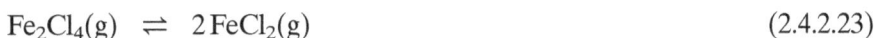

$$FeS(s) + 2\,HCl(g) \rightleftharpoons FeCl_2(g) + H_2S(g) \qquad (2.4.2.22)$$

$$Fe_2Cl_4(g) \rightleftharpoons 2\,FeCl_2(g) \qquad (2.4.2.23)$$

Starting with solid $FeS_{1.0}$ at the source, practically the same composition is obtained at the sink. A starting material FeS_x at the source with $x > 1$ and HCl as transport agent will result in an enrichment, however small, of iron in the sink's solid. Transport of FeS at its *lower phase boundary* is anticipated from thermodynamic reasons. Experimentally, transport rates up to 2 mg · h^{-1} are found. Crystals were deposited as hexagonal prisms with edge-length up to 1 mm (Kra 1976b).

Similarly, the transport of $FeS_{1.0}$ by HBr or HI can be described. Using as transport agent $GeCl_2$ and GeI_2, respectively, allows deposition of $FeS_{1.0}$, too. In these cases formation of the volatile species GeS, as sulfur carrier, besides the iron halides is crucial for the transport (Kra 1976b).

Stationary ("steady-state") transport starting from multi-phasic solids Another important application of the extended transport model lies in the theoretical treatment of transport experiments proceeding from a bi- or multi-phasic equilibrium solid at the source region. For such cases, preferred transport of one of the phases or simultaneous transport of several phases might be observed. For both scenarios a time-independent, steady state of fluxes can be anticipated. For the case of simultaneous transport the composition $AB_{x,\,sink}$ deposited at the sink determines the molar ratio $n(AB_y)/n(AB_z)$ of the phases AB_y and AB_z that are dissolved at the source. This problem has been discussed in considerable detail by *Oppermann* and co-workers for the transport behavior of vanadium oxides (Opp 1977a, Opp 1977b, Opp 1977c, Rei 1977).

In addition to the oxides containing vanadium just in a single oxidation state (VO, V_2O_3, VO_2, and V_2O_5), in the system V/O a remarkable number of phases with mixed valency are existing between V_2O_3 and VO_2, the so-called *Magnéli* phases with general composition V_nO_{2n-1}. These discrete phases are characterized by well defined compositions (V_3O_5, V_4O_7, V_5O_9, V_6O_{11}, V_7O_{13}, V_8O_{15}) and crystal structures. Their close lying chemical compositions should not be confused with the presence of a homogeneity range. The stability range of the various phases depends strongly on the oxygen partial pressure, as is visualized by the barogram of the system (Figure 2.4.2.16). Directed CVT of a given solid in the compositional regime of the *Magnéli* phases is experimentally highly demanding, due to the very close chemical compositions of the individual phases (see section 5.2.5). Of particular importance are the experimental conditions that will lead to deposition of single-phase products at the sink. Experimental parameters of influence for the composition of the crystals deposited at the sink are the starting composition of the source solid, mean transport temperature, and magnitude of the temperature gradient. Reasonably, calculations with predictive character are performed on the basis of the extended transport model.

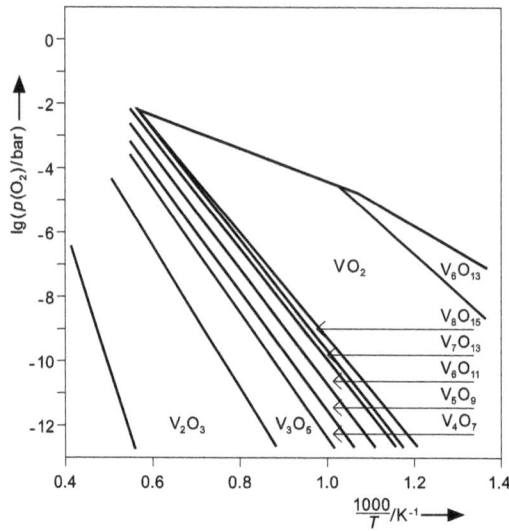

Figure 2.4.2.16 Barogram of the V/O system showing the co-existence pressures for the oxides V_2O_3 and V_2O_5 as well as for the *Magnéli* phases V_nO_{2n-1} according to *Oppermann et al.* (Opp 1977b, Opp 1977c).

Selection of the transport agent

The redox behavior of the transport agent is of particular importance for CVT in systems such as that of vanadium/oxygen. Chlorine is not suitable as a transport agent for most vanadium oxides, since these solids are irreversibly oxidized. Hydrogen chloride as well as tellurium(IV)-chloride, however, have been applied successfully. These transport agents allow the formation of gaseous oxygen carriers (e.g., H_2O, $TeOCl_2$) and thus even transport of oxides that exhibit very low oxygen co-existence pressures becomes possible (see section 5.1).

The closed thermodynamic description of the system with 4 components (V, O, Te, Cl) and 15 gas species (O_2, O, Cl_2, Cl, VCl_4, VCl_3, VCl_2, $VOCl_3$, Te_2, Te, $TeCl_4$, $TeCl_2$, $TeOCl_2$, $TeOCl_2$, TeO) requires 12 unique equilibria. Thus derived equilibrium partial pressures above solid VO_2 (Figure 2.4.2.17) and V_3O_5 (Figure 2.4.2.18) are graphically represented.

Significant pressures of $TeCl_2$ and $VOCl_3$ are present in the equilibrium gas phase above VO_2, V_3O_5, and the *Magnéli* phases. In contrast to this result (Figures 2.4.2.17 and 2.4.2.18) the partial pressures of oxygen-containing gas species of tellurium (e.g., TeO, TeO_2, $TeOCl_2$) depend strongly on the variable oxygen co-existence pressure above the *Magnéli* phases (see Figure 2.4.2.16). The particular suitability of tellurium(IV)-chloride as transport additive (note: transport additive and transport agent might not be the same!) results from its ability to form at 1000 °C over a wide range of oxygen partial pressures, namely, the range of co-existence pressures covered by the *Magnéli* phases, the oxygen transporting species $VOCl_3$, and $TeOCl_2$ with partial pressures higher than 10^{-5} bar. Thus, these species remain relevant for the transport of all *Magnéli* phases (Figure 2.4.2.19).

According to the calculated transport efficiencies (Figure 2.4.2.20) for the CVT of VO_2 by addition of $TeCl_4$, reactions (2.4.2.24a to d) (with decreasing impor-

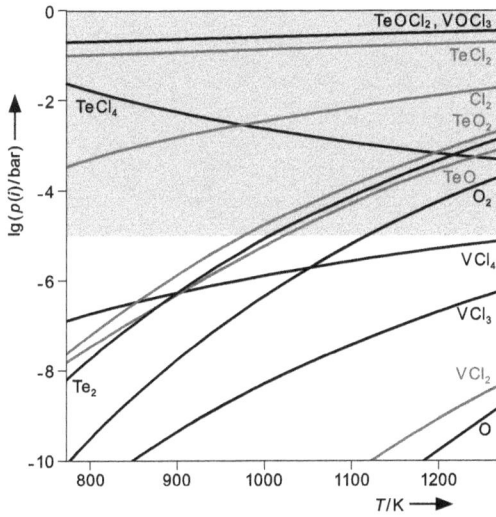

Figure 2.4.2.17 Partial pressures of the gas species involved in the dissolution of VO_2 by $TeCl_4$. Starting amounts: 0.01 mol VO_2 and $5 \cdot 10^{-3}$ mol/l $TeCl_4$.

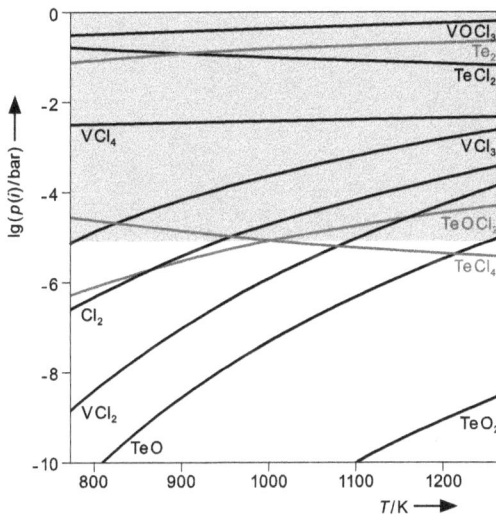

Figure 2.4.2.18 Partial pressures of the gas species involved in the dissolution of V_3O_5 by $TeCl_4$. Starting amounts: 0.01 mol V_3O_5 and $5 \cdot 10^{-3}$ mol/l $TeCl_4$.

tance) can be formulated. In the same way, one arrives for V_3O_5 (Figure 2.4.2.21) at reactions (2.4.2.25a to c). Here, different oxygen partial pressures above VO_2 and V_3O_5, respectively, cause a change in the dominating tellurium transporting species. While $Te_2(g)$ acts as the dominant gas species for the transport of V_3O_5, more oxidized species $TeOCl_2$, TeO_2, and TeO become effective, too, for the transport of VO_2, see Figure 2.4.2.19.

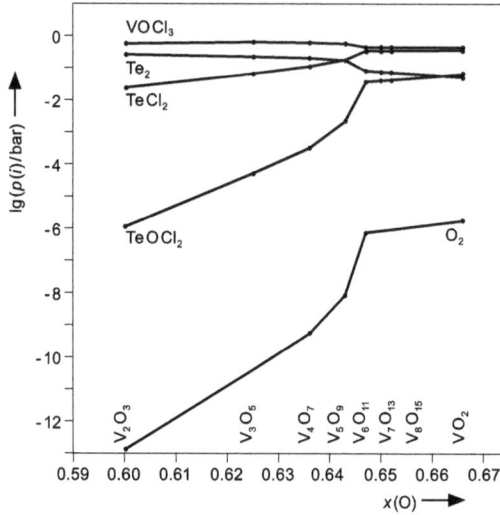

Figure 2.4.2.19 Gas-phase composition obtained for the dissolution of V_2O_3, the *Magnéli* phases, and VO_2 by $TeCl_4$ at $1000\,°C$. Starting amounts: 0.01 mol VO_x; $5 \cdot 10^{-3}$ mol/l $TeCl_4$.

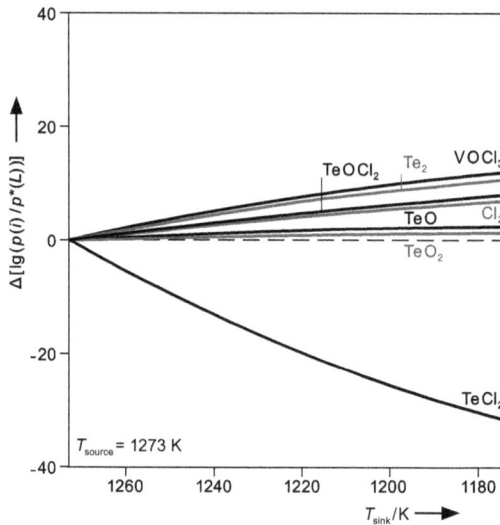

Figure 2.4.2.20 Transport efficiencies $w(i) = \Delta[p(i)/p^*(X)]$ of the gas species involved in the chemical vapor transport of VO_2 by the addition of $TeCl_4$.

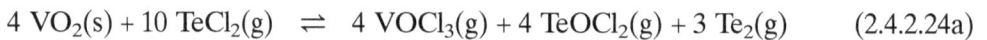

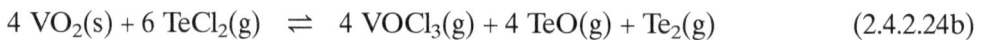

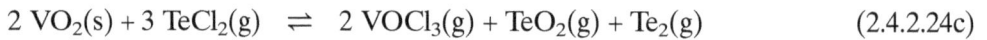

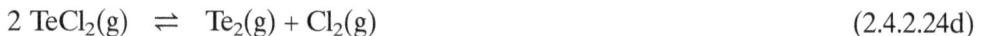

$$4\,VO_2(s) + 10\,TeCl_2(g) \;\rightleftharpoons\; 4\,VOCl_3(g) + 4\,TeOCl_2(g) + 3\,Te_2(g) \qquad (2.4.2.24a)$$

$$4\,VO_2(s) + 6\,TeCl_2(g) \;\rightleftharpoons\; 4\,VOCl_3(g) + 4\,TeO(g) + Te_2(g) \qquad (2.4.2.24b)$$

$$2\,VO_2(s) + 3\,TeCl_2(g) \;\rightleftharpoons\; 2\,VOCl_3(g) + TeO_2(g) + Te_2(g) \qquad (2.4.2.24c)$$

$$2\,TeCl_2(g) \;\rightleftharpoons\; Te_2(g) + Cl_2(g) \qquad (2.4.2.24d)$$

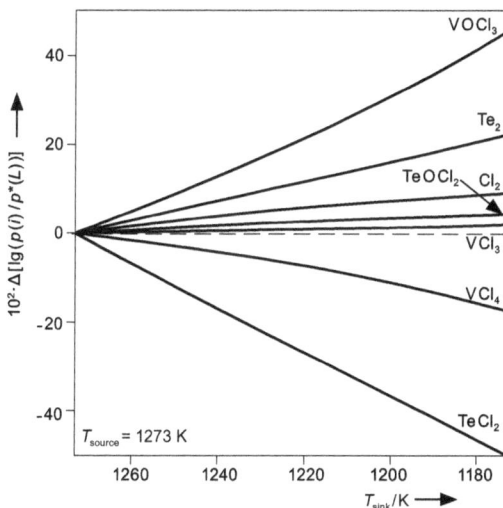

Figure 2.4.2.21 Transport efficiencies $w(i) = \Delta[p(i)/p*(X)]$ of the gas species involved in the chemical vapor transport of V_3O_5 by the addition of $TeCl_4$.

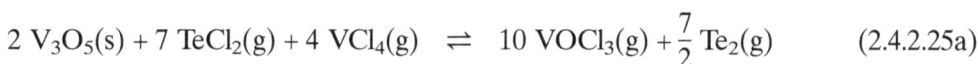

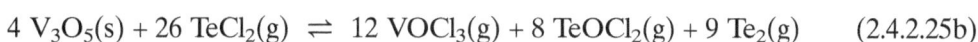

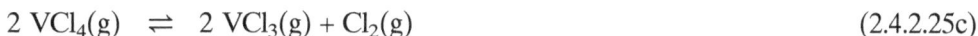

$$2\,V_3O_5(s) + 7\,TeCl_2(g) + 4\,VCl_4(g) \;\rightleftharpoons\; 10\,VOCl_3(g) + \frac{7}{2}\,Te_2(g) \qquad (2.4.2.25a)$$

$$4\,V_3O_5(s) + 26\,TeCl_2(g) \;\rightleftharpoons\; 12\,VOCl_3(g) + 8\,TeOCl_2(g) + 9\,Te_2(g) \qquad (2.4.2.25b)$$

$$2\,VCl_4(g) \;\rightleftharpoons\; 2\,VCl_3(g) + Cl_2(g) \qquad (2.4.2.25c)$$

According to the analysis of the transport efficiencies for all vanadium oxides between V_2O_3 and VO_2 CVT by addition of $TeCl_4$ should be possible in the temperature range between 900 and 1000 °C. This expectation has been confirmed by experiments (see section 5.2.5). Due to the higher oxygen co-existence pressures above oxides between VO_2 and V_2O_5, containing vanadium in oxidation states +IV and +V, CVT is only possible at lower temperatures in the range 450 to 650 °C (see 5.2.5).

According to calculations, transport of the *Magnéli* phases applying sink temperatures higher than 900 °C will lead to the deposition of oxygen-depleted solids. Starting from a two-phase equilibrium solid in the source will lead in a first stationary state to the deposition of oxide with lower oxygen content in the sink. Thus separation of two adjacent *Magnéli* phases becomes possible. Experiments have confirmed these predictions: in a temperature gradient 1000 to 900 °C one observes transport of V_2O_3 before V_3O_5, V_3O_5 before V_4O_7, … , V_8O_{15} before VO_2. Thus, at higher mean temperature the transport of the reduced vanadium oxide (lower in oxygen content) dominates vs. the co-existing oxidized vanadium oxide (e.g., V_2O_3 before V_3O_5). This results from the higher transport efficiency of Te_2, compared to $TeOCl_2$, TeO_2, and TeO.

In contrast, the transport of a two-phase mixture VO_2/V_8O_{15} in the gradient 650 to 550 °C leads to the deposition of VO_2 before V_8O_{15} (Opp 1977c), as the transport efficiency of $TeOCl_2$ rises considerably, see Figure 2.4.2.19.

2.5 Non-stationary Transport Behavior

2.5.1 Chemical Reasons for the Occurrence of Multi-phase Solids in Transport Experiments

In the preceding section the general conditions for the incongruent dissolution of a solid in the gas phase and the occurrence of multi-phase solids (more general condensed phases) in transport experiments have been described. Using rather large amounts of a solid as source material together with sufficiently short experiment duration will yield quasi-stationary (almost independent of time) transport behavior. Thus, deposition of a single-phase solid of constant composition will be possible. Experimentally, about 2 g of source solid should be charged to the ampoule, when transport rates around $1 \text{ mg} \cdot \text{h}^{-1}$ are anticipated. Thus, quasi-stationary transport behavior is accomplished, even for a source solid consisting of several compounds or of a phase exhibiting a homogeneity range. The subsequent section will describe transport experiments that show non-stationary behavior. During their course, indeed the migration of several different solids to the sink will be observed. The sequential migration and deposition of a multi-phase solid is always accompanied by a change of the gas-phase composition in the source and sink. The crystallization via CVT of solids with homogeneity range, too, might be accompanied by a variation of the composition of the sink solid over time. This time dependence can yield the formation of a concentration gradient within the deposited crystals. Experimental evidence for *non-stationary* transport behavior can be obtained from a series of transport experiments allowing for variable duration of the experiments. Much easier experimental access to non-stationary transport behavior is possible by using the so-called transport balance (see section 14.4), which allows continuous monitoring of the mass flow within a sealed ampoule. As an example, the sequential migration of $RhCl_3$ (range b) and Rh_2O_3 (range c) during the CVT of rhodium(III)-oxide by chlorine is presented in Figure 2.5.1.1.

Clearly, the description of this transport experiment has to involve two consecutive stationary states. Thus, the course of the overall experiment is non-stationary.

How a transport experiment proceeds (chemical composition and mass of deposited sink solid; sequence of deposition in the sink) is controlled by thermodynamics and is almost independent of kinetic influences. Therefore, modeling of experiments with non-stationary behavior is well suited for narrowing down the thermodynamic data of the solids involved. In this respect, the choice of a multi-phase equilibrium solid at the source, which from a preparative point of view is not desired, might even help with the thermodynamic analysis of a transport process. Thus estimated thermodynamic data will not be affected by kinetic effects, in contrast to thermodynamic data based on transport rates.

Despite charging a single-phase solid into a transport ampoule, a multi-phase equilibrium solid might form at the source region, due to the setting of chemical equilibrium at the beginning of an experiment (e.g., Figure 2.5.1.1). In the meantime, many examples for such observations have been reported in the literature, in particular in studies by *Gruehn* and co-workers (Table 2.5.1.1).

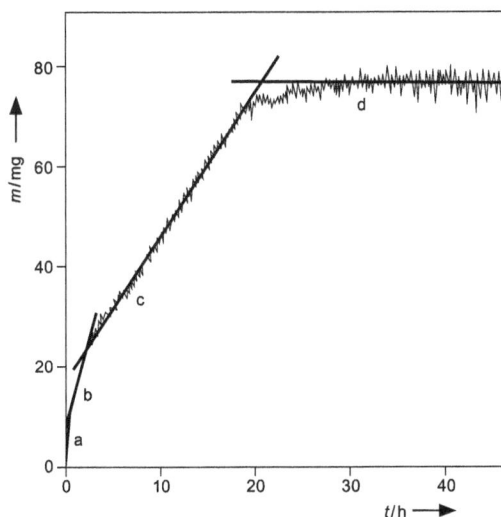

Figure 2.5.1.1 Diagram showing the course of mass vs. time for a typical transport of Rh_2O_3 by chlorine according to *Görzel* and *Glaum* (Gör 1996).
a) Setting of the chemical equilibrium between the solid and gas phase.
b) Sublimation of $RhCl_3$.
c) Transport of Rh_2O_3.
d) Constant mass achieved after complete transport of Rh_2O_3.
   ($1075 \rightarrow 975\,°C$; starting amounts: 75.6 mg Rh_2O_3, 56.7 mg Cl_2)

Formation of *multi-phase* equilibrium solids at the source region of a transport ampoule (sources solid QBK; from the German: "Quellenbodenkörper") can result from three reasons, despite charging the ampoule with a *single-phase* solid (starting material ABK; from the German: "Ausgangsbodenkörper"):

• Reaction between starting material (ABK) and transport agent.
• Thermal decomposition of the starting material (ABK) at the conditions of the transport experiment.
• Reaction between the starting material (ABK) and the ampoule material (possibly involving the transport agent).

Reaction between starting solid and transport agent   Reactions between starting solid and transport agent, which do not exclusively yield gaseous products, can not lead to CVT. Such reactions, however, will change the composition of the solid and the gas phase. For example, in transport experiments with Rh_2O_3 and chlorine (Figure 2.5.1.1) reaction 2.5.1.1 results in the formation of $RhCl_3(s)$ besides the oxide. Note: for this reason it appears appropriate and helpful to distinguish between the terms "starting solid(s) (ABK)", which has been charged into the ampoule, and "source solid(s) (QBK)", which are present in the source region after equilibration at a given time of an experiment.

Table 2.5.1.1 Examples for Non-stationary Transport Experiments Involving Multi-phase Solids.

Source solids (transport agent)	Temperature / °C	Deposition sequence in the sink	Reference
NbO, NbO$_2$ (NbCl$_5$)	1125 → 1025	I) NbO$_2$, II) NbO	(Scha 1974; Gru 1983)
NbO$_2$, Nb$_{12}$O$_{29}$ (NbCl$_5$)	1125 → 1025	I) Nb$_{12}$O$_{29}$, II) NbO$_2$	(Scha 1974; Gru 1983)
H-Nb$_2$O$_5$, FeNb$_{11}$O$_{29}$ (Cl$_2$)	1125 → 1025	I) Nb$_{12}$O$_{29}$, II) NbO$_2$	(Bru 1975; Gru 1983)
H-Nb$_2$O$_5$, AlNb$_{11}$O$_{29}$ (Cl$_2$)	1125 → 1025	I) Nb$_{12}$O$_{29}$, II) NbO$_2$	(Stu 1972; Gru 1983)
Ti$_3$O$_5$, Ti$_4$O$_7$ (HCl)	1000 → 860	I) Ti$_4$O$_7$, II) Ti$_3$O$_5$	(Sei 1984)
WO$_2$, W$_{18}$O$_{49}$ (HgI$_2$)	1060 → 980	I) WO$_2$, II) W$_{18}$O$_{49}$	(Scho 1989)
CuO (I$_2$ [a])	1000 → 900	I) CuO, II) Cu$_2$O	(Tra 1994)
CuO (I$_2$ [b])	1000 → 900	I) CuI, II) CuO	(Tra 1994)
Rh$_2$O$_3$ (Cl$_2$)	1075 → 975	I) RhCl$_3$(s), II) Rh$_2$O$_3$	(Gör 1996)
CrOCl (Cl$_2$)	600 → 500	I) CrCl$_3$, II) CrOCl	(Noc 1993a, Noc 1993b)
CrOCl (Cl$_2$)	900 → 800	I) CrCl$_3$, II) Cr$_2$O$_3$	(Noc 1993a, Noc 1993b)
CoP (I$_2$)	800 → 700	I) CoI$_2$(l), II) CoP	(Schm 1995)
Cr$_2$P$_2$O$_7$, CrP (I$_2$)	1050 → 950	I) Cr$_2$P$_2$O$_7$ + CrP [c], II) CrP	(Lit 2003)
WP$_2$O$_7$, WP (I$_2$)	1000 → 900	I) WP$_2$O$_7$ + WP [d], II) WP	(Lit 2003)
WPO$_5$, WP (I$_2$)	1000 → 900	I) WPO$_5$ + WP [e], II) WP	(Lit 2003)

[a] addition of iodine: 9 mg;
[b] addition of iodine: 140 mg;
[c] simultaneous transport of a two-phasic solid with $n(Cr_2P_2O_7) : n(CrP) \approx 3 : 8$;
[d] simultaneous transport of a two-phasic solid with $n(WP_2O_7) : n(WP) \approx 7 : 4$;
[e] simultaneous transport of a two-phasic solid with $n(WPO_5) : n(WP) \approx 7 : 3$

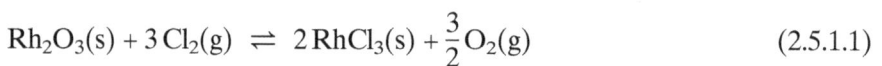

$$Rh_2O_3(s) + 3\,Cl_2(g) \rightleftharpoons 2\,RhCl_3(s) + \frac{3}{2}O_2(g) \qquad (2.5.1.1)$$

In a similar way a melt of cobalt(II)-iodide is formed besides cobalt monophosphide, when aiming at the transport of CoP and using rather high amounts of the transport agent iodine:

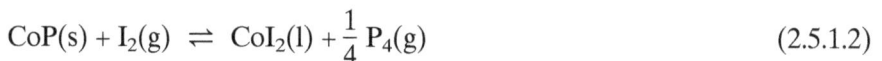

$$CoP(s) + I_2(g) \rightleftharpoons CoI_2(l) + \frac{1}{4}P_4(g) \qquad (2.5.1.2)$$

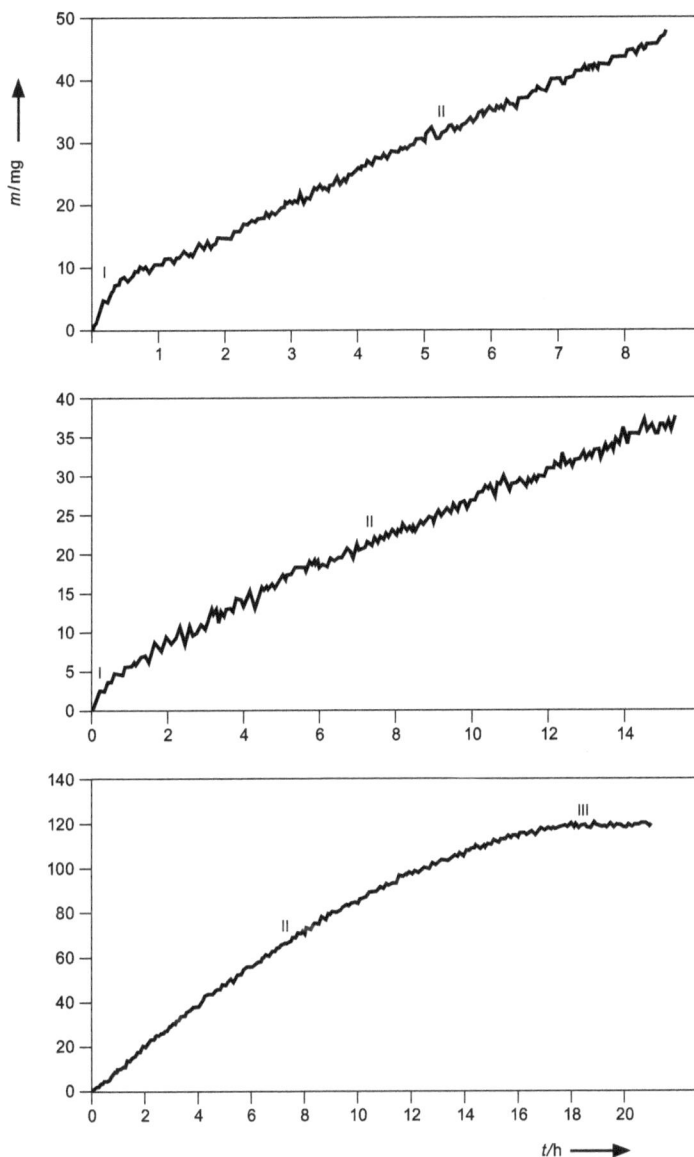

Figure 2.5.1.2 Diagram showing the course of mass vs. time for transport of CoP by iodine according to *Schmidt* and *Glaum* (Schm 1995).
(Starting amounts: 123.3 mg CoP, 300 mg iodine; $V = 20$ cm^3; $q = 2$ cm^2; diffusion length for cobalt monophosphide $s(\text{CoP}) = 9.4$ cm)
a) $700 \rightarrow 600\,°C$; b) $850 \rightarrow 750\,°C$; c) $1050 \rightarrow 950\,°C$.
I) Distillation of CoI$_2$; II) transport of CoP; III) constant mass after complete transport of CoP)

According to expectation, the amount of liquid cobalt(II)-iodide is decreasing with rising temperature due to its increasing vapor pressure (Figure 2.5.1.1).

The equilibria 2.5.1.1 and 2.5.1.2 are superimposed to the corresponding transport reactions and provide typical examples for the incongruent dissolution of solids in the gas phase. Thus, in the system Rh_2O_3/Cl_2 the oxygen content of the equilibrium gas phase is significantly higher $(p^*(O)/p^*(Rh) > 3/2)$ than expected from the congruent dissolution of the oxide in the gas phase according to equation (5.2.9.8). The situation is comparable above the source's equilibrium condensed phases $CoI_2(l)/CoP(s)$. Here we find for the ratio of the balance pressures of phosphorus and cobalt $p^*(P)/p^*(Co) > 1/1$.

Even three condensed equilibrium phases, $MI_2(l)$, $M_2P(s)$, and $MP(s)$, have been observed at the source region in experiments aiming at the transport of Fe_2P and Co_2P, respectively, by iodine:

$$M_2P(s) + I_2(g) \rightleftharpoons MI_2(l) + MP(s) \tag{2.5.1.3}$$
$$(M = Fe, Co)$$

Phosphorus is not evaporated by these reactions but kept in the solid by reaction with the metal-rich phosphides M_2P. The equilibrium partial pressures $p(P_2, P_4)$ are fixed above a solid consisting of two co-existing phosphides. These rather low phosphorus pressures $p(P_2, P_4) \ll 10^{-5}$ prevent the CVT of phosphides M_2P. Furthermore, not even transport of the monophosphides is possible under these circumstances, since the iodine is quantitatively bound as liquid MI_2 and thus not available as a transport agent. Along the same lines observations can be understood, which have been made by experiments aiming at the CVT of reduced vanadium oxides by chlorine.

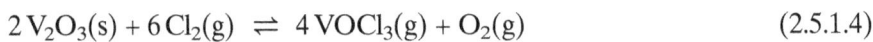

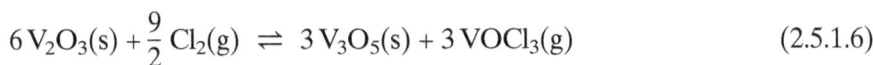

$$2\,V_2O_3(s) + 6\,Cl_2(g) \rightleftharpoons 4\,VOCl_3(g) + O_2(g) \tag{2.5.1.4}$$

$$6\,V_2O_3(s) + O_2(g) \rightleftharpoons 4\,V_3O_5(s) \tag{2.5.1.5}$$

$$6\,V_2O_3(s) + \frac{9}{2}\,Cl_2(g) \rightleftharpoons 3\,V_3O_5(s) + 3\,VOCl_3(g) \tag{2.5.1.6}$$

Chlorine, intended to act as a transport agent, is completely consumed and the gas phase consists almost exclusively of $VOCl_3$. The oxygen, formed by reaction 2.5.1.4, is completely taken away by equilibrium 2.5.1.5 and $p(O_2)$ thus remains far below the pressure necessary for transport. Transport of the *Magnéli* phase will not be possible under such conditions. Reduced titanium oxides behave in the same way (Sei 1983, Sei 1984). In contrast, non- or only weakly oxidizing transport agents (e.g., HCl, $TeCl_4$), which allow the formation of efficient gaseous oxygen carriers other than O_2 (e.g., H_2O, $TeOCl_2$, TeO, TeO_2), are well suited for the transport of these vanadium and titanium oxides (see sections 2.4.2, 5.2.4, and 5.2.5). In such experiments, if proceeding in a non-stationary way, deposition of the oxide with lower oxygen content always occurs prior to that of the oxide richer in oxygen. For transport of oxides involving molecular oxygen as oxygen carrier, the opposite deposition sequence will always be observed. The deposition sequence in these non-stationary transport experiments allows the unequivocal determination of the transport-effective oxygen carrier (Sei 1983, Sei 1984).

Thermal decomposition of the starting solid   Temperatures that lead to the quantitative decomposition of a solid are obviously not suitable for its transport. On the other hand, sometimes CVT experiments have to be carried out close to the

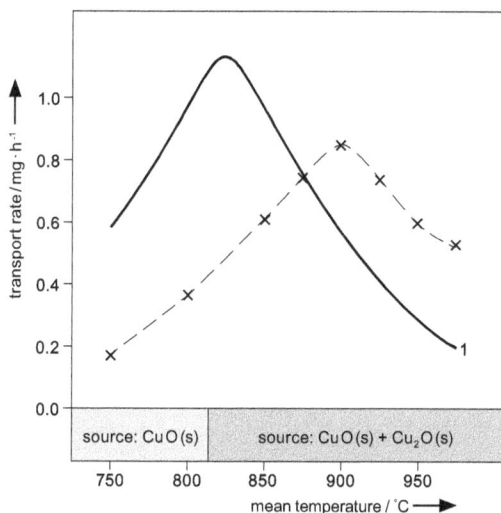

Figure 2.5.1.3 Influence of mean temperature $\bar{T}$ on experimentally observed (–×–) and calculated (——) transport rates of copper(II)-oxide and the composition of the source condensed phase according to *Trappe* (Tra 1994).
(Starting amounts: 500 mg CuO, 9.2 mg I_2, 0.01 mmol H_2O; $\Delta T = 100\,K$; ampoule: $V = 20\ cm^3$; $q = 2\ cm^2$; $s = 10\ cm$)

decomposition temperature of a solid to ensure sufficiently high partial pressures for evaporation. For these conditions, multi-phase solids frequently occur due to partial thermal decomposition. The transport behavior of copper(II)-oxide with iodine has been investigated in considerable detail (Tra 1994). Depending on the experimental conditions four different condensed phases may occur: CuO(s), Cu_2O(s), Cu(s), and CuI(l).

Four different courses have been distinguished for the transport of CuO by iodine:

1. Stationary transport of copper(II)-oxide as the only solid at $\vartheta_{source} \leq 900\,°C$ and a rather small addition of iodine.
2. Stationary transport of copper(II)-oxide and subsequent deposition of copper(I)-oxide in a second stationary state at $\vartheta_{source} \geq 900\,°C$ and a rather small addition of iodine (Figure 2.5.1.3).
3. Distillation of liquid copper(I)-iodide followed by transport of copper(II)-oxide at $\vartheta_{source} \geq 850\,°C$ and higher amounts of iodine (Figure 2.5.1.4).
4. For low starting amounts of copper(II)-oxide and iodine, at higher source temperatures formation of copper might be observed.

In addition to the heterogeneous equilibrium 2.5.1.7, which describes the CVT of copper(II)-oxide by iodine, the composition of the source's condensed phase depends on the temperature and amount of iodine added, as given by reactions 2.5.1.7 to 2.5.1.10. These reactions exhibit for the temperature range under consideration, 700 to 1100 °C, no extreme equilibrium positions, as can be shown by a quantitative thermodynamic treatment. Thereby the observed mutual interdependence of the equilibria is understood. Calculations on the basis of the model of co-operative

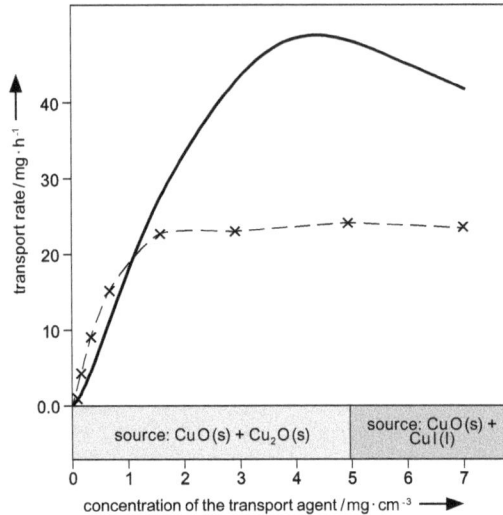

Figure 2.5.1.4 Influence of the transport agent's concentration $c(I_2)$ on experimentally observed (×) and calculated (———) transport rates of copper(II) oxide and the composition of the source condensed phase at $\overline{T} = 950\,°C$ according to *Trappe* (Tra 1994).
(500 mg CuO, variable amount of iodine, 0.01 mmol H_2O; $\Delta T = 100\,K$; ampoule: $V = 20\,cm^3$; $q = 2\,cm^2$; $s = 10\,cm$)

transport (see section 2.5.3) reproduce nicely the experimental observations (composition of condensed equilibrium phase in source and sink, deposition sequence, transport rates) (Tra 1994).

The observations made for the transport of copper(II)-oxide by iodine can serve as an example for the complex phase relations and deposition sequences encountered during the CVT of many other thermally labile compounds. Their transport behavior is characterized by partial thermal decomposition and the formation of condensed metal halides occurring besides the actual transport reaction. In this context we may point out the transport of polyphosphides. Synthesis and crystallization of FeP_4 by chemical transport experiments is always accompanied by the formation of FeP_2 and/or FeI_2 as further condensed phases (Flö 1983). An even more complicated situation is encountered during the CVT of Cu_2P_7 (Möl 1982, Öza 1993). CuP_2, $CuI(l)$, and various copper phosphide iodides, as well as adducts of phosphorus to copper(I)-iodide are observed besides the polyphosphide ($Cu_2P_3I_2$ (Möl 1986), $Cu_3I_3P_{12}$ (Pfi 1995), $Cu_2I_2P_{14}$ (Pfi 1997)). The even more phosphorus-containing phosphide Cu_2P_{20} is accessible by isothermal heating (Lan 2008). Thermodynamic modeling allows an understanding of problems associated with the synthesis of single-phase products. It shows that for directed and reproducible syntheses not only the appropriate molar ratios for the various components (copper, iodine, phosphorus) have to be kept – the absolute amounts of starting materials and the ampoule volume are crucial for successful syntheses, too – since at the synthesis temperature all components are contained at a substantial, however not equal, amount in the gas phase. According to calculations isothermal heating should be preferable over CVT if single-phase products are required. Otherwise, demixing in the temperature gradient might occur (Öza 1993).

$$CuO(s) + \frac{1}{2}I_2(g) \rightleftharpoons \frac{1}{3}Cu_3I_3(g) + \frac{1}{2}O_2(g) \qquad\qquad (2.5.1.7)$$

$$2\,CuO(s) \rightleftharpoons Cu_2O(s) + \frac{1}{2}O_2(g) \qquad\qquad (2.5.1.8)$$

$$Cu_2O(s) \rightleftharpoons 2\,Cu(s) + \frac{1}{2}O_2(g) \qquad\qquad (2.5.1.9)$$

$$CuI(l) \rightleftharpoons \frac{1}{3}Cu_3I_3(g) \qquad\qquad (2.5.1.10)$$

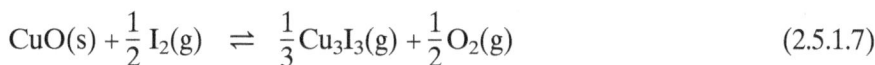

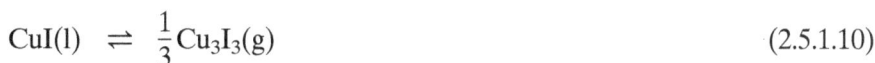

Finally, we would like to mention the chemical transport of anhydrous sulfates (see section 6.1). For this class of compounds, too, many examples are found for partial decomposition due to low thermal stability followed by non-stationary migration of multi-phase solids (e.g., $CuSO_4/Cu_2O(SO_4)/CuO$ (Bal 1983), $Fe_2(SO_4)_3/Fe_2O_3$ (Dah 1992)). The complex, non-stationary transport behavior of rhodium(III)-orthophosphate, which is characterized by the occurrence of several solids $RhPO_4$, $Rh(PO_3)_3$, $RhCl_3(s)$, and Rh, will be explained in detail in section 6.2. Transport with a sequential deposition of phases does also occur with chalkogenides co-existing either with chalkogenide halides or metal halides (see section 7.2 and 8.2).

According to the barogram of the vanadium/oxygen system (Figure 2.4.2.16) the composition of the vanadium oxide deposited in the sink, in the course of CVT using hydrogen chloride or tellurium(IV)-chloride, depends on the choice of the temperature gradient. The co-existence pressure at the source region leads at the lower temperature of the sink to the deposition of an oxide with higher oxygen content. In section 2.4.2 it has been explained that sufficiently large amounts of solid in the source and a rather short transport time will ensure that the transport system will behave in a quasi-stationary way. On the contrary, non-stationary transport behavior will follow, if the source is charged with a rather small amount of material that will be transported completely to the sink during the experiment time. For this case, the deposition at the sink of an oxide of lower oxygen content (with respect to the composition at the source) has to lead to oxygen enrichment of the sink's gas phase (and as a consequence of the source's gas phase and condensed phase, too). Eventually, the ratio of the balance pressures $p^*(O)/p^*(V)$ at the source will increase and cause deposition of the next vanadium oxide with higher oxygen content. Actually, the process might be understood as disproportionation of the source solid by migration in the temperature gradient. Not necessarily, the composition of the sink's solid after complete transport is identical to the composition of the source solid at the start of the experiment. Various effects will be of influence. First of all, a sink temperature different from that of the source might lead, due to thermodynamic reasons, to changes in the occurring condensed phases. Given that such reasons are not present, one should expect identical condensed phases. Nevertheless, synproportionation in the sink might be suppressed, if the setting of the heterogeneous equilibria at the sink is slow in comparison to the migration rate of the solid from the source to the sink. This kinetic effect may lead, in contradiction to Gibbs's phase rule, to the simultaneous occurrence of more than two binary oxides at the sink region. For the *Magnéli* phases of vanadium and titanium, setting of all equilibria appears to be rather fast. Thus, at each point of the experiments the observed solids are in agreement with thermodynamics.

Examples for the dominating influence of kinetics are found in some transport experiments with anhydrous phosphates. Thus, CVT of $Fe_5^{II}V_2^{III}(P_2O_7)_4$ leads to *dismutation* (demixing into phosphates with higher content of $Fe_2P_2O_7$ and those with higher content of $V_4(P_2O_7)_3$) and as a consequence, the simultaneous occurrence of four solid phases $Fe_2P_2O_7$, $Fe_5^{II}V_2^{III}(P_2O_7)_4$, $Fe_3^{II}V_2^{III}(P_2O_7)_3$, and $Fe^{II}V_2^{III}(P_2O_7)_2$ in the sink's solid (see section 6.2.3). Probably along the same lines, the decomposition of $In_2P_2O_7$ into $InPO_4$ and InP might be understood during CVT by iodine in a temperature gradient 800 to 750 °C (Tha 2003).

By the use of mercury(II)-bromide as transport agent, mixed Mo/W crystals can be transported in a temperature gradient 1000 to 900 °C. Apparently, for this case, the transport of tungsten is kinetically inhibited and sets in after some time, only. As a consequence, mixed crystals of Mo/W are obtained, which show a concentration gradient: The crystals' nuclei are rich in molybdenum, at their surface the concentration of tungsten increases (Ned 1996). Similar observations were made for the transport of mixtures of CoO/Ga_2O_3. Crystals deposited at the sink show a concentration gradient for cobalt that is immediately recognized due to the blue color. The crystals inside are blue and formed by the spinel $CoGa_2O_4$, while the parts closer to the surface are colorless (Ga_2O_3) (Loc 2000).

Reaction with the ampoule material A number of examples from literature show that reaction between a solid and the ampoule material might also cause the formation of additional solid phases by a transport experiment. In particular, the formation of silicates and incorporation of SiO_2 are observed. Depending on reaction conditions, SiO_2 as ampoule material might also serve as a source for silicon or oxygen, only. Experimental observations on how such reactions proceed with time are available for the formation of silicates, only (see section 6.3). It is assumed that frequently the reaction of a solid with SiO_2 from the ampoule material has to be interpreted by two simultaneous processes: short distance transport at isothermal conditions at the source, and the simultaneous, yet not coupled deposition of SiO_2 and the starting solid, at the sink of the ampoule. In any case, such reactions are determined by the kinetics of the dissolution of the ampoule material in the gas phase. Thermodynamic modeling will not allow a sufficient understanding of these cases.

2.5.2 The Time Dependence of Chemical Vapor Transport Experiments with Multi-phase Solids

In the preceding sections it has been explained why starting chemical transport experiments with multi-phase source solids can be reasonable. It has been shown that in doing so the deposition of single-phase solids with precisely defined chemical composition might be achieved (see section 7.1, transport of *Chevrel* phases $Pb_xMo_6S_y$ (Kra 1981)). Furthermore, narrowing down of thermodynamic data of compounds (condensed phases as well as gas species) involved in the transport process might be possible on the basis of observations (molar ratio of the solids, deposition sequence, and transport rate of the individual phases) related to the transport behavior of multi-phase solids. Eventually, chemical reasons have been provided for the occurrence of multi-phase equilibrium solids even in experiments where a single-phase starting solid had been charged to the transport ampoule.

For the course with time of a transport experiment involving multi-phase solids various scenarios can be imagined. These shall be visualized subsequently by several examples.

- Sequential migration by a sequence of steady states involving non-stationary behavior (changes of the gas phase with time) during the transition between the steady-state regions: e.g., CuO before Cu_2O (Tra 1994), $CoI_2(l)$ before CoP (Schm 1992), WO_2 before $W_{18}O_{49}$ (Scho 1989), CrOCl before Cr_2O_3 (Noc 1993a, Noc 1993b).
- Simultaneous (parallel) migration of solids by independent transport reactions, however, in one experiment.
- Coupled transport of two (or more) solids by a single heterogeneous reaction: e.g., $Cr_2P_2O_7$/CrP by iodine (Lit 2003), WO_2/$W_{18}O_{49}$ by $SbBr_3$ (Bur 2001).

Sequential transport   The presence of multi-phase solids at the source at the beginning of a transport experiment typically leads to sequential migration of these solids. For a transport experiment with copper(II)-oxide and iodine as transport agent the different phases of the transport with deposition of copper(II)-oxide and copper(I)-oxide are shown in Figure 2.5.2.1.

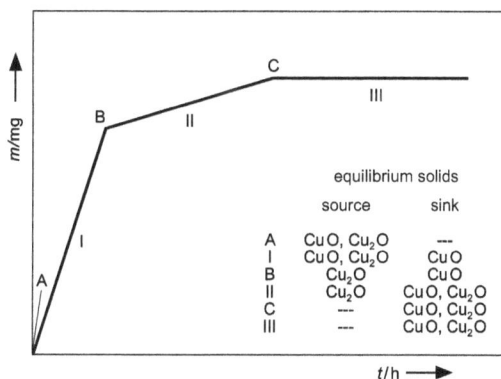

	equilibrium solids	
	source	sink
A	CuO, Cu_2O	---
I	CuO, Cu_2O	CuO
B	Cu_2O	CuO
II	Cu_2O	CuO, Cu_2O
C	---	CuO, Cu_2O
III	---	CuO, Cu_2O

Figure 2.5.2.1   Non-stationary transport behavior of the CuO/I_2 system ($1050 \rightarrow 950\,°C$; 10 mg iodine) according to *Trappe* (Tra 1994).
a) Schematic diagram of mass vs. time showing the transport of copper(II)-oxide (transport section I) and copper(I)-oxide (section II), as well as the completed transport (constant mass; section III).
b) Composition of the condensed phases at source and sink during the various sections of the transport experiment.

After equilibration, at the beginning of the experiment the source solid consists of CuO and Cu_2O in equilibrium with the gas phase. Its content of oxygen is set by equilibrium 2.5.1.7. For the ratio of the partial pressures one finds $p(O_2)_{source}/p(Cu_3I_3)_{source} > 1$. After transfer of this gas phase to the sink, cooling to the sink temperature leads to supersaturation of the gas phase, which eventually results in crystallization of the thermodynamically most stable phase, which is under the given conditions copper(II)-oxide. Dissolution of copper(II)-oxide at the source and its simultaneous deposition at the sink result in the steady state (section I), which is characterized by the following ratio of fluxes from source to sink: $J(O)/J(Cu) = 1$ and $(1/2 \cdot J(O_2))/(1/3 \cdot (Cu_3I_3)) = 1$, respectively. After complete consumption of the copper(II)-oxide at the source, copper(I)-oxide will be dissolved. For a short time, CuO still continues to be deposited at the sink (transition from section I to II). This leads to oxygen depletion from the gas phase. This results eventually in the deposition of copper(I)-oxide at the sink (section II). During section II, the deposition of Cu_2O, the ratio of the fluxes are $J(O)/J(Cu) = 1/2$ and $1/2 \cdot J(O_2))/(1/3 \cdot J(Cu_3I_3)) = 1/2$, since only $O_2(g)$ and $Cu_3I_3(g)$ are effective for the transport.

For the transition from one to next section during the course of a non-stationary transport of cobalt monophosphide by iodine provides real experimental evidence. For certain experimental conditions formation of a CoI_2 melt as an additional condensed phase (besides CoP) is observed. At high starting amounts of iodine and low temperatures $\bar{T}$ liquid cobalt(II)-iodide is already present at the source besides CoP. The reasoning for the sequential deposition given in the system $CuO/Cu_2O/I_2$ can be adapted completely. Figure 2.5.2.2, however, shows that the rate of distillation of CoI_2 seems to depend very strongly on the starting amounts of iodine. This would be very surprising, since the gas phase consists only of CoI_2 and Co_2I_4 as cobalt carriers, and chemical transport of CoI_2 via $CoI_3(g)$ can be excluded. The distillation rate of liquid cobalt iodide, however, passes through pronounced maxima when drawn against the starting amount of iodine (curve 1, curve 2) and rises strongly with $\bar{T}$ (from 650 to 800 °C, curve 1 to curve 3). Thermodynamic modeling shows, that in the ascending parts of curve 1 and curve 2 as well as in the whole range of curve 3 indeed $CoI_2(l)$ is deposited in the sink. The iodide, however, is not present as equilibrium condensed phase at the source. CoP is dissolved at the source; CoI_2 is deposited at the sink. Thus, the gas phase's phosphorus content $p*(P)_{source}/p*(Co)_{source} > 1$ is slowly increased, until at the sink deposition of CoP (instead of CoI_2) becomes more favorable. The migration rate (to avoid the in this case inappropriate expression transport rate) of CoI_2 from source to sink is mainly determined by the partial pressure difference $\Delta p(CoI_2) = p(CoI_2)_{source} - p(CoI_2)_{sink}$. The partial pressure of CoI_2 at the sink is set by the saturation pressure above liquid cobalt(II) iodide (equation 2.5.2.1). At low starting amounts of iodine and/or high temperatures no CoI_2 occurs at the source, $p(CoI_2)$ will be determined exclusively by equilibrium 2.5.2.2. Increasing the starting amount of iodine will shift 2.5.2.2 to higher pressures of $CoI_2(g)$, until eventually its saturation pressure is reached at the source, too. Then, molten iodide will occur besides CoP as the condensed phase. This situation will remain unchanged by even further increased starting amounts of iodine. However, the rate of distillation of CoI_2 will be slightly reduced by the increasing total pressure (Schm 1992).

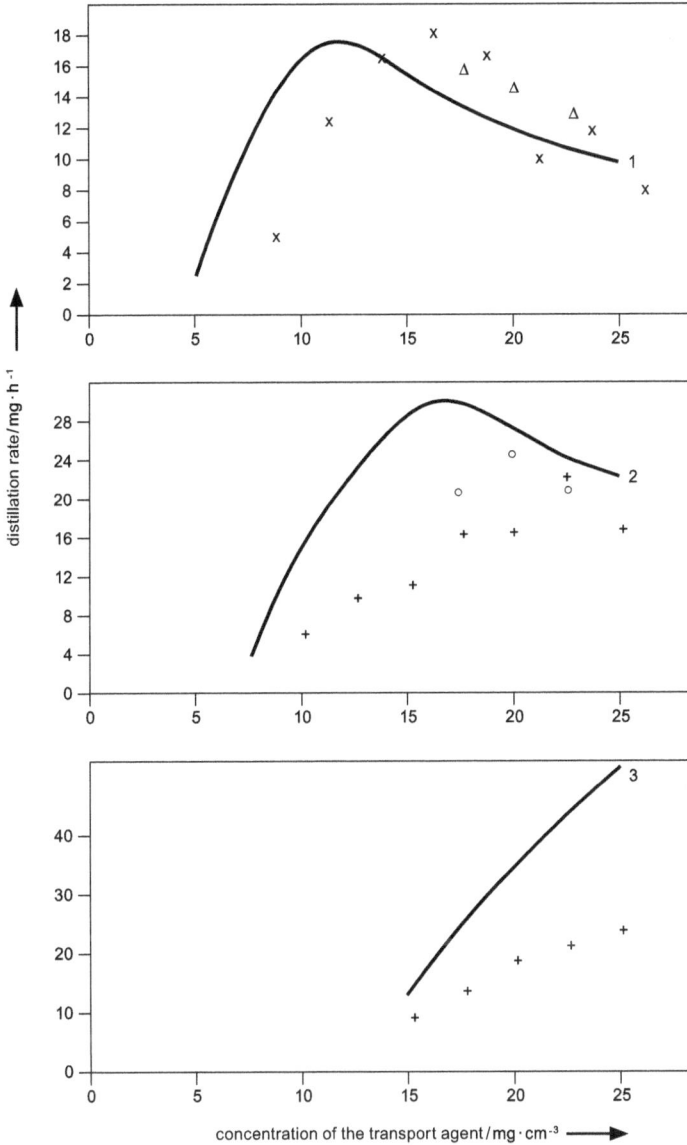

Figure 2.5.2.2 Rate of distillation of CoI_2 as a function of T and the starting amount of iodine, given as concentration of the transport agent $c(I_2)$; experimental (Δ, $\times$, $\circ$, $+$) and calculated (——) values. Graphics according to *Schmidt* (Schm 1992).
(curve 1: $T = 650\,°C$, curve 2: $T = 700\,°C$, curve 3: $T = 800\,°C$; modeling with variable starting amounts of iodine, 0.1 mmol H_2, $\Delta T = 100\,K$; ampoule: $V = 20\ cm^3$; $q = 2\ cm^2$; $s = 10\ cm$)

$$CoI_2(l) \ \rightleftharpoons \ CoI_2(g) \qquad\qquad (2.5.2.1)$$

$$CoP(s) + \frac{5}{2}\,I_2(g) \ \rightleftharpoons \ CoI_2(g) + PI_3(g) \qquad\qquad (2.5.2.2)$$

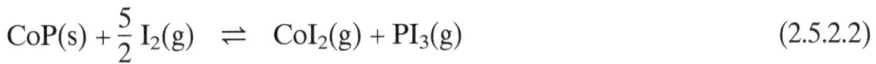

Two-phase solids of the co-existing oxides $WO_2/W_{18}O_{49}$ show remarkable differences in their transport behavior (transport agent HgI_2) upon variation of the mean

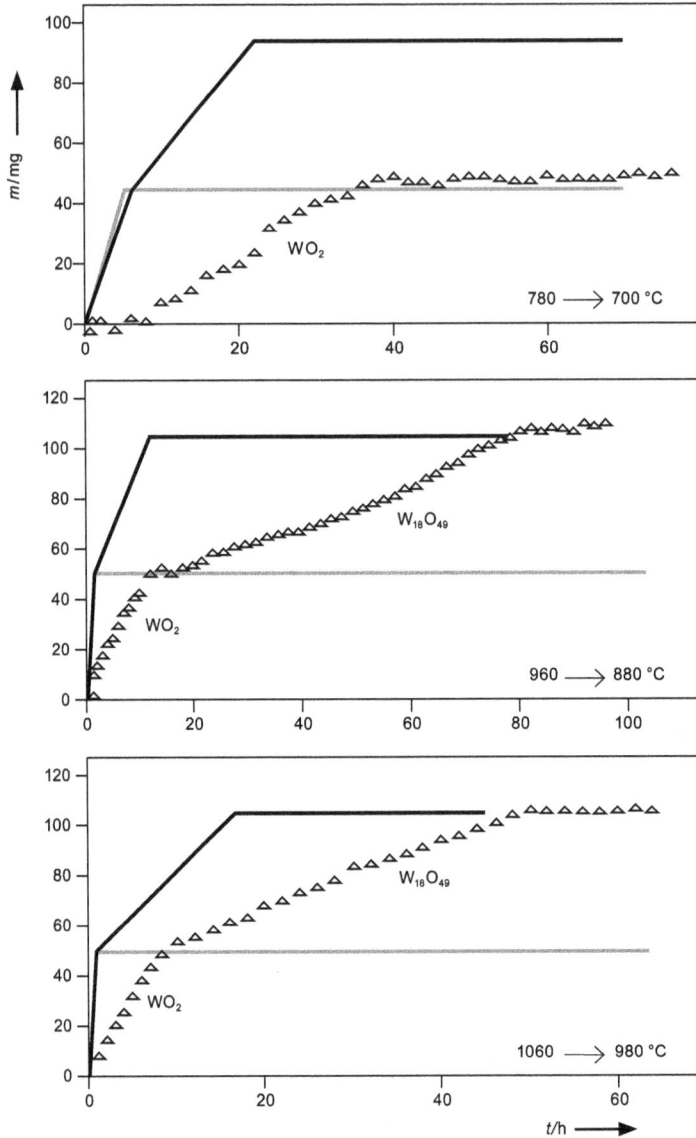

Figure 2.5.2.3 Diagrams of mass vs. time for the transport of $WO_2/W_{18}O_{49}$ by HgI_2 (HI, H_2O) in various temperature gradients according to *Schornstein* and *Gruehn* (Scho 1989); experimental ($\triangle$) and calculated courses of the transport. The latter have been calculated allowing for some moisture (0.005 mmol H_2O (——, black), and without moisture (——, gray).

temperature T and in the presence or absence of traces of moisture (Scho 1989). Experiments using the transport balance (see section 14.4) show, in agreement with thermodynamic modeling, that WO_2 migrates before $W_{18}O_{49}$ to the lower temperature of the sink. The transport rate of WO_2 increases with temperature. Equilibrium 2.5.2.3 is responsible for the transport (see section 15.1). The inhibition

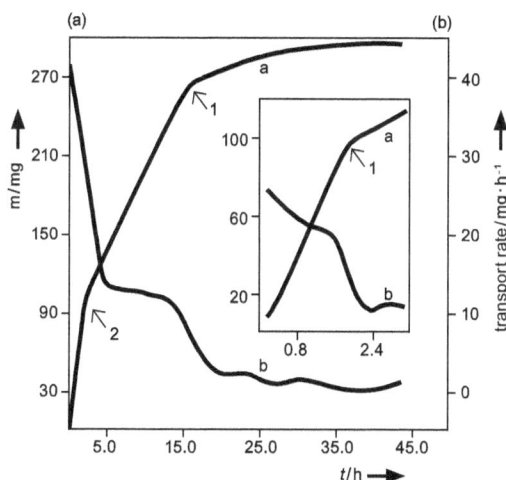

Figure 2.5.2.4 Diagram mass vs. time (a) for a transport experiment with $NiSO_4$ and addition of $PbCl_2$ according to *Plies et al.* (Pli 1989). The inset shows the steady state for deposition of $PbSO_4$ (accompanied by small amounts of $PbCl_2$) at the beginning of the experiment (until point 1). Between point 1 and 2 the transport of $NiSO_4$ takes place. Mass changes after point 2 are caused by migration of $NiSO_4$, $PbSO_4$, and $PbCl_2$ within the sink due to a small temperature gradient in this region.

of the transport of WO_2 at the beginning of the experiment at rather low temperature deserves special attention (780 → 700 °C; Figure 2.5.2.3).

$$WO_2(s) + HgI_2(g) \rightleftharpoons WO_2I_2(g) + Hg(g) \tag{2.5.2.3}$$

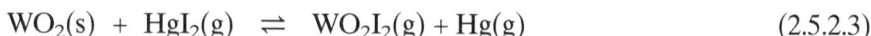

Comparison between experiment and modeling (Figure 2.5.2.3) shows that traces of moisture in the ampoule are responsible for the transport of $W_{18}O_{49}$. In anhydrous systems transport of $W_{18}O_{49}$ by HgI_2 is impossible. In the presence of water, which will be reduced by WO_2 to hydrogen, equations 2.5.2.4 and 2.5.2.5 are responsible for the transport of $W_{18}O_{49}$ in accordance with detailed modeling. Differences, in particular at lower temperatures, between the observed and calculated course of the transport of WO_2 and $W_{18}O_{49}$ are attributed by the authors to kinetic inhibition of the oxide deposition.

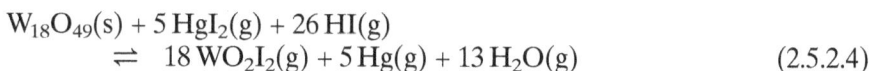

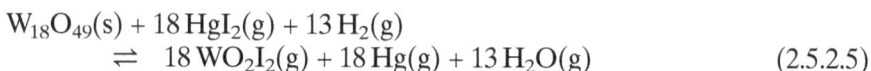

$$W_{18}O_{49}(s) + 5\,HgI_2(g) + 26\,HI(g)$$
$$\rightleftharpoons 18\,WO_2I_2(g) + 5\,Hg(g) + 13\,H_2O(g) \tag{2.5.2.4}$$

$$W_{18}O_{49}(s) + 18\,HgI_2(g) + 13\,H_2(g)$$
$$\rightleftharpoons 18\,WO_2I_2(g) + 18\,Hg(g) + 13\,H_2O(g) \tag{2.5.2.5}$$

An example of transport involving equilibrium solids of particular complex composition and sequential deposition of these solids in the sink is provided by the CVT of $NiSO_4$ with the transport additive (not the transport agent!) $PbCl_2$ (Pli 1989). Equilibration according to reaction 2.5.2.6 yields at the source $PbSO_4$ and NiO besides $NiSO_4$ and $PbCl_2$. Chlorine, formed by the same reaction, eventually acts as the transport agent according to equilibrium 2.5.2.7.

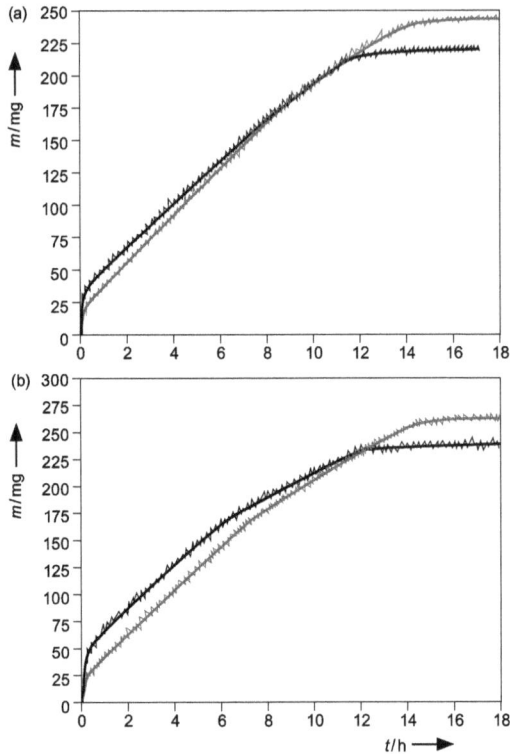

Figure 2.5.2.5 Diagram of mass vs. time for the coupled transport of $Cr_2P_2O_7$ and CrP by iodine as transport agent $(1050 \to 950\,°C)$ according to (Lit 2003).
a) Molar ratio $n(Cr_2P_2O_7) : n(CrP) = 3 : 10.53$ (two experiments)
b) $n(Cr_2P_2O_7) : n(CrP) = 3 : 4.40$ (two experiments)

$$2\,NiSO_4(s) + PbCl_2(l)$$
$$\rightleftharpoons\; PbSO_4(s) + 2\,NiO(s) + SO_2(g) + Cl_2(g) \qquad (2.5.2.6)$$

$$NiSO_4(s) + Cl_2(g)$$
$$\rightleftharpoons\; NiCl_2(g) + SO_3(g) + \frac{1}{2}O_2(g) \qquad (2.5.2.7)$$

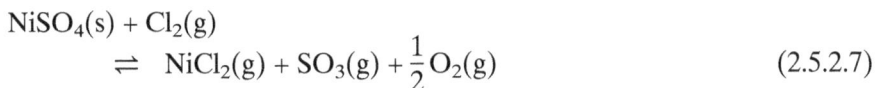

Depending on the experimental conditions, the occurrence of up to four condensed equilibrium phases ($NiSO_4$, $PbSO_4$, NiO, $PbCl_2$) can be observed (see section 6.1). Figure 2.5.2.4 shows a diagram of mass vs. time registered with transport balance for the chemical transport of $NiSO_4$ by the addition of $PbCl_2$.

Coupled transport of two and more condensed phases A peculiarity found for the transport of multi-phase solids provides transport systems with two or more solids transported simultaneously at a fixed molar ratio to the sink. This situation will be encountered if the transport of two or more condensed phases is "coupled" due to a single transport equilibrium. As an example we consider the coupled chemical transport of $Cr_2P_2O_7$ and CrP, which is caused by the synproportionation of P^{5+} and P^{3-} leading to P_4O_6 in the gas phase.

$$3\,Cr_2P_2O_7(s) + 8\,CrP(s) + 14\,I_2(g) \;\rightleftharpoons\; 14\,CrI_2(g) + \frac{7}{2}P_4O_6(g) \quad (2.5.2.8)$$

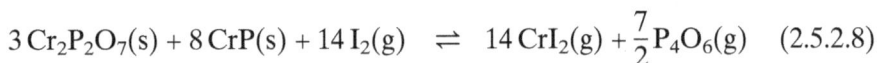

The molar ratio phosphide : phosphate in the transport reaction is determined by the synproportionation to gaseous P_4O_6. For starting amounts of $Cr_2P_2O_7$ and CrP at a molar ratio 3 : 8 both compounds should migrate in a single steady state from the source to the sink. This expectation is confirmed by the diagrams of mass vs. time of the respective transport experiments (Figure 2.5.2.5a). Starting amounts with a molar ratio $n(Cr_2P_2O_7) : n(CrP) = 3 : 4.40$ (a surplus of $Cr_2P_2O_7$ with respect to equilibrium 2.5.2.8) result in two consecutive steady states. At first, migration of $Cr_2P_2O_7$ and CrP at a ratio $n(Cr_2P_2O_7) : n(CrP) = 3 : 8$ is observed, followed by migration of the surplus of diphosphate to the sink.

Simultanenous transport It is also possible that two or more solids are independently transported in one ampoule at the same experimental conditions. The prerequisite for such a behavior is that the two solids do not react with each other; at the conditions of the transport the phase diagram should neither show any mutual miscibility nor multinary compounds. A further condition for the independent, simultaneous transport of two phases is that both solids react to a similar extent with the transport agent. According to *Emmenegger* the system NiO/SnO_2 provides an example for simultaneous, independent transport. The two binary oxides are transported by chlorine (Emm 1968).

2.5.3 The Model of Co-operative Chemical Vapor Transport

Obviously, it has to be the ultimate target of modeling to reproduce the experimental observations made during a CVT experiment as accurately as possible. For the modeling, observables that exclusively depend on the thermodynamic properties of the involved condensed phases and gas species have to be distinguished from those that are influenced or even determined by "non-thermodynamic" effects. The composition of the equilibrium solid(s) and its solubility in the gas phase belong to the former, whereas transport rates belong to the latter. These are, by the diffusion equation of *Schäfer* (see section 2.6), related to the partial pressure differences between source and sink and hence to the thermodynamic properties of the system. Nevertheless, transport rates might be significantly affected by kinetic effects as well as by mass transfer via thermal convection. Therefore, it is advisable to distinguish, when modeling transport experiments, between the calculation of the phase relations in the source and sink, which depend exclusively on thermodynamics and such calculations that, based on thermodynamics, describe the rate of the mass transfer between the equilibrium zones.

The model of co-operating equilibrium zones ("model of co-operative transport") The calculation of the equilibrium solid(s) and gas phase in source and sink uses the *model of co-operative transport* (Schw 1983a, Gru 1983) and involves minimization of the Gibbs energy according to *Eriksson* (G_{min} method (Eri 1971), see section 13.2) for the two equilibrium regions of the transport ampoule.

The experimental conditions (temperatures, ampoule volume, amounts of starting materials) as well as the thermodynamic data ($\Delta_f H_T^0$, S_T^0, $C_p^0(T)$) for all gas species and condensed phases are required as input for the calculation. This procedure bears several advantages over the more commonly known K_p method for calculation of equilibrium pressures (and solids). First of all, equilibrium states involving multiple condensed phases can be treated in a simple way without any premises. The equilibrium phases follow from the set of all possible phases of a system just from their thermodynamic stability. In addition, the G_{min} procedure can easily be automated (see section 13.2).

The main conceptual problem in modeling CVT experiments lies in the linking of the equilibrium calculations for the source and sink regions. The restriction to identical single-phase solids in both equilibrium regions (congruent transport behavior; see section 2.2 and 2.3) allows far-reaching simplification of the calculations. As side condition only Σp(source) = Σp(sink) has to be allowed for. In section 2.4.2 it has been described how the *extended transport model* can be applied to incongruent evaporation (and deposition) of solids in quasi-stationary transport experiments.

The target of the model of co-operative transport by *Schweizer* and *Gruehn* (Schw 1983a, Gru 1983) is the complete (time dependent) description of non-stationary transport behavior. For this purpose, linking the equilibrium regions to each other is crucial. In using the computer program CVTRANS, which is based on the model of co-operative transport this is done in the following way (see also Figure 2.5.3.1). In a first step, accounting for all compounds and the molar amounts of the systems components, an isothermal–isochoric equilibrium calculation (iteratively, G_{min}) for the temperature of the source and the given ampoule volume is performed. In the second step, the thus obtained equilibrium gas phase is transferred to the sink. Its element balance is used in the third step for an isothermal–isobaric equilibrium calculation for the sink temperature, allowing for the side condition Σp(source) = Σp(sink). From the so-called "sink calculation" results if and which condensed phases will be deposited. For the sink calculation, the balance of the molar amounts of the elements contained in the condensed and gas phase is equal to the corresponding element balance in the source's gas phase (see section 2.4.2). This procedure gets by without explicit balancing of the fluxes of the individual gas species, in contrast to the flux relation by *Krabbes, Oppermann,* and *Wolf* (Kra 1975, Kra 1976a, Kra 1976b, Kra 1983). As a consequence, the procedure is restricted to a mean diffusion coefficient. Individual values for the various gas species cannot be accounted for. For modeling of the transport of condensed phases without homogeneity range this slightly simpler approach leads to the same results as the application of the flux relation (section 2.4). The flow chart for the described calculation is shown in Figure 2.5.3.1.

When modeling steady-state transport of single-phase solids, the cycle of calculation steps 1 to 3 will be traversed only once. After this cycle, the system is checked for stationarity by comparison of the solid(s) in the source and sink. Stationary transport behavior is assumed if they are chemically identical. Chemically different condensed phases in the source and sink indicate non-stationary transport behavior.

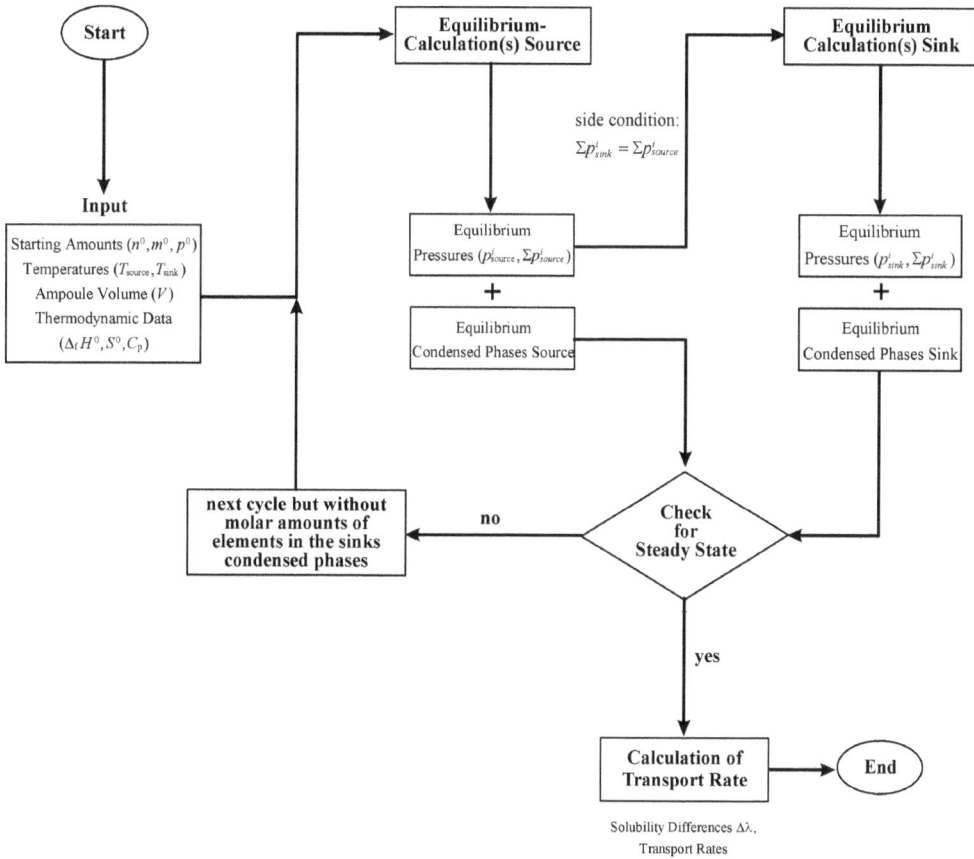

Figure 2.5.3.1 Flow-chart for calculations on the basis of the model of co-operative transport.

In a rather simple manner the model of co-operative transport offers a way to simulate the time-dependent mass transfer during a transport experiment. For this purpose, the equilibrium condensed phase(s) of the source and sink obtained by a calculation cycle are kept at these regions (Figure 2.5.3.1). The source calculation of the subsequent cycle is performed without the molar amounts of the elements deposited at the sink in the preceding cycle. The stepwise ("cyclewise") transfer of the source solid(s) to the sink is simulated by repeated calculation cycles. The calculation is finished once no condensed phase is left at the source. Alternatively, the calculation is terminated when the source solid's composition remains stable from one cycle to the next – only the molar number of the solid is decreased. According to the stationarity criterion a steady state has been reached when the gas phase (in the source and sink) remains constant from one calculation cycle to the next. With respect to the mass transfer from the source to the sink, this means that within one calculation cycle the molar numbers of the source solid's components dissolved in the gas phase and deposited at the sink are equal. If more than one condensed phase is involved in this process, we find simultaneous transport.

Eventually, for calculation of the transport rates, the diffusion equation by *Schäfer* (see section 2.6) can be applied. Solubility in the gas phase, as obtained via the model of co-operative transport is used for this purpose.

2.6 Diffusion, Stoichiometric Flow, and Transport Rate

2.6.1 Steady-state Diffusion

The subsequent considerations are confined to the simple and most frequently occurring situation where diffusion takes place along the axis of a transport ampoule. Diffusion between the axis and the wall of the tube, however, can also be easily visualized. Arrangements with a hot filament can thus be designed. For the calculation of the transport by diffusion in such an apparatus the reader is referred to the publications by *Shapiro* (Sha 1955) and *Oxley* and *Blocher* (Oxl 1961).

The shape of the theoretically ideal ampoule is shown by Figure 2.6.1.1. It is composed of the two equilibrium regions 1 and 2 and of the diffusion path between them. As an approximation a simple cylindrical ampoule can also be used either in a horizontal or in a vertical position. As already introduced in Chapter 1, the ampoule region containing the starting material is referred to as the "source region" or source, the deposition region is called the "sink".

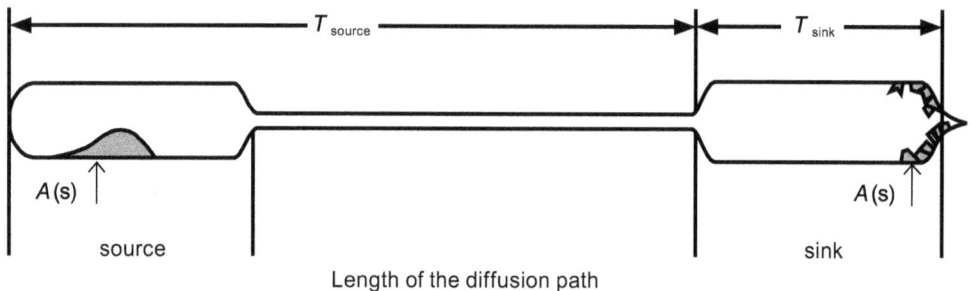

Figure 2.6.1.1 Transport of condensed (solid or liquid) substances with gas motion by diffusion in a temperature gradient $\Delta T = T_{source} - T_{sink}$. Theoretically ideal shape and horizontal position of the tube; the theoretical treatment is based on this arrangement.

2.6.2 One-dimensional Steady-state Diffusion as Rate-determining Step

The model subsequently described is based on the following assumptions (Sch 1962, Sch 1956):

- ΔT is small compared to T.
- No homogeneous reaction takes place within the diffusion path.
- The concentration (partial pressure) gradient is linear.

- The steady state is established quickly.
- The mean free path-length of the molecules is small compared to the dimensions of the apparatus (the transport ampoule).

If a simple heterogeneous reaction (2.6.2.1) proceeds with a change in the number of moles of the gaseous molecules ($k \neq j$), then the requirement of a uniform total pressure in the whole ampoule will give rise to a flow of the entire gaseous mass.

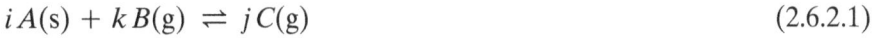

$$i A(s) + k B(g) \rightleftharpoons j C(g) \qquad (2.6.2.1)$$

This stoichiometric (laminar) flow is combined with the diffusion. The transport rate $\dot{n}$ (mol $\cdot$ sec^{-1}) as well as the transported quantity n (mol) is therefore given by equation 2.6.2.2 or 2.6.2.3. Equation 2.6.2.2 describes the transport by looking at the negative concentration gradient of the transport agent B. Equation 2.6.2.3 expresses the same facts by means of the transport-effective species C, that migrates to the deposition region.

$$\dot{n}(A) = \frac{n(A)}{t} = \frac{i \cdot n(B)}{k \cdot t}$$

$$= -\frac{i}{k} \cdot D \cdot q \cdot \frac{dc(B)}{ds} - \frac{i}{k} \cdot q \cdot c(B) \cdot W \qquad (2.6.2.2)$$

$$\dot{n}(A) = \frac{n(A)}{t} = \frac{i \cdot n(C)}{j \cdot t}$$

$$= +\frac{i}{j} \cdot D \cdot q \cdot \frac{dc(C)}{ds} + \frac{i}{j} \cdot q \cdot c(C) \cdot W \qquad (2.6.2.3)$$

i, k, j	stoichiometric coefficients
$c(B), c(C)$	concentrations
D	diffusion coefficient under experimental conditions (Σp and T are constant)
q	cross section of the diffusion path
s	length of the diffusion path
t	time
W	flow of the gas phase as a whole

The flow caused by different values of k and j ($k \neq j$) can also be expressed by diffusion, as a combination of equations 2.6.2.2 and 2.6.2.3 show:

$$W = -\frac{n(B)}{q \cdot t \cdot c(B)} - D \cdot \frac{dc(B)ds}{c(B)} = \frac{n(C)}{q \cdot t \cdot c(C)} - D \cdot \frac{dc(C)ds}{c(C)} \qquad (2.6.2.4)$$

With $\dfrac{dc(B)}{ds} = -\dfrac{dc(C)}{ds}$ eventually equations 2.6.2.5 and 2.6.2.6 result, where a term in braces { } means the "**flow factor**" F.

$$\dot{n}(A) = -\frac{i}{k} \cdot D \cdot q \cdot \frac{dc(B)}{ds} \cdot \left\{ \frac{k(c(B) + c(C))}{j \cdot c(B) + k \cdot c(C)} \right\} \tag{2.6.2.5}$$

$$\dot{n}(A) = \frac{i}{j} \cdot D \cdot q \cdot \frac{dc(C)}{ds} \cdot \left\{ \frac{j(c(B) + c(C))}{j \cdot c(B) + k \cdot c(C)} \right\} \tag{2.6.2.6}$$

The concentrations are transformed into pressures by using the ideal gas law ($p \cdot V = n \cdot R \cdot T, p = c \cdot R \cdot T$). Under the conditions of steady state, and for small ΔT, one can introduce $\Delta p/s$ instead of dp/ds, an average temperature $\overline{T}$ along the diffusion path, and an average pressure value $\overline{p} = (p_{T_1} + p_{T_2})/2$. These assumptions are simplifications. One now obtains equations 2.6.2.7 and 2.6.2.8.

$$\dot{n}(A) = -\frac{i}{k} \cdot \frac{D \cdot q}{R \cdot \overline{T}} \cdot \frac{\Delta p(B)}{s} \cdot \left\{ \frac{k(\overline{p}(B) + \overline{p}(C))}{j \cdot \overline{p}(B) + j \cdot \overline{p}(C)} \right\} \tag{2.6.2.7}$$

$$\dot{n}(A) = \frac{i}{j} \cdot \frac{D \cdot q}{R \cdot \overline{T}} \cdot \frac{\Delta p(C)}{s} \cdot \left\{ \frac{j(\overline{p}(B) + \overline{p}(C))}{j \cdot \overline{p}(B) + k \cdot \overline{p}(C)} \right\} \tag{2.6.2.8}$$

The term in braces in 2.6.2.7 and 2.6.2.8 is the *individual flow factor F*. It is determined by the stoichiometric coefficients of the transport reaction 2.6.2.1. The further variables in equations 2.6.2.7 and 2.6.2.8 are determined by the experimental conditions.

In a transport reaction without a change in the numbers of the gas molecules ($k = j$) the flow factor has the value $F = 1$. For $k \neq j$ one can take $F = 1$ as a good approximation, if one uses the component with the lower partial pressure for the calculation of the diffusion.

In case $\overline{p}(B) < \overline{p}(C)$ then $F \approx 1$ in equation 2.6.2.7   and

if $\overline{p}(B) < \overline{p}(C)$ then $F \approx 1$ in equation 2.6.2.8.

The difference of both calculations, equations 2.6.2.7 and 2.6.2.8, is determined by the ratio k/j. This shows that the influence of the flow factor is relatively small.

If one assumes $F = 1$ and in addition uses the temperature dependence of the diffusion coefficient, which has been determined semi-empirically by *Blanck* (Jel 1928):

$$D = D^0 \cdot \frac{\Sigma p^0}{\Sigma p} \cdot \left(\frac{\overline{T}}{T^0} \right)^{1.75} \tag{2.6.2.9}$$

with $\Sigma p^0 = 1$ bar and $T^0 = 273$ K, one obtains from equation 2.6.2.8 the equation 2.6.2.10. Typically, the diffusion coefficient is given with the dimensions $cm^2 \cdot s^{-1}$.

$$\dot{n}(A)' = \frac{n(A)}{t'} = \frac{i}{j} \cdot \frac{\Delta p(C)}{\Sigma p} \cdot \frac{D^0 \cdot \overline{T}^{0.75} \cdot q}{273.15^{1.75} \cdot s \cdot R} \tag{2.6.2.10}$$

Equation 2.6.2.11 is frequently referred to as *Schäfer's transport equation*. After combining all constants it describes the transport rate with the more appropriate dimension mol · h^{-1}. R has to be used with the dimension cm^3 · bar · K^{-1} · mol^{-1}.

$$\dot{n}(A) = \frac{n(A)}{t'}$$

$$= \frac{i}{j} \cdot \frac{\Delta p(C)}{\Sigma p} \cdot \frac{D^0 \cdot \overline{T}^{0.75} \cdot q}{s} \cdot 2.4 \cdot 10^{-3} \, (\text{mol} \cdot \text{h}^{-1}) \qquad (2.6.2.11)$$

$\dot{n}(A)$:   transport rate for condensed phase A
i, j: stoichiometric coefficients
$\Delta p(C)$:   difference of partial pressures for gas species C (bar)Σ
p: total pressure in the transport ampoule (bar)
D^0:   mean diffusion coefficient (0.025 cm^2 · s^{-1})
T: average temperature along the diffusion path (K)
q: cross section of the diffusion path (cm^2)
s: length of the diffusion path (cm)
t':   transport time (h)

Using $s = 10$ cm, $q = 2$ cm^2, $t' = 1$ h, $T = 1000$ K, $\Sigma p = 1$ bar, $i/j = 1$, $D^0 = 0.025$ cm^2 · sec^{-1}, and $\Delta p = 1 \cdot 10^{-3}$ bar, one calculates $\dot{n}(A) = 2.1 \cdot 10^{-6}$ mol · h^{-1} = 2.1 · 10^{-3} mmol · h^{-1}. For the molar mass $M(A) = 100$ g · mol^{-1} the calculation eventually leads to the transport rate of 2.1 · 10^{-1} ≈ 0.2 mg · h^{-1}. This can be assumed (under the conditions mentioned above) to be near the lower limit of transport effect for practical applications in the laboratory. The usefulness of equation 2.6.2.11 for estimation of the transport effect has been proven by many experiments reported in the literature. As examples one may refer to the systems NiO/HCl, Fe$_2$O$_3$/HCl, and NiFe$_2$O$_4$/HCl (Kle 1969).

Chemical transport reactions may differ in their transport rates by many orders of magnitude. The accuracy with which the transported quantities can be calculated depends on the prevailing experimental conditions, the precision of the employed thermodynamic data, and knowledge of the involved chemical equilibria. For many practical purposes it is usually sufficient to be able to calculate the correct order of magnitude of the transport effect. Errors by a factor of 2 or 3 are frequently encountered. For this reason, the use of more exact models often seems not justified, as will be seen in section 2.6.4.

2.6.3 The Use of λ for the Calculation of Transport Rates of Complex Transport Systems in Closed Ampoules

For simple systems the transport rate, which is determined by diffusion and stoichiometric flow, can be calculated from known partial pressure differences Δp (section 2.6.2). Instead of the partial pressure difference Δp, for systems of higher complexity the difference of solubilities $\Delta \lambda = \lambda_{T_{\text{source}}} - \lambda_{T_{\text{sink}}}$ (dimensionless) multiplied by the normalization pressure $\overline{p^*(L)}$ is advised (equation 2.6.3.1). In doing so, it has to be kept in mind, to define the solvent L in the same way as it was done for the definition of λ (see section 2.4).

$$\Delta p = \lambda_{T_{\text{source}}} \cdot p^*(L)_{T_{\text{source}}} - \lambda_{T_{\text{sink}}} \cdot p^*(L)_{T_{\text{sink}}}$$
$$\approx (\lambda_{T_2} - \lambda_{T_1}) \cdot \overline{p^*(L)} \tag{2.6.3.1}$$

This relation already allows for the contribution of the stoichiometric flow in systems where the transport reaction shows a change of the number of gas molecules. The use of the arithmetic average $\overline{p^*(L)}$ is an approximation, commonly used for practical calculations.

$$\overline{p^*(L)} = 0.5 \cdot (p^*(L)_{T_{\text{source}}} + p^*(L)_{T_{\text{sink}}}) \tag{2.6.3.2}$$

The value of λ comprises reversible and irreversible solubility. When calculating the solubility differences the irreversible solubility will vanish (see section 2.4).

We consider the system $Si(s)/SiI_4(g)$, $SiI_2(g)$, $I_2(g)$, $I_1(g)$, to explain the correlation between the solubility λ (of a condensed phase) in the gas phase and the balance pressure p^* introduced earlier by *Schäfer*. The amount of silicon that is dissolved in the gas phase at T_{source} and re-deposited at the sink at temperature T_{sink} depends on the gas phase concentration of SiI_2, I_2, and I. Only the fraction of SiI_2 that will not be consumed by reaction with I_2 and I (leading to SiI_4) at T_{sink} will disproportionate according to the following reaction:

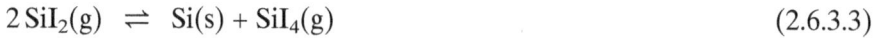

$$2\, SiI_2(g) \ \rightleftharpoons \ Si(s) + SiI_4(g) \tag{2.6.3.3}$$

Following this reasoning, the balance pressure p^* (equation 2.6.3.4) is obtained, which represents the amount of Si *reversibly* dissolved in the gas phase.

$$p^*(\text{Si})_{T_{\text{source}}} = 0.5 \cdot (p(SiI_2) - p(I_2) - 0.5\, p(I))_{T_{\text{source}}} \tag{2.6.3.4}$$

Accordingly, we arrive at equation 2.6.3.5:

$$\Delta p^*(\text{Si}) = p^*(\text{Si})_{T_{\text{source}}} - p^*(\text{Si})_{T_{\text{sink}}} = \Delta\lambda(\text{Si}) \cdot \overline{p^*(\text{I})}$$
$$= (\lambda(\text{Si})_{T_{\text{source}}} - \lambda(\text{Si})_{T_{\text{sink}}}) \cdot \overline{p^*(\text{I})} \tag{2.6.3.5}$$

Chemical vapor transport will always proceed from the region of higher solubility ("source") to that with lower solubility ("sink"). The solubility of silicon in gaseous iodine vs. temperature and pressure (Figure 2.6.3.1) passes through a maximum. Thus, at temperatures above the maximum, transport will proceed from lower to higher temperature ($T_1 \rightarrow T_2$), at temperatures below the maximum from T_2 to T_1. Such a behavior is called **reversal of transport** (direction). For temperatures of source and sink around the solubility maximum no transport will be observed. In a series of papers by *Schäfer*, e.g., on metal/iodine systems, the expression $p^*(M)$ instead of the solubility $\lambda = p^*(M)/p^*(X)$ has been used as a simplification. This is acceptable, if $p^*(X)$ is large and almost constant. For this condition one may assume $\Delta\lambda \approx \Delta p^*(M)$.

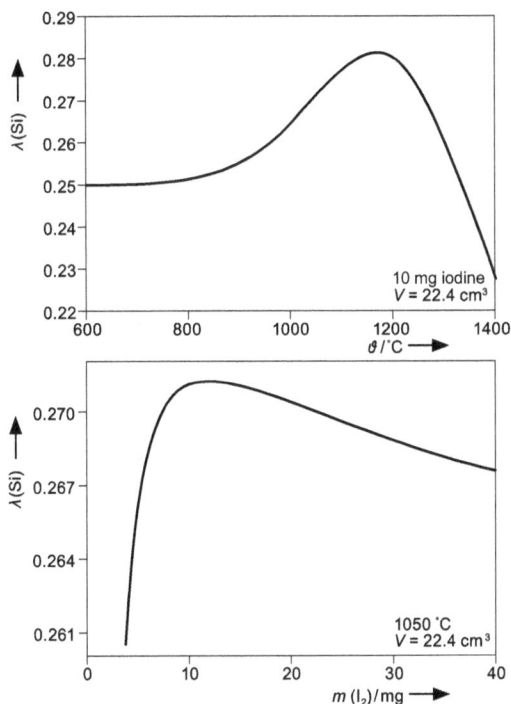

Figure 2.6.3.1 Solubility of silicon in gaseous iodine.

2.6.4 The Use of Solubility λ in Open Transport Systems

Dissolution and deposition of a condensed phase in an open system ("flow tube") are depicted in Figure 2.6.4.1. In the case of equilibrium, the transport rate is proportional to the flow rate and to the difference of the solubility λ of the condensed phase.

For open systems (in contrast to closed ampoules) the correlation between the transport rate and the solubility will be demonstrated using as example the transfer of iron(III)-oxide into the gas phase in flowing hydrogen chloride. Considering seven gas species requires four unique chemical equations according to $r_u = 7 - 4 + 1 = 4$. Equations 2.6.4.1 to 2.6.4.4 can be used for this purpose.

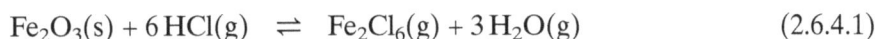

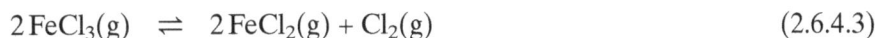

$$Fe_2O_3(s) + 6\,HCl(g) \rightleftharpoons Fe_2Cl_6(g) + 3\,H_2O(g) \tag{2.6.4.1}$$

$$Fe_2Cl_6(g) \rightleftharpoons 2\,FeCl_3(g) \tag{2.6.4.2}$$

$$2\,FeCl_3(g) \rightleftharpoons 2\,FeCl_2(g) + Cl_2(g) \tag{2.6.4.3}$$

$$Cl_2(g) \rightleftharpoons 2\,Cl(g) \tag{2.6.4.4}$$

Figure 2.6.4.1 Flow tube.

For the solubility of iron $\lambda(Fe)$ equations 2.6.4.5 and 2.6.4.6, respectively, are obtained.

$$\lambda(Fe) = \frac{2 \cdot p(Fe_2Cl_6) + p(FeCl_3) + p(FeCl_2)}{p(HCl) + 6 \cdot p(Fe_2Cl_6) + 3 \cdot p(FeCl_3) + 2 \cdot p(FeCl_2) + 2 \cdot p(Cl_2) + p(Cl)}$$

(2.6.4.5)

$$\lambda(Fe) = \frac{\Sigma(v(Fe \cdot p(Fe))}{\Sigma(v(Cl) \cdot p(Cl))} = \frac{p^*(Fe)}{p^*(Cl)} = \frac{n^*(Fe)}{n^*(Cl)}$$

(2.6.4.6)

Allowing for the flow rate $f = n^*(Cl)/t$ yields the transport rate according to 2.6.4.7:

$$\dot{n} = [\lambda(Fe)_T - \lambda(Fe)_T] \cdot f = \frac{[\lambda(Fe)_T - \lambda(Fe)_T] \cdot n^*(Cl)}{t}$$

(2.6.4.7)

There is no difference, whether $\lambda(Fe)$ or $\lambda(O)$ is used to describe the solubility of Fe_2O_3 in the gas phase. In the same way, for the calculation the choice of the solvent element (H or Cl) has no influence, as long as all calculations are carried out consistently. The equality can be seen from equations 2.6.4.8 and 2.6.4.9 for the aforementioned system Fe_2O_3/HCl.

$$\lambda'(Fe) = \frac{2 \cdot p(Fe_2Cl_6) + p(FeCl_3) + p(FeCl_2)}{p(HCl) + 2 \cdot p(H_2O)} = \frac{\Sigma(v(Fe) \cdot p(Fe))}{\Sigma(v(H) \cdot p(H))} = \frac{n^*(Fe)}{n^*(H)}$$

$$\lambda(Fe) \neq \lambda'(Fe)$$

(2.6.4.8)

For the latter case, the flow rate has to be expressed by $f' = n^*(H)/t$ and for the transport rate $\dot{n}$ follows $\dot{n} = (\lambda'(Fe)_{T_2} - \lambda'(Fe)_{T_1}) \cdot n^*(H)/t$. The solubility $\lambda(O)$ is given by equation 2.6.4.9.

$$\lambda(O) = \frac{p(H_2O)}{p(HCl) + 2 \cdot p(H_2O)} = \frac{\Sigma(v(O) \cdot p(O))}{\Sigma(v(H) \cdot p(H))} = \frac{n^*(O)}{n^*(H)}$$

(2.6.4.9)

Accordingly, with the flow rate $f' = n^*(H)/t$ the transport rate is obtained:

$$\dot{n} = \frac{(\lambda(O)_{T_2} - \lambda(O)_{T_1}) \cdot n^*(H)}{t}$$

(2.6.4.10)

2.6.5 An Example – Calculation of the Transport Rate for the Nickel/Carbon Monoxide System

The effect of the stoichiometric flow for a vertically positioned transport ampoule (T_2 at the top) can be visualized by calculation for the following reaction (Sch 1982):

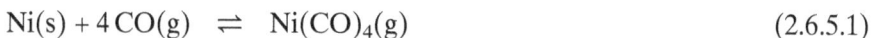

$$Ni(s) + 4\,CO(g) \rightleftharpoons Ni(CO)_4(g)$$

(2.6.5.1)

This reaction not only has extremely different values k and j (see equation 2.6.2.1) but also large differences of the gas composition at T_1 and T_2. According to *Schäfer*, we use the following data (Sch 1982):

$$\Delta_r H_{298}^0 = -146.4 \text{ kJ} \cdot \text{mol}^{-1}, \ \Delta_r S_{298}^0 = -418.4 \text{ J} \cdot \text{mol}^{-1} \cdot \text{K}^{-1},$$

$$\Delta_r C_p^0 = -34.14 + 97.53 \cdot 10^{-3} T + 7.43 \cdot 10^5 T^{-2}.$$

The sign of $\Delta_r H_{298}^0$ shows that the transport proceeds to the higher temperature.

($T_1 = 355$ K, $T_2 = 453$ K, $\Sigma p = 0.85$ bar, $p(\text{Ni (CO)}_4)_{T_1} = 1.43 \cdot 10^{-1}$ bar, $p(\text{CO})_{T_1} = 7.08 \cdot 10^{-1}$ bar, $p(\text{Ni(CO)}_4)_{T_2} = 6.06 \cdot 10^{-6}$ bar, $p(\text{CO})_{T_2} = 8.52 \cdot 10^{-1}$ bar)

As can be seen, no kind of molecule exists that has a partial pressure that is low enough at both temperatures T_1 and T_2 as required for the method of calculation described in section 2.6.2. Neglecting the flow, i.e., setting the flow factor $F = 1$, and further using $-\Delta p(\text{CO}) = \Delta p(\text{Ni(CO)}_4) = 1.43 \cdot 10^{-1}$ bar and $k = D \cdot q / R \cdot \bar{T} \cdot s$ (mol $\cdot$ s^{-1} $\cdot$ bar^{-1}) as the constant value we obtain from equation 2.6.2.7 equation 2.6.5.2, the lower limiting value of the transport effect:

$$\dot{n}(Ni) = -\frac{1}{4} \cdot (-1.43 \cdot 10^{-1}) \cdot k = 0.36 \cdot 10^{-1} \cdot k \qquad (2.6.5.2)$$

From equation 2.6.2.8 we arrive at 2.6.5.3, the upper limiting value of the transport effect:

$$\dot{n}(Ni) = \frac{1}{1} \cdot 1.43 \cdot 10^{-1} \cdot k \qquad (2.6.5.3)$$

These two different results show the influence of the stoichiometric flow that in the case of reaction 2.6.5.1 is in the direction $T_2 \rightarrow T_1$ due to the production of CO by decomposition of Ni(CO)$_4$. For the given example, the drastic reduction of the number of gas molecules is unusually high and the resulting stoichiometric flow has the opposite direction than the migration of Ni(CO)$_4$. The transport rate $\dot{n}(Ni)$ is thus reduced.

Neglecting the flow has two consequences. Firstly, the calculation of the transport rate for nickel by using $\Delta p(\text{CO})$ leads to the lower limit; and, secondly, the calculation with $\Delta p(\text{Ni (CO)}_4)$, for which the diffusion is hindered by the flow of CO, leads to the higher limit. All formulae taking into account the stoichiometric flow must lead to results within these limits. This effect is also reported for the transport of InAs by I$_2$ (Nic 1972).

Allowing for the stoichiometric flow factor F in the treatment of the chemical transport of nickel (Sch 1982) yields the following results.

$$\overline{p(B)} = \overline{p(\text{CO})} = 0.5 \cdot (p(\text{CO})_T + p(\text{CO})_{T_2})$$
$$= 7.80 \cdot 10^{-1} \text{ bar} \qquad (2.6.5.4)$$
$$\bar{p}(C) = \bar{p}(\text{Ni(CO)}_4) = 0.5 \cdot (p(\text{Ni(CO)}_4)_{T_1} + p(\text{Ni(CO)}_4)_{T_2})$$
$$= 7.17 \cdot 10^{-2} \text{ bar} \qquad (2.6.5.5)$$

Thus, we yield for equation 2.6.2.7 the result:

$$\dot{n}(Ni) = -\frac{1}{4} \cdot (-1.43 \cdot 10^{-1}) \cdot 3.19 \cdot k = 1.44 \cdot 10^{-1} \cdot k \qquad (2.6.5.6)$$

and accordingly, for equation 2.6.2.8 we arrive at:

$$\dot{n}(Ni) = \frac{1}{1} \cdot 1.43 \cdot 10^{-1} \cdot 7.98 \cdot 10^{-1} \cdot k = 1.14 \cdot 10^{-1} \cdot k \qquad (2.6.5.7)$$

Following section 2.3 we can use the solubility data λ. These exhibit the advantage that they are suitable for transport systems that are fully described by just one heterogeneous reaction, as well as for those systems where several reactions are involved in the transport process.

$$\lambda(Ni) = \frac{p(Ni(CO)_4)}{p(CO) + 4 \cdot p(Ni(CO)_4)} = \frac{p^*(Ni)}{p^*(CO)} \qquad (2.6.5.8)$$

We are using values from the literature (Sch 1982): $\lambda(Ni)_{T_1} = 1.12 \cdot 10^{-1}$, $p^*(CO)_{T_1} = 1.28$ bar, $\lambda(Ni)_{T_2} = 7.11 \cdot 10^{-6}$, $p^*(CO)_{T_2} = 8.52 \cdot 10^{-1}$ bar. Following this concept (equation 2.6.5.4) we obtain equation 2.6.5.9.

$$0.5 \cdot (p^*(CO)_{T_1} + p^*(CO)_{T_2}) = \overline{p^*(CO)} \qquad (2.6.5.9)$$

Equation 2.6.5.9 is used to relate the CO content at each place of the ampoule to the averaged amount of the solvent CO (expressed as $p(CO)$). According to equation 2.6.3.1 we get:

$$\Delta p^*(Ni) = \Delta\lambda(Ni) \cdot (0.5 \cdot (p^*(CO)_{T_1} + p^*(CO)_{T_2}))$$
$$= 1.19 \cdot 10^{-1} \text{ bar} \qquad (2.6.5.10)$$

Combining of all constant values of equations 2.6.2.9 and 2.6.2.10 leads to:

$$k = D^0 \left(\frac{\overline{T}}{T^0}\right)^{1.75} \cdot \frac{p^0}{\Sigma p} \cdot \frac{q}{s \cdot R \cdot \overline{T}}$$

$$= 0{,}06 \left(\frac{354 \text{ K}}{273.15 \text{ K}}\right)^{1.75} \cdot \frac{1 \text{ bar}}{0.8517 \text{ bar}} \cdot$$

$$\frac{2.43 \text{ cm}^2}{15 \text{ cm} \cdot 83.1415 \text{ cm}^3 \cdot \text{bar} \cdot \text{K}^{-1} \cdot \text{mol}^{-1} \cdot 354 \text{ K}} \cdot 3600 \text{ s} \cdot \text{h}^{-1}$$

$$= 2.23 \cdot 10^{-3} \text{ mol} \cdot \text{h}^{-1} \cdot \text{bar}^{-1} \qquad (2.6.5.11)$$

With this the transport rate for nickel can be written as:

$$\dot{n}(Ni) = \Delta p^*(Ni) \cdot k' = 1.1926 \cdot 10^{-1} \cdot k' \text{ mol} \cdot \text{h}^{-1} \qquad (2.6.5.12)$$

This value is close to that calculated earlier and it is also within the limits given above (section 2.6.2). As for experiments concerning the transport of Ni by means of CO and for calculations of the transport rate using formulae given in the literature (Sch 1982), see Table 2.6.5.1. The results are within the limits given by equations 2.6.2.7 and 2.6.2.8.

Table 2.6.5.1 Calculation of the transport rate for the chemical transport of nickel by carbon monoxide using different models.

Model [a]	$\dot{m}(\mathrm{Ni})/\mathrm{mg} \cdot \mathrm{h}^{-1}$
Lower limit, see equation 2.6.2.7; $F = 1$	4.69
Calculation with equations 2.6.2.7 or 2.6.2.8 using an individual flow factor F	14.98
Arizumi and *Nishinaga* (Ari 1965)	14.98
Richardson (Ric 1977, Ric 1978)	14.98
Faktor and *Garrett* (Fak 1974)	13.29
Calculation using $\lambda(\mathrm{Ni})$ (equation 2.6.5.8)	15.61
Upper limit, see equation 2.6.2.8; $F = 1$	18.76

[a] Calculations using $D^0 = 0.06 \ \mathrm{cm}^2 \cdot \mathrm{sec}^{-1}$; internal diameter of diffusion tube $d_i = 1.8$ cm, diffusion length $s = 15$ cm; $\bar{T}$(along diffusion path) $= 354$ K; $t = 3600$ sec; $\bar{p}(\mathrm{CO}) = 7.800 \cdot 10^{-1}$ bar; $\bar{p}(\mathrm{Ni}(\mathrm{CO})_4) = 7.166 \cdot 10^{-2}$ bar.

As for calculations utilizing the solubility λ see also *Gruehn* and co-workers (Schm 1981). For transport experiments under microgravity conditions, the calculations for diffusion as well as those for diffusion plus stoichiometric flow using models given by *Schäfer* (Sch 1956, Sch 1962), *Lever* (Lev 1962), *Mandel* (Man 1962), and *Faktor* (Fak 1971) lead to nearly the same results (compare also *Wiedemeier* (Wie 1976)).

At the end of this section we want to emphasize that all derivations given here, as well as by others, include various simplifications. In particular these are the following assumptions:

- Linear partial pressure gradient along the diffusion path.
- No homogeneous gas phase reaction along the diffusion path.
- The temperature of the diffusion path is assumed to be the average of the temperatures of the source and sink.
- There are no kinetic effects, in particular regarding supersaturation during nucleation.
- All gas species have the same diffusion coefficient.

Under these circumstances all efforts to calculate more precise values for transport rates are meaningless. This is especially true because the influence of the simplifications is small compared to the inaccuracies caused by errors in the thermodynamic data.

2.7 Diffusion Coefficients

2.7.1 The Binary Diffusion Coefficient D^0

General remarks D^0 is standardized for $T^0 = 273$ K and $\Sigma p^0 = 1$ bar. The interaction between the constituents of the gas mixture can be neglected. According to *Chapman* and *Cowling* (Cha 1953) the following equation holds to a first approximation for rigid spheres:

$$D^0(i, j) = \frac{3}{8 \cdot n(\sigma(i,j))^2} \cdot \sqrt{\frac{k_B \cdot T^0(m(i) + m(j))}{2 \cdot \pi \cdot m(i) \cdot m(j)}} \qquad (2.7.1.1)$$

n	means the number of molecules per cm^3 (for 1 bar and 273 K, $n = 2.6874 \cdot 10^{19}$ cm^{-3})
k_B	Boltzmann constant (= $1.38032 \cdot 10^{-23}$ J · K^{-1})
$m(i), m(j)$	are masses of molecules
$\sigma(i, j)$	means the arithmetic average diameter of the rigid elastic spheres that represent the molecules (= $0.5 \cdot [\sigma(i) + \sigma(j)]$)

Using M, the masses per mole, one gets equation 2.7.1.2:

$$D^0(i, j) = \frac{3}{8 \cdot n} \sqrt{\frac{k_B T^0}{2 \cdot \pi}} \cdot \frac{1}{j(\sigma(i,j))^2} \cdot \sqrt{\frac{(M(i) + M(j)) \cdot N_A}{M(i) \cdot M(j)}}$$

$$= 1.08 \cdot 10^{-27} \cdot \frac{1}{(\sigma(i,j))^2} \cdot \sqrt{\left(\frac{1}{M(i)} + \frac{1}{M(j)}\right) \cdot N_A} \qquad (2.7.1.2)$$

N_A is the Avogadro constant. The value for σ is relatively uncertain. Values for σ that allow a sufficiently good calculation of D^0 are given in Table 2.7.1.1.

Hastie (Has 1975) proposes empirical relations, which hold approximately:

$$\sigma = 0.841(V_c)^{\frac{1}{3}} = 1.166(V_b)^{\frac{1}{3}} = 1.221(V_m)^{\frac{1}{3}} \qquad (2.7.1.3)$$

Where V_c is the molecular volume at the critical point and V_b and V_m are the volume of the liquid at the boiling temperature and the volume of the solid at

Table 2.7.1.1 σ values determined from self-diffusion coefficients at $T = 273$ K (Lan 1950). The data given in brackets are uncertain. $\sigma(CO_2)$ is considered equal to $\sigma(N_2O)$ because the molecular weight and the structure of both molecules are identical.

Gas	He	Ne	Ar	Kr	Xe	Na	Cd
$\sigma/10^{-8}$ cm	2.5	(2.83)	(3.30)	(4.24)	(5.41)	3.7	3.0

Gas	Hg	H_2	O_2	Br_2	I_2	N_2	H_2O
$\sigma/10^{-8}$ cm	4.4	(2.91)	3.6	4.6	6.5	(3.57)	3.9

Gas	HCl	HBr	NH_3	N_2O	CO_2	CO	HCN
$\sigma/10^{-8}$ cm	(4.46)	(4.59)	3.9	4.5	(4.5)	3.7	4.1

Gas	$COCl_2$	UF_6	CH_4	CCl_4			
$\sigma/10^{-8}$ cm	4.8	(7.11)	(4.12)	6.7			

the melting temperature, respectively. Tables 2.7.1.2 to 2.7.1.5 show a collection of experimental D^0 values.

Table 2.7.1.2 Diffusion coefficients D^0 for $\Sigma p = 1$ bar and $T = 273$ K, determined experimentally for gas pairs $A + B$ with $A = H_2$ (Lan 1969, Ful 1966).

Gas B	NH_3	H_2O	N_2	O_2	HCl	CO	CH_4
D^0 (cm^2 · sec^{-1})	0.75	0.75	0.70	0.70	0.70	0.65	0.63

Gas B	CO_2	N_2O	Hg	Br_2	SO_2	CS_2	SF_6
D^0 (cm^2 · sec^{-1})	0.54	0.54	0.53	0.52	0.48	0.37	0.36

Gas B	CCl_4
D^0 (cm^2 · sec^{-1})	0.30

Table 2.7.1.3 Diffusion coefficients D^0 for $\Sigma p = 1$ bar and $T = 273$ K, determined experimentally for gas pairs $A + B$ with $A = N_2$ (also applicable for diffusion in O_2 or air) (Lan 1969, Ful 1966).

Gas B	NH_3	H_2O	NO	CH_4	CO	$COCl_2$
D^0/cm^2 · sec^{-1}	0.22	0.20	0.20	0.20	0.19	0.17

Gas B	O_2	Cd	CO_2	Hg	SO_2	Cl_2
D^0/cm^2 · sec^{-1}	0.17	0.17	0.14	0.11	0.11	0.11

Gas B	CS_2	Br_2	SF_6	I_2	CCl_4	UF_6
D^0/cm^2 · sec^{-1}	0.09	0.09	0.09	0.07	0.06	0.06

Table 2.7.1.4 Diffusion coefficients D^0 for $\Sigma p = 1$ bar and $T = 273$ K, determined experimentally for gas pairs $A + B$ with $A = CO_2$ (Lan 1969, Ful 1966).

Gas B	CH_4	H_2O	CO	N_2	I_2	Br_2	SO_2
D^0 (cm^2 · sec^{-1})	0.15	0.13	0.13	0.1	0.10	0.09	0.07

Gas B	CS_2	SF_6	$(C_2H_5)_2O$
D^0 (cm^2 · sec^{-1})	0.063	0.06	0.05

Table 2.7.1.5 Diffusion coefficients D^0 for $\Sigma p = 1$ bar and $T = 273$ K, experimentally determined for gas pairs $A + B$ (Lan 1969, Ful 1966).

Gas A	H_2O	CO	NH_3	CO	H_2O	NH_3	CO
Gas B	CH_4	NH_3	CO	CH_4	SO_2	SF_6	SF_6
D^0 (cm^2 · sec^{-1})	0.24	0.24	0.21	0.19	0.11	0.09	0.08

Empirical formulae for the ascertainment of D^0 *Andrussow* (And 1950) recommended for the calculation of D^0 the approximation 2.7.1.4.

$$D^0 = \frac{17.2 \cdot \left[1 + \sqrt{M(i) + M(j)}\right]}{\left[(V(i))^{\frac{1}{3}} + (V(j))^{\frac{1}{3}}\right]^2 \cdot \sqrt{M(i) \cdot M(j)}} \tag{2.7.1.4}$$

This formula is usable, especially for systems without H_2, when the molar volumes $V(i)$ (in cm^3) of the solid or the liquid compounds are used. These quantities may also be calculated (with less accuracy) from the volume increments given by *Biltz* (Bil 1934).

A critical collection of the values and an optimized formula based on a larger number of experimental data is given by *Fuller* (Ful 1966).

Table 2.7.1.6 Selected V increments ($cm^3 \cdot mol^{-1}$) taken from *Fuller, Schettler,* and *Giddings* (Ful 1966) (data in brackets are uncertain).

C	16.5	Cl	(19.5)	O_2	16.6	N_2O	35.9
Cl_2	(37.7)	H	1.98	S	(17.0)	Ar	16.1
NH_3	14.9	Br_2	(67.2)	O	5.48	H_2	7.07
CO	18.9	H_2O	12.7	SO_2	(41.1)	N	(5.69)
N_2	17.9	CO_2	26.9	SF_6	(69.7)		

$$D(i,j) = \frac{1.00 \cdot 10^{-3} \cdot T^{1.75}}{p \cdot \left((\Sigma V(i))^{\frac{1}{3}} + (\Sigma V(j))^{\frac{1}{3}}\right)^2} \cdot \sqrt{\frac{1}{M(i)} + \frac{1}{M(j)}} \tag{2.7.1.5}$$

$$D^0 = \frac{18.34}{\left((\Sigma V(i))^{\frac{1}{3}} + (\Sigma V(j))^{\frac{1}{3}}\right)^2} \cdot \sqrt{\frac{1}{M(i)} + \frac{1}{M(j)}} \tag{2.7.1.6}$$

The influence of Σp and T on the diffusion coefficient In the literature equation 2.7.1.7 is given to allow for the temperature dependence of D (Jel 1928).

$$D = D^0 \cdot \frac{1}{\Sigma p} \cdot \left(\frac{T}{273.15}\right)^n \tag{2.7.1.7}$$

Theoretical considerations lead to values of n that lie between 1.5 and 2. The n-values determined experimentally for many substances lie between 1.66 and 2.0 (Mül 1968). We generally use $n = 1.75$. The scattering of n-values leads at 1000 K to uncertainties of D of about 30 %.

The viability of this estimation has been shown by transport experiments in the systems $NbCl_4/NbCl_5$ and $ZrCl_4/NbCl_5$. By these experiments the diffusion coefficient for $NbCl_4$ has been determined and further on the molar mass of $NbCl_4$ (Wes 1975). But we should keep in mind that the result is strongly influenced by the vapor pressures of $NbCl_4$ and $ZrCl_4$ on which the calculation is based.

2.7.2 D^0 in Multi-compound Systems

The systems discussed in the preceding section contain the substances $A(s)$, $B(g)$, and $C(g)$. According to the phase rule in each of the equilibrium regions (source, sink) one can dispose of two degrees of freedom. They are fixed by the choice of the content of the ampoule and by the reaction temperature. Under these conditions the values of all the equilibrium pressures are readily determined because of the reasoning that, in the case of chemical transport, diffusion must occur between equilibrium regions.

If more than two gas species are present, one can for example think of a transport reaction according to the equilibrium 2.7.2.1. Then, the first two of the three disposable degrees of freedom will be fixed by the choice of the ampoule content and the temperature. The remaining degree of freedom is not fixed by the condition of equilibrium that establishes itself at the solid substance, since the equilibrium pressures are variable according to the law of mass action. The stoichiometric relation between the partial pressures of C and D will be disturbed by differences in the diffusion coefficients for C and D. **As a consequence a stationary state is established, which is dependent on diffusion.** These relations will now be examined more closely.

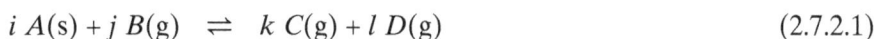

$$i\,A(s) + j\,B(g) \;\rightleftharpoons\; k\,C(g) + l\,D(g) \qquad (2.7.2.1)$$

If an ampoule is filled with the solid substance A and a known quantity of $B(g)$, the equilibrium pressures for the reaction between A and B can be calculated for the separate sub-volumes at T_1 and T_2. Conditions are chosen, such that the total pressure has the same value in both volumes. When the hypothetical wall that separates the two equilibrium regions is removed, the transport starts. In this calculation using a hypothetical wall, one has tacitly introduced the condition that there is not only equilibrium between the gas and solid phase, but also at the same time a stoichiometric relationship between the gaseous reaction products:

$$p(C) : p(D) = k : l \qquad (2.7.2.2)$$

This stoichiometric relationship can be disturbed by the diffusion, without necessarily destroying the basic premise that the gas phase and the solid phase are in equilibrium with each other. In such an ampoule a stationary state is established, which is characterized by the fact that in a heterogeneous reaction as much of a certain substance is consumed (or produced) as is introduced (or removed) by diffusion. The transport of A must proceed to the same extent, irrespective of whether $\Delta p(B)$, $\Delta p(C)$, or $\Delta p(D)$ was used in the calculations (see section 2.6). Thus, the following equation 2.7.2.3 must be valid.

$$D(B) \cdot \frac{1}{j} \cdot \Delta p(B) = D(C) \cdot \frac{1}{k} \cdot \Delta p(C) = D(D) \cdot \frac{1}{l} \cdot \Delta p(D) \qquad (2.7.2.3)$$

$D(B)$, $D(C)$, and $D(D)$ are the diffusion coefficient of the gases B, C, and D in the gas mixture. This equation shows that the presence of a molecule $X(g)$, which has a larger value for $D(X)$, must lead to a smaller value of $\Delta p(X)$. Thus, the composition of the equilibrium gas can change. This consideration only holds for reactions with no change in the molar number of the gas species: $j = k + l$. In other cases the flow of the gas mass in its entirety must be taken into account.

Usually, the individual diffusion coefficients are not available. Results were obtained, however, which were sufficiently accurate for most purposes, when one assumes that diffusion, without any change in the equilibrium pressures, starts after removal of the hypothetical wall. Therefore, **an average value for the diffusion coefficient is used**. Instead of the value $D^0 = 0.1$ cm$^2 \cdot$ s^{-1} proposed by *Schäfer*, nowadays $D^0 = 0.025$ cm$^2 \cdot$ s^{-1} is accepted. For thermodynamically accurately characterized transport systems the latter value leads to very good compliance between theory and experiment (Kra 1982, Scho 1990).

If the equilibrium pressures above the solid material and the binary diffusion coefficients are known as a function of temperature and, furthermore, if experimental conditions are chosen so that no substantial error is made by neglecting thermal convection and thermal diffusion relative to the diffusion, then it may be meaningful to refine the calculation of the transport in the case of systems with more than two gaseous components. A theoretical derivation is given by *Lever, Mandel,* and *Jona* (Man 1962, Lev 1962a, Lev 1962b, Jon 1963). For the influence of different binary diffusion coefficients in gas mixtures see also *Hugo* (Hug 1966).

In general one can say that the influence on calculated transport rates by errors of the diffusion coefficient is small compared to that of thermodynamic data.

Regarding the diffusion coefficient of one gas species in a gas mixture the approximation 2.7.2.4 by *Wilke* (Wil 1950) has been proposed (Rei 1958).

$$D(1, 2, 3, ..., n) = \frac{1 - x(1)}{x(2) \cdot \dfrac{1}{D(1, 2)} + x(3) \cdot \dfrac{1}{D(1, 3)} + x(4) \cdot \dfrac{1}{D(1, 4)} + ...} \quad (2.7.2.4)$$

$D(1, n)$ is the binary diffusion coefficient for the gases 1 and n and $x(n)$ is the molar fraction of component n in the mixture.

In order to combine these considerations with the concept of the solubility λ of a solid in the gas phase (see section 2.3) we recommend taking that molecule as the basis (gas species 1), which originates from the partial pressure difference $\Delta p = p(T_2) - p(T_1)$. The biggest solvent molecule should be taken as gas species 2. This leads to the binary diffusion coefficient $D(1, 2)$. The latter recommendation is based on the idea that the slowest step is rate determining.

In the case of a multi-compound gas mixture the following procedure also seems to be feasible (equation 2.7.2.5).

$$\lambda(A1) = \frac{p^*(A1)}{p^*(L)}; \quad \lambda(A2) = \frac{p^*(A2)}{p^*(L)}; \quad \lambda(A3) \quad \text{etc.} \quad (2.7.2.5)$$

$D(A1, A2 ... An)$, $D(A2, A1, A3 ... An)$ etc., can be evaluated with equation 2.7.2.4.

In order to get the transport rate $\dot{n}(A) = n(A)/t$ one adds the following terms, which describe the solubility (see section 2.3) and diffusion of gas species $A1$, $A2$, $A3$, etc. within the gas mixture (equation 2.7.2.6).

$$\Delta\lambda(A1) \cdot \overline{\Sigma p(L)} \cdot D\,(A1, A2 \ldots An)$$
$$+\ \Delta\lambda(A2) \cdot \overline{\Sigma p(L)} \cdot D\,(A2,\ A1,\ A3 \ldots An) + \ldots \sim n'(A) \qquad (2.7.2.6)$$

2.8 Survey of Gas Motion in Ampoules

2.8.1 General Remarks

We always assume that the equilibria between solids and gases on both sides of an ampoule are established. This is usually true at temperatures that are not too low ($\vartheta \geq 500\,°\mathrm{C}$) and at pressures higher than 0.01 bar. The homogeneous equilibria in the gas phase that may be established will not be taken into account. Thermodynamics yields the most important information, but data published by different authors on a specific subject often differ by many orders of magnitude with respect to the transport effect. The lack of reliability of thermodynamic data may lead to calculated transport rates that are substantially different from the experimental results.

Gas motion takes place by diffusion and stoichiometric (laminar) flow, if ampoules according to Figures 2.6.1.1 and 2.8.1.1 are used.

For the usually present multi-component gas mixtures one uses an average diffusion coefficient, which leads to a certain degree of inaccuracy.

For preparative purposes one may use simple ampoules in a horizontal position according to Figure 2.8.1.1, for which the model calculation allowing for the diffusion may be sufficient but only with tube diameters $d \leq 2$ cm. With larger diameters and pressures $\Sigma p > 1$ bar the transport rate becomes up to ten times larger because not only diffusion but also convection plays a role (Pao 1975). For the system $\mathrm{Ge/GeCl_4}$ *Schäfer* and *Trenkel* carried out an investigation that shows

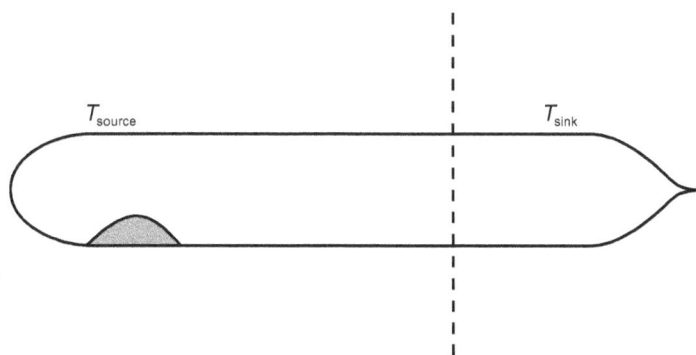

Figure 2.8.1.1 A typical transport ampoule (l = 100 ... 200 mm, q = 2 cm^2).

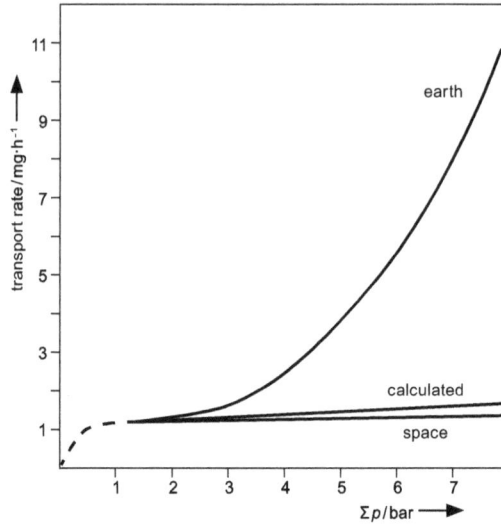

Figure 2.8.1.2 Comparison of calculated and experimental transport rates for the chemical vapor transport of Ge by iodine, according to *Oppermann* (Opp 1981).

the influence of reaction kinetics and the shape and inclination of the ampoule on transport behavior (Sch 1980a).

For the system Ge/I_2 *Oppermann* compared, using the solubility $\lambda(Ge)$, the calculated transport rate for the conditions in space and on earth (Opp 1981). He found that under the microgravity condition in space the exact transport rate for diffusion could be achieved with total pressures of 1 to 7 bar, whereas on earth with total pressures of 2 to 3 bar the transport rate starts to increase strongly due to convection (Figure 2.8.1.2).

Wiedemeier et al. observed the transport rate of GeSe and GeTe, respectively, under microgravity conditions using GeI_4 or $GeCl_4$ as transport agents. In some cases argon was added (to acheive a particular total pressure). Thus determined values for the experimental transport rates were approximately three times higher than calculated. However, experimental uncertainties seem to have been responsible in this case (Wie 1976).

Some complications Thermal convection and diffusion can take place in an ampoule simultaneously, but adjacent to the solid a boundary layer with diffusion must always be assumed. This means that for the convection that must be taken into account outside the boundary layer a shorter diffusion path length has to be considered (Wie 1981). *Klosse* and *Ullersma* propose equation 2.8.1.1 for the relation of thermal convection with diffusion (Klo 1973).

$$K = \frac{1}{[A \cdot (Sc \cdot Gr)^{-2} + B]} \tag{2.8.1.1}$$

A and B are functions of $1/d$ (length and diameter) and Sc and Gr are the *Schmidt* and *Grashof* numbers (Klo 1973).

Apart from the expansive convection the "solutional" convection (stoichiometric flow) is also of importance, e.g., when the density of the gas phase changes in the applied temperature gradient by dissociation of one of the gaseous constituents (Sch 1980b, Sch 1982b). Furthermore, the adsorption and desorption of the gas on the wall of the tube, the heat conduction of the tube material, the thermal gas demixing (thermo-diffusion), and the supersaturation on the deposited crystals are of influence.

Apart from all these influences, in the case of low total pressures ($0.01 \leq \Sigma p \leq 3$ bar) the transport rate is mainly defined by diffusion and stoichiometric flow, whereas in the case of higher pressure it is the convection that becomes more important.

Additional references give further information on the observation of the gas motion using carbon smoke (Cho 1979), on the study of natural convection in a sealed tube (Lau 1981), on numerical models to transport of Ge by GeI_4 and GeI_2 (Lau 1982), on constitutional supersaturation (Ros 1975), on numerical modeling of diffusive physical transport in cylindrical ampoules (Gre 1981, Mar 1981), and on expansion convection in vapor transport across horizontal rectangular enclosures (Iha 1982).

2.8.2 Experiments on Gas Motion – Diffusion and Convection in Closed Ampoules

The transport rate in closed cylindrical ampoules is expected to follow a scheme given by Figure 2.8.2.1. With the assumption that the equilibrium solid/gas is established at both ends of the ampoule one can calculate the transport rate $\dot{n}$ by using the concept of gas diffusion only (see section 2.6.2). This diffusion controlled value $\dot{n}_{calc}$ can be compared to the actual observed rate $\dot{n}_{obs}$ by plotting the ratio of both rates versus the total pressure (Figure 2.8.2.1).

In general three different regions can be observed, in accordance with the sequence of rate-determining processes: chemical reaction (setting of the chemical equilibrium; section I); diffusion between equilibrium regions (section II); and convection and diffusion between equilibrium regions (section III).

Section I

a) With very low amounts of transport agent and a large specific surface of the starting material, the adsorption of the gas molecules onto the surface may hinder or reduce the transport process; see MgO/HCl (Kle 1972).
b) At low total pressures the mean free path length of the molecules becomes comparable to, or even greater than, the dimensions of the transport container, so that molecular collisions will hardly take place. Let us consider as an example a hot wire arrangement for the metal transport by the iodide

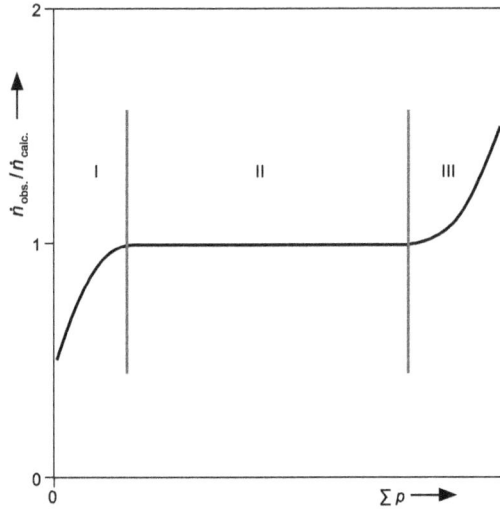

Figure 2.8.2.1 Theoretically predicted dependence of the transport rate on the total pressure (schematic representation).
($\dot{n}_{obs}$: experimentally determined transport rate; $\dot{n}_{calc}$: transport rate calculated by assuming diffusion as rate-determining step)

method (see sections 1.1 and 3.1), operated under conditions where chemical equation 2.8.2.1 can be assumed.

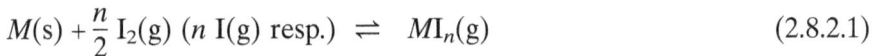

$$M(s) + \frac{n}{2} I_2(g) \ (n \ I(g) \ \text{resp.}) \ \rightleftharpoons \ MI_n(g) \tag{2.8.2.1}$$

At the surface of the crude metal the iodine is largely converted into $MI_n(g)$. The molecules MI_n, which have a long mean free path length, arrive at the hot wire where a considerable fraction of them is decomposed. The quantity of metal thus transported is proportional to the surface of the hot wire. The same applies when, besides the crude metal, solid MI_n is present whose saturation pressure is kept independent of the growing surface of the hot wire (Dör 1952, Hol 1953, Sha 1955, Loo 1959). In this case the MI_n saturation pressure can be estimated (Loo 1959, Rol 1961), if one assumes that all collisions of MI_n molecules with the hot wire lead to the deposition of the metal. In the transition range from molecular flow to diffusion, the former can be predominant in a hot wire arrangement that has a very thin wire; the quantity of deposited metal is proportional to the surface area of the hot wire. When the thickness of the wire has increased considerably, the diffusion of the gas molecules becomes the rate-determining factor in the process. In this case the quantity of deposited metal is practically independent of the surface area of the hot wire (Sha 1955).

c) It is also possible that diffusion is responsible for the migration of the solid, though the velocity of the reaction is rate determining. In this case the diffusion is so fast that the solid/gas equilibria on one or both sides of the ampoule cannot be established. This becomes true especially with small distances between source and deposit ("sandwich method", short distance transport) (Krä 1974).

Section II

This is the simplest situation of chemical transport where gas diffusion (and stoichiometric flow) occurs between equilibrium regions and where this diffusion is the rate-determining step. In general one can expect that the diffusion range will extend to lower pressures with increasing temperatures. It is also possible to expand the diffusion range by increasing the quantity of solid material or by choosing other tube dimensions, especially by introducing a narrow diffusion path.

Section III

Here we have higher pressures and thermal convection besides diffusion should play an important role. This expectation has been confirmed by transport experiments that can be described by the equilibrium 2.8.2.2 (Doe 1973, Opp 1968).

$$CrCl_3(s) + \frac{1}{2}Cl_2(g) \rightleftharpoons CrCl_4(g) \qquad (2.8.2.2)$$

Silica ampoules of 10 mm inside diameter and 200 mm length contained about 0.7 g $CrCl_3$ (prepared from electrolytic chromium and chlorine; "sublimed" in a stream of chlorine) and defined quantities of chlorine. The latter was introduced as a liquid in capillary tubes. A horizontal positioned two-zone furnace was used. The vapor pressure of the $CrCl_3$ at the required temperatures could be neglected. The experimental run lasted 20 hours, the transport distance was 15 cm, and the quantity of transported $CrCl_3$ was in the range of 3 to 30 mg.

Figure 2.8.2.2 shows the results. As comparison with Figure 2.8.2.1 shows, the trend of the experiments completely satisfies our expectation although the values of the curve $\dot{n}_{obs}/\dot{n}_{calc}$ versus Σp (section II) are too low by a factor of 0.6. The diffusion coefficient that was taken as $D^0(273 \text{ K}, 1 \text{ bar}) = 0.05 \text{ cm}^2 \cdot \text{sec}^{-1}$ seems to be responsible for this deviation. One can be certain therefore, that the transport of $CrCl_3$ in these simple ampoules at total pressures between 0.2 and 0.4 bar can be calculated on the assumption that diffusion takes place between regions where equilibrium is established. Even at a total pressure of 0.01 bar the result of the calculation is $\dot{n}_{obs}/\dot{n}_{calc} = 0.27$ and can be looked upon as an acceptable approximation. At still lower pressures (section I) the rate of the reaction becomes too small. At higher pressures (section III) the thermal convection becomes rather considerable.

Similar results have been verified for the following transport system:

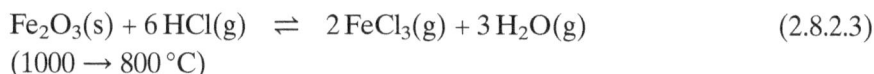

$$Fe_2O_3(s) + 6\,HCl(g) \rightleftharpoons 2\,FeCl_3(g) + 3\,H_2O(g) \qquad (2.8.2.3)$$
$$(1000 \rightarrow 800\,^\circ C)$$

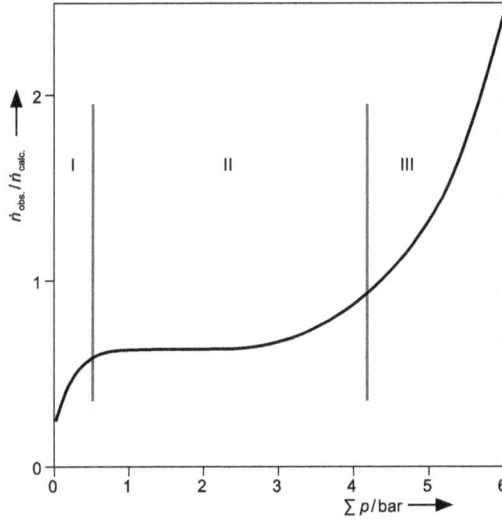

Figure 2.8.2.2   Results of $CrCl_3$ transport ($500 \rightarrow 400\,°C$) in the presence of chlorine. Here the dependence of the relative transport effect $\dot{n}_{obs}/\dot{n}_{calc}$ on the total pressure is shown (Doe 1973, Opp 1968).

The chemical transport of Fe_2O_3 carried out in sealed ampoules at total pressures between 0.04 and 0.4 bar corresponds to the calculation based on diffusion as the rate-determining step. Higher pressures have not been used in this case. At lower pressures (0.004 and 0.0004 bar), on the other hand, the observed transport rate was much smaller than calculated on the basis of a diffusion controlled transport (Sch 1956).

The diffusion range (section II) deserves special attention because with most of the transport experiments that will be described later, diffusion between the equilibrium regions (source and sink) is the rate-determining step. Obviously, this rather simple model is true for the hot-wire transport of zirconium according to the iodide method (Sha 1955, Mor 1952) as well as for transport experiments in simple ampoules with which the following transport reactions 2.8.2.4 to 2.8.2.14 were carried out in the medium pressure range (from 0.03 to 1 bar) and at medium temperatures (from 400 to 1100 °C).

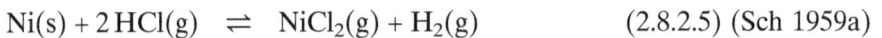

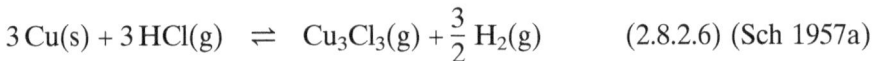

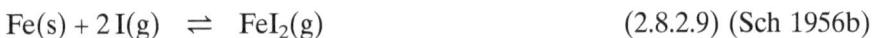

$$Fe(s) + 2\,HCl(g) \;\rightleftharpoons\; FeCl_2(g) + H_2(g) \qquad\qquad (2.8.2.4)\ (Sch\ 1959a)$$

$$Ni(s) + 2\,HCl(g) \;\rightleftharpoons\; NiCl_2(g) + H_2(g) \qquad\qquad (2.8.2.5)\ (Sch\ 1959a)$$

$$3\,Cu(s) + 3\,HCl(g) \;\rightleftharpoons\; Cu_3Cl_3(g) + \frac{3}{2}\,H_2(g) \qquad (2.8.2.6)\ (Sch\ 1957a)$$

$$Si(s) + SiCl_4(g) \;\rightleftharpoons\; 2\,SiCl_2(g) \qquad\qquad (2.8.2.7)\ (Sch\ 1956a)$$

$$Si(s) + SiI_4(g) \;\rightleftharpoons\; 2\,SiI_2(g) \qquad\qquad (2.8.2.8)\ (Sch\ 1957b)$$

$$Fe(s) + 2\,I(g) \;\rightleftharpoons\; FeI_2(g) \qquad\qquad (2.8.2.9)\ (Sch\ 1956b)$$

$$Ni(s) + 2\,Br(g) \;\rightleftharpoons\; NiBr_2(g) \qquad\qquad (2.8.2.10)\;(Sch\;1956b)$$

$$Ni(s) + 2\,I(g)/I_2(g) \;\rightleftharpoons\; NiI_2(g) \qquad\qquad (2.8.2.11)\;(Sch\;1956b)$$

$$3\,Cu_2O(s) + 6\,HCl(g) \;\rightleftharpoons\; 2\,Cu_3Cl_3(g) + 3\,H_2O(g) \quad (2.8.2.12)\;(Sch\;1957a)$$

$$Fe_2O_3(s) + 6\,HCl(g) \;\rightleftharpoons\; 2\,FeCl_3(g) + 3\,H_2O(g) \qquad (2.8.2.13)\;(Sch\;1956a)$$

$$NbCl_3(s) + NbCl_5(g) \;\rightleftharpoons\; 2\,NbCl_4(g) \qquad\qquad (2.8.2.14)\;(Sch\;1959b)$$

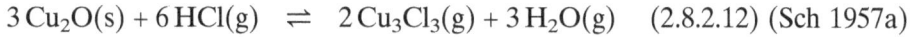

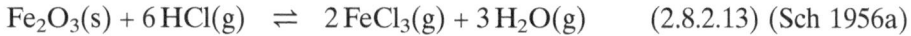

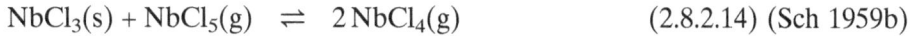

A curve like that in Figure 2.8.2.2 was also found for the system ZnS/HCl (Jon 1963). With tube diameters of 2 cm or more the diffusion section (section II) becomes smaller, that means the convection begins at lower pressure. Nevertheless the sequence of the three sections with different rate-determining steps (section I: establishing of the heterogeneous equilibrium; section II: diffusion; section III: thermal convection) will be observed (Wie 1972). Diffusion as the rate-determining step has been confirmed by several authors (Lev 1966, Dan 1973, Got 1963, Ari 1967).

In *open flow systems* at $\Sigma p = 1$ bar a similar curve as in Figure 2.8.2.2 can be observed if for the calculation of $\dot{n}_{calc}$ the total pressure is replaced by the inverse flow rate. The requirement of fast setting of the equilibrium is retained as in the case of closed systems (Sch 1949):

Section I: no equilibrium; the kinetics of the reaction between solid (condensed) phase and gas phase is rate determining.
Section II: equilibrium is established.
Section III: equilibrium is established; however, the flow of the gas is superimposed by thermal convection.

2.8.3 Experiments Concerning Thermal Convection

In section III (Figures 2.8.2.1 and 2.8.2.2) all equilibria are established. Thermal convection plays a role besides diffusion if the pressure within the tube is relatively high, e.g., several atmospheres, and especially if the main constituent is gaseous (Sch 1980b, Sch 1982b). In general, thermal convection depends on Σp, on the diameter and length of the tube, and on whether the ampoule is positioned vertically or horizontally. If one wishes to prevent thermal convection one should apply a vertical position of the transport tube with the hotter end at the top (Figure 2.8.3.1 (Wie 1982)). For $\Sigma p = 1$ bar and a horizontally positioned tube with diameter $d = 2$ cm no convection was observed. However, with the wider diameter $d = 3$ cm strong convection occurs (Pao 1975).

The combination of thermal convection and diffusion is shown in Figure 2.8.3.1. The distance between source and deposit (sink) is mainly overcome by convection. However, through a boundary layer of the gas phase adjacent to the solid, diffusion occurs as the slowest step. This means that a calculation based on diffusion only, using a length of the diffusion path that is equal to the length of the tube, leads to a transport rate that is too small.

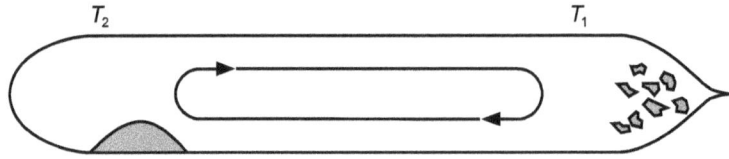

Figure 2.8.3.1 Transport by convection and diffusion. Schematic representation of gas motion in a horizontally positioned ampoule.

For growing large crystals one prefers to increase the transport rates by using conditions that favor convection. However, crystals grown under these conditions may be inhomogeneous and may have a higher number of defects (Wie 1982).

An example of the influence of convection in a dissociating gas is the chemical transport of TaS_2 with sulfur according to equation 2.8.3.1 at temperatures between 973 K and 1273 K and pressures $p(S_2)$ up to 25 bar.

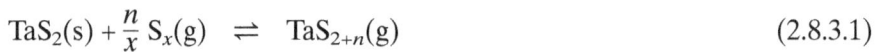

$$TaS_2(s) + \frac{n}{x} S_x(g) \rightleftharpoons TaS_{2+n}(g) \tag{2.8.3.1}$$

Sulfur molecules with $x = 2$, 3, ... 8 have to be taken into account. Therefore, thermal convection is increased by the temperature dependent dissociation of $S_8(g)$. The stoichiometric coefficient n of the gaseous species TaS_{2+n} was determined to be $n + 2 = 5$ or perhaps, allowing for the experimental uncertainties, $n + 2 = 6$ (Sch 1980b). These coefficients were found by using the transport rates for evaluating the pressures and the thermodynamic equilibria.

Combining nearly all quantities of equation 2.6.2.11 into an experimental factor $f(exp)$ the equation 2.8.17 for the transport rate $\dot{n}$ becomes:

$$\dot{n} \approx \Delta p' \cdot f(exp) \tag{2.8.3.2}$$

The factor $f(exp)$ has been determined empirically by sublimation experiments with NaCl in sulfur vapor. For these experiments the same conditions as for the transport of TaS_2 were applied, in particular, the same pressure for sulfur. When in this case the total pressure $\Sigma p(S_x)$ was increased from 1 to 25 bar, $f(exp)$ was increasing from 10 to approximately 30. The corresponding factor $f(diff)$ calculated for diffusion only was decreasing from 10 to 0.5 under these conditions. We conclude that the gas motion as a whole (diffusion, stoichiometric flow, and thermal convection) can be expressed for any system and for any total pressure by the empirically determined factor $f(exp)$. Then the calculation of the transport rate $\Delta p' = \Delta \lambda \cdot \overline{\Sigma p(S)}$ eventually yields the coefficients in equation 2.8.3.1.

Qualitative considerations concerning the influence of thermal convection can be made by means of Figure 2.8.3.2. The tube of radius r is kept at the temperature T_1 with the exception of a short section of length $l(w)$, which is heated to a higher temperature T_2. The difference in gas density between the hot and the cold part of the tube causes buoyancy. The total path the gas travels has the length l. The volume V of a gas of viscosity η that flows per time unit through the cross section of a tube of radius r is obtained from the *Hagen-Poiseuille* equation:

$$V = \frac{\pi \cdot r^4 \cdot \Delta p^*}{8 \cdot \eta \cdot l} \tag{2.8.3.3}$$

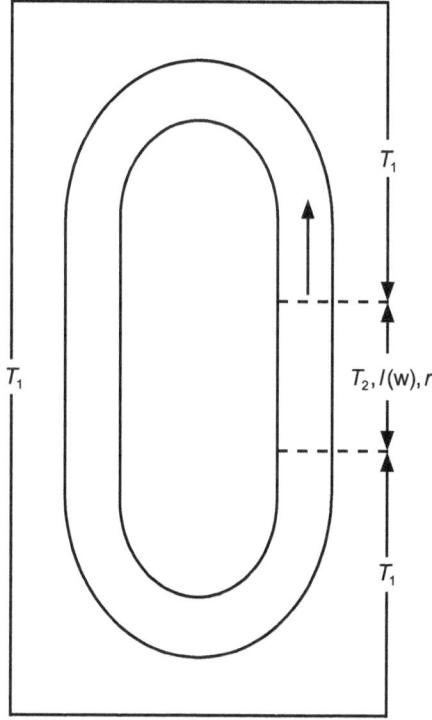

Figure 2.8.3.2 Schematic representation of gas motion by convection.

The pressure difference Δp^* follows from the difference in force per unit area between the end of the hot section of the tube and the end of an equivalent cold section of the tube, In these two sections at T_1 and T_2, m_1 and m_2 represent the quantities of the gas of molecular weight M, and p is the total pressure.

$$\Delta p^* = 981 \frac{m_1 - m_2}{r^2 \cdot \pi} \; \text{g} \cdot \text{cm}^{-1} \cdot \text{sec}^{-2} \tag{2.8.3.4}$$

Using the ideal gas law for the volume of each section one obtains:

$$m_1 - m_2 = \frac{l(\text{w}) \cdot r^2 \cdot \pi \cdot p \cdot M}{R} \cdot \left(\frac{1}{T_1} - \frac{1}{T_2} \right) \tag{2.8.3.5}$$

And further, with equations 2.8.3.3, 2.8.3.4, and 2.8.3.5, the equations 2.8.3.6 and 2.8.3.7, respectively.

$$V = \frac{981 \cdot r^4 \cdot \pi \cdot l(\text{w}) \cdot p \cdot M}{8 \cdot l \cdot R} \cdot \left(\frac{1}{T_1} - \frac{1}{T_2} \right) \tag{2.8.3.6}$$

$$V = \frac{981 \cdot 4.7 \cdot r^4 \cdot l(\text{w}) \cdot p \cdot M}{\eta \cdot l} \left(\frac{1}{T_1} - \frac{1}{T_2} \right) \tag{2.8.3.7}$$

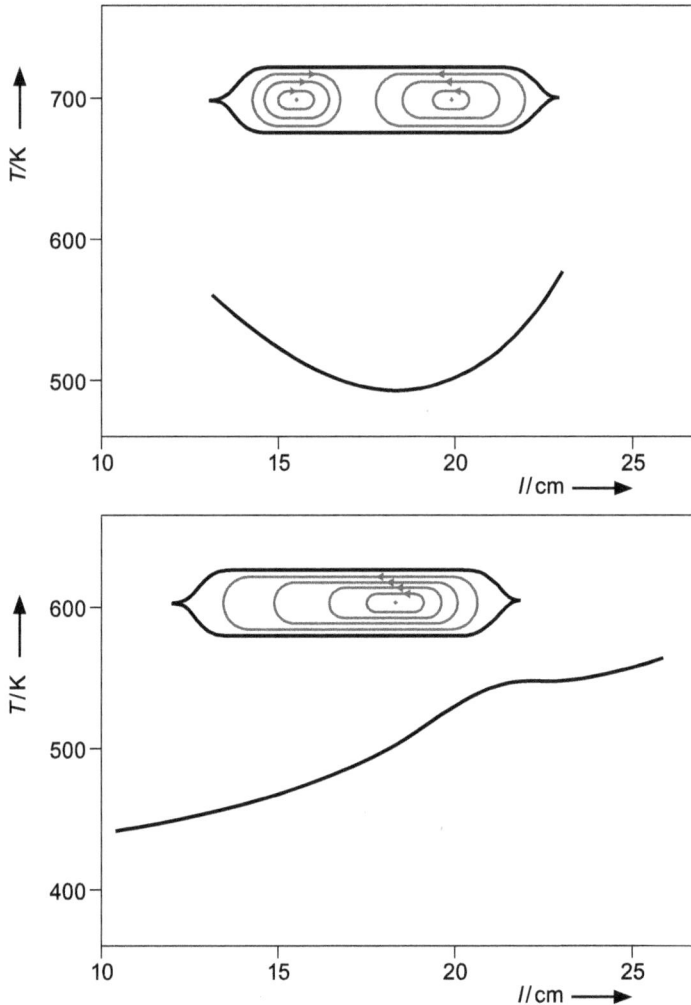

Figure 2.8.3.3 Gas motion by convection depending on the temperature profile (see (Cho 1979)). In some other cases the thermal conductivity, especially in He- or H_2-containing systems, may determine the transport rate (Schö 1980, Ind 1966).

The viscosity η is measured in g · cm^{-1}, r in cm, and p in bar (Clu 1948). V and η are related to the temperature T_1.

If one represents the number of moles of gas B that flow into the hotter zone by $n(B, \text{init.})$, one gets for t seconds with $n = p \cdot V/(R \cdot T)$:

$$n(B, \text{init.}) = (p(B, \text{init.}))^2 \cdot \left\{ \frac{4.7 \cdot 981 \cdot r^4 \cdot l(\text{w}) \cdot p \cdot M(B)}{R \cdot T_1 \cdot \eta \cdot l} \left(\frac{1}{T_1} - \frac{1}{T_2} \right) \right\}$$

$$(2.8.3.8)$$

On combining this expression with equation 2.8.3.9 that applies for flow methods (Sch 1962), one can calculate the transported quantity $n(A)$ of the condensed phase A (equation 2.8.3.10).

$$n(A) = \frac{i}{j} \cdot \frac{\Delta p(C) \cdot n(B, \text{init.})}{p(B, \text{init.})} \tag{2.8.3.9}$$

$$n(A) = \frac{i}{j} \cdot \Delta p(C) \cdot p(B, \text{ init.}) \cdot \left\{ \frac{4.7 \cdot r^4 \cdot l(w) \cdot M(B) \cdot t}{R \cdot T_1 \cdot \eta \cdot l} \left(\frac{1}{T_1} - \frac{1}{T_2} \right) \right\} \tag{2.8.3.10}$$

One should note that the transport effect by diffusion is proportional to $1/\Sigma p$, whereas with the thermal convection it is proportional to p. Consequently, material transport in closed tubes is basically caused by diffusion at low pressures, but by thermal convection at high pressures.

In practice one often works with systems that are less easy to visualize than those used in Figure 2.6.1.1, which served as the basis for the derivations related to the convection process. One would, for example, use simple ampoules as represented in Figure 2.8.1.1. In such cases it is advised to determine the braced expression in equation 2.8.3.10 empirically as a tube constant by using a known chemical transport reaction.

If one wishes to work in a pressure range where diffusion and convection contribute to about the same extent, the gas flow should be determined empirically as a function of pressure. In this way one obtains data that can be used in other transport reactions under similar conditions.

The work of *Klosse* and *Ullersma* (Klo 1973) gives one expression for the ratio of diffusion and thermal convection in a closed ampoule. Other authors have been investigating qualitatively the diffusion and convection in closed, horizontal tubes, using carbon (made by incomplete oxidation of acetylene) as an indicator for gas motion (Cho 1979). They observed different types of convection figures (Figure 2.8.3.3).

2.9 Chemical Kinetics of Heterogeneous Reactions

2.9.1 Reaction Behavior at an Atomic Level

Despite the enormous importance of reactions between solids and gases, the reaction kinetics on the solid surface are known only incompletely. We depend on model considerations. For example, for the reaction of solid Fe_2O_3 with gaseous HCl the following elementary steps may be postulated:

Forward reaction

1. Adsorption of HCl molecules on the Fe_2O_3 surface. Polarization of the bonds in the adsorbed molecules and of the bonds in the solid near the surface as well.
2. At temperatures high enough, ions (or atoms) of the solid leave the lattice and migrate into the adsorbate layer. Here the ions originating from the gas and from the solid mix in a similar way to ions in a melt.

3. Neutral molecules formed in step 2 (H_2O, $FeCl_3$) are desorbed. They move away, if a concentration gradient exists in the gas phase.

Remarks on step 2 The transfer of atoms from the bulk into the adsorbed film has been shown by means of a mass spectrometer using the example of the influence of $AlCl_3$ or I_2 on the sublimation of red phosphorus and As_2O_3 (Claudetite) (Sch 1978). This process is similar to oxide formation by the reaction of metals with oxygen, where the metal moves into the oxide (oxygen) layer on the metal surface. (Schm 1995a).

Other examples that show the similarity of the adsorbed film and a melt are the evaporation reaction of arsenic in the presence of a thallium melt (Bre 1955) and the improved decomposition of gallium nitride by an indium melt (Scho 1965).

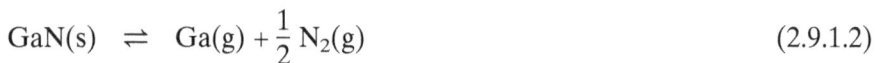

$$As(s) \; \rightleftharpoons \; \frac{1}{4} As_4(g) \tag{2.9.1.1}$$

$$GaN(s) \; \rightleftharpoons \; Ga(g) + \frac{1}{2} N_2(g) \tag{2.9.1.2}$$

Reverse reaction

The gas molecules H_2O and $FeCl_3$ are adsorbed on a solid surface. The mixing of their ions leads to the formation of nuclei and finally the growth of crystals of Fe_2O_3, and hydrogen chloride, which is desorbed.

For the formation of GaAs from gaseous As_4, GaCl, and H_2, *Cadoret* (Cad 1975) describes the following steps 2.9.1.3 to 2.9.1.5.

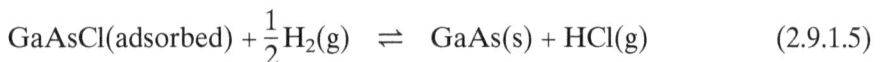

$$\frac{1}{4} As_4(g) \; \rightleftharpoons \; As(adsorbed) \tag{2.9.1.3}$$

$$As(adsorbed) + GaCl(g) \; \rightleftharpoons \; GaAsCl(adsorbed) \tag{2.9.1.4}$$

$$GaAsCl(adsorbed) + \frac{1}{2} H_2(g) \; \rightleftharpoons \; GaAs(s) + HCl(g) \tag{2.9.1.5}$$

Similar complex processes must occur, for example, during the deposition of $Nb_3O_7Cl(s)$ according to equation 2.9.1.6.

$$7 NbOCl_3(g) \; \rightleftharpoons \; Nb_3O_7Cl(s) + 4 NbCl_5(g) \tag{2.9.1.6}$$

As a further example we can look at the deposition of $Pd_5AlI_2(s)$ from gaseous $PdAl_2I_8$. This gas complex (see section 3.4) is formed by the reaction of PdI_2 and Al_2I_6. Roughly speaking the following steps should occur during deposition:

- $PdAl_2I_8$ molecules are adsorbed on the surface and dissociate.
- Pd atoms and Al atoms or PdAl molecules migrate on the surface until they reach the most favorable site where they are deposited with iodine atoms.
- The surplus of iodine and AlI_3 is released into the gas phase (Mer 1980).

$$5 PdAl_2I_8(g) \; \rightleftharpoons \; Pd_5AlI_2(s) + 9 AlI_3(g) + \frac{11}{2} I_2(g) \tag{2.9.1.7}$$

Such complex processes are hidden in the overall reaction of a chemical transport, if we consider only the equilibria between solid and gas. Many kinds of transport agents also support the migration of the components of the solid through the adsorbed film onto the surface of the growing crystal. However, they are seldomly incorporated into the solid.

2.9.2 Kinetic Influences Observed by Transport Experiments

The most important case, that for the influence of kinetic effects on the course of a chemical transport, is provided by the *Boudouard* reaction (2.9.2.1), by which the evaporation of carbon is investigated.

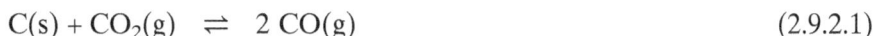

$$C(s) + CO_2(g) \;\rightleftharpoons\; 2\,CO(g) \tag{2.9.2.1}$$

(Experimental conditions: silica ampoule with 8 mm inside diameter, 200 mm length, 1 g of spectral grade carbon powder outgassed in high vacuum at 1100 °C; initial pressure $p^0(CO_2) = 0.36$ bar at 800 °C; length of the diffusion path $s = 15$ cm (Sch 1958)) Within 45 hours approximately 1 mg carbon (1000 → 600 °C) was transported.

The transport of carbon was clearly detectable under these conditions but its amount was about 100 times less than expected for a process in which diffusion is rate determining. Therefore, the rate of reaction was the slowest step. The significance of the reaction rate for the carbon transport according to the *Boudouard* reaction was demonstrated by reactions with carbon filament lamps too. Here also, the transport rate is reduced by the kinetic stability of CO.

Similar situations are encountered, as expected, with the transport of nitrides due to the high stability of the N_2-molecule. In this case transport of TiN is possible only if the temperature is high, e.g., 1500 °C (Sch 1962).

In contrast, platinum transport via $Pt(CO)_2X_2$ ($X = Cl$, Br) takes advantage of the kinetic stability of carbon monoxide (Sch 1971). To some extent this reaction is similar to the transport of platinum in a stream of metastable NO (Teb 1962). In that case, the transport reaction is superimposed by the catalytic decomposition of NO.

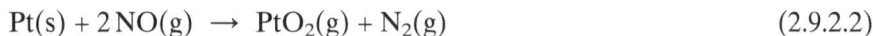

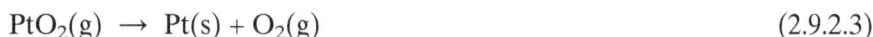

$$Pt(s) + 2\,NO(g) \;\rightarrow\; PtO_2(g) + N_2(g) \tag{2.9.2.2}$$

$$PtO_2(g) \;\rightarrow\; Pt(s) + O_2(g) \tag{2.9.2.3}$$

Arizumi and *Nishinaga* studied the transport of germanium by means of iodine in a closed system (ampoule dimensions $l = 30$ cm, $d = 1.6$ cm, $p(I_{2,298}) = 96$ mbar) (Ari 1967). They found that at 400 °C the adsorption of GeI_2 was rate determining. At 450 °C the transport rate was determined by the decomposition reaction 2.9.2.4. Above 500 °C, gas motion by diffusion (and therefore thermodynamics) was the determining factor.

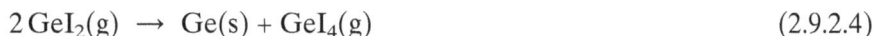

$$2\,GeI_2(g) \;\rightarrow\; Ge(s) + GeI_4(g) \tag{2.9.2.4}$$

Analogous studies on the transport behavior of germanium by iodine and bromine, respectively, in a temperature gradient 430 to 395 °C (ampoule dimensions $l = 20$ cm, $d = 0.8$ cm, $p(I_{2, 298}) = 0.1$ to 133 mbar) show that the Ge/I_2 transport was diffusion controlled. In contrast, for the system Ge/Br_2 a kinetic influence (surface-controlled process) was found (Jon 1965). Chemical kinetics and the parameter for diffusion and convection for the transports of germanium by $GeCl_4$ have been reported by *Schäfer* and *Trenkel* (Sch 1980a).

With regard to the thermodynamics a chemical transport of SiO_2 by $TeCl_4$ ($1000 \rightarrow 800$ °C) can be expected. It is quite remarkable and important for the use of silica ampoules in transport experiments that such a transport has never been observed. Here, also, the reaction rate of the solid is too small (Sch 1978). *Mizumo* and *Watanabe* studied the deposition of GaAs, GaP, InAs, and InP from a gas phase containing H_2, an inert gas, $AsCl_3$, or PCl_3. They found a region at temperature $\vartheta < 750$ °C controlled by kinetics, and another region above 750 °C where thermodynamic control of the process was observed (Miz 1975).

These results indicate exceptional cases, where the transport effect is determined by the speed of the heterogeneous reaction. These exceptions are particularly surprising for the experiments at medium and high temperatures (700 to 1000 °C). On switching to lower temperatures it is obvious that one will run into reaction inhibitions even more frequently.

The zirconium transport according to the iodide method (Sha 1955) can be cited as another example. When the temperature of the metallic starting material is only 285 °C, the transport rate of the zirconium to the hot filament depends on the surface area and the reactivity of the crude metal. The significance of the reaction rate at the crude zirconium also becomes evident from the fact that at the beginning the Zr transport rate goes through a maximum, before it passes through a minimum. This phenomenon has not been fully explained. It is plausible, however, that at the lowest temperature formation of the relatively volatile ZrI_4 is kinetically controlled. Therefore, its concentration in the gas phase increases at first with an increase in temperature. Upon further increase of the temperature the thermodynamically stable, less volatile lower iodides are formed. These compounds coat the metal surface and at the same time remove the transport agent from the gas phase, thus the transport effect is decreased. When the temperature is reached at which the lower iodides disproportionate into Zr and ZrI_4 the reactivity of the starting material and the iodide content in the gas phase will increase. As a consequence of this the transport rate increases too (Dör 1952). Similar observations were made for the transport of niobium by the iodide method.

2.9.3 Some Observations on Catalytic Effects

A well known example of the influence of catalysis on the course of a transport experiment is the transport of nickel in a temperature gradient ($80 \rightarrow 180$ °C) by means of (the metastable) carbon monoxide.

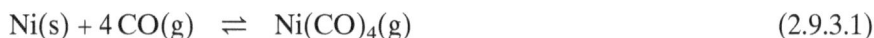

$$Ni(s) + 4 CO(g) \rightleftharpoons Ni(CO)_4(g) \tag{2.9.3.1}$$

At the low source temperature (80 °C) the reaction of the nickel foil can be observed only in the presence of sulfur as a catalyst, whereas the reverse reaction (decomposition) at 180 °C is fast enough by itself (Sch 1982). When using nickel powder of sufficiently fine particle size introduction sulfur as a catalyst is not required.

According to observations by *Gruehn* and co-workers (Red 1978, Schm 1984), a kinetic hindrance of the transport of GeO_2 by chlorine can be overcome by addition of manganese(II)-chloride or of alkali metal chlorides. For chemical transport of silica by hydrogen fluoride *Gruehn* and co-workers (Hof 1977) found that the reverse reaction can be catalyzed by adding platinum, nickel foil, and potassium fluoride, respectively. Furthermore, the transport of SiO_2 by niobium(V) chloride in a temperature gradient 1050 to 950 °C was investigated (Hib 1977). The authors observed that the reverse reaction (deposition of SiO_2) is hindered on the silica surface for low initial pressures of $NbCl_5$ ($p^0(NbCl_5, 298)$ < 2 bar).

2.9.4 Indirect Transport

An influence of kinetics may be assumed for the formation of NbO_2Cl from $Nb_2O_5/NbCl_5$ mixtures. The experiments lead to the conclusion that during the transport process initially at the sink at T_1 $NbOCl_3(s)$ is deposited (Zyl 1966), which then is decomposed according to:

$$2\,NbOCl_3(s) \; \rightleftharpoons \; NbO_2Cl(s) + NbCl_5(g) \tag{2.9.4.1}$$

A similar observation has been made during the transport of $Nb_{12}O_{29}$ by $NbCl_5$ in silica ampoules (Hib 1977). The primarily deposited $Nb_{12}O_{29}$ afterwards reacts at longer experimental time with gaseous $SiCl_4$ to form amorphous SiO_2. The latter has been deposited as pseudomorph after $Nb_{12}O_{29}$. A similar observation was made for the reaction of an iron coil with silicon(IV)- chloride. Here a coil of FeSi had been formed (Bin 2001).

At the end of this chapter on models related to CVT a "caveat" appears to be appropriate:

Each model, and in particular the various models concerning CVT, will describe the experiment reasonably well, only when the underlying data are reliable. For CVT these are the thermodynamic data for enthalpy, entropy, and specific heat of all compounds involved in the process. The accuracy of these data can be of very different quality (see Chapter 12). A critical discussion of the results of modeling of transport experiments requires as a prerequisite the assessment of the accuracy of the used (thermodynamic) data.

Bibliography

Alc 1967   C. B. Alcock, J. H. E. Jeffes, *Trans. Inst. Mining* **1967**, *76*, C246.

And 1950   L. Andrussow, *Z. Elektrochem.* **1950**, *54*, 566. Landolt-Börnstein, Table II/ 5a, **1969**.

Ari 1967   T. Arizumi, T. Nishinaga, *Crystal Growth Suppl. J. Phys. Chem. Solids* **1967**, C14.

Bal 1983   L. Bald, M. Spiess, R. Gruehn, Th. Kohlmann, *Z. Anorg. Allg. Chem.* **1983**, *498*, 153.

Bin 1996   M. Binnewies, *Chemische Gleichgewichte: Grundlagen, Berechnungen, Tabellen*, VCH, Weinheim, **1996**.

Bin 2002   M. Binnewies, E. Milke, *Thermodynamic Data of Elements and Compounds*, 2nd Ed.Wiley-VCH, Weinheim u. a., **1999**.

Bin 2001   M. Binnewies, A. Meyer, M. Schütte, *Angew. Chem.* **2001**, *113*, 3801, *Angew. Chem. Intern. Ed.* **2001**, *40*, 3688

Bre 1955   L. Brewer, J. S. Kane, *J. Phys. Chem.* **1955**, *59*, 105.

Bru 1975   H. Brunner, *Dissertation*, University of Gießen, **1975**.

Bur 1971   J. Burmeister, *Mater. Res. Bull.* **1971**, *6*, 219.

Cad 1975   R. Cadoret, L. Hollam, J. B. Loyau, M. Oberlin, A. Oberlin, *J. Cryst. Growth* **1975**, *29*, 187.

Cat 1960   E. D. Cater, E. R. Plante, P. W. Gilles, *J. Chem. Phys.* **1960**, *32*, 1269.

Cha 1953   S. Chapman, T. G. Cowling, *The Math. Theory of Non-Uniform Gases*, University Press, Cambridge, **1953**.

Cho 1979   S. Choukroun, J. C. Launay, M. Pouchard, M. Combarnous, *J. Cryst. Growth* **1979**, *46*, 644.

Clu 1948   K. Clusius, zitiert in: A. Klemenc: *Die Behandlung und Reindarstellung von Gasen*, Springer, Wien **1948**.

Cor 1983   J. D. Corbett, *Inorg. Synthesis* **1983**, *22*, 15.

Daa 1952   A. H. Daane, *Rev. Sci. Instruments* **1952**, *23*, 245.

Dah 1992   T. Dahmen, R. Gruehn *Z. Anorg. Allg. Chem.* **1992**, *609*, 139.

Dan 1973   P. N. Dangel, B. J. Wuensch, *J. Cryst. Growth* **1973**, *19*, 1.

Dör 1952   H. Döring, K. Molière, *Z. Elektrochem.* **1952**, *56*, 403.

Doe 1973   H. A. Doerner, *A. S. Bur. Mines, Tech. Paper*, 577, **1973**.

Emm 1968   F. Emmenegger, *J. Cryst. Growth* **1968**, *3/4*, 135.

Eri 1971   G. Eriksson, *Acta Chem. Scand.* **1971**, *25*, 2651.

Fak 1971   M. M. Faktor, I. Garrett, R. Heckingbottom, *J. Cryst. Growth* **1971**, *9*, 3.

Flö 1983   U. Flörke, *Z. Anorg. Allg. Chem.* **1983**, *502*, 218.

Fuh 1961   W. Fuhr, *Diplomarbeit*, University of Münster, **1961**.

Ful 1966   E. N. Fuller, P. D. Schettler, J. C. Giddings, *Inst. Enging. Chem.* **1966**, *58*, 19.

Gla 1989a   R. Glaum, R. Gruehn *Z. Anorg. Allg. Chem.* **1989**, *568*, 73.

Gla 1989b   R. Glaum, R. Gruehn, *Z. Anorg. Allg. Chem.* **1989**, *573*, 24.

Gör 1996   H. Görzel, R. Glaum, *Z. Anorg. Allg. Chem.* **1996**, *622*, 1773.

Got 1963   G. E. Gottlieb, J. F. Corboy, *R. C. A. Rev.* **1963**, *24*, 585.

Gre 1981   D. W. Greenwell, B. L. Markham, F. Rosenberger, *J. Cryst. Growth* **1981**, *51*, 413.

Gru 1983   R. Gruehn, H. J. Schweizer, *Angew. Chem.* **1983**, *95*, 80.

Gru 2000   R. Gruehn, R. Glaum, *Angew. Chem.* **2000**, *112*, 706, *Angew. Chem. Int. Ed.* **2000**, *39*, 692.

Has 1975   J. W. Hastie, *High-Temperature Vapors*, New York, **1975**.

Hib 1977   H. Hibst, R. Gruehn, *Z. Anorg. Allg. Chem.* **1977**, *434*, 63.

Hof 1977   J. Hofmann, R. Gruehn, *Z. Anorg. Allg. Chem.* **1977**, *431*, 105.

Hol 1953	R. B. Holden, B. Kopelman, *J. Electrochem. Soc.* **1953**, *100*, 120.
Hug 1966	P. Hugo, *Ber. Bunsenges. Phys. Chem.* **1966**, *70*, 44.
Iha 1982	B. S. Ihaveri, F. Rosenberger, *J. Cryst. Growth* **1982**, *57*, 57.
Ind 1966	X. Indradev, *Mater. Res. Bull.* **1966**, *1*, 173.
Jel 1928	K. Jellinek, *Lehrbuch der Physikalischen Chemie*, Enke, Stuttgart, **1928**.
Jon 1963	F. Jona, G. Mandel, *J. Chem. Phys.* **1963**, *38*, 346.
Jon 1965	F. Jona, *J. Chem. Phys.* **1965**, *42*, 1025.
Kal 1968	E. Kaldis, *J. Cryst. Growth* **1968**, *3,4*, 146.
Kem 1957	C. P. Kempter, C. Alvarez-Tostado, *Z. Anorg. Allg. Chem.* **1957**, *290*, 238.
Kle 1969	P. Kleinert, "*Chemischer Transport oxidischer Metallverbindungen, besonders Ferriten, im System Festkörper-HCl*", in: J. W. Mitchel, R. C. DeVries, R. W. Roberts, P. Cannon (Hrsg.): Proc. 6th Intl. Symp. On the Reactivity of Solids. New York, Wiley Interscience, **1969**.
Kle 1972	P. Kleinert, *Z. Anorg. Allg. Chem.* **1972**, *387*, 11.
Klo 1965a	H. Klotz, *Vakuum-Technik* **1965**, *15*, 63.
Klo 1965b	H. Klotz, *Naturwissenschaften* **1965**, *52*, 451.
Klo 1973	K. Klosse, P. Ullersma, *J. Cryst. Growth* **1973**, *18*, 167.
Klo 1975	K. Klosse, *Dissertation*, Utrecht **1973**. *J. Solid State Chem.* **1975**, *15*, 105.
Kra 1975	G. Krabbes, H. Oppermann, E. Wolf, *Z. Anorg. Allg. Chem.* **1975**, *416*, 65.
Kra 1976a	G. Krabbes, H. Oppermann, E. Wolf, *Z. Anorg. Allg. Chem.* **1976**, *421*, 111.
Kra 1976b	G. Krabbes, H. Oppermann, E. Wolf,. *Z. Anorg. Allg. Chem.* **1976**, *423*, 212.
Kra 1982	G. Krabbes, *Zur Thermodynamik heterogener Gleichgewichte bei der Abscheidung fester Verbindungen aus der Gasphase in komplexen chemischen Reaktionssystemen*, Academy of Sciences of the German Democratic Republic, Dresden, **1982**.
Kra 1983	G. Krabbes, H. Oppermann, E. Wolf, *J. Crystal Growth* **1983**, 64, 353.
Kra 2008	G. Krabbes, W. Bieger, K.-H. Sommer, T. Söhnel, U. Steiner, *Computerprogram TRAGMIN*, Version 5.0, IFW Dresden, TU Dresden, HTW Dresden, **2008**.
Krä 1974	V. Krämer, R. Nitsche, M. Schuhmacher, *J. Cryst. Growth* **1974**, *24/25*, 179.
Kub 1993	O. Kubaschewski, C. B. Alcock, P. J. Spencer, *Materials Thermochemistry*, 6. Auflage, Pergamon Press, Oxford, **1993**.
Lan 2008	S. Lange, M. Bawohl, R. Weihrich, T. Nilges, *Angew. Chem. Int. ed.* **2008**, *47*, 5654.
Lau 1981	J. C. Launay, J. Miroglio, B. Roux, *J. Cryst. Growth* **1981**, *51*, 61.
Lau 1982	J. C. Launay, B. Roux, *J. Cryst. Growth* **1982**, *58*, 354.
Len 1994	M. Lenz, R. Gruehn, *Z. Anorg. Allg. Chem.* **1994**, *620*, 867.
Len 1997	M. Lenz, R. Gruehn, *Chem. Rev.* **1997**, *97*, 2967.
Lev 1962a	R. F. Lever, G. Mandel, *J. Phys. Chem. Solids* **1962**, *23*, 599.
Lev 1962b	R. F. Lever, *J. Chem. Phys.* **1962**, *37*, 1078; **1962**, *37*, 1174; **1962**, *37*, 1177.
Lev 1966	R. F. Lever, F. Jona, *A. I. Ch. E.* **1966**, *12*, 1158.
Loc 2000	S. Locmelis, *Dissertation*, University of Hannover, **2000**.
Loo 1959	A. C. Loonam, *J. Electrochem. Soc.* **1959**, *106*, 238.
Man 1962	G. Mandel, *J. Chem. Phys.* **1962**, *37*, 1177.
Mar 1981	B. L. Markham, D. W. Greenwell, F. Rosenberger, *J. Cryst. Growth* **1981**, *51*, 426.
Mer 1980	H. B. Merker, H. Schäfer, B. Krebs, *Z. Anorg. Allg. Chem.* **1980**, *462*, 49.
Miz 1975	O. Mizumo, H. Watanabe, *J. Cryst. Growth* **1975**, *30*, 240.
Möl 1982	M. Möller, W. Jeitschko, *Z. Anorg. Allg. Chem.* **1982**, *491*, 225.
Möl 1986	M. Möller, W. Jeitschko, *J. Solid State Chem.* **1986**, *65*, 178.
Mor 1952	W. Morawietz, *Z. Elektrochem.* **1952**, *56*, 407.

Mül 1968   R. Müller, *Chem. Ing. Tech.* **1968**, *40*, 344.

Ned 1996   R. Neddermann, S. Gerighausen, M. Binnewies, *Z. Anorg. Allg. Chem.* **1996**, *622*, 21.

Nic 1972   W. Nicolaus, E. Seidowski, V. A. Voronin, *Kristall und Techn.* **1972**, *7*, 589.

Nit 1971   R. Nitsche, *J. Cryst. Growth* **1971**, *9*, 238.

Noc 1993a   K. Nocker, R. Gruehn, *Z. Anorg. Allg. Chem.* **1993**, *619*, 699.

Noc 1993b   K. Nocker, *Dissertation*, University of Gießen **1993**.

Opp 1968   H. Opperman, *Z. Anorg. Allg. Chem.* **1968**, *359*, 51.

Opp 1977a   H. Oppermann, W. Reichelt, E. Wolf, *Z. Anorg. Allg. Chem.* **1977**, *432*, 26.

Opp 1977b   H. Oppermann, W. Reichelt, G. Krabbes, E. Wolf, *Kristall Techn.* **1977**, *12*, 717.

Opp 1977c   H. Oppermann, W. Reichelt, G. Krabbes, E. Wolf, *Kristall Techn.* **1977**, *12*, 919.

Opp 1981   H. Oppermann, *Wiss. Ber. Akad. Wiss. DDR* **1981**, *22*, 51.

Oxl 1961   J. H. Oxley, J. M. Blocher, *J. Electrochem. Soc.* **1961**, *108*, 460.

Öza 1993   D. Özalp, *Dissertation*, University of Gießen, **1993**.

Pao 1975   C. Paorici, C. Pelosi, G. Attolini, G. Zuccalli, *J. Cryst. Growth* **1975**, *28*, 358.

Pau 1997   N. Pausch, J. Burggraf, R. Gruehn, *Z. Anorg. Allg. Chem.* **1997**, *623*, 1835.

Pfi 1995   A. Pfitzner, E. Freudenthaler, *Angew. Chem.* **1995**, *107*, 1784, *Angew. Chem. Int. Ed.* **1995**, *34*, 1647.

Pli 1989   V. Plies, T. Kohlmann, R. Gruehn, *Z. Anorg. Allg. Chem.* **1989**, *568*, 62.

Red 1978   W. Redlich, R. Gruehn, *Z. Anorg. Allg. Chem.* **1978**, *438*, 25.

Rei 1958   R. C. Reid, T. K. Sherwood, *The Properties of Gases and Liquids*, New York, **1958**, p. 281.

Rei 1977   W. Reichelt, Dissertation, *Academy of Sciences of the German Democratic Republic*, **1977**.

Ric 1977   M. W. Richardson, B. I. Noläng, *J. Crystal Growth* **1977**, *42*, 90.

Ros 1975   F. Rosenberger, M. C. Delong, D. W. Greenwell, J. M. Olson, G. H. Westphal, *J. Cryst. Growth* **1975**, *29*, 49.

Sae 1976   M. Saeki, *J. Crystal Growth* **1976**, *36*, 77.

Sch 1949   H. Schäfer, *Z. Anorg. Allg. Chem.* **1949**, *259*, 75.

Sch 1956a   H. Schäfer, H. Jacob, K. Etzel, *Z. Anorg. Allg. Chem.* **1956**, *286*, 27.

Sch 1956b   H. Schäfer, H. Jacob, K. Etzel, *Z. Anorg. Allg. Chem.* **1956**, *286*, 42.

Sch 1957a   H. Schäfer, K. Etzel, *Z. Anorg. Allg. Chem.* **1957**, *291*, 294.

Sch 1957b   H. Schäfer, B. Morcher, *Z. Anorg. Allg. Chem.* **1957**, *290*, 279.

Sch 1958   H. Schäfer, J. Tillack, *unpublished results*, **1958**.

Sch 1959a   H. Schäfer, K. Etzel, *Z. Anorg. Allg. Chem.* **1959**, *301*, 137.

Sch 1959b   H. Schäfer, K. D. Dohmann, *Z. Anorg. Allg. Chem.* **1959**, *300*, 1.

Sch 1962   H. Schäfer, *Chemische Transportreaktionen*, Verlag Chemie, Weinheim **1962**.

Sch 1962b   H. Schäfer, W. Fuhr, *Z. Anorg. Allg. Chem.* **1962**, *319*, 52.

Sch 1971   H. Schäfer, U. Wiese, *J. Less-Common Met.* **1971**, *24*, 55.

Sch 1972   H. Schäfer, *"Preparation of Oxides and Related Compounds by Chemical Transport"*, in: Natl. Bur. of Standards, Special Publ. *364*, **1972**, 5th Materials Res. Symposium Solid State Chemistry.

Sch 1973   H. Schäfer, *Z. Anorg. Allg. Chem.* **1973**, *400*, 242.

Sch 1978   H. Schäfer, M. Binnewies *Z. Anorg. Allg. Chem.* **1978**, *441*, 216.

Sch 1978b   H. Schäfer, M. Trenkel, *Z. Naturforsch.* **1978**, *33b*, 1318.

Sch 1980a   H. Schäfer, M. Trenkel, *Z. Anorg. Allg. Chem.* **1980**, *461*, 22.

Sch 1980b   H. Schäfer, *Z. Anorg. Allg. Chem.* **1980**, *471*, 21.

Sch 1982   H. Schäfer, *Z. Anorg. Allg. Chem.* **1982**, *493*, 17.

Sch 1982b   H. Schäfer, *Z. Anorg. Allg. Chem.* **1982**, *489*, 154.

Scha 1974	E. Schaum, *Diplomarbeit*, University of Gießen, **1974**.
Schm 1981	G. Schmidt, R. Gruehn, *J. Cryst. Growth* **1981**, *55*, 599.
Schm 1984	G. Schmidt, R. Gruehn, *Z. Anorg. Allg. Chem.* **1984**, *512*, 193.
Schm 1992	A. Schmidt, *Diplomarbeit*, University of Gießen **1992**.
Schm 1995	A. Schmidt, R. Glaum, *Z. Anorg. Allg. Chem.* **1995**, *621*, 1693.
Schm 1995a	H. Schmalzried, *Chemical Kinetics of Solids*, Wiley-VCH, Weinheim, **1995**.
Schö 1980	E. Schönherr, *Crystals, Growth, Properties, Applications*, 2, Springer, Berlin, **1980.**
Schö 2010	M. Schöneich, M. Schmidt, P. Schmidt, *Z. Anorg. Allg. Chem.* **2010**, *636*, 1810.
Scho 1965	R. C. Schoonmaker, A. Buhl, J. Lemley, *J. Phys. Chem.* **1965**, *69*, 3455.
Scho 1989	H. Schornstein, R. Gruehn, *Z. Anorg. Allg. Chem.* **1989**, *579*, 173.
Scho 1990	H. Schornstein, R. Gruehn, *Z. Anorg. Allg. Chem.* **1990**, *587*, 129.
Schw 1983a	H.-J. Schweizer, *Dissertation*, University of Gießen, **1983**.
Sei 1983	F. J. Seiwert, R. Gruehn, *Z. Anorg. Allg. Chem.* **1983**, *503*, 151.
Sei 1984	F. J. Seiwert, R. Gruehn, *Z. Anorg. Allg. Chem.* **1984**, *510*, 93.
Sha 1955	Z. M. Shapiro, cited in: B. Lustmann, F. Kerze, *The Metallurgy of Zirconium*, McGraw-Hill, New York **1955**.
Stu 1972	J. Sturm, Diplomarbeit, University of Gießen, 1972.
Teb 1962	A. Tebben, *Dissertation*, University of Münster, **1962**.
Tha 2003	H. Thauern, R. Glaum, *Z. Anorg. Allg. Chem.* **2003**, *629*, 479.
Tra 1994	O. Trappe, *Diplomarbeit*, University of Gießen, **1994**.
Tra 1999	O. Trappe, R. Glaum, R. Gruehn, *Computerprogram CVTRANS*, University of Gießen, **1999**.
Whi 1958	W. B. White, S. M. Johnson, G. B. Dantzig, *J. Chem. Phys.* **1958**, *28*, 751.
Wie 1972	H. Wiedemeier, E. A. Irene, A. K. Chaudhuri, *J. Cryst. Growth.* **1972**, *13–14*, 393.
Wie 1976	H. Wiedemeier, H. Sadeek, F. C. Klaessig, M. Norek, R. Santandrea, *E.S.A. Spec. Publ.* **1976**, *114. Mater. Sci. Space* N77 – 14066, 189.
Wie 1981	H. Wiedemeier, D. Chandra, F. C. Klaessig, *J. Cryst. Growth.* **1981**, *51*, 345.
Wil 1950	C. R. Wilke, *Chem. Eng. Progr.* **1950**, *46*, 95. Zitiert in: R. C. Reid, T. K. Sherwood, *The Properties of Gases and Liquids*, S. 281, New York, **1958**.
Zeg 1970	F. van Zeggeren, S. H. Storey, *The Computation of Chemical Equilibria*, Cambridge University Press, New York **1970**.
Zel 1968	F. J. Zeleznik, S. Gordon, *Ind. Eng. Chem.* **1968**, *60*, 27.
Zin 1935	E. Zintl, A. Harder, *Z. Elektrochem.* **1935**, *41*, 767.
Zyl 1966	L. Zylka, *Dissertation*, University of Münster, **1966**.

3 Chemical Vapor Transport of Elements

Chrom
according to *van Arkel* and *de Boer*.

H																	He
Li	Be											B	C	N	O	F	Ne
Na	Mg											Al	Si	P	S	Cl	Ar
K	Ca	Sc	Ti	V	Cr	Mn	Fe	Co	Ni	Cu	Zn	Ga	Ge	As	Se	Br	Xe
Rb	Sr	Y	Zr	Nb	Mo	Tc	Ru	Rh	Pd	Ag	Cd	In	Sn	Sb	Te	I	Kr
Cs	Ba	La	Hf	Ta	W	Re	Os	Ir	Pt	Au	Hg	Tl	Pb	Bi	Po	At	Rn

Ce	Pr	Nd	Pm	Sm	Eu	Gd	Tb	Dy	Ho	Er	Tm	Yb	Lu
Th	Pa	U	Np	Pu									

Van-Arkel–de-Boer-process

$$Zr(s) + 2\,I_2(g) \;\rightleftharpoons\; ZrI_4(g)$$

Conproportionation reaction

$$Si(s) + SiI_4(g) \;\rightleftharpoons\; 2\,SiI_2(g)$$

Transport with oxygen

$$Pt(s) + O_2(g) \;\rightleftharpoons\; PtO_2(g)$$

The CVT of elements has been studied and described in detail using metals and some semi-metals as examples. In the case of the typical non-metals phosphorus and sulfur, there is no need to increase their volatility in the sense of a CVT reaction due to their high vapor pressures. This way, those metals and semi-metals that feature high vapor pressures can also easily be transferred into the gas phase through distillation or sublimation and be deposited from it again. The following elements belong to this group: alkali and alkaline earth metals, zinc, cadmium, mercury, europium, ytterbium, arsenic, antimony, selenium, and tellurium. Some metals' melting temperatures are so low that they can be obtained in liquid form at the most. This, for example, applies for gallium, tin, and lead. Chemical vapor transport is relevant for high melting elements with low vapor pressures. These elements can be deposited from the gas phase in closed systems (ampoules), fluid systems, special reactors (hot-wire process according to *van Arkel* and *de Boer*), or through chemical vapor deposition (CVD) processes (Cho 2003). All of these processes are based on the same thermodynamic basic principles. This way, more than 40 elements can be crystallized using CVT reactions, more than 25 with iodine as the transport agent (Rol 1960, Rol 1961).

In addition to iodine as most important transport agent for elements, compounds such as aluminum(III)-chloride, gallium(III)-chloride, and iron(III)-chloride, as well as aluminum(III)-iodide and indium(III)-iodide, are described as transport effective additives (Sch 1975). These can act as halogenating and thus form gaseous halides with the transporting elements. Also, they stabilize them by forming gaseous complexes (see Chapter 11). Halogens such as fluorine, chlorine, and bromine as well as the hydrogen halides, water, the chalcogens, oxygen, sulfur, selenium, tellurium, as well as carbon monoxide are other transport agents that can be used in individual cases. Although carbon monoxide can be used for the transport of nickel only, the industrial purifying process according to *Mond* and *Langer* has found its way into chemistry textbooks, making carbon monoxide particularly prominent as a transport agent (Mon 1890).

3.1 Transport with Halogens

Transport with iodine   Due to the great importance of iodine as transport agent, the CVT of elements with iodine shall be elaborated upon. The underlying reactions can principally be applied to the transport of other halogens.

During the reaction of a solid substance with iodine with the formation of a gaseous iodide, the following cases can be distinguished:

- *The reaction is highly exothermic*: no transport effect is observed due to the fact that no back reaction can take place. Exception: during the hot-wire process according to *van Arkel* and *de Boer,* the equilibrium can be strongly shifted to the side of the formed iodides ($\Delta_r G^0$ up to $-600 \, \text{kJ} \cdot \text{mol}^{-1}$). Due to the very high temperatures, partly up to 2000 °C, the back reaction can take place anyhow.
- *The reaction is exothermic*: there is a transport from cold to hot.
- *The reaction is endothermic*: there is a transport from hot to cold.
- *The reaction is highly endothermic*: there is no transport because iodides with transport effective partial pressures are not formed.

In the case of highly exothermic reactions the elements form thermally very stable iodides, which can principally be decomposed on the deposition side. This would, however, require such a high temperature that the elements would exhibit considerable vapor pressures. In some cases the decomposition temperature of iodides is higher than the boiling temperature of the elements. This is the case for the alkali and alkaline earth metals as well as europium, manganese, zinc, cadmium, mercury, thallium, and lead.

The chemical transport of an element with iodine from T_1 to T_2 is the most frequently described transport reaction involving metals. It has extensively been studied and described among others for titanium, zirconium, hafnium, and thorium (process of *van Arkel* and *de Boer* (Ark 1925, Boe 1926, Boe 1930, Ark 1934)). Further elements that can be transported in this way are yttrium, vanadium, niobium, tantalum, chromium, iron, cobalt, nickel, copper, boron, silicon, germanium, and tin as well as uranium and protactinium (Rol 1960, Rol 1961, She 1966, Has 1967, Cue 1978, Spi 1979). This kind of transport to the hotter zone shall be illustrated with the example of the transport of zirconium with iodine. The temperatures of the dissolving side T_1 can vary between 200 and 650 °C. The most suitable temperature is between 350 and 400 °C. The temperatures of the decomposing side T_2 can be between 1100 and 2000 °C; whereby temperatures around 1400 °C are usually used. Often, a wire heated by an electrical current is the place of decomposition. The transport behavior is described by the equilibria 3.1.1 to 3.1.3.

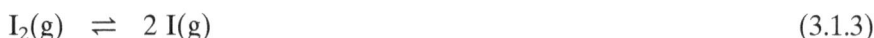

$$Zr(s) + 4\,I(g) \;\rightleftharpoons\; ZrI_4(g) \tag{3.1.1}$$

$$ZrI_2(g) + 2\,I(g) \;\rightleftharpoons\; ZrI_4(g) \tag{3.1.2}$$

$$I_2(g) \;\rightleftharpoons\; 2\,I(g) \tag{3.1.3}$$

In Figure 3.1.1 the partial pressures in the system Zr/I are illustrated as a function of temperature. Figure 3.1.2 shows the transport efficiency of the individual species. Further gas species that are to be taken into account in the system are ZrI_3 and ZrI. However, these are of minor relevance to the description of transport events (Eme 1956, Eme 1957).

The transport of uranium (850 → 950 °C) (Has 1967) and copper (420 → 920 °C) (She 1966) with iodine are two examples of exothermic reactions according to the *van Arkel–de Boer* process with relatively low temperatures on the decomposition side.

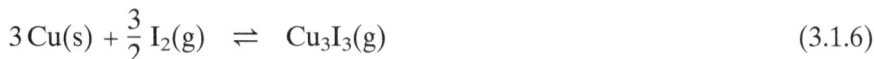

$$U(s) + \frac{3}{2}\,I_2(g) \;\rightleftharpoons\; UI_3(g) \tag{3.1.4}$$

$$U(s) + 2\,I_2(g) \;\rightleftharpoons\; UI_4(g) \tag{3.1.5}$$

$$3\,Cu(s) + \frac{3}{2}\,I_2(g) \;\rightleftharpoons\; Cu_3I_3(g) \tag{3.1.6}$$

In this temperature range (500 → 1050 °C) niobium, tantalum, and silicon can also be transferred with a carrier gas stream, which is loaded with iodine via the gas phase.

The transport of iron (800 → 1000 °C) is an example of an exothermic transport of an element with iodine in an ampoule.

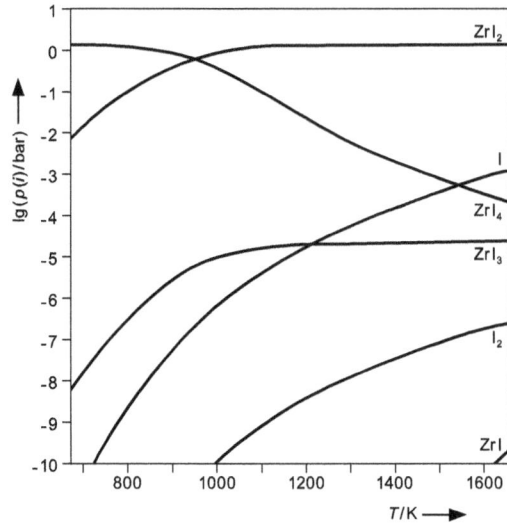

Figure 3.1.1 Partial pressures in the Zr/I system as a function of the temperature.

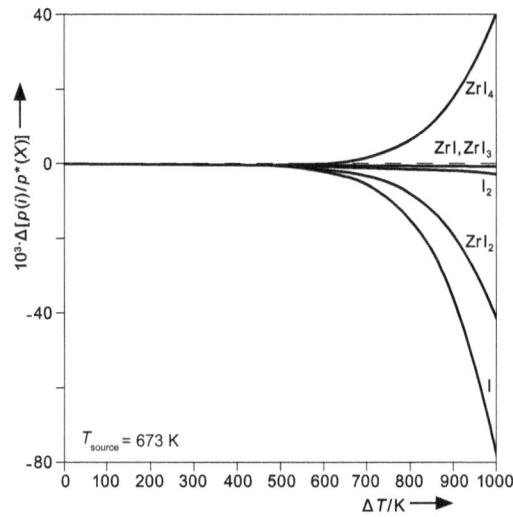

Figure 3.1.2 Transport efficiency of essential gas species in the Zr/I system.

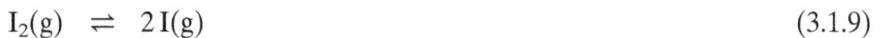

$$Fe(s) + I_2(g) \rightleftharpoons FeI_2(g) \tag{3.1.7}$$

$$2\,Fe(s) + 2\,I_2(g) \rightleftharpoons Fe_2I_4(g) \tag{3.1.8}$$

$$I_2(g) \rightleftharpoons 2\,I(g) \tag{3.1.9}$$

The first equilibrium is endothermic and thus cannot cause transport into the hotter zone. That is why the second, exothermic, equilibrium is decisive for the description of the exothermic transport behavior (see Chapter 1).

Using the example of the transport of germanium with iodine from T_2 to T_1, *Oppermann* and co-workers investigated the proportion of diffusion and convection of the gas movement at different total pressures. In comparative experiments, the transport behavior was determined at normal gravity on earth and under microgravity in space (Opp 1981). The transport rates of those experiments conducted under normal gravity increased at pressures above approximately 3 bar exponentially with the pressure. In contrast, the transport rates under microgravity did not show a dependency on the total pressure. Under microgravity conditions, the convection is negligibly small; substance transport takes place by diffusion only. This indicates that in the gravitational field of earth, gas movement above 3 bar occurs not only by diffusion, but increasingly by convection.

The knowledge gained for the exothermic transport with iodine also holds for the other halogens. However, their meaning as transport agents for elements is of little importance due to their unsuitable equilibrium position. Because the stability of halides increases from iodides to fluorides, their decomposition temperatures increase as well in this direction. Higher decomposition temperatures become necessary, which are more difficult to put into practice in experiments.

Transport with chlorine and bromine Other elements such as molybdenum, rhodium, palladium, osmium, iridium, and platinum do not form gaseous iodides with a sufficient stability to be transport effective (Rol 1961). The solid iodides decompose into non-volatile metals and elementary gaseous iodine while heating. For palladium and platinum, an endothermic transport with iodine is described, which, however, takes place with low transport rates due to the unsuitable equilibrium position. If iodine is unsuitable as a transport agent, bromine and particularly chlorine form stable and thus more transport-effective species in many cases. For example, the transport of molybdenum, palladium, rhenium, osmium, platinum, and gold with chlorine is possible. The endothermic transport of gold (700 $\rightarrow$ 500 °C) can be described with the equilibrium 3.1.10 (Sch 1975).

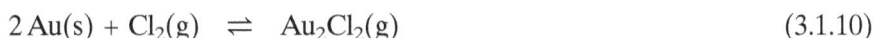

$$2\,\mathrm{Au(s)} + \mathrm{Cl_2(g)} \ \rightleftharpoons\ \mathrm{Au_2Cl_2(g)} \qquad\qquad (3.1.10)$$

The exothermic transport of osmium with chlorine (450 $\rightarrow$ 1000 °C) can be described by the following reaction (Sch 1975).

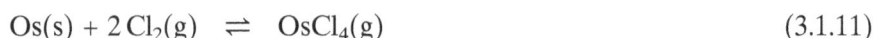

$$\mathrm{Os(s)} + 2\,\mathrm{Cl_2(g)} \ \rightleftharpoons\ \mathrm{OsCl_4(g)} \qquad\qquad (3.1.11)$$

The CVT of platinum with chlorine can be endothermic or exothermic depending on the temperature. Due to exothermic reactions, the migration (600 $\rightarrow$ 700 °C) occurs with high transport rates according to the equilibrium 3.1.12.

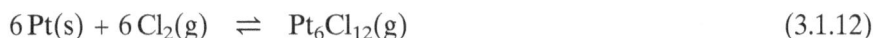

$$6\,\mathrm{Pt(s)} + 6\,\mathrm{Cl_2(g)} \ \rightleftharpoons\ \mathrm{Pt_6Cl_{12}(g)} \qquad\qquad (3.1.12)$$

The endothermic transport (1000 $\rightarrow$ 800 °C) is described by the equilibria 3.1.13 and 3.1.14.

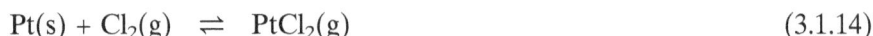

$$\mathrm{Pt(s)} + \frac{3}{2}\,\mathrm{Cl_2(g)} \ \rightleftharpoons\ \mathrm{PtCl_3(g)} \qquad\qquad (3.1.13)$$

$$\mathrm{Pt(s)} + \mathrm{Cl_2(g)} \ \rightleftharpoons\ \mathrm{PtCl_2(g)} \qquad\qquad (3.1.14)$$

One observes a reversal of the transport at a temperature of 725 °C. At this temperature, the solubility of platinum in the gas phase declines to a minimum (Sch 1974).

The transport of molybdenum and tungsten, respectively, from 400 to 1400 °C, as well as of rhenium, from T_1 to 1800 °C, are examples of the hot-wire process in which iodine is not used as a transport agent but chlorine instead.

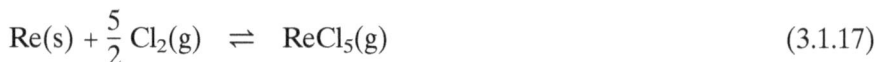

$$Mo(s) + \frac{5}{2} Cl_2(g) \;\rightleftharpoons\; MoCl_5(g) \tag{3.1.15}$$

$$W(s) + 3 Cl_2(g) \;\rightleftharpoons\; WCl_6(g) \tag{3.1.16}$$

$$Re(s) + \frac{5}{2} Cl_2(g) \;\rightleftharpoons\; ReCl_5(g) \tag{3.1.17}$$

3.2 Conproportionation Reactions

Besides the formation and decomposition of the halides, also conproportionation reactions (dissolution) and disproportionation (deposition) can be used for the CVT of elements. During such reactions, at least two halides of different composition appear in the gas phase (3.2.1 to 3.2.3).

$$Be(s) + BeX_2(g) \;\rightleftharpoons\; 2 BeX(g) \tag{3.2.1}$$
$$(X = F, Cl, 1300\,°C \rightarrow T_1)$$

$$2 B(s) + BX_3(g) \;\rightleftharpoons\; 3 BX(g) \tag{3.2.2}$$
$$(X = F, Cl, Br, I, 900 \rightarrow 600\,°C)$$

$$Si(s) + SiX_4(g) \;\rightleftharpoons\; 2 SiX_2(g) \tag{3.2.3}$$
$$(X = F, Cl, Br, I, 1100 \rightarrow 900\,°C)$$

The increase of entropy is the driving force of the endothermic formation of the low halide. The transport always takes place from T_2 to T_1. Elements that can be transported by these reactions are among others beryllium, zinc, cadmium, boron, aluminum, gallium, silicon, germanium, tin, antimony, and bismuth (Sch 1971). The respective halides can directly be used as transport agent. Instead the halogen is added in many cases. In this process, the halides are formed by a primary reaction.

Thus, boron can be transported and obtained in crystalline form with BBr_3 + Br_2 from 900 to 600 °C according to the equilibrium 3.2.2. In contrast, the transport in the temperature gradient from 900 to 400 °C with BI_3 + I_2 as transport additive leads to the deposition of amorphous boron. Due to the higher deposition temperature during the transport with iodine in the hot-wire process, crystalline boron is obtained.

The transport of aluminum, gallium, indium, antimony, bismuth, silicon, germanium, and niobium under formation of their sub-halides can be described by the equilibria 3.2.4 to 3.2.11.

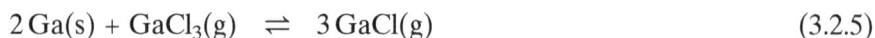

$$2 Al(s) + AlX_3(g) \;\rightleftharpoons\; 3 AlX(g) \tag{3.2.4}$$
$$(X = F, Cl, Br, I)$$

$$2 Ga(s) + GaCl_3(g) \;\rightleftharpoons\; 3 GaCl(g) \tag{3.2.5}$$

$$2 \, \text{In(s)} + \text{In}X_3(\text{g}) \; \rightleftharpoons \; 3 \, \text{In}X(\text{g}) \tag{3.2.6}$$
$(X = \text{F, Cl})$

$$2 \, \text{Sb(s)} + \text{SbCl}_3(\text{g}) \; \rightleftharpoons \; 3 \, \text{SbCl(g)} \tag{3.2.7}$$

$$2 \, \text{Bi(s)} + \text{Bi}X_3(\text{g}) \; \rightleftharpoons \; 3 \, \text{Bi}X(\text{g}) \tag{3.2.8}$$
$(X = \text{Cl, Br, I})$

$$\text{Si(s)} + \text{Si}X_4(\text{g}) \; \rightleftharpoons \; 2 \, \text{Si}X_2(\text{g}) \tag{3.2.9}$$
$(X = \text{F, Cl, Br, I})$

$$\text{Ge(s)} + \text{GeI}_4(\text{g}) \; \rightleftharpoons \; 2 \, \text{GeI}_2(\text{g}) \tag{3.2.10}$$

$$\text{Nb(s)} + 4 \, \text{NbCl}_5(\text{g}) \; \rightleftharpoons \; 5 \, \text{NbCl}_4(\text{g}) \tag{3.2.11}$$

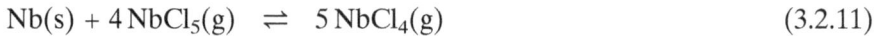

These examples can be generalized as follows:

$$M(\text{s}) + MX_2(\text{g}) \; \rightleftharpoons \; 2 \, MX(\text{g}) \tag{3.2.12}$$
$(M = \text{Be, Cd, Zn})$

$$2 \, M(\text{s}) + MX_3(\text{g}) \; \rightleftharpoons \; 3 \, MX(\text{g}) \tag{3.2.13}$$
$(M = \text{B, Al, Ga, In, Sb, Bi})$

$$M(\text{s}) + MX_4(\text{g}) \; \rightleftharpoons \; 2 \, MX_2(\text{g}) \tag{3.2.14}$$
$(M = \text{Si, Ge})$

$$M(\text{s}) + 4 \, MCl_5(\text{g}) \; \rightleftharpoons \; 5 \, MCl_4(\text{g}) \tag{3.2.15}$$
$(M = \text{Nb, Ta})$

In contrast to silicon and germanium, the transport of tin does not take place with participation of tin(IV)-chloride but via the equilibrium 3.2.16 with participation of tin(II)-chloride and tin(I)-chloride as gas species. Liquid tin is obtained at deposition temperatures between 500 and 700 °C (Spe 1972).

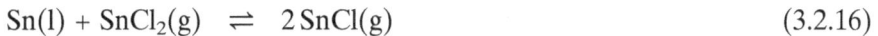

$$\text{Sn(l)} + \text{SnCl}_2(\text{g}) \; \rightleftharpoons \; 2 \, \text{SnCl(g)} \tag{3.2.16}$$

It is also possible to explain the transport of the elements titanium, zirconium, and chromium by conproportionation equilibria. In a flowing system, they are caused to react with their halides at temperatures around 1000 °C. During the reaction of titanium with titanium(IV)-chloride, gaseous TiCl_3 forms as transport agent in a previous reaction. Its partial pressure is in the range of 10^{-3} bar and thus relatively low. The reaction to titanium(II)-chloride takes place via the transport equilibrium 3.2.17 (1200 → 1000 °C).

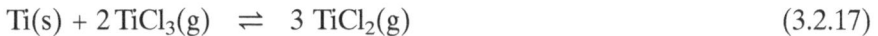

$$\text{Ti(s)} + 2 \, \text{TiCl}_3(\text{g}) \; \rightleftharpoons \; 3 \, \text{TiCl}_2(\text{g}) \tag{3.2.17}$$

The principle of conproportionation can also be used for transport reactions with chalcogenides (Sch 1979) as shown by the following examples:

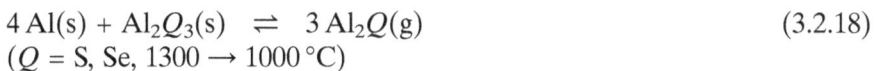

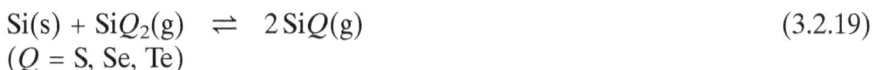

$$4 \, \text{Al(s)} + \text{Al}_2Q_3(\text{s}) \; \rightleftharpoons \; 3 \, \text{Al}_2Q(\text{g}) \tag{3.2.18}$$
$(Q = \text{S, Se}, 1300 \rightarrow 1000 \, °\text{C})$

$$\text{Si(s)} + \text{Si}Q_2(\text{g}) \; \rightleftharpoons \; 2 \, \text{Si}Q(\text{g}) \tag{3.2.19}$$
$(Q = \text{S, Se, Te})$

$$\text{Ge(s)} + \text{Ge}Q_2(\text{g}) \;\rightleftharpoons\; 2\,\text{Ge}Q(\text{g}) \tag{3.2.20}$$
$$(Q = \text{S, Se, Te})$$

3.3 Reversal of the Transport Direction

If several reactions are necessary in order to describe the transport of an element, respectively a solid, endothermic and exothermic reactions can be relevant. Which of these reactions is the dominant one and thus the direction determining is dependent on the total pressure and the temperature. Thermodynamic discussions show that the transport direction can be reversible if the transport conditions are varied.

Examination of the chemical transport of silver with iodine proved that silver can be transported exothermically from 625 to 715 °C and endothermically from 925 to 715 °C. The three independent equilibria 3.3.1 to 3.3.3 are necessary for a description of the transport (Sch 1973c).

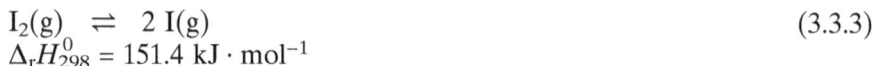

$$\text{Ag(s)} + \text{I(g)} \;\rightleftharpoons\; \text{AgI(g)} \tag{3.3.1}$$
$$\Delta_r H^0_{298} = 18.4 \text{ kJ} \cdot \text{mol}^{-1}$$

$$3\,\text{Ag(s)} + 3\,\text{I(g)} \;\rightleftharpoons\; \text{Ag}_3\text{I}_3(\text{g}) \tag{3.3.2}$$
$$\Delta_r H^0_{298} = -307 \text{ kJ} \cdot \text{mol}^{-1}$$

$$\text{I}_2(\text{g}) \;\rightleftharpoons\; 2\,\text{I(g)} \tag{3.3.3}$$
$$\Delta_r H^0_{298} = 151.4 \text{ kJ} \cdot \text{mol}^{-1}$$

If the gas-phase solubility of silver is calculated in dependency on the temperature, a minimum at approximately 680 °C will result. This indicates the change of the transport direction. Below this temperature, transport takes place from the hotter zone above to the cooler zone. This temperature is called the *reversal temperature* (turning point). The solubility of silver is illustrated in Figure 3.3.1 as a function of the temperature.

There is also a change of the transport direction during the deposition of titanium when iodine is added.

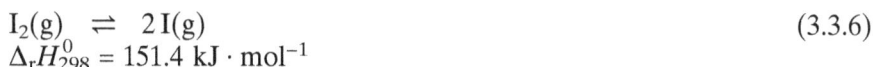

$$\text{Ti(s)} + 4\,\text{I(g)} \;\rightleftharpoons\; \text{TiI}_4(\text{g}) \tag{3.3.4}$$
$$\Delta_r H^0_{298} = -704.3 \text{ kJ} \cdot \text{mol}^{-1}$$

$$\text{Ti(s)} + \text{TiI}_4(\text{g}) \;\rightleftharpoons\; 2\,\text{TiI}_2(\text{g}) \tag{3.3.5}$$
$$\Delta_r H^0_{298} = 237.9 \text{ kJ} \cdot \text{mol}^{-1}$$

$$\text{I}_2(\text{g}) \;\rightleftharpoons\; 2\,\text{I(g)} \tag{3.3.6}$$
$$\Delta_r H^0_{298} = 151.4 \text{ kJ} \cdot \text{mol}^{-1}$$

In this case, the gas-phase solubility of titanium reaches a maximum at approximately 1000 °C. At lower temperature, the endothermic equilibrium predominates while transport at higher temperatures is determined by the exothermic equilibrium.

The transport of silicon with iodine is another, well examined example of a reversal of the transport direction. In order to describe it, the following three independent equilibria have to be formulated:

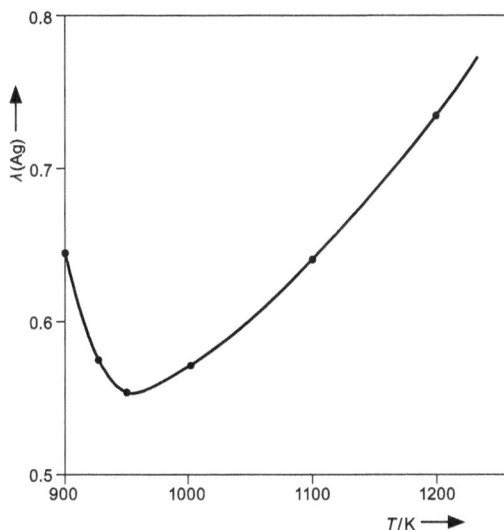

Figure 3.3.1 Gas-phase solubility of silver as a function of the temperature according to *Schäfer* (Sch 1973c).

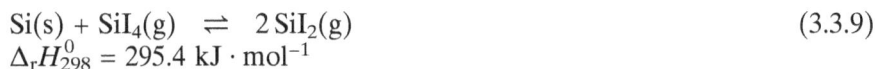

$$Si(s) + 2\,I_2(g) \;\rightleftharpoons\; SiI_4(g) \tag{3.3.7}$$
$$\Delta_r H^0_{298} = -235.3 \; kJ \cdot mol^{-1}$$

$$Si(s) + 4\,I(g) \;\rightleftharpoons\; SiI_4(g) \tag{3.3.8}$$
$$\Delta_r H^0_{298} = -537.5 \; kJ \cdot mol^{-1}$$

$$Si(s) + SiI_4(g) \;\rightleftharpoons\; 2\,SiI_2(g) \tag{3.3.9}$$
$$\Delta_r H^0_{298} = 295.4 \; kJ \cdot mol^{-1}$$

At temperatures above 600 °C, gaseous silicon(IV)-iodide is the dominating silicon-containing gas species. The formation of silicon(II)-iodide is favored with rising temperature and it becomes the dominating gas species. This is accompanied by a reversal of the transport direction. The quotient $p(SiI_4)/p(SiI_2)$ determines the transport direction. This quotient is dependent on temperature and total pressure as well (see Figure 3.3.2). Thus, silicon can be transported from 1150 to 950 °C as well as from 350 to 1000 °C (Sch 1957, Rol 1961).

The interplay of two equilibria with different algebraic signs of the reaction enthalpy can cause a reversal of the transport direction. Generally, this is not only dependent on the temperature, respectively the temperature gradient, but also on the total pressure. This is the case if the positions of the participating equilibria are dependent on the total pressure. Often, more than two equilibria are necessary to describe a transport. In these cases, it is helpful to consider the temperature dependency of the gas-phase solubility for the description of the transport and for the choice of suitable transport conditions (temperature, temperature gradient, total pressure).

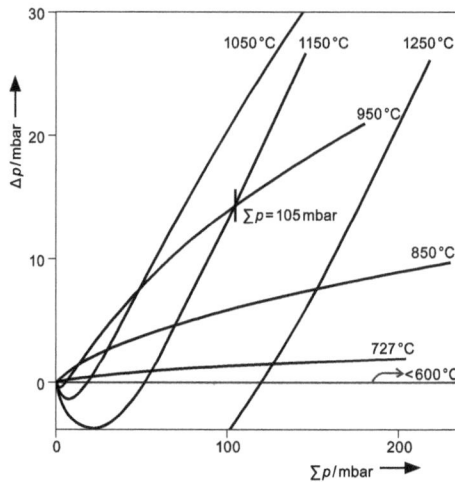

Figure 3.3.2 Influence of the total pressure in the silicon/iodine system on the solubility[1] of silicon in the gas phase according to *Schäfer* (Sch 1957).

3.4 Transport via Gas Complexes

Besides pure halogenating equilibria, halogenating equilibria in combination with complex formation equilibria are of importance for the CVT of elements (Sch 1975). In the process, the formation of gas complexes leads to an increase of the solubility of the respective element in the gas phase. Thus, the transport of gold with iodine is only possible with the participation of Al_2I_6.

AlX_3, GaX_3, InX_3, and FeX_3 (X = Cl, Br, I) are used as complexing agents whereby the chlorides are used most often. In the gas phase, the mentioned halides can be present as dimeric molecules to a considerable extent. Among others, silver, gold, cobalt, chromium, copper, nickel osmium palladium, platinum, rhodium, and ruthenium can be transported via complex formation equilibria. The composition of the different gas complexes are dealt with in Chapter 11. In many cases, in particular at low temperatures below 500 °C, the transport effective equilibria can be generally described by the following equations:

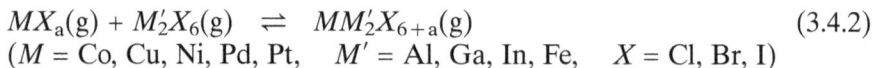

$$M^a(s) + \frac{a}{2} X_2(g) \ \rightleftharpoons \ MX_a(g) \tag{3.4.1}$$

$$MX_a(g) + M'_2X_6(g) \ \rightleftharpoons \ MM'_2X_{6+a}(g) \tag{3.4.2}$$
$(M$ = Co, Cu, Ni, Pd, Pt, M' = Al, Ga, In, Fe, X = Cl, Br, I)

The transport equation is the sum of the equilibria 3.4.1 and 3.4.2.

$$M^a(s) + \frac{a}{2} X_2(g) + M'_2X_6(g) \ \rightleftharpoons \ MM'_2X_{6+a}(g) \tag{3.4.3}$$

It should be noted that gas complexes of another composition can also be formed depending on temperature and pressure (see Chapter 11).

---

[1] The quantity of $\Delta p = p(SiI_2) + p(I_2) + 0.5 \cdot p(I)$ is considered the measure for the "amount of dissolved silicon in the gas phase" in the cited work.

The formation of gas complexes according to 3.4.2 is always endothermic and connected to a loss of enthalpy. If the formation of the gaseous halides is also exothermic according to equation 3.4.1, the addition of the complexing agent will cause an increase of the reaction enthalpy. The temperature range, which is favorable for the transport, shifts compared to the transport reaction without the addition of the complexing agent. The transport takes place from T_1 to T_2. If the formation of the gaseous halides is endothermic, the transport direction can change: If the amount of the reaction enthalpy during the halide's formation is higher than the amount of the reaction enthalpy of the gas complex formation, the transport reaction will be less endothermic but the transport direction stays the same (T_2 to T_1). If the amount of the reaction enthalpy of the exothermic gas complex formation is higher than that of the endothermic halide formation, there will be a reversal of the transport direction. The original endothermic transport from hot to cold becomes an exothermic transport from cold to hot by adding a complexing agent. In most cases, the addition of a complexing agent leads to an increase of the transport rate at low temperatures and to a decrease at higher temperatures.

At higher temperatures, the mentioned complexing agents are increasingly present in the form of monomeric molecules in the gas phase. At the same time, gas complexes of different compositions form. Other reactions take place and the thermodynamic relations change. These effects have to be discussed in more detail in the respective individual cases.

The exothermic transport of platinum with chlorine and carbon monoxide and respectively bromine and carbon monoxide from 375 to 475 °C can be explained by the formation of a gaseous complex (Sch 1970).

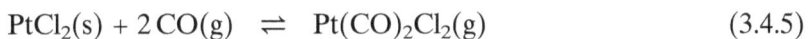

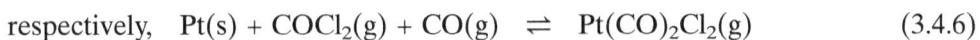

$$Pt(s) + Cl_2(g) \rightleftharpoons PtCl_2(s) \tag{3.4.4}$$

$$PtCl_2(s) + 2\,CO(g) \rightleftharpoons Pt(CO)_2Cl_2(g) \tag{3.4.5}$$

respectively, $\quad Pt(s) + COCl_2(g) + CO(g) \rightleftharpoons Pt(CO)_2Cl_2(g) \tag{3.4.6}$

In this case, the formation of a gaseous complex increases the volatility of the platinum chloride, which does not achieve transport effective partial pressures at the given temperatures without the complex formation.

By adding iodine, aluminum(III)-chloride, or gallium(III)-chloride, the volatilization of solid substances with low *vaporization coefficients* can be catalytically accelerated. The catalytic volatilization of red phosphorus is an example of this (Sch 1972, Sch 1976). Similar effects were observed during the sublimation of arsenic in the presence of thallium during which the evaporation speed is increased by the factor 100 (Bre 1955).

3.5 Transport with the Addition of Hydrogen Halides and Water

As far as the transport of metals is concerned, the hydrogen halides are of minor importance. Only chromium, iron, cobalt, nickel, and copper can be endothermically transported with hydrogen chloride. Iron can also be transported with hydrogen bromide (1020 → 900 °C) (Kot 1967). The transport equations 3.5.1 and 3.5.2 exemplarily describe the processes.

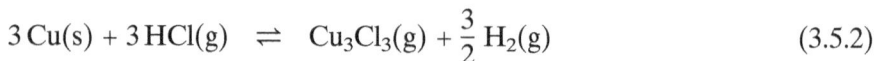

$$Ni(s) + 2\,HCl(g) \;\rightleftharpoons\; NiCl_2(g) + H_2(g) \tag{3.5.1}$$

$$3\,Cu(s) + 3\,HCl(g) \;\rightleftharpoons\; Cu_3Cl_3(g) + \frac{3}{2}\,H_2(g) \tag{3.5.2}$$

Some elements, such as molybdenum, tungsten, rhenium, gallium, germanium, tin, and antimony can be transported with water via the gas phase. The transport is based on the formation of volatile oxides respectively acids.

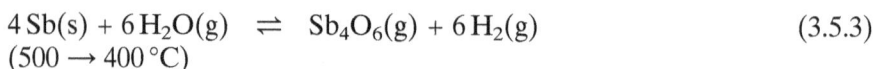

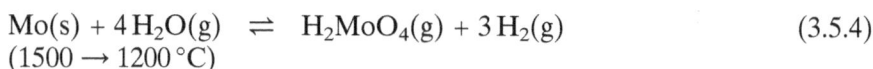

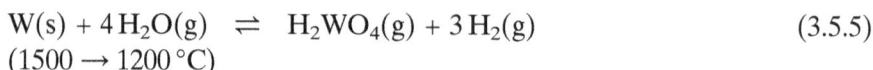

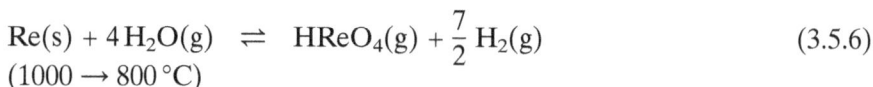

$$4\,Sb(s) + 6\,H_2O(g) \;\rightleftharpoons\; Sb_4O_6(g) + 6\,H_2(g) \tag{3.5.3}$$
$$(500 \rightarrow 400\,°C)$$

$$Mo(s) + 4\,H_2O(g) \;\rightleftharpoons\; H_2MoO_4(g) + 3\,H_2(g) \tag{3.5.4}$$
$$(1500 \rightarrow 1200\,°C)$$

$$W(s) + 4\,H_2O(g) \;\rightleftharpoons\; H_2WO_4(g) + 3\,H_2(g) \tag{3.5.5}$$
$$(1500 \rightarrow 1200\,°C)$$

$$Re(s) + 4\,H_2O(g) \;\rightleftharpoons\; HReO_4(g) + \frac{7}{2}\,H_2(g) \tag{3.5.6}$$
$$(1000 \rightarrow 800\,°C)$$

Besides the volatile acids H_2MoO_4 and respectively H_2WO_4, one has also to consider gaseous oxides as transport-effective species for the transport of molybdenum and tungsten in the given temperature range.

The CVT of germanium with water is described in particular detail (Schm 1981, Schm 1982). It takes place from 850 to 750 °C, the gas phase contains the species GeO, $(GeO)_2$, $(GeO)_3$, H_2O, and H_2.

$$Ge(s) + H_2O(g) \;\rightleftharpoons\; \frac{1}{n}\,(GeO)_n(g) + H_2(g) \quad (n = 1,\,2,\,3) \tag{3.5.7}$$

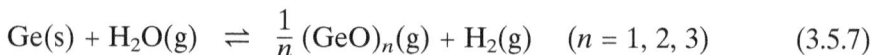

With the help of the transported amount of germanium, the moisture, which is bound reversibly in silica ampoules, was determined (Schm 1981, Schm 1982). Molybdenum and tungsten can be crystallized by adding iodine + water via exothermic chemical transport reactions (Mo: 1050 → 1150 °C; W: 800 → 1000 °C) (Det 1969, Sch 1973a). The transport processes are based on the equilibria 3.5.8 and 3.5.9.

$$Mo(s) + 2\,H_2O(g) + 3\,I_2(g) \;\rightleftharpoons\; MoO_2I_2(g) + 4\,HI(g) \tag{3.5.8}$$
$$W(s) + 2\,H_2O(g) + 3\,I_2(g) \;\rightleftharpoons\; WO_2I_2(g) + 4\,HI(g) \tag{3.5.9}$$

3.6 Oxygen as a Transport Agent

Oxygen can function as a transport agent for a series of noble metals – ruthenium, rhodium, iridium, platinum, and silver (Sch 1963, Sch 1984, Han 2005). In doing so, the transport always takes place at relatively high temperatures in strong endothermic reactions under formation of volatile oxides. Thus, platinum is transported from 1500 °C to T_1, silver from 1400 °C to T_1, iridium from 1325 to 1125 °C. The reactions 3.6.1 to 3.6.3 can be formulated as transport equilibria.

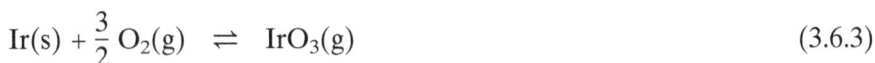

$$Pt(s) + O_2(g) \rightleftharpoons PtO_2(g) \tag{3.6.1}$$

$$Ag(s) + \frac{1}{2} O_2(g) \rightleftharpoons AgO(g) \tag{3.6.2}$$

$$Ir(s) + \frac{3}{2} O_2(g) \rightleftharpoons IrO_3(g) \tag{3.6.3}$$

In particular, the CVT of iridium takes place at low oxygen partial pressure and high transport temperatures. Under these conditions, the formation of solid iridium(IV)-oxide can be suppressed. During the transport of ruthenium with oxygen, RuO_3 and RuO_4 are the transport effective gas species. In rare cases, sulfur and selenium can become transport effective by the formation of volatile sulfides and selenides, respectively (see section 3.2). Thus, boron can be transported with sulfur via the formation of volatile BS_2. The transport of boron with selenium takes place from 900 to 800 °C via equilibrium 3.6.4.

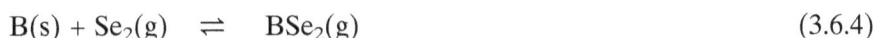

$$B(s) + Se_2(g) \rightleftharpoons BSe_2(g) \tag{3.6.4}$$

Sulfur can also increase the volatility of tellurium by the formation of gas species such as Te_2S_x ($x = 1, 2 \dots 6$).

3.7 Technical Applications

The CVT of elements is also of technical and economic interest. In particular, three practical applications should be mentioned.

- The industrial purification of nickel according to the *Mond–Langer* process.
- The deposition of carbon from the gas phase.
- Transport reactions, which are important for the technology of illuminants.

The purification of nickel according to the *Mond–Langer* process takes place via a reversible transport equilibrium during which the nickel powder to be purified reacts with carbon monoxide to form gaseous tetracarbonylnickel in a forward reaction at a partial pressure of 1 bar at temperatures between approximately 50 and 80 °C. At temperatures around 200 °C, the back reaction takes place whereby the tetracarbonylnickel is decomposed under the release of nickel to nickel granulate. The carbon monoxide that is released in this process is transferred back to the process. The essential impurities of the raw nickel are copper and cobalt. These do not form volatile carbonyls under the given conditions so that nickel can be obtained at purities of 99.9 to 99.99 %. The basic transport equation is formulated as follows (see section 2.6.5):

$$Ni(s) + 4 CO(g) \rightleftharpoons Ni(CO)_4(g) \tag{3.7.1}$$

Generally, amorphous carbon black is obtained during deposition of carbon by CVT reactions. The transport can take place via the endothermic *Boudouard* equilibrium 3.7.3.

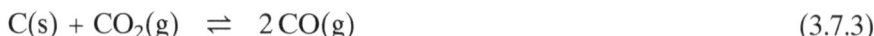

$$C(s) + O_2(g) \ \rightleftharpoons \ CO_2(g) \tag{3.7.2}$$

$$C(s) + CO_2(g) \ \rightleftharpoons \ 2\,CO(g) \tag{3.7.3}$$

The temperature of the dissolution side is above 1000 °C while the temperature on the deposition side can vary between 400 and 600 °C. The disproportionation of the carbon monoxide to carbon and carbon dioxide takes place during the back reaction under the deposition of carbon. This back reaction is kinetically inhibited at room temperature and does not take place practically. However, in the given temperature range, the equilibrium can be achieved. Analogous to the transport of carbon via the *Boudouard* equilibrium, carbon disulphide can also be used as transport agent:

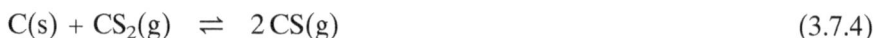

$$C(s) + CS_2(g) \ \rightleftharpoons \ 2\,CS(g) \tag{3.7.4}$$

Under special conditions, carbon can also be deposited from the gas phase in the form of diamond. However, this does not succeed by CVT reactions but with the help of a CVD-process: methane, as carbon-transferring gas species, and hydrogen are conducted over a substrate where the deposition of the diamond takes place under plasma conditions. Fine diamond powder or isotypic silicon can serve as substrate. The substrate temperatures are between 800 and 1100 °C, the total pressure between 0.1 and 0.3 bar. The composition of the gas, which was brought to reaction, varies between pure methane and gas mixtures with 2 % methane and 98 % hydrogen (Reg 2001, Ale 2003).

Currently, light bulbs with a tungsten filament are an often-used illuminant. Therefore, the transport behavior of tungsten with the different transport agents has been well examined. In particular, the works of *Neumann* (Neu 1971, Neu 1972, Neu 1974) and *Dittmer* (Dit 1977, Ditt 1981, Ditt 1983) are recommended. Light bulbs have a very low efficiency. Only approximately 5 % of the electric energy is transformed into visible light, the vast part is yielded as heat. Due to the fact that the efficiency increases with rising temperatures of the filament, high operating temperatures are required. However, the negative consequence is that the filament increasingly evaporates with rising temperatures and tungsten is deposited in the inner side of the light bulb. Thus, the filament becomes thinner with time and eventually burns out. Also, the light bulb is blackened, which decreases the lighting efficiency. This "evaporating" of the tungsten filament is additionally strengthened by the endothermic chemical transport of the tungsten with water, which is still present in the inert gas filling the light bulb (Bel 1964, Alm 1971). It is possible to back transport tungsten from the wall of the glass to the filament by exothermic CVT with halogen lamps that contain chlorine, bromine, or iodine gas mixtures respectively, which allows the filaments to be operated at higher temperatures. This increases the light efficiency, inhibits the blackening of the bulb and increases the life expectancy of the filament, because the back reaction at least partly reconstructs the filament (Bin 1986).

The "metal halide lamp" was developed recently for specific appliances (Bor 2006). In contrast to the light bulb and halogen lamps, these do not have a filament. The lamp normally does not consist of glass, respectively silica glass, but usually of aluminum oxide. Between two tungsten electrodes, a plasma is lighted in very small lamps. The lamps contain a series of substances (Na, I_2, Hg, Tl, Ar, and/or Xe), which are responsible for the light emission as plasma gases. Additionally, the lamps contain the iodide of a rare earth metal, often dysprosium as well as thallium(I)-iodide and aluminum(III)-iodide. Chemical vapor transport reactions can take place by the addition of iodide (Fis 2008).

In the process, tungsten is transported to the cooler parts of the lamp away from the electrodes. A transport of aluminum oxide is also observed. This undesired transport process negatively influences the efficiency and the life expectancy of the lamp but is not yet completely explained. It is certain that the formation of gas complexes in the lamp plays a role. Metal halide lamps are used in the headlights of cars and in video projectors.

Table 3.1 Examples of the chemical vapor transport of elements

Sink solid	Transport additive	Temperature / °C	Reference
Ag	I_2	$625 \rightarrow 715, 925 \rightarrow 715$	Sch 1973c
	$HCl + AlCl_3$	$450 \rightarrow 700$	Sch 1975
	O_2	1400	Tro 1877, War 1913
	O_2	$610 \rightarrow 720$	Sch 1986
Al	$F_2 + AlF_3, Cl_2 + AlCl_3,$ $Br_2 + AlBr_3, I_2 + AlI_3$	$1000 \rightarrow 600$	Schn 1951, Alu 1958
	$S + Al_2S_3, Se + Al_2Se_3$	$1300 \rightarrow 1000$	Kle 1948
Au	Cl_2	$1000 \rightarrow 700$	Bil 1928
	Cl_2	$300 \rightarrow 500, 700 \rightarrow 500$	Sch 1975
	I_2	$1050 \rightarrow 600, 700 \rightarrow 500$	Sch 1955, Sch 1975
	$Cl_2 + AlCl_3, Cl_2 + FeCl_3$	$500 \rightarrow 700$	Sch 1975
	$I_2 + AlI_3$	$300 \rightarrow 350, 700 \rightarrow 500$	Sch 1975
B	I_2	$870 \rightarrow 1300 \dots 1450$	Cue 1978
	$F_2 + BF_3$	not specified	Bla 1964
	$Cl_2 + BCl_3$	not specified	Gro 1949
	$Br_2 + BBr_3$	$900 \rightarrow 600$	Arm 1967
	$I_2 + BI_3$	$900 \rightarrow 400$	Arm 1967
	Se	$900 \rightarrow 800$	Bin 1981
Be	$F_2 + BeF_2, Cl_2 + BeCl_2$	$1300 \rightarrow T_1$	Gre 1963, Gre 1964, Gro 1973
	NaCl	$1000 \rightarrow T_1$	Gro 1956
Bi	$Cl_2 + BiCl_3, Br_2 + BiBr_3,$ $I_2 + BiI_3$	$T_2 \rightarrow T_1$	Cor 1957, Cub 1959, Cub 1960, Cub 1961
C	F	$1725 \rightarrow 2575$	Rie 1988

Table 3.1 (continued)

Sink solid	Transport additive	Temperature / °C	Reference
	$H_2 + CH_4$	$2045 \rightarrow 1100 \dots 700$	Ale 2003
	$H_2 + CH_4$	$2000 \dots 2500 \rightarrow T_1$	Reg 2001
	CO_2	$1000 \rightarrow 600, 1600 \rightarrow 400$	Sch 1962a
	CS_2	$T_2 \rightarrow T_1$	Sch 1958
Cd	$Br_2 + CdBr_2$	$T_2 \rightarrow T_1$	Leh 1980
Co	I_2	$800 \rightarrow 900$	Sch 1962a
	HCl	$900 \rightarrow 600$	Sch 1962a
	$GaCl_3$	$525 \rightarrow 625, 1075 \rightarrow 975$	Sch 1977
	GaI_3	$800 \rightarrow 900$	Sch 1962a
	GaI_3	$525 \rightarrow 625, 925 \rightarrow 825$	Sch 1977
	$H_2 + H_2O$	1400	Bel 1962, Bel 1967
Cr	I_2	$800 \rightarrow 1100$	Ark 1934
	I_2	$790 \rightarrow 1000$	Rol 1960, Rol 1961
	$HCl, AlCl_3$	$1050 \rightarrow 850$	Lut 1974
	$CrCl_2, CrCl_3$	1250	Lee 1958
Cu	I_2	$400 \rightarrow 900$	Ark 1934
	I_2	$360 \rightarrow 1000$	Rol 1961
	I_2	$400 \rightarrow 890$	She 1966
	I_2	not specified	She 1968, She 1968a
	I_2	$400 \rightarrow 890$	Sch 1975
	HCl	$1000 \rightarrow 500, 600 \rightarrow 500$	Sch 1975
	$HCl + AlCl_3$	$400 \rightarrow 600$	Sch 1975
Fe	Cl_2, Br_2, I_2	$800 \rightarrow 1000$	Sch 1956
	I_2	$500 \rightarrow 1100$	Ark 1934
	I_2	$550 \rightarrow 1100$	Rol 1961
	I_2	900	Nic 1973
	HCl	$1000 \rightarrow 800$	Sch 1959
	HBr	$1020 \rightarrow 900$	Kot 1967
	$GaCl_3$	$925 \rightarrow 825, 1000 \rightarrow 900$	Sch 1977
	GaI_3	$925 \rightarrow 825$	Sch 1977
	$H_2 + H_2O$	1400	Bel 1962, Bel 1967
Ga	Cl_2	$1200 \rightarrow T_1$	Gas 1962
	I_2	$500 \rightarrow 700$	Wil 1976
	$Cl_2 + GaCl_3$	$1200 \rightarrow T_1$	Gas 1962
	$I_2 + GaI_3$	$T_2 \rightarrow T_1$	Sil 1962
	$GaCl_2$	800	Lee 1958
	H_2O	$1000 \rightarrow T_1$	Coc 1962
	H_2O	$500 \rightarrow 700$	Wil 1976
Ge	Br_2	$425 \rightarrow 395$	Jon 1965

Table 3.1 (continued)

Sink solid	Transport additive	Temperature / °C	Reference
Ge	I_2	not specified	Rol 1961
	I_2	$425 \to 395$	Jon 1965
	I_2	$500 \to 400$	Ari 1966
	I_2	$545 \to 500, 880 \to 930$	Sch 1966
	I_2	$500 \to T_1, 900 \to T_1$	Sch 1980
	I_2	$T_2 \to T_1$	Fis 1981
	I_2	$895 \to 620$	Opp 1981
	$Cl_2 + GeCl_4$	$T_2 \to T_1$	Sed 1965
	$Br_2 + GeBr_4, I_2 + GeI_4$	$T_2 \to T_1$	Jon 1964
	$GaCl_3$	$500 \to 400$	Sch 1980a
	$GaCl_2$	1250	Lee 1958
	$GeCl_4$	$350 \to 250, 500 \to 400, 700 \to 600$	Sch 1980
	H_2O	$850 \to 750$	Schm 1981, Schm 1982
	H_2O	not specified	Lev 1963, Tra 1969, And 1972
	$H_2O + CO_2$	$800 \to T_1$	Rös 1954
	S	$600 \to 550$	Sch 1979
	$S + GeS_2, Se + GeSe_2$	$T_2 \to T_1$	Sch 1979
	$Te + GeTe_2$	$T_2 \to T_1$	Yel 1981
	GeI_4	$535 \to 460 \dots 485$	Lau 1982
	$BI_3 + H_2$	$590 \to 340$	Eti 1986
	BI_3	$590 \to 340$	Eti 1986
Hf	I_2	$650 \to 2000$	Ark 1925
	I_2	$T_1 \to T_2$	Boe 1926, Boe 1930
	I_2	$400 \to 1600$	Ark 1934
	I_2	not specified	Rol 1961
	I_2	$T_1 \to T_2$	Ger 1968
	I_2	not specified	She 1968, She 1968a
In	Cl_2	$1200 \to T_1$	Gas 1962
	$F_2 + InF_3$	$T_2 \to T_1$	Dim 1976
	$Cl_2 + InCl_3$	$1200 \to T_1$	Gas 1962
Ir	$Cl_2 + O_2$	not specified	Bel 1966
	$Cl_2 + AlCl_3, Cl_2 + FeCl_3$	$600 \to 800, 500 \to 700$	Sch 1975
	O_2	$1325 \to 1130$	Sch 1960, Cor 1962
	O_2	1480	Sch 1984
Mo	Cl_2	$T_1 \to T_2$	Gei 1924, Dit 1983
	Cl_2	$300 \to 1400$	Ark 1934
	Br_2	$T_1 \to T_2$	Mur 1972
	$I_2 + H_2O$	$1050 \to 1150$	Sch 1973a
	$HgBr_2$	$1000 \to 900$	Len 1993
	MoO_3	$1600 \to T_1$	Cat 1960
	$SbBr_3$	$1000 \to 900$	Pau 1994, Pau 1997
	$H_2O + H_2$	$T_2 \to T_1$	Jae 1952, Bel 1965

Table 3.1 (continued)

Sink solid	Transport additive	Temperature / °C	Reference
Nb	I_2	$240 \dots 470 \to 1200 \dots 1400$	Rol 1961
	I_2	not specified	Sch 1962
Ni	Cl_2, Br_2	$800 \to 1000$	Sch 1956
	I_2	$860 \to 1030$	Sch 1956
	HCl	$1000 \to 700$	Sch 1959
	HBr	$1000 \to 800$	Kot 1967
	$GaCl_3$, GaI_3	$1075 \to 975, 925 \to 825$	Sch 1977
	InI_3	$1000 \to 900$	Sch 1978
	$H_2 + H_2O$	1400	Bel 1962, Bel 1967
	CO	$80 \to 180$	Sch 1982
	CO	$T_1 \to T_2$	Mon 1890, Mon 1988, Cad 1990
	CO	$120 \to 180$	Las 1990
Os	Cl_2, $Cl_2 + FeCl_3$	$450 \to 1000$	Sch 1975
Pa	I_2	not specified	Spi 1979
Pd	Cl_2, $Cl_2 + AlCl_3$	$400 \to 600$	Sch 1975
	$Cl_2 + FeCl_3$	$400 \to 900$	Sch 1975
	$I_2 + AlI_3$	$375 \to 600$	Sch 1975
Pt	Cl_2	$1000 \to 800, 580 \to 650$	Sch 1974
	Cl_2	$600 \to 800$	Sch 1975
	Cl_2	$995 \to 875$	Str 1981
	Cl_2	1025, 1625	Sch 1984
	$Cl_2 + AlCl_3$, $Cl_2 + FeCl_3$	$600 \to 800$	Sch 1975
	$Cl_2 + CO$	$375 \to 475$	Sch 1970
	Br_2	1025, 1625	Sch 1984
	$Br_2 + CO$	$375 \to 475$	Sch 1970
	O_2	$1500 \to T_1$	Alc 1960
	O_2	1340	Sch 1984
	O_2	$900 \to 800$	Han 2005
Re	Cl_2	$T_1 \to 1800$	Ark 1934
	I_2	$1000 \to 900$	Sch 1973b
	$I_2 + H_2O$, H_2O	$1000 \to 800$	Sch 1973b
Rh	$Cl_2 + AlCl_3$	$600 \to 800$	Sch 1975
	O_2	$900 \to 800$	Han 2005
Ru	$Cl_2 + AlCl_3$, $Cl_2 + FeCl_3$	$450 \to 600$	Sch 1975
	O_2	$T_2 \to T_1$	Sch 1963
Sb	$I_2 + SbI_3$	$T_2 \to T_1$	Cor 1957
	$GaCl_3$, $SbCl_3$, H_2O	$500 \to 400$	Sch 1982a

Table 3.1 (continued)

Sink solid	Transport additive	Temperature / °C	Reference
Se	H_2	$160 \rightarrow T_1$	Scho 1966
Si	Cl_2	$1300 \rightarrow 1100$	Les 1961
	Br_2	1000	Nic 1973
	I_2	$1150 \rightarrow 950, 950 \rightarrow 1050$	Les 1961
	I_2	$350 \rightarrow 1000$	Rol 1961
	$F_2 + SiF_4, Cl_2 + SiCl_4,$	$1100 \rightarrow 900$	Tro 1876, Sch 1953,
	$Br_2 + SiBr_4, I_2 + SiI_4$		Tei 1966, Wol 1966
	$I_2 + H_2$	$1100 \rightarrow 800$	Gre 1961
	$AlCl_3$	$1000 \rightarrow T_1$	Lee 1958
	$CdCl_2$	not specified	Val 1967
	$S + SiS_2, Se + SiSe_2,$	$T_2 \rightarrow T_1$	Hol 1981, Dro 1969,
	$Te + SiTe_2$		Fau 1968
	$SiCl_4$	$1150 \rightarrow 950$	Sch 1957
	$SiCl_4$	1000	Lee 1958
	$SiCl_4$	$1100 \rightarrow 900$	Sch 1991
	$SiBr_4, SiI_4$	$1150 \rightarrow 950$	Sch 1957
	SiI_4	$950 \rightarrow 1150$	Sch 1957
Sn	$Cl_2 + SnCl_2$	$T_2 \rightarrow T_1$	Spe 1972
	HCl, H_2O	750	Wil 1976
Ta	I_2	$225 \dots 495 \rightarrow 1100$	Rol 1961
	I_2	not specified	Sch 1973, Gaw 1983
Te	I_2	$T_2 \rightarrow T_1$	Bur 1971
	I_2	$375 \rightarrow 325$	Sch 1991
	$AlCl_3$	$T_2 \rightarrow T_1$	Pri 1970
	S	$T_2 \rightarrow T_1$	Bin 1976
	S	$375 \rightarrow 325$	Sch 1991
Th	I_2	$650 \rightarrow 2000$	Ark 1925
	I_2	not specified	Boe 1926, Boe 1930
	I_2	$400 \rightarrow 1700$	Ark 1934
	I_2	$455 \dots 485 \rightarrow T_2$	Vei 1955
	I_2	not specified	Rol 1961
	I_2	$420 \dots 470 \rightarrow 1200 \dots 1400$	Spi 1979
Ti	Br_2	1000	Nic 1973
	I_2	$650 \rightarrow 2000$	Ark 1925
	I_2	$T_1 \rightarrow T_2$	Boe 1926, Boe 1930
	I_2	$200 \rightarrow 1400$	Ark 1934
	I_2	$525 \rightarrow 1300 \dots 1400$	Cam 1948
	I_2	$500 \rightarrow 1100$	Rol 1961
	I_2	not specified	She 1968, She 1968a
	$F_2 + TiF_3, Cl_2 + TiCl_3$	$1200 \rightarrow 1000$	Gro 1952, Sch 1958a
	$TiCl_3, TiCl_4$	1250, 1300	Lee 1958

Table 3.1 (continued)

Sink solid	Transport additive	Temperature / °C	Reference
Ti	NaCl	$1000 \rightarrow T_1$	Gro 1956
U	I_2	$800 \rightarrow 1000$	Has 1967
	NaCl	$1000 \rightarrow T_1$	Gro 1956
	$I_2 + H_2$	not specified	Ber 1954
V	I_2	$800 \rightarrow 1200$	Ark 1934
	I_2	$800 \rightarrow 1300$	Car 1961
	I_2	not specified	Rol 1961
	NaCl	$1000 \rightarrow T_1$	Gro 1956
W	F_2	$T_1 \rightarrow T_2$	Neu 1971a, Neu 1971b, Dit 1977
	Cl_2	$T_1 \rightarrow T_2$	Lan 1915, Ark 1923, Rie 1960, Wei 1970, Neu 1971a, Neu 1971b, Neu 1972, Neu 1973, Smi 1986
	Cl_2	$300 \rightarrow 1400$	Ark 1934
	Br_2	$T_1 \rightarrow T_2$	Neu 1971a, Neu 1971b, Yan 1972
	Br_2	1000	Nic 1973
	Br_2	$775 \rightarrow 1125$	Wei 1977
	$HgBr_2$	$1000 \rightarrow 900$	Len 1994
	I_2	not specified	Rol 1961
	I_2	$T_1 \rightarrow T_2$	Neu 1971a
	HgI_2	$T_2 \rightarrow T_1$	Scho 1991
	$F_2 + H_2$	$T_1 \rightarrow T_2$	Neu 1973d
	$F_2 + Br_2, F_2 + O_2, F_2 + Br_2 + O_2$	not specified	Rie 1987
	$F_2 + O_2$	$T_1 \rightarrow T_2$	Neu 1971c, Neu 1973b, Har 1976, Dit 1981
	$Cl_2 + H_2$	$T_1 \rightarrow T_2$	Neu 1973c
	$Cl_2 + O_2, Br_2 + O_2, I_2 + O_2$	$T_1 \rightarrow T_2$	Neu 1971c, Neu 1973b, Dit 1981
	$F_2 + H_2 + O_2$	$T_1 \rightarrow T_2$	Neu 1973f
	$F_2 + H_2 + O_2 + C$	$T_1 \rightarrow T_2$	Neu 1973g
	$Cl_2 + H_2$	$T_1 \rightarrow T_2$	Neu 1973d
	$Cl_2 + H_2 + O_2$	$T_1 \rightarrow T_2$	Neu 1973f, Neu 1973g, Det 1974
	$Br_2 + H_2$	$T_1 \rightarrow T_2$	Neu 1973d
	$Br_2 + H_2 + O_2$	$T_1 \rightarrow T_2$	Neu 1973f
	$Br_2 + O_2 + C$	$T_1 \rightarrow T_2$	Neu 1974
	$Br_2 + H_2 + O_2 + C$	$T_1 \rightarrow T_2$	Neu 1973g, Neu 1974a
	$Br_2 + H_2O$	$T_1 \rightarrow T_2$	Det 1974
	$I_2 + H_2 + O_2$	$T_1 \rightarrow T_2$	Neu 1973f, Det 1974
	$I_2 + H_2O$	$800 \rightarrow 1000$	Det 1969, Sch 1973a

Table 3.1 (continued)

Sink solid	Transport additive	Temperature / °C	Reference
W	BF_3, SiF_4	not specified	Rie 1987, Rie 1987a
	$BF_3 + BBr_3$, $SiF_4 + SiBr_4$	not specified	Rie 1987
	CF_4, SF_6, WF_6	not specified	Rie 1987a
	BF_3, NF_3, PF_3, SiF_4, SF_6, WF_6	$T_1 \rightarrow T_2$	Dit 1977
	$PSCl_3$, $PSBr_3$	$T_1 \rightarrow T_2$	Neu 1972a
	$P + N_2 + Cl_2$, $P + N_2 + Br_2$	$T_1 \rightarrow T_2$	Neu 1972b
	H_2O	$T_2 \rightarrow T_1$	Lan 1913, Smi 1921, Alt 1924, Mil 1957, Bel 1964, Hof 1964, Wie 1970, Sch 1973a, Pra 1974
	H_2O	$2400 \rightarrow T_1$	Smi 1952
	H_2O	$2525 \rightarrow 1225$	Alm 1971
	H_2O	$1100 \rightarrow 900$	Sch 1973a
	$O_2 + H_2$	$T_2 \rightarrow T_1$	Neu 1973a
	O_2	not specified	Neu 1971
	CH_2Br_2	$T_1 \rightarrow T_2$	Neu 1974b
Zn	$Cl_2 + ZnCl_2$	$T_2 \rightarrow T_1$	Gai 1964
Zr	I_2	$650 \rightarrow 2000$	Ark 1925
	I_2	$T_1 \rightarrow T_2$	Boe 1926, Boe 1930, Eme 1956, Eme 1957
	I_2	$200 \rightarrow 1400$	Ark 1934
	I_2	not specified	Rol 1961, She 1968, She 1968a
	$ZrCl_4$	1225	Lee 1958
Y	I_2	$560 \rightarrow 1200$	Rol 1960

Bibliography

Alc 1960 — C. B. Alcock, G. W. Hooper, *Proc. Roy. Soc.* **1960**, *A254*, 551.

Ale 2003 — V. D. Aleksandrov, I. V. Sel'skaya, *Inorg. Mater.* **2003**, *39*, 455.

Alm 1971 — F. H. R. Almer, P. Wiedijk, *Z. Anorg. Allg. Chem.* **1971**, *385*, 312.

Alt 1924 — H. Alterthum, *Z. Phys. Chem.* **1924**, *110*, 1.

Alu 1958 — Aluminium Laboratories Limited, Montreal, *US Patent 2914398 (DE1130605)* **1958**.

And 1972 — R. Andrade, E. Butter, *Krist. Tech.* **1972**, *7*, 581.

Ari 1966 — T. Arizumi, T. Nishinaga, *Crystal Growth (Suppl. J. Phys. Chem. Solids), Conference of Crystal Growth, Boston* **1966**, C 14.

Ark 1923 — A. E. van Arkel, *Physica* **1923**, *3*, 76.

Ark 1925   A. E. van Arkel, J. H. de Boer, *Z. Anorg. Allg. Chem.* **1925**, *148*, 345.

Ark 1934   A. E. van Arkel, *Metallwirtschaft* **1934**, *13*, 405.

Bel 1967   G. R. Belton, A. S. Jordan, *J. Phys. Chem.* **1965**, *71*, 4114.

Ber 1954   G. Berge, G. P. Monet, *US Patent.* US2743113 **1954**.

Bil 1928   W. Biltz, W. Fischer, R. Juza, *Z. Anorg. Allg. Chem.* **1928**, *176*, 121.

Bin 1976   M. Binnewies, *Z. Anorg. Allg. Chem.* **1976**, *422*, 43.

Bin 1981   M. Binnewies, **1981**, *unpublished results.*

Bin 1986   M. Binnewies, *Chem. unserer Zeit* **1986**, *5*, 141.

Bla 1964   J. Blauer, M. A. Greenbaum, M. Farber, *J. Phys. Chem.* **1964**, *68*, 2332.

Boe 1926   J. H. de Boer, J. D. Fast, *Z. Anorg. Allg. Chem.* **1926**, *153*, 1.

Boe 1930   J. H. de Boer, J. D. Fast, *Z. Anorg. Allg. Chem.* **1930**, *187*, 193.

Bor 2006   M. Born, T. Jüstel, *Chem. unserer Zeit* **2006**, *40*, 294.

Bur 1971   J. Burmeister, *Mater. Res. Bull.* **1971**, *6*, 219.

Cad 1990   R. Cadoret, *Ann. Chim. Fr.* **1990**, *15*, 156.

Cam 1948   I. E. Campbell, R. I. Jaffee, J. M. Blocher, J. Gurland, B. W. Gonser, *Trans. Electrochem. Soc.* **1948**, *93*, 271.

Car 1961   D. N. Carlson, C. W. Owen, *J. Electrochem. Soc.* **1961**, *108*, 88.

Cat 1960   E. D. Cater, E. R. Plante, P. W. Gilles, *J. Chem. Phys.* **1960**, *32*, 1269.

Cho 2003   K. L. Choy, *Progress in Materials Science* **2003**, *48*, 57.

Coc 1962   C. N. Cochran, L. M. Foster, *J. Electrochem. Soc.* **1962**, *109*, 149.

Cor 1957   J. D. Corbett, S. v. Winbush, F. C. Albers, *J. Anorg. Chem. Soc.* **1957**, *79*, 3020.

Cor 1962   E. H. P. Cordfunke, G. Meyer, *Rec. Trav. Chim.* **1962**, *81*, 495.

Cub 1958   D. Cubicciotti, F. J. Keneshea, G. M. Kelley, *J. Phys. Chem.* **1958**, *62*, 463.

Cub 1959   D. Cubicciotti, F. J. Keneshea, *J. Phys. Chem.* **1959**, *63*, 295.

Cub 1961   D. Cubicciotti, *J. Phys. Chem.* **1961**, *65*, 521.

Cue 1978   J. Cueilleron, J. C. Viala, *J. Less-Common Met.* **1978**, *58*, 123.

Det 1969   J. H. Dettingmeijer, J. Tillack, H. Schäfer, *Z. Anorg. Allg. Chem.* **1969**, *369*, 161.

Det 1974   J. H. Dettingmeijer, B. Meinders, L. M. Nijland, *J. Less-Common Met.* **1974**, *35*, 159.

Dim 1976   V. S. Dimitriev, V. A. Smirnov, *Zh. Fiz. Khim.* **1976**, *50*, 2445.

Dit 1977   G. Dittmer, A. Klopfer, J. Schröder, *Philips Res. Repts.* **1977**, *32*, 341.

Dit 1981   G. Dittmer, U. Niemann, *Philips J. Res.* **1981**, *36*, 87.

Dit 1983   G. Dittmer, U. Niemann, *Mater. Res. Bull.* **1983**, *18*, 355.

Dro 1969   M. I. Dronyuk, *Vestnik, L'rovskogo Politechniceskogo Intituta* **1969**, *34*, 11.

Eme 1956   V. S. Emelyanov, P. D. Bystrov, A. I. Evstyukhin, *J. Nucl. Energy* **1956**, *3*, 121.

Eme 1957   V. S. Emelyanov, P. D. Bystrov, A. I. Evstyukhin, *J. Nucl. Energy* **1957**, *4*, 253.

Eti 1986   D. Étienne, N. Archagui, G. Bougnot, *J. Cryst. Growth* **1986**, *74*, 145.

Fau 1968   J. W. Faust, H. F. John, C. Pritchard, *J. Cryst. Growth* **1968**, *3*, 321.

Fis 1981   H. J. Fischer, R. Kuhl, H. Oppermann, R. Herrmann, K. Hilbert, A. S. Okhotin, G. E. Ignatjev, V. T. Khrjapov, E. M. Markov, I. V. Barmin, *Adv. Space Res.* **1981**, *1*, 111.

Fis 2008   S. Fischer, U. Niemann, T. Markus, *Appl. Phys.* **2008**, *41*, 144015.

Gai 1964   B. Gaiek, F. Proshek, *Russ. J. Inorg. Chem.* **1964**, *9*, 256.

Gas 1962   E. Gastinger, *Z. Anorg. Allg. Chem.* **1962**, *316*, 161.

Gaw 1983   I. I. Gawrilow, A. I. Ewstjuschin, A. Schulow, M. M. Koslow, *Zh. Fiz. Khim.* **1983**, *57*, 1347.

Ger 1968   J. Gerlach, J. P. Krumme, F. Pawlek, *J. Less-Common Met.* **1968**, *15*, 303.

Gre 1961   E. S. Greiner, J. A. Gutowski, W. C. Ellis, *J. Appl. Phys.* **1961**, *32*, 2489.

Gre 1963   M. A. Greenbaum, R. E. Yates, M. L. Arin, M. Arshadi, J. Weiher, M. Far-ber, *J. Phys. Chem.* **1963**, *67*, 703.

Gre 1964   M. A. Greenbaum, M. L. Arin, M. Wong, M. Farber, *J. Phys. Chem.* **1964**, *68*, 791.

Gro 1949   P. Groß, *French Patent* **1949**, 960785.

Gro 1952   P. Groß, *Austr. Patent* **1952**, 161020.

Gro 1956   P. Groß, D. L. Levi, *Extr. Refining Rarer Metals, Proc. Sympos.* London **1956**, 337.

Gro 1973   P. Groß, R. H. Levin, *Chem. Soc. Faraday Division, Sympos. High Temp. Studies in Chem.,* London, **1973**.

Han 2005   L. Hannevold, O. Nilsen, A. Kjekshus, H. Fjellvåg, *J. Cryst. Growth* **2005**, *279*, 206.

Har 1976   G. Hartel, H. G. Kloss, *Z. Phys. Chem.* **1976**, *257*, 873.

Has 1967   T. Hashino, T. Kawai, *Trans. Faraday Soc.* **1967**, *63*, 3088.

Hof 1964   T. W. Hoffmann, J. Nikliborc, *Acta Phys. Pol.* **1964**, *25*, 633.

Hol 1981   C. Holm, E. Sirtl, *J. Cryst. Growth* **1981**, *54*, 253.

Jae 1952   G. Jaeger, R. Krasemann, *Werkst. Korros.* **1952**, *3*, 401.

Jon 1964   F. Jona, R. F. Lever, H. R. Wendt, *J. Electrochem. Soc.* **1964**, *111*, 413.

Jon 1965   F. Jona, *J. Chem. Phys.* **1965**, *42*, 1025.

Kan 1964   A. S. Kanaan, J. L. Margrave, *Inorg. Chem.* **1964**, *3*, 1037.

Kem 1957   C. P. Kempter, C. Alvarez-Tostado, *Z. Anorg. Allg. Chem.* **1957**, *290*, 238.

Kle 1948   W. Klemm, K. Geiersberger, B. Schaeler, H. Mindt, *Z. Anorg. Allg. Chem.* **1948**, *255*, 287.

Kot 1967   M. Kotrbová, Z. Hauptman, *Krist. Tech.* **1967**, *2*, 505.

Lan 1913   I. Langmuir, *Proc. AIEE* **1913**, *32*, 1894, 1913.

Lan 1915   I. Langmuir, *J. Amer. Chem. Soc.* **1915**, *137*, 1139.

Las 1990   J. Laskowski, M. Oledzka, *Ann. Chim. Fr.* **1990**, *15*, 153.

Lau 1982   J. C. Launay, *J. Cryst. Growth* **1982**, *60*, 185.

Lee 1958   M. F. Lee, *J. Phys. Chem.* **1958**, *62*, 877.

Leh 1980   G. Lehmann, *private communication* **1980**.

Len 1993   M. Lenz, R. Gruehn, *Z. Anorg. Allg. Chem.* **1993**, *619*, 731.

Len 1994   M. Lenz, R. Gruehn, *Z. Anorg. Allg. Chem.* **1994**, *620*, 867.

Les 1961   R. Lesser, E. Erben, *Z. Anorg. Allg. Chem.* **1961**, *309*, 297.

Lev 1963   R. F. Lever, F. Jona, *J. Appl. Phys.* **1963**, *34*, 3139.

Lut 1974   H. D. Lutz, K. H. Bertram, G. Wrobel, M. Ridder, *Mh. Chem.* **1974**, *105*, 849.

Mil 1957   T. Millner, *Acta Techn. Acad. Sci. Hung.* **1957**, *17*, 67.

Mon 1890   L. Mond, C. Langer, F. Quincke, *J. Chem. Soc.* **1890**, 749.

Mon 1988   Y. Monteil, P. Raffin, J. Bouix, *Spectrochim. Acta* **1988**, *44*, 429.

Mor 1962   C. R. Morelock, *Acta Metall. Mater.* **1962**, *10*, 161.

Mur 1972   J. J. Murray, J. B. Taylor, L. Asner, *J. Cryst. Growth* **1972**, *15*, 231.

Mur 1995   K. Murase, K. Machida, G. Adachi, *J. Alloys Compd.* **1995**, *217*, 218.

Neu 1971   G. M. Neumann, G. Gottschalk, *Z. Naturforsch.* **1971**, *26a*, 882.

Neu 1971a   G. M. Neumann, W. Knatz, *Z. Naturforsch.* **1971**, *26a*, 863.

Neu 1971b   G. M. Neumann, G. Gottschalk, *Z. Naturforsch.* **1971**, *26a*, 870.

Neu 1971c   G. M. Neumann, G. Gottschalk, *Z. Naturforsch.* **1971**, *26a*, 1046.

Neu 1972   G. M. Neumann, U. Müller, *J. Less-Common Met.* **1972**, *26*, 391.

Neu 1972a   G. M. Neumann, *Thermochim. Acta* **1972**, *5*, 25.

Neu 1972b   G. M. Neumann, *Thermochim. Acta* **1972**, *4*, 73.

Neu 1973   G. M. Neumann, *J. Less-Common Met.* **1974**, *35*, 45.

Neu 1973a   G. M. Neumann, *Z. Metallkde.* **1973**, *64*, 193.
Neu 1973b   G. M. Neumann, *Z. Metallkde.* **1973**, *64*, 26.
Neu 1973c   G. M. Neumann, D. Schmidt, *J. Less-Common Met.* **1973**, *33*, 209.
Neu 1973d   G. M. Neumann, *Z. Metallkde.* **1973**, *64*, 117.
Neu 1973f   G. M. Neumann, *Z. Metallkde.* **1973**, *64*, 379.
Neu 1973g   G. M. Neumann, *Z. Metallkde.* **1973**, *64*, 444.
Neu 1974   G. M. Neumann, *J. Less-Common Met.* **1974**, *35*, 51.
Neu 1974a   G. M. Neumann, *Thermochim. Acta* **1974**, *8*, 369.
Neu 1974b   G. M. Neumann, *Z. Naturforsch.* **1974**, *29a*, 1471.
Nic 1971a   J. J. Nickl, J. D. Koukoussas, *J. Less-Common Met.* **1971**, *23*, 73.
Nic 1973   J. J. Nickl, J. D. Koukoussas, A. Mühlratzer, *J. Less-Common Met.* **1973**, *32*, 243.
Opp 1981   H. Oppermann, A. S. Okhotin, *Adv. Space Res.* **1981**, *1*, 51.
Pau 1994   N. Pausch, *Diplomarbeit,* University of Giessen **1994**.
Pau 1997   N. Pausch, J. Burggraf, R. Gruehn, *Z. Anorg. Allg. Chem.* **1997**, *623*, 1835.
Pra 1974   M. Prager, *J. Crystal Growth* **1974**, *22*, 6.
Pri 1970   D. J. Prince, J. D. Corbett, B. Garbisch, *Inorg. Chem.* **1970**, *9*, 2731.
Reg 2001   L. Regel, W. Wilcox, *Acta Astronaut.* **2001**, *48*, 129.
Rie 1960   G. D. Rieck, H. A. C. M. Bruning, *Acta Metallurg.* **1960**, *8*, 97.
Rie 1987   L. Riesel, A. Dimitrov, P. Szillat, *Z. Anorg. Allg. Chem.* **1987**, *547*, 205.
Rie 1987a   L. Riesel, A. Dimitrov, P. Szillat, H. Vogt, *Z. Anorg. Allg. Chem.* **1987**, *547*, 216.
Rie 1988   L. Riesel, K.- H. Rietze, *Z. Anorg. Allg. Chem.* **1988**, *557*, 191.
Rol 1959   R. F. Rolsten, *J. Electrochem. Soc.* **1959**, *106*, 975.
Rol 1960   R. F. Rolsten, *Z. Anorg. Allg. Chem.* **1960**, *305*, 25.
Rol 1961   R. F. Rolsten, *Iodide Metals and Metal Iodides,* J. Wiley, New York **1961**.
Rös 1954   O. Rösner, *Patent. DE976701* **1954**.
Sch 1953   H. Schäfer, J. Nickl, *Z. Anorg. Allg. Chem.* **1953**, *274*, 250.
Sch 1955   H. Schäfer, B. Morcher, **1955**, *unpublished results.*
Sch 1956   H. Schäfer, H. Jacob, K. Etzel, *Z. Anorg. Allg. Chem.* **1956**, *286*, 42.
Sch 1956a   H. Schäfer, H. Jacob, K. Etzel, *Z. Anorg. Allg. Chem.* **1956**, *286*, 27.
Sch 1957   H. Schäfer, B. Morcher, *Z. Anorg. Allg. Chem.* **1957**, *290, 279.*
Sch 1958   H. Schäfer, H. Wiedemeier, *Z. Anorg. Allg. Chem.* **1958**, *296*, 241.
Sch 1958a   H. Schäfer, F. Wartenpfuhl, **1958**, *unpublished results.*
Sch 1959   H. Schäfer, K. Etzel, *Z. Anorg. Allg. Chem.* **1959**, *301*, 137.
Sch 1960   H. Schäfer, H. J. Heitland, *Z. Anorg. Allg. Chem.* **1960**, *304, 249.*
Sch 1962   H. Schäfer, M. Hüsker, *Z. Anorg. Allg. Chem.* **1962**, *317*, 321.
Sch 1962a H. Schäfer, *Chemische Transportreaktionen,* VCH, Weinheim, **1962**.
Sch 1963   H. Schäfer, A. Tebben, W. Gerhardt, *Z. Anorg. Allg. Chem.* **1963**, *321,* 41.
Sch 1966   H. Schäfer, H. Odenbach, *Z. Anorg. Allg. Chem.* **1966**, *346, 127.*
Sch 1970   H. Schäfer, U. Wiese, *J. Less-Common Met.* **1971**, *24*, 55.
Sch 1971   H. Schäfer, *J. Cryst. Growth* **1971**, *9*, 17.
Sch 1972   H. Schäfer, M. Trenkel, *Z. Anorg. Allg. Chem.* **1972**, *391,* 11.
Sch 1973   H. Schäfer, *J. Less-Common Met.* **1973**, *30*, 141.
Sch 1973a   H. Schäfer, T. Grofe, M. Trenkel, *J. Sol. State. Chem.* **1973**, *8*, 14.
Sch 1973b   H. Schäfer, *Z. Anorg. Allg. Chem.* **1973**, *400*, 253.
Sch 1973c H. Schäfer, *Z. Anorg. Allg. Chem.* **1973**, *401*, 227.
Sch 1974   H. Schäfer, *Z. Anorg. Allg. Chem.* **1974**, *410, 269.*
Sch 1975   H. Schäfer, M. Trenkel, *Z. Anorg. Allg. Chem.* **1975**, *414*, 137.
Sch 1976   H. Schäfer, M. Trenkel, *Z. Anorg. Allg. Chem.* **1976**, *420, 261.*
Sch 1977   H. Schäfer, J. Nowitzki, *Z. Anorg. Allg. Chem.* **1977**, *435*, 49.

Sch 1978	H. Schäfer, J. Nowitzki, *Z. Anorg. Allg. Chem.* **1978**, *439*, 80.
Sch 1978a	H. Schäfer, M. Binnewies, *Z. Anorg. Allg. Chem.* **1978**, *441,* 216.
Sch 1979	H. Schäfer, M. Trenkel, *Z. Anorg. Allg. Chem.* **1979**, *458,* 234.
Sch 1980	H. Schäfer, M. Trenkel, *Z. Anorg. Allg. Chem.* **1980**, *461*, 22.
Sch 1980a	H. Schäfer, M. Trenkel, *Z. Anorg. Allg. Chem.* **1980**, *461*, 29.
Sch 1982	H. Schäfer, *Z. Anorg. Allg. Chem.* **1982**, *493*, 17.
Sch 1982a	H. Schäfer, *Z. Anorg. Allg. Chem.* **1982**, *489*, 154.
Sch 1983a	H. Schäfer, T. Grofe, M. Trenkel, *J. Sol. State Chem.* **1973**, *8*, 14.
Sch 1984	H. Schäfer, W. Gerhardt, *Z. Anorg. Allg. Chem.* **1984**, *512,* 79.
Sch 1986	H. Schäfer, W. Kluy, *Z. Anorg. Allg. Chem.* **1986**, *536,* 53.
Sch 1991	H. Schäfer, C. Brendel, *Z. Anorg. Allg. Chem.* **1991**, *598*, 293.
Schm 1981	G. Schmidt, R. Gruehn, *J. Cryst. Growth* **1981**, *55,* 599.
Schm 1982	G. Schmidt, R. Gruehn, *J. Cryst. Growth* **1982**, *58,* 623.
Schn 1951	A. Schneider, W. Schmidt, *Z. Metallkunde* **1951**, *42,* 43.
Scho 1966	H. Scholz, **1966**, *unpublished results.*
Scho 1991	H. Schornstein, *Dissertation,* University of Giessen, **1991**.
Schr 1975	J. Schroeder, *Philips Tech. Rev.* **1975**, *35*, 332.
Sed 1965	T. O. Sedgwick, *J. Electrochem. Soc.* **1965**, *112,* 496.
She 1966	R. A. J. Shelton, *Trans. Faraday Soc.* **1966**, *62*, 222.
She 1968	R. A. J. Shelton, *Trans. Inst. Mining and Metallurgy* **1968**, *77,* C 32.
She 1968a	R. A. J. Shelton, *Trans. Inst. Mining and Metallurgy* **1968**, *77,* C 113.
Sil 1962	V. J. Silvestri, V. J. Lions, *J. Electrochem. Soc.* **1962**, *109,* 963.
Smi 1921	C. J. Smithells, *Trans. Faraday Soc.* **1921**, *17*, 485.
Smi 1986	V. P. Smirnov, Y. I. Sidorov, V. P. Yanchur, *Poverkhnost* **1986**, *4*, 123.
Spe 1972	D. M. Speros, R. M. Caldwell, W. E. Smyser, *High Temp. Sci.* **1972**, *4,* 99.
Spi 1979	H. C. Spirlet, *J. Phys. Colloq.* **1979**, *40*, 87.
Str 1981	P. Strobel, Y. Le Page, *J. Cryst. Growth* **1981***, 54*, 345.
Tei 1966	R. Teichmann, E. Wolf, *Z. Anorg. Allg. Chem.* **1966**, *347*, 145.
Tra 1969	R. F. Tramposch, *J. Electrochem. Soc.* **1969**, *116*, 654.
Tro 1876	L. Troost, O. Hautefeuille, *Ann. Chem. Phys.* **1876**, *5,* 452.
Tro 1877	L. Troost, O. Hautefeuille, *C. R. Acad. Sci.* **1877**, *84*, 946.
Val 1967	J. A. Valov, R. L. Plečko, *Krist. Tech.* **1967**, *2*, 535.
Vei 1955	N. D. Veigel, E. M. Sherwood, I. E. Campbell, *J. Electrochem. Soc.* **1955**, *102*, 687.
War 1913	H. von Wartenberg, *Z. Elektrochem.* **1913**, *19,* 482.
Wei 1970	G. Weise, G. Owsian, *J. Less-Comm. Met.* **1970**, *22*, 99.
Wei 1977	G. Weise, W. Richter, *Growth of Crystals* **1977**, *12*, 34.
Wie 1970	G. Wiese, R. Günther, *Krist. Tech.* **1970**, *5*, 323.
Wil 1976	M. Wilhelm, S. Frohmader, G. Ziegler, *Mater. Res. Bull.* **1976**, *11*, 491.
Wol 1966	E. Wolf, C. Herbst, *Z. Anorg. Allg. Chem.* **1966**, *347*, 113.
Yan 1972	L. N. Yannopoulos, *J. Appl. Phys.* **1972**, *43*, 2435.
Yel 1981	N. Yellin, G. Gafni, *J. Cryst. Growth* **1981***, 53*, 409.

4 Chemical Vapor Transport of Metal Halides

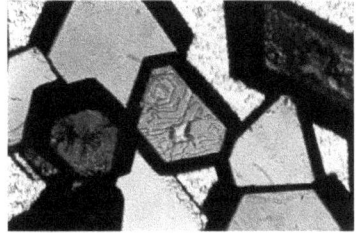

CoCl$_2$

H																	He
Li	Be											B	C	N	O	F	Ne
Na	Mg											Al	Si	P	S	Cl	Ar
K	Ca	Sc	Ti	V	Cr	Mn	Fe	Co	Ni	Cu	Zn	Ga	Ge	As	Se	Br	Xe
Rb	Sr	Y	Zr	Nb	Mo	Tc	Ru	Rh	Pd	Ag	Cd	In	Sn	Sb	Te	I	Kr
Cs	Ba	La	Hf	Ta	W	Re	Os	Ir	Pt	Au	Hg	Tl	Pb	Bi	Po	At	Rn

Ce	Pr	Nd	Pm	Sm	Eu	Gd	Tb	Dy	Ho	Er	Tm	Yb	Lu
Th	Pa	U	Np	Pu									

Formation of higher halides

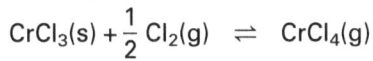

$$CrCl_3(s) + \frac{1}{2}\,Cl_2(g) \;\rightleftharpoons\; CrCl_4(g)$$

Conproportionation reactions

$$MoCl_3(s) + MoCl_5(g) \;\rightleftharpoons\; 2\,MoCl_4(g)$$

Formation of gas complexes

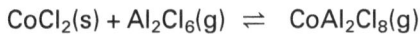

$$CoCl_2(s) + Al_2Cl_6(g) \;\rightleftharpoons\; CoAl_2Cl_8(g)$$

Halogen transfer reactions

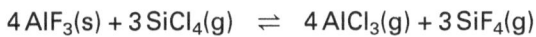

$$4\,AlF_3(s) + 3\,SiCl_4(g) \;\rightleftharpoons\; 4\,AlCl_3(g) + 3\,SiF_4(g)$$

Most metal halides are sufficiently stable in thermodynamic terms to evaporate them undecomposed. They can also be volatilized by distillation or sublimation and deposited again at lower temperatures. Some metal halides decompose at higher temperatures either to the elements or to a low halide and the according halogen. Thus, platinum(II)-chloride decomposes notably above 500 °C to platinum and chlorine. Copper(II)-chloride decomposes above 300 °C under the formation of copper(I)-chloride and chlorine. The tendency of decomposing generally increases from the fluorides to the iodides. Some metal halides disproportionate while heating: molybdenum(III)-chloride essentially dissociates above 600 °C under the formation of solid molybdenum(II)-chloride and gaseous molybdenum(IV)-chloride.

Many metal halides can be obtained by CVT reactions. In the process, four different types of solid/gas reactions play a central role. These are discussed in the following.

- Halogens as transport agents under the formation of higher halides.
$$CrCl_3(s) + \frac{1}{2} Cl_2(g) \; \rightleftharpoons \; CrCl_4(g)$$
- Conproportionation reactions.
$$MoCl_3(s) + MoCl_5(g) \; \rightleftharpoons \; 2\,MoCl_4(g)$$
- Formation of gas complexes.
$$CoCl_2(s) + Al_2Cl_6(g) \; \rightleftharpoons \; CoAl_2Cl_8(g)$$
- Halogen transfer reactions.
$$4\,AlF_3(s) + 3\,SiCl_4(g) \; \rightleftharpoons \; 4\,AlCl_3(g) + 3\,SiF_4(g)$$

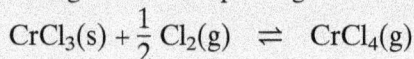

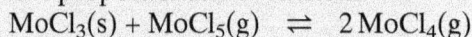

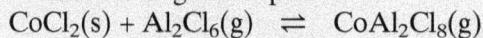

Additionally, further reactions of different kinds are known, which can be used for the transport of metal halides. However, their application is limited so far. The original literature can help. *Oppermann* provides an overview on the CVT of halides (Opp 1990).

4.1 Formation of Higher Halides

If the boiling temperatures of the halides of a metal in different oxidation numbers are considered, one will find decreasing boiling temperatures for increasing oxidation numbers. The reason for this effect is the covalence of the metal/halogen-compound, which increases with higher oxidation number. If a metal halide reacts with a halogen, high-volatile halides may be formed in which the metal atom has a higher oxidation number than in the solid. This is particularly often observed with halides of the transition metals. However, the halides of an element tend to decompose with higher oxidation numbers at the same time.

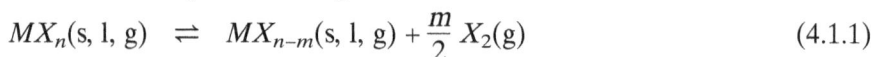

$$MX_n(s, l, g) \; \rightleftharpoons \; MX_{n-m}(s, l, g) + \frac{m}{2}\,X_2(g) \qquad (4.1.1)$$

Thus, often a high partial pressure of the halogen is necessary to form a sufficiently high transport effective partial pressure. Gaseous ruthenium(IV)-bromide is formed during the transport of ruthenium(III)-bromide with bromine (Bro 1968a). However, a high bromine pressure of 15 bar is necessary to cause a sufficient transport effect.

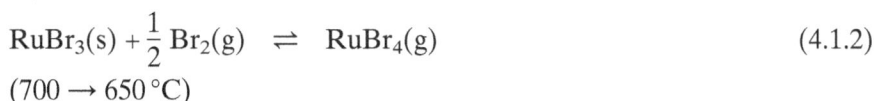

$$RuBr_3(s) + \frac{1}{2} Br_2(g) \quad \rightleftharpoons \quad RuBr_4(g) \qquad (4.1.2)$$
$$(700 \rightarrow 650\,°C)$$

Generally, the halogen is used as a transport agent, which is also contained in the solid. Sometimes, however, another halogen is used as the transport agent (McC 1964).

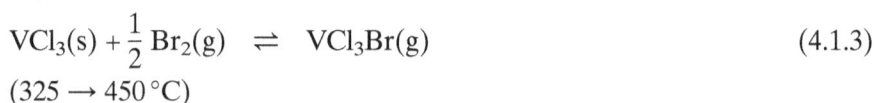

$$VCl_3(s) + \frac{1}{2} Br_2(g) \quad \rightleftharpoons \quad VCl_3Br(g) \qquad (4.1.3)$$
$$(325 \rightarrow 450\,°C)$$

During the deposition, a small amount of bromide is introduced. A solid of the composition $VCl_{2.97}Br_{0.03}$ forms in the sink.

4.2 Conproportionating Reactions

Often, the transition metals can appear in their binary halides in more than two oxidation numbers, which are stable under transport conditions. This particularly applies for metals of group 5 and 6. This circumstance can be used in order to transport a solid metal halide in which the metal has a low oxidation number with a gaseous metal halide in which the metal has an oxidation number that is higher by two units or more. The transport of niobium(III)-chloride with niobium(V)-chloride as transport agent is an example of this (Sch 1962).

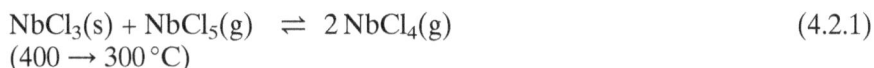

$$NbCl_3(s) + NbCl_5(g) \quad \rightleftharpoons \quad 2\,NbCl_4(g) \qquad (4.2.1)$$
$$(400 \rightarrow 300\,°C)$$

Gaseous niobium(IV)-chloride is formed, which disproportionates to solid niobium(III)-chloride and gaseous niobium(V)-chloride. Frequently, the according halogen is used as transport agent instead of the one that was formulated in the transport equation. The actually effective transport agent forms by the reaction of the transport additive with the solid. Additional examinations and/or thermodynamic model calculations are necessary in order to decide whether the added halogen or a higher halide that is formed by the halogen is the actual transport agent.

4.3 Formation of Gas Complexes

The term gas complex refers to a gaseous metal/halogen compound in which several metal atoms are connected with each other by halogen bridges. Gas complexes with several identical metal atoms, such as Al_2Cl_6, are called *homeo complexes*. Those with different metal atoms, such as $NaAlCl_4$, *hetero complexes* (Sch 1976, Sch 1983). In Chapter 11, there is a short overview of common types of gas complexes and their stability. Gas complexes are important for many chemical vapor transports (Sch 1975), in particular for the CVT of halides: In these cases, the solid to be transported is a metal halide with a high boiling temperature, the transport agent is a high volatility halide, particularly often an aluminum halide. The aluminum halides have low boiling temperatures and form stable gas complexes with a number of metal halides. Gallium(III)-halides, indium(III)-halides, and iron(III)-halides are used as transport agents as well. Exceptionally, the addition of other metal halides, e.g., titanium(IV)-chloride or tantalum(V)-chloride, cause a transport effect (Sch 1981a).

Monohalides, such as alkali metal halides, MX, form gas complexes of the composition $MAlX_4$ with the aluminum halides, AlX_3. These are characterized by an extremely high stability. However, the solid and liquid ternary halides of these compositions are so stable that it does not work to transport alkali metal halides with aluminum halides: in the sink, not the alkali metal halides but a different ternary phase is always deposited. Accordingly, this also applies when gallium(III)-halides, indium(III)-halides, and iron(III)-halides are used as transport agents.

Dihalides, MX_2, and aluminum halides, gallium halides, and iron halides essentially form gas complexes of the composition MAl_2X_8 and $MAlX_5$. Furthermore, the formation of gas complexes of the composition MAl_3Cl_{11} has been reported in some cases (Sch 1980a, Kra 1987b). At relatively low temperatures around 300 to 400 °C, the transport of dihalides with aluminum(III)-chloride takes place via MAl_2X_8 as transport-effective species.

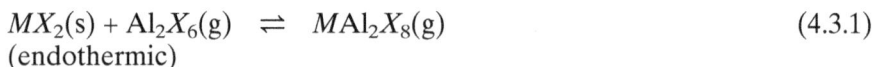

$$MX_2(s) + Al_2X_6(g) \rightleftharpoons MAl_2X_8(g) \qquad (4.3.1)$$
(endothermic)

These reactions are always endothermic. The reaction enthalpies are between 30 and 60 kJ · mol^{-1}. The transport equation 4.3.1 can formally be split into the reaction equations 4.3.2 and 4.3.3.

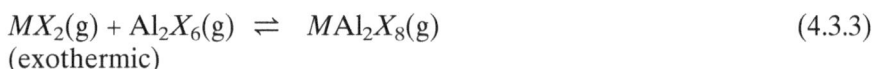

$$MX_2(s) \rightleftharpoons MX_2(g) \qquad (4.3.2)$$
(endothermic)

$$MX_2(g) + Al_2X_6(g) \rightleftharpoons MAl_2X_8(g) \qquad (4.3.3)$$
(exothermic)

Apparently, the enthalpy increase by the formation of the complex does not entirely compensate for the sublimation enthalpy so that the transport reaction 4.3.1 is always endothermic. If, however, the transport reaction is conducted at conditions under which the transport agent aluminum(III)-chloride is present as monomeric $AlCl_3(g)$, the transport direction can change. In this case, the transport occurs according to equation 4.3.4. *Lange* was able to prove this behavior with the example of the transport of $EuCl_2$ (Lan 1993).

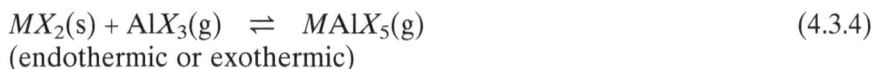

$$MX_2(s) + AlX_3(g) \rightleftharpoons MAlX_5(g) \qquad (4.3.4)$$
(endothermic or exothermic)

Besides $EuAl_2Cl_8$, the gas complexes $EuAl_3Cl_{11}$ and $EuAl_4Cl_{14}$ play a certain role. This way, a number of dihalides can already be obtained in crystalline form at low temperatures. Typical transport temperatures are 400 to 300 °C. Gallium(III)-halides, indium(III)-halides, and iron(III)-halides react in an analogous way. At temperatures above 500 °C, the mentioned transport agents are increasingly present in monomeric form. Additionally, complexes of the composition $MAlX_5$ become more and more important. The transport then takes place to the hotter zone and can be described by the transport equation 4.3.5.

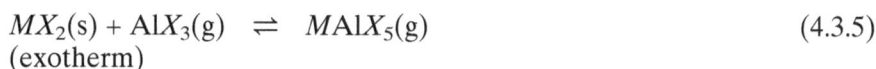

$$MX_2(s) + AlX_3(g) \; \rightleftharpoons \; MAlX_5(g) \qquad\qquad (4.3.5)$$
(exotherm)

Krauße and *Oppermann* discussed in detail the thermodynamic relations that lead to the reverse of the transport direction with the help of the transport example of manganese(II)-chloride with aluminum-chloride, gallium-chloride, and indium-chloride (Kra 1987).

Trihalides, MX_3, and aluminum halides, gallium halides, and iron halides can form gas complexes of the composition MAl_3X_{12} and MAl_4X_{15} (Oye 1969, Gru 1967). The formation of these gas complexes can be used to transport heavy volatile trihalides (Las 1974, Sch 1974). The transport is endothermic, typical transport temperatures are 500 to 400 °C. This way, chromium(III)-chloride (Las 1971) and the trihalides of the lanthanoid metals (Gun 1987) can be obtained in crystalline form with aluminum(III)-chloride as transport agent.

The formation of UAl_2Cl_{10} and, respectively, $ThAl_2Cl_{10}$ is assumed during the transport of the tetra halides UCl_4 and $ThCl_4$ with aluminum(III)-chloride (Sch 1974). These reactions are also endothermic (UCl_4: 350 → 250 °C; $ThCl_4$: 500 → 400 °C).

Pentahalides of metals are not very common. Their vapor pressure is relatively high so that they can be sublimed without any problems and are of minor interest for transport reactions.

The formation of gas complexes can be useful during the synthesis of metal halides from the elements: metals generally react with halogens in exothermic reactions under the formation of a metal halide. If these have a low vapor pressure under the synthesis conditions, a covering metal halide layer will form on the surface of the metal. This covering layer slows down the further reaction. If these syntheses are conducted in the presence of aluminum halides, the halides can get to the gas phase under formation of a gas complex. The formation of the covering layer can be avoided and the synthesis takes place a lot faster than without the aluminum halide. This was shown with the help of the examples of the synthesis of some transition metal halides (Sch 1978, Sch 1980c).

The formation of gas complexes has also been used for the separation of the halides of the lanthanoids. In the sense of a *transpiration experiment*, a flow of aluminum(III)-chloride is conducted by a mixture of lanthanoid halides. At a suitable temperature, the individual lanthanoid halides form gas complexes of different stability. These complexes decompose again under the back formation of the halides LnX_3 at different places. This way, the halides of the lanthanoids can be separated from each other (Shi 1990, Ada 1991, Mur 1992, Mur 1993, Oza 1997, Yu 1997, Oza 1998a, Oza 1998b, Oza 1999, Wan 1999, Wan 2000, Yan 2003). During separations by such a "fractionalized chemical vapor transport" a solid consisting of lanthanoids can be used. This solid is mixed with carbon and brought to reaction with a chlorine-containing carrier gas. At approximately 1000 °C, a reaction according to the following equation takes place:

$$Ln_2O_3(s) + 3\,C(s) + 3\,Cl_2(g) \; \rightleftharpoons \; 2\,LnCl_3(g) + 3\,CO(g) \tag{4.3.6}$$

The so-formed lanthanoid trichlorides can be transferred with aluminum(III)-chloride to the gas phase and can be separated as described. In the presence of alkali metal chlorides, volatile gas-phase complexes of the composition $MLnCl_4$ (M = Li, Na, K, Rb, Cs) form (Sun 2002, Sun 2004). Aluminum halides are unsuitable for the CVT of almost all oxido compounds because a very low oxygen partial pressure virtually always leads to the deposition of the particularly stable aluminum(III)-oxide (Sch 1978). Only the lanthanoid oxides are more stable than the aluminum oxide. This can be seen by the fact that in the so-called metal halide lamps ("xenon lamps"), the iodides of selected lanthanoid metals, which serve as the filling, react with the aluminum oxide of the lamp bulb. In the process, the aluminum oxide is partly transferred to aluminum iodide under the formation of the lanthanoid oxide. This leads to an undesired corrosion of the lamp bulb, on the one hand, and to a change in the gas-phase composition in the lamp, on the other hand, which can be negative for the emission behavior of the lamp (see Chapter 3).

It is important to note that the addition of the very moisture sensitive aluminum(III)-chloride as transport agent requires special experimental care. The same applies for iron(III)-halides and indium(III)-halides, and in particular for gallium(III)-halides. It is not advisable to fill the transport ampoules with commercially available substances under suitable precautions because the degree of purity of these substances is not always sufficient. It is better to synthesize the halides *in situ* from the elements and to sublime them to the transport ampoule (see section 14.2); the stoichiometric use of the respective metal with the according halogen is recommended.

4.4 Halogen Transfer Reactions

In contrast to the semi- and non-metals, the fluorides of the metals have essentially higher boiling temperatures than the chlorides, bromides, and iodides: The boiling temperatures of aluminum fluoride is 1275 °C, those of the other halides 181 °C ($AlCl_3$), 254 °C ($AlBr_3$), and 374 °C (AlI_3). Often, metal fluorides cannot be obtained in crystalline form by sublimation. Due to this circumstance, transport reactions come into focus as a preparative method. In a few cases, transport with silicon(IV)-chloride succeeded (Bon 1978, Red 1983).

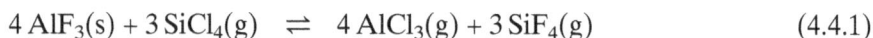

$$4\,AlF_3(s) + 3\,SiCl_4(g) \; \rightleftharpoons \; 4\,AlCl_3(g) + 3\,SiF_4(g) \tag{4.4.1}$$

The fact that silicon(IV)-fluoride as well as silicon(IV)-chloride are very highly volatile compounds is used in this case.

4.5 Formation of Interhalogen Compounds

The CVT of fluorides with halogens as transport agent is not possible via equilibria such as 4.5.1, due to their unfavorable position. The release of fluorine, which occurs during the reaction, is thermodynamically unfavorable. Nevertheless, magnesium fluoride can be crystallized with iodine as transport agent in the temperature gradient $1000 \rightarrow 900\ °C$ (Zen 1999a). Thermodynamic model calculations with data for the gaseous iodine fluorides IF_n ($n = 1, 3, 5, 7$) from the *ab initio* calculations (Dix 2008) reflect the observed transport effect. These calculations suggest a significant participation of IF_5 in the transport process according to 4.5.2 (Gla 2008). It has not yet been examined whether the transport reaction 4.5.2 can also be applied for other fluorides.

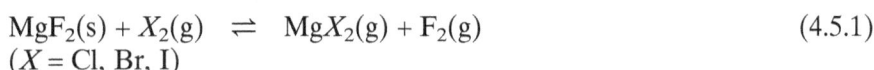

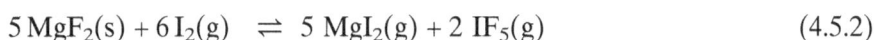

$$MgF_2(s) + X_2(g) \rightleftharpoons MgX_2(g) + F_2(g) \qquad (4.5.1)$$
$$(X = Cl, Br, I)$$

$$5\,MgF_2(s) + 6\,I_2(g) \rightleftharpoons 5\,MgI_2(g) + 2\,IF_5(g) \qquad (4.5.2)$$

Table 4.1 Examples of the chemical vapor transport of halides.

Sink solid	Transport additive	Temperature / °C	Reference
AlF_3	$SiCl_4$	$600 \ldots 800 \rightarrow 400 \ldots 650$	Bon 1978
$AuCl$	Cl_2	$235 \rightarrow 247$	Jan 1974
$BaCl_2$	$FeCl_3$	$780 \rightarrow 680$	Emm 1977
$BaBr_2$	$AlCl_3$	$500 \rightarrow T_1$	Zva 1979
$CaCl_2$	$GaCl_3$	$390 \ldots 455 \rightarrow 495 \ldots 525$	Sch 1977
	$FeCl_3$	$495 \rightarrow 325 \ldots 400$	Emm 1977
	$UCl_5 + Cl_2$	$500 \rightarrow 400$	Sch 1981b
$CeCl_3$	$AlCl_3$	$500 \rightarrow 400$	Gun 1987
	$AlCl_3$	$400 \rightarrow 180$	Yin 2000
$CeBr_3$	$AlBr_3$	$500 \rightarrow T_1$	Zva 1979
$CeBr_{3-x}Cl_x$	$AlCl_3 + AlBr_3$	$500 \rightarrow 400$	Schu 1991
$CoCl_2$	$AlCl_3$	$400 \rightarrow 300$	Sch 1974
	$GaCl_3$	$350 \rightarrow 300$	Sch 1974
	$AlCl_3$	$360 \rightarrow 240$	Del 1975
	$GaCl_3$	$410 \ldots 440 \rightarrow 500 \ldots 590$	Sch 1977
	$AlCl_3$	$400 \rightarrow 350$	Sch 1978
	$AlCl_3$	$390 \rightarrow 310$	Sch 1979
	$AlCl_3$	$400 \rightarrow 350$	Sch 1981a
	$GaCl_3$	$400 \rightarrow 350$	Sch 1981a
	$FeCl_3$	$400 \rightarrow 350$	Sch 1981a
	$UCl_5 + Cl_2$	$500 \rightarrow 400$	Sch 1981a
$CoBr_2$	$AlBr_3$	$345 \rightarrow 245$	Gee 1975

Table 4.1 (continued)

Sink solid	Transport additive	Temperature / °C	Reference
$CrCl_2$	$AlCl_3$	$T_2 \rightarrow 200$	Las 1972
	$GaCl_3$	$420 \ldots 430 \rightarrow 520 \ldots 530$	Sch 1977
$CrCl_3$	Cl_2	$T_2 \rightarrow 400$	Opp 1968
	Cl_2	$400 \ldots 700 \rightarrow T_1$	Ban 1969
	$AlCl_3$	$500 \rightarrow 400$	Las 1971
	$NbCl_5$	$600 \rightarrow 500$	Sch 1974
	CCl_4	$550 \rightarrow 400$	Ahm 1989
$CrBr_3$	Br_2	$600 \rightarrow T_1$	Sch 1962
	Br_2	$625 \ldots 875 \rightarrow T_1 (\Delta T = 50)$	Noc 1994
	Br_2	$640 \rightarrow 580$	Bel 1966
CrI_2	AlI_3	$450 \rightarrow 400$	Sch 1978
$CuCl_2$	$AlCl_3$	$450 \rightarrow 350$	Sch 1974
	$FeCl_3 + Cl_2$	$400 \rightarrow 350$	Sch 1974
	$GaCl_3 + Cl_2$	$425 \rightarrow 515$	Sch 1977
	$AlCl_3$	$400 \rightarrow 350$	Sch 1978
	$AlCl_3$	$390 \rightarrow 310$	Sch 1979
	$AlCl_3$	$400 \rightarrow 350$	Sch 1981a
	$GaCl_3$	$400 \rightarrow 350$	Sch 1981a
	$FeCl_3$	$400 \rightarrow 350$	Sch 1981a
$CsNb_4Cl_{11}$	$NbCl_5$	$730 \rightarrow 700$	Bro 1969
	$NbBr_5$	$630 \rightarrow 610$	Bro 1969
$DyCl_3$	$AlCl_3$	$400 \rightarrow 180$	Yin 2000
$DyAl_3Cl_{12}$	$AlCl_3$	$250 \rightarrow 160$	Hak 1990
$EuCl_2$	$AlCl_3$	$510 \rightarrow 570, 400 \rightarrow 300$	Lan 1993
	$AlCl_3$	$400 \rightarrow 180$	Yin 2000
Eu_5Cl_{11}	$AlCl_3$	$480 \rightarrow 545$	Lan 1993
$Eu_{14}Cl_{33}$ *	$AlCl_3$	$550 \rightarrow 575$	Lan 1993
$Eu(AlCl_4)_2$	$AlCl_3$	$320 \rightarrow 200$	Lan 1993
$Eu(AlCl_4)_3$	$AlCl_3$	$190 \rightarrow 160$	Lan 1993
$EuBr_3$	$AlCl_3$	$500 \rightarrow T_1$	Zva 1979
FeF_3	$SiCl_4$	$580 \rightarrow 360$	Bon 1978
$FeCl_2$	$AlCl_3$	$350 \rightarrow 250$	Sch 1974
	$FeCl_3$	$350 \rightarrow 250$	Sch 1974
	$GaCl_3$	$500 \rightarrow 400$	Sch 1977
	$AlCl_3$	$400 \rightarrow 350$	Sch 1978
FeI_2	I_2	$500 \rightarrow T_1$	Sch 1962
	AlI_3	$450 \rightarrow 400$	Sch 1978
$GdCl_3$	$AlCl_3$	$400 \rightarrow 310$	Gun 1987
	$AlCl_3$	$400 \rightarrow 180$	Yin 2000
$GdBr_3$	$AlCl_3$	$500 \rightarrow T_1$	Zva 1979
$GdBr_{3-x}Cl_x$	$AlCl_3 + AlBr_3$	$500 \rightarrow 400$	Schu 1991

Table 4.1 (continued)

Sink solid	Transport additive	Temperature / °C	Reference
$Ge_{4,06}I$	I_2	$1070 \rightarrow 625$	Nes 1986
HfNBr	NH_4Br	$760 \rightarrow 860$	Oro 2002
$HoCl_3$	$AlCl_3$	$400 \rightarrow 180$	Yin 2000
$HoAl_3Cl_{12}$	$AlCl_3$	$250 \rightarrow 160$	Hak 1990
$IrBr_3$	Br_2	$900 \rightarrow 450$	Bro 1968b
IrI_3	I_2	$1000 \rightarrow 400$	Bro 1968b
$LaCl_3$	$AlCl_3$	$750 \rightarrow 650$	Sch 1974
	$AlCl_3$	$500 \rightarrow 400$	Gun 1987
	$AlCl_3$	$400 \dots 750 \rightarrow 350$	Opp 1999
	$AlCl_3$	$400 \rightarrow 180$	Yin 2000
$LaBr_3$	$AlBr_3$	$350 \dots 600 \rightarrow T_1,$ $700 \dots 900 \rightarrow T_1$	Opp 2001
$LuCl_3$	$AlCl_3$	$400 \rightarrow 180$	Yon 1999
	$AlCl_3$	$400 \rightarrow 180$	Yin 2000
$LuBr_3$	$AlCl_3$	$500 \rightarrow T_1$	Zva 1979
	$AlBr_3$	$400 \rightarrow 600$	Jia 2005
MgF_2	I_2	$1000 \rightarrow 900$	Zen 1999
$MgCl_2$	$FeCl_3$	$575 \rightarrow 525$	Emm 1977
$MnCl_2$	$AlCl_3$	$400 \rightarrow 350$	Sch 1978
	$AlCl_3$	$580 \rightarrow T_1$	Kra 1987
	$GaCl_3$	$380 \rightarrow T_1, 380 \rightarrow T_2$	Kra 1987
	$InCl_3$	$730 \rightarrow T_1$	Kra 1987
$MoCl_2$ (Mo_6Cl_{12})	$MoCl_4$	$950 \rightarrow 850$	Sch 1967
$MoCl_3$	$MoCl_5$	$400 \rightarrow 375$	Sch 1967
	$MoCl_5$	$450 \rightarrow 375$	Opp 1972
	$AlCl_3$	$450 \rightarrow 350$	Sch 1974
	$MoCl_5$	$445 \rightarrow 405$	Sch 1980a
$MoBr_3$	Br_2	$450 \rightarrow 350$	Opp 1973
MoI_2 (Mo_6I_{12})	I_2	$800 \rightarrow 1000$	Ali 1981
NaCl	$NbCl_5$	$600 \rightarrow 500$	Sch 1974
$NbF_{2.5}$ (Nb_6F_{15})	NbF_5	$700 \rightarrow T_1$	Sch 1965a
$NbCl_3$	$NbCl_5$	$390 \rightarrow 355$	Sch 1962
$NbCl_{2.67}$ (Nb_3Cl_8)	$NbCl_5$	$T_2 \rightarrow 355$	Sch 1955
	$NbCl_5$	$600 \rightarrow 580$	Sch 1959

Table 4.1 (continued)

Sink solid	Transport additive	Temperature / °C	Reference
$NbCl_{2.33}$ (Nb_6Cl_{14})	$NbCl_5$	$840 \rightarrow 830$	Sim 1965
$NbCl_x$ ($x = 2.7 \ldots 4$)	$NbCl_5$	$T_2 \rightarrow 350$	Sae 1972
$NbBr_{2.67}$ (Nb_3Br_8)	$NbBr_5$	$800 \rightarrow 760$	Sim 1966
$NbBr_3$	$NbBr_5$	$450 \rightarrow 400$	Sch 1961
Nb_3I_8	NbI_5	$800 \rightarrow 760$	Sim 1966
$NdCl_3$	$AlCl_3$	$500 \rightarrow 400$	Gun 1987
	$AlCl_3$	$400 \rightarrow 180$	Yin 2000
$NdBr_{3-x}Cl_x$	$AlCl_3 + AlBr_3$	$500 \rightarrow 400$	Schu 1991
$NiCl_2$	$AlCl_3$	$350 \rightarrow 300$	Sch 1974
	$GaCl_3$	$450 \rightarrow 400$	Sch 1974
	$GaCl_3$	$505 \rightarrow 405$	Sch 1977
	$AlCl_3$	$400 \rightarrow 350$	Sch 1978
	$AlCl_3$	$465 \rightarrow 405$	Sch 1979
$NiBr_2$	$AlBr_3$	$345 \rightarrow 245$	Gee 1975
NiI_2	AlI_3	$450 \rightarrow 400$	Sch 1978
$OsCl_{3,5}$	$OsCl_4$	$500 \rightarrow 480$	Hun 1986
$OsCl_4$	Cl_2	$470 \rightarrow 420$	Hun 1986
$PdCl_2$	$AlCl_3$	$350 \rightarrow 300$	Sch 1978
	$AlCl_3$	$350 \rightarrow 300$	Sch 1974
	$GaCl_3 + Cl_2$	$445 \rightarrow 520$	Sch 1977
$PmBr_3$	$AlCl_3$	$500 \rightarrow T_1$	Zva 1979
$PrCl_3$	$AlCl_3$	$500 \rightarrow 400$	Gun 1987
	$AlCl_3$	$400 \rightarrow 180$	Yin 2000
$PrBr_{3-x}Cl_x$	$AlCl_3 + AlBr_3$	$500 \rightarrow 400$	Schu 1991
$PtCl_2$	Cl_2	$650 \rightarrow 550$	Sch 1970
	$AlCl_3$	$350 \rightarrow 300$	Sch 1974
$PtCl_3$	Cl_2	$600 \rightarrow 400$	Sch 1970
$PuCl_3$	Cl_2	$400 \ldots 750 \rightarrow T_1$	Ben 1962
$RbNb_4Cl_{11}$	$NbCl_5$	$730 \rightarrow 700$	Bro 1969
$RhCl_3$	$AlCl_3$	$300 \rightarrow 220$	Bog 1998
$RuCl_3$	Cl_2	$500 \rightarrow T_1$	Sch 1962
$RuBr_3$	Br_2	$700 \rightarrow 600$	vSc 1966
	Br_2	$750 \rightarrow 650$	Bro 1968a

Table 4.1 (continued)

Sink solid	Transport additive	Temperature / °C	Reference
ScF_3	$SiCl_4$, $GeCl_4$	$900 \rightarrow 850$	Red 1983
$ScCl_3$	$AlCl_3$	$400 \rightarrow 180$	Yin 2000
Sc_7Cl_{10}	$ScCl_3$	$880 \rightarrow 900$	Poe 1977
$ScBr_3$	$AlCl_3$	$500 \rightarrow T_1$	Yin 2000
$SmCl_2$	$AlCl_3$	$460 \rightarrow 500$	Rud 1997
$Sm_{14}Cl_{32}$	$AlCl_3$	$500 \rightarrow 520$	Rud 1997
$SmCl_3$	$AlCl_3$	$400 \rightarrow 310$	Gun 1987
	$AlCl_3$	$400 \rightarrow 180$	Yin 2000
$SmBr_{3-x}Cl_x$	$AlCl_3 + AlBr_3$	$500 \rightarrow 400$	Schu 1991
$SrCl_2$	$FeCl_3$	$790 \rightarrow 690$	Emm 1977
$SrBr_2$	$AlCl_3$	$500 \rightarrow T_1$	Zva 1979
$TaCl_3$	$TaCl_5$	$600 \ldots 620 \rightarrow 365$	Sch 1964
$TaCl_{2.5}$ (Ta_6Cl_{15})	$TaCl_5$	$630 \rightarrow 470$	Sch 1964
$TaBr_3$	$TaBr_5$	$620 \rightarrow 380$	Sch 1965a
$TaCl_{2.5}$ (Ta_6Br_{15})	$TaBr_5$	$620 \rightarrow 450$	Sch 1965a
$TaI_{2.33}$ (Ta_6I_{14})	TaI_5	$650 \rightarrow 350 \ldots 510$	Bau 1965
$TbCl_3$	$AlCl_3$	$500 \rightarrow 400$, $400 \rightarrow 310$	Gun 1987
$TbCl_3$	$AlCl_3$	$400 \rightarrow 180$	Yin 2000
$TbAl_3Cl_{12}$	$AlCl_3$	$250 \rightarrow 160$	Hak 1990
Te_3AlCl_4	$AlCl_3$	$290 \rightarrow 225$	Pri 1970
$ThCl_4$	$AlCl_3$	$500 \rightarrow 400$	Sch 1974
TiF_3	$SiCl_4$	$600 \rightarrow 400$	Bon 1978
$TiCl_3$	$AlCl_3$	$300 \rightarrow 250$	Sch 1981c
Ti_7Cl_{16}	$AlCl_3$	$400 \rightarrow 350$	Sch 1981c
$TiBr_3$	$AlBr_3$	$350 \rightarrow 250$	Sch 1981c
Ti_7Br_{16}	Br_2	$350 \rightarrow 250$	Sch 1981c
$TiNCl$	NH_4Cl	$400 \rightarrow 440$	Yam 2009
$TmCl_3$	$AlCl_3$	$400 \rightarrow 180$	Yin 2000
$TmBr_3$	$AlCl_3$	$500 \rightarrow T_1$	Zva 1979
UF_5	UF_6	not specified	Lei 1980
UCl_3	I_2	$770 \rightarrow T_1$	Bar 1951
UCl_4	Cl_2	$650 \ldots 750 \rightarrow T_1$	Lel 1914
	$AlCl_3$	$350 \rightarrow 250$	Sch 1974
VCl_2	I_2	$350 \rightarrow 400$	McC 1964
VCl_3	Cl_2	$300 \rightarrow 250$	Sch 1962
	Br_2	$325 \rightarrow 400$	McC 1964
	$AlCl_3$	$400 \rightarrow 350$	Sch 1981b

Table 4.1 (continued)

Sink solid	Transport additive	Temperature / °C	Reference
VCl$_3$	I$_2$	$T_2 \rightarrow 260 \dots 280$	Cor 1980
VI$_2$	I$_2$	$580 \rightarrow 520$	Juz 1969
	I$_2$	$500 \rightarrow 400, 800 \rightarrow 700$	Lam 1980
VI$_3$	I$_2$	$400 \rightarrow 320$	Juz 1969
WI$_3$	I$_2$	$450 \rightarrow 350$	Sch 1984
YbCl$_3$	AlCl$_3$	$400 \rightarrow 180$	Yin 2000
YbBr$_3$	AlCl$_3$	$500 \rightarrow T_1$	Zva 1979
YCl$_3$	AlCl$_3$	$600 \rightarrow 510$	Sch 1974
	FeCl$_3$ + Cl$_2$	$600 \rightarrow 500$	Sch 1974
	AlCl$_3$	$550 \rightarrow 300$	Opp 1995
	AlCl$_3$	$400 \rightarrow 180$	Yin 2000
YBr$_3$	AlBr$_3$	$550 \dots 800 \rightarrow 500$	Opp 1998
YI$_3$	AlI$_3$	$550 \dots 800 \rightarrow 500$	Opp 1998
ZrCl$_2$ (Zr$_6$Cl$_{12}$)	ZrCl$_4$	$610 \rightarrow 700$	Imo 1981

Bibliography

Ada 1991	G.-Y. Adachi, K. Shinozaki, Y. Hirashima, K.-I. Machida, *J. Less-Common Met.* **1991**, *169*, L1.
Ada 1992	G.-Y. Adachi, K. Murase, K. Shinozaki, K.-I. Machida, *Chem. Lett. (Chemical Soc. of Japan)* **1998**, 511.
Ahm 1989	A. U. Ahmed, *J. Bagladesh Acad. Sci.* **1989**, *13*, 181.
Ali 1981	Z. G. Aliev, L. A. Klinkova, I. V. Dubrovnin, L. O. Atovmyan, *Zh. Neorg. Khim.* **1981**, *26*, 1964.
Ban 1969	J. S. Bandorawolla, V. A. Altekar, *Trans. Indian. Inst. Metals* **1969**, 34.
Bar 1951	C. H. Barkelew, in: J. J. Katz, E. Rabinowitch (Ed's), *The Chemistry of Uranium* McGraw Hill, New York, National Nuclear Energy Series, **1951**.
Bau 1965	D. Bauer, H. G. v. Schnering, H. Schäfer, *J. Less-Common Met.* **1965**, *8*, 388.
Bel 1966	L. M. Belyaev, V. A. Lyakhovitdkaya, V. D. Spytsyna, *Izvest. Akad. Nauk. SSR Neorg. Mater.* **1966**, *2*, 2074.
Ben 1962	R. Benz, *J. Inorg. Nucl. Chem.* **1962**, *24*, 1191.
Bog 1998	S. Boghosian, G. D. Zissi, *Electrochem. Soc. Proc.* **1998**, *98*, 377.
Bon 1978	C. Bonnamy, J. C. Launay, M. Pouchard, *Rev. Chim. Minéral.* **1978**, *15*, 178.
Bro 1968a	K. Brodersen, H. K. Breitbach, G. Thiele, *Z. Anorg. Allg. Chem.* **1968**, *357*, 162.
Bro 1968b	K. Brodersen, *Angew. Chem.* **1968**, *80*, 155.
Bro 1969	A. Broll, A. Simon, H. G. v. Schnering, H. Schäfer, *Z. Anorg. Allg. Chem.* **1969**, *367*, 1.
Cor 1980	P. Corradini, A. Rovello, A. Sirigu, *Congr. Naz. Chim. Inorg. (Atti)* 13th **1980**, 185.

Del 1975   A. Dell'Ánna, F. P. Emmenegger, *Helv. Chim. Acta* **1975**, *58*, 1145.

Dix 2008   D. A. Dixon, D. J. Grant, K. O. Christe, K. A. Peterson, *Inorg. Chem.* **2008**, *47*, 5485.

Emm 1977   F. P. Emmenegger, *Inorg. Chem.* **1977**, *16*, 343.

Gee 1975   R. Gee, R. A. J. Sheldon, *J. Less-Common Met.* **1975**, *40*, 351.

Gla 2008   R. Glaum, *unpublished results*, University of Bonn, **2008**.

Gru 1967   D. M. Gruen, H. A. Oye, *Inorg. Nucl. Chem. Letters* **1967**, *3*, 453.

Gun 1987   H. Gunsilius, W. Urland, R. Kremer, *Z. Anorg. Allg. Chem.* **1987**, *550*, 35.

Hak 1990   D. Hake, W. Urland, *Z. Anorg. Allg. Chem.* **1990**, *586*, 99.

Hun 1986   K.-H. Huneke, H. Schäfer, *Z. Anorg. Allg. Chem.* **1980**, *534*, 216.

Imo 1981   H. Imoto, J. D. Corbett, A. Cisar, *Inorg. Chem.* **1981**, *20*, 145.

Jan 1974   E. M. W. Janssen, J. C. W. Folmer, G. A. Wiegers, *J. Less-Common Met.* **1974**, *38*, 71.

Jia 2005   J.-H. Jiang, X.-R- Xiao, T. Kuang, *Appl. Chem. Ind.* (China) **2005**, *34*, 317.

Juz 1969   D. Juza, D. Giegling, H. Schäfer, *Z. Anorg. Allg. Chem.* **1969**, *366*, 121.

Kra 1987   R. Krausze, H. Oppermann, *Z. Anorg. Allg. Chem.* **1987**, *550*, 123.

Lam 1980   G. Lamprecht, E. Schönherr, *J. Cryst. Growth* **1980**, *49*, 415.

Lan 1993   F. Th. Lange, H. Bärnighausen, *Z. Anorg. Allg. Chem.* **1993**, *619*, 1747.

Las 1971   K. Lascelles, H. Schäfer, *Z. Anorg. Allg. Chem.* **1971**, *382*, 249.

Las 1972   K. Lascelles, R. A. J. Shelton, H. Schäfer, *J. Less-Common Met.* **1972**, *29*, 109.

Lei 1980   J. M. Leitnaker, *High Temp. Sci.* **1980**, *12*, 289.

Lel 1914   D. Lely, L. Hamburger, *Z. Anorg. Allg. Chem.* **1914**, *87*, 209.

McC 1964   R. E. Mc Carley, J. W. Roddy, K. O. Berry, *Inorg. Chem.* **1964**, *3*, 50.

Mur 1992   K. Murase, K. Shinozaki, K.-I. Machida, G.-Y. Adachi, *Bull. Chem. Soc. Jpn.* **1992**, *65*, 2724.

Mur 1993   K. Murase, K. Shinozaki, Y. Hirashima, K.-I. Machida, G.-Y. Adachi, *J. Alloys Compd.* **1993**, *198*, 31.

Nes 1986   R. Nesper, J. Curda, H.-G. v. Schnering, *Angew. Chem.* **1986**, *98*, 369, *Angew. Chem. Int. Ed.* **1986**, *25*, 350.

Noc 1994   K. Nocker, R. Gruehn, *Z. Anorg. Allg. Chem.* **1994**, *620*, 73.

Opp 1968   H. Oppermann, *Z. Anorg. Allg. Chem.* **1968**, *359*, 51.

Opp 1972   H. Oppermann, G. Stöver, *Z. Anorg. Allg. Chem.* **1972**, *387*, 218.

Opp 1973   H. Oppermann, *Z. Anorg. Allg. Chem.* **1973**, *395*, 249.

Opp 1990   H. Oppermann, *Solid State Ionics* **1990**, *39*, 17.

Opp 1995   H. Oppermann, D. Q. Huong, *Z. Anorg. Allg. Chem.* **1995**, *621*, 659.

Opp 1998   H. Oppermann, S. Herrera, D. Q. Huong, *Z. Naturforsch.* **1998**, *53 b*, 361.

Opp 1999   H. Oppermann, H. D. Quoc, A. Morgenstern, *Z. Naturforsch.* **1999**, *54 b*, 1410.

Opp 2001   H. Oppermann, H. D. Quoc, M. Zhang-Preße, *Z. Naturforsch.* **2001**, *56 b*, 908.

Oro 2002   J. Oró-Solé, M. Vlassov, D. Beltrán-Porter, M. T. Caldés, V. Primo, A. Fuertes, *Solid States Sci.* **2002**, *4*, 475.

Oye 1969   H. A. Oye, D. M. Gruen, *J. Amer. Chem. Soc.* **1969**, *91*, 2229.

Oza 1998a   T. Ozaki, K.-I. Machida, G.-Y. Adachi, *Rare Earths* **1988**, *32*, 154.

Oza 1998b   T. Ozaki, J. Jiang, K. Murase, K.-I. Machida, G.-Y. Adachi, *J. Alloys Compd.* **1998**, *265*, 125.

Oza 1999   T. Ozaki, K.-I. Machida, G.-Y. Adachi, *Metall. Mater. Trans.* **1999**, *30 B*, 45.

Poe 1977   K. R. Poeppelmeier, J. D. Corbett, *Inorg. Chem*, **1977**, *16*, 1107.

Pri 1970   D. J. Prince, J. D. Corbett, B. Garbisch, *Inorg. Chem*, **1970**, *9*, 2731.

Red 1983   W. Redlich, T. Petzel, *Rev. Chim. Min.* **1983**, *20*, 54.

Rud 1997   M. Rudolph, W. Urland, *Z. Anorg. Allg. Chem.* **1997**, *623*, 1349.

Sae 1972   Y. Saeki, M. Yanai, A. Sofue, *Denki Kagagu*, **1972**, *40*, 816.
Sch 1955   H. Schäfer, *Angew. Chem.* **1955**, *67*, 748.
Sch 1959   H. Schäfer, K. D. Dohmann, *Z. Anorg. Allg. Chem.* **1959**, *300*, 1.
Sch 1961   H. Schäfer, K. D. Dohmann, *Z. Anorg. Allg. Chem.* **1961**, *311*, 134.
Sch 1962   H. Schäfer, *Chemische Transportreaktionen*, Verlag Chemie, Weinheim, **1962**.
Sch 1964   H. Schäfer, H. Scholz, R. Gehrken, *Z. Anorg. Allg. Chem.* **1964**, *331*, 154.
Sch 1965a   H. Schäfer, R. Gehrken, H. Scholz, *Z. Anorg. Allg. Chem.* **1965**, *335*, 96.
Sch 1965b   H. Schäfer, H. G. v. Schnering, K. J. Niehues, H. G. Nieder-Vahrenholz, *J. Less-Common Met.* **1965**, *9*, 95.
Sch 1967   H. Schäfer, H. G. v. Schnering, J. Tillack, F. Kuhnen, H. Wöhrle, H. Baumann, *Z. Anorg. Allg. Chem.* **1967**, *353*, 281.
Sch 1970   U. Wiese, H. Schäfer, H. G. v. Schnering, C. Brendel, *Angew. Chem.* **1970**, *82*, 135.
Sch 1974   H. Schäfer, M. Binnewies, W. Domke, J. Karbinski, *Z. Anorg. Allg. Chem.* **1974**, *403*, 116.
Sch 1975   H. Schäfer, *J. Cryst. Growth* **1975**, *31*, 31.
Sch 1976   H. Schäfer, *Angew. Chem.* **1976**, *88*, 775, *Angew. Chem. Int. Ed.* **1976**, *15*, 713.
Sch 1977   H. Schäfer, M. Trenkel, *Z. Anorg. Allg. Chem.* **1977**, *437*, 10.
Sch 1978   H. Schäfer, J. Nowitzki, *J. Less-Common Met.* **1978**, *61*, 47.
Sch 1979   H. Schäfer, J. Nowitzki, *Z. Anorg. Allg. Chem.* **1979**, *457*, 13.
Sch 1980   H. Schäfer, U. Wiese, C. Brendel, J. Nowitzki, *J. Less-Common Met.* **1980**, *76*, 63.
Sch 1980a   H. Schäfer, *Z. Anorg. Allg. Chem.* **1980**, *469*, 123.
Sch 1980b   H. Schäfer, U. Flörke, *Z. Anorg. Allg. Chem.* **1980**, *469*, 172.
Sch 1981a   H. Schäfer, *Z. Anorg. Allg. Chem.* **1981**, *479*, 105.
Sch 1981b   H. Schäfer, U. Flörke, M. Trenkel, *Z. Anorg. Allg. Chem*, **1981**, *478*, 191.
Sch 1981c   H. Schäfer, R. Laumanns, *Z. Anorg. Allg. Chem*, **1981**, *474*, 135.
Sch 1983   H. Schäfer, *Adv. Inorg. Nucl. Chem.* **1983**, *26*, 201.
Sch 1984   H. Schäfer, H. G. Schulz, *Z. Anorg. Allg. Chem.* **1984**, *516*, 196.
Schu 1991   M. Schulze, W. Urland, *Eur. J. Solid State Inorg. Chem.* **1991**, *28*, 571.
Shi 1975   C. F. Shieh, N. W. Gregory, *J. Phys. Chem.* **1975**, *79*, 828.
Shi 1990   K. Shinozaki, Y. Hirashima, G. Adachi, *Trans. Nonferrous Met. Soc. China* **1990**, *16*, 42.
Sim 1965   A. Simon, H. G. v. Schnering, H. Wöhrle, H. Schäfer, *Z. Anorg. Allg. Chem.* **1965**, *339*, 155.
Sim 1966   A. Simon, H. G. v. Schnering, *J. Less-Common Met.* **1966**, *11*, 31.
Sun 2002   Y.-H.- Sun, L. Zhang, P. X. Lei, Z.-C. Wang, L. Guo, *J. Alloys Compd.* **2002**, *335*, 196.
Sun 2004   Y.-H. Sun, Z.-F. Chen, Z.-C. Wang, *Trans. Nonferrous Met. Soc. China* **2004**, *14*, 412.
vSc 1966   H. G. v. Schnering, K. Brodersen, F. Moers, H. Breitbach and G. Thiele, *J. Less-Common Met.* **1966**, *11*, 288.
Wan 1999   Z.-C. Wang, Y.-H. Sun, L. Guo, *J. Alloys Compd.* **1999**, *287*, 109.
Wan 2000   Y. H. Wang, L. S. Wang, *J. Chinese Rare Earth Soc.* **2000**, *18*, 74.
Yam 2009   S. Yamanaka, T. Yasunaga, K. Yamaguchi, M. Tagawa, *J. Mater. Chem.* **2009**, *19*, 2573.
Yan 2003   D. Yang, J. Yu, J. Jiang, Z. Wang, *J. Mater. and Metall.* **2003**, *2*, 113.
Yin 2000   L. Yin, W. Linshan, W. Yuhong, *Rare Met.* **2000**, *19*, 157.
Yon 2000   S. Yonbo, W. Linshan, W. Yuhong, *Rare Met.* **1999**, *18*, 217.
Yu 1997   J. Yu, Y. Yu, Z. Wang, *Jinshu Xuebao (Acta Metallurgica Sinica)* **1997**, *3*, 391.
Zva 1979   T. S. Zvarova, *Radiokhirniya* **1979**, *21*, 627.
Zen 1999   L.-P. Zenser, *Dissertation*, University of Gießen, **1999**.

5 Chemical Vapor Transport of Binary and Multinary Oxides

CeTa$_3$O$_9$

H																	He
Li	Be											B	C	N	O	F	Ne
Na	Mg											Al	Si	P	S	Cl	Ar
K	Ca	Sc	Ti	V	Cr	Mn	Fe	Co	Ni	Cu	Zn	Ga	Ge	As	Se	Br	Xe
Rb	Sr	Y	Zr	Nb	Mo	Tc	Ru	Rh	Pd	Ag	Cd	In	Sn	Sb	Te	I	Kr
Cs	Ba	La	Hf	Ta	W	Re	Os	Ir	Pt	Au	Hg	Tl	Pb	Bi	Po	At	Rn

Ce	Pr	Nd	Pm	Sm	Eu	Gd	Tb	Dy	Ho	Er	Tm	Yb	Lu
Th	Pa	U	Np	Pu									

Chlorine as transport agent

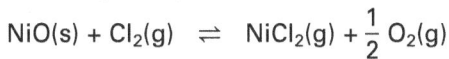

$$NiO(s) + Cl_2(g) \;\rightleftharpoons\; NiCl_2(g) + \frac{1}{2} O_2(g)$$

Tellurium(IV)-chloride as transport agent

$$ZrO_2(s) + TeCl_4(g) \;\rightleftharpoons\; ZrCl_4(g) + TeO_2(g)$$

Hydrogen chloride as transport agent

$$NiMoO_4(s) + 4\,HCl(g) \;\rightleftharpoons\; NiCl_2(g) + MoO_2Cl_2(g) + 2\,H_2O(g)$$

Auto transport

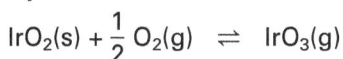

$$IrO_2(s) + \frac{1}{2} O_2(g) \;\rightleftharpoons\; IrO_3(g)$$

The oxides are the largest group among all compounds, and have been crystalli-zed by chemical transport reactions with more than 600 examples. Among these are binary oxides, such as iron(III)-oxide, oxides with complex anions, such as phosphates or sulfates, as well as oxides with several cations, such as $ZnFe_2O_4$ or $Co_{1-x}Ni_xO$. Ternary oxides, to which belong most of the compounds with complex anions, are most common. There are more than 100 examples of binary oxides, most of them containing transition metals. The smallest group is the one of the quaternary and multinary oxides, with slightly more than 50 examples.

The question whether an oxide is transportable is not only dependent on *its* thermodynamic stability but also on the stabilities of the gas species that are formed. Thus, zirconium(IV)-oxide, which is thermodynamically very stable, is suited for transport reactions in contrast to the less stable rubidium oxide. Metal oxides are thermodynamically very stable compounds. However, only a few of them evaporate undecomposed; among them are CrO_3, MoO_3, WO_3, Re_2O_7, GeO, SnO, PbO, and TeO_2. Most of the metal oxides decompose at higher tempe-ratures. In these cases, oxygen and the respective metal or a metal-rich oxide are formed. The latter can be present as condensed phase or gas. The following three examples show the different thermal behavior of metal oxides:

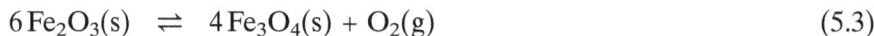

$$2\,ZnO(s) \;\rightleftharpoons\; 2\,Zn(g) + O_2(g) \tag{5.1}$$

$$2\,SiO_2(s) \;\rightleftharpoons\; 2\,SiO(g) + O_2(g) \tag{5.2}$$

$$6\,Fe_2O_3(s) \;\rightleftharpoons\; 4\,Fe_3O_4(s) + O_2(g) \tag{5.3}$$

The oxygen partial pressure that is built up during these decompositions is called the **decomposition pressure**. It is the leading determinant of the transport beha-vior and the composition of the transported solid. This can be explained with the example of Fe_2O_3. If the oxygen partial pressure in the system is higher than the decomposition pressure of Fe_2O_3, this oxide is stable as a solid. If it is lower Fe_3O_4, a compound with lower oxygen content, will be formed. If the oxygen partial pressure is identical with the decomposition pressure at a certain tempera-ture, the two solid phases will co-exist. This is called the **co-existence decomposi-tion pressure**. The respective oxygen co-existence decomposition pressure of an oxide phase results from its decomposition equilibrium, where the oxygen-rich solid phase is converted to the co-existing, oxygen-poorer solid phase (in border cases the metal) and gaseous oxygen.

In order to compare quantitatively different decomposition reactions, the re-action equation and the corresponding law of mass action is formulated in a way that the equilibrium constants K_p have the same unit, e. g., bar.

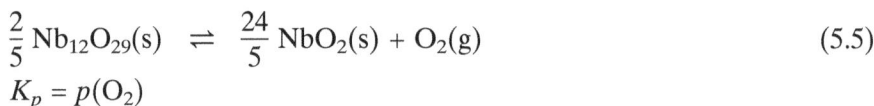

$$2\,Ag_2O(s) \;\rightleftharpoons\; 4\,Ag(s) + O_2(g) \tag{5.4}$$
$$K_p = p(O_2)$$

$$\frac{2}{5}\,Nb_{12}O_{29}(s) \;\rightleftharpoons\; \frac{24}{5}\,NbO_2(s) + O_2(g) \tag{5.5}$$
$$K_p = p(O_2)$$

$$\frac{2}{3} HgO(s) \; \rightleftharpoons \; \frac{2}{3} Hg(g) + \frac{1}{3} O_2(g) \qquad\qquad (5.6)$$
$$K_p = p^{2/3}(Hg) \cdot p^{1/3}(O_2)$$

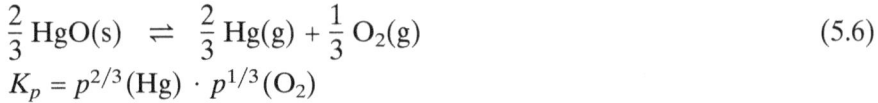

The equilibrium constant and the oxygen partial pressure can be calculated from the thermodynamic data of a decomposition reaction (see section 15.6). The co-existence decomposition pressure of different oxides of an element increases with temperature. At a given temperature, it increases from the more oxygen-poor to the more oxygen-rich phase. If the logarithm of a co-existence decomposition pressure is plotted against the reciprocal temperature for different oxides of one metal, the **phase barogram** of the system will be obtained. The linear dependency of $\lg p(O_2)$ versus T^{-1} defines the co-existence decomposition line. It separates the respective existence ranges of the two phases. Figure 5.1 shows schematically a hypothetical phase diagram (T/x), in which the phases MO, M_2O_3 as well as a solid with a homogeneity range $MO_{2\pm\delta}$ and M_2O_5 appear. In Figure 5.2 the co-existence decomposition pressures are illustrated over the mentioned phases as function of temperature. Figure 5.3 shows the same fact in another way of visualization ($\lg p$ versus T^{-1}). Both illustrations can be understood as phase diagrams because they show the existence ranges of the different phases as a function of temperature and oxygen pressure.

Due to the fact that the oxygen partial pressure is decisive for the composition of the deposited solid, important criteria for the transport conditions can be derived from the phase barogram. In the following, eight different possibilities of the endothermic transport from T_2 to T_1 in a binary system are discussed with the help of Figures 5.1 and 5.3.

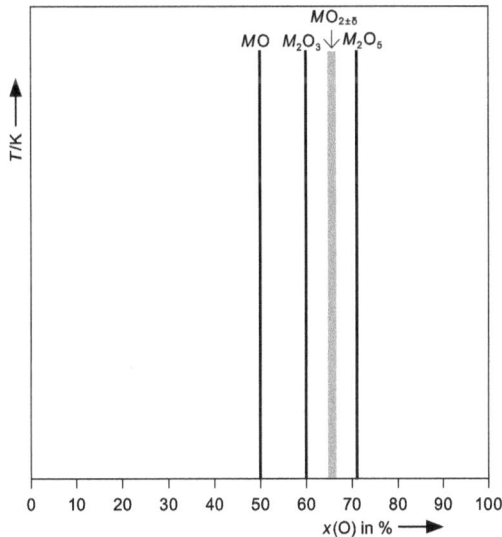

Figure 5.1 Schematic illustration of a hypothetical phase diagram for a metal/oxygen system.

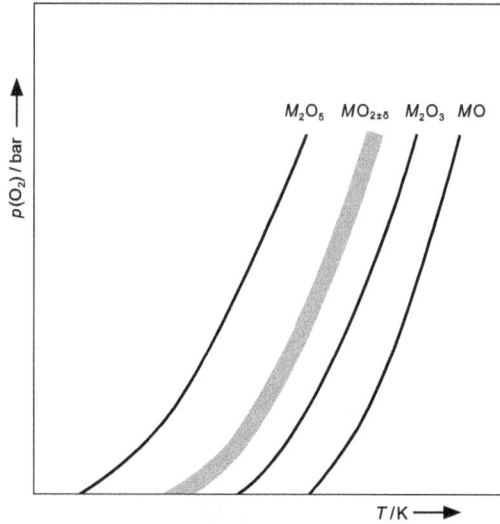

Figure 5.2 Schematic illustration of the temperature dependency of the oxygen partial pressure above the different phases (p versus T).

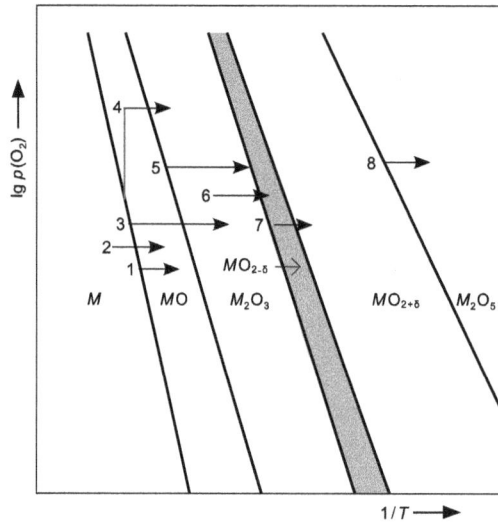

Figure 5.3 Schematic illustration of the temperature dependency of the oxygen partial pressures above the different phases ($\lg p$ versus T^{-1}). Arrows indicate the transport behavior of the oxides.

Case 1: MO is the source solid at T_2. The chosen temperature gradient between T_2 and T_1 is smaller than the temperature difference at which MO and M_2O_3 have the same oxygen co-existence pressure (see Figure 5.3). Thus, *stationary transport* takes place in the existence range of MO, which is deposited as the sink solid.

Case 2: The source consists of two equilibrium solids at T_2, M and MO. The chosen temperature gradient is, as in case 1, smaller than the temperature difference at which MO and M_2O_3 have the same oxygen co-existence pressure (see Figure 5.3). MO is formed as the sink solid.

Case 3: MO is provided as the source solid at T_2. In this case, the chosen temperature gradient between T_2 and T_1 is larger than the temperature difference at which MO and M_2O_3 have the same oxygen co-existence pressure (see Figure 5.3). At first, the deposition of M_2O_3 at T_1 is observed. By this, the gas phase as well as the source solid are depleted of oxygen. The consequence is a new stationary state, which can be described as case 2. Afterwards MO is deposited. Hence, a *non-stationary transport behavior* is expected.

Case 4: As in case 1, the source solid consists of MO. However, the oxygen partial pressure is increased by the transport agent, residual gas, or traces of moisture. Due to this, the co-existence line of M_2O_3 is crossed despite the same temperature gradient as in case 1; MO is not deposited but M_2O_3. The resulting behavior is the same as in case 3.

Case 5: M_2O_3 is used as the source solid at T_2. The temperature gradient between T_2 and T_1 is chosen in a way that it reaches into the homogeneity range of the phase $MO_{2-\delta}$. In this case, the non-stationary deposition of $MO_{2-\delta}$ takes place at T_1. A time dependent variation of δ is observed.

Case 6: In this case, a two-phase solid, composed of M_2O_3 and MO_2, is used. The temperature gradient between T_2 and T_1 is chosen in a way, as in case 5, that it reaches into the homogeneity range of the phase $MO_{2\pm\delta}$. The deposition takes place in the homogeneity range between $MO_{2-\delta}$ and $MO_{2+\delta}$. Depending on the ratio of $n(M_2O_3) : n(MO_2)$ the transport can be quasi-stationary.

Case 7: The transport occurs out of the homogeneity range. The temperature gradient is slightly larger than the temperature difference at which $MO_{2-\delta}$ and $MO_{2+\delta}$ have the same oxygen co-existence pressure. $MO_{2+\delta}$ is deposited.

Case 8: The most oxygen-rich compound M_2O_5 is provided. Due to the fact that there is no compound that is more oxygen-rich in the system, the oxygen partial pressure is free according to *Gibbs's* phase rule. Thus, the temperature gradient is not limited. M_2O_5 is always deposited as the sink solid.

These different examples show that the oxygen partial pressure in oxide systems is decisive for the directed deposition of a certain compound. Likewise, the choice of the transport agent is dependent on the oxygen partial pressure. If the oxygen partial pressure over solid phases is low, the transport agent must be able to form oxygen-transferring species. There are many examples of this behavior in the literature. If, for example, $TeCl_4$ is used as the transport agent for oxides, $TeOCl_2$ and/or TeO_2 can be effective as oxygen-transferring species. If gaseous metal oxide halides are formed, the situation is different. These species can transfer simultaneously both metal and oxygen atoms from the source to the sink.

5.1 Transport Agents

Chlorinating equilibria proved suitable for the CVT of oxides. Apart from chlorine and hydrogen chloride, tellurium(IV)-chloride is an important transport agent. Tellurium(IV)-chloride is used especially when the oxygen partial pressure in the system varies, and the setting of the oxygen partial pressure is of essential importance for the transport. Some other chlorinating additives include phosphorus(V)-chloride, niobium(V)-chloride, selenium(IV)-chloride, and tetrachloromethane as well as mixtures of sulfur/chlorine, vanadium(III)-chloride/chlorine, and chromium(III)-chloride/chlorine. Due to the generally unfavorable equilibrium position, brominating and iodinating equilibria are of minor importance. Transport agents or transport effective additives, respectively, are: bromine and iodine, hydrogen bromide and hydrogen iodide, phosphorus(V)-bromide, niobium(V)-bromide and niobium(V)-iodide as well as sulfur + iodine. Iodine as transport agent and iodinating equilibria are of interest if chlorine is too oxidizing or if, as is the case with rare-earth metal oxides, stable solid oxide chlorides form. Thus the transport agent is taken from the gas phase. Some further transport agents or transport effective additives, respectively, are hydrogen, oxygen, water, carbon monoxide and in special cases, fluorine or hydrogen fluoride. In some cases, the solid oxides can form gaseous oxide halides: transport-effective species, which contain both oxygen and halogen atoms. There are many metal/oxygen systems in which the oxygen partial pressure varies depending on the composition of the solid phase. The transport agent must allow for this fact.

Transports with halogens The process of dissolution of an oxide in the gas phase, for example by the reaction with a halogen, can be split into two partial reactions:

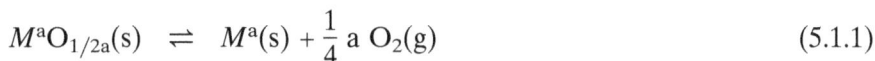

$$M^aO_{1/2a}(s) \; \rightleftharpoons \; M^a(s) + \frac{1}{4} a \; O_2(g) \tag{5.1.1}$$

(a = oxidation number of the metal)

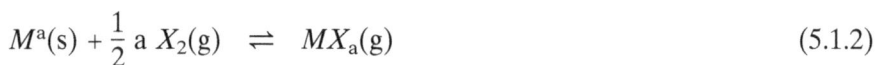

$$M^a(s) + \frac{1}{2} a \; X_2(g) \; \rightleftharpoons \; MX_a(g) \tag{5.1.2}$$

(X = Cl, Br, I).

Generally, the first reaction is endergonic; the second one is exergonic in almost all cases. The difference of the Gibbs energy of both reactions is decisive for the transportability of an oxide. The transport reaction results as a sum of these two reactions if no gaseous oxide halides are formed. The Gibbs energy of a transport reaction should not be extreme (see Chapter 1 and 2), usually between -100 kJ $\cdot$ mol^{-1} and $+100$ kJ $\cdot$ mol^{-1}. In oxide systems, the absolute value of the heat of reaction of the first partial reaction is generally higher than that of the second one so that an endothermic transport from T_2 to T_1 is expected. The absolute value of the Gibbs energy of the second partial reaction decreases from chlorine via bromine to iodine. Due to the higher stability of the chloride compared to the bromide and the resulting equilibrium position, mostly chlorine is used as the transport agent for the CVT of oxides. In the process, sufficiently stable gas species are formed with adequately high partial pressures ($p > 10^{-5}$ bar) and sufficiently high partial pressure differences along the temperature gradient. The generalized transport equation of the transport of a binary oxide is:

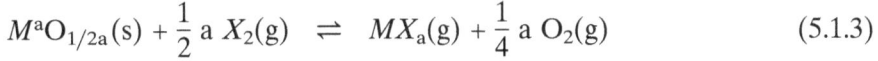

$$M^a O_{1/2a}(s) + \frac{1}{2} a\, X_2(g) \;\rightleftharpoons\; MX_a(g) + \frac{1}{4} a\, O_2(g) \tag{5.1.3}$$

The following equation applies for a ternary oxide:

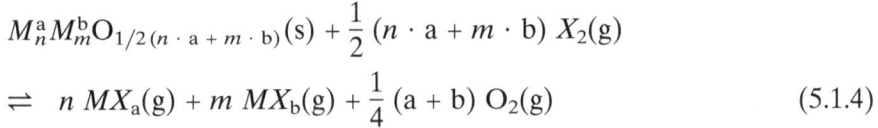

$$M_n^a M_m^b O_{1/2(n\cdot a + m\cdot b)}(s) + \frac{1}{2}(n\cdot a + m\cdot b)\, X_2(g)$$

$$\rightleftharpoons\; n\, MX_a(g) + m\, MX_b(g) + \frac{1}{4}(a+b)\, O_2(g) \tag{5.1.4}$$

The transports of Fe_2O_3 and $NiGa_2O_4$ with chlorine shall be used as examples.

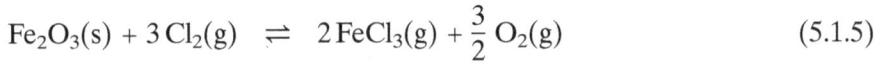

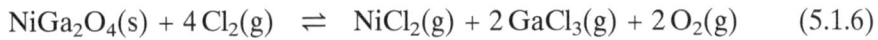

$$Fe_2O_3(s) + 3\, Cl_2(g) \;\rightleftharpoons\; 2\, FeCl_3(g) + \frac{3}{2}\, O_2(g) \tag{5.1.5}$$

$$NiGa_2O_4(s) + 4\, Cl_2(g) \;\rightleftharpoons\; NiCl_2(g) + 2\, GaCl_3(g) + 2\, O_2(g) \tag{5.1.6}$$

Halogens are also suited as transport agents for oxides when gaseous oxide halides are formed. This way, for example, the transport of molybdenum(VI)-oxide with chlorine succeeds:

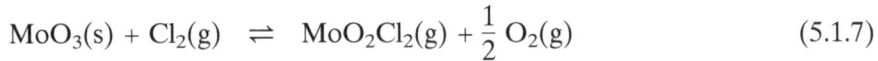

$$MoO_3(s) + Cl_2(g) \;\rightleftharpoons\; MoO_2Cl_2(g) + \frac{1}{2}\, O_2(g) \tag{5.1.7}$$

Instead of introducing pure halogens, decomposition of less stable halides, such as PtX_2 (X = Cl, Br, I), can be used to form halogens. If mercury halides are employed as transport agents, the equilibrium position of the transport reaction will shift compared to elemental halogen. By decomposing gaseous mercury halides to the elements (equation 5.1.8), additional gas species are formed. There will be a change of the entropy balance shifting the equilibrium to the side of the reaction products.

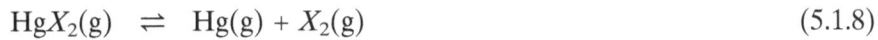

$$HgX_2(g) \;\rightleftharpoons\; Hg(g) + X_2(g) \tag{5.1.8}$$

Transport with hydrogen halides Hydrogen chloride, hydrogen bromide, and hydrogen iodide are often used and are effective transport agents for the transport of oxides. Sometimes also hydrogen fluoride is used as a transport agent, especially for silicates. During the transport of a binary oxide with a hydrogen halide, a gaseous metal halide is formed besides water. The general transport equation is:

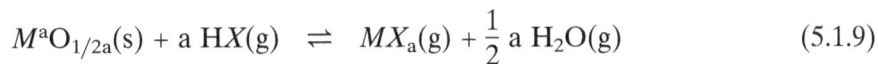

$$M^a O_{1/2a}(s) + a\, HX(g) \;\rightleftharpoons\; MX_a(g) + \frac{1}{2} a\, H_2O(g) \tag{5.1.9}$$

The transport of zinc oxide with hydrogen chloride is an example:

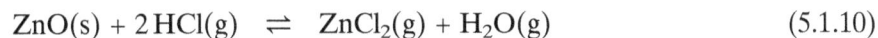

$$ZnO(s) + 2\, HCl(g) \;\rightleftharpoons\; ZnCl_2(g) + H_2O(g) \tag{5.1.10}$$

The following generalizing transport equation applies for ternary oxides:

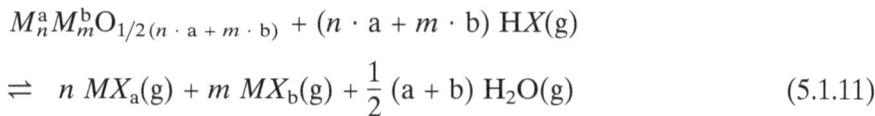

$$M_n^a M_m^b O_{1/2(n \cdot a + m \cdot b)} + (n \cdot a + m \cdot b)\, HX(g)$$

$$\rightleftharpoons\ n\, MX_a(g) + m\, MX_b(g) + \frac{1}{2}(a + b)\, H_2O(g) \tag{5.1.11}$$

These equations only apply if no volatile acid, such as $H_2MoO_4(g)$, hydroxide, and oxide halide, respectively, is formed. In some cases, a more favorable equilibrium position can be achieved by using hydrogen halides instead of halogens. Often ammonium halides (NH_4X, $X = Cl$, Br, I) are used as the hydrogen halide source. These solids are easy to handle and to dose (see Chapter 14). They decompose to ammonia and hydrogen halide at increased temperature. However, the formation of ammonia leads to an increase of the total pressure in the system and thus to lower transport rates. It is important to consider that ammonia and the hydrogen that is formed by decomposition above 600 °C, create a reducing atmosphere. This can lead to a reduction of the gas species and/or the solid phase.

In some cases, in which the transport of oxides or sulfides with moisture-sensitive halides, such as aluminum(III)-chloride or tellurium(IV)-chloride, is described, hydrogen chloride can be expected as the transport agent. Traces of water, which can never be excluded completely, cause the formation of hydrogen halide.

Tellurium(IV)-halides as transport agents Tellurium(IV)-chloride is a flexible transport agent, which can especially be used for oxides of the transition metals and compounds with complex anions as the works of *Oppermann* and co-workers show (Opp 1975, Mer 1981). Let us consider the transport of a dioxide above 700 °C as an example. The following simplified transport equation can be assumed:

$$MO_2(s) + TeCl_4(g) \ \rightleftharpoons\ MCl_4(g) + TeO_2(g) \tag{5.1.12}$$

In this simplification, however, the equilibria 5.1.13 to 5.1.18 in the system Te/O/Cl are not considered. *Reichelt* discussed in detail the complex reaction behavior of tellurium(IV)-chloride (Rei 1977).

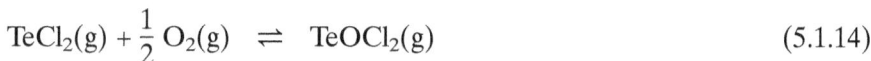

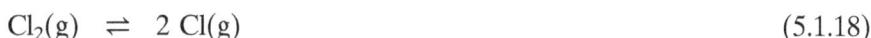

$$TeCl_4(g) \ \rightleftharpoons\ TeCl_2(g) + Cl_2(g) \tag{5.1.13}$$

$$TeCl_2(g) + \frac{1}{2} O_2(g) \ \rightleftharpoons\ TeOCl_2(g) \tag{5.1.14}$$

$$TeOCl_2(g) \ \rightleftharpoons\ TeO(g) + Cl_2(g) \tag{5.1.15}$$

$$TeO_2(g) \ \rightleftharpoons\ \frac{1}{2} Te_2(g) + O_2(g) \tag{5.1.16}$$

$$Te_2(g) \ \rightleftharpoons\ 2\, Te(g) \tag{5.1.17}$$

$$Cl_2(g) \ \rightleftharpoons\ 2\, Cl(g) \tag{5.1.18}$$

Tellurium(IV)-chloride is specially suited as a transport additive for oxide systems with a wide range of oxygen partial pressures between 10^{-25} and 1 bar. This is because it creates a complex redox system, which is described by the equations 5.1.12 to 5.1.18. At low oxygen partial pressures, the oxygen-free gas species

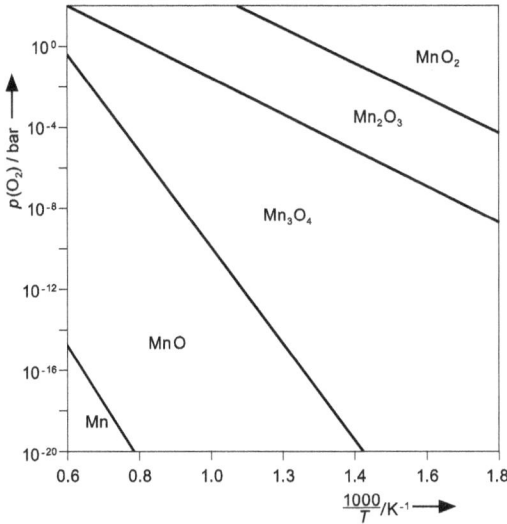

Figure 5.1.1 Phase barogram of the Mn/O system according to *Rossberg* (Ros 1987).

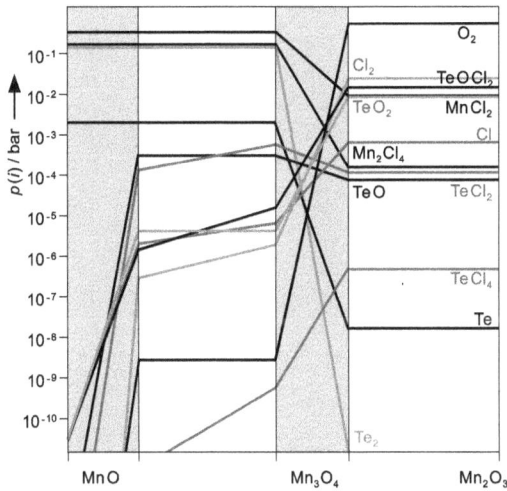

Figure 5.1.2 Gas-phase composition over the different manganese oxides in equilibrium with tellurium(IV)-chloride at 1200 K according to *Rossberg* (Ros 1988).

$TeCl_4$, $TeCl_2$, Te_2, Te, Cl_2, and Cl dominate, for example during the transport of Mn_3O_4 with tellurium(IV)-chloride as transport agent (for a phase barogram of the system Mn/O, see Figure 5.1.1). The gas phase consists of the gas species $MnCl_2$, Te_2, Mn_2Cl_4, Te, TeO, and $TeCl_2$ in the temperature range around 1000 °C; the partial pressure of the only oxygen-transferring species TeO is only $5 \cdot 10^{-4}$ bar (see Figures 5.1.2 and 5.1.3). The partial pressures of the other oxygen-containing gas species TeO_2 and $TeOCl_2$ are clearly below 10^{-5} bar (Ros 1988). At medium or high oxygen partial pressures, the amount of oxygen-containing gas species TeO_2, TeO, and $TeOCl_2$ becomes significantly higher, for example during the transport of the more oxygen-rich manganese(III)-oxide. Here, at

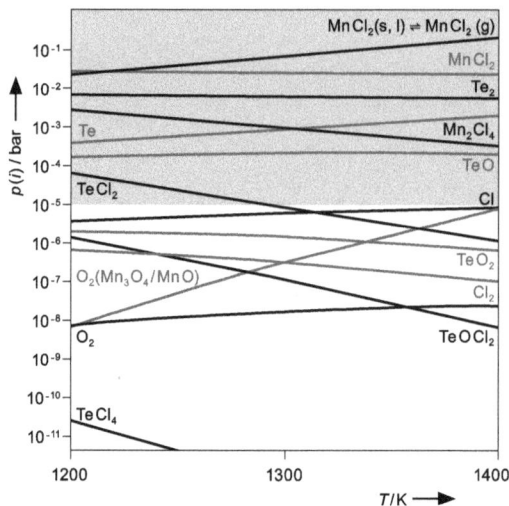

Figure 5.1.3 Gas-phase composition of the $Mn_3O_4/TeCl_4$ system as a function of the temperature according to *Rossberg* (Ros 1988).

1000 °C, the gas phase contains the species O_2, Cl_2, $TeOCl_2$, TeO_2, $MnCl_2$, Cl, $TeCl_2$, and TeO in the pressure range between 1 and 10^{-5} bar (Ros 1988).

Tellurium(IV)-chloride proves an ideal transport additive for those oxides that differ only slightly in their composition and stability and thus are thermodynamically stable only in narrow ranges of the oxygen partial pressure, such as *Magnéli*-phases. *Oppermann* and his co-workers succeeded in depositing crystalline phases V_nO_{2n-1} ($n = 2 \ldots 8$) by CVT with tellurium(IV)-chloride (see section 2.4) (Opp 1977, Opp 1977c, Rei 1977).

Tellurium(IV)-chloride is also suitable for the transport of oxide phases that show homogeneity ranges that are dependent on the oxygen partial pressure, such as "VO_2" (Opp 1975, Opp 1977a) and for oxides of transition metals that have similar stabilities, such as MnO and Mn_3O_4 (Ros 1988, Sch 1977). Similar redox systems form when tellurium(IV)-bromide (Opp 1978a) and tellurium(IV)-iodide (Opp 1980) are used as transport agents. These compounds, however, are less effective as transport agents and therefore of minor importance.

Combination of transport additives The combination of two transport additives is often used, for example the combination of sulfur + chlorine, sulfur + bromine, sulfur + iodine, selenium + chlorine, selenium + bromine, carbon + chlorine, and carbon + bromine. These also form complex redox systems and can be treated in a similar way to tellurium(IV)-chloride. The mechanism of the combination of sulfur + iodine is described exemplarily by the following equation:

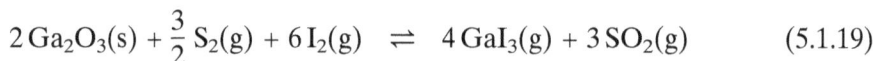

$$2\,Ga_2O_3(s) + \frac{3}{2}\,S_2(g) + 6\,I_2(g) \ \rightleftharpoons\ 4\,GaI_3(g) + 3\,SO_2(g) \tag{5.1.19}$$

Sulfur transfers the oxygen; iodine transfers the gallium. The transport agent combinations carbon + chlorine and carbon + bromine are introduced in the form of CCl_4 and CBr_4, respectively.

$$Y_2O_3(s) + 3\,CCl_4(g) \;\rightleftharpoons\; 2\,YCl_3(g) + 3\,CO(g) + 3\,Cl_2(g) \qquad (5.1.20)$$

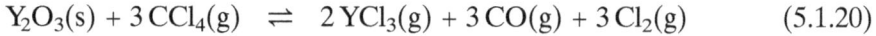

The formation of gaseous SO_2 and CO, respectively, causes a more balanced equilibrium position compared to the cases in which oxygen is formed. The reaction equilibrium is shifted to the side of the gaseous reaction products.

Other transport additives In some cases, a transport agent combination consisting of a halide and a halogen is applied, for example for the transport of SiO_2 with $CrCl_4 + Cl_2$. In this process, the surplus of halogen leads to a shift of the equilibrium position to the side of the reaction products.

$$SiO_2(s) + CrCl_4(g) + Cl_2(g) \;\rightleftharpoons\; SiCl_4(g) + CrO_2Cl_2(g) \qquad (5.1.21)$$

Further transport agents for the CVT of oxides are some metal halides and nonmetal halides, e. g., $NbCl_5$ or $TaCl_5$, and PCl_5 or PBr_5, respectively. They have a halogenating effect on the metal and are transport effective for the oxygen.

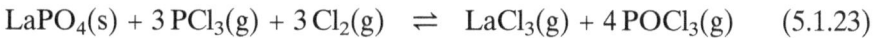

$$Nb_2O_5(s) + 3\,NbCl_5(g) \;\rightleftharpoons\; 5\,NbOCl_3(g) \qquad (5.1.22)$$

$$LaPO_4(s) + 3\,PCl_3(g) + 3\,Cl_2(g) \;\rightleftharpoons\; LaCl_3(g) + 4\,POCl_3(g) \qquad (5.1.23)$$

Aluminum(III)-chloride is not suited for the transport of oxides because aluminum oxide is formed. Observed transport effects can most often be traced back to the formation of hydrogen chloride.

Apart from the described transport agents, which have halogenating effects in different ways, also water (Gle 1963) and oxygen are used for the transport in some oxide systems. If water is used as the transport agent, a volatile hydroxide and a gaseous acid (e. g. H_2MoO_4), respectively, will be formed; for example during the transport of beryllium oxide with water:

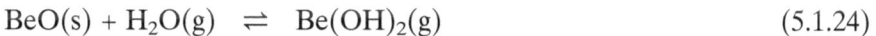

$$BeO(s) + H_2O(g) \;\rightleftharpoons\; Be(OH)_2(g) \qquad (5.1.24)$$

If oxygen is used as the transport agent, the transport will generally occur via a higher, more volatile, oxide, which is formed in the transport equilibrium.

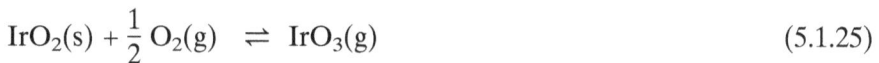

$$IrO_2(s) + \frac{1}{2}\,O_2(g) \;\rightleftharpoons\; IrO_3(g) \qquad (5.1.25)$$

Equilibrium position for different transport agents The following equations describe the transport of magnesium oxide. The given numerical values of the Gibbs energy provide an overview of the equilibrium positions for the different transport agents:

Transport with halogens

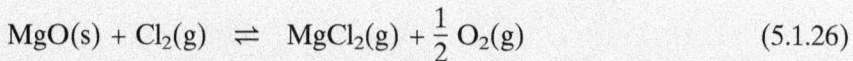

$$MgO(s) + Cl_2(g) \;\rightleftharpoons\; MgCl_2(g) + \frac{1}{2}\,O_2(g) \qquad (5.1.26)$$

$$\Delta_rG^0_{1300} = 42\ kJ \cdot mol^{-1}$$

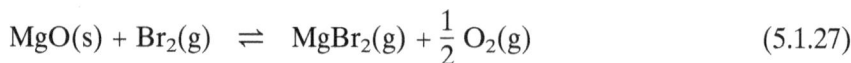

$$MgO(s) + Br_2(g) \rightleftharpoons MgBr_2(g) + \frac{1}{2}O_2(g) \qquad (5.1.27)$$
$$\Delta_r G^0_{1300} = 102 \text{ kJ} \cdot \text{mol}^{-1}$$

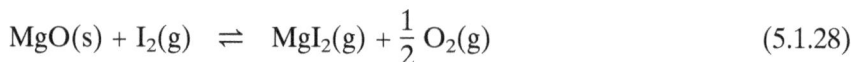

$$MgO(s) + I_2(g) \rightleftharpoons MgI_2(g) + \frac{1}{2}O_2(g) \qquad (5.1.28)$$
$$\Delta_r G^0_{1300} = 211 \text{ kJ} \cdot \text{mol}^{-1}$$

Transport with hydrogen halides

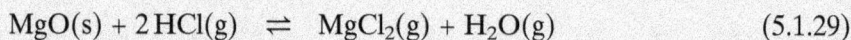

$$MgO(s) + 2\,HCl(g) \rightleftharpoons MgCl_2(g) + H_2O(g) \qquad (5.1.29)$$
$$\Delta_r G^0_{1300} = 71 \text{ kJ} \cdot \text{mol}^{-1}$$

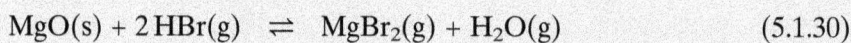

$$MgO(s) + 2\,HBr(g) \rightleftharpoons MgBr_2(g) + H_2O(g) \qquad (5.1.30)$$
$$\Delta_r G^0_{1300} = 51 \text{ kJ} \cdot \text{mol}^{-1}$$

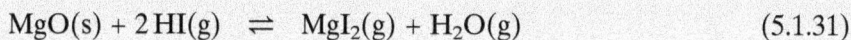

$$MgO(s) + 2\,HI(g) \rightleftharpoons MgI_2(g) + H_2O(g) \qquad (5.1.31)$$
$$\Delta_r G^0_{1300} = 68 \text{ kJ} \cdot \text{mol}^{-1}$$

Transport with halogen compounds

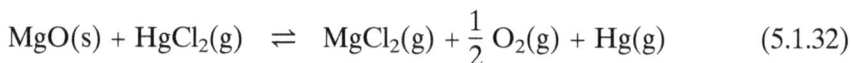

$$MgO(s) + HgCl_2(g) \rightleftharpoons MgCl_2(g) + \frac{1}{2}O_2(g) + Hg(g) \qquad (5.1.32)$$
$$\Delta_r G^0_{1300} = 119 \text{ kJ} \cdot \text{mol}^{-1}$$

$$MgO(s) + PCl_3(g) + Cl_2(g) \rightleftharpoons MgCl_2(g) + POCl_3(g) \qquad (5.1.33)$$
$$\Delta_r G^0_{1300} = -113 \text{ kJ} \cdot \text{mol}^{-1}$$

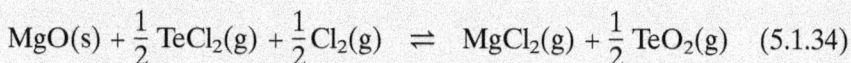

$$MgO(s) + \frac{1}{2}TeCl_2(g) + \frac{1}{2}Cl_2(g) \rightleftharpoons MgCl_2(g) + \frac{1}{2}TeO_2(g) \qquad (5.1.34)$$
$$\Delta_r G^0_{1300} = 79 \text{ kJ} \cdot \text{mol}^{-1}$$

Transport with two transport additives

$$MgO(s) + Cl_2(g) + \frac{1}{4}S_2(g) \rightleftharpoons MgCl_2(g) + \frac{1}{2}SO_2(g) \qquad (5.1.35)$$
$$\Delta_r G^0_{1300} = -92 \text{ kJ} \cdot \text{mol}^{-1}$$

$$MgO(s) + Cl_2(g) + \frac{1}{2}C(s) \rightleftharpoons MgCl_2(g) + \frac{1}{2}CO_2(g) \qquad (5.1.36)$$
$$\Delta_r G^0_{1300} = -146 \text{ kJ} \cdot \text{mol}^{-1}$$

$$MgO(s) + Cl_2(g) + CO(g) \rightleftharpoons MgCl_2(g) + CO_2(g) \qquad (5.1.37)$$
$$\Delta_r G^0_{1300} = -128 \text{ kJ} \cdot \text{mol}^{-1}$$

$$MgO(s) + Cl_2(g) + \frac{1}{4}Se_2(g) \rightleftharpoons MgCl_2(g) + \frac{1}{2}SeO_2(g) \qquad (5.1.38)$$
$$\Delta_r G^0_{1300} = -5 \text{ kJ} \cdot \text{mol}^{-1}$$

$$MgO(s) + \frac{1}{2}CrCl_4(g) + \frac{1}{2}Cl_2(g) \rightleftharpoons MgCl_2(g) + \frac{1}{2}CrO_2Cl_2(g)$$
$$\Delta_r G^0_{1300} = -3 \text{ kJ} \cdot \text{mol}^{-1} \qquad (5.1.39)$$

$$MgO(s) + VCl_4(g) + \frac{1}{2}Cl_2(g) \rightleftharpoons MgCl_2(g) + VOCl_3(g) \qquad (5.1.40)$$
$$\Delta_r G^0_{1300} = -103 \text{ kJ} \cdot \text{mol}^{-1}$$

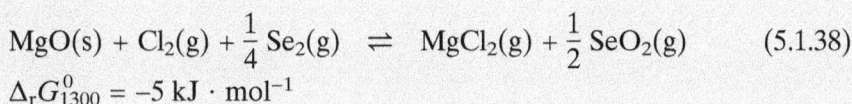

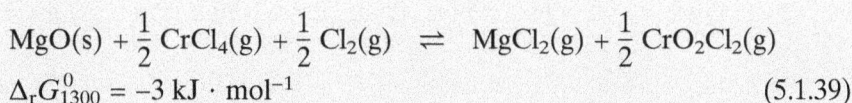

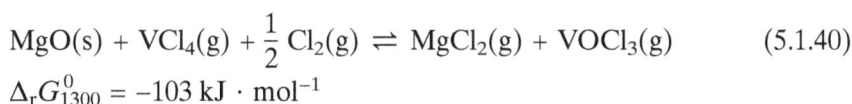

The equilibrium position is highly important for the choice of a suitable transport agent and transport additive, respectively. The positions of the various halogenating equilibria for the reaction of one mole of magnesium oxide differ immensely. Thus, a value of the Gibbs energy $\Delta_r G^0_{1300} = 211$ kJ $\cdot$ mol^{-1} results for the reaction with iodine, $\Delta_r G^0_{1300} = -113$ kJ $\cdot$ mol^{-1} for PCl$_3$/Cl$_2$. Considering the transport equilibria of magnesium oxide with the halogens, only the transport with chlorine appears promising. It should also be possible to transport magnesium oxide with all hydrogen halides. Further transport agents, which allow us to expect a transport of magnesium oxide, are tellurium(IV)-chloride and the combinations selenium/chlorine as well as chromium(IV)-chloride/chlorine. In many cases, conclusions from analogy with the already described transport reactions help choosing the transport agent and the transport conditions. Table 5.1 provides a detailed composition of known transport reactions.

5.2 Solids

In this section, the CVT of oxides is discussed with the help of selected examples. The periodic system serves as the principle classification for the extensive material. The allocation of certain compounds to a group of the periodic system underlies a certain degree of arbitrariness. Hence, the transport of MgGeO$_3$, for example, can be dealt with the same justification when oxides of group 2 or group 14 are treated. The decision made here orients on the usual notation of the first mentioned atom.

5.2.1 Group 1

The CVT of oxides of group 1 is a special case. Due to the unfavorable equilibria, which are far on the side of the halides, a transport of binary alkali metal oxides with the abovementioned transport agents is not possible. However, the volatility of lithium oxide (Li$_2$O) in the presence of water vapor is described in the literature. Accordingly, lithium oxide can be transported with water in an endothermic transport from 1000 °C to the cooler zone T_1 according to equation 5.2.1.1 (Ark 1955, Ber 1963).

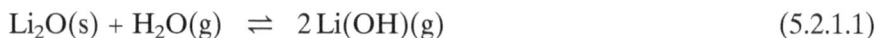

$$Li_2O(s) + H_2O(g) \rightleftharpoons 2\,Li(OH)(g) \qquad (5.2.1.1)$$

Besides the monomeric gaseous lithium hydroxide, LiOH, the existence of a dimer, $Li_2(OH)_2$, has been proven as well (Ber 1960).

There are only very few examples of the transport of ternary or multinary oxides that contain alkali metal atoms. One of the few exceptions is the CVT of lithium niobate, $LiNbO_3$, with sulfur, which is possible in a temperature gradient from 1000 to 900 °C with added niobium(V)-oxide in the presence of a high sulfur pressure (Sch 1988). Other exceptions are the tungsten and niobium bronzes of the composition M_xWO_3 and $M_xNb_yW_{1-y}O_3$, respectively (M = Li, K, Rb, Cs, e. g., $K_{0.25}WO_3$) as well as the molybdenum bronzes Li_xMoO_3. Here endothermic transport succeeds with relatively high transport rates with the mercury(II)-halide $HgCl_2$ or $HgBr_2$ in the temperature gradient from 800 to 750 °C (Hus 1991, Hus 1994, Hus 1997). The lithium-containing tungsten bronzes and molybdenum bronzes, respectively, are endothermically transported with tellurium(IV)-chloride (Schm 2008). Here, the exact process is still unexplained.

5.2.2 Group 2

Among the oxides of the elements of group 2 the CVT of magnesium oxide with different transport agents has been most intensively investigated. Furthermore, there are a number of examples of the CVT of ternary oxides of magnesium, such as $MgWO_4$. Beryllium oxide and calcium oxide can become volatile by the reactions with water vapor. Compounds of the heavy alkaline earth metals are hardly (Ca, Sr) and not at all (Ba), respectively, crystallized by CVT due to the low volatility of their halides.

Magnesium oxide can be transported with chlorine (1200 → 1000 °C) (Bay 1985) as well as with hydrogen chloride (1000 → 800 °C) (Klei 1972). The transport equilibria, which become effective in the process, correspond to the above given general transport equation for the transport with halogens and hydrogen halides, respectively. Additionally, magnesium oxide can be transported with the help of the transport additive hydrogen, carbon, or carbon monoxide under the formation of magnesium vapor:

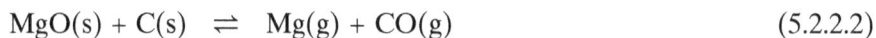

$$MgO(s) + H_2(g) \rightleftharpoons Mg(g) + H_2O(g) \qquad (5.2.2.1)$$

$$MgO(s) + C(s) \rightleftharpoons Mg(g) + CO(g) \qquad (5.2.2.2)$$

$$MgO(s) + CO(g) \rightleftharpoons Mg(g) + CO_2(g) \qquad (5.2.2.3)$$

Based on the transport behavior of MgO, mixed-crystals can be obtained by CVT with hydrogen chloride in an endothermic reaction in the systems MnO/MgO, CoO/MgO, and NiO/MgO (Skv 2000).

A volatility due to the formation of gaseous hydroxides in the temperature range around 1500 °C is described for beryllium oxide (Bud 1966, Stu 1964, You 1960) as well as for calcium oxide (Mat 1981a). By the reaction, BeO is deposited in the form of fine, fibrous crystals at the lower temperature T_1. The composition of the gaseous hydroxide (possibly $Be(OH)_2$) that was formed in the process, is not known for sure. The detection of Be_3O_3- and Be_4O_4-molecules in the gas phase is seen as an indication that $Be_nO_n \cdot H_2O$-molecules were formed as transport-effective species by the reaction with the water vapor (Chu 1959). Also, the transport of beryllium oxide with hydrogen chloride from 1100 to 800 °C is possible according to the following equilibrium (Spi 1930):

$$BeO(s) + 2\,HCl(g) \;\rightleftharpoons\; BeCl_2(g) + H_2O(g) \tag{5.2.2.4}$$

Typical examples of the transport of ternary and quaternary magnesium/transition metal oxides, respectively, are $MgFe_2O_4$, $MgTiO_3$, $MgWO_4$, $MgNb_2O_6$, $MgTa_2O_6$, and $Mg_{0.5}Mn_{0.5}Fe_2O_4$. They are transportable with chlorine and hydrogen chloride, respectively, each from 1000 °C to T_1; the transport takes place via magnesium chloride. Transport with different bromine compounds as transport agent is described for MgV_2O_4 (Pic 1973b). $Mg_2Mo_3O_8$ can be transported with chlorine as well as with bromine (Ste 2003a). *Steiner* assumes that hydrogen chloride and hydrogen bromide, respectively, functions as a transport agent that is formed under the influence of traces of water. According to equilibrium calculations the transport process can be described by the equations 5.2.2.5 to 5.2.2.7.

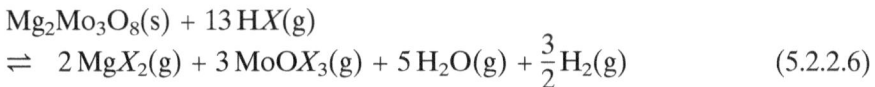

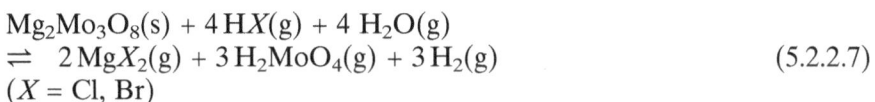

$$\begin{aligned} Mg_2Mo_3O_8(s) + 10\,HX(g) \\ \rightleftharpoons\; 2\,MgX_2(g) + 3\,MoO_2X_2(g) + 2\,H_2O(g) + 3\,H_2(g) \end{aligned} \tag{5.2.2.5}$$

$$\begin{aligned} Mg_2Mo_3O_8(s) + 13\,HX(g) \\ \rightleftharpoons\; 2\,MgX_2(g) + 3\,MoOX_3(g) + 5\,H_2O(g) + \frac{3}{2}H_2(g) \end{aligned} \tag{5.2.2.6}$$

$$\begin{aligned} Mg_2Mo_3O_8(s) + 4\,HX(g) + 4\,H_2O(g) \\ \rightleftharpoons\; 2\,MgX_2(g) + 3\,H_2MoO_4(g) + 3\,H_2(g) \\ (X = Cl, Br) \end{aligned} \tag{5.2.2.7}$$

When hydrogen chloride is used as the transport agent, the description of the transport behavior takes place by all three equations; when hydrogen bromide is used, by 5.2.2.5 and 5.2.2.7.

The ternary calcium compounds $CaMoO_4$, $CaMo_5O_8$, $CaNb_2O_6$, and $CaWO_4$ are endothermically transportable via chlorination. While the CVT of strontium oxide is not described, there are reports on the transport of three strontium-containing ternary compounds $SrMoO_4$, $SrMo_5O_8$ (Ste 2006), and $SrWO_4$ (Ste 2005a). These are transportable with chlorine (1150 → 1050 °C). The transport of $CaMoO_4$ and $SrMoO_4$ can be described well with the following equation:

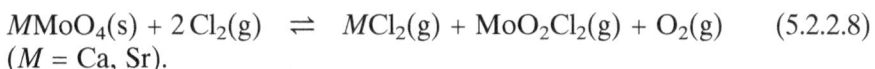

$$MMoO_4(s) + 2\,Cl_2(g) \;\rightleftharpoons\; MCl_2(g) + MoO_2Cl_2(g) + O_2(g) \tag{5.2.2.8}$$
$$(M = Ca, Sr).$$

The transport of alkaline earth metal tungstates can be described in an analogous way (Ste 2005a).

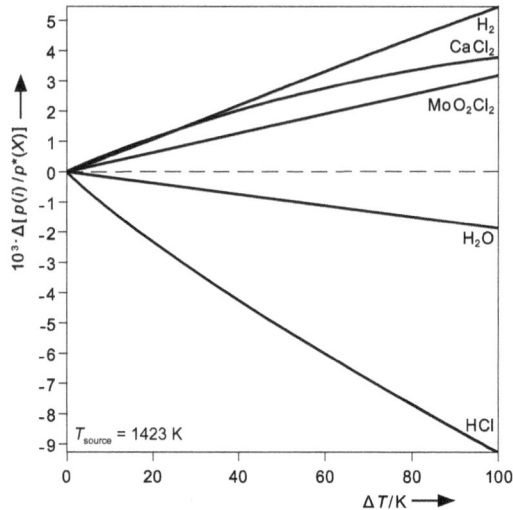

Figure 5.2.2.1 Transport efficiency of the essential gas species of the chemical vapor transport of $CaMoO_4$ when chlorine is added in presence of water according to *Steiner* and *Reichelt* (Ste 2006) ($n(Cl) = 2.5 \cdot 10^{-5}$ mol, $n(H_2O) = 2.5 \cdot 10^{-5}$ mol).

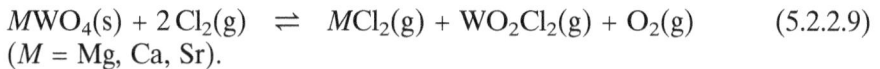

$$MWO_4(s) + 2\,Cl_2(g) \;\rightleftharpoons\; MCl_2(g) + WO_2Cl_2(g) + O_2(g) \qquad (5.2.2.9)$$
$$(M = Mg,\ Ca,\ Sr).$$

It is important to note that the transport rate of tungstates decreases systematically from magnesium (0.7 mg $\cdot$ h^{-1}) via calcium to strontium (0.1 mg $\cdot$ h^{-1}).

The transport behavior of the binary and ternary oxide compounds of the alkaline earth metals is limited by the high thermodynamic stability of these compounds. The transport reactions are marked by extreme equilibrium positions. The transport efficient alkaline earth metal halides show low partial pressures so that a deposition of the oxide compounds is only possible at temperatures above 1000 °C. The transport rates are low in every case. It has proven necessary to add the transport agent only in very small concentrations (0.05 mg $\cdot$ cm^{-3}) to avoid a condensation of the alkaline earth metal halides.

5.2.3 Group 3, Lanthanoids and Actinoids

Binary oxides In the following section, the transport behavior of the binary, ternary, and multinary oxides of the rare earth metals (*RE)* is described. This includes the elements scandium, yttrium, and lanthanum as well as those 14 elements that follow lanthanum up to lutetium. The mentioned elements have similar chemical characteristics. Seen in the total number of publications on the CVT of oxides, the transport of oxide compounds of rare earth metals is rather a rarity.

Little is known about the transport of binary oxides of rare earth metals. The transport reactions of scandium(III)-oxide, yttrium(III)-oxide, and cerium(IV)-oxide are described. The transport of the other rare earth metal oxides does not succeed with the usual working techniques. The transport of gadolinium(III)-

oxide is a special case because the crystallization was examined and described as transport at very high temperatures (1800 °C → T_1) (Kal 1971).

The special problems during the CVT of oxide compounds of rare earth metals are based on the following characteristics:

- The oxides of the rare earth metals show an extraordinary high thermodynamic stability.
- The halides of the rare earth metals have low vapor pressures.
- The solid oxide halides of the rare earth metals $REOX$ show a special stability.

The very high stability of the binary oxides RE_2O_3 becomes clear on comparison of some standard enthalpies of formation with that of aluminum(III)-oxide:

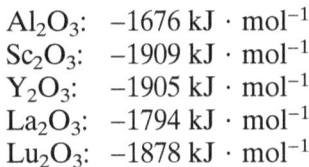

Al_2O_3: -1676 kJ $\cdot$ mol^{-1}
Sc_2O_3: -1909 kJ $\cdot$ mol^{-1}
Y_2O_3: -1905 kJ $\cdot$ mol^{-1}
La_2O_3: -1794 kJ $\cdot$ mol^{-1}
Lu_2O_3: -1878 kJ $\cdot$ mol^{-1}

Due to the equilibrium position, only the chlorinating equilibria and thus the rare earth metal(III)-chlorides play a role for the CVT of rare earth metal oxido compounds. The saturation pressures of the trichlorides are at the border of transport efficiency. If there is a higher partial pressure for a transport equilibrium, there will be a condensation of the trichloride and the transport will not take place.

The rare earth metal(III)-halides get to the gas phase in the form of monomeric molecules due to their high thermal stability. Europium and samarium trihalides are exceptions; they show a low thermal stability. The equilibrium partial pressures of rare earth metal(III)-chlorides principally increases from lanthanum to lutetium. Lanthanum(III)-chloride reaches an equilibrium partial pressure of $6 \cdot 10^{-4}$ bar at 1000 °C, gadolinium(III)-chloride of $1.5 \cdot 10^{-3}$ bar, and ytterbium(III)-chloride of $2 \cdot 10^{-2}$ bar (Opp 2005a).

If, for example, lanthanum(III)-oxide reacts with chlorine, the solid oxide chloride LaOCl will form:

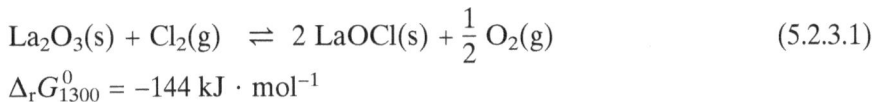

$$La_2O_3(s) + Cl_2(g) \rightleftharpoons 2\,LaOCl(s) + \frac{1}{2}\,O_2(g) \qquad (5.2.3.1)$$
$$\Delta_r G^0_{1300} = -144 \text{ kJ} \cdot \text{mol}^{-1}$$

The decomposition of LaOCl occurs according to the following equilibrium:

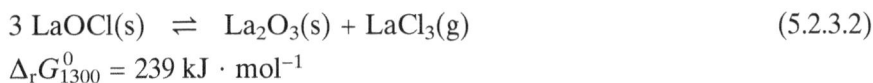

$$3\,LaOCl(s) \rightleftharpoons La_2O_3(s) + LaCl_3(g) \qquad (5.2.3.2)$$
$$\Delta_r G^0_{1300} = 239 \text{ kJ} \cdot \text{mol}^{-1}$$

The partial pressure of the lanthanum(III)-chloride, which results from it, is $2 \cdot 10^{-10}$ bar. Thus, it is far below the transport effective range of $p \geq 10^{-5}$ bar.

The CVT of scandium(III)-oxide, yttrium(III)-oxide, cerium(IV)-oxide, and gadolinium(III)-oxide shall be explained in more detail. Scandium(III)-oxide can be transported with chlorine as the transport agent from 1100 to 1000 °C (Ros 1990a). The transport of scandium(III)-oxide with chlorine is possible in contrast to the chemically related lanthanum(III)-oxide because scandium oxide chloride allows a transport effective scandium(III)-chloride partial pressure due to its lower stability.

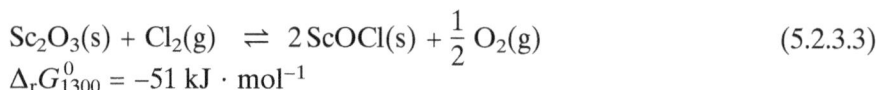

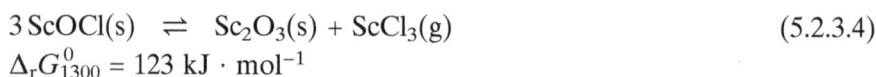

$$Sc_2O_3(s) + Cl_2(g) \rightleftharpoons 2\,ScOCl(s) + \frac{1}{2}\,O_2(g)$$ (5.2.3.3)
$$\Delta_r G^0_{1300} = -51 \text{ kJ} \cdot \text{mol}^{-1}$$

$$3\,ScOCl(s) \rightleftharpoons Sc_2O_3(s) + ScCl_3(g)$$ (5.2.3.4)
$$\Delta_r G^0_{1300} = 123 \text{ kJ} \cdot \text{mol}^{-1}$$

The partial pressure of scandium(III)-chloride over ScOCl is 10^{-5} bar at 1300 K. Therefore, the transport can take place according to the following equation:

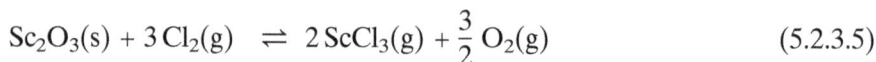

$$Sc_2O_3(s) + 3\,Cl_2(g) \rightleftharpoons 2\,ScCl_3(g) + \frac{3}{2}\,O_2(g)$$ (5.2.3.5)

Cerium(IV)-oxide is also endothermically transportable with chlorine from 1100 to 1000 °C (Scha 1986). It reacts like scandium(III)-oxide (but in contrast to lanthanum(III)-oxide) to form gaseous trichloride and not by forming the solid oxide chloride. The following transport equation is assumed:

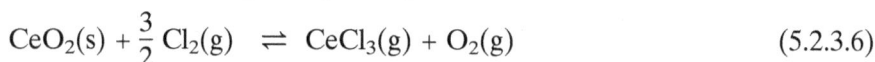

$$CeO_2(s) + \frac{3}{2}\,Cl_2(g) \rightleftharpoons CeCl_3(g) + O_2(g)$$ (5.2.3.6)

Adding carbon leads to the formation of CO and CO_2, respectively. By this process, more favorable equilibrium positions for chlorinating equilibria of the oxides of rare earth metals RE_2O_3 and thus higher transport rates are achieved.

Matsumoto and co-workers (Mat 1983) succeeded in transporting yttrium(III)-oxide from 1160 to 1100 °C by using bromine + carbon monoxide as transport agent. Analogue transport experiments that used bromine exclusively (5.2.3.7) did not cause a transport. The equilibrium 5.2.3.8 describes the observed transport behavior.

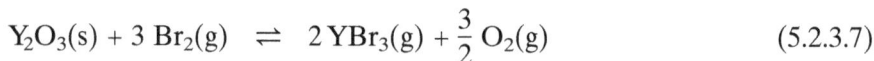

$$Y_2O_3(s) + 3\,Br_2(g) \rightleftharpoons 2\,YBr_3(g) + \frac{3}{2}\,O_2(g)$$ (5.2.3.7)

$$Y_2O_3(s) + 3\,Br_2(g) + 3\,CO(g) \rightleftharpoons 2\,YBr_3(g) + 3\,CO_2(g)$$ (5.2.3.8)

The Gibbs energy $\Delta_r G^0_{1300}$ for 5.2.3.8 is $= -166 \text{ kJ} \cdot \text{mol}^{-1}$. This value is still unfavorable for transport reactions. Due to the fact that yttrium(III)-oxide is transportable under the given conditions, the other rare earth metal(III)-oxides should be deposited this way, too; e. g., gadolinium(III)-oxide ($\Delta_r G^0_{1300} = -105 \text{ kJ} \cdot \text{mol}^{-1}$) or erbium(III)-oxide ($\Delta_r G^0_{1300} = -67 \text{ kJ} \cdot \text{mol}^{-1}$).

Yttrium(III)-oxide crystals that have been doped with europium can be transported from 1190 to 1090 °C by endothermic transport reactions with the transport agents bromine, carbon monoxide, hydrogen bromide, and the transport agent combination carbon monoxide + bromine (Mat 1982). Gadolinium(III)-

oxide crystallizes during the transport from 1800 °C to T_1 when hydrogen chloride is used as the transport agent (Kal 1971).

Orlovskii (Orl 1985) made extensive thermodynamic considerations on the transport behavior of binary oxides of the rare earth metals. Model calculations of the systems RE_2O_3/Cl_2, $/Br_2$, $/HBr$, $/BBr_3$, $/CO + Br_2$, and $/S + Br_2$ for the temperature range between 900 and 1150 °C prove the limiting influence of the formation of solid oxide halides on the CVT of rare earth metals. It can be shown that higher transport rates are to be expected for the transport of Sc_2O_3 and Y_2O_3, as well as CeO_2, and that the combinations carbon monoxide + bromine and sulfur + bromine, respectively, should be the most effective transport additives. Transport also seems possible when hydrogen bromide is added; however, with a transport rate that is decreased by a factor of 10 compared to the use of sulfur + bromine. The formation of solid boron(III)-oxide besides the respective rare earth metal bromides suggests boron(III)-bromide to be an unsuitable transport additive.

Multinary oxido compounds *Phosphates, arsenates, antimonates:* Comparing the transport behavior of the oxido compounds of the rare earth metals, one sees that compounds with the elements lanthanum to europium are generally transportable with higher transport rates than the corresponding compounds of the heavier lathanoids from gadolinium to lutetium. This particularly applies for compounds with complex anions, such as $REPO_4$, $REAsO_4$, $RESbO_4$, and $REVO_4$. In the following, thermodynamic basic principles of the transport of multinary oxido compounds are explained. The transport equation results from detailed thermodynamic considerations of the transport of a rare earth metal arsenate with chlorine, which is used as an example.

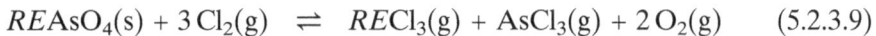

$$REAsO_4(s) + 3\,Cl_2(g) \;\rightleftharpoons\; RECl_3(g) + AsCl_3(g) + 2\,O_2(g) \qquad (5.2.3.9)$$

This equation can be formally split into three partial equations:

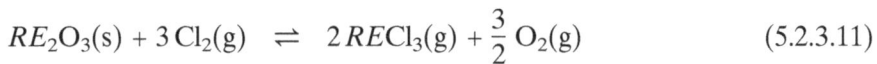

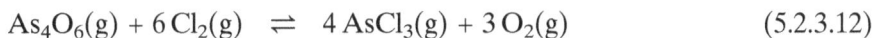

$$2\,REAsO_4(s) \;\rightleftharpoons\; RE_2O_3(s) + \frac{1}{2}\,As_4O_6(g) + O_2(g) \qquad (5.2.3.10)$$

$$RE_2O_3(s) + 3\,Cl_2(g) \;\rightleftharpoons\; 2\,RECl_3(g) + \frac{3}{2}\,O_2(g) \qquad (5.2.3.11)$$

$$As_4O_6(g) + 6\,Cl_2(g) \;\rightleftharpoons\; 4\,AsCl_3(g) + 3\,O_2(g) \qquad (5.2.3.12)$$

The following can be laid down for the series of rare earth metals concerning the above formulated partial equations 5.2.310 to 5.2.3.12:

- The amounts of the Gibbs energies of the first partial reaction are close to each other.
- The Gibbs energies of the second partial reaction can be clearly distinguished.
- The Gibbs energies of the third partial reaction are identical.

The Gibbs energies of the transport process can be traced back to the difference between the Gibbs energy of the gaseous rare earth metal(III)-chloride and of the rare earth metal(III)-oxide. These differences are exemplarily shown in Table 5.2.3.1 for 1300 K.

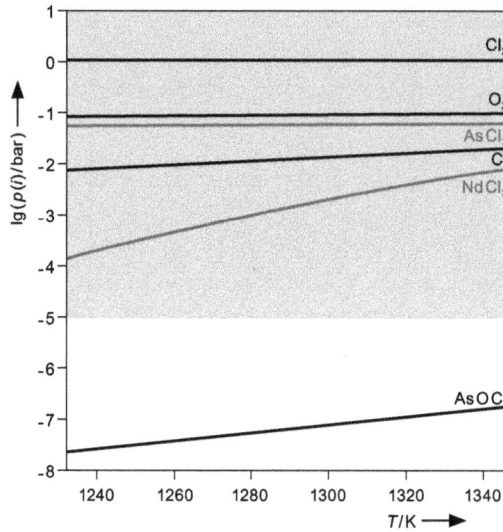

Figure 5.2.3.1 Gas-phase composition of the chemical vapor transport of NdAsO$_4$ with chlorine.

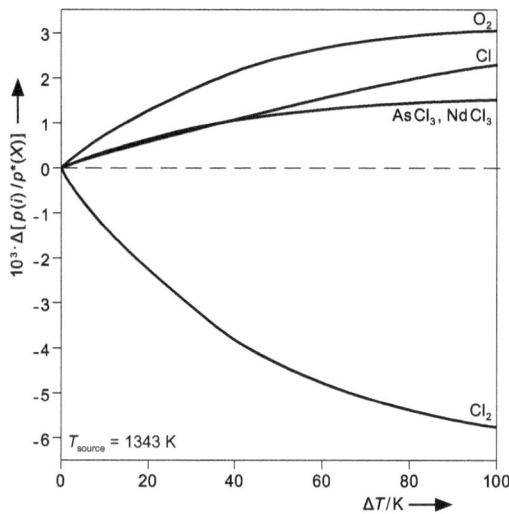

Figure 5.2.3.2 Transport efficiency of the essential gas species during the chemical vapor transport of NdAsO$_4$ with chlorine.

This compilation shows three groups with different values of the Gibbs energy of the transport reactions: group 1 (values around −450 kJ · mol^{-1}) with relatively high transport rates; group 2 (values in the range from −350 to −300 kJ · mol^{-1}) with medium to low transport rates; and group 3 (values between −300 and −190 kJ · mol^{-1}) with low transport rates for transport reactions with chlorine or chlorine-containing transport agents from 1100 to 1000 °C. There is no transport with chlorine or chlorine-containing transport agents for some compound classes, such as the rare earth metal vanadates(V) and some rare earth metal antimona-

Table 5.2.3.1 Thermodynamic data of the transport of rare earth metal(III)-oxido compounds.

Numerical values: $2 \cdot \Delta G^0_{1300} (RECl_3(g)) - \Delta G^0_{1300} (RE_2O_3(s))/kJ \cdot mol^{-1}$

Group I		Group II		Group III	
La	−472	Gd	−376	Er	−264
Ce	−447	Tb	−322	Tm	−267
Pr	−454	Dy	−304	Yb	−285
Nd	−432	Ho	−294	Lu	−190
Sm	–				
Eu	−457	Y	−318	Sc	−224

tes(V), with rare earth metals of group 3. Transport effects are observed when the Gibbs energy is in the range between −100 and 100 kJ · mol^{-1}. If this range is compared to the rare earth metal dependent amount of the Gibbs energy of the transport reaction a variance of 280 kJ · mol^{-1} results. This explains the different transport behavior of these oxido compounds of the rare earth metals. If one wants to transport an entire series of rare earth metal oxide compounds (La...Lu), a change of the transport agent and/or the transport conditions will be necessary (position and range of the temperature gradient). For some compound classes, such as rare earth metal phosphates ($REPO_4$, RE = Dy ... Lu), a change of the equilibrium position, and thus a transport, can be achieved by swapping to bromine and bromine-containing transport agents, respectively.

In the following, the transport of phosphates, arsenates, and antimonates shall be dealt with individually. The following equation applies for all given examples of transport with chlorine:

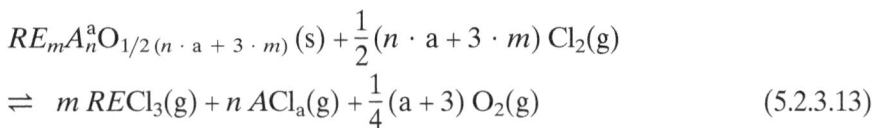

$$RE_m A^a_n O_{1/2 (n \cdot a + 3 \cdot m)} (s) + \frac{1}{2}(n \cdot a + 3 \cdot m) Cl_2(g)$$

$$\rightleftharpoons m\, RECl_3(g) + n\, ACl_a(g) + \frac{1}{4}(a+3) O_2(g) \qquad (5.2.3.13)$$

(A = Central atom of the complex anion, e. g., Si, Ge, P, As, Sb, Ti, V, Nb, or Ta)

The rare earth metal phosphate can be endothermically transported via different chlorinating and brominating equilibria, respectively, along different temperature gradients. The following transport additives are used: phosphorus(V)-chloride, chlorine + carbon monoxide, phosphorus(V)-bromide, bromine + carbon, bromine + carbon monoxide, and ammonium bromide. The rare earth metal phosphates of the composition $REPO_4$ (RE = Dy, Tm, Yb, Lu) can only be transported by brominating equilibria. The transport of considerable amounts of silicon(IV)-oxide, which deposits as cristobalite adjacent to the rare earth metal phosphates, is a problem; especially if phosphorus(V)-bromide is used as the transport additive (Orl 1971).

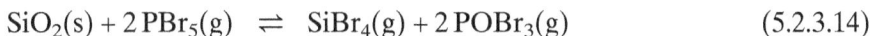

$$SiO_2(s) + 2\, PBr_5(g) \rightleftharpoons SiBr_4(g) + 2\, POBr_3(g) \qquad (5.2.3.14)$$

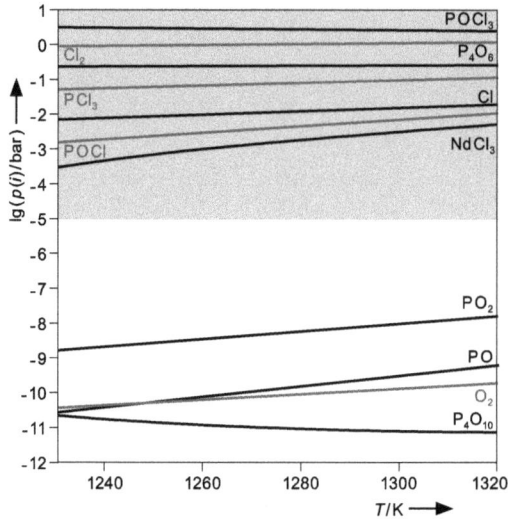

Figure 5.2.3.3 Gas-phase composition of the chemical vapor transport of $NdPO_4$ with PCl_5.

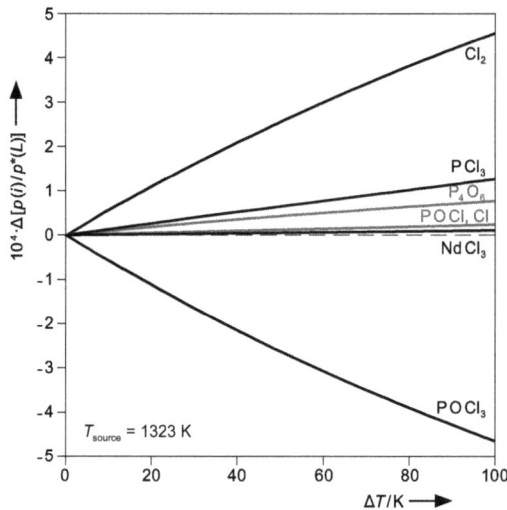

Figure 5.2.3.4 Transport efficiency of the essential gas species during the chemical vapor transport of $NdPO_4$ with PCl_5.

The CVT of lanthanum phosphate can take place with phosphorus(V)-chloride as the transport additive. In the literature, the transport is described by the following equation (Orl 1971):

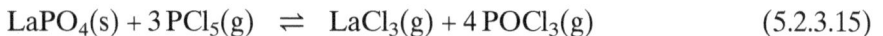

$$LaPO_4(s) + 3\,PCl_5(g) \; \rightleftharpoons \; LaCl_3(g) + 4\,POCl_3(g) \tag{5.2.3.15}$$

Model calculations for the system $NdPO_4/PCl_5$ show that $POCl_3$ functions as the transport agent.

Further examples of the transport of rare earth metal phosphates are compiled in Table 5.1. *Schäfer* made a detailed analysis of the CVT of lanthanum phosphate, $LaPO_4$, with bromine-containing transport agents (Sch 1972). In doing so, the transport behavior of lanthanum phosphate with the transport additives bromine, hydrogen bromide and phosphorus(V)-bromide ($PBr_3 + Br_2$, respectively) as well as bromine + carbon (CO, respectively) was thermodynamically analyzed. Experiments and thermodynamic calculations showed that lanthanum phosphate can be transported with all the mentioned transport agents except bromine. The transport took place endothermically from 1130 to 930 °C in every case; the highest transport rate of 13.4 mg · h^{-1} was achieved with phosphorus(V)-bromide and $PBr_3 + Br_2$, respectively, as transport agent. The gas species, on which the calculation was based, are $LaBr_3$, HBr, Br, Br_2, H_2O, O_2, CO, CO_2, PBr_5, PBr_3, $POBr_3$, and P_4O_{10}. The transport of lanthanum phosphate can be described by the two following equilibria:

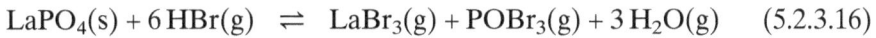

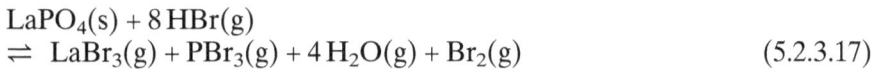

$$LaPO_4(s) + 6\,HBr(g) \;\rightleftharpoons\; LaBr_3(g) + POBr_3(g) + 3\,H_2O(g) \qquad (5.2.3.16)$$

$$LaPO_4(s) + 8\,HBr(g)$$
$$\rightleftharpoons LaBr_3(g) + PBr_3(g) + 4\,H_2O(g) + Br_2(g) \qquad\qquad (5.2.3.17)$$

The second equilibrium provides the decisive contribution to the transport process. The following equilibrium can be given for the transport with phosphorus(V)-bromide:

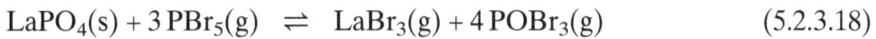

$$LaPO_4(s) + 3\,PBr_5(g) \;\rightleftharpoons\; LaBr_3(g) + 4\,POBr_3(g) \qquad (5.2.3.18)$$

If the transport agent combination bromine + carbon and bromine + carbon monoxide, respectively, is used, the following transport effective equilibrium can be assumed:

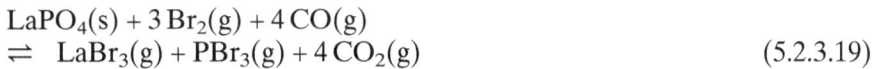

$$LaPO_4(s) + 3\,Br_2(g) + 4\,CO(g)$$
$$\rightleftharpoons LaBr_3(g) + PBr_3(g) + 4\,CO_2(g) \qquad\qquad (5.2.3.19)$$

Based on these assumptions, *Orlovskii* was able to conduct successfully the transport of a rare earth metal phosphate with one of the heavy rare earth metals (lutetium phosphate, $LuPO_4$), which had not been described until then (Orl 1974, Orl 1978). The transport took place from 1200 to 1000 °C with phosphorus(V)-bromide (1.9 mg · cm^{-3}) and bromine + carbon monoxide. The transport experiments were conducted in ampoules with radii between approx. 8 and 35 mm, whereby the transport rate drastically increased with radii above 20 mm, which could be traced back to the massively increasing convective amount. The endothermic transport of rare earth metal phosphate(V)-chlorides of the composition $REPO_4$ (RE = Ce, Pr, Nd, Gd, Ho, Er) was described with phosphorus(V)-chloride as the transport agent as well as with phosphorus(V)-bromide for $REPO_4$ (RE = Dy, Ho, Er, Tm, Yb, Lu) (Mül 2004). What is more, the transport of rare earth metal phosphate mixed-crystals of the composition $Nd_{1-x}Pr_xPO_4$, $Nd_{1-x}Sm_xPO_4$, and $Sm_{1-x}Gd_xPO_4$ succeeded as well. The transport reactions took place endothermically with phosphorus(V)-chloride as the transport agent.

Beside the transport behavior of the rare earth metal phosphates, the one of the less stable, higher homologues, the rare earth metal arsenates(V) ($REAsO_4$) and the rare earth metal antimonates(V) ($RESbO_4$) were investigated, too (Schm 2005, Ger 2007). The rare earth metal arsenates $REAsO_4$ (RE = Sc, Y, La ... Nd, Sm ... Lu) can be crystallized by endothermic CVT with tellurium(IV)-chloride. Furthermore, the example of neodymium arsenate shows that the transport additives tellurium(IV)-bromide, phosphorus(V)-chloride, phosphorus(V)-bromide, mercu-

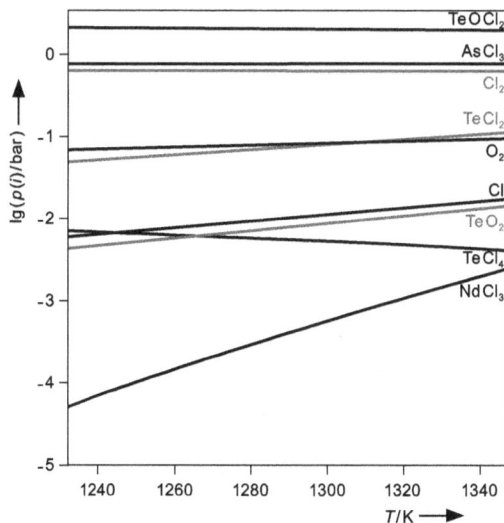

Figure 5.2.3.5 Gas-phase composition of the chemical vapor transport of $NdAsO_4$ with $TeCl_4$.

ry(II)-chloride, sulfur + chlorine, arsenic + chlorine, and arsenic + bromine are also well suited. Experiments with different iodine-containing transport agents did not work. Additionally, a series of differently composed rare earth metal arsenate mixed-crystals of the general formula $RE_{1-x}RE'_xAsO_4$ was deposited in crystalline form with tellurium(IV)-chloride in the given temperature range (see Table 5.1). The following dominating transport equation was derived from thermodynamic calculations:

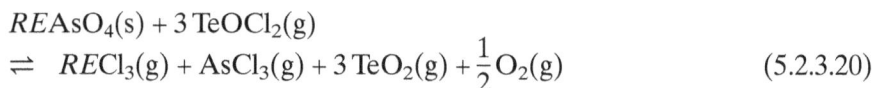

$$REAsO_4(s) + 3\,TeOCl_2(g)$$
$$\rightleftharpoons\ RECl_3(g) + AsCl_3(g) + 3\,TeO_2(g) + \frac{1}{2}O_2(g) \qquad (5.2.3.20)$$

The rare earth metal-transferring species is solely the rare earth metal(III)-chloride with partial pressures in the range of 10^{-3} bar. Arsenic(III)-chloride becomes transport effective as arsenic-transferring species. The gas species As_4O_6 and $AsOCl$, which were also considered, are at the border of transport efficiency with partial pressures of approx. 10^{-5} bar and 10^{-6} bar, respectively (see Figure 5.2.3.5). Comparing the different rare earth metal arsenates(V) shows that the transport rate of the arsenates of the light rare earth metals (La … Nd, Sm, Eu) is higher than that of the arsenates of the heavy rare earth metals (Gd … Lu).

Apart from the rare earth metal arsenates (V), the more unstable rare earth metal antimonates(V) (RE = La, Pr, Nd, Sm, Eu, Gd, Tb, Dy, Ho, Er, Lu) can also be deposited in crystalline form by CVT with tellurium(IV)-chloride. Transport experiments with praseodymium antimonate utilizing tellurium(IV)-bromide, tellurium(IV)-iodide, phosphorus(V)-chloride, phosphorus(V)-bromide, chlorine, and bromine showed that a crystallization of $PrSbO_4$ is only possible with chlorine-containing transport agents. The generally low transport rates of the rare earth metal antimonates(V) of 0.05 to 0.1 mg $\cdot$ h^{-1} are achieved with unusually high concentrations of the transport agent. Transport is no longer observed with transport agent

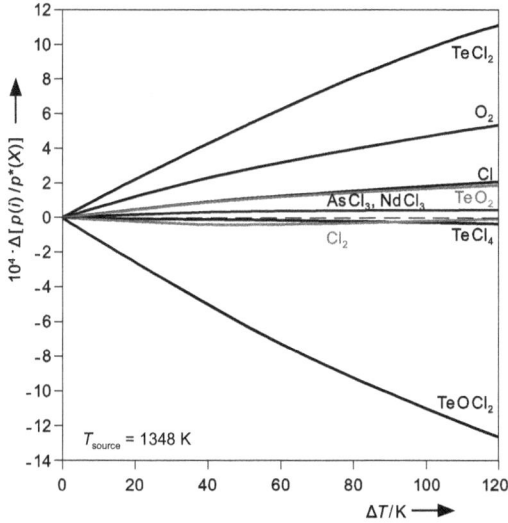

Figure 5.2.3.6 Transport efficiency of the essential gas species during the chemical vapor transport of $NdAsO_4$ with $TeCl_4$.

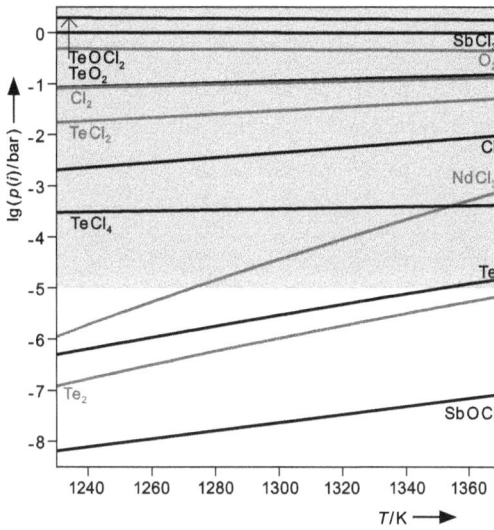

Figure 5.2.3.7 Gas-phase composition of the chemical vapor transport of $NdSbO_4$ with $TeCl_4$.

concentrations below $2.5 \, \text{mg} \cdot \text{cm}^{-3}$. Transport agent concentrations in the range of 5 to $10 \, \text{mg} \cdot \text{cm}^{-3}$ lead to a clear increase of the transport rate. This increase is due to the rising total pressure in the ampoule of up to 6.7 bar so that the convective part predominates. The following dominating transport equation was derived from the calculated transport efficiency (see Figure 5.2.3.8):

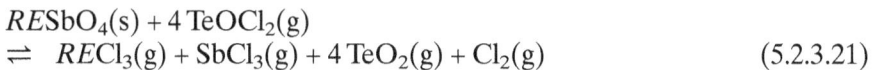

$$RESbO_4(s) + 4\,TeOCl_2(g)$$
$$\rightleftharpoons RECl_3(g) + SbCl_3(g) + 4\,TeO_2(g) + Cl_2(g) \qquad (5.2.3.21)$$

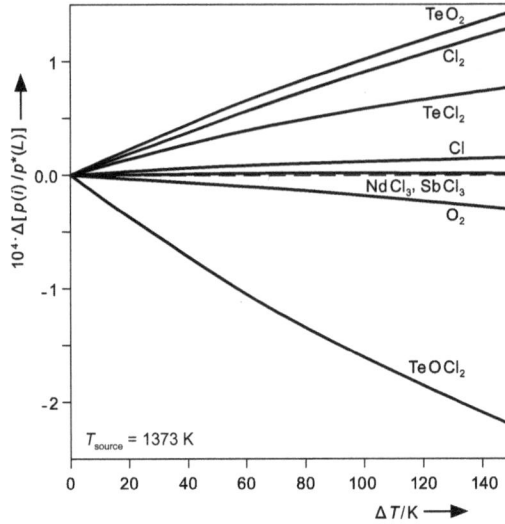

Figure 5.2.3.8 Transport efficiency of the essential gas species during the chemical vapor transport of $NdSbO_4$ with $TeCl_4$.

Nine independent equilibria (r_u) corresponding to $r_u = s + 1 - k$ are necessary for a complete description of the complex five components (k) containing transport system (RE, Sb, O, Te, Cl). Apart from the condensed phases $RESbO_4$, RE_2O_3, $REOCl$, $RECl_3$, 13 gas species (s) have to be considered during thermodynamic modelings: $TeOCl_2$, $SbCl_3$, O_2, TeO_2, Cl_2, $TeCl_2$, Cl, $RECl_3$, $TeCl_4$, Te, Te_2, $SbCl$, $SbOCl$.

Further rare earth metal oxido metallates: Apart from the rare earth metal compounds with complex anions (PO_4^{3-}, AsO_4^{3-}, SbO_4^{3-}), a series of ternary and a series of quaternary oxides of the rare earth metal oxides can be deposited by CVT reactions. Examinations on the CVT in the systems RE_2O_3/TiO_2 show that titanates of the composition $RE_2Ti_2O_7$ (RE = Nd, Sm ... Lu) can be transported endothermically (e. g., $1050 \rightarrow 950\,°C$) with chlorine as the transport agent (Hüb 1992):

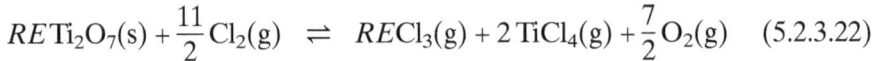

$$RE\,Ti_2O_7(s) + \frac{11}{2}Cl_2(g) \;\rightleftharpoons\; RECl_3(g) + 2\,TiCl_4(g) + \frac{7}{2}O_2(g) \quad (5.2.3.22)$$

A sub-reaction with the transport agent chlorine can lead to the formation of rare earth metal oxide halides $REOCl$; thus disturbing the transport reactions of oxide rare earth metal-containing compounds:

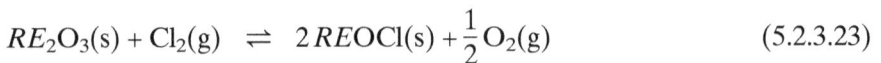

$$RE_2O_3(s) + Cl_2(g) \;\rightleftharpoons\; 2\,REOCl(s) + \frac{1}{2}O_2(g) \qquad\qquad (5.2.3.23)$$

In the system RE_2O_3/TiO_2, the following reaction can take place:

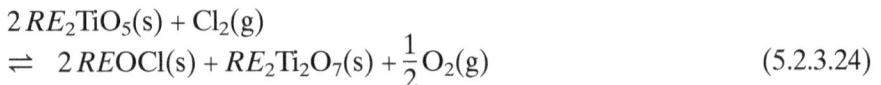

$$2\,RE_2TiO_5(s) + Cl_2(g)$$
$$\rightleftharpoons\; 2\,REOCl(s) + RE_2Ti_2O_7(s) + \frac{1}{2}O_2(g) \qquad (5.2.3.24)$$

(RE = La, Pr, Nd, Sm ... Gd).

However, also multinary oxide chlorides, e. g., of the composition $RE_2Ti_3O_8Cl_2$, adjacent to $REOCl$ can be deposited at T_1 (see Chapter 8). If the rare earth metal : titanium ratio is ≤ 1, these sub-reactions are not observed. Besides chlorine, mercury(II)-chloride can be used as transport agent (Zen 1999a). The analogue, oxidation-sensitive cerium(III)-compound $Ce_2Ti_2O_7$ can neither be transported with chlorine nor with hydrogen chloride. Here, a transport with ammonium chloride or with mercury(II)-chloride to the cooler zone is possible (Pre 1996). Furthermore, the neodymium titanate $Nd_4Ti_9O_{24}$ can be transported from T_2 to T_1 with chlorine (Hüb 1992a). In short distance transport experiments, the titanates $Nd_2Ti_4O_{11}$ (Hüb 1992b) as well as $Pr_4Ti_9O_{24}$ (Zen 1999a) crystallize. When ammonium chloride is added, cerium(III)-silicate-titanate ($Ce_2Ti_2SiO_9$) is transported endothermically from 1050 to 900 °C (Zen 1999).

Besides the rare earth metal titanates, a number of rare earth metal vanadates, rare earth metal niobates, and rare earth metal tantalates can be obtained in crystalline form by CVT reactions. Yttrium vanadate(V), YVO_4, which was doped with europium, crystallizes in the presence of tellurium(IV)-chloride in the cooler zone (Mat 1981). The endothermic transport of rare earth metal vanadates(V) $REVO_4$ (RE = Y, La ... Nd, Sm ... Ho) as well as of mixed rare earth metal vanadates(V), $Pr_{1-x}Nd_xVO_4$ and $Sm_{1-x}Eu_xVO_4$, is possible with tellurium(IV)-chloride as the transport agent (Schm 2005a). Based on the equilibrium calculations on the transport efficiency of the individual gas species, the transport of the rare earth metal vanadates(V) can be described in good approximation, with the help of the following transport equation:

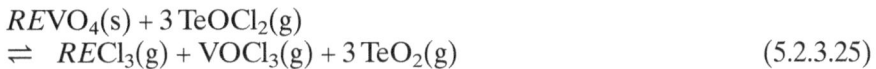

$$REVO_4(s) + 3\,TeOCl_2(g)$$
$$\rightleftharpoons RECl_3(g) + VOCl_3(g) + 3\,TeO_2(g) \qquad (5.2.3.25)$$

In this case, it is not the added tellurium(IV)-chloride but $TeOCl_2$, which develops from the complex reaction, that functions as the actual transport agent.

Rare earth metal niobates can be obtained by endothermic CVT with chlorine as the transport agent (Scha 1991, Gru 2000) (see Table 5.1). The transport additive ammonium chloride or ammonium bromide is also possible for some of the mentioned compounds. The transport of $LaNbO_4$ with chlorine is described as follows:

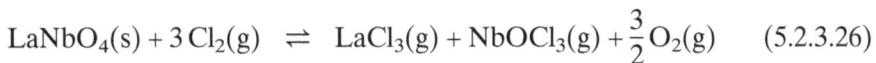

$$LaNbO_4(s) + 3\,Cl_2(g) \;\rightleftharpoons\; LaCl_3(g) + NbOCl_3(g) + \frac{3}{2}O_2(g) \qquad (5.2.3.26)$$

This is exemplary for the transport equilibria of depositing the other rare earth metal niobates (Gru 2000). Apart from the rare earth metal niobates, the transport of a series of rare earth metal tantalates is published (see Table 5.1). It takes place endothermically with chlorine as the transport agent, whereby the temperatures are 1100 °C on the dissolution side and 1000 °C on the deposition side. Generally, the following equilibrium can be given for the transport of rare earth metal tantalate(V) with chlorine:

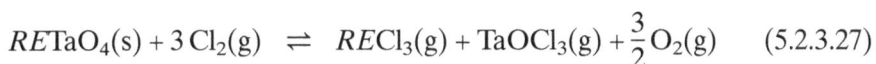

$$RETaO_4(s) + 3\,Cl_2(g) \;\rightleftharpoons\; RECl_3(g) + TaOCl_3(g) + \frac{3}{2}O_2(g) \qquad (5.2.3.27)$$

$CeTaO_4$ is an exception (Scha 1989). It is transported exothermically from 1000 to 1100 °C with added carbon monoxide and tetrabromomethane. In the process a reducing gas phase forms, which is composed of carbon monoxide and bromine. Bromine is the actual transport agent. During the transport of the rare earth metal tantalates, a similar tendency of forming multinary oxide halides as is the case with the rare earth metal titanates is observed. The reaction takes place with participation of traces of water and the transport agent chlorine is exhausted completely:

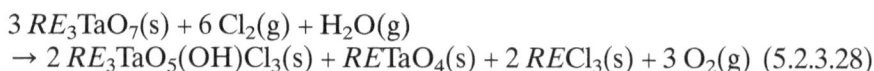

$$3\,RE_3TaO_7(s) + 6\,Cl_2(g) + H_2O(g)$$
$$\rightarrow 2\,RE_3TaO_5(OH)Cl_3(s) + RETaO_4(s) + 2\,RECl_3(s) + 3\,O_2(g) \quad (5.2.3.28)$$

The CVT of garnet crystals is topic of many publications (see Table 5.1). In particular, it is dealt with the deposition of the yttrium-iron-garnet, $Y_3Fe_5O_{12}$. Apart from that, the transport of a gadolinium-iron-garnet (Klei 1977) and a garnet of the composition $Gd_{2.66}Tb_{0.34}Fe_5O_{12}$ is described (Gib 1973). Yttrium-iron-garnet can be transported endothermically with different chlorine-containing transport additives (HCl, HCl + FeCl_3, GdCl_3, YCl_3, CCl_4, FeCl_3). Among other things, the transport behavior of the yttrium-iron-garnet with hydrogen chloride is described and analyzed with the help of the thermodynamic calculations (Weh 1970). The following transport equation is given:

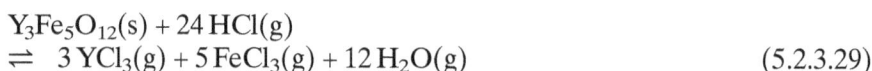

$$Y_3Fe_5O_{12}(s) + 24\,HCl(g)$$
$$\rightleftharpoons 3\,YCl_3(g) + 5\,FeCl_3(g) + 12\,H_2O(g) \quad (5.2.3.29)$$

The experiments show that $Y_3Fe_5O_{12}$ can only be transported with hydrogen chloride from 1140 to 1045 °C when iron(III)-chloride is additionally used. In the given temperature range, the transport experiments with hydrogen chloride + iron(III)-chloride lead to up to 5 mm size yttrium-iron-garnet crystals within seven days. Without adding iron(III)-chloride, only the deposition of iron(III)-oxide takes place. If iron(III)-chloride is added, there is also a simultaneous transport of yttrium-iron-garnet and iron(III)-oxide observed. *Piekarczyk* uses tetrachloromethane as an effective transport agent for $Y_3Fe_5O_{12}$ in the temperature range from 1100 to 950 °C (Pie 1981). He assumes the solid phases $Y_3Fe_5O_{12}$, $YFeO_3$, Fe_2O_3, Y_2O_3, and YOCl and the gas species CCl_4, YCl_3, $FeCl_3$, $FeCl_2$, Fe_2Cl_6, Cl_2, CO_2, CO, and O_2 for a thermodynamic description. The following equation describes the transport reaction:

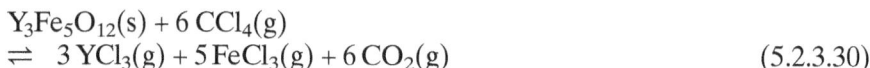

$$Y_3Fe_5O_{12}(s) + 6\,CCl_4(g)$$
$$\rightleftharpoons 3\,YCl_3(g) + 5\,FeCl_3(g) + 6\,CO_2(g) \quad (5.2.3.30)$$

Yttrium-iron-garnet and gadolinium-iron-garnet can be obtained when yttrium(III)-chloride or gadolinium(III)-chloride, respectively, is used (Klei 1977, Klei 1977a). The transport takes place from T_2 to T_1, during which the yttrium-iron-garnet can be transported almost single-phased. The deposition of $Gd_3Fe_5O_{12}$ adjacent to Fe_2O_3 takes place in the temperature gradient from 1165 to 1050 °C. *Gibart* publishes the CVT of the yttrium-iron-garnet with chlorine as well as the one with chlorine + iron(II)-chloride as transport agent, besides the one of $Gd_{2.34}Tb_{0.66}Fe_5O_{12}$ with chlorine and iron(II)-chloride. Adjacent to iron(III)-oxide, the deposition of garnet crystals is observed during the experiments. The following transport equation is assumed:

$$Y_3Fe_5O_{12}(s) + \frac{19}{2}Cl_2(g)$$
$$\rightleftharpoons\ 3\,YCl_3(g) + 5\,FeCl_2(g) + 6\,O_2(g) \tag{5.2.3.31}$$

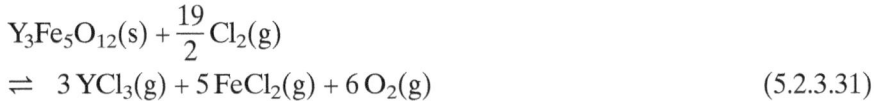

Calculations show that, in contrast to the claims of some authors, $FeCl_2(g)$, and not $FeCl_3(g)$, is the dominating gas species in the temperature range of about 1100 °C, which also applies especially when tetrachloromethane and hydrogen chloride are used (Gib 1973).

A considerable number of ternary and quaternary oxido compounds of the rare earth elements can be obtained by CVT reactions. Ternary compounds can be transported better than the binary rare earth metal oxides. The reason for this different transport behavior is that the formation of solid oxide chlorides, $REOCl$, does not take place because other reactions are thermodynamically preferred. The ternary oxido compounds of the rare earth metals $(RE_2O_3)_x(MO_n)_y$ differ from the binary rare earth metal oxide in thermodynamic terms because the thermodynamic activity of the RE_2O_3 is less than 1. Due to this, the equilibrium position of all reactions shifts to the left side. In doing so, the formation of oxide chlorides fails to appear and the transport equilibrium is more balanced. Principally, the activity decreases when the quotient x/y gets smaller. The rare earth metal-poor oxido compounds can generally be transported better than rare earth metal-rich oxido compounds.

Oxido compounds of the actinoids   The transport behavior is examined with the help of the binary oxides thorium(IV)-oxide (ThO_2), uranium(IV)-oxide (UO_2), uranium(IV,VI)-oxide (U_3O_8), and uranium(IV, VI)-oxide (U_4O_9), as well as neptunium(IV)-oxide (NpO_2). Examination of thorium silicate, thorium titanate, thorium zirconate, thorium niobates, and thorium tantalates, as well as uranium niobates and uranium tantalates, have also been published.

Thorium(IV)-oxide can be transported endothermically over the chlorinating equilibria with the transport agents tellurium(IV)-chloride (Kor 1989), and chlorine and ammonium chloride, respectively, (as HCl source) (Schm 1991a). The following transport equations become effective:

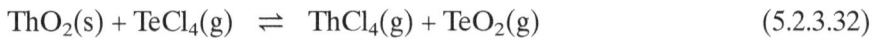

$$ThO_2(s) + TeCl_4(g)\ \rightleftharpoons\ ThCl_4(g) + TeO_2(g) \tag{5.2.3.32}$$

$$ThO_2(s) + 2\,Cl_2(g)\ \rightleftharpoons\ ThCl_4(g) + O_2(g) \tag{5.2.3.33}$$

$$ThO_2(s) + 4\,HCl(g)\ \rightleftharpoons\ ThCl_4(g) + 2\,H_2O(g) \tag{5.2.3.34}$$

Uranium(IV)-oxide can be transported endothermically by different halogenating equilibria with the transport agents chlorine, bromine, hydrogen chloride, and especially tellurium(IV)-chloride, as well as with bromine + selenium and bromine + sulfur. *Oppermann* and co-workers describe the transport of UO_2 with tellurium(IV)-chloride from 1100 to 900 °C with high transport rates. During this process, well formed crystals of up to 5 mm edge length were obtained (Opp 1975). The transport equation is as follows:

$$UO_2(s) + TeCl_4(g) \; \rightleftharpoons \; UCl_4(g) + TeO_2(g) \qquad (5.2.3.35)$$

If halogens are used as the transport agent, the following transport equation can be given (Sin 1974):

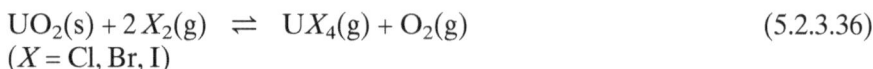

$$UO_2(s) + 2\,X_2(g) \; \rightleftharpoons \; UX_4(g) + O_2(g) \qquad (5.2.3.36)$$
$$(X = Cl, Br, I)$$

Uranium(IV)-oxide can be transported endothermically with the transport agents chlorine, bromine, hydrogen chloride, and bromine + sulfur (Nai 1971). The transport equations are given:

$$UO_2(s) + 4\,HCl(g) \; \rightleftharpoons \; UCl_4(g) + 2\,H_2O(g) \qquad (5.2.3.37)$$

$$UO_2(s) + 2\,Br_2(g) + \frac{1}{2}S_2(g) \; \rightleftharpoons \; UBr_4(g) + SO_2(g) \qquad (5.2.3.38)$$

This way, well formed crystals are obtained with high transport rates. Even higher transport rates are achieved if chlorine is used. In this process, the deposition of U_4O_9 takes place first and not, as expected, the one of UO_2 (Nai 1971). The reason for this non-stationary transport behavior is assumed to be a change in the oxygen partial pressure, which occurs during the course of the experiment.

The mixed-valent uranium oxide U_4O_9 (Nei 1971, Opp 1977d, Nom 1981) can be deposited, as U_3O_8 (Nei 1971, Nom 1981), during congruent transport with the mentioned transport agents. Also, the transport of U_3O_8 and U_4O_9 is described with hydrogen chloride (Nom 1981). Here, the participation of UCl_4, UCl_5, and UCl_6 besides the gaseous uranium oxide chlorides of the composition $UOCl_2$ and UO_2Cl_2 is discussed. However, the transport efficiency is not quantifiable due to unsure thermodynamic data. The consideration that during the CVT of uranium oxides and uranium oxide-containing compounds, respectively, gaseous uranium oxide chlorides become transport effective, is supported by mass spectrometric investigation on the existence of UO_2Cl_2 in the gas phase (Schl 1999). Investigations on the thermal behavior of solid UO_2F_2 show that it sublimes partly undecomposed. However, a part of it decomposes with the formation of solid U_3O_8, and gaseous UF_6 and oxygen. Because of this, it can be assumed that U_3O_8 can be transported with oxygen and uranium(VI)-fluoride.

$$2\,U_3O_8(s) + 3\,UF_6(g) + O_2(g) \; \rightleftharpoons \; 9\,UO_2F_2(g) \qquad (5.2.3.39)$$

Besides the binary oxides, the transport of mixed-crystals $U_{1-x}Th_xO_2$ ($x \le 0.69$) (Kam 1978) is also possible. The endothermic transport takes place from 1100 to 950 °C with hydrogen chloride as the transport agent. The gas species UCl_4 and $ThCl_4$ are considered to be transport effective:

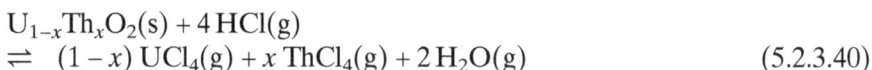

$$\begin{aligned} &U_{1-x}Th_xO_2(s) + 4\,HCl(g) \\ &\rightleftharpoons \; (1-x)\,UCl_4(g) + x\,ThCl_4(g) + 2\,H_2O(g) \end{aligned} \qquad (5.2.3.40)$$

The crystallization of thorium(IV)-orthosilicate $ThSiO_4$ is observed during transport experiments with thorium(IV)-oxide in silica glass ampoules. This behavior is more defined when pure thorium(IV)-oxide is used than when thorium oxide-containing compounds are used. This can be traced back to the following reaction:

$$ThCl_4(g) + SiO_2(s) \rightleftharpoons ThO_2(s) + SiCl_4(g) \qquad (5.2.3.41)$$

$$SiCl_4(g) + ThCl_4(g) + 4\,H_2O(g) \rightleftharpoons ThSiO_4(s) + 8\,HCl(g) \qquad (5.2.3.42)$$

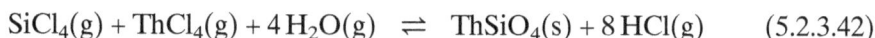

The transport of thorium(IV)-orthosilicate (Kam 1979, Schm 1991a) is dealt with in Chapter 6 under the topic of CVT of silicates.

Another example of the transport of an actinoid oxide is that of neptunium(IV)-oxide from 1050 to 960 °C with $TeCl_4$. In this case, relatively high transport agent concentrations of 5 mg $\cdot$ cm^{-3} are necessary (Spi 1980). This high transport agent concentration is also necessary for the successful transport of some rare earth metal-containing oxido compounds. Beside uranium(IV)-oxide and neptunium(IV)-oxide, thorium(IV)-oxide and plutonium(IV)-oxide are also transportable endothermically with tellurium(IV)-chloride; hereby a strong corrosion in the silica glass ampoule occurs in the process.

Transport experiments with thorium titanates, thorium niobates, and thorium tantalates prove that the formation of thorium(IV)-orthosilicate, which takes place at the same time as the transport reaction and leads to the strong corrosion of the silica glass ampoule, is supressed. This is due to the high stability of these compounds and the reduced activity of the thorium oxide. $ThNb_2O_7$, $ThNb_4O_{12}$, $Th_2Nb_2O_9$, $ThTa_2O_7$, $ThTa_4O_{12}$, $ThTa_8O_{22}$, $Th_2Ta_2O_9$, and $Th_2Ta_6O_{19}$ can be transported endothermically with chlorine or ammonium chloride. $ThTa_6O_{17}$ can be transported by adding chlorine and small amounts of vanadium or tantalum (Schm 1991a). The endothermic transport of $Th_4Ta_{18}O_{53}$ with chlorine and added tantalum(V)-chloride is similar. When chlorine and sulfur are added, the thorium tantalate $ThTi_2O_6$ crystallizes in the coldest side of the ampoule. The ternary thorium(IV)/zirconium(IV)-oxide $Th_{1-x}Zr_xO_2$ can be crystallized in the temperature gradient from 1000 to 980 °C with zirconium(IV)-chloride as transport agent (Bus 1996). In an analogous way, the crystallization of $Th_2Ta_6O_{19}$ is possible with zirconium(IV)- and hafnium(IV)-chloride, respectively, as transport additive.

Beside the ternary thorium compounds, different uranium niobates and uranium tantalates could be obtained by CVT reactions. In all cases, the endothermic transport takes place via the chlorinating equilibria. The uranium niobates $UNbO_5$ and UNb_2O_7 can be transported with chlorine and added niobium(V)-chloride. UNb_6O_{16} as well as the uranium tantalates UTa_3O_{10} and UTa_6O_{17} can be deposited with added ammonium chloride. These compounds as well as $U_4Ta_{18}O_{53}$ are obtained by CVT reactions with chlorine and the addition of tantalum(V)-chloride.

Mass spectrometric examinations of the transport behavior of $UTaO_5$ with chlorine prove that in the gas phase, the molecules UO_2Cl_2 and $TaOCl_3$ are present (Schl 1999). From this, the following transport equation can be derived:

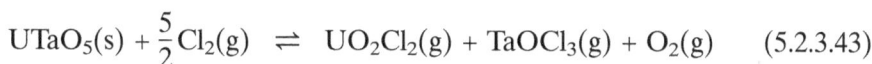

$$UTaO_5(s) + \frac{5}{2}Cl_2(g) \rightleftharpoons UO_2Cl_2(g) + TaOCl_3(g) + O_2(g) \qquad (5.2.3.43)$$

Apart from the transport of the pure oxides of actinoid metales, further examples of the transport of oxido compounds are known. First, there is the example of the thorium(IV)-oxide sulfide ThOS, which crystallizes in the presence of iodine in the cooler part of the ampoule. Furthermore, there is uranium telluride oxide, UTeO, which is obtained during the exothermic transport of U_2Te_3 with bromine from 900 to 950 °C as an overgrowth on the U_2Te_3 crystals (Shl 1995).

5.2.4 Group 4

Titanium oxides Group 4 offers one of the best and most extensively investigated systems as far as CVT reactions are concerned, the titanium/oxygen system. Besides titanium(IV)-oxide and titanium(III)-oxide the CVT of a number of titanium oxides of the composition Ti_nO_{2n-1} (*Magnéli* phases), which exist between TiO_2 and Ti_3O_5, with different transport agents are described extensively and in-depth by numerous authors (among others Far 1955, Hau 1967, Wäs 1972, Mer 1973, Ban 1981, Hon 1982, Str 1982a, Sei 1984, Kra 1987, Mul 2004). So far, Titanium(II)-oxide has not been deposited by CVT reactions.

The following transport agents and transport effective additives, respectively, can be used for transport reactions of the titanium oxides: chlorine, mercury(II)-chloride, hydrogen chloride, ammonium chloride, tetrachloromethane, chlorine + sulfur, selenium(IV)-chloride, tellurium(IV)-chloride, ammonium bromide, tellurium(IV)-chloride + sulfur, as well as iodine + sulfur. The transport of all titanium/oxygen compounds always takes place endothermically (see Table 5.1). Extensive thermodynamic model calculations, made by a series of authors, on the solid/solid and solid/gaseous equilibria showed that $TiCl_4$ is the essential, transport-effective and titanium-transferring gas species for chlorinating equilibria (Sei 1983, Sei 1984, Kra 1987, Kra 1988). Other gas species, which have to be considered, are $TiCl_3$, Ti_2Cl_6, $TiCl_2$, $TiCl$, $TiOCl$, and $TiOCl_2$, of which $TiCl_3$ is present in clearly smaller amounts than $TiCl_4$ but is still in the transport-effective range. The other mentioned gas species are necessary for a complete description of the gas phase; however, they are of minor importance for the transport due to their very low partial pressures. Due to the fact that there are numbers of discrete phases in the titanium/oxygen system of which each one only exists in very narrow oxygen co-existence pressure ranges, the setting of the oxygen partial pressure for the directed deposition of certain phases is decisive. The oxygen partial pressure can only be set freely for the most oxygen-rich compound, TiO_2. All other titanium oxides can only be deposited from the gas phase in a narrow region of the oxygen co-existence pressure ranges. The following general co-existence decomposition equilibrium can be formulated for the titanium-oxygen phases:

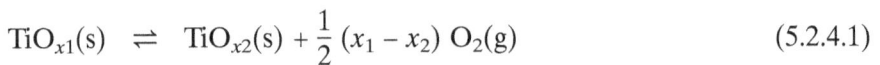

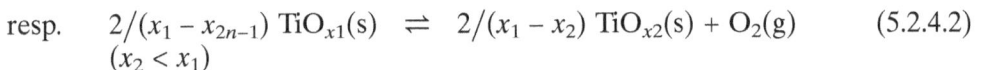

$$TiO_{x1}(s) \rightleftharpoons TiO_{x2}(s) + \frac{1}{2}(x_1 - x_2)\,O_2(g) \tag{5.2.4.1}$$

resp. $\quad 2/(x_1 - x_{2n-1})\,TiO_{x1}(s) \rightleftharpoons 2/(x_1 - x_2)\,TiO_{x2}(s) + O_2(g) \tag{5.2.4.2}$
$\quad (x_2 < x_1)$

The equilibrium constant of 5.2.4.2 results from the Gibbs energy of this reaction. From it, the oxygen partial pressure over the gas phase results (see section 15.6).

$$\Delta_r G_T^0 = -R \cdot T \cdot \ln K \tag{5.2.4.3}$$

$$K_p = p(O_2)/bar \tag{5.2.4.4}$$

The oxygen partial pressures at 1000 °C, which result from it, are in the range of 10^{-17} bar as the upper limit for Ti_nO_{2n-1} and in the range of $6 \cdot 10^{-23}$ bar as the lower limit for Ti_2O_3. The oxygen partial pressures over the *Magnéli*-phases of the composition Ti_nO_{2n-1} are in the respective narrow co-existence pressure ranges between these values. Due to the fact that there is no upper limit for the oxygen partial pressure of the CVT of TiO_2, there are no limitations on the choice of transport agent. Because of this, all transport agents and transport agent combinations that are mentioned above are suitable. If chlorine is used as the transport agent for the transport of TiO_2 (Mer 1982, Str 1982, Str 1982b, Schm 1983a, Mon 1984, Schm 1984, Kra 1987), the transport equation can be formulated as follows:

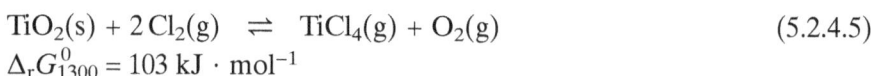

$$TiO_2(s) + 2\,Cl_2(g) \;\rightleftharpoons\; TiCl_4(g) + O_2(g) \qquad\qquad (5.2.4.5)$$
$$\Delta_r G^0_{1300} = 103 \text{ kJ} \cdot \text{mol}^{-1}$$

The CVT of TiO_2 with hydrogen chloride can be described by the following equilibrium (Far 1955):

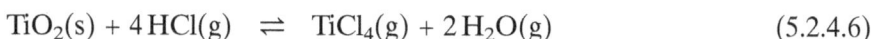

$$TiO_2(s) + 4\,HCl(g) \;\rightleftharpoons\; TiCl_4(g) + 2\,H_2O(g) \qquad\qquad (5.2.4.6)$$

In an analogous way, the transport can be described with added ammonium chloride (Wäs 1972, Izu 1979, Ban 1981] or ammonium bromide (Izu 1979), respectively, as source for the hydrogen chloride or hydrogen bromide, respectively, which act as the actual transport agent. The mode of action of the transport agent combination iodine + sulfur for the CVT of binary oxides can be explained with the example TiO_2. The following transport equation is given (Nit 1967a):

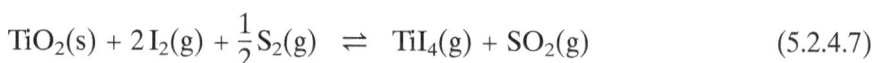

$$TiO_2(s) + 2\,I_2(g) + \frac{1}{2}S_2(g) \;\rightleftharpoons\; TiI_4(g) + SO_2(g) \qquad\qquad (5.2.4.7)$$

In contrast to chlorine, iodine is not suited as a transport agent for the transport of TiO_2 due to the equilibrium position. By adding sulfur, which allows the formation of SO_2, there is, however, a shift of the equilibrium constant to the more favorable values of the transport reactions.

The transport of other binary oxides, which exist in the titanium/oxygen system, differs generally from the TiO_2-transport due to the dependency on the partial pressure and the formation of redox systems between the oxide that is to be transported and the transport agent. However, CVT reactions are suitable for crystallizing these oxides as well. In the process, certain requirements apply for the transport agent that is to be chosen. A suitable transport agent must form stable gaseous oxygen compounds because of the low oxygen partial pressures over the solid phases and the low stability of titanium oxide halides. The following transport agents can be used (oxygen-transferring gas species in ( )): HCl, HBr, NH_4Cl or NH_4Br (H_2O), $TeCl_4$ (TeO_2, $TeOCl_2$, TeO), $SeCl_4$ (SeO_2, $SeOCl_2$, SeO), CCl_4 (CO_2, CO).

The transport agent that is used most often in the titanium/oxygen system is tellurium(IV)-chloride. It allows the CVT of oxides in a wide oxygen partial pressure range from 1 to 10^{-25} bar, in which titanium oxides are able to exist as well (Mer 1973, Mer 1973a, Mer 1977, Opp 1975, Fou 1977, Wes 1980, Mer 1982,

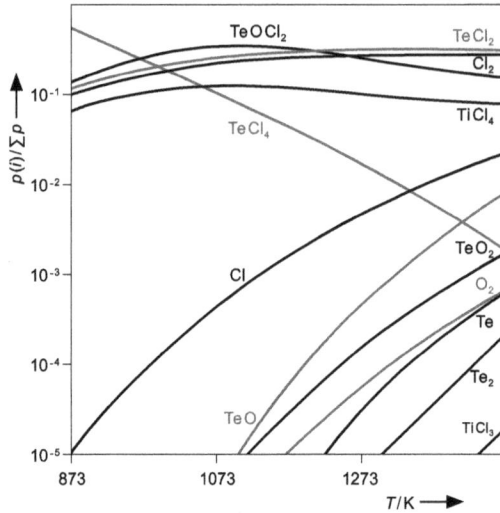

Figure 5.2.4.1 Gas-phase composition of the chemical vapor transport of TiO_2 with $TeCl_4$ according to *Krabbes et al.* (Kra 1987).

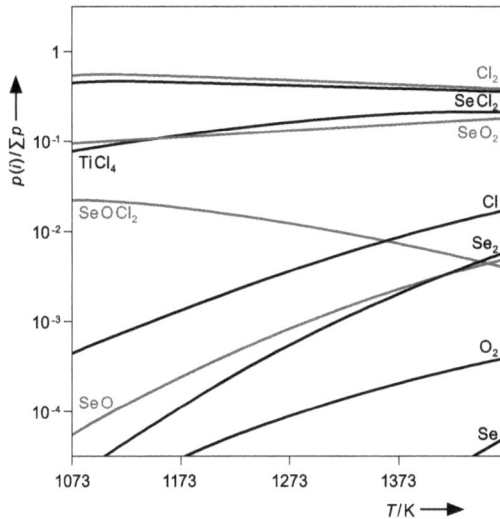

Figure 5.2.4.2 Gas-phase composition of the chemical vapor transport of TiO_2 with $SeCl_4$ according to *Krabbes et al.* (Kra 1987).

Hon 1982, Sei 1983, Kra 1987). The essential titanium-transferring gas species is titanium(IV)-chloride, the oxygen-transferring species are essentially $TeOCl_2(g)$ and with subordinated participation $TeO(g)$ and $TeO_2(g)$ (Kra 1987, Wes 1980).

Compared to the vanadium/oxygen or niobium/oxygen systems, the calculations show an unexpected gas-phase composition for the more oxygen-poor titanium oxides TiO_{2-x}. The titanium-transferring gas species $TiCl_4$ and $TiCl_3$ as well as the species Te_2, Te, and $TeCl_2$ do have transport effective partial pressures; however, the partial pressure of the oxygen-carrier of the possible gas species

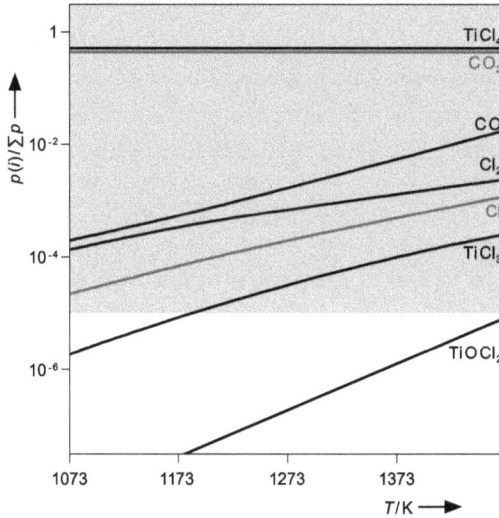

Figure 5.2.4.3 Gas-phase composition of the chemical vapor transport of TiO_2 with CCl_4 according to *Krabbes et al.* (Kra 1987).

TeO_2, $TeOCl_2$, and TeO are clearly below 10^{-5} bar in the temperature range between 600 and 1100 °C. Thus, $TeCl_4$ is not suited as a transport agent for the transport of titanium oxides TiO_{2-x}.

Seiwert and *Gruehn* analyzed the CVT of the binary oxides that exist in the titanium/oxygen system, with tellurium(IV)-chloride by thermodynamic model calculations (Sei 1983, Sei 1984). In doing so, the transport of the titanium oxides is explained in detail by comparison to the efficiency of mercury(II)-chloride, ammonium chloride, hydrogen chloride, and tellurium(IV)-chloride as transport agents, as well as the influence of water in the system. The calculations of the equilibrium show that titanium oxides of the composition TiO_{2-x} are not transportable with tellurium(IV)-chloride and mercury chloride because there are no suitable oxygen-transferring gas species with sufficiently high partial pressures. Observed transport effects for Ti_3O_5, Ti_4O_7, and Ti_5O_9 can be explained by the presence of water. Already traces of water lead to the formation of hydrogen chloride by hydrolysis of tellurium(IV)-chloride. The hydrogen chloride then becomes effective as the essential transport agent.

The reducing ability of the titanium compounds, which is higher than those of the compounds VO_{2-x} and NbO_{2-x}, is the reason why TeO_2 and $TeOCl_2$ are not transport effective. Dependent on the substance amount and the ratio of solid/transport agent, the additive tellurium(IV)-chloride leads to the oxidation of titanium to TiO_2 and $TiCl_4$ and thus to the formation and eventually deposition of elemental tellurium. Only a single-phase TiO_2-solid allows the CVT with tellurium(IV)-chloride with high transport rates without the participation of water. In this process, $TiCl_4$, $TeCl_4$, $TeOCl_2$, $TeCl_2$ and, Cl_2 are the dominant gas species. In contrast to the vanadium/oxygen and niobium/oxygen systems, only the most oxygen-rich oxide can be transported under exclusion of water in the titanium/oxygen system. This is due to the lower stability of the gaseous oxide chloride

$TiOCl_2$ compared to $VOCl_3$ and $NbOCl_3$ which are essentially involved in the transport in the respective systems. In the titanium/oxygen system, no transport effective $TiOCl_2$-partial pressures can form.

Analogous to the transport of the titanium oxides with tellurium(IV)-chloride under exclusion of water, the transport behavior with selenium(IV)-chloride can be described. Here too, TiO_2 can be transported with selenium(IV)-chloride without any problems. During the transport the gas species $TiCl_4$, SeO_2, $SeCl_2$, $SeOCl_2$, SeO, Cl_2, Cl, and Se_2 determine the transport. If the source solid is a titanium-rich oxide TiO_{2-x}, the gas phase, which is in equilibrium with the solid, contains no oxygen-transferring species with a partial pressure above 10^{-5} bar (Kra 1987). In contrast to this, transport with tetrachloromethane or chlorine + carbon, respectively, is possible. Investigations and thermodynamic calculations show that TiO_2 as well as the more titanium-rich phases TiO_{2-x} can be endothermically deposited with tetrachloromethane as transport agent (Kra 1987, Kra 1988). The transport determining gas species are $TiCl_4$ and CO_2 as well as $TiCl_3$, CO, and Cl_2, which also show transport effective partial pressures. The following simplified transport equation can be formulated:

$$TiO_2(s) + CCl_4(g) \rightleftharpoons TiCl_4(g) + CO_2(g) \qquad (5.2.4.8)$$

With this transport equilibrium, the transport of titanium oxides of the composition TiO_x ($x \geq 1.5$) is possible. In the case of more titanium-rich oxides ($x < 1.5$), there is a reaction with the SiO_2 of the transport ampoule, and through this Ti_5Si_3 is deposited as well as Ti_2O_3.

Transport with hydrogen chloride and ammonium chloride, respectively, is described by different authors (Wäs 1972, Mer 1977, Izu 1979, Ban 1981, Mul 2004) and extensively analyzed by thermodynamic model calculations (Sei 1984). If ammonium chloride (hydrogen chloride source) is used as the transport additive, there will be a higher hydrogen partial pressure in the system than when hydrogen chloride is used due to the decomposition of the initially formed ammonia. Hence, ammonium chloride has a stronger reducing effect than hydrogen chloride. In order to understand the CVT of the titanium oxides with the mentioned transport agents, the equilibrium state in the dissolution and deposition chamber have to be considered separately from each other. The source solid Ti_nO_{2n-1}, which develops from the hydrogen chloride in the equilibrium, reacts in a redox equilibrium under the formation of a heterogeneous, poly-phase solid of the composition Ti_nO_{2n-1} and $Ti_{n+1}O_{2n}$ and a gas phase, which contains hydrogen and titanium(IV)-chloride besides hydrogen chloride. After this equilibration, the reaction of the newly formed heterogeneous solid with the actual transport agents titanium(IV)-chloride and hydrogen takes place according to the following equation:

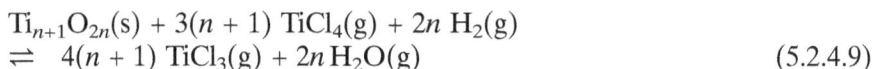

$$Ti_{n+1}O_{2n}(s) + 3(n + 1)\, TiCl_4(g) + 2n\, H_2(g)$$
$$\rightleftharpoons 4(n + 1)\, TiCl_3(g) + 2n\, H_2O(g) \qquad (5.2.4.9)$$

In this way, not the original source solid Ti_nO_{2n-1} but the phase $Ti_{n+1}O_{2n}$ that formed in the redox equilibrium between the source solid and transport agent, crystallizes on the deposition side at T_1. The composition of the solid deposited is determined by the temperature of the source solid, its composition, the amount of the transport agent, and the temperature gradient. Small temperature gradients ($\Delta T \leq 50$ K) in combination with small amounts of the transport agent generally lead to the deposition of the solid, in the phase diagram, adjacent to more oxygen-rich phases. Higher temperature gradients and/or higher transport agent concentrations lead to the crossing of several oxygen co-existence pressure lines in the phase barograph, and thus always to the transport of the more oxygen-rich phases. The oxygen/titanium ratio increases with increasing temperature gradients and transport agent concentrations so that TiO_2 can be deposited as the most oxygen-rich phase, too.

Besides the extensive examinations on the CVT of binary titanium oxide, some works are published that describe the transport of ternary titanium-containing oxides. On the one hand, there are the already-mentioned rare earth metal titanates of the composition $RE_2Ti_2O_7$, the thorium titanate $ThTi_2O_6$, and the quaternary cerium(III)-silicate-titanate $Ce_2Ti_2SiO_9$. On the other hand, these works deal with the mixed-crystals $Mo_{1-x}Ti_xO_2$ (Mer 1982), $Ti_{1-x}Ru_xO_2$ (Tri 1983, Kra 1991), and $V_{1-x}Ti_xO_2$ (Hör 1976), nickel titanate $NiTiO_3$ (Emm 1968b, Bie 1990), and iron titanate Fe_2TiO_5, (Mer 1980) as well as TiO_2 doped with aluminum(III)-oxide, iron(III)-oxide, gallium(III)-oxide, indium(III)-oxide, magnesium(II)-oxide, niobium(IV)-oxide (Izu 1979, Raz 1981, Mul 2004). In these compounds, titanium is present in the highest oxidation number IV. The only ternary compound with titanium in a lower oxidation number whose transport was described, is $(Ti_{1-x}V_x)_4O_7$ (Mer 1980). The endothermic CVT of differently composed molybdenum(IV)/titanium(IV)-mixed-oxides with tellurium(IV)-chloride can take place from 900 to 850 °C and with tellurium(IV)-chloride/sulfur from 745 to 685 °C (Mer 1982). Based on the binary oxides TiO_2 and RuO_2, the CVT of mixed-crystals of the composition $Ti_{1-x}Ru_xO_2$ succeeds with the transport agents HCl, $TeCl_4$, CBr_4, and Br_2. Hereby, the chlorinating transport agents are more effective than the brominating. Also, the deposition of iridium-, indium-, niobium-, or tantalum-doped rutile-crystals from 1050 to 1000 °C can be observed. In these cases too, TiO_2 and the respective binary oxide of the doping element are assumed as the initial materials (Tri 1983).

Krabbes and co-workers also analyzed the ruthenium/titanium/oxygen system with tellurium(IV)-chloride and tetrachloromethane as suitable transport agents with the help of thermodynamic considerations while modeling and interpreting CVT experiments of binary solid solutions (Kra 1991). The calculation of the gas-phase composition shows that $TiCl_4$ is the essential titanium-transferring species and that the transport of ruthenium takes place via the oxide chloride RuOCl as well as via the chlorides $RuCl_3$ and $RuCl_4$. The calculated oxygen partial pressures are in the range of over 10^{-5} bar at 1075 °C. However, the essential oxygen-transferring species is carbon dioxide. Analogous to titanium(IV)/ruthenium(IV)-mixed-oxide, the transport of vanadium(IV)/titanium(IV)-mixed-oxide with tellurium(IV)-chloride is possible from 1080 to 970 °C (Hör 1976).

Nickel titanate, $NiTiO_3$, can be obtained via the gas phase with chlorine or selenium(IV)-chloride, respectively, as transport agent (Emm 1968b, Bie 1990). The transport behavior of $NiTiO_3$ with $SeCl_4$ was analyzed with extensive thermodynamic calculations (Bie 1990). In the temperature range from 1050 to 1000 °C, the following dominating transport equilibrium was given:

$$NiTiO_3(s) + 3\,SeCl_2(g) \;\rightleftharpoons\; NiCl_2(g) + TiCl_4(g) + 3\,SeO(g) \quad (5.2.4.10)$$

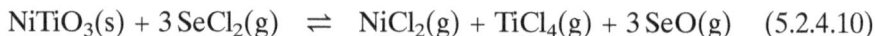

The equilibrium calculation show that selenium(II)-chloride functions as the actual transport agent and gaseous selenium(II)-oxide as the oxygen-transferring species. Doped TiO_2 can be obtained by CVT (Izu 1979, Raz 1981). For the purpose of doping, the elements aluminum, iron, gallium, indium, and magnesium in the form of their oxides are used in fractions between 1 and 10 mol %. Ammonium chloride and ammonium bromide as well as tellurium(IV)-chloride are used as transport additives. Titanium(IV)-oxide, which is doped with niobium, is transported endothermically when ammonium chloride is added in the form of up to 5 mm large mono-crystals (Mul 2004).

Zirconium oxides and hafnium oxides Besides titanium(IV)-oxides, the dioxide of the heavy homologues of group 4, ZrO_2 and HfO_2, can be obtained by CVT as well. *Oppermann* and *Ritschel* describe the CVT of transition metal dioxides with tellurium halides (Opp 1975). Accordingly, zirconium(IV)-oxide can be deposited from 1100 to 900 °C and hafnium(IV)-oxide from 1100 to 1000 °C with tellurium(IV)-chloride as the transport agent. Transport rates of around $1\ mg \cdot h^{-1}$ are achieved. Neither of the two oxides can be transported with tellurium(IV)-bromide. In the given temperature range, the transport essentially takes place via the following equilibrium:

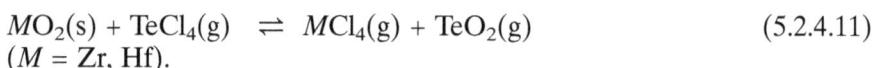

$$MO_2(s) + TeCl_4(g) \;\rightleftharpoons\; MCl_4(g) + TeO_2(g) \qquad\qquad (5.2.4.11)$$
$$(M = Zr, Hf).$$

The tellurium(IV)-chloride, which was added as transport agent, dissociates to $TeCl_2(g)$ and $Cl_2(g)$ at transport conditions around 1000 °C. Examination of the transport behavior of zirconium(IV)-oxide and hafnium(IV)-oxide with the transport agents Cl_2 and $TeCl_4$ show that both oxides can be transported endothermically with chlorine as well as with tellurium(IV)-chloride from 1100 to 1000 °C. In the process, $TeCl_4$ is the more effective transport agent with a transport rate that is up to ten times higher ($2\ mg \cdot h^{-1}$) (Dag 1992). The following equilibrium is given for transport with chlorine:

$$ZrO_2(s) + 2\,Cl_2(g) \;\rightleftharpoons\; ZrCl_4(g) + O_2(g) \qquad\qquad (5.2.4.12)$$

Zirconium(IV)-oxide is also transportable with iodine + sulfur from 1050 to 1000 °C (Nit 1966, Nit 1967a). Among the ternary zirconium oxides and hafnium oxides, the deposition of their orthosilicates, $ZrSiO_4$ and $HfSiO_4$ respectively, via the gas phase is patented (Hul 1968). The halogens chlorine, bromine, and iodine, or the corresponding tetrahalides, $ZrCl_4$, $ZrBr_4$, and ZrI_4 and $HfCl_4$, $HfBr_4$, and HfI_4 respectively, function as transport agents. The transport is endothermic at relatively high temperatures from T_2 (1150 ... 1250 °C) to T_1 (1050 ... 1150 °C).

5.2.5 Group 5

Vanadium oxides Vanadium is the lightest element of this group and with oxygen forms a number of discrete phases, which, in part, differ only slightly in the ratio $n(V)/n(O)$ (see Figure 5.2.5.1). Some of these compounds exhibit homogeneity ranges of different extents (Brü 1983).

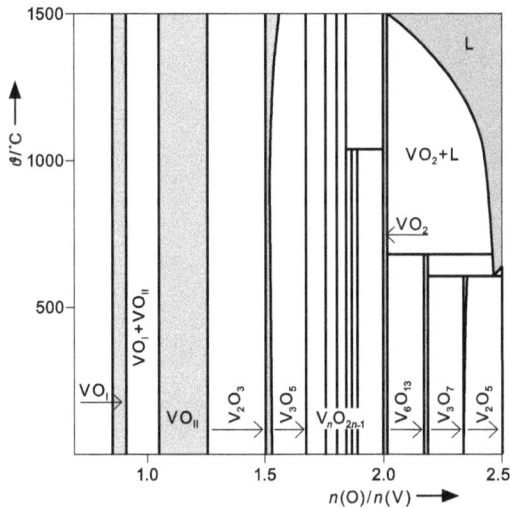

Figure 5.2.5.1 Phase diagram of the vanadium/oxygen system according to *Oppermann et al.* (Opp 1977).

The vanadium/oxygen system is one of the best investigated systems as far as CVT reactions are concerned. Large numbers of experiments with different transport agents have been conducted. Additionally, gas phase compositions have been determined spectrometrically and thermodynamic calculations made. The following transport agents and transport agent combinations were used: Cl_2, I_2, $HgCl_2$, $HgBr_2$, HgI_2, HCl, HBr (NH_4Cl and NH_4Br, respectively), $NH_4Cl + H_2O$, $NH_4Br + H_2O$, $I_2 + S$, $I_2 + H_2O$, and especially $TeCl_4$. *Nagasawa* (Nag 1972), *Oppermann* and co-workers (Opp 1977, Opp 1977a, Opp 1977c, Rei 1977), *Bando* (Ban 1978), *Brückner* (Brü 1983), *Ohtani* (Oht 1986) as well as *Gruehn* and co-workers (Wen 1989, Wen 1991, Hac 1998) provide important publications and reviews on the topic.

The oxides that exist in the vanadium/oxygen system, V_2O_5, V_3O_7, V_6O_{13}, VO_2, V_nO_{2n-1} ($4 \leq n \leq 9$), V_3O_5, and V_2O_3, could be deposited by CVT from T_2 to T_1. Only the transport of the vanadium(II)-oxide, VO has not been described yet. The temperatures vary between 500 and 1050 °C on the dissolution side and between 450 and 950 °C on the deposition side, depending on the transported compound and the transport agent. The exothermic transport of VO_2 with hydrogen chloride and with $HCl + Cl_2$, which takes place from 450 to 600 °C and from 530 to 680 °C respectively, is an exception (Opp 1977, Opp 1977a, Opp 1977c, Rei 1977).

The vanadium/oxygen system is an example of a binary system A/B with a number of phases AB_x that are in equilibrium with a gas phase, which contains one of the components in elemental form (Opp 1975a). The oxygen co-existence pressure is determined by the following decomposition equilibrium (see sections 5.2.4 and 15.6):

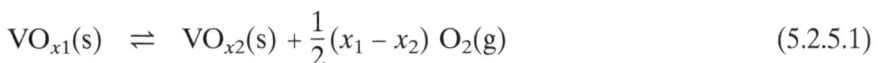

$$VO_{x1}(s) \; \rightleftharpoons \; VO_{x2}(s) + \frac{1}{2}(x_1 - x_2)\, O_2(g) \tag{5.2.5.1}$$

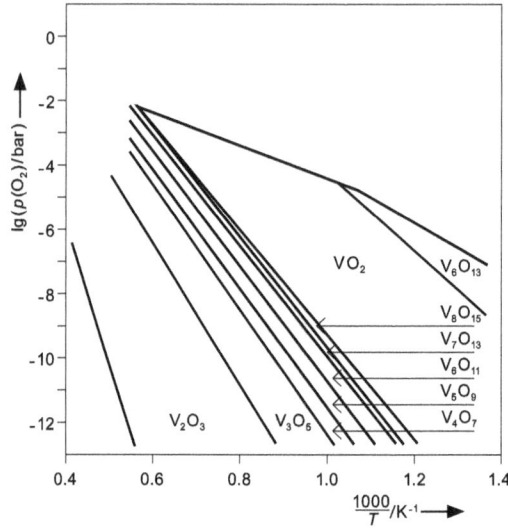

Figure. 5.2.5.2 Phase barogram of the vanadium/oxygen system according to *Oppermann et al.* (Opp 1977c).

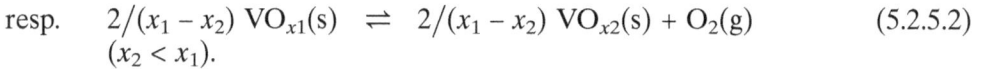

resp. $2/(x_1 - x_2) \, VO_{x1}(s) \;\rightleftharpoons\; 2/(x_1 - x_2) \, VO_{x2}(s) + O_2(g)$ (5.2.5.2)
$(x_2 < x_1).$

In this system, $p(O_2)$ ranges from 10^{-3} to 10^{-20} bar at 1000 °C. Tellurium(IV)-chloride allows the transport of oxides over a wide oxygen partial pressure range from 10^{-25} to 1 bar. Due to this, it is well suited as a transport agent in the vanadium/oxygen system.

The experimental results prove that all binary vanadium oxides can be transported with tellurium(IV)-chloride with the single exception of VO (see Table 5.1). The endothermic transport of the most oxygen-rich phase, V_2O_5, was described with the transport additives $TeCl_4$, H_2O, $I_2 + H_2O$, NH_4Cl, $NH_4Cl + Cl_2$, $NH_4Cl + H_2O$, and $NH_4Br + H_2O$. The oxygen partial pressure of this phase is not limited. Generally, the temperature of the source T_2 is close to the melting temperature of V_2O_5 (676 °C). Nevertheless, two publications mention a temperature of the source that is above the melting temperature (Rei 1975, Mar 1999).

Investigations of the transport of V_2O_5 from 580 to 480 °C with added H_2O, $I_2 + H_2O$, NH_4Br, $NH_4Br + H_2O$, NH_4Cl, $NH_4Cl + Cl_2$, or $NH_4Cl + H_2O$ show that, besides water, ammonium chloride as well as combinations that contain ammonium chloride are well suited for the transport of V_2O_5. The transport with iodine + water is less suited. The use of ammonium bromide and water always leads to the deposition of V_3O_7 besides V_2O_5. This reducing effect is enhanced when ammonium bromide is used solely. Starting from V_2O_5 as the source solid, phase pure V_3O_7 is transported in this case (Wen 1989).

In order to examine the influence of the deposition temperature on the phase range of the V_2O_5 crystals that are obtained by CVT, the temperatures were varied between 550 and 660 °C at the source and between 450 and 620 °C at the sink. In every case, tellurium(IV)-chloride was used as the transport agent. It was discovered that the phase range of V_2O_5 is temperature dependent. It increases significantly at temperatures above 550 °C (Kir 1994).

The role of water as an effective transport agent for V_2O_5 was examined with the help of the Knudsen cell-mass spectrometry and thermodynamic modeling. The detection of the gas species $V_2O_3(OH)_4$ succeeded so that the endothermic transport can be formulated by the following transport equation:

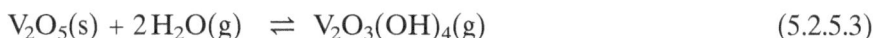

$$V_2O_5(s) + 2\,H_2O(g) \;\rightleftharpoons\; V_2O_3(OH)_4(g) \tag{5.2.5.3}$$

Furthermore, it could be shown by mass spectrometry as well as by thermodynamic modeling that the gaseous vanadium oxides V_4O_{10} and V_4O_8 have partial pressures that are too low to be important for a transport effect (Hac 1998).

In particular tellurium(IV)-chloride proved a universal transport agent for compounds that differ only slightly from the adjacent phases in the molar ratio $n(V)/n(O)$ and do only exist in a narrow oxygen co-existence pressure range (Nag 1972, Opp 1977c). The compounds V_3O_7 and V_6O_{13} can also be transported with NH_4Br, NH_4Cl, H_2O, $NH_4Cl + H_2O$, and $I_2 + H_2O$ (Wen 1989). In some cases, the transport with chlorine (Wen 1991) or hydrogen chloride (Gra 1986) was tested; usually mercury(II)-chloride was used as the chlorine source. Hence, the endothermic transport of the compounds V_3O_5, V_4O_7, V_5O_9, V_6O_{11}, V_7O_{13}, V_8O_{15}, and V_9O_{17} is described in detail for transports from 900 to 840 °C with the additive mercury(II)-chloride (Wen 1991). The authors usually introduced a two-phase mixture as source solid, which contained the phase that was to be transported and the co-existing more oxygen-poor phase.

The transport of VO_2 is examined and described in detail (see also section 2.4). VO_2 can be transported endothermically with $TeCl_4$, I_2, NH_4Cl, and NH_4Br as well as with Cl_2 ($HgCl_2$) and Br_2 ($HgBr_2$) in a wide temperature range from 1100 to 450 °C. The exothermic VO_2 transport to the hotter zone with hydrogen chloride or hydrogen chloride + chlorine, respectively, is an exception. The investigation shows that VO_2 at the upper phase boundary is transported with the mentioned transport agents from 530 to 680 °C. The following transport equation is given:

$$VO_2(s) + 2\,HCl(g) + \frac{1}{2}Cl_2(g) \;\rightleftharpoons\; VOCl_3(g) + H_2O(g) \tag{5.2.5.4}$$

The deposition at the lower phase boundary $VO_{2-\delta}$ is impossible under these conditions (Opp 1975a, Opp 1977a). However, the deposition of VO_2 at the upper as well as at the lower phase boundary is possible with tellurium(IV)-chloride (1000 → 900 °C) (Brü 1983). Furthermore, tellurium(IV)-bromide can be used as the transport agent, although the transport rates are low in comparison to transport with $TeCl_4$ (Opp 1975).

Vanadium(III)-oxide is the most oxygen-poor transportable compound in the vanadium/oxygen system. The transport is always endothermic. Different suitable transport additives are described, for example $I_2 + S$ (Pic 1973), HCl (Pou1973), $TeCl_4$ (Nag 1971, Lau 1976, Opp 1977c, Gra 1986), Cl_2 (Bli 1977, Pes 1980), as well as HgX_2 ($X = Cl$, Br, I) (Wen 1991). Instead of V_2O_3, one of the more oxygen-rich phases can be deposited if chlorine is used, which is an oxidizing transport agent.

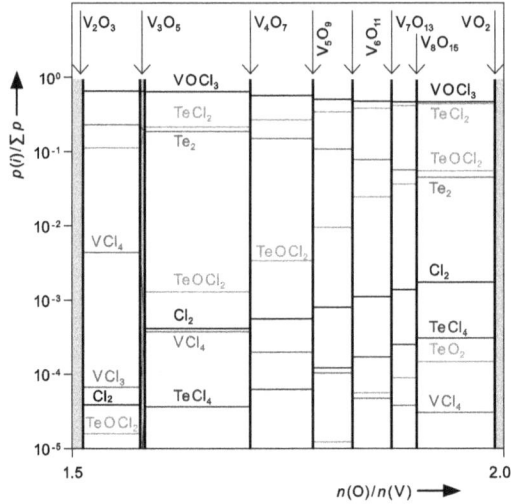

Figure 5.2.5.3 Gas-phase composition over the different vanadium oxides in equilibrium with TeCl$_4$ at 900 °C according to *Oppermann et al.* (Opp 1977c).

Tellurium(IV)-chloride is the most efficient and most flexible transport agent for vanadium oxides. In the following, vapor transport with TeCl$_4$ is used as an example for a binary system with a number of discrete phases, which show only small differences in their composition. Therefore it will be explained in more detail (see also section 2.4). According to *Oppermann* and co-workers, the following ten independent reaction equilibria are necessary ($r_u = 13 - 4 + 1$) for a complete thermodynamic description of the four-component system V/O/Te/Cl, which forms thirteen gas species that are essential for the transport (VOCl$_3$, VCl$_4$, VCl$_3$, VCl$_2$, TeCl$_4$, TeCl$_2$, TeOCl$_2$, TeO$_2$, Te$_2$, Te, Cl$_2$, Cl, and O$_2$) (Rei 1977):

$$2\,VO_x(s) + 3\,Cl_2(g) \;\rightleftharpoons\; 2\,VOCl_3(g) + (x-1)\,O_2(g) \tag{5.2.5.5}$$

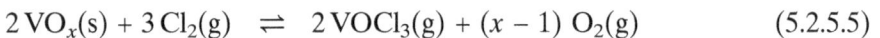

$$TeCl_4(g) \;\rightleftharpoons\; TeCl_2(g) + Cl_2(g) \tag{5.2.5.6}$$

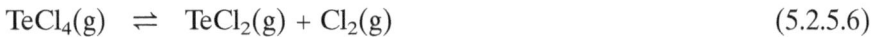

$$TeCl_2(g) + \frac{1}{2}O_2(g) \;\rightleftharpoons\; TeOCl_2(g) \tag{5.2.5.7}$$

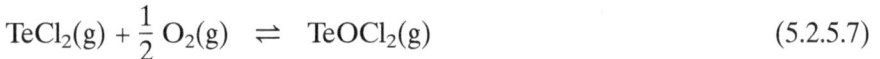

$$TeCl_2(g) + O_2(g) \;\rightleftharpoons\; TeO_2(g) + Cl_2(g) \tag{5.2.5.8}$$

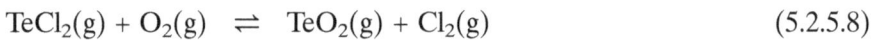

$$VOCl_3(g) + \frac{1}{2}Cl_2(g) \;\rightleftharpoons\; VCl_4(g) + \frac{1}{2}O_2(g) \tag{5.2.5.9}$$

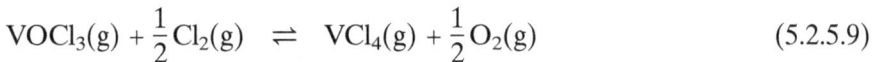

$$TeO_2(g) \;\rightleftharpoons\; \frac{1}{2}Te_2(g) + O_2(g) \tag{5.2.5.10}$$

$$Te_2(g) \;\rightleftharpoons\; 2\,Te(g) \tag{5.2.5.11}$$

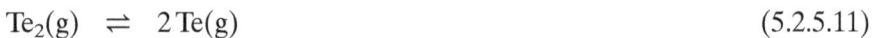

$$2\,VCl_4(g) \;\rightleftharpoons\; 2\,VCl_3(g) + Cl_2(g) \tag{5.2.5.12}$$

$$2\,VCl_3(g) \;\rightleftharpoons\; VCl_4(g) + VCl_2(g) \tag{5.2.5.13}$$

$$Cl_2(g) \;\rightleftharpoons\; 2\,Cl(g) \tag{5.2.5.14}$$

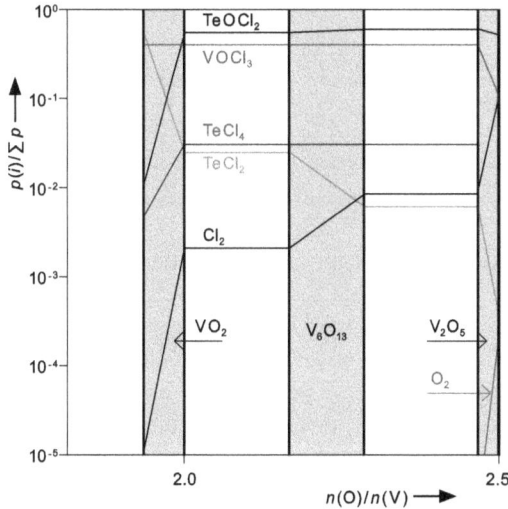

Figure 5.2.5.4   Gas-phase composition over VO_2, V_6O_{13}, and V_2O_5 in equilibrium with $TeCl_4$ at 500 °C according to *Oppermann et al.* (Opp 1977c).

The oxygen co-existence decomposition pressure decreases by more than 15 orders of magnitude, in going from the most oxygen-riche phas V_2O_5 to the most oxygen-poor phase V_2O_3, which is still transportable. Accordingly, the gas phase compositions are remarkably different. Above 675 °C, $VOCl_3$ appears as the dominating gas species over solids in the composition range between V_2O_3 and V_2O_5 (Figure 5.2.5.4). In contrast, the partial pressures of the oxygen-transferring, tellurium-containing species $TeOCl_2$ and TeO increase with rising oxygen partial pressure in the system. In the same way, the partial pressures of $TeCl_4$ and Cl_2 increase; the partial pressure of $TeCl_2$ increases as well but more slightly. The partial pressures of VCl_4, VCl_3, and VCl_2 massively increase with decreasing oxygen partial pressure (from VO_2 to V_2O_3). At temperatures below 675 °C, the gas phase over the solid V_2O_5 is dominated by $VOCl_3$ and $TeOCl_2$. The partial pressure of TeO_2 massively decreases with decreasing oxygen partial pressure, the $TeCl_2$-partial pressure, on the other hand, increases (Opp 1977, Opp 1977c).

Besides the mentioned works, a series of articles was published that deal with the CVT of doped vanadium oxides (V_2O_3:Cr, V_2O_5:Na, V_6O_{13}:Fe), in particular of mixed-crystals of the composition $V_{1-x}M_xO_2$ (M = Al, Ga, Ti, Fe, Nb, Mo, Ru, W, Os). The mixed-crystals are deposited endothermically in all cases. Generally, tellurium(IV)-chloride is used as the transport agent (Hör 1972, Lau 1973, Hör 1973, Brü 1976, Rit 1977, Kra 1991, Fec 1993). The transport of mixed-phases $V_{1-x}Ru_xO_2$ and $V_{1-x}Os_xO$ occurred with mercury(II)-chloride as an effective transport additive (Arn 2008). Detailed examinations on the ternary system vanadium/niobium/oxygen have been published (Fec 1993, Fec 1993a, Woe 1997). They include the thermodynamic description of the phase relations and the transport behavior of the ternary phases in the system. The transport agents tellurium(IV)-chloride, hydrogen chloride, and chlorine have been used. Among others, the transport of VNb_9O_{25} with $TeCl_4$ and with HCl is described. Further ternary compounds, which are accessible by CVT with tellurium(IV)-chloride, are $VTeO_4$ and $V_6Te_6O_{25}$ (Rei 1979).

Niobium oxides The transport of the niobium oxides, in particular niobium(IV)-oxide and niobium(V)-oxide, are equally well examined. In many regards, the niobium/oxygen system is similar to the vanadium/oxygen system. The phase diagram is illustrated in Figure 5.2.5.5.

All phases that appear in the system Nb/O (Nb_2O_5, $Nb_{53}O_{132}$, $Nb_{25}O_{62}$, $Nb_{47}O_{116}$, $Nb_{22}O_{54}$, $Nb_{12}O_{29}$, NbO_2, and NbO) can be crystallized by CVT. The transport takes place endothermically with the exception of NbO. NbO is one of the few examples of an *exothermic* CVT of an oxide. It can be transported with iodine as well as with the addition of the ammonium halides NH_4X (X = Cl, Br, I) from 920 ... 990 °C to 1080 ... 1150 °C. The change of the transport direction in the niobium/oxygen system is described as dependent on the oxygen partial pressure for the transport with iodine (Sch 1962). The "oxygen-poorest" phase of the system, elemental niobium, can be transported exothermically from 725 to 1000 °C according to the following equilibrium:

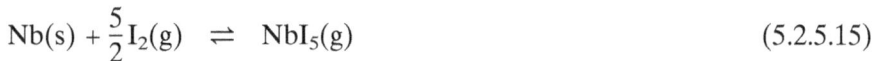

$$Nb(s) + \frac{5}{2}I_2(g) \; \rightleftharpoons \; NbI_5(g) \tag{5.2.5.15}$$

The most oxygen poor niobium oxide NbO can also be transported exothermically with iodine from 950 to 1100 °C. $NbOI_3(g)$ is both the niobium- and oxygen-transferring species. The following transport equation describes the process:

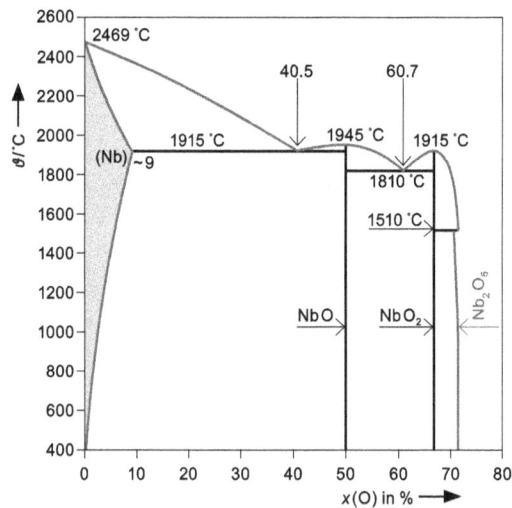

$$NbO(s) + \frac{3}{2}I_2(g) \; \rightleftharpoons \; NbOI_3(g) \tag{5.2.5.16}$$

Figure 5.2.5.5 Phase diagram of the niobium/oxygen system according to *Massalski* (Mas 1990).

However, iodine is not an effective transport agent for the more oxygen-rich phase NbO_2, because the reaction enthalpy of the exothermic effective transport equilibrium (5.2.5.17) is close to zero.

$$NbO_2(s) + \frac{1}{2}I_2(g) + NbI_5(g) \rightleftharpoons 2\,NbOI_3(g) \qquad (5.2.5.17)$$

However, crystals of NbO_2 can only be prepared with iodine from a starting mixture of NbO_2 and elemental niobium in a molar ratio of 7 : 5 at 970 °C. The NbO that forms in the process is transported to 1100 °C; at 970 °C NbO_2 crystals remain. The most oxygen-rich phase in the system, Nb_2O_5, can be transported endothermically with a transport agent combination I_2/NbI_5 from 740 to 650 °C.

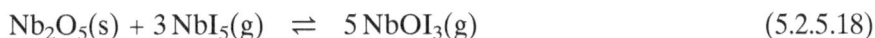

$$Nb_2O_5(s) + 3\,NbI_5(g) \rightleftharpoons 5\,NbOI_3(g) \qquad (5.2.5.18)$$

From a starting mixture of elemental niobium and niobium(V)-oxide, NbO was crystallized in a gradient from 990 to 1150 °C with ammonium halides NH_4X (X = Cl, Br, I) as additives (Kod 1976). If the temperature gradient was reversed, the crystallization of NbO_2 could be observed at T_1 (980 °C). The essential transport equilibria are as follows:

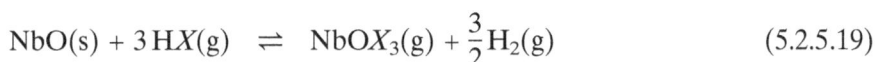

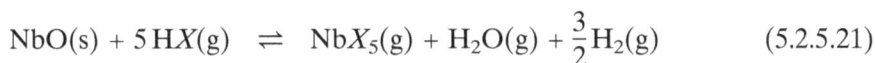

$$NbO(s) + 3\,HX(g) \rightleftharpoons NbOX_3(g) + \frac{3}{2}H_2(g) \qquad (5.2.5.19)$$

$$NbO(s) + 4\,HX(g) \rightleftharpoons NbX_4(g) + H_2O(g) + H_2(g) \qquad (5.2.5.20)$$

$$NbO(s) + 5\,HX(g) \rightleftharpoons NbX_5(g) + H_2O(g) + \frac{3}{2}H_2(g) \qquad (5.2.5.21)$$

The endothermic transport of niobium(IV)-oxide can take place with iodine, tellurium(IV)-chloride, tellurium(IV)-bromide, the ammonium halides, as well as niobium(V)-chloride (Sak 1972, Kod 1975, Kod 1975a, Opp 1975, Kod 1976, Schw 1982). According to systematic investigations on the transport behavior of NbO_2 with a variety of transport agents, the transport efficiency decreases from ammonium chloride via ammonium bromide, ammonium iodide, niobium(V)-chloride to iodine (Kod 1975a).

Between niobium(IV)- and niobium(V)-oxide a series of phases with homogeneity ranges of different extent exist. The close compositional proximity of these phases makes the crystallization of NbO_2 more difficult. Therefore the transport conditions have to be chosen in a way that the deposition takes place within the existence area of NbO_2. In some cases, simultaneous transport of the co-existing more oxygen-rich phase $Nb_{12}O_{29}$ is observed when tellurium(IV)-chloride is used as the transport agent. The reason for this is the very low oxygen co-existence pressure of NbO_2 and the oxidizing character of the transport agent tellurium(IV)-chloride. This causes the formation of $Nb_{12}O_{29}$ besides NbO_2 in the source. From this two-phase mixture in the source at first, the transport of $Nb_{12}O_{29}$ to the deposition zone takes place. After this, the actual transport of NbO_2 is observed (Rit 1978).

Ritschel and *Oppermann* (Rit 1978) as well as other authors (Hus 1986) described the crystallization of the phases with compositions between NbO_2 and Nb_2O_5: $Nb_{12}O_{29}$, $Nb_{22}O_{54}$, $Nb_{47}O_{116}$, $Nb_{25}O_{62}$, and $Nb_{53}O_{132}$. The deposition of two co-existing niobium oxides is observed if $TeCl_4$ is used as the transport agent. The transport of the mentioned compounds with mercury(II)-chloride as transport additive succeeded at 1250 to 1200 °C (Hus 1986).

The most oxygen-rich niobium oxide, Nb_2O_5, has a considerable phase range. It can be deposited with different transport agents in endothermic reactions. It crystallizes in a number of different modifications whose occurrences are both temperature and pressure dependent. Their compositions vary in the range from $NbO_{2.4}$ to $NbO_{2.5}$ (Sch 1966, Hib 1978). Nb_2O_5 can be transported with the niobium(V)-halides NbX_5 (X = Cl, Br, I), the niobium(V)-halides + halogen, as well as with tellurium(IV)-chloride, hydrogen halides, sulfur, and also with $NbOF_3$. If the transport takes place via the halogenating equilibria with niobium(V)-halides and/or halogens (Lave 1964, Sch 1964, Emm 1968, Hib 1978), the following transport equations can be formulated:

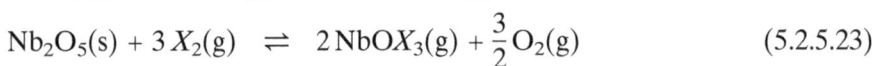

$$Nb_2O_5(s) + 3\,NbX_5(g) \;\rightleftharpoons\; 5\,NbOX_3(g) \tag{5.2.5.22}$$

$$Nb_2O_5(s) + 3\,X_2(g) \;\rightleftharpoons\; 2\,NbOX_3(g) + \frac{3}{2}O_2(g) \tag{5.2.5.23}$$

The additional use of chlorine together with niobium(V)-chloride prevents the formation of reduced phases Nb_2O_{5-x} and leads to the deposition of the upper boundary composition Nb_2O_5. *Gruehn* and co-workers described the formation of Nb_2O_{5-x} at the lower phase boundary by CVT from 1080 to 980 °C with $NbCl_5$ + $NbOCl_3$ in the presence of the oxygen buffer CO_2/CO in the ratio $1:1$ (Gru 1967). Crystals of the composition of the lower phase boundary can be obtained by CVT with sulfur from 1000 to 900 °C (Sch 1988), too. During investigations in the Nb_2O_5/Nb_3O_7F system, the transport of Nb_2O_5 was observed. The transport equation was given (Gru 1973):

$$Nb_2O_5(s) + 3\,NbF_5(g) \;\rightleftharpoons\; 5\,NbOF_3(g) \tag{5.2.5.24}$$

Oppermann and *Ritschel* discussed the way tellurium(IV)-chloride works as a transport agent for Nb_2O_5 on the basis of thermodynamic calculations (Rit 1978, Rit 1978a). Niobium(V)-oxide can be transported with $TeCl_4$ in a wide temperature range from T_2 (600 to 1000 °C) to T_1. The comparison of the gas-phase composition over Nb_2O_5 at the upper phase boundary and the lower phase boundary shows clear differences: in both cases, $NbOCl_3$ is the dominating, niobium-transferring gas species. The partial pressure of $NbCl_5$ is still in the transport effective range ($p > 10^{-5}$ bar) in both cases. The pressure of $NbCl_4$ is in the range of 10^{-3} bar for the lower phase boundary and thus transport effective; for the upper phase boundary it is $5 \cdot 10^{-7}$ bar. The difference is more distinct when the tellurium-containing species are considered: The gas phase over Nb_2O_5 of the lower phase boundary is defined by Te_2, Te, and $TeCl_2$. The partial pressure of $TeOCl_2$ is below 10^{-8} bar. In contrast, the gas phase over Nb_2O_5 at the upper phase boundary is dominated by $TeOCl_2$ with a partial pressure of 10^{-1} bar. Also, the partial pressures of TeO_2 and TeO are in the transport effective range. These differences in the gas-phase composition reflect the large change of the oxygen partial pressure within the phase range of Nb_2O_5. This is an illustrative example of how small changes in the composition of a solid can decisively influence the gas phase composition.

Mass spectrometric investigations in a temperature range between 1100 and 830 °C provide experimental evidence for the gas-phase composition during CVT reactions of NbO_2 with $TeCl_4$, as well as its change in dependency on the composition of the solid and the content of residual moisture (Kob 1981). In the same context, the transports of NbO_2 and $Nb_{12}O_{29}$ with tellurium(IV)-chloride based on different compositions of the source solid and the influence of HCl as transport agent are examined and discussed.

Heterogeneous equilibria between solids and the gas phase can massively increase the speed of isothermal solid state reactions. So, at 900 °C, there is virtually no reaction between elemental niobium and niobium(V)-oxide. Adding hydrogen (1 bar) enables this reaction and the formation of NbO takes place. The necessary transfer of the oxygen can be described by the following equilibria:

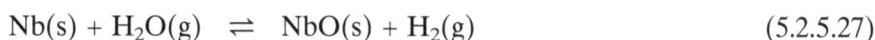

$$Nb_2O_5(s) + H_2(g) \rightleftharpoons 2\,NbO_2(s) + H_2O(g) \tag{5.2.5.25}$$

$$NbO_2(s) + H_2(g) \rightleftharpoons NbO(s) + H_2O(g) \tag{5.2.5.26}$$

$$Nb(s) + H_2O(g) \rightleftharpoons NbO(s) + H_2(g) \tag{5.2.5.27}$$

Tantalum oxide Ta_2O_5 is the only tantalum oxide. It can be deposited endothermically from the source at temperatures of 800 and 1000 °C, to the sink at temperatures of 700 and 800 °C, respectively, via chlorinating equilibria with tantalum(V)-chloride, ammonium chloride, chlorine, chromium(III)-chloride + chlorine, as well as sulfur as transport agent. Exothermic transport with tetrabromomethane from 900 to 1000 °C is also possible (Scha 1989). The different efficiencies of the transport equilibria are shown in comparison.

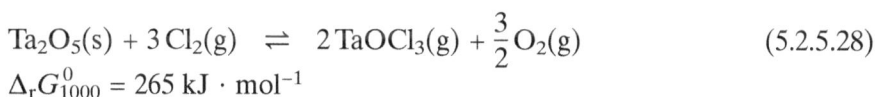

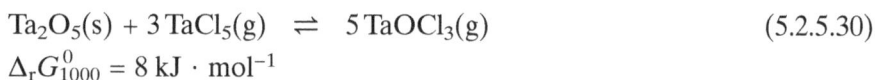

$$Ta_2O_5(s) + 3\,Cl_2(g) \rightleftharpoons 2\,TaOCl_3(g) + \frac{3}{2}O_2(g) \tag{5.2.5.28}$$
$$\Delta_r G^0_{1000} = 265 \text{ kJ} \cdot \text{mol}^{-1}$$

$$Ta_2O_5(s) + \frac{3}{2}CrCl_4(g) + \frac{3}{2}Cl_2(g)$$
$$\rightleftharpoons 2\,TaOCl_3(g) + \frac{3}{2}CrO_2Cl_2(g) \tag{5.2.5.29}$$
$$\Delta_r G^0_{1000} = 124 \text{ kJ} \cdot \text{mol}^{-1}$$

$$Ta_2O_5(s) + 3\,TaCl_5(g) \rightleftharpoons 5\,TaOCl_3(g) \tag{5.2.5.30}$$
$$\Delta_r G^0_{1000} = 8 \text{ kJ} \cdot \text{mol}^{-1}$$

Accordingly, $TaCl_5$ is a suitable transport agent for tantalum(V)-oxide because the *Gibbs* energy is close to zero and thus in an optimal range for transport reactions. For this reaction the highest transport rates are expected. In all chlorinating equilibria, $TaOCl_3$ is the decisive transport effective tantalum carrier. Other gaseous tantalum oxide chlorides, e. g., TaO_2Cl, $Ta_2O_4Cl_2$, or binary tantalum chlorides have very low partial pressures, out of the transport-effective range. $TaOCl_3$ is also the oxygen-transferring species.

The deposition of tantalum(V)-oxide provides an example of sulfur alone being the transport agent. The transport of Ta_2O_5 is possible from 1000 to 900 °C because sulfur forms stable volatile compounds with tantalum, presumable TaS_5, as well as with oxygen, SO_2. The following transport equation can be formulated (Sch 1980):

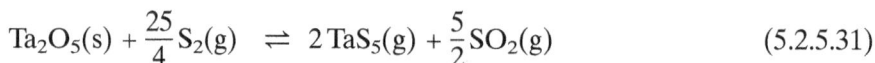

$$Ta_2O_5(s) + \frac{25}{4}S_2(g) \rightleftharpoons 2\,TaS_5(g) + \frac{5}{2}SO_2(g) \tag{5.2.5.31}$$

On the one hand, the transport of oxides of group 5 is defined by the fact that the oxygen co-existence pressures over the binary oxides extend over a wide range. Some of the oxides are thermodynamically stable only in narrow co-existence ranges. This applies in particular for the vanadium/oxygen and niobium/oxygen systems in which there are a number of phases that differ only slightly in their ratio $n(M)/n(O)$, although some of these phases have widely different homogeneity ranges. On the other hand, it can be stated that in all systems, the respective metal oxide halides of the composition MOX_3 (M = V, Nb, Ta; X = F, Cl, Br, I) are the metal-transferring species, and, in many cases, also the oxygen carriers when halogenating transport agents are used. The stability of these compounds decreases from the fluorides via the chlorides and bromides to the iodides, and increases from vanadium via niobium to tantalum. This is shown by the comparison of the standard heats of formation:

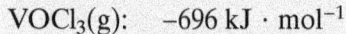

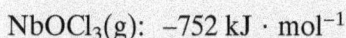

$VOCl_3(g)$: $-696\ kJ \cdot mol^{-1}$

$NbOCl_3(g)$: $-752\ kJ \cdot mol^{-1}$

$TaOCl_3(g)$: $-783\ kJ \cdot mol^{-1}$

Besides publications on the transport behavior of the binary oxides of group 5, there is a large body of work on the CVT of ternary and quaternary vanadium-, niobium-, and tantalum-containing oxidic compounds, respectively. These are described in the chapters about the respective metal vanadates, niobates, and tantalates.

5.2.6 Group 6

Chromium oxides   Among the binary chromium oxides, only chromium(III)-oxide is transportable. Chromium(III)-oxide is volatile at temperatures above 1000 °C in the presence of oxygen (Grim 1961, Cap 1961). Transport reactions also allow the crystallization of chromium(III)-oxide during endothermic transports via chlorinating and brominating equilibria. Transport reactions are generally possible with the transport agents Cl_2, $Cl_2 + O_2$, $HgCl_2$, $TeCl_4$, Br_2 and $CrBr_3 + Br_2$ from 1000 to 900 °C. *Nocker* and *Gruehn* deal extensively with transport in the chromium/oxygen/chlorine system and contribute to the understanding of the transport of chromium(III)-oxide with chlorine and mercury(II)-chloride, respectively, with experiments and thermodynamic calculations (Noc 1993a, Noc 1993b). The calculations have shown that the following equilibrium plays an essential role:

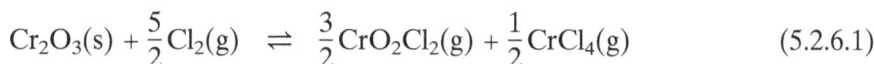

$$Cr_2O_3(s) + \frac{5}{2}Cl_2(g) \rightleftharpoons \frac{3}{2}CrO_2Cl_2(g) + \frac{1}{2}CrCl_4(g) \qquad (5.2.6.1)$$

In the presence of additional oxygen in the system, another transport reaction becomes more important:

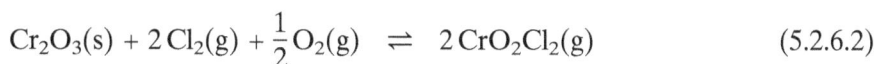

$$Cr_2O_3(s) + 2\,Cl_2(g) + \frac{1}{2}O_2(g) \rightleftharpoons 2\,CrO_2Cl_2(g) \qquad (5.2.6.2)$$

If water is present in the system, hydrogen chloride is formed additionally:

$$Cr_2O_3(s) + 3\,Cl_2(g) + H_2O(g) \rightleftharpoons 2\,CrO_2Cl_2(g) + 2\,HCl(g) \qquad (5.2.6.3)$$

The transport behavior of chromium(III)-oxide with mercury(II)-chloride can be described in an according way. The transport of chromium(III)-oxide with bromine and chromium(III)-bromide + bromine, respectively, was investigated as well (Noc 1994). The endothermic transport with bromine can be formulated via analogous equilibria. If chromium(III)-bromide + bromine is used, the system is less oxidized so that $CrOBr_2(g)$ and not $CrO_2Br_2(g)$ is the transport-effective species, although $CrBr_4(g)$ is the actual transport agent. Thus, the following transport equation can be formulated:

$$Cr_2O_3(s) + 3\,CrBr_4(g) \rightleftharpoons 3\,CrOBr_2(g) + 2\,CrBr_3(g) \qquad (5.2.6.4)$$

In the literature, transport reactions of a series of ternary and quaternary chromium-containing oxido compounds are also published. With the exception of the transport of $Cr_2BP_3O_{12}$ with iodine and phosphorus + iodine (Schm 2002), respectively, only endothermic chlorinating equilibria are mentioned. Examples are the transport of $Cr_{0.18}In_{1.82}GeO_7$ (Pfe 2002b) and of $CrGa_2O_4$ (Pat 2000b) with chlorine. The chromium-containing oxido compounds $CrTaO_4$ and $CrNbO_4$ are transportable with chlorine and niobium(V)-chloride + chlorine, respectively (Emm 1968b, Emm 1968c). The transport of a chromium/niobium-mixed oxide, $(Cr, Nb)_{12}O_{29}$, is examined with chlorine as well as with niobium(V)-chloride (Hof 1994).

Molybdenum oxides Molybdenum(IV)-oxide (MoO_2), molybdenum(VI)-oxide (MoO_3), as well as the mixed-valent oxides Mo_4O_{11}, Mo_8O_{23}, and Mo_9O_{26} can be obtained by CVT. The endothermic transport of all mentioned oxides succeeds with tellurium(IV)-chloride as the transport agent (Ban 1976). *Monteil et al.* (Mon 1984) described and analyzed the thermodynamics of the molybdenum(IV)-oxide transport with tellurium(IV)-chloride and determined experimentally the gas-phase composition by Raman spectroscopy. In doing so, it became clear that, in the lower temperature range between 400 and 500 °C, transport takes place according to the following transport equation:

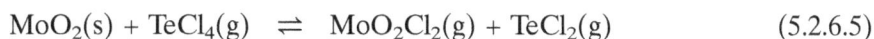

$$MoO_2(s) + TeCl_4(g) \rightleftharpoons MoO_2Cl_2(g) + TeCl_2(g) \qquad (5.2.6.5)$$

Above these temperatures, the dissociation of $TeCl_4$ leads to the formation of tellurium(II)-chloride, which becomes effective as a transport agent.

$$MoO_2(s) + TeCl_2(g) \rightleftharpoons MoO_2Cl_2(g) + \frac{1}{2}Te_2(g) \tag{5.2.6.6}$$

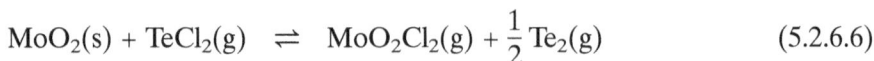

Monteil confirmed the data of *Ritschel* and *Oppermann* (Rit 1980), which analyzed the transport of molybdenum(IV)-oxide with tellurium(IV)-chloride as well as tellurium(IV)-bromide in thermodynamic terms in the temperature range around 1000 °C. It could be shown that $MoO_2X_2(g)$ (X = Cl, Br) is both the essential molybdenum-transferring and the oxygen-transferring species. The partial pressures of the species $MoOX_3$, $MoOX_4$ as well as MoX_4 (X = Cl, Br) are at the boundary of the transport effective range. This also applies for $TeOX_2$ (X = Cl, Br). The calculations show that the tellurium(II)-halides are the real transport agents.

The transport behavior of MoO_2 and Mo_4O_{11} with mercury(II)-chloride was investigated with the help of experiments and model calculations. In doing so, the influence of the source solid, the transport agent concentration, as well as the range and position of the temperature gradient were analyzed (Scho 1990b). The transport of MoO_2 can be described by the following equilibrium:

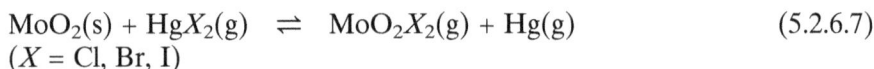

$$MoO_2(s) + HgX_2(g) \rightleftharpoons MoO_2X_2(g) + Hg(g) \tag{5.2.6.7}$$
$$(X = Cl, Br, I)$$

The calculated gas-phase composition shows that the gas species MoO_2Cl_2, $MoOCl_3$, and $MoCl_4$ become transport effective for solids Mo/MoO_2 at 800 °C (Figure 5.2.6.1). Besides MoO_2Cl_2, also the volatile molybdenum oxides Mo_3O_9, Mo_4O_{12}, and Mo_5O_{15} become transport effective over a solid that is composed of MoO_2/MoO_3. These oxides are also the dominating gas species above 950 °C. If water is present in the system, the formation of molybdenum acid H_2MoO_4 with transport effective partial pressures is possible for oxygen-rich solids.

Besides the halogenating equilibria that have been described so far, the CVT of molybdenum(IV)-oxide is also published with iodine from 1000 to 800 °C. Transport takes place according to the following reaction equation:

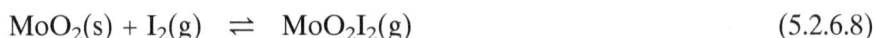

$$MoO_2(s) + I_2(g) \rightleftharpoons MoO_2I_2(g) \tag{5.2.6.8}$$

The description of the CVT of molybdenum(IV)-oxide with iodine also includes a comparison of the thermodynamic stabilities of the gaseous metal oxide halides of group 6 of the composition MO_2X_2 (X = Cl, Br, I) (Opp 1971). These species prove essential for all transports of molybdenum oxido compounds with halogenating transport agents.

Many publications deal with the transport of a series of molybdates(IV) (Ste 1983, Rei 1994, Ste 1996, Söh 1997, Ste 2000, Ste 2003, Ste 2004a, Ste 2005b, Ste 2006) as well as of molybdenum(IV)-oxide-mixed phases $Mo_{1-x}Ti_xO_2$ (Mer 1982), $Mo_{1-x}V_xO_2$ (Brü 1977, Rit 1977), $Mo_{1-x}Ru_xO_2$ (Nic 1993), and $Mo_{1-x}Re_xO_2$ (Fel 1998a). The transport takes place endothermically with tellurium(IV)-chloride as the transport agent. During transport experiments in the molybdenum/rhenium/oxygen system, mixed-crystals of the composition $Mo_{1-x}Re_xO_2$ ($0 < x < 0.42$) can be deposited with I_2 ($1100 \rightarrow 1000$ °C) and $TeCl_4$ ($1000 \rightarrow 900$ °C) as transport agent (Fel 1998a). Compared to the transport with tellurium(IV)-chloride, the crystals that were transported with iodine have a superior shape. Fur-

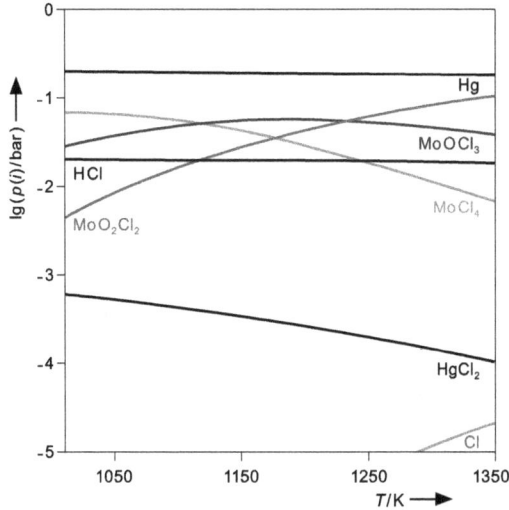

Figure 5.2.6.1 Gas-phase composition over Mo/MoO_2 in equilibrium with $HgCl_2$ in the presence of water according to *Schornstein* (Scho 1992).

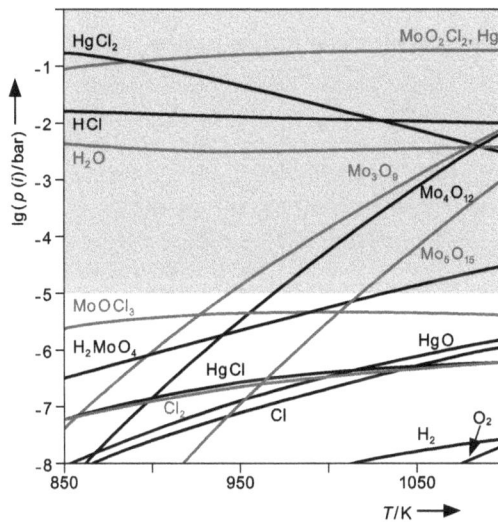

Figure 5.2.6.2 Gas-phase composition over MoO_2/Mo_4O_{11} in equilibrium with $HgCl_2$ in the presence of water according to *Schornstein* (Scho 1992).

thermore a phase of the composition Mo_3ReO_{11} can be transported with $TeCl_4$ and $HgCl_2$ from 740 to 700 °C. Thermodynamic calculations show that the transport of mixed-phases with iodine takes place via the molybdenum-containing species $MoO_2I_2(g)$ and via the rhenium-containing species $ReO_3I(g)$. If tellurium(IV)-chloride is used as the transport agent, the composition of the gas phase is essentially defined by the species Re_2O_7, ReO_3Cl, $TeOCl_2$, $TeCl_2$, and MoO_2Cl_2.

The most oxygen-rich binary molybdenum oxide MoO_3 can be deposited in crystalline form at temperatures above 780 °C by sublimation. In the process, trimers, tetramers, and pentamers are observed in the gas phase. The formation of the molybdenum acid is the reason that molybdenum(VI)-oxide can be transferred to the gas phase with the help of water in the temperature range between 600 and 700 °C (Gle 1962). The following equilibrium is formulated:

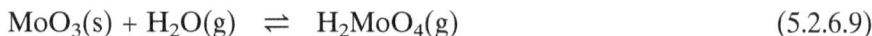

$$MoO_3(s) + H_2O(g) \;\rightleftharpoons\; H_2MoO_4(g) \qquad (5.2.6.9)$$

During transport experiments with the aim of maintaining *Magnéli* phases by transport with tellurium(IV)-chloride in the molybdenum/oxygen system, Mo_nO_{3n-1}, crystals of the compound Mo_5TeO_{16} were deposited in the temperature range from 680 to 690 °C (Neg 2000). The following formation reaction was given:

$$6\,MoO_3(s) + MoO_2(s) + TeCl_4(g)$$
$$\rightarrow \; Mo_5TeO_{16}(s) + 2\,MoO_2Cl_2(g) \qquad (5.2.6.10)$$

Tungsten oxides   The CVT of tungsten oxides is verified by a number of publications over a long period. The binary oxides tungsten(IV)-oxide (WO_2), tungsten(VI)-oxide (WO_3), as well as the mixed-valent oxides $W_{18}O_{49}$, and $W_{20}O_{58}$ can be obtained by CVT with different transport agents; all transport equilibria are endothermic. The mercury(II)-halides HgX_2 (X = Cl, Br, I), tellurium(IV)-chloride, and iodine proved particularly effective as transport agents for the transport of WO_2. *Schornstein* and *Gruehn* describe WO_2 transport with the mercury(II)-halides (Scho 1988, Scho 1989, Scho 1990a). Extensive thermodynamic calculations explain the transport behavior in detail (see section 15.1). The influence of the range and position of the temperature gradient and dependency on the transport agent used, as well as the influence of moistures traces in the system, is discussed as well. Transport with all three transport agents takes place via the following equilibrium:

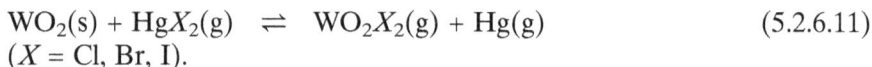

$$WO_2(s) + HgX_2(g) \;\rightleftharpoons\; WO_2X_2(g) + Hg(g) \qquad (5.2.6.11)$$
$$(X = Cl, Br, I).$$

In the presence of water, another equilibrium becomes transport effective:

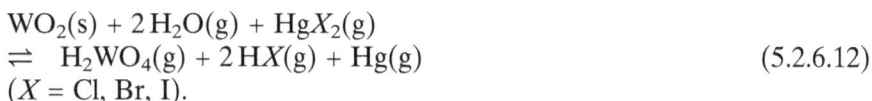

$$WO_2(s) + 2\,H_2O(g) + HgX_2(g)$$
$$\rightleftharpoons \; H_2WO_4(g) + 2\,HX(g) + Hg(g) \qquad (5.2.6.12)$$
$$(X = Cl, Br, I).$$

The formed tungsten acid H_2WO_4 becomes transport effective in the temperature range around 1000 °C. In this process, its partial pressure is approximately two magnitudes lower than the partial pressure of the essential tungsten-transferring species WO_2X_2. Other gaseous tungsten oxide halides WOX_3 and WOX_4 do not have transport effective partial pressures. The transport of tungsten(IV)-oxide with the mercury(II)-halides is an illustrative example of the interaction of the position of the temperature gradient and the transport agent used. If mercury(II)-chloride is used, maximal transport rates are achieved during the transport from

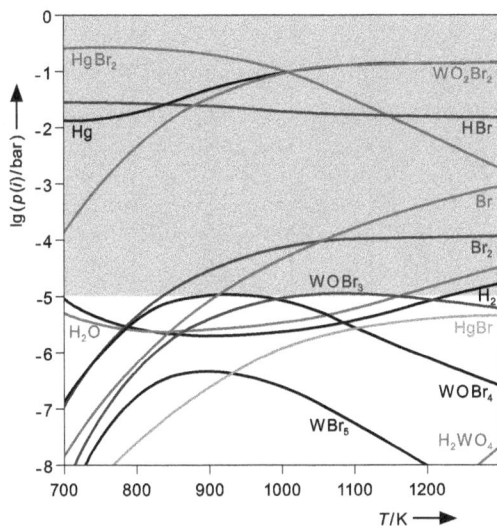

Figure 5.2.6.3 Gas-phase composition over $WO_2/W_{18}O_{49}$ in equilibrium with $HgCl_2$ in the presence of water according to *Schornstein* (Scho 1992).

700 to 600 °C. If mercury(II)-bromide is used, they are achieved from 800 to 700 °C, and if mercury(II)-iodide is used, this is the case from 1060 to 980 °C.

Iodine and tellurium(IV)-chloride are also suitable. Furthermore, tungsten(IV)-oxide can be transported with water (Mil 1949). The endothermic transport takes place via the following equilibrium:

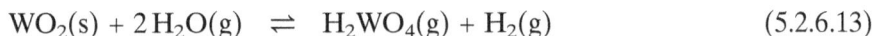

$$WO_2(s) + 2\,H_2O(g) \;\rightleftharpoons\; H_2WO_4(g) + H_2(g) \tag{5.2.6.13}$$

In an analogous way, the deposition of tungsten(VI)-oxide with water via the gas phase is also possible (Mil 1949, Glen 1962, Heu 1981, Sah 1983).

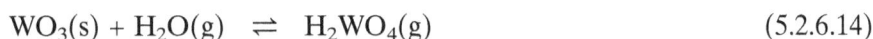

$$WO_3(s) + H_2O(g) \;\rightleftharpoons\; H_2WO_4(g) \tag{5.2.6.14}$$

Tungsten(VI)-oxide can be transported endothermically via different chlorinating equilibria in a wide temperature range (KIe 1966, Opp 1985a, Scho 1990a). The phases $W_{18}O_{49}$ and $W_{20}O_{58}$, which are positioned between tungsten(IV)-oxide and tungsten(VI)-oxide, can also be obtained by CVT. Like WO_3, $W_{18}O_{49}$ can be transported effectively with mercury(II)-halides (Scho 1988, Scho 1989). Tellurium(IV)-chloride is a suitable transport agent for $W_{18}O_{49}$ as well as for $W_{20}O_{58}$ (Opp 1985a).

There are further examples of the CVT of tungsten-containing oxido compounds. Besides the mixed-crystals $V_{1-x}W_xO_2$ (Hör 1972, Rit 1977, Brü 1977) and $Ru_{1-x}W_xO_2$ (Nic 1993), these are the tungstates $CaWO_4$ (Ste 2005a), $CdWO_4$ (Ste 2005a), $CuWO_4$ (Yu 1993, Ste 2005a), $FeWO_4$ (Emm 1968b, Sie 1982, Yu 1993), $MnWO_4$ (Emm 1968b, Ste 2005), $MgWO_4$ (Emm 1968b, Ste 2005a), $SrWO_4$ (Ste 2005a), and $ZnWO_4$ (Cur 1965, Emm 1968b, Ste 2005), as well as the tungsten bronzes M_xWO_3 (M = Li, K, Rb, Cs, In) (Hus 1994, Hus 1997, Schm 2008, Rüs 2008, Ste 2008). The transport behavior of these compounds is explained in the context of the first mentioned elements.

Above all, the transport behavior of the oxide compounds of group 6 is marked by the formation of volatile oxide halides of the composition MO_2X_2 (M = Cr, Mo, W; X = Cl, Br, I), which, as transport-effective species, are essentially responsible for the transfer of the metal atoms. These are independent of the oxidation number in the solid, and always transferred to the oxidation number VI so that the transport process is marked by redox equilibria (except for the metal(VI)-oxides). These equilibria determine the choice of suitable transport agent as well as the composition of the deposited solid. Other gas species, such as molybdenum and tungsten acid H_2MO_4 (M = Mo, W), respectively, and other oxide halides than MO_2X_2 are formed but generally do not reach transport-effective pressure.

5.2.7 Group 7

The CVT of binary, ternary, and quaternary oxides of manganese and rhenium is known. There are no details about the CVT of technetium-containing oxido compounds.

Manganese oxide   In the manganese/oxygen system, there are the binary oxides MnO_2, Mn_2O_3, Mn_3O_4, and MnO, all of which can be deposited by endothermic transports except for MnO_2. The transport of Mn_2O_3 is possible with the transport agents chlorine, bromine, hydrogen chloride, hydrogen bromide, selenium(IV)-chloride, and tellurium(IV)-chloride from T_2 to T_1 (Ros 1987, Ros 1988). Mn_3O_4 can be transported endothermically with the same transport agents. Chlorine and bromine, however, are only partially suitable because only Mn_3O_4 of the upper phase boundary can be deposited when these are used (Klei 1963, Klei 1972a, Yam 1972, Ros 1987, Ros 1988). In order to deposit Mn_3O_4 of the lower phase boundary as well as MnO, selenium(IV)-chloride and tellurium(IV)-chloride are particularly suitable (Ros 1988). Manganese oxide transport with chlorine and bromine can essentially be described by the following three equilibria:

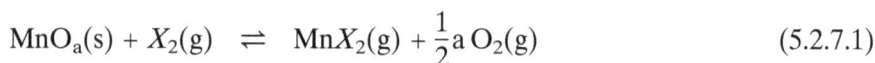

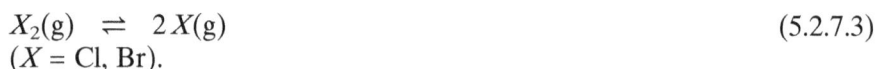

$$MnO_a(s) + X_2(g) \; \rightleftharpoons \; MnX_2(g) + \frac{1}{2}a\,O_2(g) \qquad (5.2.7.1)$$

$$2\,MnX_2(g) \; \rightleftharpoons \; Mn_2X_4(g) \qquad (5.2.7.2)$$

$$X_2(g) \; \rightleftharpoons \; 2\,X(g) \qquad (5.2.7.3)$$
$$(X = \text{Cl, Br}).$$

The transport effective gas species MnX_2 (X = Cl, Br) that have to be considered, their respective dimers Mn_2X_4, and oxygen result from these equations. Due to the fact that manganese does not form volatile oxide halides and that there are no further species that can transfer the oxygen in the $Mn/O/X$ system (X = Cl, Br), the oxygen partial pressure determines the transportability of the individual

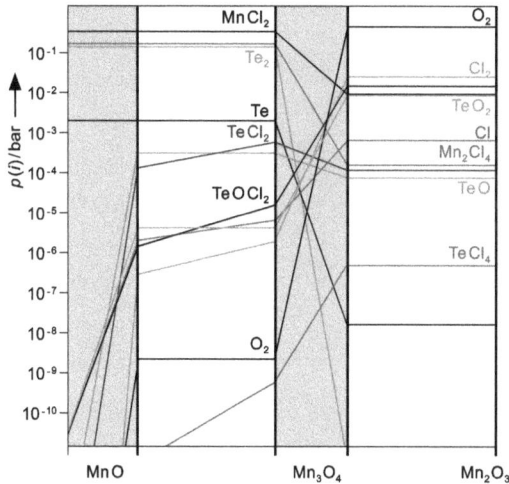

Figure 5.2.7.1 Gas-phase composition over the different manganese oxides inequilibrium with TeCl$_4$ at 1200 K according to *Rossberg* (Ros 1988).

manganese oxides. The oxygen partial pressure decreases massively with decreasing oxygen : manganese ratio. It crosses the transport effective limit of 10^{-5} bar at 1000 °C in the homogeneity area of Mn$_3$O$_4$ so that Mn$_3$O$_4$ of the lower phase boundary as well as MnO are not transportable with halogens. Hydrogen halides allow the transport of MnO as well as of Mn$_3$O$_4$ of the lower phase boundary by forming water as an additional, oxygen-transferring species. If the respective hydrogen halides HCl and HBr are used as the transport agent, the following transport effective equilibrium is assumed:

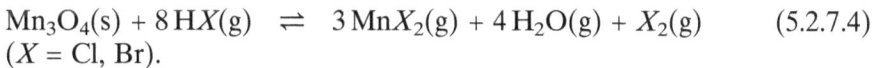

$$\text{Mn}_3\text{O}_4(s) + 8\,\text{H}X(g) \; \rightleftharpoons \; 3\,\text{Mn}X_2(g) + 4\,\text{H}_2\text{O}(g) + X_2(g) \qquad (5.2.7.4)$$
$$(X = \text{Cl, Br}).$$

The transport agents selenium(IV)-chloride and tellurium(IV)-chloride, respectively, which are able to form oxygen-transferring species, work in a similar way. During the transport of Mn$_2$O$_3$ and Mn$_3$O$_4$ of the upper phase boundary, MnCl$_2$ and oxygen are the transport effective gas species. During the transport of Mn$_3$O$_4$ of the lower phase boundary and MnO, these are MnCl$_2$ and TeO, MnCl$_2$, and SeO and SeO$_2$, respectively (Ros 1987, Ros 1988).

There are no experimental indications on the CVT of MnO$_2$ in the literature. Model calculations on the transport behavior of MnO$_2$ with the transport agents AlCl$_3$, GaCl$_3$, and InCl$_3$ cover the formation of gas complexes of the composition MnM_2Cl$_8$ in particular (Kra 1988b). If the requirement that the oxygen partial pressure over MnO$_2$ is 1 bar at 527 °C is met, one can assume that the trichlorides of aluminum and gallium will be transferred virtually completely to the oxide according to reaction 5.2.7.5.

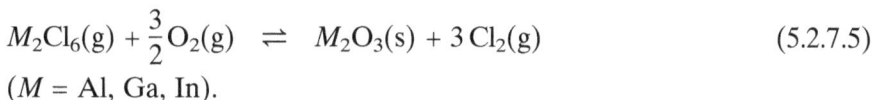

$$M_2\text{Cl}_6(g) + \frac{3}{2}\text{O}_2(g) \; \rightleftharpoons \; M_2\text{O}_3(s) + 3\,\text{Cl}_2(g) \qquad (5.2.7.5)$$
$$(M = \text{Al, Ga, In}).$$

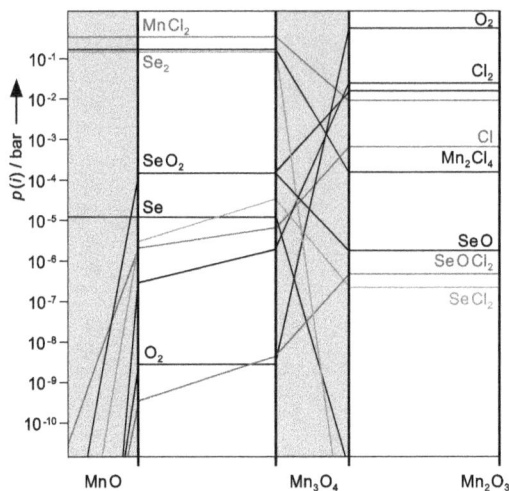

Figure 5.2.7.2 Gas-phase composition over the different manganese oxides in equilibrium with $SeCl_4$ at 1200 K according to *Rossberg* (Ros 1988).

Hence, the trichlorides of aluminum and gallium are disqualified as transport agents; indium(III)-chloride remains as a possible transport agent. By additionally adding chlorine adjacent to the transport agent indium(III)-chloride, the equilibrium is shifted to the side of the source solids and thus the formation of indium(III)-oxide is counteracted. The calculated transport rates for the transport of MnO_2 with indium(III)-chloride/chlorine from 530 to 480 °C are in the range between 10^{-3} and 10^{-2} mg $\cdot$ h^{-1}. Hence, a minimal transport effect is expected. The transport experiments are in agreement with the calculations insofar as no transport of MnO_2 is observed throughout duration of the experiment of 700 hours.

There are numbers of examples of the transport of ternary and quaternary manganese-containing oxido compounds in which manganese always appears in the oxidation number II. Among others, the transport of spinels is published during which chlorinating equilibria become exclusively effective and chlorine, hydrogen chloride and ammonium chloride, respectively, are used as transport additives (Klei 1963, Klei 1964, Mem 1968b, Trö 1972, Klei 1972a, Klei 1973). Further examples of the transport of ternary manganese oxides are $MnMoO_4$ and $Mn_2Mo_3O_8$: their transport behavior with the transport agents chlorine, hydrogen chloride, tellurium(IV)-chloride, and selenium(IV)-chloride and with chlorine, iodine, hydrogen chloride, hydrogen iodide, and selenium(IV)-chloride, respectively, were examined in detail with thermodynamic calculations and experiments (Rei 1994). The transport behavior and the gas-phase composition are defined by the same characteristics as the transport of binary oxides. $MnCl_2$ and Mn_2Cl_4 are the transport effective, manganese-transferring species. MoO_2Cl_2 is responsible for the transfer of molybdenum.

$MnNb_2O_6$ can be endothermically deposited with chlorine/ammonium chloride as well as with niobium(V)-chloride/chlorine (Emm 1968c, Ros 1991a). The following transport effective equilibria describe the reaction process:

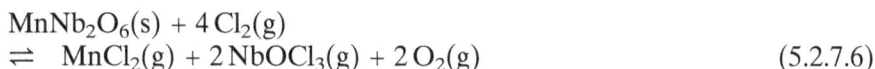

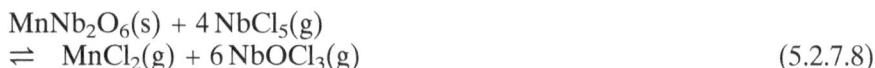

$MnNb_2O_6(s) + 4\,Cl_2(g)$
$\rightleftharpoons\ MnCl_2(g) + 2\,NbOCl_3(g) + 2\,O_2(g)$ (5.2.7.6)

$MnNb_2O_6(s) + 8\,HCl(g)$
$\rightleftharpoons\ MnCl_2(g) + 2\,NbOCl_3(g) + 4\,H_2O(g)$ (5.2.7.7)

$MnNb_2O_6(s) + 4\,NbCl_5(g)$
$\rightleftharpoons\ MnCl_2(g) + 6\,NbOCl_3(g)$ (5.2.7.8)

Garnets of the composition $Mn_3Cr_2Ge_3O_{12}$, $Mn_3Fe_2Ge_3O_{12}$, and $Mn_3Ga_2Ge_3O_{12}$ can be transported with added chlorine, sulfur + chlorine, tellurium(IV)-chloride, or tetrachloromethane from T_2 to T_1 (Paj 1985, Paj 1986a). Transport takes place according to the following equilibria:

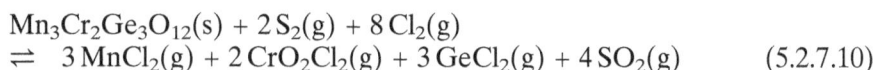

$Mn_3Cr_2Ge_3O_{12}(s) + 8\,Cl_2(g)$
$\rightleftharpoons\ 3\,MnCl_2(g) + 2\,CrO_2Cl_2(g) + 3\,GeCl_2(g) + 4\,O_2(g)$ (5.2.7.9)

$Mn_3Cr_2Ge_3O_{12}(s) + 2\,S_2(g) + 8\,Cl_2(g)$
$\rightleftharpoons\ 3\,MnCl_2(g) + 2\,CrO_2Cl_2(g) + 3\,GeCl_2(g) + 4\,SO_2(g)$ (5.2.7.10)

$Mn_3Cr_2Ge_3O_{12}(s) + 4\,TeCl_4(g)$
$\rightleftharpoons\ 3\,MnCl_2(g) + 2\,CrO_2Cl_2(g) + 3\,GeCl_2(g) + 4\,TeO_2(g)$ (5.2.7.11)

$Mn_3Cr_2Ge_3O_{12}(s) + 4\,CCl_4(g)$
$\rightleftharpoons\ 3\,MnCl_2(g) + 2\,CrO_2Cl_2(g) + 3\,GeCl_2(g) + 4\,CO_2(g)$ (5.2.7.12)

In the process, the tetrachloromethane proved the most effective transport agent because it allows the highest transport rate besides chlorine, but does not lead to the undesired non-stationary transport of chromium(III)-oxide adjacent to the $Mn_3Cr_2Ge_3O_{12}$. Moreover, crystals that have been transported with tetrachloromethane are the biggest and the best formed.

The manganese-containing germanates $Mn_{1-x}Co_xGeO_3$, $Mn_{1-x}Zn_xGeO_3$, and $(Mn_{1-x}Zn_x)_2GeO_4$ can be deposited by endothermic transport with hydrogen chloride from 900 to 700 °C and from 1050 to 900 °C, respectively (Pfe 2002, Pfe 2002c). The transport of $Mn_{1-x}Mg_xO$-mixed-crystals is also possible with hydrogen chloride (Skv 2000).

Rhenium oxides Rhenium(IV)-oxide, ReO_2, and rhenium(VI)-oxide, ReO_3, can be crystallized by CVT. Rhenium(VII)-oxide, Re_2O_7, is transferable undecomposed via the gas phase. In doing so, the thermal stability of the different rhenium oxides is decisive for the temperature range in which the transport reactions shall take place (Opp 1985b). The compound ReO_2, which is the most thermally stable, can be transported in a wide temperature range. The temperatures of the dissolution side varies between 700 and 1100 °C; the temperatures on the deposition side, between 600 and 1000 °C. ReO_3-transport, which is also endothermic, takes place in a temperature range from 385 to 600 °C towards temperatures from 350 to 500 °C. The thermally relatively stable, volatile compound Re_2O_7 can be sublimed at temperatures below 200 °C. The gas species Re_2O_7 was detected by mass spectrometry (Rin 1967). Adding water or water and oxygen increases the volatility in the sense of a CVT during which the following equilibrium becomes transport effective:

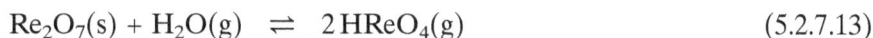

$Re_2O_7(s) + H_2O(g)\ \rightleftharpoons\ 2\,HReO_4(g)$ (5.2.7.13)

This effect can be used in closed as well as in flowing systems for depositing Re_2O_7 (Gle 1964a, Mül 1965).

Rhenium(VI)-oxide ReO_3 can be obtained in an analogous way to the oxides Re_2O_7 and ReO_2 with water as the transport agent, by forming gaseous perrhenic acid, $HReO_4$, from 400 to 350 °C. Furthermore, halogens, mercury halides, tellurium(IV)-chloride, and hydrogen chloride are used as transport additives. Mercury(II)-chloride, which causes unusually high transport rates during endothermic transports of up to 26 mg $\cdot$ h^{-1} from 600 to 500 °C (see section 15.3), is particularly well suited. If bromine or iodine is used as transport agent, the transport rate is slightly lower. Adding tellurium(IV)-chloride results in transport rates of approximately 0.1 mg $\cdot$ h^{-1}. The gas phase is essentially defined by the gas species Re_2O_7 and ReO_3X (X = Cl, Br, I). Re_2O_7 forms above 300 °C by disproportionation of ReO_3 and simultaneous formation of solid ReO_2. If there are traces of water in the system, $HReO_4$ is formed additionally. In all cases, the transport-effective species are the gaseous rhenium oxide halides, ReO_3X, so that the following transport equation can be formulated:

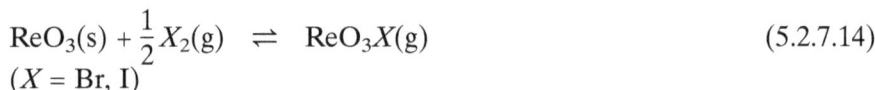

$$ReO_3(s) + \frac{1}{2}X_2(g) \ \rightleftharpoons \ ReO_3X(g) \tag{5.2.7.14}$$
$$(X = Br, I)$$

Besides the mentioned gas species, ReX_3 and ReX_5 appear in the Re/O/X system (X = Cl, Br, I), and ReO_2Cl_3 and $ReOCl_5$ appear additionally in the Re/O/Cl system (Fel 1998).

In many cases, the morphology of the deposited crystals is dependent on the temperature, the temperature gradient, as well as on the transport agent and its amount. Hence, rhenium(VI)-oxide can be deposited in the form of cubes, flakes, or needles.

Rhenium(IV)-oxide can be transported via different halogenating equilibria as well as with water. The CVT of ReO_2 with added water as transport agent takes place from 700 to 600 °C via the following equilibrium:

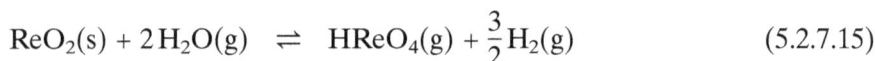

$$ReO_2(s) + 2H_2O(g) \ \rightleftharpoons \ HReO_4(g) + \frac{3}{2}H_2(g) \tag{5.2.7.15}$$

The gaseous perrhenic acid is transport effective during the transport of Re_2O_7 too (Sch 1973b). The halogens as well as tellurium(IV)-chloride are very effective transport agents for the transport of ReO_2. Also, the mercury(II)-halides HgX_2 (X = Cl, Br, I) proved to be good transport additives. High transport rates are achieved with the mentioned transport agents. Rhenium oxide halides of the composition ReO_3X (X = Cl, Br, I) are the essential transport effective gas species with a sufficiently high partial pressure difference between the dissolution and deposition side. The high Re_2O_7-equilibrium pressure increases with decreasing temperature (see Figure 5.2.7.4). Thus, Re_2O_7(g) does not contribute to the endothermic transport of ReO_2. Rhenium and oxygen are transferred via ReO_3Cl(g). Thus, the following essential transport equation can be formulated:

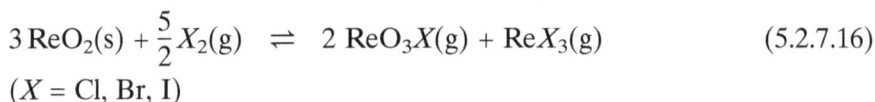

$$3\,ReO_2(s) + \frac{5}{2}X_2(g) \ \rightleftharpoons \ 2\,ReO_3X(g) + ReX_3(g) \tag{5.2.7.16}$$
$$(X = Cl, Br, I)$$

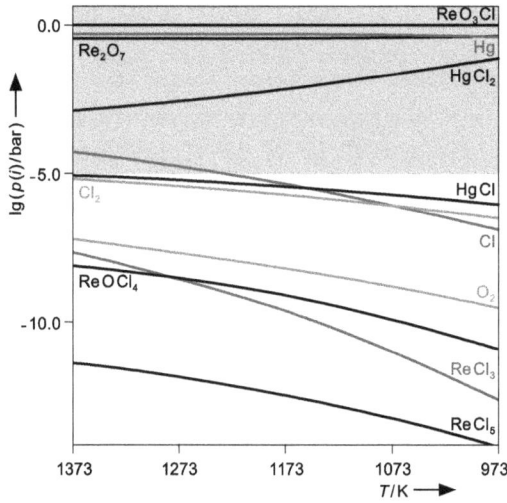

Figure 5.2.7.3 Gas-phase composition over ReO_3 in equilibrium with $HgCl_2$ according to *Feller et al.* (Fel 1998).

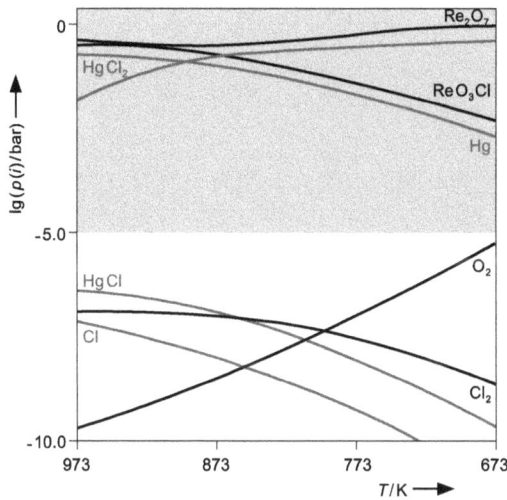

Figure 5.2.7.4 Gas-phase composition over ReO_2 in equilibrium with $HgCl_2$ according to *Feller et al.* (Fel 1998).

Higher transport rates can be achieved if there is an additional oxygen-containing species in the system, which can be realized, for example, by a surplus of rhenium(VII)-oxide. In this case, the transport takes place via the following equilibrium:

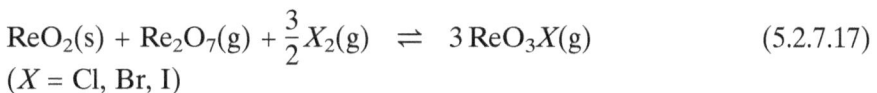

$$ReO_2(s) + Re_2O_7(g) + \frac{3}{2}X_2(g) \rightleftharpoons 3\,ReO_3X(g) \qquad (5.2.7.17)$$
$$(X = Cl, Br, I)$$

The transport of the mixed-phase $Re_{1-x}Mo_xO_2$ succeeds under the derived transport conditions of ReO_2. The mixed-phase can be transported with iodine from 1100 to 1000 °C (Fel 1998a).

If the thermochemical characteristics of the binary manganese oxides and rhenium oxides are compared to those of the technetium oxides Tc_2O_7 and TcO_2, the transport of the technetium oxides should be possible. This is shown by a compilation of the standard enthalpies of formation.

MnO_2(s): $-520\,kJ \cdot mol^{-1}$
TcO_2(s): $-433\,kJ \cdot mol^{-1}$
ReO_2(s): $-449\,kJ \cdot mol^{-1}$
Tc_2O_7(s): $-1115\,kJ \cdot mol^{-1}$
Re_2O_7(s): $-1263\,kJ \cdot mol^{-1}$

Technetium(VII)-oxide already melts at 120 °C and can be obtained in crystalline form at 300 °C by sublimation. At the melting temperature, the Tc_2O_7-partial pressure is approximately 10^{-3} bar. In particular, technetium(IV)-oxide should be transportable over the chlorinating equilibria because the existence of gaseous technetium oxide chlorides TcO_3Cl and $TcOCl_3$ is known. $HTcO_4$ as well as volatile oxides could be further possible transport effective gas species. At high temperatures, the following gas species were detected by mass spectrometry: Tc_2O_6, Tc_2O_5, Tc_2O_4, TcO_3, TcO_2, and TcO (Gue 1972, Rar 2005).

5.2.8 Group 8

Iron oxides Iron forms the binary oxides iron(II)-oxide, $Fe_{1-x}O$, iron(II, III)-oxide, Fe_3O_4, and iron(III)-oxide, Fe_2O_3. All three of them can be obtained by CVT. The transport of iron(II)-oxide is described in only one publication. The transport of the wustite, which is only thermodynamically stable above 570 °C, took place on a magnesium(II)-oxide substrate from 725 to 700 °C with hydrogen chloride as the transport agent. The relatively low transport agent pressures varied in the range between 0.001 and 0.05 bar (Bow 1974). The thermodynamic calculations for the analysis of the CVT in the iron/oxygen system under consideration of the oxygen co-existence pressure show that the transport of $Fe_{1-x}O$ should only be possible with hydrogen chloride, if possible at all. Due to the very low oxygen co-existence pressure in the co-existence area $Fe_{1-x}O$ /Fe_3O_4, no transport effective oxygen partial pressure develops in the system. In order to transfer oxygen, the formation of an oxygen-containing, gaseous *compound* is necessary. If hydrogen chloride is used as the transport agent, this compound will be water. Even though the calculations predict the transport of a phase sequence Fe_3O_4 prior to $Fe_{1-x}O$, only Fe_3O_4 is detected in most experiments (Ger 1977b). The small existence area of $Fe_{1-x}O$ as far as the oxygen partial pressure is concerned always leads to the deposition of the more oxygen-rich phase Fe_3O_4. This is already the case in application of small temperature gradients.

The phase Fe_3O_4, which co-exists with $Fe_{1-x}O$, can be endothermically transported with the chlorinating transport agents hydrogen chloride and tellurium(IV)-chloride from 1000 to 800 °C. Here, the formation of an oxygen-transferring gas species is decisive, too, due to the low oxygen partial pressure. If $TeCl_4$ is used as transport agent, $TeOCl_2$ is transport effective. The iron-transferring species are $FeCl_2$, Fe_2Cl_4 as well as $FeCl_3$ and Fe_2Cl_6 (Hau 1962, Klei 1972a, Mer 1973a, Ger 1977b, Bli 1977, Pes 1984).

Due to its relatively high oxygen partial pressure, iron(III)-oxide, which is the most oxygen-rich binary oxide, can be endothermically transported with the halogens chlorine and iodine (presumably also bromine) as well as with the halogenating transport agents hydrogen chloride and tellurium(IV)-chloride (e. g., 1000 → 800 °C); with hydrogen chloride, it can also be transported exothermically (e. g., 300 °C → 400 °C) (Klei 1966) (see Table 5.1). At 1000 °C, the transport of iron(III)-oxide with hydrogen chloride is essentially described by the following endothermic equilibrium:

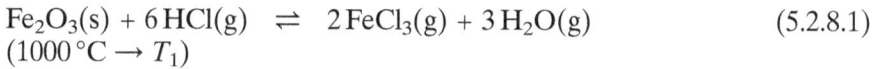

$$Fe_2O_3(s) + 6\,HCl(g) \; \rightleftharpoons \; 2\,FeCl_3(g) + 3\,H_2O(g) \qquad (5.2.8.1)$$
$$(1000\,°C \rightarrow T_1)$$

Besides the transport effective gas species $FeCl_3$, Fe_2Cl_6 also appears to be an iron-transferring species with a negligibly small partial pressure at 1000 °C. Due to the temperature dependent dimerization equilibrium, the Fe_2Cl_6-partial pressure increases at low temperatures (300 to 400 °C) in an order that Fe_2Cl_6 becomes the essential iron-transferring gas species. In this case, the following transport equation applies:

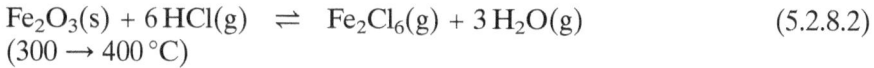

$$Fe_2O_3(s) + 6\,HCl(g) \; \rightleftharpoons \; Fe_2Cl_6(g) + 3\,H_2O(g) \qquad (5.2.8.2)$$
$$(300 \rightarrow 400\,°C)$$

This transport equilibrium is exothermic so that a change of the transport direction results from different transport effective gas species during the transport of Fe_2O_3 with hydrogen chloride. Two equilibria are additionally important for the reaction process:

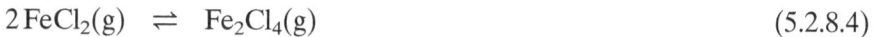

$$FeCl_3(g) \; \rightleftharpoons \; FeCl_2(g) + \frac{1}{2}Cl_2(g) \qquad (5.2.8.3)$$

$$2\,FeCl_2(g) \; \rightleftharpoons \; Fe_2Cl_4(g) \qquad (5.2.8.4)$$

This formation of iron(II)-ions causes the incorporation of Fe^{2+} in the Fe_2O_3 lattice (Klei 1970, Sch 1971). The amount of $FeCl_2$ formation increases with temperature, which is why considerably less Fe^{2+} is introduced during the exothermic transport of Fe_2O_3 from 300 to 400 °C than during the endothermic transport from 1000 to 800 °C. Other possibilities to avoid the incorporation of Fe^{2+} are the use of chlorine or hydrogen chloride + chlorine as transport agent (Sch 1971), or the use of hydrogen chloride + oxygen (Klei 1970).

Krabbes, Gerlach and *Reichelt* examined the change of the isotope relation $^{16}O/^{18}O$ during the transport process with the example of the transport of Fe_2O_3 with chlorine and traced the occuring change back to the influence of different diffusion coefficients of the isotopomer oxygen molecules. In order to ensure pure diffusive transport, the transport was realized under microgravitation in an orbit in space. Additionally, the transport agent concentration was varied. These results are published with model calculations as well as with results of mixed diffusive/convective gas movements under the influence of the Earth's gravitation as references (Kra 2005).

Besides the binary oxides, the transport of a series of ternary and quaternary oxides, which contain iron as a component, is known. Among others, this includes the extensive group of the ferrites whose transport is topic of many publications (see Table 5.1). Most of these compounds can be obtained by endothermic CVT via the chlorinating equilibrium. Furthermore, the transport of iron(III)/chromium(III)-oxide mixed-crystals and iron(III)/gallium(III)-oxide mixed-crystals is described with tellurium(IV)-chloride as the transport agent from 1000 to 900 °C (Pes 1984, Pat 2000c). Thermodynamic calculations show that the transport of iron(III)/chromium(III)-oxide mixed phases essentially takes place via $FeCl_3(g)$ and $CrO_2Cl_2(g)$ as metal-transferring gas species. All in all, 24 gas species, besides the solids, are necessary for an exact description of the system. Among others, this included the gas complexes $CrFe_2Cl_8$ and $CrFe_3Cl_{12}$ (Pes 1984). Iron(II, III)-arsenate $Fe_3^{II}Fe_4^{III}(AsO_4)_6$ as a mixed-valent iron compound with a complex anion, can be obtained by CVT with ammonium chloride from 900 to 800 °C (Wei 2004b).

Ruthenium oxides and osmium oxides   The transport of ruthenium oxido compounds is described for ruthenium(IV)-oxide, RuO_2, as well as for some ruthenium(IV)-metal(IV)-oxide mixed-crystals. Ruthenium(IV)-oxide can be endothermically transported with oxygen, chlorine, and tellurium(IV)-chloride. Oxygen can also be used as a transport agent for ruthenium(IV)-oxide in regard of the deposition of well formed RuO_2 crystals (Par 1982). The transport is based on the equilibria 2.5.8.2 and 2.5.8.3 (Sch 1963, Sch 1963a). According to the following equations, the auto transport of RuO_2 is also possible under its own decomposition pressure (Opp 2005).

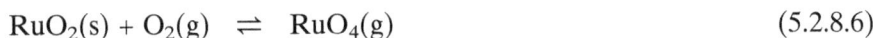

$$RuO_2(s) + \frac{1}{2}O_2(g) \;\rightleftharpoons\; RuO_3(g) \tag{5.2.8.5}$$

$$RuO_2(s) + O_2(g) \;\rightleftharpoons\; RuO_4(g) \tag{5.2.8.6}$$

If the transport takes place via the chlorinating equilibria with chlorine or tellurium(IV)-chloride as transport agents, the ruthenium-transferring species are $RuOCl$, $RuCl_3$, and $RuCl_4$ (Rei 1991). The transport of $Ru_{1-x}V_xO_2$ mixed-crystals with tellurium(IV)-chloride can be described this way, too, although the essential metal-transferring species are $VOCl_3$ and $RuOCl$. The partial pressures of the ruthenium chlorides $RuCl_3$ and $RuCl_4$ are at the limit of transport effectiveness. The partial pressures of the vanadium chlorides VCl_4, VCl_3, and VCl_2 as well as of the volatile ruthenium oxides RuO_3 and RuO_4 are not in the transport effective range (Kra 1986). Besides the $Ru_{1-x}V_xO_2$ mixed-phase, other mixed-phases, $Ru_{1-x}Ti_xO_2$, $Ru_{1-x}Mo_xO_2$, $Ru_{1-x}W_xO_2$, and $Ru_{1-x}Sn_xO_2$, can be obtained with chlorine or tellurium(IV)-chloride from 1100 to 1000 °C (Rei 1991, Kra 1991, Nic 1993). $Ru_{1-x}Ir_xO_2$ mixed-crystals are preparatively accessible over the gas phase when oxygen is added as the transport agent. This is due to the fact that iridium forms, analogous to ruthenium, volatile oxides of the composition $IrO_3(g)$ and $IrO_4(g)$ (Geo 1982, Tri 1982).

Osmium(IV)-oxide can be crystallized with oxygen as well as with tellurium(IV)-chloride as well as by auto transport. Transport with oxygen can be explained by the formation of gaseous oxides OsO_3 and OsO_4 (Opp 1998). Well formed osmium(IV)-oxide crystals are obtained, for example, in the temperature gradient from 930 to 900 °C (Yen 2004). The following transport equation is assumed:

$$OsO_2(s) + \frac{1}{2}O_2(g) \rightleftharpoons OsO_3(g) \tag{5.2.8.7}$$

Osmium(IV)-oxide can be deposited without the addition of a transport agent in the gradient from 900 to 600 °C and thus belongs to the few examples of an auto transport of oxides. Osmium(IV)-oxide has an oxygen decomposition pressure of approx. 1 bar at 1400 °C. However, the total pressure over osmium(IV)-oxide already achieves 1 bar at approx. 900 °C: that is why, OsO_4 and OsO_3 appear in the vapor over osmium(IV)-oxide; here the OsO_4 partial pressure is one and a half magnitudes higher than that of OsO_3. This thermal decomposition provides the transport agent OsO_4, which allows the auto transport according to the following equilibrium (Opp 2005):

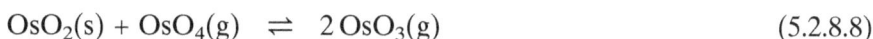

$$OsO_2(s) + OsO_4(g) \rightleftharpoons 2\,OsO_3(g) \tag{5.2.8.8}$$

The transport of osmium(IV)-oxide with tellurium(IV)-chloride is possible from T_2 to T_1 (Opp 1975). In this process, especially $OsCl_4$ as well as $OsCl_3$, $OsCl_2$, $OsCl$, OsO_2Cl_2, and $OsOCl_4$ can be assumed as osmium-transferring species (Hun 1986). The following transport equation describes the process in approximation:

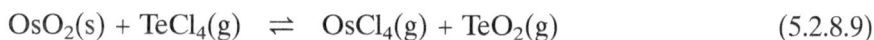

$$OsO_2(s) + TeCl_4(g) \rightleftharpoons OsCl_4(g) + TeO_2(g) \tag{5.2.8.9}$$

The highly volatile osmium(VIII)-oxide, OsO_4, can be obtained by sublimation at 35 to maximum 40 °C. Besides the transport of osmium(IV)-oxide, the CVT of only one more osmium oxido compound is published in the literature. This is the transport of a vanadium(IV)/osmium(IV) mixed-phase from 900 to 800 °C with mercury(II)-chloride as the transport effective additive, as published in the literature (Arn 2008).

5.2.9 Group 9

The transport reaction of numerous oxido compounds of the elements of group 9, cobalt, rhodium, and iridium is described. The binary cobalt oxides, cobalt(II)-oxide (Kle 1966a, Emm 1968c) and cobalt(II, III)-oxide, Co_3O_4, (Kle 1963, Emm 1968c, Klei 1972a, Tar 1984, Pat 2000b) can be transported endothermically with chlorine or hydrogen chloride. The transport equilibria can be formulated as follows.

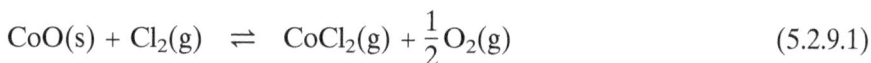

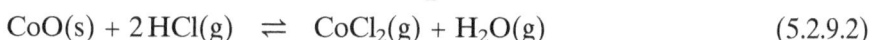

$$CoO(s) + Cl_2(g) \rightleftharpoons CoCl_2(g) + \frac{1}{2}O_2(g) \tag{5.2.9.1}$$

$$CoO(s) + 2\,HCl(g) \rightleftharpoons CoCl_2(g) + H_2O(g) \tag{5.2.9.2}$$

$$Co_3O_4(s) + 3\,Cl_2(g) \; \rightleftharpoons \; 3\,CoCl_2(g) + 2\,O_2(g) \tag{5.2.9.3}$$

$$Co_3O_4(s) + 6\,HCl(g) \; \rightleftharpoons \; 3\,CoCl_2(g) + 3\,H_2O(g) + \frac{1}{2}O_2(g) \tag{5.2.9.4}$$

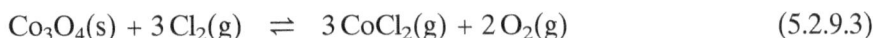

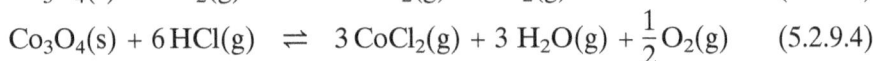

Cobalt(II)-chloride is the transport effective cobalt-containing gas species for all cobalt-containing oxido compounds. Adjacent to it, Co_2Cl_4 and $CoCl_3$ appear with considerably lower partial pressures.

Ternary cobalt-containing oxido compounds, whose transport is described and partially analyzed in detail with the help of extensive thermodynamic calculations, are compiled in Table 5.1. They are transported endothermically via chlorinating equilibria with the transport additives chlorine, mercury(II)-chloride, hydrogen chloride, ammonium chloride, or tellurium(IV)-chloride. Many of these compounds are spinels or mixed-spinels. During their transport, the transport agents Cl_2, HCl, or $TeCl_4$ lead to well formed crystals. If mercury(II)-chloride and ammonium chloride are used, well formed crystals appear as well; however, they have low transport rates due to the unfavorable equilibrium position. Besides the chlorine-containing transport agents, bromine and ammonium bromide can be used for molybdates $Co_2Mo_3O_8$ and $Co_{2-x}Zn_xMo_3O_8$ (Ste 2004a, Ste 2005), respectively. Tellurium(IV)-chloride (Str 1981) and hydrogen fluoride (Schm 1964a) can be used as transport additives for cobalt(II)-silicate, Co_2SiO_4. During the transport of ternary or quaternary cobalt-containing oxido compounds, one has also to consider the existence of gas complexes, which can appear for example as $CoFe_2Cl_8$, $CoFeCl_5$, $CoNiCl_4$, $CoGa_2Cl_8$, or $CoGaCl_5$, in order to describe the CVT (see section 11.1).

The transport reactions of $CoNb_2O_6$ with chlorine, mercury(II)-chloride, or ammonium chloride are examples of the transport of ternary oxides, which combine the characteristics of the transport of the respective binary oxide CoO and Nb_2O_5. Transport from 1020 to 960 °C can be described by the equilibria 5.2.9.5 to 5.2.9.7:

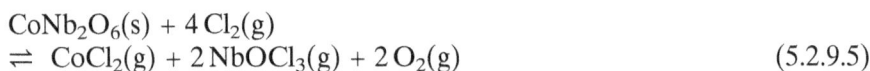

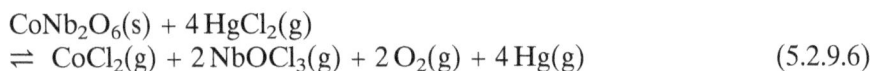

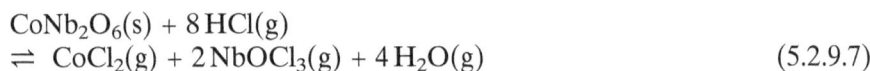

$$CoNb_2O_6(s) + 4\,Cl_2(g)$$
$$\rightleftharpoons CoCl_2(g) + 2\,NbOCl_3(g) + 2\,O_2(g) \tag{5.2.9.5}$$

$$CoNb_2O_6(s) + 4\,HgCl_2(g)$$
$$\rightleftharpoons CoCl_2(g) + 2\,NbOCl_3(g) + 2\,O_2(g) + 4\,Hg(g) \tag{5.2.9.6}$$

$$CoNb_2O_6(s) + 8\,HCl(g)$$
$$\rightleftharpoons CoCl_2(g) + 2\,NbOCl_3(g) + 4\,H_2O(g) \tag{5.2.9.7}$$

In contrast to cobalt, niobium as an element of group 5 forms volatile oxide chlorides so that $NbOCl_3$ is the transport-effective species for niobium. The transport effective oxygen-containing species are elemental oxygen or water besides $NbOCl_3$ (Ros 1992). Further examples of the transport of cobalt-containing oxido compounds are cobalt(II)-diarsenate $Co_2As_2O_7$, with chlorine (Wei 2005) on the one hand and cobalt(II)-selenate(IV), $CoSeO_3$, with tellurium(IV)-chloride on the other.

Rhodium(III)-oxide, as the only binary oxide of rhodium, can be transported. The transport is endothermic with chlorine from 1050 to 950 °C and with hydrogen chloride from 1000 to 850 °C (Poe 1981, Goe 1996). The transport with chlorine is described by the following equilibrium:

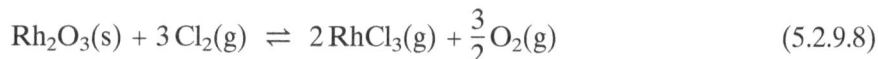

$$Rh_2O_3(s) + 3\,Cl_2(g) \;\rightleftharpoons\; 2\,RhCl_3(g) + \frac{3}{2}O_2(g) \tag{5.2.9.8}$$

Model calculations show that $RhCl_3$ is the essential rhodium-transferring gas species; the other gas species $RhCl_2$, $RhCl_4$, and $RhOCl_2$ are of minor importance (Goe 1996).

Rhodium(III)-arsenate(V) as well as rhodium(III)-niobate(V) and rhodium(III)-tantalate(V) can be deposited in crystalline form with chlorine as the transport agent. Furthermore, $RhVO_4$, $RhNbO_4$, and $RhTaO_4$ are transportable with tellurium(IV)-chloride (1000 → 900 °C). The spinel phase $CuRh_2O_4$ can be obtained as a further rhodium oxido compound by an endothermic CVT with chlorine as well as with tellurium(IV)-chloride (Jen 2009).

Iridium is chemically similar to ruthenium and can be transported like ruthenium and osmium via volatile oxides. Thus, iridium(IV)-oxide IrO_2 can be transported, analogous to ruthenium(IV)-oxide RuO_2, with oxygen, chlorine, and tellurium(IV)-chloride to the lower temperature zone. Transport with oxygen takes place from 1100–1200 to 1000–1050 °C by forming the volatile oxide IrO_3 (Sch 1960, Cor 1962, Geo 1982, Tri 1982, Opp 2005). The following transport equation describes the process:

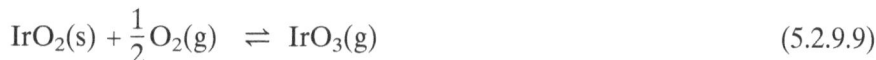

$$IrO_2(s) + \frac{1}{2}O_2(g) \;\rightleftharpoons\; IrO_3(g) \tag{5.2.9.9}$$

If tellurium(IV)-chloride is used as the transport agent, the endothermic transport from 1100 to 1000 °C can be described by the following equilibrium (Opp 1975):

$$IrO_2(s) + TeCl_4(g) \;\rightleftharpoons\; IrCl_4(g) + TeO_2(g) \tag{5.2.9.10}$$

Furthermore, IrO_2 can be obtained by auto transport from 1050 to 850 °C in well formed crystals. In doing so, part of the iridium(IV)-oxide decomposes incongruently at approximately 1050 °C to elemental iridium and oxygen with the formation of a two-phase source solid (IrO_2/Ir). According to equilibrium 5.2.9.9, the released oxygen reacts as a transport agent with the IrO_2 that remains in the solid, to form the transport effective gas species IrO_3. At 850 °C, the back reaction takes place; in the process IrO_2 is deposited.

5.2.10 Group 10

For nickel, palladium, and platinum, only the binary oxides NiO and PdO as well as a series of ternary and quaternary nickel oxido compounds can be transported. Nickel(II)-oxide can be transported with chlorine and bromine as well as with hydrogen chloride and hydrogen bromide from 1000 to 900 °C. The following transport equations can be formulated (Sto 1966, Emm 1968c):

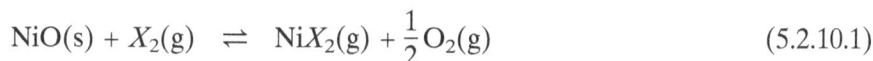

$$NiO(s) + X_2(g) \;\rightleftharpoons\; NiX_2(g) + \frac{1}{2}O_2(g) \tag{5.2.10.1}$$

$$NiO(s) + 2\,HX(g) \;\rightleftharpoons\; NiX_2(g) + H_2O(g) \qquad\qquad (5.2.10.2)$$
$$(X = Cl,\ Br)$$

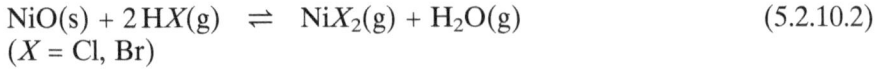

Beside the transport of the binary nickel(II)-oxide, the transports of nickel-containing mixed-crystals are described, for example the transport of $Ni_{1-x}Co_xO$. The binary oxides NiO and CoO are isotypic (Bow 1972). Transport takes place with hydrogen chloride as transport agent from 900 to 800 °C; the following transport equation is given:

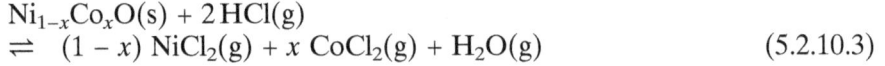

$$Ni_{1-x}Co_xO(s) + 2\,HCl(g)$$
$$\rightleftharpoons\; (1-x)\,NiCl_2(g) + x\,CoCl_2(g) + H_2O(g) \qquad (5.2.10.3)$$

Furthermore, the formation of mixed-crystals in the system NiO/ZnO has been examined. These oxides crystallize in different structure types, NiO in the rock-salt type and ZnO in the wurtzite type. Transport takes place with hydrogen chloride as transport agent from 900 to 750 °C. In the process, nickel oxide-rich as well as zinc oxide-rich mixed-crystals can be transported. Thermodynamic calculations show that $NiCl_2$, Ni_2Cl_4 as well as the gas complex $ZnNiCl_4$ are the nickel-transferring gas species and $ZnCl_2$, $ZnNiCl_4$ as well as Zn_2Cl_4 are the zinc-transferring gas species (Loc 1999a).

The transport of nickel-containing spinels takes place endothermically with chlorine or hydrogen chloride as transport agent, e. g., $Ni_{0.5}Zn_{0.5}Fe_2O_4$ (Klei 1973), $NiCr_2O_4$ (Emm 1968b), $NiFe_2O_4$ (Klei 1964, Bli 1977, Pes 1978), and $NiGa_2O_4$ (Paj 1990). Hence, the transport of $NiGa_2O_4$ with chlorine as the transport agent can be described by the following transport equation:

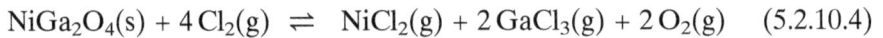

$$NiGa_2O_4(s) + 4\,Cl_2(g) \;\rightleftharpoons\; NiCl_2(g) + 2\,GaCl_3(g) + 2\,O_2(g) \quad (5.2.10.4)$$

Further equilibria determine the composition of the gas phase:

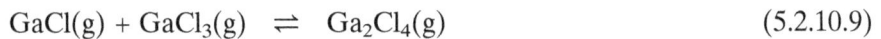

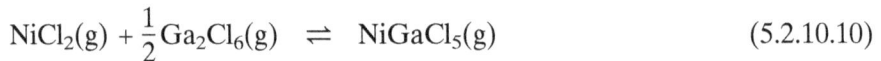

$$2\,NiCl_2(g) \;\rightleftharpoons\; Ni_2Cl_4(g) \qquad\qquad\qquad (5.2.10.5)$$

$$2\,GaCl_3(g) \;\rightleftharpoons\; Ga_2Cl_6(g) \qquad\qquad\qquad (5.2.10.6)$$

$$GaCl_3(g) \;\rightleftharpoons\; GaCl(g) + Cl_2(g) \qquad\qquad (5.2.10.7)$$

$$2\,GaCl(g) \;\rightleftharpoons\; Ga_2Cl_2(g) \qquad\qquad\qquad (5.2.10.8)$$

$$GaCl(g) + GaCl_3(g) \;\rightleftharpoons\; Ga_2Cl_4(g) \qquad\qquad (5.2.10.9)$$

$$NiCl_2(g) + \frac{1}{2}Ga_2Cl_6(g) \;\rightleftharpoons\; NiGaCl_5(g) \qquad\qquad (5.2.10.10)$$

$$NiCl_2(g) + 2\,GaCl_3(g) \;\rightleftharpoons\; NiGa_2Cl_8(g) \qquad\qquad (5.2.10.11)$$

All of the partial pressures of the gas species that appear in the mentioned equilibria are in the transport relevant range except for GaCl. However, the essential, nickel-transferring and gallium-transferring gas species are $NiCl_2$ and $GaCl_3$ respectively. This example shows that the transport of ternary or complex compounds is essentially defined by the characteristics of the transport of the respective binary oxides on the one hand. During the transport of ternary or complex compounds, on the other hand, one has to consider that ternary compounds are generally more stable in thermodynamic terms than the totality of the binary oxides from which they are composed. Also, the formation of gas complexes is possible, which influences the equilibrium position. Thus, the transport of ternary and complex oxides cannot only be considered as simultaneous transport of the respective binary oxides. This applies if the complex oxide is provided as the source solid, and also if stoichiometric mixtures of the binary oxides are used. According to the solid/solid and solid/gaseous equilibria a one- or multi-phase solid forms, which is thermodynamically stable at these temperatures and the respective partial pressures of the species develop that determine the transport of the compounds. Different gas phase solubilities of the binary oxides can lead to enrichment or depletion processes during the transport. This phenomenon applies especially during the transport of mixed-phases (Kra 1979, Kra 1984, Kra 1991).

Nickel(II)-niobate(V) can be transported with chlorine as well as with ammonium chloride from T_2 to T_1. Thermodynamic model calculations show the transport via endothermic equilibria:

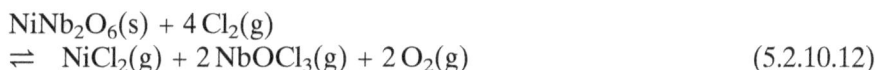

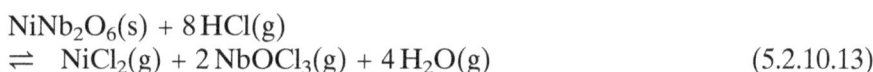

$$\begin{aligned} &\text{NiNb}_2\text{O}_6(s) + 4\,\text{Cl}_2(g) \\ &\rightleftharpoons \ \text{NiCl}_2(g) + 2\,\text{NbOCl}_3(g) + 2\,\text{O}_2(g) \end{aligned} \qquad (5.2.10.12)$$

$$\begin{aligned} &\text{NiNb}_2\text{O}_6(s) + 8\,\text{HCl}(g) \\ &\rightleftharpoons \ \text{NiCl}_2(g) + 2\,\text{NbOCl}_3(g) + 4\,\text{H}_2\text{O}(g) \end{aligned} \qquad (5.2.10.13)$$

The partial pressures of Ni_2Cl_4 and NbCl_5 are at the border of transport efficiency (Ros 1992b). Besides nickel(II)-niobate(V), nickel(II)-tantalate(V) can be obtained with tantalum(V)-chloride/chlorine from 1000 to 900 °C by CVT.

Examination of the CVT in the ternary system Ni/Mo/O shows that the compounds NiMoO_4 and $\text{Ni}_2\text{Mo}_3\text{O}_8$ can be deposited with chlorine and bromine as transport agent (Ste 2006a). The following transport equation was given:

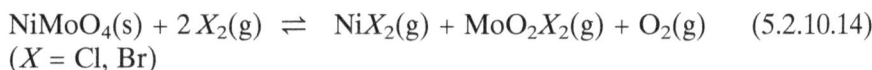

$$\text{NiMoO}_4(s) + 2\,X_2(g) \ \rightleftharpoons \ \text{Ni}X_2(g) + \text{MoO}_2X_2(g) + \text{O}_2(g) \qquad (5.2.10.14)$$
$$(X = \text{Cl, Br})$$

Thermodynamic model calculations prove that hydrogen chloride, formed by traces of moisture, plays a decisive role for the transport of NiMoO_4, and especially for that of $\text{Ni}_2\text{Mo}_3\text{O}_8$:

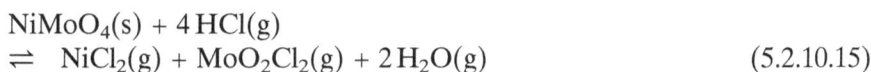

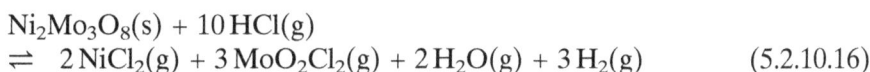

$$\begin{aligned} &\text{NiMoO}_4(s) + 4\,\text{HCl}(g) \\ &\rightleftharpoons \ \text{NiCl}_2(g) + \text{MoO}_2\text{Cl}_2(g) + 2\,\text{H}_2\text{O}(g) \end{aligned} \qquad (5.2.10.15)$$

$$\begin{aligned} &\text{Ni}_2\text{Mo}_3\text{O}_8(s) + 10\,\text{HCl}(g) \\ &\rightleftharpoons \ 2\,\text{NiCl}_2(g) + 3\,\text{MoO}_2\text{Cl}_2(g) + 2\,\text{H}_2\text{O}(g) + 3\,\text{H}_2(g) \end{aligned} \qquad (5.2.10.16)$$

In the Ni/Mo/O system, it is often favorable to start with a multi-phased solid, instead of a one-phase solid, that contains the target phase that is to be transported, and a co-existing compound. Nickel(II)-tungstate(VI), NiWO_4, can be transported with chlorine (Emm 1968b, Ste 2005a). Further nickel compounds that can be transported with chlorine are compiled in Table 5.1.

Besides chlorine and hydrogen chloride, tellurium(IV)-chloride functions as a transport agent in exceptional cases during the transport of nickel-containing

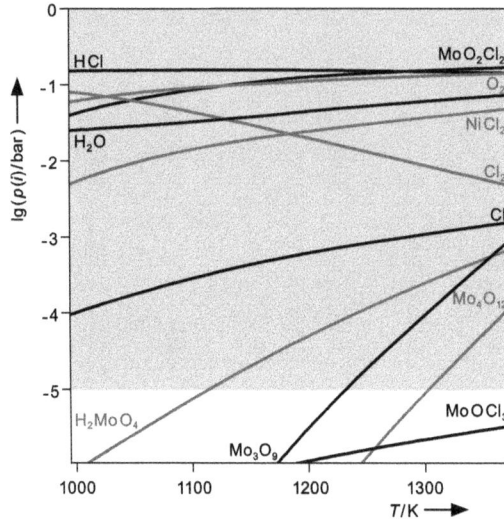

Figure 5.2.10.1 Gas-phase composition over $NiMoO_4$ in equilibrium with Cl_2 in the presence of water according to *Steiner* and *Reichelt* (Ste 2006a).

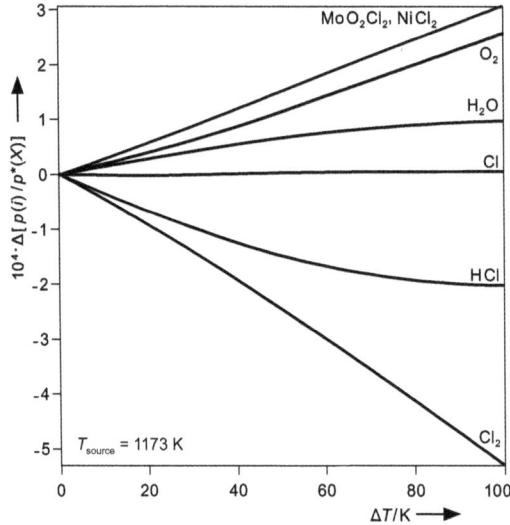

Figure 5.2.10.2 Transport efficiency of the essential gas species of the transport of $NiMoO_4$ with Cl_2 in the presence of water according to S*teiner* and *Reichelt* (Ste 2006a).

spinels. The transport of $NiFe_2O_4$ with tellurium(IV)-chloride from 980 to 880 °C is well described by equation 5.2.10.17 (Pes 1978):

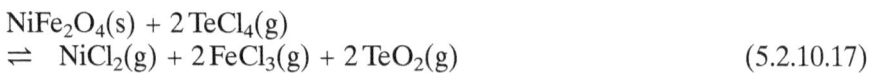

$$NiFe_2O_4(s) + 2\,TeCl_4(g)$$
$$\rightleftharpoons\; NiCl_2(g) + 2\,FeCl_3(g) + 2\,TeO_2(g) \qquad (5.2.10.17)$$

Another ternary nickel oxido compound, whose transport behavior with selenium(IV)-chloride was described and thermodynamically examined in detail, is $NiTiO_3$ (Bie 1990). Selenium(II)-chloride is the transport agent:

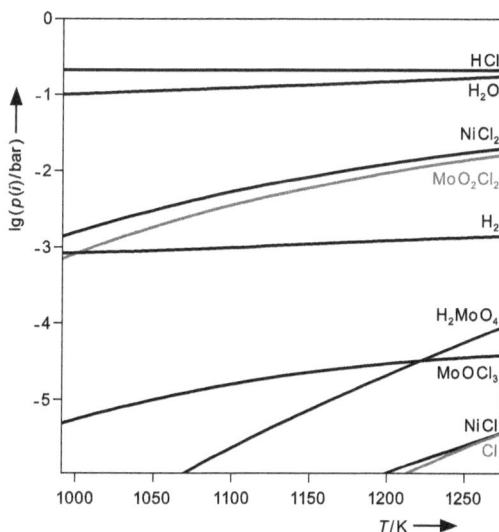

Figure 5.2.10.3 Gas-phase composition over $Ni_2Mo_3O_8/MoO_2$ in equilibrium with Cl_2 in the presence of water according to *Steiner* and *Reichelt* (Ste 2006a).

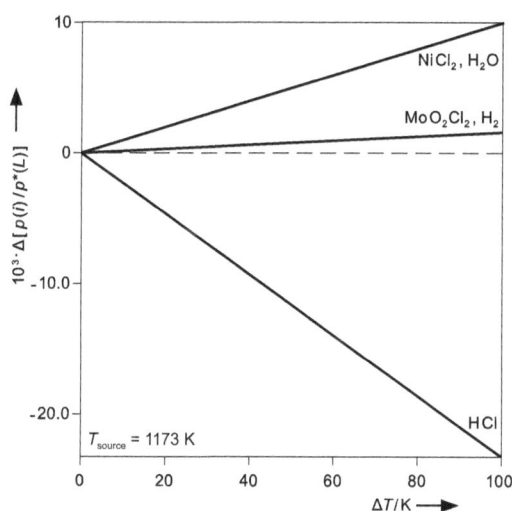

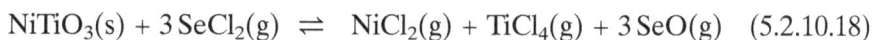

Figure 5.2.10.4 Transport efficiency of the essential gas species of the transport of $Ni_2Mo_3O_8$ with Cl_2 in the presence of water according to *Steiner* and *Reichelt* (Ste 2006a).

$$NiTiO_3(s) + 3\,SeCl_2(g) \;\rightleftharpoons\; NiCl_2(g) + TiCl_4(g) + 3\,SeO(g) \quad (5.2.10.18)$$

In order to describe the thermodynamic situation during the transport of $NiTiO_3$ with selenium(IV)-chloride completely, altogether 18 gas species have to be considered besides the solid phases $NiTiO_3$, NiO, TiO_2, Ni, and $NiCl_2$. The vast majority of the respective partial pressure are in the relevant range for transport reactions. Nickel(II)-orthosilicate, Ni_2SiO_4, can be transported with SiF_4. Based on a NiO-solid that was provided in a platinum crucible in a silica ampoule that simultaneously serves as SiO_2-source, Ni_2SiO_4 is formed in the temperature gradient from 1200 to 1030 °C. In this process, the formation of Ni_2SiO_4-crystals can be observed at T_1 as well as at T_2 in the platinum crucible. It can be concluded from this that the nickel-containing part is transported endothermically while the silicon-containing part migrates exothermically. The transport of nickel and silicon via the gas phase with different transport directions can be described as an exchange reaction.

$$2\,NiO(s) + SiF_4(g) \; \rightleftharpoons \; 2\,NiF_2(g) + SiO_2(s) \qquad (5.2.10.19)$$

Due to the fact that in this case there is no oxygen transport via the gas phase, one can only speak of a "partial transport". Traces of water also contribute to the explanation of the crystallization of Ni_2SiO_4 via the gas phase (Hof 1977a). The transport agent hydrogen fluoride, which was formed by hydrolysis, allows the transport of endothermic nickel oxide as well as the transport of exothermic silicon(IV)-oxide according to the following equilibria:

$$NiO(s) + 2\,HF(g) \; \rightleftharpoons \; NiF_2(g) + H_2O(g) \qquad (5.2.10.20)$$

$$SiO_2(g) + 4\,HF(g) \; \rightleftharpoons \; SiF_4(g) + 2\,H_2O(g) \qquad (5.2.10.21)$$

Also, the formation of Ni_2SiO_4 is possible in a medium temperature zone in the presence of traces of water:

$$2\,NiF_2(g) + SiF_4(g) + 4\,H_2O(g) \; \rightleftharpoons \; Ni_2SiO_4(s) + 8\,HF(g) \qquad (5.2.10.22)$$

Only the transport of one oxido compound of palladium is described, that of palladium(II)-oxide. The exothermic transport succeeds with chlorine as transport agent from 800 to 900 °C (Rog 1971).

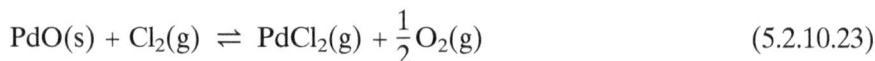

$$PdO(s) + Cl_2(g) \; \rightleftharpoons \; PdCl_2(g) + \frac{1}{2}O_2(g) \qquad (5.2.10.23)$$

5.2.11 Group 11

Regarding the elements copper, silver, and gold, only the CVT of the two binary copper oxides CuO and Cu_2O as well as a series of ternary copper oxido compounds has been published. The binary oxides of the elements silver and gold cannot be obtained by CVT due to their low thermal stability – Ag_2O decomposes at 420 °C, Au_2O_3 already at 310 °C – additionally the partial pressure of the halides are low at these temperatures. Silver and gold halides are the prime metal-transferring species. Volatile oxides or oxide halides are not formed.

Copper(II)-oxide can be transported essentially via the chlorinating equilibria when Cl_2 (Kle 1970, Bal 1985), $HgCl_2$ (Bal 1985), HCl (Sch 1976, Kra 1977, Des 1998), and NH_4Cl (Bal 1985) as well as $TeCl_4$ (Des 1989) and CuCl (Bal 1985) are added. Also, I_2 (Bal 1985), HI (Kle 1970), and HBr (Kle 1970) can be used as transport agents. All reactions occurs endothermically; T_2 is in the range from 800 to 1000 °C and T_1 between 700 and 900 °C. The peculiarity during the transport of Cu(II)-compounds is that no copper(II)-halides appear in the gas phase.

Only copper(I)-halides are formed, which predominantly are monomeric and/or trimeric. The CVT of copper(II)-oxide with chlorine and with mercury(II)-chloride can be formulated with the following equilibria:

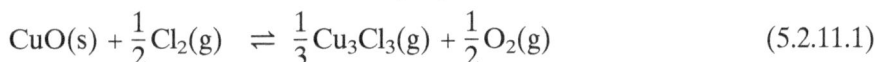

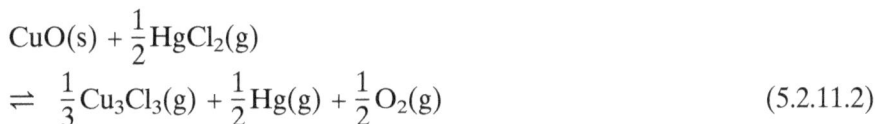

$$CuO(s) + \frac{1}{2}Cl_2(g) \;\rightleftharpoons\; \frac{1}{3}Cu_3Cl_3(g) + \frac{1}{2}O_2(g) \qquad (5.2.11.1)$$

$$CuO(s) + \frac{1}{2}HgCl_2(g)$$
$$\rightleftharpoons \frac{1}{3}Cu_3Cl_3(g) + \frac{1}{2}Hg(g) + \frac{1}{2}O_2(g) \qquad (5.2.11.2)$$

If hydrogen halides or ammonium halides are used as transport additives, the transport equation is as follows:

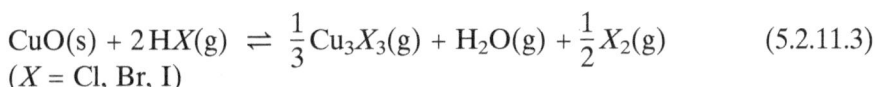

$$CuO(s) + 2\,HX(g) \;\rightleftharpoons\; \frac{1}{3}Cu_3X_3(g) + H_2O(g) + \frac{1}{2}X_2(g) \qquad (5.2.11.3)$$
$$(X = Cl, Br, I)$$

The transport rate with mercury(II)-chloride is the highest. Compared to this, the rate with chlorine, hydrogen halides, or ammonium halides, as well as with copper(I)-chloride, are relatively low (Bal 1985). In this process, the trimeric copper(I)-halides, Cu_3X_3, (X = Cl, Br, I), are the essential copper-transferring species. The monomer CuX and the tetramer Cu_4X_4 are clearly less effective than the trimer as far as the transport efficiency is concerned. However, they always have to be considered in thermodynamic calculations because their partial pressures are, at least in chlorine-containing systems, above 10^{-5} bar. The according transport reactions are always connected to the simultaneous redox reaction because the transfer of copper always takes place as copper(I)-species. This is particularly important when ternary or higher copper oxido compounds are to be transported, which have high oxygen partial pressures. These can counteract the reduction.

The exothermic transport of one-phased copper(I)-oxide is possible only if hydrogen chloride or ammonium chloride is used as the transport additive in a wide temperature range from 650–900 °C to 730–1000 °C (Sch 1957, Kra 1977, Bal 1985, Mar 1999). If hydrogen bromide and particularly hydrogen iodide are used as transport agents, elemental copper is deposited adjacent to Cu_2O (Kra 1977). The following characteristic transport equation can be formulated:

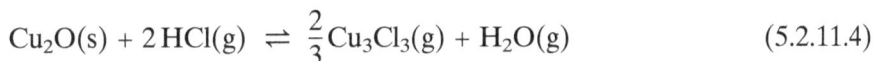

$$Cu_2O(s) + 2\,HCl(g) \;\rightleftharpoons\; \frac{2}{3}Cu_3Cl_3(g) + H_2O(g) \qquad (5.2.11.4)$$

Additionally transport effective partial pressures are achieved by the gas species Cu_4Cl_4 and $CuCl$ in the given temperature range. Due to the low saturation vapor pressure of liquid CuCl, the condensation of the chlorinating transport agents is often observed in the lower temperature zone, which can already appear at low transport agent pressure (e. g., 0.04 bar at 850 °C) (Kra 1977).

Krabbes and *Oppermann* (Kra 1977) as well as *Bald* and *Gruehn* (Bal 1985) have published results on the transport behavior of the binary system copper/oxygen. These deal mainly with thermodynamic aspects. Furthermore, works have been published that deal with the CVT of mixed-crystals in the system

CuO/ZnO as well as of copper oxides with complex anions. In the CuO/ZnO system the binary oxides crystallize with different structures. The copper oxide-rich and zinc oxide-rich mixed-crystals can be deposited by CVT with chlorine from T_2 to T_1. The transport effective gas species essentially are $ZnCl_2$, Cu_3Cl_3, Cu_4Cl_4, and O_2 (Loc 1999c).

Extensive examinations are known for the copper/molybdenum/oxygen system (Ste 1996). The copper(II)-molybdates $CuMoO_4$ and $Cu_3Mo_2O_9$ can be transported endothermically with the transport agents chlorine, bromine, hydrogen chloride, and tellurium(IV)-chloride. Thermodynamic model calculations show that MoO_2Cl_2 is the predominant, molybdenum-transferring species when chlorinating transport agents are used. Primarily, the copper-transferring species are Cu_3Cl_3 and Cu_4Cl_4, the oxygen-transferring species are elemental oxygen besides MoO_2Cl_2 as well as H_2O, TeO_2, and $TeOCl_2$ depending on the transport agent. Apart from copper(II)-molybdate, $CuMoO_4$, a mixed phase $Cu_{1-x}Zn_xMoO_4$ can be obtained by CVT with chlorine and bromine as well as with ammonium chloride (Rei 2000, Ste 2003). The essential transport effective gas species are MoO_2X_2, Cu_3X_3, and ZnX_2 (X = Cl, Br). Copper(II)-tungstate, $CuWO_4$, can be transported endothermically with chlorine and tellurium(IV)-chloride (Yu 1993, Ste 2005a). Besides the mentioned ternary and quaternary copper oxide compounds, $CuSb_2O_6$ (Pro 2003), $Cu_2V_2O_7$ (Pro 2001), and $CuTe_2O_5$ as well as Cu_2SeO_4 (Meu 1976), $CuSeO_3$ (Meu 1976), and $CuSe_2O_5$ (Jan 2009) can be transported to the lower temperature zone with tellurium(IV)-chloride. Transport of the compound $CuSe_2O_5$ with tellurium(IV)-chloride as transport agent can be described in good approximation from 380 to 280 °C.

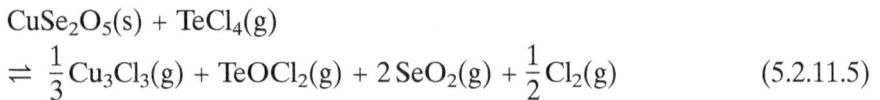

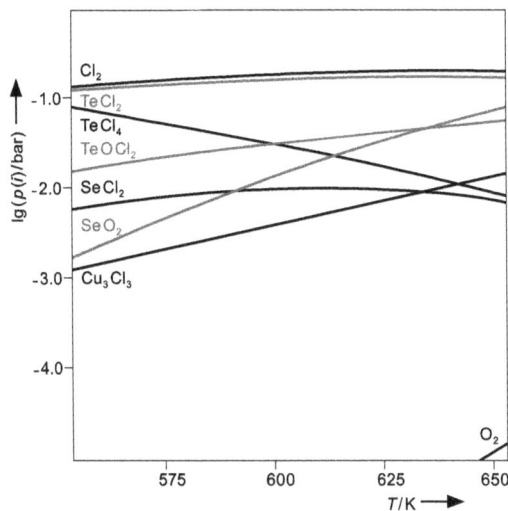

$$CuSe_2O_5(s) + TeCl_4(g)$$
$$\rightleftharpoons \frac{1}{3}Cu_3Cl_3(g) + TeOCl_2(g) + 2\,SeO_2(g) + \frac{1}{2}Cl_2(g) \qquad (5.2.11.5)$$

Figure 5.2.11.1 Gas-phase composition over $CuSe_2O_5$ in equilibrium with $TeCl_4$.

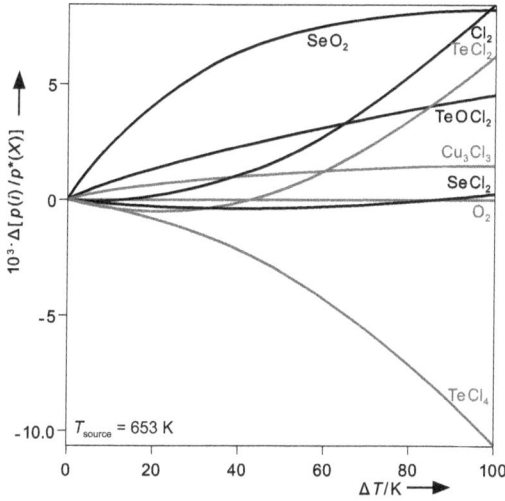

Figure 5.2.11.2 Transport efficiency of the essential gas species in the transport of $CuSe_2O_5$ with $TeCl_4$.

Here, the transport effective gas species is also Cu_3Cl_3 as far as copper is concerned. In the given temperature range, the oxygen-transferring gas species are $TeOCl_2$ and SeO_2.

The copper(II)-compounds $Cu_2As_2O_7$, $CuSb_2O_6$, and $CuTe_2O_5$ can be transported with chlorine. A dominating transport equation can be formulated:

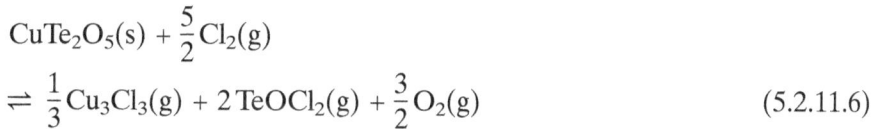

$$CuTe_2O_5(s) + \frac{5}{2}Cl_2(g)$$

$$\rightleftharpoons \frac{1}{3}Cu_3Cl_3(g) + 2\,TeOCl_2(g) + \frac{3}{2}O_2(g) \qquad (5.2.11.6)$$

Furthermore, $CuTe_2O_5$ can be transported with $TeCl_4$ with high transport rates from 590 to 490 °C. Due to the fact that only copper(I)-halides play a role as copper-transferring gas species, it is often problematic or not possible to crystallize ternary or multinary copper(II)-oxido compounds, respectively, by CVT reactions at high temperatures around 1000 °C and with low oxygen partial pressures in the system. Another problem is the formation of stable solid quaternary oxide halides such as in the copper/niobium/oxygen/chlorine system, which bind the transport agent in the solid phase and thus disrupt the transport.

5.2.12 Group 12

The CVT of numerous oxido compounds of the elements of group 12 is described in the literature. Zinc oxide and cadmium oxide can be transported in contrast to mercury(II)-oxide. With regard to mercury, however, the transport of the two ternary compounds $HgAs_2O_6$ (Wei 2000) and $(Hg_2)_2As_2O_7$ (Wei 2003) is known.

ZnO, as the only existing zinc oxide, and can be transported endothermically with many transport agents at temperatures from 1000–1100 °C to 800–1000 °C. Thus, ZnO can be transported with the halogens chlorine (Kle 1966b, Opp 1984, Pat 1999), bromine (Wid 1971, Opp 1984), and iodine (Sch 1966), and the hydrogen halides hydrogen chloride (Kle 1966a, Opp 1984), hydrogen bromide (Opp 1984), and hydrogen iodide (Mat 1988) when the relevant ammonium halides are added. Other transport additives that are suitable for the transport of zinc oxide are hydrogen (Shi 1971), carbon (Pal 2006), carbon monoxide (Pal 2006), water (Gle 1957, Pal 2006), and ammonia (Shi 1971) as well as mercury(II)-chloride (Sch 1972a, Shi 1973) and zinc(II)-chloride (Mat 1985, Mat 1988). Zinc oxide has a small but clear phase range; the choice of the transport agent is decisive for directed depositions at the upper or lower phase boundary (Opp 1985). In order to deposit ZnO_{1-x} at the lower phase boundary, hydrogen bromide and ammonium bromide are particularly suitable. The deposition at the upper phase boundary is possible with the transport agents chlorine and bromine. The following transport effective equilibria for the different transport agents can be formulated ($X = $ Cl, Br, I) (see section 2.4):

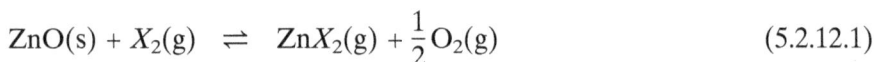

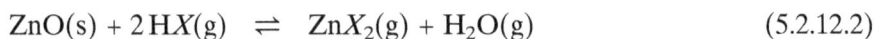

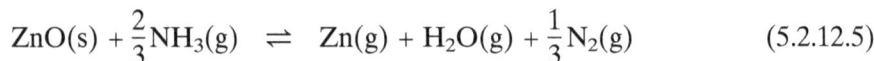

$$ZnO(s) + X_2(g) \rightleftharpoons ZnX_2(g) + \frac{1}{2}O_2(g) \qquad (5.2.12.1)$$

$$ZnO(s) + 2\,HX(g) \rightleftharpoons ZnX_2(g) + H_2O(g) \qquad (5.2.12.2)$$

$$ZnO(s) + H_2(g) \rightleftharpoons Zn(g) + H_2O(g) \qquad (5.2.12.3)$$

$$ZnO(s) + CO(g) \rightleftharpoons Zn(g) + CO_2(g) \qquad (5.2.12.4)$$

$$ZnO(s) + \frac{2}{3}NH_3(g) \rightleftharpoons Zn(g) + H_2O(g) + \frac{1}{3}N_2(g) \qquad (5.2.12.5)$$

Solid solutions of iron(II)-oxide, manganese(II)-oxide, and cobalt(II)-oxide in zinc(II)-oxide can also be transported endothermically with hydrogen chloride or ammonium chloride (Loc 1999b, Kra 1984). Furthermore, there are numbers of ternary and quaternary zinc oxido compounds that can be obtained by endothermic CVT via halogenating equilibria (see Table 5.1).

The transfer of zinc takes place via $ZnCl_2$ and Zn_2Cl_4 in all cases. The specific transport behavior of the individual compounds is not only determined by zinc but also by the other metal atoms. In order to describe the gas phase exactly, the formation of gas complexes, e.g., $FeZnCl_4$ has to be considered (see Chapter 11). The influence of different transport agents on the transport behavior in the ternary $Zn/Mo/O$ systems is discussed with regard to thermodynamics (Söh 1997).

Cadmium oxide can be deposited with bromine as well as with iodine according to the following transport equation (Emm 1978c):

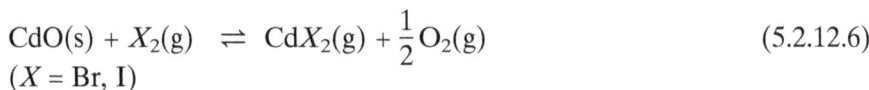

$$CdO(s) + X_2(g) \rightleftharpoons CdX_2(g) + \frac{1}{2}O_2(g) \qquad (5.2.12.6)$$
$$(X = \text{Br, I})$$

Cadmium niobate(V), $CdNb_2O_6$, can be transported with mercury(II)-chloride and mercury(II)-bromide as well as with hydrogen halides when ammonium chloride and ammonium bromide, respectively, are added (Kru 1987). Also, the cadmium molybdate $CdMoO_4$ can be transported with chlorine, bromine, and iodine, and $Cd_2Mo_3O_8$ with chlorine and bromine from T_2 to T_1 (Ste 2000). The cadmium(II)/iron(III)-spinel $CdFe_2O_4$ can be endothermically transported with hydrogen chloride. The compounds $CdAs_2O_6$ (Wei 2001), $Cd_2As_2O_7$ (Wei 2001a), $CdTe_2O_5$, and $CdWO_4$ (Ste 2005a) can be obtained by endothermic transport with chlorine. $CdTe_2O_5$ can also be transported with tellurium(IV)-chloride from 600 to 500 °C. Thermodynamic model calculations show that the respective cadmium(II)-halides are the transport-effective species. The cadmium(I)-halides, which were considered in the calculations as well, do not achieve transport effective partial pressures.

The cadmium molybdates are exemplary for the transport of ternary cadmium oxido compounds. *Steiner* (Ste 2000) examined and described the transport behavior of $CdMoO_4$ and $Cd_2Mo_3O_8$ with the halogens chlorine, bromine, and iodine as well as with the respective hydrogen halides. The observed transport behavior is compared to the results of the thermodynamic model calculations and discussed in detail. In doing so, the emphasis is on the interpretation of the model calculations with regard to the co-existence relations of the solid phases, the gas phase composition, and the transport efficiency of the individual gas species. In particular, the meaning of the transport efficiency for the interpretation of the substance flows is in the focus of this work.

5.2.13 Group 13

The binary oxides of group 13, aluminum(III)-oxide, gallium(III)-oxide, and indium(III)-oxide, can be transported to the lower temperature. In particular transport with different transport agents is described for gallium(III)-oxide and indium(III)-oxide. Boron(III)-oxide and thallium(III)-oxide can be sublimed above 1200 °C (B_2O_3) and above 500 °C (Tl_2O_3) respectively. In the process, thallium(III)-oxide forms gaseous thallium(I)-oxide and oxygen in the vapor phase. Thallium(III)-oxide can be obtained in form of up to 2 mm sized crystals by sublimation in the oxygen flow at 900 °C (Sle 1970).

Binary oxides *Aluminum oxide* can be transported with very low transport rates at very high temperatures corresponding to the following chlorinating equilibria:

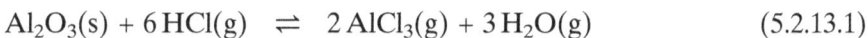

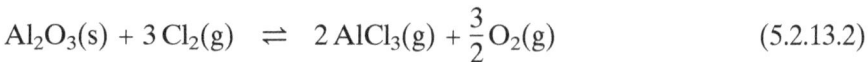

$$Al_2O_3(s) + 6\,HCl(g) \rightleftharpoons 2\,AlCl_3(g) + 3\,H_2O(g) \tag{5.2.13.1}$$

$$Al_2O_3(s) + 3\,Cl_2(g) \rightleftharpoons 2\,AlCl_3(g) + \frac{3}{2}O_2(g) \tag{5.2.13.2}$$

The transport also succeeds by adding metal-fluorides (e.g., PbF_2) (Tsu 1966, Whi 1974) from T_2 to T_1. Furthermore, it is possible to deposit aluminum oxide, especially in the form of whiskers, with the help of moist hydrogen via the gas phase (Dev 1959, Sea 1963). Generally, temperatures in the range of 1700 to 2000 °C are required. The deposition occurs at T_1:

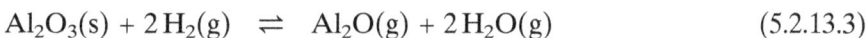

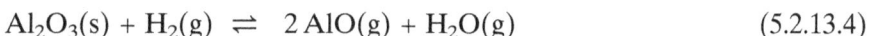

$$Al_2O_3(s) + 2\,H_2(g) \rightleftharpoons Al_2O(g) + 2\,H_2O(g) \tag{5.2.13.3}$$

$$Al_2O_3(s) + H_2(g) \rightleftharpoons 2\,AlO(g) + H_2O(g) \tag{5.2.13.4}$$

Apart from the sub-oxides, aluminum atoms were detected in the gas phase (Mar 1959). Due to the extreme conditions during the transport of aluminum oxide, other crystallization procedures are preferable (e.g., the *Verneuil* or *Czochralski* processes).The endothermic transport of *gallium(III)-oxide* is described with different chlorine- and iodine-containing transport agents. Often, the transport agent combinations sulfur + chlorine and sulfur + iodine, respectively, are used. By adding sulfur, $SO_2(g)$ forms as the oxygen-transferring species. This formation of SO_2 shifts the equilibrium position to the side of the reaction products. According to the thermodynamic calculations of *Juskowiak* (Jus 1988), the following transport equation can be derived for the transport of gallium(III)-oxide with sulfur and chlorine:

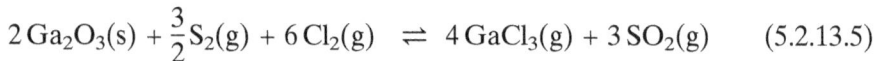

$$2\,Ga_2O_3(s) + \frac{3}{2}S_2(g) + 6\,Cl_2(g) \;\rightleftharpoons\; 4\,GaCl_3(g) + 3\,SO_2(g) \qquad (5.2.13.5)$$

At lower temperatures, gallium(III)-oxide shows an exothermic transport behavior with sulfur and chlorine as transport agent from 550 to 750 °C (Jus 1988). The following transport equation (Nit 1966, Aga 1985] is given for the transport of gallium(III)-oxide with sulfur and iodine (Nit 1966, Aga 1985):

$$2\,Ga_2O_3(s) + \frac{3}{2}S_2(g) + 6\,I_2(g) \;\rightleftharpoons\; 4\,GaI_3(g) + 3\,SO_2(g) \qquad (5.2.13.6)$$

Extensive examinations on the CVT of gallium(III)-oxide with tellurium(IV)-chloride have been published (Ger 1977a). Transport occurs endothermically in a temperature range from 750 to 900 °C at a temperature gradient of 100 K. The calculations of the gas-phase composition over solid gallium(III)-oxide, which is in equilibrium with tellurium(IV)-chloride, indicate that the gas species $GaCl_3$ and Ga_2Cl_6 are essential for the transfer of gallium; the existence of gaseous $GaCl_2$ is controversial. The oxygen transport occurs primarily via $TeOCl_2$ at low temperatures; at higher temperatures the amount of TeO_2 and O_2 increases. The following transport equation describes the process:

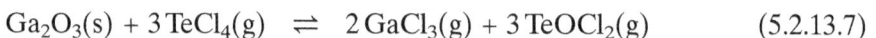

$$Ga_2O_3(s) + 3\,TeCl_4(g) \;\rightleftharpoons\; 2\,GaCl_3(g) + 3\,TeOCl_2(g) \qquad (5.2.13.7)$$

At higher temperatures, the transport essentially takes place via $GaCl_3$ and TeO_2.

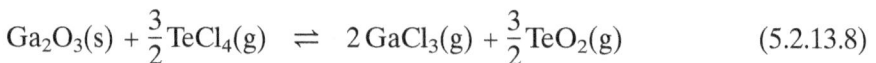

$$Ga_2O_3(s) + \frac{3}{2}TeCl_4(g) \;\rightleftharpoons\; 2\,GaCl_3(g) + \frac{3}{2}TeO_2(g) \qquad (5.2.13.8)$$

The given transport equations are simplified. An exact description is only possible if further equilibria are considered as well (Ger 1977a).

If tellurium(IV)-chloride is used, there is no reverse of the transport direction whereas this is possible in the case of sulfur + chlorine.

Indium(III)-oxide can be transported endothermically only with different halogenating transport agents, such as chlorine and hydrogen chloride, and with sulfur + iodine. *Werner* offers extensive thermodynamic calculations on the transport behavior of indium(III)-oxide with chlorine as well as with sulfur + iodine (Wer 1996). Thermodynamic calculations indicate that sulfur binds the oxygen with the formation of SO_2. The presence of iodine leads to the formation of different indium iodides. If chlorine is added during the transport, indium(III)-chloride is

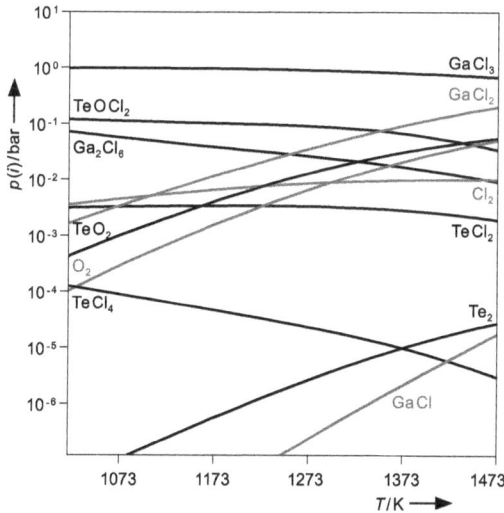

Figure 5.2.13.1 Gas-phase composition over Ga_2O_3 in equilibrium with $TeCl_4$ according to *Gerlach* and *Oppermann* (Ger 1977a).

discussed as the actual transport agent (Wer 1996). Like gallium(III)-oxide and indium(III)-oxide, the mixed phase $Ga_{2-x}In_xO_3$ can also be deposited by CVT reactions with high transport rates (Pat 2000).

Ternary compounds *Boron* forms many ternary and multinary oxido compounds. Among them, the CVT of $FeBO_3$, Fe_3BO_6, BPO_4, and $Cr_2BP_3O_{12}$ have been published (the transport of boracites is treated in section 8.1). The compounds $TiBO_3$, VBO_3, and $CrBO_3$ crystallize via the gas phase with the help of titanium(II)-iodide, vanadium(II)-chloride, and chromium(II)-chloride each in combination with water (Schm 1964). If chlorine or hydrogen chloride are used as transport agent, the transport of $FeBO_3$ will occur from 760 to 670 °C via the chlorinating equilibria with the formation of BCl_3 and $FeCl_3$, as well as O_2 and H_2O as gas species (Die 1975). Fe_3BO_6 can be transported with hydrogen chloride (Die 1976). The source solids $FeBO_3$ and Fe_3BO_6 were provided in a platinum crucible to minimize reactions with the silica tube. The deposition of the crystals is endothermic; the temperature of the crystallization zone is in the range of 800 to 905 °C. The following transport equation was given:

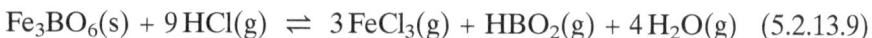

$$Fe_3BO_6(s) + 9\,HCl(g) \rightleftharpoons 3\,FeCl_3(g) + HBO_2(g) + 4\,H_2O(g) \quad (5.2.13.9)$$

The boron phosphate BPO_4 could be deposited via the following equilibrium when phosphorus(V)-chloride was added (Schm 2004):

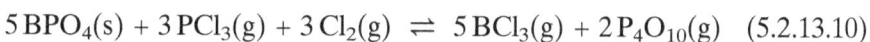

$$5\,BPO_4(s) + 3\,PCl_3(g) + 3\,Cl_2(g) \rightleftharpoons 5\,BCl_3(g) + 2\,P_4O_{10}(g) \quad (5.2.13.10)$$

The deposition partly took place on a target of glassy carbon, which was introduced into the ampoule in order to avoid the overgrowing of the boron phosphate on the silica tube. This way, especially, pure single-crystals were obtained for further examination.

The CVT of ternary oxides with *aluminum* as one component is only known by the examples of Al_2SiO_5, $Al_2Ge_2O_7$ and, $Al_2Ti_7O_{15}$. The transport of Al_2SiO_5 is endothermic with the help of Na_3AlF_6 (Nov 1966a). The similar endothermic transport of $Al_2Ge_2O_7$ with aluminum(III)-chloride as transport additive succeeds from 1000 to 900 °C. Experiments with the transport agent tellurium(IV)-chloride, hydrogen chloride, as well as sulfur + iodine were not successful (Aga 1985a). If vanadium(IV)-oxide is to be doped with aluminum via the gas phase only a small amount of aluminum is used when tellurium(IV)-chloride is used as transport agent ($V_{1-x}Al_xO_2$; $x \approx 0.007$) (Brü 1976).

A series of examples of ternary oxido compounds that contain *gallium* and transition metal atoms is known from the literature. They can be endothermically transported via the chlorinating equilibrium in temperature ranges between 1000 and 800 °C with elemental chlorine or hydrogen chloride as transport agents. Equilibrium calculations show that gallium is transferred essentially as gallium(III)-chloride (Lec 1991, Pat 1999). The dimer $Ga_2Cl_6(g)$ only plays a minor role. The GaCl-partial pressure is not in the transport relevant range even at high temperatures above 1000 °C. The transfer of the transition metal usually takes place as dichloride. In the case of iron, the dimer Fe_2Cl_4 becomes transport effective next to the monomer $FeCl_2$. Furthermore, calculations (Lec 1991) indicate that the gas phase has a considerable amount of $FeCl_3$ over solid $FeGa_2O_4$ that is in equilibrium with chlorine. Equilibrium calculations in the $ZnO/Ga_2O_3/Cl_2$ system and the $ZnO/Ga_2O_3/HCl$ system demonstrate that the gas complexes $ZnGa_2Cl_8$ and $ZnGaCl_5$ are formed but do not become transport effective; $GaCl_3$ and $ZnCl_2$ are transport effective (Pat 1999).

There is also literature about the transport of ternary oxido compounds of indium. With the help of examples of ternary indium molybdates and the In_2O_3/SnO_2 mixed-phase, the transport behavior shall be explained. *Steiner et al.* describe and explain the transport behavior of the molybdate $In_2Mo_3O_{12}$ with the transport agents chlorine and bromine, as well as $InMo_4O_6$ with water in detail with the help of thermodynamic calculations (Ste 2005b). The compound $In_2Mo_3O_{12}$ can be transported with chlorine according to the following equilibrium:

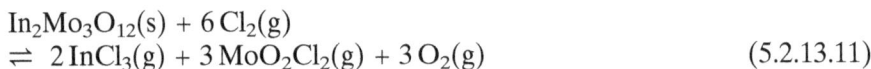

$$In_2Mo_3O_{12}(s) + 6\,Cl_2(g)$$
$$\rightleftharpoons 2\,InCl_3(g) + 3\,MoO_2Cl_2(g) + 3\,O_2(g) \tag{5.2.13.11}$$

This transport equation results from the thermodynamic model calculations (see Figures 5.2.13.2 and 5.2.13.3).

If traces of water are taken into account, another equilibrium is transport effective.

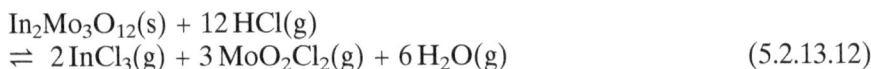

$$In_2Mo_3O_{12}(s) + 12\,HCl(g)$$
$$\rightleftharpoons 2\,InCl_3(g) + 3\,MoO_2Cl_2(g) + 6\,H_2O(g) \tag{5.2.13.12}$$

Transport with bromine as transport agent can be understood in an analogous way, whereby $InBr_2$ was also considered as the indium-transferring species besides $InBr_3$. The transfer of molybdenum takes place via MoO_2Br_2. The essential difference to the transport with chlorine is that HBr does not become transport effective in the presence of water when bromine is used as transport agent. Fur-

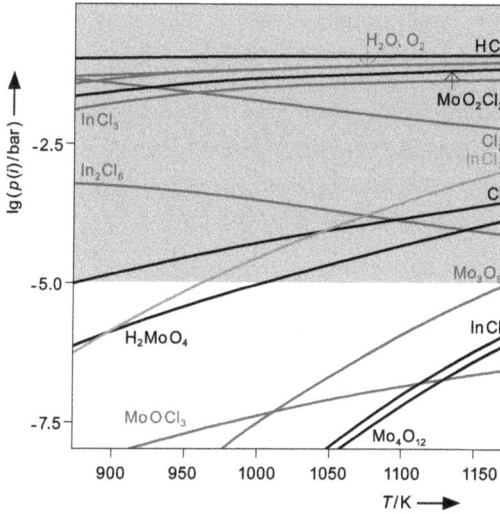

Figure 5.2.13.2 Gas-phase composition over $In_2Mo_3O_{12}$ in equilibrium with Cl_2 in the presence of water according to *Steiner* and *Reichelt* (Ste 2005b).

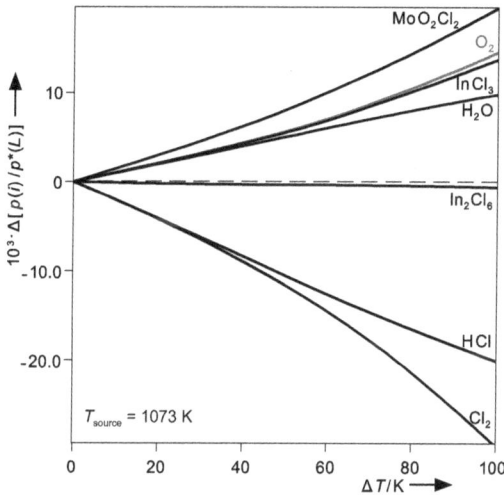

Figure 5.2.13.3 Transport efficiency of the essential gas species for the chemical vapor transport of $In_2Mo_3O_{12}$ with Cl_2 in the presence of water according to *Steiner* and *Reichelt* (Ste 2005b).

ther calculations show that the transport of $InMo_4O_6$ cannot take place via the known gaseous molybdenum oxide halides MoO_2X_2 (X = Cl, Br, I) or the molybdenum acid H_2MoO_4 because their partial pressures are clearly below 10^{-6} bar (Ste 2008). The gaseous species In_2MoO_4 is responsible for the transport effect (Kap 1985). The transport occurs endothermically from 1000 to 900 °C in this case, too. During the CVT of ternary oxides in the $In/W/O$ system, the compounds $In_2W_3O_{12}$ and In_6WO_{12} were deposited in crystalline form in the temperature gradient from 800 to 700 °C when chlorine is used as transport agent. The tungsten bronze In_xWO_3 crystallizes during the endothermic transport from 900 to 800 °C with the aid of ammonium chloride (Ste 2008).

The transport of In_2O_3/SnO_2-mixed-crystals (ITO) was published with the transport agents chlorine (Wer 1996, Pat 2000a) and sulfur + iodine (Wer 1996) as an endothermic transport. The different chlorides $InCl_3$ and $SnCl_4$ as well as $InCl_2$ were described as transport-effective species. The transport can be described assuming the following transport equation:

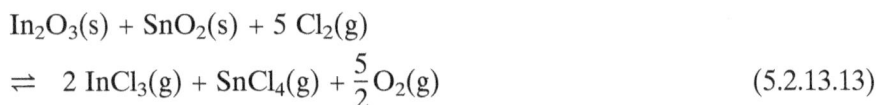

$$In_2O_3(s) + SnO_2(s) + 5\ Cl_2(g)$$
$$\rightleftharpoons\ 2\ InCl_3(g) + SnCl_4(g) + \frac{5}{2}O_2(g) \tag{5.2.13.13}$$

When sulfur + iodine are used as transport additive, sulfur dioxide is formed; additionally, tin iodides and indium iodides appear in low oxidation numbers.

Let us consider the only example of the transport of a *thallium*/oxygen compound; that of thallium ruthenate $Tl_2Ru_2O_7$ with oxygen as the transport agent (Sle 1972). Chemical vapor transport reactions can also be used to dope transition metal oxides, such as titanium(IV)-oxide (Izu 1979) or vanadium(IV)-oxide (Brü 1976a), with aluminum, gallium, or indium.

5.2.14 Group 14

Binary oxides   The transport behavior of the binary oxides silicon dioxide, germanium dioxide, and tin dioxide is described with many examples and different transport additives by a lot of authors.

In several ways, it is important to understand the CVT of *silicon dioxide*. On the one hand, silicon dioxide is in close relation with the silicates. On the other hand, almost all transport reactions are conducted in silica tubes so that silicon dioxide can be involved in the transport reactions due to the ampoule material. Depending on the transport agent used, silicon dioxide can be transported both endothermically and exothermically. Among other things, water can cause crystallization over the gas phase. In the gas phase, the appearance of different silicic acids of the general formula $Si_nO_{2n-x}(OH)_{2x}$ is discussed (Bra 1953). Compounds of this kind are also important during hydrothermal syntheses (Rab 1985). In the process, the transport of the substance takes place by thermal convection and not by diffusion.

Besides water, hydrogen can serve as the transport agent for silicon dioxide by the following endothermic equilibrium:

$$SiO_2(s) + H_2(g)\ \rightleftharpoons\ SiO(g) + H_2O(g) \tag{5.2.14.1}$$

Furthermore, the transport of silicon dioxide is possible via a number of halogenating equilibria in the temperature range from 1200 to 900 °C. Described transport additives are, among others, phosphorus(III)-chloride + chlorine, phosphorus(V)-chloride, phosphorus(III)-bromide + bromine, chromium(III)-chloride + chlorine, tantalum(V)-chloride, niobium(V)-chloride, titanium(II)-oxide/titanium(II)-chloride, and silicon + iodine. One always has to consider a reaction of the silicon dioxide of the ampoule with the transport agents during experiments in the mentioned temperature range. Besides the corrosion of the ampoule, depositions of silicon dioxide or silicates might occur. So, the transport with phosphorus(III)-chloride + chlorine takes place from 900 to 1100 °C (Orl 1976):

$$SiO_2(s) + PCl_3(g) + 2\,Cl_2(g) \rightleftharpoons SiCl_4(g) + 2\,POCl_3(g) \qquad (5.2.14.2)$$

Transport with niobium(V)-chloride can be described as follows:

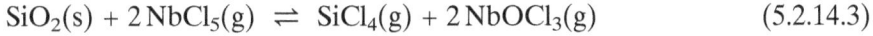

$$SiO_2(s) + 2\,NbCl_5(g) \rightleftharpoons SiCl_4(g) + 2\,NbOCl_3(g) \qquad (5.2.14.3)$$

Hibst found that the deposition of silicon dioxide crystals is inhibited on a silica glass surface (Hib 1977). In contrast, there is no inhibition of the silicon dioxide deposition on a niobium(V)-oxide substrate. Remarkably, the deposited silicon dioxide crystals show the form of the substrate crystals. The formation of these crystals can be seen as an exchange equilibrium between silicon dioxide and Nb_2O_5. Also, silicon dioxide can be transported with chromium(IV)-chloride and chlorine from 1100 to 900 °C:

$$SiO_2(s) + CrCl_4(g) + Cl_2(g) \rightleftharpoons SiCl_4(g) + CrO_2Cl_2(g) \qquad (5.2.14.4)$$

By forming gaseous chromium(VI)-oxide chloride, the equilibrium shifts further to the side of those reaction products that form free oxygen. The surplus of chlorine stabilizes the chromium(IV)-chloride. The deposition of SiO_2 via the gas phase can also take place with added silicon and iodine (Sch 1957a). Silicon dioxide migrates from T_2 (1270 °C) to T_1 (1000 °C). At T_2, solid silicon and silicon dioxide react with the formation of gaseous SiO, which migrates to the lower temperature. There it reacts back to solid silicon dioxide. The following equilibria describe the transport:

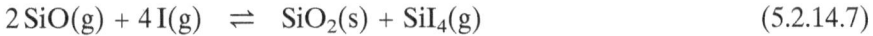

$$SiO_2(s) + Si(s) \rightleftharpoons 2\,SiO(g) \qquad (5.2.14.5)$$

$$2\,SiO(g) + SiI_4(g) \rightleftharpoons SiO_2(s) + 2\,SiI_2(g) \qquad (5.2.14.6)$$

$$2\,SiO(g) + 4\,I(g) \rightleftharpoons SiO_2(s) + SiI_4(g) \qquad (5.2.14.7)$$

In an analogous way, silicon dioxide transport can be conducted with the help of the transport agents silicon(IV)-chloride or silicon(IV)-bromide from 1100 to 900 °C (Sch 1957a). Furthermore, the deposition of silicon dioxide with fibrous form in a flow of hydrogen chloride is described.

$$SiO_2(s) + 4\,HCl(g) \rightleftharpoons SiCl_4(g) + 2\,H_2O(g) \qquad (5.2.14.8)$$

The exothermic transport of silicon dioxide with hydrogen fluoride can be traced back to the following reaction for low temperatures (150 → 500 °C) (Chu 1965) as well as for high temperatures (600 → 1100 °C) (Hof 1977a):

$$SiO_2(s) + 4\,HF(g) \rightleftharpoons SiF_4(g) + 2\,H_2O(g) \qquad (5.2.14.9)$$

Silicon dioxide can also be deposited in an exothermic transport reaction via the following equilibrium:

$$SiO_2(s) + 3\,SiF_4(g) \rightleftharpoons 2\,Si_2OF_6(g) \qquad (5.2.14.10)$$

If hydrogen fluoride is used as the transport agent, transport ampoules made of platinum should be used.

Like silicon dioxide, *germanium dioxide* can be transported with the help of hydrogen, water as well as halogenating transport agents, such as germanium(IV)-chloride, tellurium(IV)-chloride, chlorine, hydrogen chloride, hydrogen fluoride, and sulfur + iodine. The following exothermic transport reaction is assumed for hydrogen chloride and hydrogen fluoride (Schu 1964):

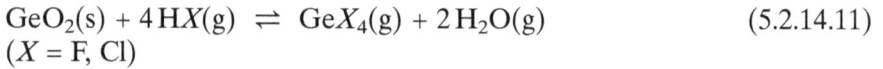

$$GeO_2(s) + 4\,HX(g) \;\rightleftharpoons\; GeX_4(g) + 2\,H_2O(g) \qquad (5.2.14.11)$$
$$(X = F,\ Cl)$$

Extensive work on the transport of germanium dioxide with hydrogen contain numerous thermodynamic calculations (Schm 1981, Schm 1981a, Schm 1981b, Schm 1983). The transport from T_2 to T_1 can be described according to the following equilibrium:

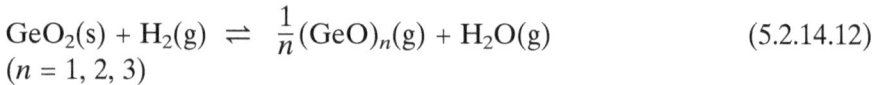

$$GeO_2(s) + H_2(g) \;\rightleftharpoons\; \frac{1}{n}(GeO)_n(g) + H_2O(g) \qquad (5.2.14.12)$$
$$(n = 1,\ 2,\ 3)$$

Examinations on the CVT of germanium dioxide with chlorine have been published by different authors (Red 1978, Schm 1985). Accordingly, the endothermic transport from 900 to 850 °C takes place with the deposition of well formed, columnar crystals of 2 to 3 mm length. Adding NaCl, KCl, and MnO counteracts the reaction inhibition and favors the formation of germanium dioxide-crystals in the rutile modification. The reaction takes place as follows:

$$GeO_2(s) + 2\,Cl_2(g) \;\rightleftharpoons\; GeCl_4(g) + O_2(g) \qquad (5.2.14.13)$$

Redlich and *Gruehn* also examined the transport efficiency of the gaseous germanium oxide chlorides $GeOCl_2$ and Ge_2OCl_6. They concluded that the efficiency of Ge_2OCl_6 can be neglected. The fraction of $GeOCl_2$ in the gas phase is in the percentage range.

Agafonov et al. described the CVT of germanium dioxide with tellurium(IV)-chloride from 1000 to 900 °C (Aga 1984) as well as with sulfur + iodine from 1100 to 1000 °C (Aga 1985). In the chosen temperature range, transport with tellurium(IV)-chloride can be described approximately as follows:

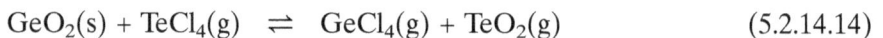

$$GeO_2(s) + TeCl_4(g) \;\rightleftharpoons\; GeCl_4(g) + TeO_2(g) \qquad (5.2.14.14)$$

As mentioned above, other equilibria are negligible in the transport system germanium dioxide/tellurium(IV)-chloride. In this process, among others, $TeOCl_2$ is formed but this is of minor importance for oxygen transport in the chosen temperature range. If sulfur + iodine is used as transport additive, transport will take place via the following equilibrium:

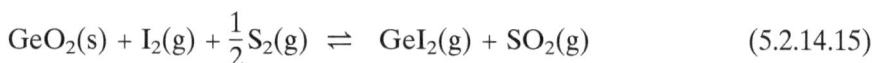

$$GeO_2(s) + I_2(g) + \frac{1}{2}S_2(g) \;\rightleftharpoons\; GeI_2(g) + SO_2(g) \qquad (5.2.14.15)$$

Due to the lower stability of the tetraiodides, compared to the chlorides, germanium is transferred by the gaseous diiodide GeI_2 in the given temperature range.

Two unusual transport reactions of germanium dioxide, which are not based on halogenating equilibria, shall be mentioned. *Ito* published (Ito 1978) the transport of germanium dioxide with carbon as transport agent. A mixture of pulverized germanium dioxide and graphite powder was brought to reaction in a nitrogen carrier gas stream at 900 °C. The gaseous GeO formed in the process was carried away from the dissolution side by the gas stream and reacted with air at the deposition side at approximately 750 °C. The oxidation caused by the aerial oxygen led to the deposition of needle-shaped GeO_2 crystals.

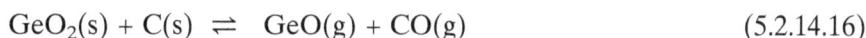

$$GeO_2(s) + C(s) \rightleftharpoons GeO(g) + CO(g) \tag{5.2.14.16}$$

$$GeO(g) + \frac{1}{2}O_2(g) \rightarrow GeO_2(s) \tag{5.2.14.17}$$

The volatility of germanium dioxide in the presence of tungsten dioxide was analyzed by *Gruehn* and co-workers (Pli 1983). The experimental result of the migration of GeO_2 in the presence of WO_2 in the temperature gradient from 930 to 830 °C is explained and justified by thermodynamic calculations: the system contains water traces that are released by the ampoule walls. Water reacts with the tungsten(IV)-oxide in the solid with the formation of hydrogen. By this, germanium(IV)-oxide is reduced to the highly volatile germanium(II)-oxide. Gas species that contribute essentially to the transport behavior are GeO, Ge_2O_2, Ge_3O_3, H_2, and H_2O. The tungsten bronze $Ge_{0.25}WO_3$, which is also deposited during this reaction, is transferred by the gas species $GeWO_4$. Gaseous GeW_2O_7 was also detected by mass spectrometry (Pli 1982). The transfer of $Ge_{0.25}WO_3$ via the gas phase is called a "conproportionative sublimation", because no transport agent in the true sense is involved in the transport (Pli 1983).

Tin dioxide can be transported with different halogenating transport additives, such as Cl_2, $TeCl_4$, CCl_4, and HBr as well as with water, hydrogen, or carbon monoxide. All transport equilibria that result from this are endothermic. *Toshev* examined the transport of tin(IV)-oxide with chlorine and tetrachloromethane (Tos 1988). The deposition (e. g., from 1010 → 970 °C) can be described by the following equilibria:

$$SnO_2(s) + 2Cl_2(g) \rightleftharpoons SnCl_4(g) + O_2(g) \tag{5.2.14.18}$$

$$SnO_2(s) + CCl_4(g) \rightleftharpoons SnCl_4(g) + CO_2(g) \tag{5.2.14.19}$$

The CVT of tin(IV)-oxide with sulfur and iodine takes place via the following transport equilibrium (Mat 1977):

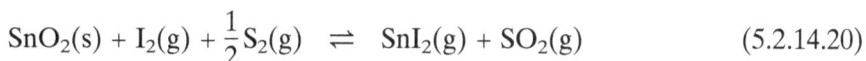

$$SnO_2(s) + I_2(g) + \frac{1}{2}S_2(g) \rightleftharpoons SnI_2(g) + SO_2(g) \tag{5.2.14.20}$$

Also, tin(IV)-oxide can be transported by reaction with hydrogen and carbon monoxide as reducing transport agents and with carbon as the transport effective additive. In this process, tin is transferred via the gas species SnO, which is formed according to the equilibria 5.2.14.21 to 5.2.14.23:

$$SnO_2(s) + H_2(g) \rightleftharpoons SnO(g) + H_2O(g) \tag{5.2.14.21}$$

$$SnO_2(s) + CO(g) \rightleftharpoons SnO(g) + CO_2(g) \tag{5.2.14.22}$$

$$SnO_2(s) + \frac{1}{2}C\,(s) \;\rightleftharpoons\; SnO(g) + \frac{1}{2}CO_2(g) \qquad (5.2.14.23)$$

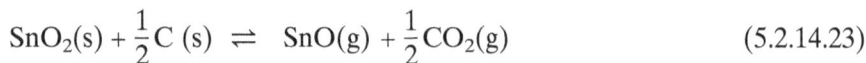

Besides the described binary oxides, a number of examples of transport reactions of ternary and quaternary oxides of silicon, germanium, and tin are known. In the following section, some examples of their transport behavior are discussed. Further examples of the transport of these compounds can be found in other section of this chapter and in Table 5.1.

Ternary compounds The crystallization of ortho*silicates* with participation of the gas phase is shown with the help of the examples Be_2SiO_4, Co_2SiO_4, Ni_2SiO_4, Zn_2SiO_4, $ThSiO_4$, $HfSiO_4$, and Eu_2SiO_4 (see section 6.3). Often fluorinating equilibria with hydrogen fluoride or metal fluorides, such as SiF_4, NaF, Li_2BeF_4, Na_2BeF_4, Na_3AlF_6, or $LiZnF_3$ are used. Thus, beryllium-orthosilicate, Be_2SiO_4, can be transported endothermically from 1290–900 °C to 1240–850 °C via the following equilibrium (among other equilibria) (Nov 1967):

$$Be_2SiO_4(s) + 3\,SiF_4(g) + 2\,NaF(g)$$
$$\rightleftharpoons\; 2\,NaBeF_3(g) + 4\,SiOF_2(g) \qquad (5.2.14.24)$$

Schmid described the transport of Co_2SiO_4 under the influence of hydrogen fluoride (Schm 1964a):

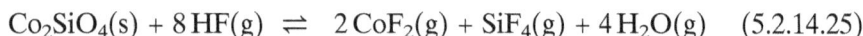

$$Co_2SiO_4(s) + 8\,HF(g) \;\rightleftharpoons\; 2\,CoF_2(g) + SiF_4(g) + 4\,H_2O(g) \quad (5.2.14.25)$$

Strobel et al. analyzed the formation of Co_2SiO_4 crystals, which is undesirable in thermodynamic terms, during the CVT of cobalt-manganese-spinels with tellurium(IV)-chloride in silica ampoules (Str 1981a). During the endothermic transport, the deposition of Co_2SiO_4 crystals takes place in the sink. In the process, the cobalt-containing solid reacts with the transport agent and forms gaseous $CoCl_2$ which reacts with the silicon dioxide of the ampoule wall. Apparently, deposition of Co_2SiO_4 is preferred compared to that of Co_3O_4.

The transport of $ThSiO_4$ is an example of the crystallization of a very stable silicate via chlorinating equilibria (Kam 1979, Schm 1991a). The endothermic transport reactions can be described as follows:

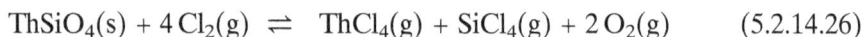

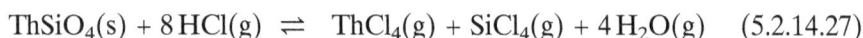

$$ThSiO_4(s) + 4\,Cl_2(g) \;\rightleftharpoons\; ThCl_4(g) + SiCl_4(g) + 2\,O_2(g) \qquad (5.2.14.26)$$

$$ThSiO_4(s) + 8\,HCl(g) \;\rightleftharpoons\; ThCl_4(g) + SiCl_4(g) + 4\,H_2O(g) \qquad (5.2.14.27)$$

In contrast to this, $HfSiO_4$ can be obtained with chromium(III)-chloride + selenium in the gradient from 920 to 980 °C (Fuh 1986).

The transport of Eu_2SiO_4 with hydrogen chloride or iodine, which was described by *Kaldis*, is a good example of a transport reaction in a metal ampoule at very high temperatures (Kal 1970, Kal 1971). The endothermic transport takes place from 1980 to 1920 °C in vertical molybdenum ampoules. Substance transport occurs mainly by convection. Presumably the following reaction is essential:

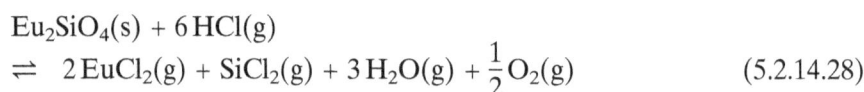

$$Eu_2SiO_4(s) + 6\,HCl(g)$$
$$\rightleftharpoons\; 2\,EuCl_2(g) + SiCl_2(g) + 3\,H_2O(g) + \frac{1}{2}O_2(g) \qquad (5.2.14.28)$$

The formation of oxygen is in accordance with the observation that Eu_2SiO_5 crystallizes during some experiments.

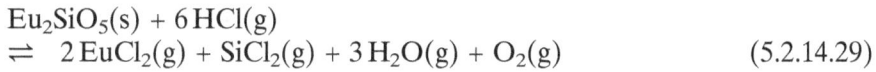

$$Eu_2SiO_5(s) + 6\,HCl(g)$$
$$\rightleftharpoons\ 2\,EuCl_2(g) + SiCl_2(g) + 3\,H_2O(g) + O_2(g) \quad\quad (5.2.14.29)$$

In both reactions, the transfer of silicon takes place via the gas species $SiCl_2$, and not $SiCl_4$ as in the previous examples. The reason for this is the unusually high temperature, which shifts the following equilibrium to the side of $SiCl_2$:

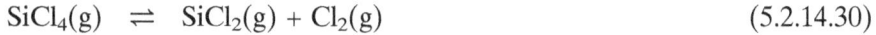

$$SiCl_4(g)\ \rightleftharpoons\ SiCl_2(g) + Cl_2(g) \quad\quad (5.2.14.30)$$

Several authors deal with the CVT of ternary and quaternary *germanates* with main group and transition metal atoms, for example magnesium germanate, $MgGeO_3$, with chlorine from 1100 to 1000 °C (Kru 1986a). The thermodynamic calculations show that $MgGeO_3$ is transferred by the gas species $MgCl_2$ and $GeCl_2$. Besides the $MgGeO_3$ crystals in the sink, crystals of Mg_2GeO_4 were observed in the source solid. The crystallization of other meta germanates is described for $MnGeO_3$, $FeGeO_3$, and $CoGeO_3$ (Roy 1963, Roy 1963a). The transport takes place with hydrogen chloride, or ammonium chloride as the hydrogen chloride source, from T_2 to T_1.

Transport in the germanium(IV)-oxide/gallium(III)-oxide system was examined with the transport agents hydrogen chloride and sulfur/iodine. In doing so, the compounds Ga_2GeO_5 and Ga_4GeO_8 can be transported endothermically with hydrogen chloride as well as with sulfur + iodine. In order to describe the system completely, the transport experiments were realized with the binary phases GeO_2 and Ga_2O_3. They were transported endothermically from 1100 to 1000 °C and from 1000 to 900 °C with sulfur + iodine (Aga 1985). In contrast to chlorinating equilibria, where germanium is generally transported via the gas phase as $GeCl_4$, the transfer takes place as GeI_2 when iodine-containing transport agents are used. The transport is described by the following equation:

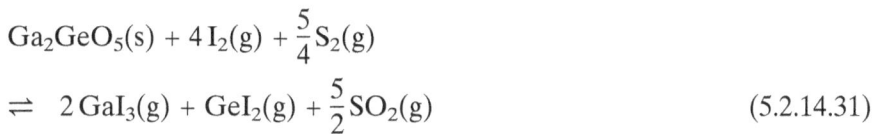

$$Ga_2GeO_5(s) + 4\,I_2(g) + \frac{5}{4}S_2(g)$$

$$\rightleftharpoons\ 2\,GaI_3(g) + GeI_2(g) + \frac{5}{2}SO_2(g) \quad\quad (5.2.14.31)$$

Additionally, the following equations have to be considered:

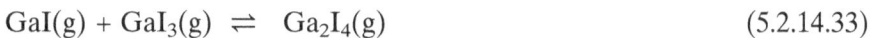

$$GaI_3(g)\ \rightleftharpoons\ GaI(g) + I_2(g) \quad\quad (5.2.14.32)$$

$$GaI(g) + GaI_3(g)\ \rightleftharpoons\ Ga_2I_4(g) \quad\quad (5.2.14.33)$$

Pfeifer and *Binnewies* describe the CVT of ternary and quaternary germanates by endothermic reactions with hydrogen chloride and chlorine as transport agents (Pfe 2002, Pfe 2002a, Pfe 2002b, Pfe 2002c). Cobalt meta germanate, $CoGeO_3$, and cobalt ortho germanate, Co_2GeO_4, were transported with hydrogen chloride. It could be revealed in the process that the ratio of cobalt and germanium in the source solid, and not the transport temperatures, is decisive for the deposition of the respective germanate. If equimolar mixtures of CoO and GeO_2 are used, $CoGeO_3$ will be deposited in a wide temperature range while a small surplus of

Co_3O_4 leads to the deposition of Co_2GeO_4. Transport takes place via gaseous $CoCl_2$, $GeCl_4$, and H_2O.

Nickel germanate, Ni_2GeO_4, can be transported with chlorine as well as with hydrogen chloride from 1050 to 900 °C. Ammonium chloride is not suitable as the transport agent or as the hydrogen chloride source due to the reducing effect. Its use leads to the deposition of metallic nickel, nickel(II)-oxide, and germanium dioxide in the sink. Transport with chlorine can be described with the gas species $NiCl_2$ and $GeCl_4$. The formation of a nickel germanate, such as $NiGeO_3$, could be observed (Pfe 2002a). Examination of the transport of $Ni_{1-x}Co_xGeO_3$-mixed-crystals in the temperature range from 900 to 700 °C with hydrogen chloride as the transport agent showed that a mixed-crystal with x between 0 and 0.4 exists and can be deposited in crystalline form as the sink solid. Higher contents of nickel led to the deposition of an ortho germanate $(Ni_{1-x}Co_x)_2GeO_4$. Furthermore, *Pfeifer* was able to prove that a transport of mixed-phases $(Ni_{1-x}Co_x)_2GeO_4$ is possible with hydrogen chloride in the range of $0 \leq x \leq 1$ from 1000 to 900 °C and that the transport takes place almost congruently (Pfe 2002a). Analogous examinations of the iron(II)/cobalt(II)-germanates system and manganese(II)-germanates system were made by *Pfeifer* (Pfe 2002). All transport reactions with hydrogen chloride are endothermic. In the process, $FeGeO_3$, Fe_2GeO_4, $CoGeO_3$, Co_2GeO_4, and $MnGeO_3$ could be deposited in crystalline form. Also, numerous mixed-crystals from the series $Fe_{1-x}Co_xGeO_3$ and $Fe_{2-x}Co_xGeO_4$ as well as different meta germanates of the composition $Mn_{1-x}Co_xGeO_3$ were transported under the given conditions. Co_2GeO_4 can also be transported to the lower temperature with tellurium(IV)-chloride (Hos 2007). Other examples of the CVT of germanates in the quaternary systems $Cr_2O_3/In_2O_3/GeO_2$, $Ga_2O_3/In_2O_3/GeO_2$, $Mn_2O_3/In_2O_3/GeO_2$, and $Fe_2O_3/In_2O_3/GeO_2$ have been published (Pfe 2002b). These examinations prove that it is possible to directly influence the composition of the deposited crystals in multi-component systems as well. It could be proven for the $Fe_2O_3/In_2O_3/GeO_2$ system that the composition of the deposited mixed-phases is predominantly influenced by variation of the composition of the source solid. All transport reactions occur endothermically via the chlorinating equilibria; hydrogen chloride was used as the transport agent in the $Mn_2O_3/In_2O_3/GeO_2$ system and chlorine in the other systems. The transport effective equilibria can be described by the formation of gaseous trichloride and germanium(IV)-chloride. The transport of $Mn_3Cr_2Ge_3O_{12}$, which was obtained with different transport agents (Cl_2, $S + Cl_2$, $TeCl_4$, CCl_4) by endothermic transport, is another example of the CVT of a quaternary germanate (Paj 1985). If chlorine is used as the transport agent, Cr_2O_3 was deposited in the sink adjacent to $Mn_3Cr_2Ge_3O_{12}$.

In contrast to the numerous examples of the transport of silicates and germanates, the transport of *stannates* is rarely verified. Examinations on the crystallization of mixed-phases in the In_2O_3/SnO_2 system are known (Wer 1996, Pat 2000a) and have already been discussed in section 5.2.13. Additionally, the transport of the cobalt ortho stannate Co_2SnO_4 with chlorine from 1030 to 1010 °C (Emm 1968b) and with ammonium bromide from 950 to 800 °C (Trö 1972) has been described. The following transport equations can be formulated:

$$Co_2SnO_4(s) + 4\,Cl_2(g) \;\rightleftharpoons\; 2\,CoCl_2(g) + SnCl_4(g) + 2\,O_2(g) \quad (5.2.14.34)$$

$$Co_2SnO_4(s) + 8\,HCl(g)$$
$$\rightleftharpoons\; 2\,CoCl_2(g) + SnCl_4(g) + 4\,H_2O(g) \qquad (5.2.14.35)$$

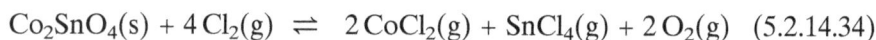

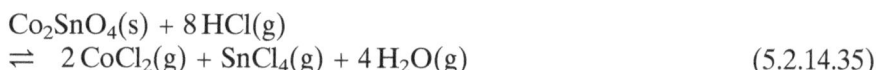

The endothermic transport of $Ru_{1-x}Sn_xO_2$ mixed-crystals is described with chlorine from 1100 to 1000 °C (Nic 1993).

Transport effective halides of the elements of group 14 If a solid contains an element of group 14, a transport effect can be achieved via halogenating equilibria in most cases. Chlorine, hydrogen chloride, and ammonium chloride prove suitable as well as the combination of sulfur with iodine. It has to be considered that ammonium chloride cannot be substituted for hydrogen chloride in every case because of its reducing effect. Also, tellurium(IV)-chloride and – in particular for the transport of silicon dioxide – a series of transition metalhalides, such as $NbCl_5$, $TaCl_5$, or a combination of $CrCl_3 + Cl_2$ play a role. If transport takes place via the chlorides, one can generally assume that the gas species $SiCl_4$ becomes transport effective in the case of silicon. Only in the temperature range about 2000 °C $SiCl_2$ does play a role. The dihalides $GeCl_2$ and $SnCl_2$ have to be considered for germanium and tin, respectively, at temperatures of 1000 or 850 °C, respectively. If iodides are formed by the reaction with iodine-containing transport agents, the tetraiodides SiI_4 (up to approximately 1000 °C) as well as GeI_4 and SnI_4 (up to approximately 750 °C) are transport effective. Above these temperatures, diiodides are increasingly formed, which have to be taken into account for equilibrium considerations. In the process, the tendency to form dihalides increases from silicon to tin and from fluoride to iodide. These coherences can be proven by the comparison of the free reaction enthalpies for the formation of some dihalides as follows:

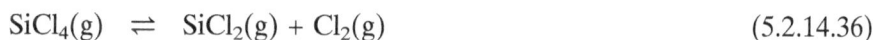

$$SiCl_4(g) \;\rightleftharpoons\; SiCl_2(g) + Cl_2(g) \qquad\qquad (5.2.14.36)$$
$$\Delta_r G_{1000}^0 = 271 \text{ kJ} \cdot \text{mol}^{-1}$$

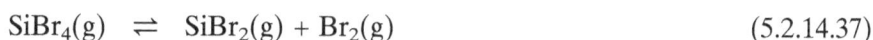

$$SiBr_4(g) \;\rightleftharpoons\; SiBr_2(g) + Br_2(g) \qquad\qquad (5.2.14.37)$$
$$\Delta_r G_{1000}^0 = 228 \text{ kJ} \cdot \text{mol}^{-1}$$

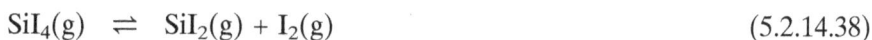

$$SiI_4(g) \;\rightleftharpoons\; SiI_2(g) + I_2(g) \qquad\qquad (5.2.14.38)$$
$$\Delta_r G_{1000}^0 = 106 \text{ kJ} \cdot \text{mol}^{-1}$$

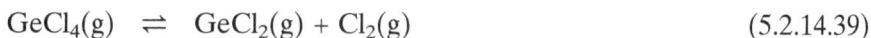

$$GeCl_4(g) \;\rightleftharpoons\; GeCl_2(g) + Cl_2(g) \qquad\qquad (5.2.14.39)$$
$$\Delta_r G_{1000}^0 = 110 \text{ kJ} \cdot \text{mol}^{-1}$$

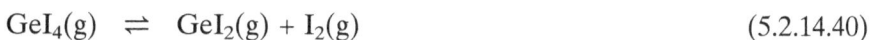

$$GeI_4(g) \;\rightleftharpoons\; GeI_2(g) + I_2(g) \qquad\qquad (5.2.14.40)$$
$$\Delta_r G_{1000}^0 = 43 \text{ kJ} \cdot \text{mol}^{-1}$$

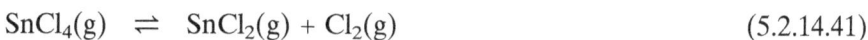

$$SnCl_4(g) \;\rightleftharpoons\; SnCl_2(g) + Cl_2(g) \qquad\qquad (5.2.14.41)$$
$$\Delta_r G_{1000}^0 = 61 \text{ kJ} \cdot \text{mol}^{-1}$$

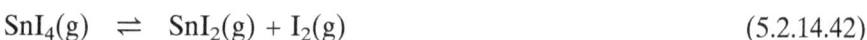

$$SnI_4(g) \;\rightleftharpoons\; SnI_2(g) + I_2(g) \qquad\qquad (5.2.14.42)$$
$$\Delta_r G_{1000}^0 = 36 \text{ kJ} \cdot \text{mol}^{-1}$$

5.2.15 Group 15

Binary oxides Only the binary oxides of antimony, Sb_2O_3 and Sb_2O_4, can be deposited by CVT. The oxides P_4O_6, P_4O_{10}, and As_2O_3 have vapor pressures that are high enough to sublimate them at relatively low temperatures. As_2O_5 and Sb_2O_5 already decompose at approximately 400 and 350 °C, respectively. In the process oxygen is released. Sb_2O_4 can be transported endothermically with tellurium(IV)-iodide (Dem 1980). Transport takes place from 950 to 930 °C according to the following transport equation.

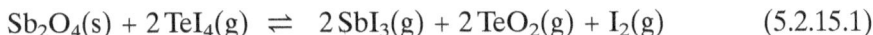

$$Sb_2O_4(s) + 2\,TeI_4(g) \;\rightleftharpoons\; 2\,SbI_3(g) + 2\,TeO_2(g) + I_2(g) \qquad (5.2.15.1)$$

The decomposition of SbI_3 is of importance, too:

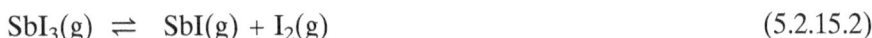

$$SbI_3(g) \;\rightleftharpoons\; SbI(g) + I_2(g) \qquad (5.2.15.2)$$

In the lower temperature range, $TeOI_2$ has also to be considered as an oxygen-transferring gas species (Opp 1980). Besides tellurium(IV)-iodide, tellurium(IV)-chloride can also be used for the transport of Sb_2O_4. The endothermic transport takes place from 1100 to 950 °C with high transport rates.

Numerous investigations prove that the CVT of bismuth(III)-oxide does not work. Different experiments with halogenating transport agents, as well as with water and with transport agent combinations of halogenating transport agent + water, showed that bismuth oxide halides instead of Bi_2O_3 are deposited as the sink solid (see section 8.1).

Ternary compounds *Oppermann* and co-workers (Opp 1999, Opp 2002, Rad 2000, Rad 2001) and *Schmidt et. al.* (Schm 1999, Schm 2000) describe the transport of ternary compounds of the Bi/Se/O system with different transport agents. The phase diagram of the Bi/Se/O system is illustrated in Figure 5.2.15.1. The compounds $Bi_2Se_4O_{11}$ and $Bi_2Se_3O_9$ Bi_2SeO_5 and $Bi_2O_2Se^1$ are the only bismuth-containing oxido compounds whose transport reactions are described besides Bi_2TeO_5 and $Bi_2Te_4O_{11}$ (Schm 1997). Two further bismuth oxides, $BiReO_4$ and $BiRe_2O_6$, can be transported by endothermic transports with bismuth(III)-iodide and with ammonium iodide in a wide temperature range. For the transport of Bi_2O_2Se thermodynamic calculations show that the gas species BiI, BiSe, SeO_2, Se_2, and SeO are transport effective and that SeI_2 functions as the transport agent. SeI_2 is formed by the reaction of the solid with bismuth(III)-iodide:

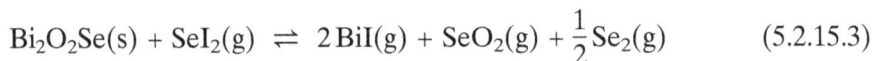

$$Bi_2O_2Se(s) + SeI_2(g) \;\rightleftharpoons\; 2\,BiI(g) + SeO_2(g) + \frac{1}{2}Se_2(g) \qquad (5.2.15.3)$$

Bi_2SeO_5 and $Bi_2Se_3O_9$ can be deposited endothermically with different bromine- and iodine-containing transport agents. The transport equation for the transport of Bi_2SeO_5 with BiI_3 can be formulated as follows (Opp 2002):

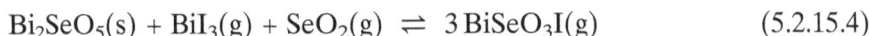

$$Bi_2SeO_5(s) + BiI_3(g) + SeO_2(g) \;\rightleftharpoons\; 3\,BiSeO_3I(g) \qquad (5.2.15.4)$$

---

[1] This formula indicates that this compound is a selenide oxide

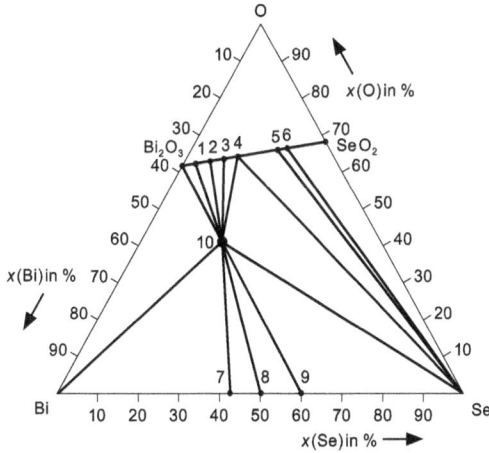

Figure 5.2.15.1 Figure 5.2.15.1 Phase diagram of the Bi/Se/O system according to *Oppermann et al.* (Opp 1999). 1: $Bi_{12}SeO_{20}$, 2: $Bi_{10}Se_2O_{19}$, 3: $Bi_{16}Se_5O_{34}$, 4: Bi_2SeO_5, 5: Bi_2SeO_9, 6: $Bi_2Se_4O_{11}$, 7: Bi_4Se_3, 8: $BiSe$, 9: Bi_2Se_3, 10: Bi_2O_2Se.

It is noteworthy that the gas species formed, $BiSeO_3I(g)$, is transport effective for bismuth, selenium, and oxygen at the same time. The exact conditions for the transport with BiI_3 + SeO_2 result from the phase barogram of the total $Bi_2O_3/SeO_2/BiI_3$ system (see section 8.1). Bi_2SeO_5 can also be transported with iodine with a clearly lower transport rate. The following equilibria describe the process:

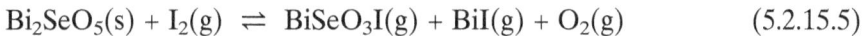

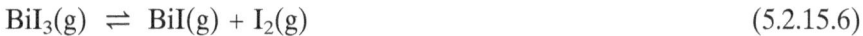

$$Bi_2SeO_5(s) + I_2(g) \rightleftharpoons BiSeO_3I(g) + BiI(g) + O_2(g) \tag{5.2.15.5}$$

$$BiI_3(g) \rightleftharpoons BiI(g) + I_2(g) \tag{5.2.15.6}$$

For equilibrium calculations, the dissociation of BiI_3 has to be considered for temperatures above 750 °C as well as the formation of iodine atoms above 900 °C. Analogous to Bi_2SeO_5, $Bi_2Se_3O_9$ can be effectively transported with bismuth(III)-iodide from 500 °C to T_1. The transport is limited above 500 ° by the decomposition of $Bi_2Se_3O_9$ to Bi_2SeO_5 and gaseous SeO_2. The most selenium(IV)-oxide-rich compound $Bi_2Se_4O_{11}$, which exists on the quasi-binary Bi_2O_3/SeO_2 section can also be transported with bismuth(III)-iodide; however, at lower temperatures from 300 to 250 °C. Above 300 °C, this is not possible due to the considerable SeO_2 decomposition pressure of $Bi_2Se_4O_{11}$.

The transport of Bi_2TeO_5 and $Bi_2Te_4O_{11}$ is possible from 600 to 500 °C with the addition of ammonium chloride (Schm 1997). Essentially, the transport can be described by the equilibrium 5.2.15.7.

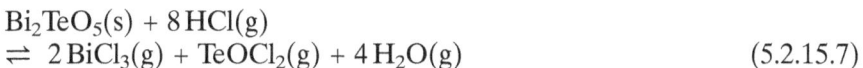

$$Bi_2TeO_5(s) + 8\,HCl(g)$$
$$\rightleftharpoons 2\,BiCl_3(g) + TeOCl_2(g) + 4\,H_2O(g) \tag{5.2.15.7}$$

The gas species that is transport effective for bismuth is $BiCl_3$ because the formation of $BiCl$ can be neglected in the given temperature range. The tellurium-transferring species is $TeOCl_2$. The oxygen is also transferred by TeO_2 as well as by H_2O, which always appears when hydrogen chloride is used as the transport agent in oxide systems.

Metal *arsenates* and metal *antimonates* are characterized structurally by oxido anions. Due to this, the transport of these compounds is dealt with in section 6.2 together with other compounds that also contain oxidido anions. The majority of the CVTs of the arsenates and antimonates are rare earth metal arsenates(V) and rare earth metal antimonates(V), respectively. Because the transport behavior of these rare earth metal oxido compounds is defined by the rare earth metal, it is treated in section 5.2.3.

Transport effective gas species of the elements of group 15   Solids that contain an element of group 15 are transported via the halogenating equilibria in most cases. The pentahalides do not play a role as transport-effective species due to their low stability. If halides are transport effective, the trichlorides are stable up to high temperatures ($BiCl_3$ to 1000 °C). As for the tribromides and in particular the tri-iodides, one has to expect the formation of mono-halides at medium temperatures already (BiI above 750 °C). The stability of the trihalides decreases from phosphorus to bismuth and from chloride to iodide:

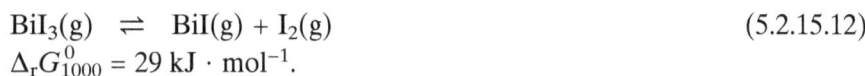

$$PCl_3(g) \;\rightleftharpoons\; PCl(g) + Cl_2(g) \tag{5.2.15.8}$$
$$\Delta_r G^0_{1000} = 218 \; kJ \cdot mol^{-1}$$

$$SbCl_3(g) \;\rightleftharpoons\; SbCl(g) + Cl_2(g) \tag{5.2.15.9}$$
$$\Delta_r G^0_{1000} = 120 \; kJ \cdot mol^{-1}$$

$$BiCl_3(g) \;\rightleftharpoons\; BiCl(g) + Cl_2(g) \tag{5.2.15.10}$$
$$\Delta_r G^0_{1000} = 120 \; kJ \cdot mol^{-1}$$

$$BiBr_3(g) \;\rightleftharpoons\; BiBr(g) + Br_2(g) \tag{5.2.15.11}$$
$$\Delta_r G^0_{1000} = 117 \; kJ \cdot mol^{-1}$$

$$BiI_3(g) \;\rightleftharpoons\; BiI(g) + I_2(g) \tag{5.2.15.12}$$
$$\Delta_r G^0_{1000} = 29 \; kJ \cdot mol^{-1}.$$

During the transport of compounds that contain phosphorus and oxygen, P_4O_{10} and/or P_4O_6 forms in the gas phase. If chlorine-containing transport agents are used, $POCl_3$, and PO_2Cl as well as PCl_3 are formed; however, not with transport effective partial pressures (see section 6.2).

The gas-phase composition over oxido compounds of aresenic and antimomy, which are in equilibrium with chlorinating transport agents, is characterized by the gas species $AsCl_3$ and $SbCl_3$, respectively. The volatile oxide halides $AsOCl$ and $SbOCl$ do not become transport effective due to their low partial pressures. Volatile oxide halides of bismuth are not described. In contrast to phosphorus, the only known gaseous oxide of arsenic is As_4O_6. Its partial pressure does not contribute to the transport. The high saturation pressure, however, explains the sublimation of arsenic(III)-oxide above 250 °C in a vacuum.

5.2.16 Group 16

Binary oxides   As far as the CVT is concerned, only one binary oxide tellurium(IV)-oxide plays an essential role because the binary oxides of sulfur and

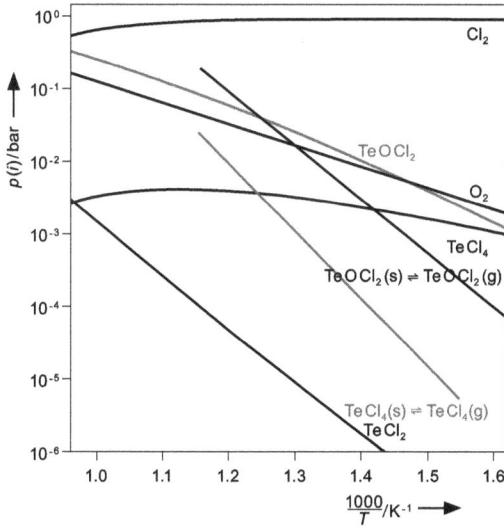

Figure 5.2.16.1 Gas-phase composition over TeO_2 in equilibrium with Cl_2 according to *Oppermann et al.* (Opp 1977b).

Figure 5.2.16.2 Gas-phase composition over TeO_2 in equilibrium with $TeCl_4$ according to *Oppermann et al.* (Opp 1977b).

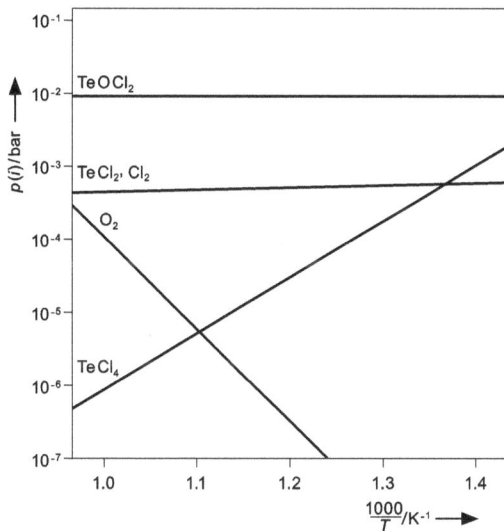

selenium have low boiling temperatures. Thus, SeO_2 can be deposited from the gas phase by sublimation in the oxygen flow around 420 °C.

The groups of *Oppermann* (Opp 1977b) and *Schäfer* (Sch 1977) dealt with the CVT of tellurium(IV)-oxide by chlorinating equilibria. During the transport of tellurium(IV)-chloride from 600 to 400 °C, the following equilibria play a role:

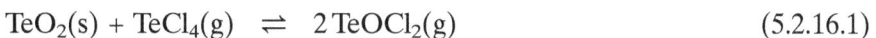

$$TeO_2(s) + TeCl_4(g) \rightleftharpoons 2\,TeOCl_2(g) \tag{5.2.16.1}$$

$$TeCl_4(g) \; \rightleftharpoons \; TeCl_2(g) + Cl_2(g) \tag{5.2.16.2}$$

$$TeO_2(s) + Cl_2(g) \; \rightleftharpoons \; TeOCl_2(g) + \frac{1}{2}O_2(g) \tag{5.2.16.3}$$

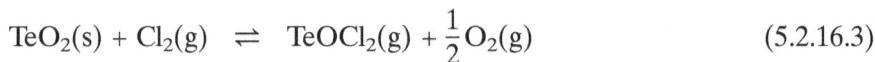

Equilibrium 5.2.16.3 can also be used for the description of the transport of tellurium(IV)-oxide with chlorine. If certain experimental conditions are fulfilled, instead of TeO_2, $Te_6O_{11}Cl_2$ will be deposited. $Te_6O_{11}Cl_2$ is in equilibrium with the solid TeO_2 and the gaseous $TeOCl_2$. If $Te_6O_{11}Cl_2$ is the solid, its decomposition pressure will determine the total pressure in the transport ampoule. The decomposition pressure and the deposition temperature are important in order to find out which of the two solids will be deposited. If the partial pressure of $TeOCl_2$ is, according to the transport equilibrium, higher than the co-existence decomposition pressure, $Te_6O_{11}Cl_2$ will be deposited; if it is lower, TeO_2 will be formed (Opp 1977b) (see section 8.1).

Tellurium(IV)-oxide can also be transported with bromine or tellurium(IV)-bromide. The description is analogous to the $Te/O/Cl$ system. $TeOBr_2$ appears in the gas phase. Solid $Te_6O_{11}Br_2$ can be formed under certain circumstances in an analogous way to $Te_6O_{11}Cl_2$ (Opp 1978a).

Oppermann and co-workers examined the CVT in the $Te/O/I$ system in the same way. Accordingly, tellurium(IV)-oxide can be transported with iodine from 700 to 450 °C, with tellurium(IV)-iodide from 600 to 400 °C as well as with tellurium(IV)-iodide and water from 520 to 420 °C. The gas phase is described in an analogous way to the $Te/O/Cl$ and $Te/O/Br$ systems. In this process, the gas species $TeOI_2$ is involved. Its existence is concluded from the transport behavior. In contrast to the chlorine- and bromine-containing systems, no solid ternary compound of the composition $Te_6O_{11}I_2$ appears in the $Te/O/I$ system (Opp 1980).

Tellurium(IV)-oxide can also be transported with hydrogen chloride (600 → 400 °C) and with hydrogen bromide (600 → 500 °C). If hydrogen halides are used, a fourth independent equilibrium will become important, which additionally considers the gas species water:

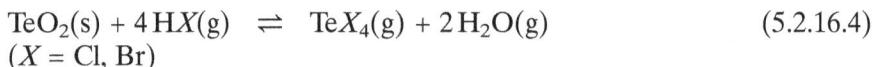

$$TeO_2(s) + 4\,HX(g) \; \rightleftharpoons \; TeX_4(g) + 2\,H_2O(g) \tag{5.2.16.4}$$
$$(X = Cl,\, Br)$$

Ternary compounds Sulfur and selenium appear in ternary and multinary oxido compounds almost exclusively in complex anions. Chapter 6 deals with the transport of oxido compounds with complex anions. The ternary oxido compounds of tellurium are discussed when the relevant metals are discussed. Due to the fact that the transport behavior of selenates and tellurates is similar, some examples of the transport of selenates are also dealt with in this chapter.

5.3 Overview of the Transport of Oxides

The majority of the binary oxides can be obtained via a CVT. There are two reasons for the few exceptions that exist.

- The oxides (e.g., Au_2O_3) are thermally unstable and already decompose at temperatures at which potential metal-transferring gas species do not reach transport effective partial pressures.
- The transport is theoretically possible, but for thermodynamic reasons only at very high temperatures (e.g., for BaO and most of the rare earth metal(III)-oxides). Silica ampoules can be used up to a maximum temperature of 1200 °C. Higher temperatures require alternative ampoule materials, which must be inert toward the oxide solid phases as well as toward the gas phase (molybdenum, platinum, corundum). Handling these materials requires great effort and is generally avoided.

The numerous examples of the transport of binary, ternary, and quaternary oxides can be the basis of planning transport experiments of oxido compounds and crystallization of oxides. The conditions for the transport of oxido compounds can be found on the basis of thermodynamic considerations or by conclusion from analogy with the help of the described examples.

As far as transport experiments of crystallizing multinary compounds are concerned, one can apply polycrystalline powder materials that already contain the target compounds, on the one hand; or on the other hand, one can apply a mixture of binary oxides. Using powdered materials in which the multinary compound already exists proved more suitable. In doing so, one should not use a single-phased source solid but rather apply a multi-phased solid that already contains the target compound and a mix of the binary oxides. The temperature range in which the deposition can take place is limited by the thermodynamic stability of the target compound. The deposition of the target compound can generally only take place in a temperature range in which it is thermodynamically stable. The transport of a multinary compound is not possible in all cases, not even if the binary compounds can be transported. In these cases, the equilibrium situation for the target compound has to be re-considered with the help of the Gibbs energy in the temperature range that is transport relevant. In doing so, the Gibbs energy should be between $-100 \text{ kJ} \cdot \text{mol}^{-1}$ and $+100 \text{ kJ} \cdot \text{mol}^{-1}$. Due to errors of estimating the thermodynamic data of unknown compounds, or by neglecting certain gas species, these values should only be seen as a rough orientation. The transport effective gas species during the transport of multinary oxides are basically the same as those that determine the transport of binary oxides. In some cases, however, the formation of gas complexes may also occur. If a source solid is to be transported for the first time, the data that are required for the calculation of the Gibbs energy can be estimated with the help of the rules described in Chapter 12. Experience from the the transport of binary oxides can be used to aid the choice of a suitable transport agent. The great experimental potential of chemical transport reactions creates a number of opportunities for the crystallization of new oxide compounds.

Table 5.1 Examples of the chemical vapor transport of oxides.

Sink solid	Transport additive	Temperature / °C	Reference
Al_2O_3	H_2	$2000 \rightarrow T_1$	Dev 1959
	H_2	$2000 \rightarrow T_1$	Sea 1963
	Cl_2, HCl	$1240 \rightarrow T_1, 1100 \rightarrow T_1$	Fis 1932
	HCl	$1150 \rightarrow T_1$	Ker 1963
	HCl	$1650 \rightarrow T_1$	Kal 1971
	PbF_2	$1300 \ldots 1100 \rightarrow T_1$	Tsu 1966
	PbF_2	$1380 \ldots 1260 \rightarrow T_1$	Whi 1974
	$H_2 + C$	$2000 \rightarrow T_1$	Isa 1973
	$PbO + F_2O$	$1250 \rightarrow T_1$	Tim 1964
$Al_2Ge_2O_7$	$AlCl_3$	$1000 \rightarrow 900$	Aga 1985a
Al_2SiO_5	Na_3AlF_6	$1100 \ldots 1240 \rightarrow 1050 \ldots 1190$	Nov 1966a
$Al_2Ti_7O_{15}$	$TeCl_4$	$950 \rightarrow 910$	Rem 1988
$Al_{1-x}V_xO_2$	$TeCl_4$	$1100 \rightarrow 1000$	Brü 1976
BPO_4	PCl_5	$800 \rightarrow 700$	Schm 2004
$BaAl_{12}O_{19}:Fe^{3+}$	PbF_2	$1300 \rightarrow T_1$	Tsu 1966
$BaTa_4O_{11}$	Cl_2	$1255 \rightarrow 1155$	Bay 1983
BeO	HCl	$1100 \rightarrow 800$	Spi 1930
	H_2O	$1100 \rightarrow 800$	You 1960
	H_2O	$1380 \rightarrow 1065$	Stu 1964
	H_2O	$1800 \rightarrow 1600$	Bud 1966
Be_2SiO_4	NaF, BeF_2,	$1200 \rightarrow 1100$,	Nov 1967
	Na_2BeF_4	$1100 \rightarrow 1000$	
		$950 \rightarrow 850$	
	Li_2BeF_4	$1290 \rightarrow 1240$	Nov 1966a
	Na_2BeF_4	$1000 \rightarrow 850$	Nov 1964
	Na_2BeF_4	$900 \rightarrow 850$	Nov 1966
Bi_2O_2Se	BiI_3	$800 \rightarrow 750$	Schm 1999
	BiI_3	$800 \rightarrow 750$	Schm 2000
	NH_4I	$750 \rightarrow 650, 900 \rightarrow 800$	Opp 1999
Bi_2SeO_5	Br_2, $SeOBr_2$	$750 \rightarrow 600$	Rad 2001
	I_2, BiI_3,	$700 \rightarrow 500, 600 \rightarrow 500$,	Opp 2002
	$SeO_2 + BiI_3$, SeO_2	$600 \rightarrow 500, 600 \rightarrow 500$	
$Bi_2Se_3O_9$	Br_2, $BiBr_3$	$500 \rightarrow 400$	Rad 2000
	BiI_3	$500 \rightarrow 300$	Opp 2002
$Bi_2Se_4O_{11}$	BiI_3	$300 \rightarrow 250$	Opp 2002
$Bi_4Si_3O_{12}$	NaF, BiF_3	$T_2 \rightarrow T_1$	Nov 1967
Bi_2TeO_5	NH_4Cl	$600 \rightarrow 500$	Schm 1997
$Bi_2Te_4O_{11}$	NH_4Cl	$600 \rightarrow 500$	Schm 1997
CaO	Br_2, HCl	$1400 \rightarrow T_1$	Vei 1967
	H_2O	$1745 \rightarrow 1405$	Mat 1981a
$CaMoO_4$	Cl_2	$1150 \rightarrow 1050$	Ste 2006
$CaMo_5O_8$	Cl_2	$1150 \rightarrow 1050$	Ste 2006

Table 5.1 (continued)

Sink solid	Transport additive	Temperature / °C	Reference
$CaNb_2O_6$	Cl_2	$1020 \rightarrow 980$	Emm 1968a
	Cl_2	$1020 \rightarrow 980$	Emm 1968d
	HCl	$1010 \rightarrow 980$	Emm 1968c
$CaTa_4O_{11}$	Cl_2, $TeCl_4$	$1100 \rightarrow 1000$,	Bay 1983
	$Cl_2 + FeCl_3$	$1015 \rightarrow 915$	
$CaWO_4$	Cl_2	$1150 \rightarrow 1050$	Ste 2005a
CdO	H_2	$720 \rightarrow 530$	Fuh 1964
	Br_2, I_2	$760 \rightarrow 700$	Emm 1968c
	I_2	$665 \rightarrow 595$	Fuh 1964
$CdAs_2O_6$	$PtCl_2$	$720 \rightarrow 680$	Wei 2001
$Cd_2As_2O_7$	$PtCl_2$	$650 \rightarrow 600$	Wei 2001a
$CdFe_2O_4$	HCl	$1000 \rightarrow 800$	Klei 1964
$CdMoO_4$	Cl_2, Br_2, I_2	$700 \rightarrow 600$	Ste 2000
$Cd_2Mo_3O_8$	Cl_2, Br_2	$700 \rightarrow 600$	Ste 2000
$CdNb_2O_6$	HCl, NH_4Cl, NH_4Br, $HgCl_2$, $HgBr_2$	$1000 \rightarrow 900$	Kru 1987
$CdSiO_3$	Br_2	$750 \rightarrow 600$	Fuh 1964
$CdWO_4$	Cl_2	$900 \rightarrow 800$	Ste 2005a
CeO_2	Cl_2, $Cl_2 + C$	$1100 \rightarrow 1000$	Scha 1989
$CeAsO_4$	$TeCl_4$	$1050 \rightarrow 950$	Schm 2005
$CeNbO_4$	Cl_2	$1000 \rightarrow 900$	Hof 1993
$CeNb_3O_9$	Cl_2	$950 \rightarrow 900$	Stu 1975
	Cl_2	$950 \rightarrow 900$, $1100 \rightarrow 1000$	Hof 1993
$CeNb_5O_{14}$	Cl_2, NH_4Br	$1100 \rightarrow 1000$	Hof 1993
$CeNb_7O_{19}$	Cl_2	$800 \rightarrow 780$	Hof 1991
	Cl_2	$850 \rightarrow 800$	Hof 1993
$CePO_4$	PCl_5	$1100 \rightarrow 1000$	Orl 1971
	PCl_5	$1050 \rightarrow 950$	Mül 2004
$CeTaO_4$	Cl_2, $CBr_4 + CO_2$	$1000 \rightarrow 1100$, $1100 \rightarrow 1000$	Scha 1989
$CeTa_3O_9$	Cl_2	$1090 \rightarrow 1000$	Scha 1988
	Cl_2, NH_4Br	$1100 \rightarrow 1000$	Scha 1989
	Cl_2	$1100 \rightarrow 1000$	Scha 1990
$CeTa_7O_{19}$	Cl_2	$1100 \rightarrow 1000$	Scha 1990
	Cl_2, NH_4Br	$1100 \rightarrow 1000$	Scha 1991
$Ce_2Ti_2O_7$	$HgCl_2$, NH_4Cl	$1050 \rightarrow 950$	Pre 1996
$Ce_2Ti_2SiO_9$	NH_4Cl	$1050 \rightarrow 900$	Zen 1999
$CeVO_4$	$TeCl_4$	$1100 \rightarrow 1000$	Schm 2005a
CoO	Cl_2, HCl	$1000 \rightarrow 900$, $970 \rightarrow 900$	Kle 1966a
	Cl_2, HCl	$1000 \rightarrow 900$, $970 \rightarrow 900$	Emm 1968c
	Cl_2	$950 \rightarrow 850$	Kru 1986
$Co_{1-x}Mg_xO$	HCl	$925 \rightarrow 825$	Skv 2000
$Co_{1-x}Ni_xO$	HCl	$800 \rightarrow 650$	Loc 1999
$Co_{1-x}Zn_xO$	HCl	$800 \rightarrow 650$	Loc 1999b
	Br_2	$950 \rightarrow 900$	Kra 1984

Table 5.1 (continued)

Sink solid	Transport additive	Temperature / °C	Reference
Co_3O_4	Cl_2	$980 \rightarrow 860$	Klei 1963
	HCl	$900 \rightarrow 700$	
	HCl	$1000 \rightarrow 800$	Klei 1972a
	Cl_2	$980 \rightarrow 860, 1000 \rightarrow 900$	Emm 1968c
	Cl_2	$975 \rightarrow 855$	Tar 1984
	Cl_2	$980 \rightarrow 860$	Pat 2000b
	HCl	$980 \rightarrow 860$	Tar 1984
$Co_{3-x}Fe_xO_4$	HCl	$1000 \rightarrow 900$	Szy 1970
$Co_2As_2O_7$	Cl_2	$880 \rightarrow 800$	Wei 2004c
	Cl_2	$880 \rightarrow 800$	Wei 2005
$CoCr_2O_4$	Cl_2	$900 \rightarrow 800$	Emm 1968b
	Cl_2	$1045 \rightarrow 945$	Pes 1982
$Co(Cr_{1-x}Fe_x)_2O_4$	$TeCl_4$	$1000 \rightarrow 900$	Pat 2000, Pat 2000c
$CoFe_2O_4$	HCl	$1000 \rightarrow 800$	Klei 1963
	HCl	$800 \ldots 1200 \rightarrow T_1$ $(\Delta T = 50 \ldots 100)$	Cur 1965
	HCl	$1000 \rightarrow 800$	Klei 1972a
	$TeCl_4$	$1000 \rightarrow 900$	Pat 2000c
$Co(Fe_{1-x}Ga_x)_2O_4$	Cl_2	$1000 \rightarrow 900$	Pat 2000, Pat 2000c
$CoGa_2O_4$	Cl_2	$980 \rightarrow 860$	Pat 2000b
$Co(Ga_{1-x}Co_x)_2O_4$	Cl_2	$980 \rightarrow 860$	Pat 2000b
$CoGeO_3$	Cl_2, NH_4Cl	$950 \rightarrow 870, 870 \rightarrow 770$	Kru 1986
	HCl	$950 \rightarrow 650$	Roy 1963
	HCl	$900 \rightarrow 700, 1000 \rightarrow 900$	Pfe 2002a
	NH_4Cl	$1000 \rightarrow 700$	Roy 1963a
Co_2GeO_4	Cl_2, NH_4Cl	$920 \rightarrow 830, 850 \rightarrow 770$	Kru 1986
	HCl	$900 \rightarrow 700, 1000 \rightarrow 900$	Pfe 2002a
	$TeCl_4$	$850 \rightarrow T_1$	Hos 2007
$Co_{1-x}Fe_xGeO_3$	HCl	$900 \rightarrow 700$	Pfe 2002
$CoMn_2O_4$	PbF_2	$1150 \rightarrow T_1$	Tsu 1966
$Co_{1-x}Mn_xGeO_3$	HCl	$900 \rightarrow 700$	Pfe 2002
$CoMoO_4$	Cl_2	$1020 \rightarrow 975$	Emm 1968b
	Cl_2, Br_2	$900 \rightarrow 800$	Ste 2004a
$Co_2Mo_3O_8$	Cl_2, Br_2	$900 \rightarrow 800$	Ste 2004a
	$HCl, TeCl_4$	$935 \rightarrow 815, 965 \rightarrow 815$	Str 1983
$Co_{2-x}Zn_xMo_3O_8$	NH_4Cl, NH_4Br	$1000 \rightarrow 900$	Ste 2005
$CoNb_2O_6$	Cl_2	$1010 \rightarrow 970$	Emm 1968c
	$HgCl_2, PtCl_2, NH_4Cl$	$1020 \rightarrow 960$	Ros 1992
	$TeCl_4$	$1000 \rightarrow 900$	Sch 1978
$CoTa_2O_6$	Cl_2	$880 \rightarrow 850$	Emm 1968b
$CoWO_4$	Cl_2	$1000 \rightarrow 905$	Emm 1968b
	Cl_2	$900 \rightarrow 800$	Ste 2005a
Co_2SiO_4	HF	$1000 \rightarrow T_1$	Schm 1964a
	$TeCl_4$	$925 \rightarrow 705$	Str 1981

Table 5.1 (continued)

Sink solid	Transport additive	Temperature / °C	Reference
Co_2SnO_4	Cl_2	$1030 \rightarrow 1010$	Emm 1968b
	NH_4Cl	$950 \rightarrow 800$	Trö 1972
Cr_2O_3	Cl_2	$980 \rightarrow 860$	Emm 1968c
	Cl_2	$1050 \rightarrow 850$	Pes 1973a
	Cl_2	$1050 \rightarrow 850$	Bli 1977
	Cl_2	$960 \rightarrow 870$	Kru 1986
	$Cl_2, HgCl_2$	$1000 \rightarrow 900$	Noc 1993
	$Br_2, Br_2 + CrBr_3$	$1000 \rightarrow 900$	Noc 1994
	O_2	$1785 \rightarrow 1565$	Grim 1961
	$O_2, O_2 + H_2O$	$1200 \rightarrow 1000$	Cap 1961
	$O_2, O_2 + H_2O$	$1400 \rightarrow T_1$	Kim 1974
	$Cl_2 + O_2$	$725 \rightarrow 625$	San 1974
	$TeCl_4$	$900 \rightarrow 850$	Pes 1984
$CrBO_3$	$CrCl_2 + H_2O$	$1000 \rightarrow 900$	Schm 1964
$Cr_2BP_3O_{12}$	$I_2, I_2 + P$	$1100 \rightarrow 1000$	Schm 2002
$CrGa_2O_4$	Cl_2	$980 \rightarrow 860$	Pat 2000b
Cr_2GeO_5	Cl_2	$1080 \rightarrow 980$	Kru 1986
$Cr_{0.18}In_{1.82}Ge_2O_7$	Cl_2	$950 \rightarrow 850$	Pfe 2002b
$(Cr,Nb)_{12}O_{29}$	$Cl_2, NbCl_5$	$1040 \rightarrow 1000$	Hof 1994
$CrNbO_4$	Cl_2	$1020 \rightarrow 960$	Ros 1990a
	$NbCl_5 + Cl_2$	$980 \rightarrow 860$	Emm 1968c
$CrTaO_4$	Cl_2	$1010 \rightarrow 950$	Emm 1968b
$CrWO_4$	$TeCl_4$	$980 \rightarrow 820$	Vla 1976
$Cs_xNb_yW_{1-y}O_3$	$HgCl_2$	$850 \rightarrow 800$	Hus 1994
Cs_xWO_3	$HgCl_2$	$850 \rightarrow 800$	Hus 1994
	$HgCl_2, HgBr_2$	$800 \rightarrow 700$	Hus 1997
Cu_2O	HCl	$900 \rightarrow 1000$	Sch 1957
	HCl	$740 \rightarrow 760$	Jag 1966
	HCl	$850 \rightarrow 950$	Kra 1977
	NH_4Cl	$1100 \rightarrow 950, 800 \rightarrow 950$	Bal 1985
	NH_4Cl, NH_4Br	$650 \rightarrow 730, 680 \rightarrow 750$	Mar 1999
	I_2	$1000 \rightarrow 900$	Tra 1994
CuO	Cl_2	$870 \rightarrow 800$	Kle 1970
	$Cl_2, I_2,$	$805 \rightarrow 725, 855 \rightarrow 705,$	Bal 1985
	$HgCl_2$	$805 \rightarrow 705$	
	HCl	$790 \rightarrow 710$	Jag 1966
	HCl	$1000 \rightarrow 900$	Yam 1973
	HCl	$800 \rightarrow 700$	Sch 1976
	HCl	$800 \rightarrow 700$	Kra 1977
	$HCl, TeCl_4$	$890 \rightarrow 830, 810 \rightarrow 750$	Des 1989
	$NH_4Cl, CuCl$	$825 \rightarrow 745, 805 \rightarrow 705$	Bal 1985
	NH_4Cl	$900 \rightarrow 865$	Mil 1990
	$BaO_2 + CuI$	$900 \rightarrow 800$	Zhe 1998
	HBr, HI	$900 \rightarrow 700, 900 \rightarrow 800$	Kle 1970
	I_2	$1000 \rightarrow 900$	Tra 1994

Table 5.1 (continued)

Sink solid	Transport additive	Temperature / °C	Reference
$Cu_{1-x}Zn_xO$	Cl_2	$900 \rightarrow 800$	Loc 1999c
$CuO \cdot CuSO_4$	I_2	$800 \rightarrow 720$	Mar 1998
$Cu_2As_2O_7$	Cl_2	$880 \rightarrow 800$	Wei 2004, Wei 2004a, Wei 2004b, Wei 2004c
	Cl_2	$880 \rightarrow 800$	Wei 2005
$CuGeO_3$	I_2	$920 \rightarrow 1010$	Red 1976
$CuMoO_4$	Cl_2, Br_2, HCl, $TeCl_4$	$750 \rightarrow 700$	Ste 1996
$Cu_3Mo_2O_9$	Cl_2, Br_2, HCl, $TeCl_4$	$750 \rightarrow 600$	Ste 1996
$CuRh_2O_4$	$TeCl_4$	$1050 \rightarrow 850$	Jen 2009
$CuSb_2O_6$	Cl_2, $TeCl_4$	$920 \rightarrow 800$, $920 \rightarrow 880$	Pro 2003
$CuSeO_3$	$TeCl_4$	$400 \rightarrow T_1$	Meu 1976
$CuSe_2O_5$	$TeCl_4$	$380 \rightarrow 280$	Jan 2009
Cu_2SeO_4	$TeCl_4$	$400 \rightarrow T_1$	Meu 1976
$CuTe_2O_5$	$PtCl_2$, $TeCl_4$	$590 \rightarrow 490$	Jan 2009
	$TeCl_4$	$790 \rightarrow 680$	Yu 1993
$Cu_2V_2O_7$	$TeCl_4$	$620 \rightarrow 560$	Pro 2001
CuV_2O_6	$TeCl_4$	$600 \rightarrow 500$	Jen 2009
$CuWO_4$	Cl_2	$900 \rightarrow 800$	Ste 2005a
$Cu_{3(1-x)}Zn_{3x}Mo_2O_9$	Cl_2	$700 \rightarrow 600$	Rei 2005
$DyAsO_4$	$TeCl_4$	$1050 \rightarrow 950$	Schm 2005
$DyPO_4$	PBr_5	$1000 \rightarrow 930$	Mül 2004
$DySbO_4$	$TeCl_4$	$1100 \rightarrow 950$	Ger 2007
$Dy_2Ti_2O_7$	Cl_2	$1050 \rightarrow 950$	Hüb 1992
$DyVO_4$	$TeCl_4$	$1100 \rightarrow 1000$	Schm 2005a
$ErAsO_4$	$TeCl_4$	$1050 \rightarrow 950$	Schm 2005
$Er_{1-x}Dy_xAsO_4$	$TeCl_4$	$1075 \rightarrow 975$	Schm 2003
$Er_{1-x}La_xAsO_4$	$TeCl_4$	$1075 \rightarrow 975$	Schm 2003
$ErPO_4$	PCl_5, PBr_5	$1050 \rightarrow 950$, $1075 \rightarrow 960$	Mül 2004
$Er_2Ti_2O_7$	Cl_2	$1050 \rightarrow 950$	Hüb 1992
$EuAsO_4$	$TeCl_4$	$1050 \rightarrow 950$	Schm 2005
$EuPO_4$	PCl_5	$1100 \rightarrow 1000$	Orl 1971
	PCl_5	$1000 \rightarrow 900$, $1100 \rightarrow 1000$	Rep 1971
$EuSbO_4$	$TeCl_4$	$1100 \rightarrow 950$	Ger 2007
Eu_2SiO_4	I_2, HCl	$1980 \rightarrow 1920$	Kal 1970
	HCl	$2000 \rightarrow 1940$	Kal 1969
Eu_2SiO_5	HCl	$1980 \rightarrow 1920$	Kal 1970
$EuVO_4$	$TeCl_4$	$1100 \rightarrow 1000$	Schm 2005a
$Eu_2Ti_2O_7$	Cl_2	$1050 \rightarrow 950$	Hüb 1992
$Eu_{1-x}Sm_xVO_4$	$TeCl_4$	$1100 \rightarrow 1000$	Schm 2005a

Table 5.1 (continued)

Sink solid	Transport additive	Temperature / °C	Reference
"FeO"	HCl	725 → 700	Bow 1974
$Fe_{1-x}Zn_xO$	NH_4Cl	900 → 750	Loc 1999b
Fe_2O_3	Cl_2, I_2	980 → 860, 980 → 860	Emm 1968c
	Cl_2	1050 → 1000	Pes 1974
	Cl_2, HCl	1000 → 800	Ger 1977b
	$TeCl_4$	1000 → 800, 1000 → 900	
	$Cl_2, TeCl_4$	1000 → 950, 970 → 920	Bli 1977
	Cl_2	1100 → 800, T_2 → 300	Kle 1966a
	HCl	1000 → 800	Sch 1956
	HCl	1000 → 800	Klei 1964
	HCl	1100 → 800	Kle 1966a
	HCl	300 → 400	Klei 1966
	HCl	1000 → 800	Klei 1970
	HCl	1000 → 800	Klei 1972
	YCl_3	1170 → 1050	Klei 1977
	$TeCl_4$	1000 → 900	Aga 1984
	$TeCl_4$	700 → 800, 1100 → 900	Ger 1977
	$TeCl_4$	970 → 820	Pes 1975
	$TeCl_4$	800 → 850	Pes 1984
$Fe_2(^{16}O_{1-x}{}^{18}O_x)_3$	Cl_2	850 → 750	Kra 2005
$Fe_{2-x}Cr_xO_3$	$TeCl_4$	900 → 850	Pes 1984
$(Fe_{1-x}Cr_x)_2O_3$	$Cl_2, FeCl_3$	1070 → 770	Hay 1980
	$TeCl_4$	1000 → 900	Pat 2000c
$(Fe_{1-x}Ga_x)_2O_3$	$TeCl_4$	1000 → 900	Pat 2000c
Fe_3O_4	HCl	1000 → 800	Hau 1962
	HCl	1000 → 800	Klei 1972a
	HCl	1000 → 800	Mer 1973a
	HCl, $TeCl_4$	1000 → 800	Ger 1977b
	$TeCl_4$	970 → 820	Bli 1977
	$TeCl_4$	900 → 850	Pes 1984
$Fe_{2-x}Co_xGeO_4$	HCl	900 → 700	Pfe 2002
$Fe_{3-x}V_xO_4$	HCl	925 → 825, 1000 → 900	Bab 1987
$Fe_7(AsO_4)_6$	NH_4Cl	900 → 800	Wei 2004b
$FeBO_3$	HCl	760 → 670	Die 1975
Fe_3BO_6	HCl	875 → 835	Die 1976
$FeGa_2O_4$	Cl_2	880 → 850	Lec 1991
	Cl_2	980 → 860	Pat 2000b
$FeGeO_3$	HCl	900 → 700, 950 → 650	Roy 1963
	NH_4Cl	1000 → 700	Roy 1962
	NH_4Cl	1000 → 700	Roy 1963a
	Cl_2	880 → 820	Kru 1986
Fe_2GeO_4	Cl_2	880 → 820	Kru 1986
	HCl	900 → 700	Pfe 2002
	$TeCl_4$	920 → 760	Str 1980
Fe_2GeO_5	$TeCl_4$,	1050 → 950,	Aga 1984
	$TeCl_4$ + HCl	800 → 1050	
$Fe_{3.2}Ge_{1.8}O_8$	Cl_2	880 → 820	Kru 1986

Table 5.1 (continued)

Sink solid	Transport additive	Temperature / °C	Reference
$Fe_8Ge_3O_{18}$	$TeCl_4$,	$1000 \to 900$,	Aga 1984
	$TeCl_4 + HCl$	$800 \to 1050$	
	Cl_2	$1060 \to 980$	Kru 1986
$Fe_{15}Ge_8O_{36}$	Cl_2	$880 \to 820$	Kru 1986
$Fe_{1-x}Co_xGeO_3$	HCl	$900 \to 700$	Pfe 2002
$Fe_{1-x}In_xGe_2O_7$	Cl_2	$840 \to 780$	Pfe 2002b
$Fe_{1-x}Mn_xWO_4$	$TeCl_4$	$985 \to 900$	Sie 1983
$Fe_2Mo_3O_8$	$HCl, TeCl_4$	$1070 \to 1025, 960 \to 855$	Str 1982a
	$TeCl_4$	$955 \to 865$	Str 1983
	$Cl_2, HCl, TeCl_4$	$970 \to 800$	Str 1983a
$FeNbO_4$	Cl_2	$1000 \to 900$	Bru 1976a
	$Cl_2 + NbCl_5$	$920 \to 750$	Emm 1968c
$FeNb_2O_6$	$Cl_2, NH_4Cl,$	$1020 \to 960, 1020 \to 960,$	Emm 1968c
	$Cl_2 + NbCl_5$	$1005 \to 935$	
	Cl_2, NH_4Cl	$1020 \to 960$	Ros 1990a
$FeTaO_4$	$Cl_2 + TaCl_5$	$1000 \to 900$	Emm 1968b
Fe_2TiO_5	$TeCl_4$	$1100 \to 900$	Pie 1978
	$TeCl_4$	$T_2 \to T_1$	Mer 1980
FeV_2O_4	HCl	$925 \to 825$	Bab 1987
$FeWO_4$	Cl_2	$1010 \to 980$	Emm 1968b
	$TeCl_4$	$985 \to 900$	Sie 1982
	$TeCl_4$	$985 \to 900$	Yu 1993
$Fe_{1-x}V_xO_2$	$TeCl_4$	$1100 \to 1000$	Brü 1976
Ga_2O_3	Cl_2	$880 \to 800$	Red 1976
	Cl_2	$945 \to 895$	Jus 1986
	Cl_2	$1000 \to 800$	Pat 2000
	HCl	$1000 \to 800$	Pat 1999
	NH_4Cl	$945 \to 895$	Paj 1986
	$TeCl_4$	$1000 \to 900$	Ger 1977a
	C, CO, CH_4	$1100 \to T_1$	Fos 1960
	$Cl_2 + S$	$545 \to 745, 945 \to 845$	Jus 1988
	$I_2 + S$	$1150 \to 1100$	Nit 1966
	$I_2 + S$	$1150 \to 1100$	Nit 1967a
	$I_2 + S$	$1000 \to 900$	Aga 1985
Ga_2GeO_5	$Cl_2, GeCl_4$	$890 \to 820, 900 \to 860$	Red 1976
	$I_2 + S$	$1000 \to 900, 1050 \to 1000,$	Aga 1985
	HCl	$1100 \to 1000$	
$Ga_{2-x}In_xO_3$	Cl_2	$1000 \to 800, 1000 \to 900$	Pat 2000
$Ga_{2-x}V_xO_3$	$TeCl_4$	$1100 \to 900$	Kra 1991
Ga_4GeO_8	$HCl, I_2 + S$	$1100 \to 1050, 1000 \to 900$	Aga 1985
$(Ga_{0.6}In_{1.4})_2Ge_2O_7$	Cl_2	$1050 \to 950$	Pfe 2002b
$(Ga_{1.9}In_{0.1})_2Ge_2O_7$	Cl_2	$1050 \to 950$	Pfe 2002b
$Ga_{1-x}V_xO_y$	$TeCl_4$	$1000 \ldots 1050 \to 850 \ldots 900$	Kra 1991
$Ga_{1-x}V_xO_2$	$TeCl_4$	$1100 \to 900$	Brü 1976a

Table 5.1 (continued)

Sink solid	Transport additive	Temperature / °C	Reference
Gd_2O_3	HCl	$1800 \rightarrow T_1$	Kal 1971
$GdAsO_4$	$TeCl_4$	$1050 \rightarrow 950$	Schm 2005
$Gd_3Fe_5O_{12}$	$GdCl_3$	$1165 \rightarrow 1050$	Klei 1972
	$GdCl_3$	$1165 \rightarrow 1050$	Klei 1974
	$GdCl_3$	$1165 \rightarrow 1050$	Klei 1977, Klei 1977a
$GdPO_4$	PCl_5	$1100 \rightarrow 1000$	Orl 1971
	PCl_5	$1000 \rightarrow 900, 1100 \rightarrow 1000$	Rep 1971
	PCl_5	$1050 \rightarrow 950$	Mül 2004
$Gd_{1-x}Sm_xPO_4$	PCl_5	$1050 \rightarrow 950$	Mül 2004
$GdSbO_4$	$TeCl_4$	$1100 \rightarrow 950$	Ger 2007
$Gd_{2.66}Tb_{0.34}Fe_5O_{12}$	$Cl_2 + FeCl_2$	$1145 \rightarrow 1085$	Gib 1973
$Gd_2Ti_2O_7$	Cl_2	$1050 \rightarrow 950$	Hüb 1992
$GdVO_4$	$TeCl_4$	$1100 \rightarrow 1000$	Schm 2005a
GeO_2	H_2	$850 \rightarrow 750$	Schm 1981, Schm1981a
	H_2	$850 \rightarrow 750$	Schm 1983
	Cl_2	$900 \rightarrow 850$	Red 1978
	Cl_2	$950 \rightarrow 850$	Schm 1985
	$Cl_2 + MCl (M = Li,$ Na, K, Rb, Cs), $Cl_2 + (MnO, CuO, Fe_2O_3)$	$900 \rightarrow 850$	Red 1978
	HCl	$200 \rightarrow 500$	Chu 1964
	$GeCl_4$	$1000 \rightarrow 900$	Klei 1982
	$TeCl_4$	$1000 \rightarrow 900$	Aga 1984
	$I_2 + S$	$1100 \rightarrow 1000$	Aga 1985
	$H_2 + H_2O$	$745 \rightarrow 555$	Fak 1965
	$H_2O + WO_2$	$925 \rightarrow 825$	Pli 1983
HfO_2	$Cl_2, TeCl_4$	$1100 \rightarrow 1000$	Dag 1992
	$TeCl_4$	$1100 \rightarrow 1000$	Opp 1975
$HfSiO_4$	$Cl_2, HfCl_4, Br_2,$ $HfBr_4,$ I_2, HfI_4	$1150 \dots 1250 \rightarrow 1050 \dots 1150$ $(\Delta T = 100)$	Hul 1968
	$CrCl_3 + Se$	$920 \rightarrow 980$	Fuh 1986
$HgAs_2O_6$	$HgCl_2$	$650 \rightarrow 550$	Wei 2000
$(Hg_2)_2As_2O_7$	$HgCl_2$	$550 \rightarrow 500$	Wei 2003
$HoAsO_4$	$TeCl_4$	$1050 \rightarrow 950$	Schm 2005
$HoPO_4$	PCl_5	$1070 \rightarrow 930$	Mül 2004
	$Br_2 + PBr_3$	$1025 \rightarrow 925, 1125 \rightarrow 925$	Orl 1974
$Ho_2Ti_2O_7$	Cl_2	$1050 \rightarrow 950$	Hüb 1992
$HoVO_4$	$TeCl_4$	$1100 \rightarrow 1000$	Schm 2005a

Table 5.1 (continued)

Sink solid	Transport additive	Temperature / °C	Reference
In_2O_3	Cl_2, NH_4Cl	$1000 \rightarrow 900$, $900 \rightarrow 800$	Kru 1986
	Cl_2	$950 \rightarrow 900$	Joz 1987
	Cl_2	$1000 \rightarrow 800$	Pat 2000
	HCl	$950 \rightarrow 720$	Wit 1971
	$I_2 + S$	$1150 \rightarrow 1100$	Nit 1966
	$I_2 + S$	$1150 \rightarrow 1100$	Nit 1967a
In_2O_3:SnO_2	Cl_2, $I_2 + S$	$975 \rightarrow 925$, $1025 \rightarrow 925$	Wer 1996
	Cl_2	$1050 \rightarrow 900$	Pat 2000a
$(In_{1.9}Mn_{0.1})_2Ge_2O_7$	Cl_2	$1000 \rightarrow 800$	Pfe 2002b
$In_2Ge_2O_7$	Cl_2, NH_4Cl	$840 \rightarrow 780$, $820 \rightarrow 720$	Kru 1986
$InMo_4O_6$	H_2O	$1000 \rightarrow 900$	Ste 2005b
$In_2Mo_3O_{12}$	Cl_2, Br_2	$700 \rightarrow 600$, $900 \rightarrow 800$	Ste 2005b
In_xWO_3	NH_4Cl	$900 \rightarrow 800$	Ste 2008
$In_2W_3O_{12}$	Cl_2	$800 \rightarrow 700$	Ste 2008
In_6WO_{12}	$TeCl_4$	$1000 \rightarrow 835$	Gae 1993
IrO_2	Cl_2	$1100 \rightarrow 900$	Bel 1966
	O_2	$1205 \rightarrow 1035$	Sch 1960
	O_2	$1090 \rightarrow 1010$	Cor 1962
	O_2	$1230 \rightarrow 1050$	Geo 1982
	O_2	$1230 \rightarrow 1050$	Tri 1982
	O_2	$1150 \rightarrow 1000$	Rea 1976
	$TeCl_4$	$1100 \rightarrow 1000$	Opp 1975
$Ir_{1-x}Ru_xO_2$	O_2	$1230 \rightarrow 1050$	Geo 1982
	O_2	$1230 \rightarrow 1050$	Tri 1982
$K_xNb_{1-y}W_yO_3$	$HgCl_2$	$850 \rightarrow 800$	Hus 1994
$K_xP_4W_8O_{32}$		$1200 \rightarrow 1000$	Rou 1997
$K_{0.25}WO_3$	$HgCl_2$, $HgBr_2$	$800 \rightarrow 750$	Hus 1991
K_xWO_3	Cl_2, I_2	$900 \rightarrow T_1$, $1000 \rightarrow 900$	Scho 1992
	$HgCl_2$	$850 \rightarrow 800$	Hus 1994
	$HgCl_2$, $HgBr_2$, HgI_2, $PtCl_2$	$800 \rightarrow 750$	Hus 1997
$LaAsO_4$	$TeCl_4$	$1050 \rightarrow 950$	Schm 2005
$LaNbO_4$	Cl_2	$1090 \rightarrow 1000$	Stu 1976
	HCl	$800 \ldots 1200 \rightarrow T_1$ $(\Delta T = 50 \ldots 100)$	Cur 1965
	NH_4Br	$1090 \rightarrow 1000$	Stu 1976
$LaNb_3O_9$	Cl_2	$1100 \rightarrow 1000$	Stu 1975, Stu 1976
$LaNb_5O_{14}$	Cl_2, NH_4Br	$1050 \rightarrow 950$	Hof 1990
$LaNb_7O_{19}$	Cl_2	$800 \rightarrow 780$	Hof 1991
	Cl_2	$900 \rightarrow 800$	Bus 1996

Table 5.1 (continued)

Sink solid	Transport additive	Temperature / °C	Reference
$LaPO_4$	PCl_5	$1100 \rightarrow 1000$	Tan 1968
	PCl_5	$1100 \rightarrow 1000$	Orl 1971
	NH_4Br, $Br_2 + C$, $Br_2 + CO$, $Br_2 + PBr_3$	$1125 \rightarrow 925$	Sch 1972
$LaSbO_4$	$TeCl_4$	$1100 \rightarrow 950$	Ger 2007
$LaTaO_4$	Cl_2, NH_4Br	$1050 \rightarrow 950$	Stu 1976
	Cl_2, NH_4Br	$1050 \rightarrow 950$	Lan 1986b
	Cl_2	$1100 \rightarrow 1000$	Scha 1990
$LaTa_3O_9$	Cl_2, NH_4Br	$1090 \rightarrow 1000$	Stu 1976
	Cl_2	$1090 \rightarrow 1000$	Lan 1987
	Cl_2	$1100 \rightarrow 1000$	Scha 1990
"$LaTa_5O_{14}$" ($La_{4.67}Ta_{22}O_{62}$)	Cl_2, NH_4Br	$1390 \rightarrow 1300$	Scha 1989
"$LaTa_5O_{14}$" ($La_{4.67}Ta_{22}O_{62}$)	Cl_2, NH_4Br	$1390 \rightarrow 1300$	Scha 1989a
$LaTa_7O_{19}$	Cl_2, NH_4Br	$1120 \rightarrow 1050$	Lan 1986
	Cl_2, NH_4Br	$1120 \rightarrow 1050$	Lan 1986b
	Cl_2	$1100 \rightarrow 1000$	Scha 1990
$LaVO_4$	$TeCl_4$	$1100 \rightarrow 1000$	Schm 2005a
Li_2O	H_2O	$1000 \rightarrow T_1$	Ark 1955
	H_2O	$870 \rightarrow 820$	Ber 1963
$LiNbO_3$	S	$1000 \rightarrow 900$	Sch 1988
Li_xMoO_3	$TeCl_4$	$850 \rightarrow 750$	Schm 2008
Li_xWO_3	$TeCl_4$	$850 \rightarrow 750$	Schm 2008
	$HgCl_2$	$800 \rightarrow 700$	Rüs 2008
$LuAsO_4$	$TeCl_4$	$1050 \rightarrow 950$	Schm 2005
$LuPO_4$	$Br_2 + PBr_3$	$1025 \rightarrow 925$, $1125 \rightarrow 925$	Orl 1974
	$Br_2 + CO$	$1125 \rightarrow 925$, $1200 \rightarrow 1000$	Orl 1978
	$Br_2 + PBr_3$	$1125 \rightarrow 925$, $1200 \rightarrow 1000$	
	PBr_5	$1050 \rightarrow 975$	Mül 2004
$LuSbO_4$	$TeCl_4$	$1100 \rightarrow 950$	Ger 2007
$Lu_2Ti_2O_7$	Cl_2	$1050 \rightarrow 950$	Hüb 1992
MgO	H_2, C	$1550 \rightarrow 950$, $1525 \rightarrow 1350$	Wol 1965
	Cl_2	$1200 \rightarrow 1000$	Bay 1985
	HCl	$1000 \rightarrow 800$	Klei 1972
	HCl	$T_2 \rightarrow 1000$	Gru 1973
	HCl	$T_2 \rightarrow 1000$	Lib 1994
	H_2O	$1735 \rightarrow 1505$	Ale 1963
	CO	$1600 \rightarrow 1400$	Bud 1967
$Mg_xFe_yMn_zO_4$	HCl	$1000 \rightarrow 800 \dots 900$	Klei 1973

Table 5.1 (continued)

Sink solid	Transport additive	Temperature / °C	Reference
$Mg_{0.5}Fe_2Mn_{0.5}O_4$	HCl	$1000 \to 800$	Klei 1973
$MgFe_2O_4$	HCl	$1000 \to 800$	Klei 1963
	HCl	$1000 \to 800$	Klei 1972a
$MgGeO_3$	Cl_2	$1100 \to 1000$	Kru 1986a
$MgMoO_4$	Cl_2	$1060 \to 990$	Emm 1968b
	Cl_2, Br_2	$1000 \to 900$	Ste 2003a
$Mg_2Mo_3O_8$	Cl_2, Br_2	$1000 \to 900$	Ste 2003a
$MgNb_2O_6$	HCl	$1005 \to 935$	Emm 1968c
$MgTa_2O_6$	Cl_2	$1060 \to 980$	Emm 1968b
	Cl_2, $TeCl_4$	$1100 \to 1020, 1100 \to 1015,$	Bay 1983
	$Cl_2 + FeCl_3$	$1025 \to 965$	
$MgTiO_3$	Cl_2	$1060 \to 980$	Emm 1968b
MgV_2O_4	I_2, HBr, $MgBr_2$,	$800 \to 600$	Pic 1973b
	$MgBr_2 + I_2$,		
	$MgBr_2 + S$		
$Mg_{1+x}V_{2-x}O_4$	$MgBr_2 + I_2$	$800 \to 600$	Pic 1973,
			Pic 1973b
$MgWO_4$	Cl_2	$1060 \to 980$	Emm 1968b
	Cl_2	$1000 \to 900$	Ste 2005a
	HCl	$800 \dots 1200 \to T_1$	Cur 1965
		$(\Delta T = 50 \dots 100)$	
MnO	Cl_2	$980 \to 900$	Kle 1966a
	Cl_2	$980 \to 900$	Emm 1968c
	Br_2	$1000 \to 900$	Loc 2005
	HCl	$1025 \to 925$	Ros 1987
	$SeCl_4$, $TeCl_4$	$1025 \to 925$	Ros 1988
$Mn_{1-x}Zn_xO$	NH_4Cl	$900 \to 750$	Loc 1999b
$Mn_{1-x}Mg_xO$	HCl	$925 \to 825$	Skv 2000
Mn_2O_3	Cl_2	$980 \to 860$	Emm 1968c
	Cl_2, Br_2	$1025 \to 925, 1125 \to 1025$	Ros 1987
	HCl, HBr	$1025 \to 925, 1025 \to 925$	
	$AlCl_3$	$790 \to 660$	Mar 1999
	$SeCl_4$, $TeCl_4$	$1025 \to 925$	Ros 1988
Mn_3O_4	Cl_2, Br_2	$1025 \to 925$	Ros 1987
	HCl	$1000 \to 800$	Klei 1963
	HCl	$1000 \to 800$	Klei 1972a
	HCl	$1050 \to 950$	Yam 1972
	HCl, HBr	$1025 \to 925$	Ros 1987
	$SeCl_4$, $TeCl_4$	$1025 \to 925$	Ros 1988
$Mn_{1.286}Fe_{1.714}O_4$	HCl	$1000 \to 800$	Klei 1973
$MnFe_{2-x}Mn_xO_4$	HCl	$1000 \to 800$	Klei 1964
$Mn_{0.5}Zn_{0.5}Fe_2O_4$	HCl	$1100 \to 1000$	Klei 1973
$Mn_{0.5}Zn_{0.45}Fe_{2.05}O_4$	HCl	$1100 \to 1000$	Klei 1973
$Mn_{0.75}Fe_{2.25}O_4$	HCl	$1000 \to 800$	Klei 1973
$Mn_{1-x}Zn_xCr_2O_4$	Cl_2	$1030 \to 950$	Lec 1993

Table 5.1 (continued)

Sink solid	Transport additive	Temperature / °C	Reference
$Mn_2As_2O_7$	Cl_2	$880 \rightarrow 800$	Wei 2004c
	Cl_2	$880 \rightarrow 800$	Wei 2005
$MnCr_2O_4$	Cl_2	$980 \rightarrow 860$	Emm 1968b
$MnFe_2O_4$	HCl	$1000 \rightarrow 800$	Klei 1963
	HCl	$800 \dots 1200 \rightarrow T_1$ $(\Delta T = 50 \dots 100)$	Cur 1965
	HCl	$1000 \rightarrow 800$	Klei 1972a
$Mn(Cr_{2-x}Fe_x)O_4$	Cl_2	$1000 \rightarrow 900$	Emm 1968b
$MnGeO_3$	$GeCl_4$	$1010 \rightarrow 930$	Red 1976
	HCl	$900 \rightarrow 700$	Pfe 2002
	HCl	$950 \rightarrow 650$	Roy 1963
	NH_4Cl	$1000 \rightarrow 700$	Roy 1963a
$Mn_{1-x}Zn_1GeO_3$	HCl	$1050 \rightarrow 900$	Pfe 2002c
$(Mn_{1-x}Zn_x)_2GeO_4$	HCl	$1050 \rightarrow 900$	Pfe 2002c
$MnInGe_2O_7$	Cl_2	$1000 \rightarrow 800$	Pfe 2002b
$Mn_3Cr_2Ge_3O_{11}$	Cl_2, CCl_4	$1290 \rightarrow 1220$	Paj 1985
$Mn_3Cr_2Ge_3O_{12}$	CCl_4, $TeCl_4$	$T_2 \rightarrow T_1$, $980 \rightarrow 910$	Paj 1985
	Cl_2, $Cl_2 + S$	$900 \rightarrow 750$, $950 \rightarrow 880$	
	Cl_2, CCl_4	$1000 \rightarrow 950$	Paj 1986a
$Mn_{6.5}In_{0.5}GeO_{12}$	Cl_2	$1000 \rightarrow 800$	Pfe 2002b
$Mn_{1-x}Co_xGeO_3$	HCl	$900 \rightarrow 700$	Pfe 2002
$MnMoO_4$	Cl_2	$905 \rightarrow 870$	Emm 1968b
	Cl_2, HCl, $SeCl_4$, $TeCl_4$	$1100 \rightarrow 1000$	Rei 1994
$Mn_2Mo_3O_8$	HCl, $TeCl_4$	$940 \rightarrow 820$, $945 \rightarrow 845$	Str 1983
	Cl_2, I_2, HCl, HI, $SeCl_4$	$1100 \rightarrow 1000$	Rei 1994
	Cl_2, HCl, $TeCl_4$	$970 \rightarrow 800$	Str 1983a
$MnNb_2O_6$	$Cl_2 + NbCl_5$	$1010 \rightarrow 970$	Emm 1968c
	$PtCl_2$, NH_4Cl	$1020 \rightarrow 960$	Ros 1991a
$MnTa_2O_6$	Cl_2	$1010 \rightarrow 950$	Emm 1968b
$MnWO_4$	Cl_2	$1000 \rightarrow 900$	Emm 1968b
	Cl_2	$900 \rightarrow 800$	Ste 2005a
Mn_2SnO_4	NH_4Cl	$950 \rightarrow 800$	Trö 1972
MoO_2	I_2	$900 \rightarrow 700$	Ben 1969
	I_2	$1000 \rightarrow 800$	Opp 1971
	I_2	$1000 \rightarrow 800$, $730 \rightarrow 650$	Sch 1973a
	I_2	$950 \rightarrow 750$	Ben 1974
	I_2	$1100 \rightarrow 1000$	Schu 1971
	$HgCl_2$	$660 \rightarrow 580$	Scho 1990
	$HgCl_2$, HgI_2	$660 \rightarrow 580$, $980 \rightarrow 900$	Scho 1992
	I_2, $HgCl_2$, $HgBr_2$, HgI_2	$1100 \rightarrow 1000$	Fel 1996
	$HgBr_2$	$820 \rightarrow 740$	Scho 1992
	$TeCl_4$	$1100 \rightarrow 1000$	Opp 1975

Table 5.1 (continued)

Sink solid	Transport additive	Temperature / °C	Reference
MoO_2	$TeCl_4$	$700 \rightarrow 630$	Ban 1976
	$TeCl_4$, $TeCl_4 + S$	$750 \rightarrow 700$, $670 \rightarrow 620$	Mer 1982
	$TeCl_4$, $TeCl_4 + S$	$T_2 \rightarrow T_1$ [1]	Mon 1984
	$TeBr_4$	$1150 \rightarrow 950$	Rit 1980
	CuI, H_2O	$1000 \rightarrow 800$	Sch 1973a
MoO_2:Ni	I_2	$950 \rightarrow 750$	Ben 1974
$Mo_xRe_{1-x}O_2$	I_2	$1100 \rightarrow 1000$	Fel 1998a
$Mo_{1-x}Ru_xO_2$	$TeCl_4$	$1100 \rightarrow 1000$	Nic 1993
$Mo_{1-x}V_xO_2$	$TeCl_4$	$1075 \rightarrow 965$	Hör 1973
	$TeCl_4$	$1100 \rightarrow 900$	Brü 1977
	$TeCl_4$	$1100 \rightarrow 900$	Rit 1977
Mo_4O_{11}	I_2, $TeCl_4$	$560 \rightarrow 510$, $740 \rightarrow 680$	Koy 1988, Ino 1988
	$MoCl_5 + TeCl_4$	$540 \rightarrow 525$	Fou 1984
	$TeCl_4$	$570 \rightarrow 550$, $690 \rightarrow 650$	
	I_2, $HgCl_2$, $HgBr_2$, HgI_2	$740 \rightarrow 700$	Fel 1996
	$HgCl_2$, $HgBr_2$, HgI_2	$780 \rightarrow 700$	Scho 1992
	$TeCl_4$	$690 \rightarrow 640$, $650 \rightarrow 560$	Ban 1976
	$TeCl_4$	$660 \rightarrow 510$	Neg 1994
Mo_8O_{23}	I_2, $HgCl_2$, $HgBr_2$, HgI_2	$740 \rightarrow 700$	Fel 1996
	$TeCl_4$	$750 \rightarrow 730$	Ban 1976
Mo_9O_{26}	I_2	$675 \rightarrow T_1$	Roh 1994
	I_2, $HgCl_2$, $HgBr_2$, HgI_2	$740 \rightarrow 700$	Fel 1996
	$TeCl_4$	$770 \rightarrow 745$	Ban 1976
MoO_3	Cl_2, $HgCl_2$	$T_2 \rightarrow T_1$	Sch 1985
	I_2, $HgCl_2$, $HgBr_2$, HgI_2	$740 \rightarrow 700$	Fel 1996
	HCl	$205 \ldots 365$	Hul 1956
	$TeCl_4$	$650 \rightarrow 600$	Ban 1976
	$TeCl_4$	$750 \rightarrow 700$	Fou 1979
	H_2O	$600 \ldots 690$	Gle 1962
	$MoCl_5$	$680 \rightarrow 600$	Fou 1984
Mo_3ReO_{11}	$TeCl_4$, $HgCl_2$	$740 \rightarrow 700$	Fel 1998a
Mo_5TeO_{16}	$TeCl_4$	$690 \rightarrow 680$	Neg 2000
	$TeCl_4$	$600 \rightarrow 575$	Fou 1984
NbO	I_2	$950 \rightarrow 1100$	Sch 1962
	I_2	$920 \rightarrow 1080$	Hib 1978
	NH_4Cl, NH_4Br, NH_4I	$990 \rightarrow 1150$	Kod 1976
NbO_2	Cl_2	$1125 \rightarrow 1025$	Stu 1972, Bru 1975, Schw 1983

[1] Gas phase analysis by Raman spectroscopy and thermodynamic calculations.

Table 5.1 (continued)

Sink solid	Transport additive	Temperature / °C	Reference
NbO_2	$HgCl_2$	$1125 \rightarrow 1025$	Schw 1982
	NH_4Br, $TeBr_4$	$1100 \rightarrow 950$, $1105 \rightarrow 990$	Kod 1975
	I_2, NH_4Cl,	$1105 \rightarrow 990$, $1110 \rightarrow 980$	Kod 1975a
	NH_4Br, NH_4I,	$1100 \rightarrow 1000$, $1110 \rightarrow 990$	
	$NbCl_5$	$1140 \rightarrow 980$	
	NH_4Cl, NH_4Br	$1135 \rightarrow 980$, $1140 \rightarrow 1000$	Kod 1976
	$TeCl_4$	$1100 \rightarrow 920$	Sak 1972
	$TeCl_4$	$1100 \rightarrow 900$	Opp 1975
	$TeCl_4$	$1100 \rightarrow T_1$	Rit 1978
	$NbCl_5$, Nb_3O_7Cl	$1100 \rightarrow 1000$, $1225 \rightarrow 1025$	Schw 1982
	$NbCl_5$	$1125 \rightarrow 1025$	Schw 1983
$NbO_{2.417}$	$TeCl_4$	$1000 \rightarrow 900$	Rit 1978a
$NbO_{2.42}$	I_2	$740 \rightarrow 650$	Sch 1962
$NbO_{2.464}$	$NbCl_5$, $NbOCl_3$	$1200 \rightarrow 1160$	Gru 1969
$NbO_{2.483}$	$NbCl_5$,	$1080 \rightarrow 980$	Gru 1967
	$NbOCl_3 + CO_2/CO$		
$Nb_{12}O_{29}$	Cl_2	$1125 \rightarrow 1025$	Stu 1972, Bru 1975, Schw 1983
	$TeCl_4$	$950 \rightarrow 900$	Sch 1962
	$TeCl_4$	$1100 \rightarrow 780$, $1100 \rightarrow 920$	Sak 1972
	NH_4Br, $TeBr_4$	$1100 \rightarrow 950$, $1105 \rightarrow 990$	Kod 1975
	I_2	$750 \rightarrow 650$	Hib 1978
	$TeCl_4$	$950 \rightarrow 900$	Rit 1978
	$NbCl_5$	$1125 \rightarrow 1025$	Scha 1974
$Nb_{22}O_{54}$	$HgCl_2$	$1250 \rightarrow 1200$	Hus 1986
$Nb_{25}O_{62}$	$HgCl_2$	$1250 \rightarrow 1200$	Hus 1986
$Nb_{47}O_{116}$	$HgCl_2$	$1250 \rightarrow 1200$	Hus 1986
$Nb_{53}O_{132}$	$HgCl_2$	$1250 \rightarrow 1200$	Hus 1986
Nb_2O_5	$NbCl_5$, NbI_5	$1000 \rightarrow 700$, $700 \rightarrow 550$	Lav 1964
	$NbCl_5$	$1050 \rightarrow 950$	Hib 1978
	$NbCl_5$, $NbBr_5$,	$850 \rightarrow 750$	Sch 1964
	NbI_5,		
	$Cl_2 + NbCl_5$,		
	$TeCl_4$	$780 \rightarrow 740$	Tor 1976
	$TeCl_4$	$900 \rightarrow 800$, $1000 ... 600 \rightarrow T_1$ ($\Delta T = 100$)	Rit 1978a
	$TeCl_4$	$1000 \rightarrow 900$	Sch 1978
	$Cl_2 + H_2O$, HCl,	$800 \rightarrow 600$	Gru 1966
	$HCl + H_2O$,		
	$NbOCl_3$		
	$Cl_2 + NbCl_5$	$890 \rightarrow 835$	Sch 1966
	Cl_2, $Cl_2 + NbCl_5$,	$850 \rightarrow 750$, $1050 \rightarrow 980$,	Emm 1968
	$Cl_2 + NbCl_5 + H_2O$	$700 \rightarrow 600$,	
	$Cl_2 + NbCl_5 +$	$970 \rightarrow 860$	
	SnO_2		

Table 5.1 (continued)

Sink solid	Transport additive	Temperature / °C	Reference
Nb_2O_5	$Cl_2 + NbCl_5$	$800 \rightarrow T_1$	Kod 1972
	$I_2 + NbI_5,$	$790 \rightarrow 710,$	Emm 1968c
	$NbCl_5 + HCl$	$750 \rightarrow 720$	
	$NbF_5, NbOF_3$	1270	Gru 1973
	S	$1000 \rightarrow 900$	Sch 1980
$(Nb,W)O_x$	$HgCl_2$	$1000 \rightarrow 925$	Hus 1989
$Nb_{18}As_2O_{50}$	Cl_2	$1020 \rightarrow 975$	Emm 1968b
$Nb_xCr_{1-x}O_2$	$Cl_2, NbCl_5 + Cl_2$	$980 \rightarrow 860$	Ben 1978
$Nb_{18}GeO_{47}$	Cl_2	$1000 \rightarrow 900$	Emm 1968b
$Nb_{18}V_2O_{50}$	$Cl_2 + NbCl_5$	$980 \rightarrow 880$	Emm 1968b
$Nb_{1-x}V_xO_2$	$TeCl_4$	$1100 \rightarrow 900$	Brü 1977
	$TeCl_4$	$1100 \rightarrow 900$	Rit 1977
	$Cl_2, HCl, TeCl_4$	$900 \rightarrow 800$	Fec 1993
	$TeCl_4$	$1050 \rightarrow 940$	Lau 1973
	$TeCl_4$	$1000 \rightarrow 900$	Woe 1997
$NdAsO_4$	$HgCl_2, TeCl_4,$	$1075 \rightarrow 950$	Schm 2005
	$TeBr_4, PCl_5, PBr_5,$		
	$As + Cl_2,$		
	$As + PtCl_2,$		
	$As + PtBr_2,$		
	$S + PtCl_2$		
$Nd_{0.5}Ln_{0.5}AsO_4$ $Ln = Sm \dots Yb$	$TeCl_4$	$1075 \rightarrow 975$	Schm 2003
$Nd_{1-x}Pr_xAsO_4$	$TeCl_4$	$1075 \rightarrow 975$	Schm 2005
$NdNbO_4$	Cl_2	$1090 \rightarrow 1000$	Gru 2000
$NdNb_7O_{19}$	Cl_2	$800 \rightarrow 750$	Hof 1993
$NdPO_4$	PCl_5	$1100 \rightarrow 1000$	Orl 1971
	PCl_5	$1000 \rightarrow 900, 1100 \rightarrow 1000$	Rep 1971
	PCl_5	$1050 \rightarrow 950$	Mül 2004
$NdSbO_4$	$TeCl_4$	$1100 \rightarrow 950$	Ger 2007
$NdTaO_4$	Cl_2	$980 \rightarrow 880, 1100 \rightarrow 1000$	Scha 1989
	NH_4Br	$980 \rightarrow 880, 1100 \rightarrow 1000$	
$NdTaO_4$	Cl_2	$1100 \rightarrow 1000$	Scha 1990
$NdTa_3O_9$	Cl_2	$1100 \rightarrow 1000$	Scha 1988a
	Cl_2, NH_4Br	$1100 \rightarrow 1000$	Scha 1989
$NdTa_7O_{19}$	Cl_2	$1100 \rightarrow 1000$	Scha 1990
	Cl_2, NH_4Br	$1100 \rightarrow 1000$	Scha 1991
$Nd_2Ti_2O_7$	Cl_2	$1050 \rightarrow 950$	Hüb 1992
$Nd_4Ti_9O_{24}$	Cl_2	$1000 \rightarrow 900$	Hüb 1992a
$Nd_{1-x}Pr_xVO_4$	$TeCl_4$	$1100 \rightarrow 1000$	Schm 2005a
$NdVO_4$	$TeCl_4$	$1100 \rightarrow 1000$	Schm 2005a
NiO	Cl_2, HCl	$980 \rightarrow 860$	Kle 1966a
	Cl_2	$980 \rightarrow 860$	Emm 1968c
	$Br_2,$	$950 \rightarrow 920$	Sto 1966
	$HCl,$	$950 \rightarrow 920, 980 \rightarrow 890$	
	HBr	$940 \rightarrow 910$	

Table 5.1 (continued)

Sink solid	Transport additive	Temperature / °C	Reference
NiO	HCl	$1000 \rightarrow 800$	Klei 1964
	HCl	$1000 \rightarrow 800$	Klei 1972
	HCl	$1050 \rightarrow 1000$	Kur 1972
	HCl	$1050 \rightarrow 1000$	Kur 1975
	HCl	$1100 \rightarrow 1050$	Chu 1995
$Ni_{1-x}Co_xO$	HCl	$900 \rightarrow 800$	Bow 1972
	HCl	$800 \rightarrow 650$	Loc 1999
$Ni_{1-x}Mg_xO$	HCl	$925 \rightarrow 825$	Skv 2000
$Ni_{1-x}Zn_xO$	HCl	$900 \rightarrow 750$	Loc 1999a
$Ni_xCo_{3-x}O_4$	Cl_2, HCl	$1055 \rightarrow 855$	Tar 1984
$Ni_2As_2O_7$	Cl_2	$880 \rightarrow 800$	Wei 2004c
	Cl_2	$880 \rightarrow 800$	Wei 2005
$NiCr_2O_4$	Cl_2	$950 \rightarrow 800$	Emm 1968b
	Cl_2	$1025 \rightarrow 925$	Pes 1982
$Ni(Fe_{2-x}Cr_x)O_4$	Cl_2	$1000 \rightarrow 900$	Emm 1968b
$NiFe_2O_4$	Cl_2, $TeCl_4$	$1000 \rightarrow 950$, $980 \rightarrow 880$	Bli 1977
	HCl	$1000 \rightarrow 800$	Klei 1963a
	HCl	$1000 \rightarrow 800$	Klei 1964
	HCl	$800 \dots 1200 \rightarrow T_1$	Cur 1965
		$(\Delta T = 50 \dots 100)$	
	HCl	$900 \rightarrow 800$	Klei 1965
	HCl	$1220 \rightarrow 1190$	Klei 1967
	HCl	$1000 \rightarrow 800$	Klei 1972a
	$TeCl_4$	$980 \rightarrow 880$	Pes 1978
$(Ni_{1-x}Fe_x)Fe_2O_4$	HCl	$1000 \rightarrow 800$	Klei 1964
$Ni_{0.5}Zn_{0.5}Fe_2O_4$	HCl	$1100 \rightarrow 1000$	Klei 1973
$Ni_{0.8}Fe_{2.2}O_4$	HCl	$1000 \rightarrow 800$	Klei 1973
$(Ni_{1-x}Co_x)_2O_4$ $(Fe_{1-x}Cr_x)_2O_4$	HCl	$1000 \rightarrow 800$	Pat 2000
$NiGa_2O_4$	Cl_2	$1030 \rightarrow 900$	Paj 1990
Ni_2GeO_4	Cl_2	$1050 \rightarrow 950$	Kru 1986
	Cl_2	$1050 \rightarrow 900$,	Pfe 2002a
	HCl	$900 \rightarrow 700$, $1000 \rightarrow 900$	
$Ni_{1-x}Co_xGeO_3$	HCl	$900 \rightarrow 700$	Pfe 2002a
$(Ni_{1-x}Co_x)_2GeO_4$	HCl	$1000 \rightarrow 900$	Pfe 2002a
$NiMoO_4$	Cl_2	$905 \rightarrow 870$	Emm 1968b
	Cl_2, Br_2	$900 \rightarrow 800$	Ste 2006a
$Ni_2Mo_3O_8$	Cl_2, Br_2	$900 \rightarrow 800$	Ste 2006a
	$TeCl_4$	$965 \rightarrow 815$	Str 1983
	Cl_2, HCl, $TeCl_4$	$970 \rightarrow 800$	Str 1983a
$NiNb_2O_6$	Cl_2	$1010 \rightarrow 970$	Emm 1968c
	$PtCl_2$, NH_4Cl	$1020 \rightarrow 960$, $1020 \rightarrow 960$	Ros 1992a
$NiTa_2O_6$	$TaCl_5 + Cl_2$	$1000 \rightarrow 900$	Emm 1968b
$NiTiO_3$	Cl_2	$1030 \rightarrow 960$	Emm 1968b
	$SeCl_4$	$1050 \rightarrow 1000$	Bie 1990
Ni_2SiO_4	SiF_4	$1190 \rightarrow 1040$	Hof 1977

Table 5.1 (continued)

Sink solid	Transport additive	Temperature / °C	Reference
NiWO$_4$	Cl$_2$	1040 → 1010	Emm 1968b
	Cl$_2$	900 → 800	Ste 2005a
NpO$_2$	TeCl$_4$	1075 → 975	Spi 1979
	TeCl$_4$	1050 → 960	Spi 1980
OsO$_2$	O$_2$	not specified	Sch 1964
	O$_2$	900 → 600	Opp 1998
	O$_2$	930 → 900	Yen 2004
	O$_2$, OsO$_4$	960 → 900	Gre 1968
	TeCl$_4$	1100 → 1000	Opp 1975
	NaClO$_3$	960 → 900	Rog 1969
	O$_2$	920 → 900	Yen 2004
Os$_{1-x}$V$_x$O$_2$	HgCl$_2$	900 → 800	Arn 2008
PdO	Cl$_2$	800 → 900	Rog 1969
	PdCl$_2$	825 → 900	Rog 1971
PrAsO$_4$	TeCl$_4$	1050 → 950	Schm 2005
Pr$_{1-x}$La$_x$AsO$_4$	TeCl$_4$	1075 → 950	Schm 2005
Pr$_{1-x}$Nd$_x$AsO$_4$	TeCl$_4$	1075 → 975	Schm 2003
PrNb$_3$O$_9$	Cl$_2$	950 → 900	Hof 1993
PrNb$_7$O$_{19}$	Cl$_2$	800 → 750	Hof 1993
PrPO$_4$	PCl$_5$	1100 → 1000	Orl 1971
	PCl$_5$	1000 → 900, 1100 → 1000	Rep 1971
	PCl$_5$	1050 → 950	Mül 2004
Pr$_{1-x}$Nd$_x$PO$_4$	PCl$_5$	1050 → 950	Mül 2004
PrSbO$_4$	PtCl$_2$, PCl$_5$, TeCl$_4$	1100 → 950	Ger 2007
PrTaO$_4$	Cl$_2$, NH$_4$Br	1120 → 1020	Stei 1987
	Cl$_2$	1100 → 1000	Scha 1990
PrTa$_3$O$_9$	Cl$_2$, NH$_4$Br	1100 → 1000	Stei 1987
	Cl$_2$	1100 → 1020	Scha 1988a
	Cl$_2$	1100 → 1000	Scha 1990
	Cl$_2$	1100 → 1020	Stei 1990
PrTa$_7$O$_{19}$	Cl$_2$, NH$_4$Br	1120 → 1020	Stei 1987
	Cl$_2$	1100 → 1000	Scha 1990
PrVO$_4$	TeCl$_4$	1100 → 1000	Schm 2005a
PuO$_2$	TeCl$_4$	1075 → 975	Spi 1979
Rb$_x$Nb$_{1-y}$W$_y$O$_3$	HgCl$_2$	850 → 800	Hus 1994
Rb$_x$WO$_3$	HgCl$_2$	850 → 800	Hus 1994
	HgCl$_2$, HgBr$_2$	800 → 700	Hus 1997

Table 5.1 (continued)

Sink solid	Transport additive	Temperature / °C	Reference
ReO_2	I_2	$850 \rightarrow 825$	Rog 1969
	I_2, H_2O	$700 \rightarrow 600, 850 \rightarrow 825$	Sch 1973b
	$I_2 + H_2O$	$700 \rightarrow 600$	
	$I_2, HgCl_2, HgBr_2,$	$1100 \rightarrow 1000$	Fel 1998
	$HgI_2, TeCl_4$		
	$TeCl_4$	$1100 \rightarrow 1000$	Opp 1975
$Re_{1-x}Mo_xO_2$	$I_2, TeCl_4$	$1100 \rightarrow 1000, 1000 \rightarrow 900$	Fel 1998a
ReO_3	I_2	$400 \rightarrow 370$	Fer 1965
	I_2	$380 \rightarrow 360$	Qui 1970
	$I_2,$	$380 \rightarrow 370$	Pea 1973
	HCl	$650 \rightarrow 425, 600 \rightarrow 550$	
	$I_2,$	$400 \rightarrow 370,$	Sch 1973
	H_2O, Re_2O_7	$400 \rightarrow 350$	
	I_2, H_2O	$385 \rightarrow 370, 400 \rightarrow 350$	Sch 1973b
	$I_2, HgCl_2, HgBr_2,$	$600 \rightarrow 500$	Fel 1998
	$HgI_2, TeCl_4$		
Re_2O_7	H_2O	$220 \rightarrow 165$	Gle 1964a
	H_2O, O_2	$180 \rightarrow T_1$	Mül 1965
Rh_2O_3	Cl_2	$1050 \rightarrow 950$	Goe 1996
	HCl	$1000 \rightarrow 850$	Poe 1981
$RhAsO_4$	Cl_2	$850 \rightarrow 750$	Goe 1996
$RhNbO_4$	Cl_2	$1100 \rightarrow 1000$	Goe 1996
	$TeCl_4$	$1050 \rightarrow 850$	Jen 2009
$RhTaO_4$	Cl_2	$1100 \rightarrow 1000$	Goe 1996
	$TeCl_4$	$1050 \rightarrow 850$	Jen 2009
$RhVO_4$	$TeCl_4$	$1000 \rightarrow 900$	Jen 2009
RuO_2	$Cl_2, TeCl_4,$	$1100 \rightarrow 1000,$	Rei 1991
	O_2	$1100 \rightarrow 1000, 1300 \rightarrow 1150$	
	$HgCl_2, TeCl_4$	$1100 \rightarrow 1000$	Fel 1996
	$TeCl_4$	$1100 \rightarrow 1000$	Opp 1975
	O_2	$1170 \rightarrow 1070$	Sch 1963
	O_2	$1205 \rightarrow 760$	Sch 1963a
	O_2	$1250 \rightarrow T_1$	But 1971
	O_2	$1230 \rightarrow 920$	Sha 1979
	O_2	$1230 \rightarrow 1050$	Geo 1982
	O_2	$1350 \rightarrow 1100$	Par 1982
	O_2	$1230 \rightarrow 1050$	Tri 1982
$Ru_{1-x}Mo_xO_2$	$Cl_2, TeCl_4$	$1100 \rightarrow 1000$	Rei 1991
$Ru_{1-x}Ti_xO_2$	$Br_2, HCl, CBr_4,$	$1050 \rightarrow 1000$	Tri 1983
	$TeCl_4$		
$Ru_{1-x}V_xO_2$	$Cl_2, TeCl_4$	$1100 \rightarrow 1000$	Rei 1991
	$TeCl_4$	$1000 \rightarrow 900$	Kra 1986
	$CCl_4, TeCl_4$	$1000 \rightarrow 850$	Kra 1991
	$HgCl_2$	$1000 \rightarrow 900$	Arn 2008
	$TeCl_4$	$1100 \rightarrow 900$	Rit 1977

Table 5.1 (continued)

Sink solid	Transport additive	Temperature / °C	Reference
Sb_2O_4	$TeCl_4$	$1100 \to 950$	Gol 2011
	TeI_4	$650 \to 630, 950 \to 930$	Dem 1980
Sc_2O_3	Cl_2	$1100 \to 1000$	Ros 1990a
$ScAsO_4$	$TeCl_4$	$1080 \to 950$	Schm 2005
$ScNbO_4$	$Cl_2, Cl_2 + NbCl_5$	$1100 \to 1000$	Ros 1990
$Sc_{11}Nb_3O_{24}$	Cl_2	$1100 \to 1000$	Ros 1990a
$ScPO_4$	PCl_5	$900 \to 800, 1100 \to 1000$	Rep 1971
	PBr_5	$1050 \to 975$	Mül 2004
$ScSbO_4$	$TeCl_4$	$1100 \to 950$	Ger 2007
SiO_2	H_2	$< 1700 \to T_1$	Flö 1963
	HF	$150 \to 500$	Chu 1965
	HF	$600 \to 1100$	Hof 1977a
	HF	$1000 \to T_1$	Schm 1964a
	HCl	$1200 \to T_1$	Spi 1930
	H_2O	$1200 \to 600$	Gre 1933
	H_2O	not specified	Bra 1953, Neu 1956
	$NbCl_5$	$1050 \to 950$	Hib 1977
	$PCl_5,$	$900 \to 1100,$	Orl 1976
	$TaCl_5, Cl_2 + CrCl_3$	$1190 \to 1000, 1100 \to 900$	
	$SiCl_4, SiBr_4,$	$1100 \to 900, 1270 \to 1000$	Sch 1957a
	$I_2 + Si, TiO_2 + TiCl_4$	$1190 \to 1000$	
	$Cl_2 + CrCl_4$	$1100 \to 900$	Sch 1962a
	$Cl_2 + PCl_3$	$1100 \to 1000$	Orl 1971
$SmAsO_4$	$TeCl_4$	$1050 \to 950$	Schm 2005
$Sm_3BSi_2O_{10}$	NH_4Cl	$1000 \to 920$	Lis 1996
$Sm_{1-x}La_xAsO_4$	$TeCl_4$	$1075 \to 975$	Schm 2003
$Sm_{1-x}Nd_xAsO_4$	$TeCl_4$	$1075 \to 975$	Schm 2005
$SmPO_4$	PCl_5	$1100 \to 1000$	Orl 1971
$Sm_{1-x}Nd_xPO_4$	PCl_5	$1050 \to 950$	Mül 2004
$SmSbO_4$	$TeCl_4$	$1100 \to 950$	Ger 2007
$Sm_2Ti_2O_7$	Cl_2	$1050 \to 950$	Hüb 1992
$SmVO_4$	$TeCl_4$	$1100 \to 1000$	Schm 2005a
SnO_2	H_2	$900 \to T_1$	Sch 1962a
	Cl_2, CCl_4	$1010 \to 970, 1070 \to 950$	Tos 1988
	HBr	$1025 \to 825$	Nol 1976
	$TeCl_4$	$700 \to 600$	Mar 1999
	CO	$900 \to T_1$	Sch 1962a
	$I_2 + S$	$1100 \to 900$	Mat 1977
	$CO (H_2)$	$1300 \to T_1$	Gha 1974
	O_2	$1475 \to T_1$	Rea 1976
	O_2	$1540 \to 1400$	Mur 1976

Table 5.1 (continued)

Sink solid	Transport additive	Temperature / °C	Reference
$Sn_{1-x}Ru_xO_2$	Cl_2	$1100 \rightarrow 1000$	Nic 1993
$SnO_2:IrO_2$	O_2	$1475 \rightarrow 1050$	Rea 1976
$SrMoO_4$	Cl_2	$1150 \rightarrow 1050$	Ste 2006
$SrMo_5O_8$	Cl_2	$1150 \rightarrow 1050$	Ste 2006
$SrTa_4O_{11}$	Cl_2, $TeCl_4$	$1100 \rightarrow 1000, 1225 \rightarrow 1100$	Bay 1983
$SrWO_4$	Cl_2	$1150 \rightarrow 1050$	Ste 2005a
Ta_2O_5	$TaCl_5$	$800 \rightarrow 700$	Sch 1960a
	Cl_2, $Cl_2 + CrCl_3$	$900 \rightarrow 700$	Sch 1962a
	NH_4Cl	$1000 \rightarrow 800$	Hum 1992
	CBr_4	$900 \rightarrow 1000$	Scha 1989
	S	$1000 \rightarrow 900$	Sch 1980, Sch 1988
TaON	NH_4Cl	$1100 \rightarrow 1000, 1000 \rightarrow 900$	Bus 1969
$Ta_{1-x}Ce_xO_2$	$Cl_2 + C$	$1100 \rightarrow 1000$	Scha 1989
$TbAsO_4$	$TeCl_4$	$1050 \rightarrow 950$	Schm 2005
$TbPO_4$	PCl_5	$1100 \rightarrow 1000$	Orl 1971
	PCl_5	$1100 \rightarrow 1000, 1000 \rightarrow 900$	Rep 1971
$TbSbO_4$	$TeCl_4$	$1100 \rightarrow 950$	Ger 2007
$Tb_2Ti_2O_7$	Cl_2	$1050 \rightarrow 950$	Hüb 1992
$TbVO_4$	$TeCl_4$	$1100 \rightarrow 1000$	Schm 2005a
TeO_2	Cl_2, HCl, $TeCl_4$	$600 \rightarrow 400$	Opp 1977b
	Br_2, HBr, $TeBr_4$	$700 \rightarrow 600$	Opp 1978a
	I_2, TeI_4,	$700 \rightarrow 450, 600 \rightarrow 400$	Opp 1980
	$TeI_4 + H_2O$	$520 \rightarrow 420$	
	NH_4Cl	$600 \rightarrow 500$	Schm 1997
	$TeCl_4$	$500 \rightarrow 450$	Sch 1977
	H_2O	$700 \dots 450$	Gle 1964
ThO_2	$TeCl_4$	$1075 \rightarrow 975$	Spi 1979
	Cl_2, NH_4Cl	$1050 \rightarrow 950$	Schm 1991a
	$TeCl_4$	$1100 \rightarrow 1000$	Kor 1989
ThCuAsO	I_2	$800 \rightarrow 900$	Alb 1996
$ThCu_{1-x}PO$	I_2	$900 \rightarrow 1000$	Alb 1996
$Th_{1-x}U_xO_2$	HCl	$1100 \rightarrow 950$	Kam 1978
$ThNb_2O_7$	Cl_2, NH_4Cl	$1050 \rightarrow 950$	Schm 1990, Schm 1991
$ThNb_4O_{12}$	NH_4Cl	$1050 \rightarrow 950$	Schm 1991a
$Th_2Nb_2O_9$	Cl_2	$1050 \rightarrow 900$	Schm 1991a
$ThSiO_4$	Cl_2, HCl	$1050 \rightarrow 950$	Kam 1979
	Cl_2, NH_4Cl	$1050 \rightarrow 950$	Schm 1991a
$ThTa_2O_7$	Cl_2, NH_4Cl	$1100 \rightarrow 900, 1100 \rightarrow 1050$	Schm 1990a, Schm 1991a
$ThTa_4O_{12}$	Cl_2, $Cl_2 + Ta$, $Cl_2 + V$, NH_4Cl	$1050 \rightarrow 1000$	Schm 1991a

Table 5.1 (continued)

Sink solid	Transport additive	Temperature / °C	Reference
$ThTa_6O_{17}$	$Ta + Cl_2$, $V + Cl_2$	$1050 \rightarrow 1000$	Schm 1991a
$ThTa_8O_{22}$	Cl_2	$1100 \rightarrow 900$	Schm 1991a
$Th_2Ta_2O_9$	Cl_2, NH_4Cl	$1100 \rightarrow 1000$, $1050 \rightarrow 950$	Schm 1989
	Cl_2,	$950 \rightarrow 900$, $1100 \rightarrow 1000$	Schm 1991a
	NH_4Cl	$1050 \rightarrow 950$	
$Th_2Ta_6O_{19}$	Cl_2, $ZrCl_4$, $HfCl_4$	$1000 \rightarrow 980$	Bus 1996
$Th_4Ta_{18}O_{53}$	$Cl_2 + TaCl_5$	$1100 \rightarrow T_1$	Bus 1996
$ThTi_2O_6$	$Cl_2 + S$	$1100 \rightarrow T_1$	Gru 2000
Ti_2O_3	Cl_2, $TeCl_4$	$1045 \rightarrow 950$, $1045 \rightarrow 935$	Str 1982b
	HCl	$1000 \rightarrow 900$	Hau 1967
	CCl_4	$1000 \rightarrow 900$	Kra 1988
	$TeCl_4$	$1050 \rightarrow 950$	Pes 1975a
	$TeCl_4$	$1050 \rightarrow 950$	Bli 1982
	$TeCl_4$	$900 \rightarrow 880$	Hon 1982
	$TiCl_4$	$1100 \rightarrow 950$	Fou 1977
Ti_3O_5	Cl_2	$995 \rightarrow 980$	Str 1982b
	HCl	$1000 \rightarrow 860$	Sei 1984
	$TeCl_4$	$1100 \rightarrow 1080$	Mer 1973
	$TeCl_4$	$T_2 \rightarrow T_1$	Mer 1973a
	$TeCl_4$	$900 \rightarrow 880$	Hon 1982
	$TiCl_4$	$1100 \rightarrow 950$	Fou 1977
Ti_4O_7	Cl_2	$995 \rightarrow 980$	Str 1982b
	HCl	$1000 \rightarrow 860$	Sei 1984
	$HgCl_2$	$1075 \rightarrow 1005$	Sei 1983
	$TeCl_4$	$1100 \rightarrow 1080$	Mer 1973
	$TeCl_4$	$T_2 \rightarrow T_1$	Mer 1973a
	$TeCl_4$	$960 \rightarrow 910$	Hon 1982
	$TiCl_4$	$1100 \rightarrow 950$	Fou 1977
$(Ti_{1-x}V_x)_4O_7$	$TeCl_4$	$T_2 \rightarrow T_1$	Mer 1980
Ti_5O_9	$HgCl_2$	$1045 \rightarrow 985$	Sei 1983
	$TeCl_4$	$1100 \rightarrow 1080$	Mer 1973
Ti_6O_{11}	Cl_2	$1070 \rightarrow 1040$	Str 1982b
	$HgCl_2$	$1035 \rightarrow 985$	Sei 1983
Ti_9O_{17}	Cl_2	$1015 \rightarrow 960$	Str 1982
	Cl_2	$1045 \rightarrow 1005$	Str 1982b
Ti_xO_{2x-1}	HCl	$1100 \rightarrow 1070$	Mer 1973
	Cl_2, HCl, CCl_4	$1000 \rightarrow 900$	Kra 1988
	NH_4Cl, $TeCl_4$	$1000 \rightarrow T_1$	Mer 1977
	NH_4Cl, $TeCl_4$	$1050 \rightarrow 1000$	Ban 1981
	HCl, NH_4Cl	$1010 \rightarrow 1000$, $1080 \rightarrow 1040$	Sei 1984
	$SeCl_4$, $TeCl_4$	$T_2 \rightarrow T_1$	Kra 1987
Ti_xO_{2x-1} $(x = 2 \ldots 9)$	Cl_2	$1070 \rightarrow 1040$	Str 1982
	Cl_2	$1070 \rightarrow 1040$	Str 1982b
Ti_xO_{2x-1} $(x < 4)$	HCl	$1000 \rightarrow 900$	Kra 1988
Ti_xO_{2x-1} $(x = 16 \ldots 20)$	NH_4Cl	$1050 \rightarrow 1000$	Ban 1981

Table 5.1 (continued)

Sink solid	Transport additive	Temperature / °C	Reference
TiO_2	Cl_2	$820 \to 750$	Mer 1982
	Cl_2	$1000 \to 850$	Schm 1983a
	Cl_2	$1000 \to 850$	Schm 1984
	Cl_2, Cl_2 + S,	$T_2 \to T_1$ [3]	Mon 1984
	$TeCl_4$, $TiCl_4$,		
	$TeCl_4$ + S		
	Cl_2	$1000 \to 900$	Kra 1987
	I_2 + S	$1150 \to 1100$	Nit 1967a
	HCl	$1125 \to 525$	Far 1955
	HCl, NH_4Cl	$930 \to 780$	Wäs 1972
	NH_4Cl, NH_4Br	$855 \to 750$	Izu 1979
	NH_4Cl	$750 \ldots 800 \to 650 \ldots 700$	Hos 1997
	NH_4Cl	$750 \ldots 800 \to 650 \ldots 700$	Sek 2000
	CCl_4, $SeCl_4$,	$1000 \to 900$	Kra 1987
	$TeCl_4$	$850 \to 800$	
	CCl_4	$1000 \to 900$	Kra 1988
	$TeCl_4$	$1100 \to 900$	Nie 1967
	$TeCl_4$	not specified	Lon 1973
	$TeCl_4$	$1100 \to 900$	Opp 1975
	$TeCl_4$	$1125 \to 725$	Wes 1980
	$TeCl_4$,	$840 \to 750$, $1000 \to 800$,	Mer 1982
	$TiCl_4$	$820 \to 720$	
	$TeCl_4$	$680 \to 780$	Kav 1996
	$TiCl_4$	$1100 \to 950$	Fou 1977
$TiBO_3$	TiI_2 + H_2O	$900 \to T_1$	Schm 1964
$TiNb_{14}O_{37}$	Cl_2, $NbCl_5$, $TiCl_4$	$850 \to 750$	Bru 1976
TiO_2:Ir, In, Nb, Ta	HCl, $TeCl_4$	$1050 \to 1000$	Tri 1983
TiO_2:Al_2O_3	NH_4Cl, NH_4Br	$860 \to 755$	Izu 1979
	$TeCl_4$	$700 \to 640$	Raz 1981
TiO_2:Fe_2O_3	$TeCl_4$	$700 \to 640$	Raz 1981
TiO_2:Ga_2O_3	NH_4Cl, NH_4Br	$860 \to 800$, $850 \to 740$	Izu 1979
TiO_2:In_2O_3	NH_4Cl, NH_4Br	$850 \to 745$, $860 \to 750$	Izu 1979
TiO_2:MgO	$TeCl_4$	$700 \to 640$	Raz 1981
TiO_2:Nb	NH_4Cl	$800 \to 700$	Mul 2004
$Ti_{1-x}Mo_xO_2$	$TeCl_4$, $TeCl_4$ + S	$900 \to 850$, $745 \to 685$	Mer 1982
$Ti_{1-x}Ru_xO_2$	CCl_4, $TeCl_4$	$1075 \to 1000$	Kra 1991
	$TeCl_4$	$1050 \to 1000$	Tri 1985
$Ti_{1-x}V_xO_2$	$TeCl_4$	$1075 \to 965$	Hör 1976
$Tl_2Ru_2O_7$	O_2	$950 \to T_1$	Sle 1971
$TmAsO_4$	$TeCl_4$	$1050 \to 950$	Schm 2005
$TmPO_4$	PBr_3 + Br_2	$1025 \to 925$, $1125 \to 925$	Orl 1974
	PBr_5	$1050 \to 975$	Mül 2004
$Tm_2Ti_2O_7$	Cl_2	$1050 \to 950$	Hüb 1992

[3] Gas phase analysis by Raman spectroscopy and thermodynamic calculations.

Table 5.1 (continued)

Sink solid	Transport additive	Temperature / °C	Reference
UO_2	Cl_2, Br_2, I_2, $Br_2 + S$, $Br_2 + Se$, HCl	$1000 \to 850$	Nai 1971
	Cl_2	$970 \to 950$	Sin 1974
	$TeCl_4$	$1075 \to 975$	Spi 1979
	HCl	$1000 \to 850$	Nom 1981
	$TeCl_4$	$1100 \to 900$	Opp 1975
	$TeCl_4$	$800 \to 600$	Opp 1977d
	$TeCl_4$	$1050 \to 950$	Fai 1978
	$TeCl_4$	$1100 \to 900$	Paj 1986a
U_3O_8	Cl_2, Br_2, HCl, $Br_2 + Se$	$1000 \to 850$	Nai 1971
	HCl	$1000 \to 850$	Nom 1981
	O_2, $O_2 + H_2O$	$1100 \ldots 1350 \to T_1$	Dha 1974
	$O_2 + UF_6$	$800 \to 760$	Kna 1969
U_4O_9	Cl_2, Br_2, HCl, $Br_2 + Se$,	$1000 \to 850$	Nai 1971
	HCl	$1000 \to 850$	Nom 1981
	$TeCl_4$	$1050 \to 1000$	Opp 1977d
$U_{1-x}Hf_xO_2$	HCl	$1000 \to 950$	Schl 1999
$U_{1-x}Zr_xO_2$	HCl	$1000 \to 950$	Schl 1999
$UNbO_5$	$NbCl_5$, Cl_2	$1000 \to 980$	Schl 1999
UNb_2O_7	$NbCl_5$, Cl_2	$1000 \to 990$	Bus 1994
UNb_6O_{16}	NH_4Cl	$1000 \to 990$	Bus 1994
$U_{1-x}Pu_xO_2$	$TeCl_4$	$1050 \to 950$	Kol 2002
UOTe	Br_2	$900 \to 950$	Shl 1995
$UTaO_5$	Cl_2	$T_2 \to T_1$	Schl 1999a
UTa_3O_{10}	NH_4Cl, $TaCl_5$	$1050 \to 1000$	Schm 1991a
UTa_6O_{17}	NH_4Cl, $TaCl_5$	$1050 \to 1000$	Schm 1991a
$U_2Ta_2O_9$	Cl_2	$1040 \to 925$	Schm 1991a
$U_2Ta_6O_{19}$	HCl	$1000 \to 950$	Schl 2000
$U_4Ta_{18}O_{53}$	$Cl_2 + TaCl_5$	$T_2 \to T_1$	Bus 1996
V_2O_3	Cl_2	$1050 \to T_1$ ($\Delta T = 50 \ldots 300$)	Pes 1973
	Cl_2	$1050 \to 950$	Bli 1977
	Cl_2	$1125 \to 525$	Pes 1980
	$HgCl_2$, $HgBr_2$, HgI_2	$900 \to 840$, $960 \to 900$	Wen 1991
	HBr, $I_2 + S$	$800 \to 600$	Pic 1973, Pic 1973b
	HCl, $TeCl_4$	$1050 \to 930$, $990 \to 900$	Pou 1973
	HCl, $TeCl_4$	$1000 \to 900$	Gra 1986
	$TeCl_4$	$1050 \to 950$	Nag 1970
	$TeCl_4$	$1050 \to 950$	Nag 1971
	$TeCl_4$	$1050 \to 950$	Nag 1972
	$TeCl_4$	$990 \to 890$, $1040 \to 910$	Lau 1976
	$TeCl_4$	$1000 \to 900$	Opp 1977, Opp 1977c

Table 5.1 (continued)

Sink solid	Transport additive	Temperature / °C	Reference
V_2O_3:Cr	$TeCl_4$	$1050 \rightarrow 950$	Kuw 1980
V_3O_5	Cl_2	$1125 \rightarrow 525$	Pes 1980
	$HgCl_2$	$900 \rightarrow 840$	Wen 1991
	HCl, $TeCl_4$	$1000 \rightarrow 900$	Gra 1986
	$TeCl_4$	$1050 \rightarrow 950$	Nag 1972
	$TeCl_4$	$1100 \rightarrow 900$	Ter 1976
	$TeCl_4$	$1000 \rightarrow 900$	Opp 1977c
	$TeCl_4$	$1040 \rightarrow 920$	Nag 1969
V_3O_7	$TeCl_4$	$700 \rightarrow 500$	Rei 1975
	NH_4Cl, NH_4Br, H_2O, $I_2 + H_2O$	$580 \rightarrow 480$	Wen 1989
	$HgCl_2$	$900 \rightarrow 840$	Wen 1991
	$TeCl_4$	$1050 \rightarrow 950$	Nag 1972
	$TeCl_4$	$1000 \rightarrow 900$	Opp 1977c
	NH_4Cl	$400 \ldots 620 \rightarrow 350 \ldots 580$	Li 2009
V_5O_9	$TeCl_4$	$1050 \rightarrow 950$	Nag 1970
	$TeCl_4$	$1050 \rightarrow 950$	Nag 1972
	$TeCl_4$	$1000 \rightarrow 900$	Opp 1977c
	$HgCl_2$	$900 \rightarrow 840$	Wen 1991
V_6O_{11}	$HgCl_2$	$900 \rightarrow 840$	Wen 1991
	$TeCl_4$	$1040 \rightarrow 920$	Nag 1969
	$TeCl_4$	$1050 \rightarrow 950$	Nag 1972
	$TeCl_4$	$1000 \rightarrow 900$	Opp 1977c
$V_{6-x}Fe_xO_{13}$	$TeCl_4$	$600 \rightarrow 550$	Gree 1982
V_6O_{13}	NH_4Cl, NH_4Br, H_2O, $I_2 + H_2O$ $NH_4Cl + H_2O$, $NH_4Br + H_2O$	$580 \rightarrow 480$	Wen 1989
	$TeCl_4$	$670 \rightarrow 650$	Sae 1973
	$TeCl_4$	$600 \rightarrow 550$	Kaw 1974
	$TeCl_4$	$700 \rightarrow 500$	Rei 1975
	$TeCl_4$	$500 \rightarrow 450$	Opp 1977, Opp 1977c
	$HgCl_2$	$900 \rightarrow 840$	Wen 1991
	$TeCl_4$	$1050 \rightarrow 950$	Nag 1972
	$TeCl_4$	$1000 \rightarrow 900$	Opp 1977c
V_8O_{15}	$HgCl_2$	$900 \rightarrow 840$	Wen 1991
	$TeCl_4$	$1050 \rightarrow 950$	Nag 1970
	$TeCl_4$	$1050 \rightarrow 950$	Nag 1972
	$TeCl_4$	$1000 \rightarrow 900$	Opp 1977, Opp 1977c
V_9O_{17}	$HgCl_2$	$900 \rightarrow 840$	Wen 1991
	$TeCl_4$	$1050 \rightarrow 950$	Nag 1972
	$TeCl_4$	$1025 \rightarrow 955$	Kuw 1981
V_xO_{2x-1} $(x = 3 \ldots 8)$	$TeCl_4$	$1050 \rightarrow 950$	Nag 1969
	$TeCl_4$	$1050 \rightarrow 950$	Nag 1971

Table 5.1 (continued)

Sink solid	Transport additive	Temperature / °C	Reference
V_xO_{2x-1} ($x = 2 \dots 8$)	$TeCl_4$	$1050 \rightarrow 950$	Nag 1972
V_xO_{2x-1}	$TeCl_4$	$1000 \rightarrow 900$	Opp 1977, Opp 1977c
VO_2	HCl, $Cl_2 + HCl$	$500 \rightarrow 800$, $530 \rightarrow 680$	Opp 1977a
	$HgCl_2$, $HgBr_2$, NH_4Cl, NH_4Br	$900 \rightarrow 840$	Wen 1991
	I_2	$1100 \rightarrow 1000$	Kli 1978
	HCl, $TeCl_4$	$500 \rightarrow 800$, $1000 \rightarrow 900$, $650 \rightarrow 450$	Opp 1975a
	HCl,	$450 \rightarrow 650$	Opp 1977
	$TeCl_4$	$650 \rightarrow 550$, $1000 \rightarrow 900$	Opp 1977c
	NH_4Cl	$780 \rightarrow 680$	Wen 1989
	$TeCl_4$	$600 \rightarrow 500$, $1000 \rightarrow 850$	Ban 1978
	$TeCl_4$	$1050 \rightarrow 950$	Nag 1971
	$TeCl_4$	$1040 \rightarrow 920$	Nag 1972
	$TeCl_4$	$1050 \rightarrow 1000$	Prz 1972
	$TeCl_4$	$700 \rightarrow 500$	Rei 1975
	$TeCl_4$	$1100 \rightarrow 900$	Opp 1975
	$TeCl_4$	$1100 \rightarrow 900$	Opp 1975b
	$TeCl_4$	$1000 \rightarrow 900$, $500 \rightarrow 450$	Opp 1977
	$TeCl_4$	$650 \rightarrow 550$, $1000 \rightarrow 900$	Opp 1977c
V_2O_5	NH_4Cl, H_2O, $I_2 + H_2O$, $NH_4Cl + PtCl_2$, $NH_4Cl + H_2O$, $NH_4Br + H_2O$	$580 \rightarrow 480$	Wen 1989
	NH_4Cl	$640 \rightarrow 540$	Wen 1990
	$TeCl_4$	$700 \rightarrow 500$	Rei 1975
	$TeCl_4$	$600 \rightarrow 580$	Vol 1976
	$TeCl_4$	$500 \rightarrow 450$	Opp 1977c
	$TeCl_4$	$550 \rightarrow 450$, $650 \rightarrow 550$	Kir 1994
	$TeCl_4$	$700 \rightarrow 600$	Mar 1999
	H_2O	$580 \rightarrow 480$	Hac 1998
$VMoO_5$	$TeCl_4$	$650 \rightarrow 450$, $560 \rightarrow 510$	Shi 1998
VNb_9O_{25}	HCl, $TeCl_4$	$900 \rightarrow 800$	Fec 1993
$VTeO_4$	$TeCl_4$	$600 \rightarrow 500$	Lau 1974
	$TeCl_4$	$500 \rightarrow 450$	Rei 1979
V_2O_5:Na	$TeCl_4$	$530 \rightarrow 500$	Roh 1994
$V_3Nb_9O_{29}$	$TeCl_4$	$900 \rightarrow 800$	Fec 1993c
$V_6Te_6O_{25}$	$TeCl_4$	$600 \rightarrow 400$	Rei 1979
$V_8Nb_5O_{29}$	HCl, $TeCl_4$	$900 \rightarrow 800$	Fec 1993
$V_{1-x}Ga_xO_2$	$TeCl_4$	$1000 \rightarrow 850$	Kra 1991
V_xO_{2x-1} ($x = 3 \dots 8$)	$TeCl_4$	$1050 \rightarrow 950$	Nag 1969
	$TeCl_4$	$1050 \rightarrow 950$	Nag 1971
V_xO_{2x-1} ($x = 2 \dots 8$)	$TeCl_4$	$1050 \rightarrow 950$	Nag 1972

Table 5.1 (continued)

Sink solid	Transport additive	Temperature / °C	Reference
V_xO_{2x-1}	$TeCl_4$	$1000 \rightarrow 900$	Opp 1977, Opp 1977c
VBO_3	$VCl_2 + H_2O$	$900 \rightarrow T_1$	Schm 1964
$V_{1-x}Ga_xO_y$	$TeCl_4$	$1000 \ldots 1050 \rightarrow 850 \ldots 900$	Kra 1991
$VNbO_4$	Cl_2, NH_4Cl	$1020 \rightarrow 880$	Ros 1990a
VTe_2O_9	$TeCl_4$	$375 \rightarrow 340$	Kho 1981
$V_{4-x}Ti_xO_7$	NH_4Cl, $TeCl_4$	$1050 \rightarrow 950$	Cal 2005
$V_{5-x}Ti_xO_9$	NH_4Cl, $TeCl_4$	$1050 \rightarrow 950$	Cal 2005
WO_2	Cl_2, HCl	$1000 \rightarrow 900$, $950 \rightarrow 900$	Kle 1968
	I_2	$900 \rightarrow 800$	Ben 1969
	I_2	$1000 \rightarrow 800$	Det 1969
	I_2	$1000 \rightarrow 960$	Rog 1969
	I_2, $I_2 + H_2O$	$1000 \rightarrow 800$	Sch 1973a
	I_2	$950 \rightarrow 850$	Ben 1974
	I_2	$1200 \rightarrow 1100$	Bab 1977
	HCl	$1000 \rightarrow 900$	Kle 1968
	Br_2, NH_4Cl, $HgCl_2$, $HgBr_2$, HgI_2	$840 \rightarrow 760$, $1025 \rightarrow 925$, $690 \rightarrow 610$, $840 \rightarrow 760$, $1060 \rightarrow 980$	Scho 1992
	$HgCl_2$, $HgBr_2$	$950 \rightarrow 850$, $840 \rightarrow 760$	Scho 1988
	HgI_2	$900 \rightarrow 820$, $840 \rightarrow 760$ $1060 \rightarrow 890$	Scho 1989
	$HgBr_2$	$1000 \rightarrow 600$ $(\Delta T = 100)$	Len 1994
	$TeCl_4$, $TeBr_4$	$1100 \rightarrow 1000$, $1000 \rightarrow 900$	Opp 1975
	$TeCl_4$	$1000 \rightarrow T_1$	Opp 1978
	$TeCl_4$	$1000 \rightarrow 900$	Wol 1978
	$TeCl_4$	$1000 \rightarrow 900$	Opp 1985, Opp 1985a
	$H_2O + H_2$	$1000 \rightarrow T_1$	Mil 1949
WO_2:Ni	I_2	$950 \rightarrow 850$	Ben 1974
$WO_{2.72}$	$TeCl_4$	$1200 \rightarrow 1150$	Wol 1978
$W_{18}O_{49}$	I_2	$950 \rightarrow 850$	Bab 1977
	$HgCl_2$, $HgBr_2$	$950 \rightarrow 850$, $840 \rightarrow 760$	Scho 1988
	$HgCl_2$, HgI_2	$690 \rightarrow 610$, $1060 \rightarrow 980$	Scho 1992
	HgI_2	$900 \rightarrow 820$, $840 \rightarrow 760$ $1060 \rightarrow 890$	Scho 1989
	$TeCl_4$	$1000 \rightarrow T_1$	Opp 1978
	$TeCl_4$	$1000 \rightarrow 900$	Opp 1985
	$TeCl_4$	$1100 \rightarrow 1000$	Opp 1985a
	H_2O	$1000 \rightarrow T_1$	Mil 1949
	H_2O	not specified	Ahm 1966
$W_{20}O_{58}$	Cl_2, HCl	$1000 \rightarrow 950$, $950 \rightarrow 900$	Kle 1968
	I_2	$950 \rightarrow 850$	Bab 1977
	$TeCl_4$	$1000 \rightarrow T_1$	Opp 1978
	$TeCl_4$	$1000 \rightarrow 900$	Opp 1985

Table 5.1 (continued)

Sink solid	Transport additive	Temperature / °C	Reference
$W_{20}O_{58}$	$TeCl_4$	$900 \to 800$	Opp 1985a
	H_2O	not specified	Ahm 1966
WO_3	Cl_2, HCl,	$800 \to 750, 1100 \to 1050,$	Kle 1966
	CCl_4	$700 \to 600$	
	Cl_2, Br_2	$925 \to T_1$	Vei 1979
	Br_2, HCl	$625 \to T_1$	Vei 1979
	HCl	$1000 \to 950$	Bih 1970
	$HgCl_2$	$1025 \to 925$	Scho 1990a
	$AlCl_3$	$880 \to 750$	Mar 1999
	$TeCl_4$	$1000 \to T_1$	Opp 1978
	$TeCl_4$	$1000 \to 900$	Opp 1985,
			Opp 1985a
	H_2O	$1100 \to T_1$	Mil 1949
	H_2O	$1100 \to 900$	Gle 1962
	H_2O	$1050 \to 900$	Sah 1983
"$W_2Nb_{18}O_{51}$" [4]	Cl_2	$900 \to 800$	Heu 1979
"$W_2Nb_{34}O_{91}$" [4]	Cl_2	$900 \to 800$	Heu 1979
"$W_3Nb_{30}O_{84}$" [4]	Cl_2	$900 \to 800$	Heu 1979
"$W_4Nb_{52}O_{142}$" [4]	Cl_2	$900 \to 800$	Heu 1979
$W_{1-x}Ru_xO_2$	Cl_2	$1100 \to 1000$	Nich 1993
$W_{1-x}V_xO_2$	$TeCl_4$	$1075 \to 950$	Hör 1972
	$TeCl_4$	$1050 \to 950$	Rit 1977
	$TeCl_4$	$1100 \to 900$	Brü 1977
Y_2O_3	Cl_2	$T_2 \to T_1$	Red 1976
	$Br_2 + CO$	$1160 \to 1100$	Mat 1983
Y_2O_3:Eu	Br_2, HBr,	$1160 \to 1100, 1190 \to 1090,$	Mat 1982
	$Br_2 + CO$	$1190 \to 1090$	
$YAsO_4$	$TeCl_4$	$1080 \to 950$	Schm 2005
$YFeO_3$	$GdCl_3$	$1200 \to 1090$	Klei 1972
	YCl_3	$1165 \to 1050$	Klei 1974
	YCl_3	$1165 \to 1050$	Klei 1977
$Y_2Ge_2O_7$	Cl_2, $Cl_2 + GeCl_4$	$T_2 \to T_1$	Red 1976
$Y_3Fe_5O_{12}$	Cl_2, $Cl_2 + FeCl_2$	$1155 \to 1095, 1145 \to 1095$	Gib 1973
	HCl	$800 \dots 1200 \to T_1$	Cur 1965
		$(\Delta T = 50 \dots 100)$	
	$HCl + FeCl_3$	$1140 \to 1045$	Weh 1970
	$HCl + FeCl_3$	$1100 \to 1000$	Lau 1972
	CCl_4	$1100 \to 950$	Pie 1981
	$FeCl_3$,	$1150 \to 1050$	Pie 1982
	$FeCl_3 + YFeO_3$		
	$GdCl_3$	$1200 \to 1090$	Klei 1972
	YCl_3	$1170 \to 1065$	Klei 1974
	YCl_3	$1170 \to 1065$	Klei 1977,
			Klei 1977a
YPO_4	$PBr_3 + Br_2$	$1025 \to 925$	Orl 1974

[4] Block structures; the given composition is the composition of individual blocks.

Table 5.1 (continued)

Sink solid	Transport additive	Temperature / °C	Reference
YVO$_4$	TeCl$_4$	1100 → 950	Mat 1981
	TeCl$_4$	1100 → 1000	Schm 2005a
YbAsO$_4$	TeCl$_4$	1050 → 950	Schm 2005
YbPO$_4$	PBr$_5$	1075 → 960	Mül 2004
Yb$_2$Ti$_2$O$_7$	Cl$_2$	1050 → 950	Hüb 1992
Yb$_x$Er$_{1-x}$AsO$_4$	TeCl$_4$	1075 → 975	Schm 2005
ZnO	H$_2$, Cl$_2$, HCl, NH$_4$Cl, HgCl$_2$, NH$_3$	1020 → 925	Shi 1971
	Cl$_2$	900 → 820	Kle 1966b
	Cl$_2$, Br$_2$, I$_2$	1150 ... 700 → T_1 ($\Delta T = 150$)	Pie 1972
	Cl$_2$, Br$_2$, HCl, HBr, NH$_4$Cl, NH$_4$Br	1025 → 825	Opp 1984
	Cl$_2$, Br$_2$, HCl, HBr, NH$_4$Cl, NH$_4$Br	1000 → 900	Opp 1985
	Cl$_2$, C	1000 → 900	Nte 1999
	Cl$_2$	1000 → 900	Pat 1999
	Br$_2$	1010 → 990	Wid 1971
	HCl	1005 → 935	Kle 1966a
	NH$_4$Cl, NH$_4$Br, NH$_4$I, ZnCl$_2$	1000 → 800	Mat 1988
	C	1050 → 1020	Mik 2005
	C, CO, H$_2$O	1000 → 960, 1050 → 960	Pal 2006
	HCl + H$_2$	890 → 630	Quo 1975
	ZnCl$_2$, ZnCl$_2$ + Zn	1000 → 900	Mat 1985
	H$_2$O	1350 → 1300	Gle 1957
	CO$_2$ + Zn	1015 → 1000	Mik 2007
	H$_2$ + C + H$_2$O, N$_2$ + C + H$_2$O	1150 → 1100	Myc 2004a
ZnO	HgCl$_2$ + Zn, HgCl$_2$ + Al, HgCl$_2$ + In	1000 → T_1	Mat 1991
ZnO	C + O$_2$	1000 → 900	Mun 2005
	$[-CH_2(CHOH)-]_n$	1100 → 1090	Udo 2008
	C	1000 → T_1	Wei 2008
	C	1000 → 880	Jok 2009
	C	970 → 965	Hon 2009
ZnO:Mn	H$_2$ + C + H$_2$O, N$_2$ + C + H$_2$O	1100 → T_1	Myc 2004

Table 5.1 (continued)

Sink solid	Transport additive	Temperature / °C	Reference
$Zn_{1-x}Mn_xO$	C	$1050 \rightarrow 1010$	Sav 2007
$Zn_{1-x}Ni_xO$	HCl	$900 \rightarrow 750$	Loc 1999a
$Zn_2As_2O_7$	Cl_2	$880 \rightarrow 800$	Wei 2004c, Wei 2005
$Zn_xCo_{3-x}O_4$	Cl_2	$1000 \rightarrow 900$, $910 \rightarrow 830$	Pie 1988
$ZnCr_2O_4$	Cl_2	$975 \rightarrow 915$	Paj 1981
$ZnFe_2O_4$	HCl	$1000 \rightarrow 800$	Klei 1964
	$TeCl_4$	$950 \rightarrow 750$	Pes 1976
	$TeCl_4$	$950 \rightarrow 750$	Bli 1977
$ZnGa_2O_4$	Cl_2, HCl	$1000 \rightarrow 900$, $1000 \rightarrow 800$	Pat 1999
Zn_2GeO_4	Cl_2, HCl	$850 \rightarrow 750$, $1050 \rightarrow 900$	Pfe 2002c
	C	$1000 \rightarrow 400 \dots 500$	Yan 2009
$Zn_{2-x}Co_xGeO_4$	Cl_2	$850 \rightarrow 750$	Pfe 2002c
$Zn_{3-x}Co_xGeO_4$	Cl_2	$910 \rightarrow 820$	Pie 1988
$ZnMn_2O_4$	PbF_2	$1100 \rightarrow T_1$	Tse 1966
$Zn_{1-x}Mn_xCr_2O_4$	Cl_2	$1030 \rightarrow 950$	Lec 1993a
$ZnMoO_4$	Cl_2, Br_2, I_2, HCl, HBr	$950 \rightarrow 850$	Söh 1997
$Zn_{1-x}Cu_xMoO_4$	Cl_2, Br_2, NH_4Cl	$700 \rightarrow 600$	Ste 2003
	Cl_2, NH_4Cl	$750 \rightarrow 600$	Rei 2000
$Zn_2Mo_3O_8$	I_2, HCl	$950 \rightarrow 850$	Söh 1997
$Zn_3Mo_2O_9$	Br_2, I_2, HCl	$950 \rightarrow 850$	Söh 1997
$(Zn_{1-x}Nb_x)_{12}O_{29}$	Cl_2, NH_4Cl	$1000 \rightarrow 900$	Kru 1987
$ZnNb_2O_6$	Cl_2, HCl, NH_4Cl, $NbCl_5 + Cl_2$, $HgCl_2$, $HgBr_2$	$1000 \rightarrow 900$	Kru 1987
	HCl	$1020 \rightarrow 975$	Emm 1968c
	$TeCl_4$	$1050 \rightarrow 850$	Jen 2009
$Zn_{1+x}V_{2-x}O_4$	$TeCl_4$	$1000 \rightarrow 800$	Pic 1973a
Zn_2SiO_4	BeF_2, NaF, Na_2BeF_4	$1200 \rightarrow T_1$	Sob 1960
	$LiZnF_3$	$1200 \rightarrow T_1$	Nov 1961
$(Zn_{1-x}Be_x)_2$ SiO_4:Mn	$LiZnF_3$	$1200 \rightarrow T_1$	Nov 1961
$ZnWO_4$	HCl	$800 \dots 1200 \rightarrow T_1$ $(\Delta T = 50 \dots 100)$	Cur 1965
	Cl_2	$1075 \rightarrow 1040$	Emm 1968b
	Cl_2	$1050 \rightarrow 1000$	Wid 1971
	Cl_2	$900 \rightarrow 800$	Ste 2005a
ZrO_2	Cl_2, $TeCl_4$	$1100 \rightarrow 1000$	Dag 1992
	$TeCl_4$	$1100 \rightarrow 900$	Opp 1975
	$I_2 + S$	$1050 \rightarrow 1000$	Nit 1966
	$I_2 + S$	$1050 \rightarrow 1000$	Nit 1967a
$Zr_{1-x}Th_xO_2$	$ZrCl_4$	$1000 \rightarrow 980$	Bus 1996
$ZrSiO_4$	Cl_2, $ZrCl_4$, Br_2, $ZrBr_4$, I_2, ZrI_4	$1150 \dots 1250 \rightarrow 1050 \dots 1150$ $(\Delta T = 100)$	Hul 1968

Bibliography

Aga 1984	V. Agafonov, D. Michel, M. Perez y Jorba, M. Fedoroff, *Mater. Res. Bull.* **1984**, *19*, 233.
Aga 1985	V. Agafonov, D. Michel, A. Kahn, M. Perez Y Jorba, *J. Cryst. Growth* **1985**, *71*, 12.
Aga 1985a	V. Agafonov, A. Kahn, D. Michel, M. Perez Y Jorba, M. Fedoroff, *J. Cryst. Growth* **1985**, *71*, 256.
Ahm 1966	I. Ahmad, G. P. Capsimalis, *Intern. Conf. Cryst. Growth,* Boston **1966**.
Alb 1996	J. H. Albering, W. Jeitschko, *Z. Naturforsch.* **1996**, *51b*, 257.
Ale 1963	C. A. Alexander, J. S. Ogden, A. Levy, *J. Chem. Phys.* **1963**, *39*, 3057.
Ark 1955	A. E. van Arkel, U. Spitsbergen, R. D. Heyding, *Can. J. Chem.* **1955**, *33*, 446.
Arn 2008	J. Arnold, J. Feller, U. Steiner, *Z. Anorg. Allg. Chem.* **2008**, *634*, 2026.
Bab 1977	A. V. Babushkin, L. A. Klinkova, E. D. Skrebkova, *Izv. Akad. Nauk SSSR Neorg. Mater.* **1977**, *13*, 2114.
Bab 1987	E. V. Babkin, A. A. Charyev, *Izv. Akad. Nauk SSSR, Neorg. Mater.* **1987**, *236*, 996.
Bag 1976	A. M. Bagamadova, S. A. Semiletov, R. A. Rabadanov, *C. A.* **1976**, *84*, 10985.
Bal 1985	L. Bald, R. Gruehn, *Z. Anorg. Allg. Chem.* **1985,** *521*, 97.
Ban 1976	Y. Bando, Y. Kato, T. Takada, *Bull. Inst. Chem. Res. Kyoto Univ.* **1976**, *54*, 330.
Ban 1978	Y. Bando, M. Kyoto, T. Takada, S. Muranaka, *J. Cryst. Growth* **1978**, *45*, 20.
Ban 1981	Y. Bando, S. Muranaka, Y. Shimada, M. Kyoto, T. Takada, *J. Cryst. Growth* **1981**, *53*, 443.
Bay 1983	E. Bayer, *Dissertation,* University of Gießen, **1983**.
Bay 1985	E. Bayer, R. Gruehn, *J. Cryst. Growth* **1985**, *71*, 817.
Bel 1966	W. E. Bell, M. Tagami, *J. Phys. Chem.* **1966**, *703*, 640.
Ben 1969	L. Ben- Dor, L. E. Conroy, *Isr. J. Chem.* **1969**, *7*, 713.
Ben 1974	L. Ben- Dor, Y. Shimony, *Mater. Res. Bull.* **1974**, *9*, 837.
Ben 1978	L. Ben- Dor, Y. Shimony, *J. Cryst. Growth* **1978**, *43*, 1.
Ber 1960	J. Berkowitz, D. J. Meschi, W. A. Chupka, *J. Chem. Phys.* **1960***, 33*, 533.
Ber 1963	J. B. Berkowitz-Mattuck, A. Büchler, *J. Phys. Chem.* **1963**, *67*, 1386.
Ber 1989	O. Bertrand, N. Floquet, D. Jaquot, *J. Cryst. Growth* **1989**, *96*, 708.
Bern 1981	C. Bernard, G. Constant, R. Feurer, *J. Electrochem. Soc.* **1981**, *128*, 2447.
Bie 1990	W. Bieger, W. Piekarczyk, G. Krabbes, G. Stöver, N. v. Hai, *Cryst. Res. Technol.* **1990**, *25*, 375.
Bih 1970	R. Le Bihan, C. Vacherand, *C. A.* **1970**, *72*, 71459.
Bli 1977	G. Bliznakov, P. Peshev, *Russ. J. Inorg. Chem.* **1977**, *22*, 1603.
Bow 1972	H. K. Bowen, W. D. Kingery, M. Kinoshita, C. A. Goodwin, *J. Cryst. Growth* **1972**, *13/14*, 402.
Bow 1974	H. K. Bowen, W. D. Kingery, *J. Cryst. Growth* **1974**, *21*, 69.
Bra 1953	E. L. Brady, *J. Phys. Chem.* **1953**, *57*, 706.
Bru 1975	H. Brunner, *Dissertation,* University of Gießen, **1975**.
Bru 1976	H. Brunner, R. Gruehn, W. Mertin, *Z. Naturforsch. B* **1976**, *31*, 549.
Bru 1976a	H. Brunner, R. Gruehn, *Z. Naturforsch. B* **1976**, *31*, 318.
Brü 1976	W. Brückner, U. Gerlach, W. Moldenhauer, H.-P. Brückner, B. Thuss, H. Oppermann, E. Wolf, I. Storbeck, *J. Phys-Paris* **1976**, *10*, C4–63l.
Brü 1976a	W. Brückner, U. Gerlach, W. Moldenhauer, H.-P. Brückner, N. Mattern, H. Oppermann, E. Wolf, *Phys. Status Solidi* **1976**, *38*, 93.
Brü 1977	W. Brückner, U. Gerlach, H. P. Brückner, W. Moldenhauer, H. Oppermann, *Phys. Status Solidi a* **1977**, *42*, 295.

Brü 1983	W. Brückner, H. Oppermann, W. Reichelt, J. I. Terukow, F. A. Tschudnowski, E. Wolf, *Vanadiumoxide. Darstellung, Eigenschaften, Anwendung, Akademie-Verlag*, Berlin **1983**.
Bud 1966	P. P. Budnikov, V. I. Stusakovskij, D. B. Sandalov, F. P. Butra, *Izv. Akad. Nauk. SSSR, Neorg. Mater.* **1966**, *2*, 829.
Bud 1967	P. P. Budnikov, D. B. Sandulov, *Krist. Tech.* **1967**, *2*, 549.
Bus 1969	Y. A. Buslaev, G. M. Safronov, V. I. Pachomov, M. A. Gluschkova, V. P. Repko, M. M. Erschova, A. N. Zhukov, T. A. Zhdanova, *Izv. Akad. Nauk. SSSR, Neorg. Mater.* **1969**, *5*, 45.
Bus 1994	J. Busch, R. Gruehn, *Z. Anorg. Allg. Chem.* **1994**, *620*, 1066.
Bus 1996	J. Busch, R. Hofmann, R. Gruehn, *Z. Anorg. Allg. Chem.* **1996**, *622*, 67.
But 1971	S. R. Butler, G. L. Gillson, *Mater. Res. Bull.* **1971**, *6*, 81.
Cal 2005	D. Calestani, F. Licci, E. Kopnin, G. Calestani, F. Bolzoni, T. Besagni, V. Boffa, M. Marezio, *Cryst. Res. Technol.* **2005**, *40*, 1067.
Cap 1961	D. Caplan, M. Cohen, *J. Electrochem. Soc.* **1961**, *108*, 438.
Chu 1959	W. A. Chupka, J. Berkowitz, C. F. Giese, *J. Chem. Phys.* **1959**, *30,* 827.
Chu 1964	T. L. Chu, J. R. Gavaler, G. A. Gruber, Y. C. Kao, *J. Electrochem. Soc.* **1964**, *111*, 1433.
Chu 1965	T. L. Chu, G. A. Gruber, *Trans. Metal Soc. AIME* **1965**, *233*, 568.
Chu 1995	Y. Chung, B. J. Wuensch, *Mat. Res. Soc. Symp. Proc.* **1995**, *357*, 139.
Cor 1962	E. H. P. Cordfunke, G. Meyer, *Rec. Trav. Chim.* **1962**, *81*, 495.
Cur 1965	B. J. Curtis, J. A. Wilkinson, *J. Am. Ceram. Soc.* **1965**, *48*, 49.
Dag 1992	F. Dageförde, R. Gruehn, *Z. Anorg. Allg. Chem.* **1992**, *611*, 103.
Dem 1980	L. A. Demina, V. A. Dolgikh, S. Yu. Stefanovich, S. A. Okonenko, B. A. Popovkin, *Izv. Akad. Nauk SSSR, Neorg. Mater.* **1980**, *16*, 470.
DeS 1989	W. DeSisto, B. T. Collins, R. Kershaw, K. Dwight, A. Wold, *Mater. Res. Bull.* **1989**, *24*, 1005.
Det 1969	J. H. Dettingmeijer, J. Tillack, H. Schäfer, *Z. Anorg. Allg. Chem.* **1969**, *369*, 161.
Dev 1959	R. C. DeVries, G. W. Sears, *J. Chem. Phys.* **1959**, *31*, 1256.
Dha 1974	D. R. Dharwadkar, S. N. Tripathi, M. D. Karkhanavala, M. S. Chandrasek-Havaisch, *Thermodyn. Nucl. Mater. Symp.* **1974**, *2*, 455.
Die 1975	R. Diehl, A. Räuber, F. Friedrich, *J. Cryst. Growth* **1975**, *29*, 225.
Die 1976	R. Diehl, F. Friedrich, *J. Cryst. Growth* **1976**, *36*, 263.
Emm 1968	F. P. Emmenegger, M. L. A. Robinson, *J. Phys. Chem. Solids* **1968**, *29*, 1673.
Emm 1968a	F. Emmenegger, *J. Cryst. Growth* **1968**, *2*, 109.
Emm 1968b	F. Emmenegger, *J. Cryst. Growth* **1968**, *3/4*, 135.
Emm 1968c	F. Emmenegger, A. Petermann, *J. Cryst. Growth* **1968**, *2*, 33.
Fai 1978	S. P. Faile, *J. Cryst. Growth* **1978**, *43*, 133.
Fak 1965	M. M. Faktor, J. I. Carasso, *J. Electrochem. Soc.* **1965**, *112*, 817.
Far 1955	M. Farber, A. J. Darnell, *J. Chem. Phys.* **1955**, *23*, 1460.
Fec 1993	St. Fechter, S. Krüger, H. Oppermann, *Z. Anorg. Allg. Chem.* **1993**, *619*, 424.
Fel 1996	J. Feller, *Dissertation*, University of Dresden, **1996**.
Fel 1998	J. Feller, H. Oppermann, M. Binnewies, E. Milke, *Z. Naturforsch.* **1998**, *53b*, 184.
Fel 1998a	J. Feller, H. Oppermann, R. Kucharkowski, S. Däbritz, *Z. Naturforsch.* **1998**, *53b*, 397.
Fer 1965	A. Ferretti, D. B. Rogers, J. B. Goodenough, *J. Phys. Chem. Solids* **1965**, *26*, 2007.
Fis 1932	W. Fischer, R. Gewehr, *Z. Anorg. Allg. Chem.* **1932**, *209*, 17.
Flö 1963	O. W. Flörke, *Z. Kristallogr.* **1963**, *118*, 470.

Fos 1960   L. M. Foster, G. Long, *U.S. Pat. 2962370*, **1960**.
Fou 1977   G. Fourcaudot, J. Dumas, J. Devenyi, J. Mercier, *J. Cryst. Growth* **1977**, *40*, 257.
Fou 1979   G. Fourcaudot, M. Gourmala, J. Mercier, *J. Cryst. Growth* **1979**, *46*, 132.
Fou 1984   G. Fourcaudot, J. Mercier, H. Guyot, *J. Cryst. Growth* **1984**, *66*, 679.
Fuh 1964   W. Fuhr, *Dissertation*, University of Münster, **1964**.
Fuh 1986   J. Fuhrmann, J. Pickardt, *Z. Anorg. Allg. Chem.* **1986**, *532*, 171.
Gae 1993   T. Gaewdang, J. P. Chaminade, A. Garcia, C. Fouassier, M. Pouchard, P. Hagenmuller, B. Jaquier, *Mater. Lett.* **1993**, *18*, 64.
Geo 1982   C. A. Georg, P. Triggs, F. Lévy, *Mater. Res. Bull.* **1982**, *17*, 105.
Ger 1977   U. Gerlach, H. Oppermann, *Z. Anorg. Allg. Chem.* **1977**, *429*, 25.
Ger 1977a   U. Gerlach, H. Oppermann, *Z. Anorg. Allg. Chem.* **1977**, *432*, 17.
Ger 1977b   U. Gerlach, G. Krabbes, H. Oppermann, *Z. Anorg. Allg. Chem.* **1977**, *436*, 253.
Ger 2007   S. Gerlach, R. Cardoso Gil, E. Milke, M. Schmidt, *Z. Anorg. Allg. Chem.* **2007**, *633*, 83.
Gha 1974   D. B. Ghare, *J. Cryst. Growth* **1974**, *23*, 157.
Gib 1973   P. Gibart, *J. Cryst. Growth* **1973**, *18*, 129.
Gle 1957   O. Glemser, H. G. Völz, B. Meyer, *Z. Anorg. Allg. Chem.* **1957**, *292*, 311.
Gle 1962   O. Glemser, R. v. Haeseler, *Z. Anorg. Allg. Chem.* **1962**, *316*, 168.
Gle 1963   O. Glemser, *Österr. Chem. Ztg.* **1963**, *64*, 301.
Gru 1966   R. Gruehn, *J. Less-Common Met.* **1966**, *11*, 119.
Gru 1967   R. Gruehn, R. Norin, *Z. Anorg. Allg. Chem.* **1967**, *355*, 176.
Gru 1969   R. Gruehn, R. Norin, *Z. Anorg. Allg. Chem.* **1969**, *367*, 209.
Gru 1973   R. Gruehn, *Z. Anorg. Allg. Chem.* **1973**, *395*, 181.
Gru 1983   R. Gruehn, H. J. Schweizer, *Angew. Chem.* **1983**, *95*, 80.
Gru 1991   R. Gruehn, U. Schaffrath, N. Hübner, R. Hofmann, *Eur. J. Solid State Inorg. Chem.* **1991**, *28*, 495.
Gru 2000   R. Gruehn, R. Glaum, *Angew. Chemie* **2000**, *112*, 706, *Angew. Chem. Int. Ed.* **2000**, *39*, 692.
Grub 1973   P. E. Gruber, *J. Cryst. Growth* **1973**, *18*, 94.
Gue 1972   A. Guest, C. J. L. Lock, *Can. J. Chem.* **1972**, *50*, 1807.
Hac 1998   A. Hackert, R. Gruehn, *Z. Anorg. Allg. Chem.* **1998**, *624*, 1756.
Hau 1962   Z. Hauptmann, *Czechoslov. J. Phys. B* **1962**, *12*, 148.
Hau 1967   Z. Hauptmann, D. Schmidt, S. K. Banerjee, *Collect. Czechoslov. Chem. Commun.* **1967**, *32*, 2421.
Hay 1980   K. Hayashi, A. S. Bhalla, R. E. Newnham, L. E. Cross, *J. Cryst. Growth* **1980**, *49*, 687.
Heu 1979   G. Heurung, R. Gruehn, *Z. Naturforsch.* **1979**, *34b*, 1377.
Hib 1977   H. Hibst, R. Gruehn, *Z. Anorg. Allg. Chem.* **1977**, *434*, 63.
Hib 1978   H. Hibst, R. Gruehn, *Z. Anorg. Allg. Chem.* **1978**, *440*, 137.
Hof 1977   J. Hofmann, R. Gruehn, *J. Cryst. Growth* **1977**, *37*, 155.
Hof 1977a   J. Hofmann, R. Gruehn, *Z. Anorg. Allg. Chem.* **1977**, *431*, 105.
Hof 1990   R. Hofmann, R. Gruehn, *Z. Anorg. Allg. Chem.* **1990**, *590*, 81.
Hof 1991   R. Hofmann, R. Gruehn, *Z. Anorg. Allg. Chem.* **1991**, *602*, 105.
Hof 1993   R. Hofmann, *Dissertation*, University of Gießen, **1993**.
Hof 1994   G. Hoffmann, R. Roß, R. Gruehn, *Z. Anorg. Allg. Chem.* **1994**, *620*, 839.
Hon 1982   S. H. Hong, *Acta Chem. Scand A* **1982**, *36*, 207.
Hon 2009   S.-H. Hong, M. Mikami, K. Mimura, M. Uchikoshi, A. Yauo, S. Abe, K. Masumoto, M. Isshiki, *J. Cryst. Growth* **2009**, *311*, 3609.
Hör 1972   T. Hörlin, T. Niklewski, M. Nygren, *Mater. Res. Bull.* **1972**, *7*, 1515.

Hör 1973	T. Hörlin, T. Niklewski, M. Nygren, *Mater. Res. Bull.* **1973**, *8*, 179.
Hör 1976	T. Hörlin, T. Niklewski, M. Nygren, *Acta Chem. Scand. A* **1976**, *30*, 619.
Hos 1997	N. Hosaka, T. Sekiya, C. Satoko, S. Kurita, *J. Phy. Soc. Jpn.* **1997**, *66*, 877.
Hos 2007	T. Hoshi, H. Aruga Katori, M. Kosaka, H. Takagi, *J. Magn.. Mater.* **2007**, *310*, 448.
Hüb 1992	N. Hübner, *Dissertation*, University of Gießen, **1992**.
Hüb 1992a	N. Hübner, R. Gruehn, *Z. Anorg. Allg. Chem.* **1992**, *616*, 86.
Hüb 1992b	N. Hübner, R. Gruehn, *J. Alloys Compd.* **1992**, *183*, 85.
Hul 1956	N. Hultgren, L. Brewer, *J. Phys. Chem.* **1956**, *60*, 947.
Hul 1968	F. Hulliger, *US Patent 3515508*, **1968**.
Hum 1992	H.-U. Hummel, R. Fackler, P. Remmert, *Chem. Ber.* **1992**, *125*, 551.
Hun 1986	K.-H. Huneke, H. Schäfer, *Z. Anorg. Allg. Chem.* **1986**, *534*, 216.
Hus 1986	A. Hussain, B. Reitz, R. Gruehn, *Z. Anorg. Allg. Chem.* **1986**, *535*, 186.
Hus 1989	A. Hussain, R. Gruehn, *Z. Anorg. Allg. Chem.* **1989**, *571*, 91.
Hus 1991	A. Hussain, R. Gruehn, *J. Cryst. Growth* **1991**, *108*, 831.
Hus 1994	A. Hussain, L. Permér, L. Kihlborg, *Eur. J. Solid State Chem.* **1994**, *31*, 879.
Hus 1997	A. Hussain, R. Gruehn, C. H. Rüscher, *J. Alloys Compd.* **1997**, *246*, 51.
Isa 1973	A. S. Isaikin, V. N. Gribkov, B. V. Shchetanov, V. A. Silaev, M. K. Levinskaya, *Tiz. Khim. Obrat. Mater.* **1973**, 112.
Ito 1978	S. Ito, K. Kodaira, T. Matsushita, *Mater. Res. Bul.* **1978**, *13*, 97.
Izu 1979	F. Izumi, H. Kodama, A. Ono, *J. Cryst. Growth* **1979**, *47*, 139.
Jag 1966	W. Jagusch, *Dissertation*, University of Münster, **1966**.
Jan 2009	O. Janson, W. Schnelle, M. Schmidt, Yu. Prots, S.-L. Drexler, S. H. Filatov, H. Rosner, *New J. Phys.* **2009**, *11*, 113034.
Jen 2009	J. Jentsch, *Diplomarbeit*, HTW Dresden, **2009**.
Jia 1997	Jianzhuang Jiang, Tesuya Ozaki, Ken-ichi Machida, Gin-ya Adachi, *J. Alloys Comp.* **1997**, *260*, 222.
Jok 2009	S. J. Jokela, M. C. Tarun, M. D. McCluskey, *Phys. B* **2009**, *404*, 4810.
Joz 1987	M. Józefowicz, W. Piekarczyk, *Mater. Res. Bull.* **1987**, *22*, 775.
Jus 1986	H. Juskowiak, A. Pajaczkowska, *J. Mater. Sci.* **1986**, *21*, 3430.
Kal 1969	E. Kaldis, *Referate der Kurzvortäge der 10. Diskussionstagung der Sektion für Kristallkunde der DMG* **1969**, 444.
Kal 1970	E. Kaldis, R. Verreault, *J. Less-Common Met.* **1970**, *20*, 177.
Kal 1971	E. Kaldis, *J. Cryst. Growth* **1971**, *9*, 281.
Kam 1978	N. Kamegashira, K. Ohta, K. Naito, *J. Cryst. Growth* **1978**, *44*, 1.
Kam 1979	N. Kamegashira, K. Ohta, K. Naito, *J. Mater. Sci.* **1979**, *14*, 505.
Käm 1998	H. Kämmerer, R. Hofmann, R. Gruehn, *Z. Anorg. Allg. Chem.* **1998**, *624*, 1533.
Kap 1985	O. Kaposi, L. Lelik, G. A. Semenov, E.N. Nikolaev, *Acta Chim. Hung.* **1985**, *120*, 79.
Kav 1996	L. Kavan, M. Grätzel, S. E. Gilbert, C. Klemenz, H. J. Scheel, *J. Am. Chem. Soc.* **1996**, *118*, 6716.
Kaw 1974	K. Kawashima, Y. Ueda, K. Kosuge, S. Kachi, *J. Cryst. Growth* **1974**, *26*, 321.
Ker 1963	J. V. Kerrigan, *J. Appl. Phys.* **1963**, *34*, 3408.
Kho 1981	I. A. Khodyakova, *Deposited Doc.* **1981**, *VINITI 575–82*, 241.
Kim 1974	Y. W. Kim, G. R. Belton, *Met. Trans.* **1974**, *5*, 1811.
Kir 1994	L. Kirsten, H. Oppermann, *Z. Anorg. Allg. Chem.* **1994**, *620*, 1476.
Kle 1966	W. Kleber, M. Hähnert, R. Müller, *Z. Anorg. Allg. Chem.* **1966**, *346*, 113.
Kle 1966a	W. Kleber, J. Noak, H. Berger, *Krist. Tech.* **1966**, *1*, 7.
Kle 1966b	W. Kleber, R. Mlodoch, *Krist. Tech.* **1966**, *1*, 249.
Kle 1968	W. Kleber, H. Raidt, U. Dehlwes, *Krist. Tech.* **1968**, *3*, 153.

Kle 1970	W. Kleber, H. Raidt, R. Klein, *Krist. Tech.* **1970**, *5*, 479.
Klei 1963	P. Kleinert, *Z. Chem.* **1963**, *3*, 353.
Klei 1964	P. Kleinert, *Z. Chem.* **1964**, *4*, 434.
Klei 1965	P. Kleinert, E. Glauche, *Z. Chem.* **1965**, *5*, 30.
Klei 1966	P. Kleinert, D. Schmidt, *Z. Anorg. Allg. Chem.* **1966**, *348*, 142.
Klei 1967	P. Kleinert, D. Schmidt, E. Glauche, *Z. Chem.* **1967**, *7*, 33.
Klei 1970	P. Kleinert, *Z. Anorg. Allg. Chem.* **1970**, *378*, 71.
Klei 1972	P. Kleinert, *Z. Anorg. Allg. Chem.* **1972**, *387*, 11.
Klei 1972a	P. Kleinert, *Z. Anorg. Allg. Chem.* **1972**, *387*, 129.
Klei 1973	P. Kleinert, D. Schmidt, *Z. Anorg. Allg. Chem.* **1973**, *396*, 308.
Klei 1974	P. Kleinert, J. Kirchhof, *Krist. Tech.* **1974**, *9*, 165.
Klei 1977	P. Kleinert, J. Kirchhof, *Z. Anorg. Allg. Chem.* **1977**, *429*, 137.
Klei 1977a	P. Kleinert, J. Kirchhof, *Z. Anorg. Allg. Chem.* **1977**, *429*, 147.
Klei 1982	P. Kleinert, D. Schmidt, H.-J. Laukner, *Z. Anorg. Allg. Chem.* **1982**, *495*, 157.
Kli 1978	L. A. Klinkova, E. D. Skrebkova, *Izv. Akad. Nauk. SSSR, Neorg. Mater.* **1978**, *14*, 373.
Kna 1969	O. Knacke, G. Lossmann, F. Müller, *Z. Anorg. Allg. Chem.* **1969**, *371*, 32.
Kob 1981	Y. Kobayashi, S. Muranaka, Y. Bando, T. Takada, *Bull. Inst. Chem. Res. Kyoto Univ.* **1981**, *59*, 248.
Kod 1972	H. Kodama, T. Kikuchi, M. Goto, *J. Less-Common Met.* **1972**, *29*, 415.
Kod 1975	H. Kodama, M. Goto, *J. Cryst. Growth* **1975**, *29*, 77.
Kod 1975a	H. Kodama, M. Goto, *J. Cryst. Growth* **1975**, *29*, 222.
Kod 1976	H. Kodama, H. Komatsu, *J. Cryst. Growth* **1976**, *36*, 121.
Kol 2002	D. Kolberg, F. Wastin, J. Rebizant, P. Boulet, G. H. Lander, J. Schoenes, *Phys. Rev. B* **2002**, *66*, 214418.
Kor 1989	V. O. Kordyukevich, V. I. Kuznetsov, M. V. Razumeenko, O. A. Yuminov, *Radiochim.* **1989**, *31*, 147.
Kra 1977	G. Krabbes, H. Oppermann, *Krist. Tech.* **1977**, *12*, 929.
Kra 1979	G. Krabbes, H. Oppermann, E. Wolf, *Z. Anorg. Allg. Chem.* **1979**, *450*, 21.
Kra 1983	G. Krabbes, H. Oppermann, E. Wolf, *J. Cryst. Growth* **1983**, *64*, 353.
Kra 1984	G. Krabbes, J. Klosowski, H. Oppermann, H. Mai, *Cryst. Res. Technol.* **1984**, *19*, 491.
Kra 1986	G. Krabbes, U. Gerlach, E. Wolf, W. Reichelt, H. Oppermann, J. C. Launay, Proc 6th *Symposium on Material Sciences under microgravity Conditions, Bordeaux* **1986**.
Kra 1987	G. Krabbes, D. V. Hoanh, N. Van Hai, H. Oppermann, S. Velichkow, P. Peshev, *J. Cryst. Growth* **1987**, *82*, 477.
Kra 1988a	G. Krabbes, D. V. Hoanh, *Z. Anorg. Allg. Chem.* **1988**, *562*, 62.
Kra 1988b	R. Krausze, H. Oppermann, *Z. Anorg. Allg. Chem.* **1988**, *558*, 46.
Kra 1991	G. Krabbes, W. Bieger, K.- H. Sommer, E. Wolf, *J. Cryst. Growth* **1991**, *110*, 433.
Kra 2005	G. Krabbes, U. Gerlach, W. Reichelt, *Z. Anorg. Allg. Chem.* **2005**, *631*, 375.
Kru 1987	F. Krumeich, R. Gruehn, *Z. Anorg. Allg. Chem.* **1987**, *554*, 14.
Kru 1986	B. Krug, *Dissertation,* University of Gießen, **1986**.
Kru 1986a	B. Krug, R. Gruehn, *J. Less-Common Met.* **1986**, *116*, 105.
Kub 1983	O. Kubaschewski, C. B. Alcock, *Metallurgical Thermochemistry*, 5.th Ed., Pergamon Press, Oxford, **1983**.
Kur 1972	K. Kurosawa, S. Saito, S. Takemoto, *Jpn. J. Appl. Phys.* **1972**, *11*, 1230.
Kur 1975	K. Kurosawa, S. Saito, S. Takemoto, *Jpn. J. Appl. Phys.* **1975**, *14*, 887.
Kuw 1980	H. Kuwamoto, J. M. Honig, *J. Solid State Chem.* **1980**, *32*, 335.
Kuw 1981	H. Kuwamoto, N. Otsuka, H. Sato, *J. Solid State Chem.* **1981**, *36*, 133.

Kyo 1977   M. Kyoto, Y. Bando, T. Takada, *Chem. Lett. Jpn.* **1977**, 595.

Lan 1986   B. Langenbach- Kuttert, J. Sturm, R. Gruehn, *Z. Anorg. Allg. Chem.* **1986**, *543*, 117.

Lan 1986b   B. Langenbach- Kuttert, *Dissertation,* University of Gießen, **1986**.

Lan 1987   B. Langenbach- Kuttert, J. Sturm, R. Gruehn, *Z. Anorg. Allg. Chem.* **1987**, *548*, 33.

Lau 1972   J. C. Launay, M. Onillon, M. Pouchard, *Rev. Chim. Miner.* **1972**, *9*, 41.

Lau 1973   J. C. Launay, G. Villeneuve, M. Pouchard, *Mater. Res. Bull.* **1973**, *8*, 997.

Lau 1974   J.-C. Launay, M. Pouchard, *J. Cryst. Growth* **1974**, *23*, 85.

Lau 1976   J.- C. Launay, M. Pouchard, R. Ayroles, *J. Cryst. Growth* **1976**, *36*, 297.

Lav 1964   F. Laves, R. Moser, W. Petter, *Naturwissenschaften* **1964**, *51*, 356.

Lec 1991   F. Leccabue, R. Panizzieri, B. E. Watts, D. Fiorani, E. Agostinelli, A. Testa, E. Paparazzo, *J. Cryst. Growth* **1991**, *112*, 644.

Lec 1993   F. Leccabue, B. E. Watts, C. Pelosi, D. Fiorani, A. M. Testa, A. Pajączkowska, G. Bocelli, G. Calestani, *J. Cryst. Growth* **1993**, *128*, 859.

Lec 1993a   F. Leccabue, B. E. Watts, D. Fiorani, A. M . Testa, J. Alvarez, V. Sagredo, G. Bocelli, *J. Mater. Sci.* **1993**, *28*, 3945.

Len 1994   M. Lenz, R. Gruehn, *J. Cryst. Growth* **1994**, *137*, 499.

Len 1997   M. Lenz, R. Gruehn, *Chem. Rev.* **1997**, *97*, 2967.

Li 2009   C. Li, M. Isobe, H. Ueda, Y. Matsuhita, Y. Ueda, *J. Solid State Chem.* **2009**, *182*, 3222.

Lib 1994   M. Liberatore, B. J. Wuensch, I. G. Solorzano, J. B. Vander Sande, *Mater. Res. Soc. Symp. Proc.* **1994**, *318*, 637.

Lis 1996   Lisheng Chi, Huayang Chen, Shuiquan Deng, Honghui Zhuang, Jinshun Huang, *J. Alloy. Compd.* **1996**, *242*, 1.

Loc 1999   S. Locmelis, M. Binnewies, *Z. Anorg. Allg. Chem.* **1999**, *625*, 294.

Loc 1999a   S. Locmelis, R. Wartchow, G. Patzke, M. Binnewies, *Z. Anorg. Allg. Chem.* **1999**, *625*, 661.

Loc 1999b   S. Locmelis, M. Binnewies, *Z. Anorg. Allg. Chem.* **1999**, *625*, 1573.

Loc 2005   S. Locmelis, U. Hotje, M. Binnewies, *Z. Anorg. Allg. Chem.* **2005**, *631*, 3080.

Lon 1973   M. C. De Long, *U. S. Nat. Tech. Inform. Serv. A. D. Rep.* **1973**, Nr. 760733.

Mar 1959   G. de Maria, J. Drowart, M. E. Inghram, *J. Chem. Phys.* **1959**, *30*, 318.

Mar 1999   K. Mariolacos, *N. Jb. Miner. Mh.* **1999**, *9*, 415.

Mas 1990   H. Massalski, *Binary Alloy Phase Diagrams*, 2nd Ed. ASN International, **1990**.

Mat 1977   K. Matsumoto, S. Kaneko, K. Takagi, *J. Cryst. Growth* **1977**, *40*, 291.

Mat 1981   K. Matsumoto, T. Kawanishi, K. Takagi, *J. Cryst. Growth* **1981**, *55*, 376.

Mat 1981a   K. Matsumoto, T. Sata, *Bull. Chem. Soc. Jpn.* **1981**, *54*, 674.

Mat 1982   K. Matsumoto, T. Kawanishi, K. Takagi, S. Kaneko, *J. Cryst. Growth* **1982**, *58*, 653.

Mat 1983   K. Matsumoto, S. Kaneko, K. Takagi, S. Kawanishi, *J. Electrochem. Soc.* **1983**, *130*, 530.

Mat 1985   K. Matsumoto, K. Konemura, G. Shimaoka, *J. Cryst. Growth* **1985**, *71*, 99.

Mat 1988   K. Matsumoto, G. Shimaoka, *J. Cryst. Growth* **1988**, *86*, 410.

Mat 1991   K. Matsumoto, K. Noda, *J. Cryst. Growth* **1991**, *109*, 309.

Mer 1973   J. Mercier, S. Lakkis, *J. Cryst. Growth* **1973**, *20*, 195.

Mer 1973a   J. Mercier, *Bull. Soc. Sci., Bretagne* **1973**, *48*, 135.

Mer 1977   J. Mercier, J. J. Since, G. Fourcaudot, J. Dumas, J. Devenyi, *J. Cryst. Growth* **1977**, *42*, 583.

Mer 1982   J. Mercier, G. Fourcaudot, Y. Monteil, C. Bec, R. Hillel, *J. Cryst. Growth* **1982**, *59*, 599.

Mer 1982a   J. Mercier, *J. Cryst. Growth* **1982**, *56*, 235.
Meu 1976   G. Meunier, M. Bertaud, *J. Appl. Cryst.* **1976**, *9*, 364.
Mik 2005   M. Mikami, T. Eko, J. Wang, Y. Masa, M. Isshiki, *J. Cryst. Growth* **2005**, *276*, 389.
Mik 2007   M. Mikami, S.- H- Hong, T. Sato, S. Abe, J. Wang, K. Matsumoto, Y. Masa, M. Isshiki, *J. Cryst. Growth* **2007**, *304*, 37.
Mil 1949   T. Millner, J. Neugebauer, *Nature* **1949**, *163*, 601.
Mil 1990   E. C. Milliken, J. F. Cordaro, *J. Mater. Res.* **1990**, *5*, 53.
Mon 1984   Y. Monteil, C. Bec, R. Hillel, J. Bouix, C. Bernard, J. Mercier, *J. Cryst. Growth* **1984**, *67*, 595.
Mül 1965   A. Müller, B. Krebs, O. Glemser, *Naturwissenschaften* **1965**, *52*, 55.
Mul 2004   D. D. Mulmi, T. Sekiya, N. Kamiya, S. Kurita, Y. Murakami, T. Kodaira, *J. Phys. Chem. Solids* **2004**, 1181.
Mül 2004   U. Müller, *Diplomarbeit,* HTW Dresden, **2004**.
Mun 2005   V. Munoz-Sanjosé, R. Tena-Zaera, C. Martínez-Tomás, J. Zúniga-Pérez, S. Hassani, R. Triboulet, *Phys. Stat. Sol.* **2005**, *2*, 1106.
Mur 1976   M. de Murcia, J. P. Fillard, *Mater. Res. Bull.* **1976**, *11*, 189.
Mur 1995   K. Murase, K. Machida, G. Adachi, *J. Alloys Compd.* **1995**, *217*, 218.
Myc 2004   A. Mycielski, A. Szadkowski, L. Kowalczyk, B. Witkowska, W. Kaliszek, B. Chwalisz, A. Wysmołek, R. Stępniewski, J. M. Baranowski, M. Potemski, A. M. Witowski, R. Jakieła, A. Barcz, P.Aleshkevych, M. Jouanne, W. Szuszkiewicz, A. Suchocki, E. Łusakowska, E. Kamińska, W. Dobrowolski, *Phys. Status Solidi C* **2004**, *1*, 884.
Myc 2004a   A. Mycielski, L. Kowalczyk, A. Szadkowski, B. Chwalisz, A. Wysmołek, R. Stępniewski, J. M. Baranowski, M. Potemski, A. M. Witowski, R. Jakieła, A. Barcz, B. Witkowska, W. Kaliszek, A. Jędrzejczak, A. Suchocki, E. Łusakowska, E. Kamińska, *J. Alloys Compd.* **2004**, *371*, 150.
Nag 1969   K. Nagasawa, Y. Bando, T. Takada, *Jpn. J. Appl. Phys.* **1969**, *8*, 1262.
Nag 1970   K. Nagasawa, Y. Bando, T. Takada, *Jpn. J. Appl. Phys.* **1970**, *9*, 407.
Nag 1971   K. Nagasawa, *Mater. Res. Bull.* **1971**, *6*, 853.
Nag 1972   K. Nagasawa, Y. Bando, T. Takada, *J. Cryst. Growth* **1972**, *17*, 143.
Nai 1971   K. Naito, N. Kamegashira, Y. Nomura, *J. Cryst. Growth* **1971**, *8*, 219.
Neg 1994   H. Negishi, T. Miyahara, M. Inoue, *J. Cryst. Growth* **1994**, *144*, 320.
Neg 2000   H. Negishi, S. Negishi, M. Sasaki, M. Inoue, *Jpn. J. Appl. Phys.* **2000**, *39*, 505.
Neu 1956   A. Neuhaus, *Chem. Ing. Tech.* **1956**, *28*, 350.
Nic 1993   G. Nichterwitz, *Dissertation,* University of Dresden, **1993**.
Nie 1967   T. Niemyski, W. Piekarczyk, *J. Cryst. Growth* **1967**, *1*, 177.
Nit 1966   R. Nitsche, *Intern. Conference on Crystal Growth, Boston* **1966**.
Nit 1967   R. Nitsche, *Fortschr. Miner.* **1967**, *442*, 231.
Nit 1967a   R. Nitsche, *J. Phys. Chem. Solids Suppl.* **1967**, *1*, 215.
Noc 1993   K. Nocker, R. Gruehn, *Z. Anorg. Allg. Chem.* **1993**, *619*, 1530.
Noc 1993a   K. Nocker, R. Gruehn, *Z. Anorg. Allg. Chem.* **1993**, *619*, 699.
Noc 1994   K. Nocker, R. Gruehn, *Z. Anorg. Allg. Chem.* **1994**, *620*, 266.
Nol 1976   B. I. Noläng, M. W. Richardson, *J. Cryst. Growth* **1976**, *34*, 205.
Nom 1981   Y. Nomura, N. Kamegashira, K. Naito, *J. Cryst. Growth* **1981**, *52*, 279.
Nov 1961   A. V. Novoselova, V. N. Babin, B. P. Sobolev, *Russ. J. Inorg. Chem.* **1961**, *6*, 113.
Nov 1964   A. V. Novoselova, U. K. Orlova, B. P. Sobolev, L. N. Sidorov, *Dokl. Akad. Nauk SSSR* **1964**, *159*, 1338.
Nov 1965   A. V. Novoselova, *Izv. Akad. Nauk. SSSR, Neorg. Mater.* **1965**, *1*, 1010.

Nov 1966	A. V. Novoselova, Y. V. Azhikina, *Izv. Akad. Nauk SSSR, Neorg. Mater.* **1966**, *2*, 1604.
Nov 1966a	A. V. Novoselova, V. N. Babin, B. P. Sobolev, *Kristallografija* **1966**, *11*, 477.
Nov 1967	A. V. Novoselova, *Krist. Tech.* **1967**, *2*, 511.
Nte 1999	J.- M. Ntep, S. Said Hassani, A. Lusson, A. Tromson-Carli, D. Ballutaud, G. Didier, R. Triboulet, *J. Cryst. Growth* **1999**, *207*, 30.
Oht 1986	T. Ohtani, T. Yamaoka, K. Shimamura, *Chem. Lett.* **1986**, 947.
Opp 1971	H. Oppermann, *Z. Anorg. Allg. Chem.* **1971**, *383*, 285.
Opp 1975	H. Oppermann, M. Ritschel, *Krist. Tech.* **1975**, *10*, 485.
Opp 1975a	H. Oppermann, W. Reichelt, E. Wolf, *J. Cryst. Growth* **1975**, *31*, 49.
Opp 1975b	H. Oppermann, W. Reichelt, U. Gerlach, E. Wolf, W. Brückner, W. Moldenhauer, H. Wich, *Physica Status Solidi* **1975**, *28*, 439.
Opp 1977	H. Oppermann, W. Reichelt, G. Krabbes, E. Wolf, *Krist. Tech.* **1977**, *12*, 717.
Opp 1977a	H. Oppermann, W. Reichelt, E. Wolf, *Z. Anorg. Allg. Chem.* **1977**, *432*, 26.
Opp 1977b	H. Oppermann, E. Wolf, *Z. Anorg. Allg. Chem.* **1977**, *437*, 33.
Opp 1977c	H. Oppermann, W. Reichelt, G. Krabbes, E. Wolf, *Krist. Tech.* **1977**, *12*, 919.
Opp 1977d	H. Oppermann, *Proc. 6th Intern. Conf. on Rare Metals, Pécs*, **1977**, *1*, 166.
Opp 1978	H. Oppermann, G. Stöver, M. Ritschel, *Kurzreferate Kristallsynthesen der 13. Jahrestagung der Vereinigung für Kristallographie in der Gesellschaft für Geologische Wissenschaften der DDR* **1978**, P 3.12, P 3.13.
Opp 1978a	H. Oppermann, V. A. Titov, G. Kunze, G. A. Kokovin, E. Wolf, *Z. Anorg. Allg. Chem.* **1978**, *439*, 13.
Opp 1980	H. Oppermann, G. Kunze, E. Wolf, G. A. Kokovin, I. M. Sitschova, G. E. Osigova, *Z. Anorg. Allg. Chem.* **1980**, *461*, 165.
Opp 1984	H. Oppermann, G. Stöver, *Z. Anorg. Allg. Chem.* **1984**, *511*, 57.
Opp 1985	H. Oppermann, G. Stöver, A. Heinrich, K. Teske, E. Ziegler, *Acta Phys. Hung.* **1985**, *57*, 213.
Opp 1985a	H. Oppermann, G. Stöver, E. Wolf, *Cryst. Res. Technol.* **1985**, *20*, 883.
Opp 1985b	H. Oppermann, *Z. Anorg. Allg. Chem.* **1985**, *523*, 135.
Opp 1990	H. Oppermann, *Solid State Ionics* **1990**, *39*, 17.
Opp 1998	H. Oppermann, B. Marklein, *Z. Naturforsch.* **1998**, *53b*, 1352.
Opp 1999	H. Oppermann, H. Göbel, P. Schmidt, H. Schadow, V. Vassilev, I. Markova-Deneva, *Z. Naturforsch.* **1999**, *54b*, 261.
Opp 2002	H. Oppermann, D. Q. Huong, P. Schmidt, *Z. Anorg. Allg. Chem.* **2002**, *628*, 2509.
Opp 2005	H. Oppermann, M. Schmidt, P. Schmidt, *Z. Anorg. Allg. Chem.* **2005**, *631*, 197.
Opp 2005a	H. Oppermann, P. Schmidt, *Z. Anorg. Allg. Chem.* **2005**, *631*, 1309.
Orl 1971	V. P. Orlovskii, H. Schäfer, V. P. Repko, G. M. Safronov, I. V. Tananaev, *Izv. Akad. Nauk. SSSR, Neorg. Mater.* **1971**, *7*, 971.
Orl 1974	V. P. Orlovskii, T. V. Belyaevskaya , V. I. Bugakov, B. S. Khalikov, *Izv. Akad. Nauk. SSSR, Neorg. Mater.* **1977**, *13*, 1489.
Orl 1976	V. P. Orlovskii, V. P. Repko, T. V. Belyaevskaya, *Izv. Akad. Nauk. SSSR, Neorg. Mater.* **1976**, *13*, 1526.
Orl 1978	V. P.Orlovskii, B. Khalikov, Kh. M. Kurbanov, V. I. Bugakov, L.N. Kargareteli, *Zh. Neorg. Khim.* **1978**, *232*, 316.
Orl 1985	V. P. Orlovskii, V. V. Nechaev, A. I. Mironenko, *Neorg. Materialy* **1985**, *21*, 664.
Paj 1981	A. Pajączkowska, W. Piekarczyk, P. Peshev, A. Toshev, *Mater. Res. Bull.* **1981**, *16*, 1091.
Paj 1985	A. Pajączkowska, K. Machjer, *J. Cryst. Growth* **1985**, 71, 810.

Paj 1986   A. Pajączkowska, H. Juskowiak, *J. Cryst. Growth* **1986**, 79, 421.
Paj 1986a   A. Pajączkowska, G. Jasiołek, K. Majcher, *J. Cryst. Growth* **1986**, 79, 417.
Paj 1990   A. Pajączkowska, *J. Cryst. Growth* **1990**, *104*, 498.
Pal 2006   W. Palosz, *J. Cryst. Growth* **2006**, *286*, 42.
Par 1982   H. L. Park, *J. Korean. Phys. Soc.* **1982**, *15*, 51.
Pat 1999   G. R. Patzke, S. Locmelis, R. Wartchow, M. Binnewies, *J. Cryst. Growth* **1999**, *203*, 141.
Pat 2000   G. Patzke, M. Binnewies, *Solid State Sci.* **2000**, *2*, 689.
Pat 2000a   G. R. Patzke, M. Binnewies, U. Nigge, H.-D. Wiemhöfer, *Z. Anorg. Allg. Chem.* **2000**, *626*, 2340.
Pat 2000b   G. R. Patzke, J. Koepke, M. Binnewies, *Z. Anorg. Allg. Chem.* **2000**, *626*, 1482.
Pat 2000c   G. R. Patzke, M. Binnewies, *Z. Naturforsch.* **2000**, *55b*, 26.
Pea 1973   T. Pearsall, *J. Cryst. Growth* **1973**, *20*, 192.
Pes 1973   P. Peshev, G. Bliznakov, G. Gyurov, M. Ivanova, *Mater. Res. Bull* **1973**, *8*, 915.
Pes 1973a   P. Peshev, G. Bliznakov, G. Gyurov, M. Ivanova, *Mater. Res. Bull.* **1973**, *8*, 1011.
Pes 1974   P. Peshev, A. Toshev, *Mater. Res. Bull.* **1974**, *9*, 873.
Pes 1975   P. Peshev, A. Toshev, *Mater. Res. Bull.* **1975**, *10*, 1335.
Pes 1975a   P. Peshev, M. S. Ivanova, *Phys. Status Solidi* **1975**, *28*, K1.
Pes 1976   P. Peshev, A. Tosehv, *Mater. Res. Bull.* **1976**, *11*, 1433.
Pes 1978   P. Peshev, A. Toshev, *J. Mater. Sci.* **1978**, *13*, 143.
Pes 1980   P. Peshev, I. Z. Babievskaja, V. A. Krenev, *J. Mater. Sci.* **1980**, *15*, 2942.
Pes 1982   P. Peshev, A. Toshev, *Mater. Res. Bull.* **1982**, *17*, 1413.
Pes 1984   P. Peshev, A. Toshev, G. Krabbes, U. Gerlach, H. Oppermann, *J. Cryst. Growth* **1984**, *66*, 147.
Pfe 2002   A. Pfeifer, M. Binnewies, *Z. Anorg. Allg. Chem.* **2002**, *628*, 1678.
Pfe 2002a   A. Pfeifer, M. Binnewies, *Z. Anorg. Allg. Chem.* **2002**, *628*, 1091.
Pfe 2002b   A. Pfeifer, M. Binnewies, *Z. Anorg. Allg. Chem.* **2002**, *628*, 2605.
Pfe 2002c   A. Pfeifer, M. Binnewies, *Z. Anorg. Allg. Chem.* **2002**, *628*, 2273.
Pic 1973   J. Pickardt, B. Reuter, J. Söchtig, *Z. Anorg. Allg. Chem.* **1973**, *401*, 21.
Pic 1973a   J. Pickardt, B. Reuter, *Z. Anorg. Allg. Chem.* **1973**, *401*, 37.
Pie 1972   W. Piekarczyk, S. Gazda, T. Niemyski, *J. Cryst. Growth* **1972**, *12*, 272.
Pie 1978   W. Piekarczyk, P. Peshev, A. Toshev, A. Pajączkowska, *Mater. Res. Bull.* **1978**, *13*, 587.
Pie 1981   W. Piekarczyk, *J. Cryst. Growth* **1981**, *55*, 543.
Pie 1982   W. Piekarczyk, *J. Cryst. Growth* **1982**, *60*, 166.
Pie 1987   W. Piekarczyk, *J. Cryst. Growth* **1987**, *82*, 367.
Pie 1988   W. Piekarczyk, *J. Cryst. Growth* **1988**, *89*, 267.
Pli 1982   V. Plies, *Z. Anorg. Allg. Chem.* **1982**, *484*, 165.
Pli 1983   V. Plies, W. Redlich, R. Gruehn, *Z. Anorg. Allg. Chem.* **1983**, *503*, 141.
Poe 1981   K. R. Poeppelmeier, G. B. Ansell, *J. Cryst. Growth* **1981**, *51*, 587.
Pou 1973   M. Pouchard, J. C. Launay, *Mater. Res. Bull.* **1973**, *8*, 95.
Pov 1998   V. G. Povarov, A. G. Ivanov, V. M. Smirnov, *Inorg. Mater.* **1998**, *34*, 800.
Pre 1996   A. Preuß, *Dissertation,* University of Gießen, **1996.**
Pro 2001   A. V. Prokofiev, R. K. Kremer, W. Assmus, *J. Cryst. Growth* **2001**, *231*, 498.
Pro 2003   A. V. Prokofiev, F. Ritter, W. Assmus, B. J. Gibson, R. K. Kremer, *J. Cryst. Growth* **2003**, *247*, 457.
Prz 1972   J. Przedmojski, B. Pura, W. Piekarczyk, S. Gazda, *Phys. Status Solidi* **1972**, *11*, K1.
Qui 1970   R. H. Quinn, P. G. Neiswander, *Mater. Res. Bull.* **1970**, *5*, 329.

Quo 1975   H. H. Quon, D. P. Malanda, *Mater. Res. Bull.* **1975**, *10*, 349.
Rab 1985   A. Rabenau, *Angew. Chem.* **1985**, *97*, 1017.
Rad 2000   O. Rademacher, H. Goebel, H. Oppermann, *Z. Kristallogr.- New Cryst. Struct.* **2000**, *215*.
Rad 2001   O. Rademacher, H. Goebel, M. Ruck, H. Oppermann, *Z. Kristallogr.- New Cryst. Struct.* **2001**, *216*, 29.
Rar 2005   J. A. Rard, *J. Nucl. Radiochem. Sci.* **2005**, *6*, 197.
Raz 1981   M. V. Razumeenko, V. S. Grunin, A. A. Boitsov, *Sov. Phys. Crystallogr.* **1981**, *26*, 371.
Rea 1976   F. M. Reames, *Mater. Res. Bull.* **1976**, *11*, 1091.
Red 1976   W. Redlich, *Dissertation,* University of Gießen, **1976**.
Red 1978   W. Redlich, R. Gruehn, *Z. Anorg. Allg. Chem.* **1978**, *438*, 25.
Rei 1975   W. Reichelt, H. Oppermann, E. Wolf, *Jahresbericht Zentralinstitut für Festkörperphysik und Werkstoffforschung*, Dresden, **1975**.
Rei 1977 W. Reichelt, Dissertation, *Akademie der Wissenschaften der DDR*, **1977**.
Rei 1979   W. Reichelt, H. Oppermann, E. Wolf, *Z. Anorg. Allg. Chem.* **1979**, *452*, 96.
Rei 1991   W. Reichelt, *Habilitationschrift,* Technische University of Dresden, **1991**.
Rei 1994   W. Reichelt, H. Oppermann, *Z. Anorg. Allg. Chem.* **1994**, *620*, 1463.
Rei 2000   W. Reichelt, T. Weber, T. Söhnel, S. Däbritz, *Z. Anorg. Allg. Chem.* **2000**, *626*, 2020.
Rei 2005   W. Reichelt, U. Steiner, T. Söhnel, O. Oeckler, V. Duppel, L. Kienle, *Z. Anorg. Allg. Chem.* **2005**, *631*, 596.
Reis 1971   A. Reisman, J. E. Landstein, *J. Electrochem. Soc.* **1971**, *118*, 1479.
Rem 1982   F. Remy, O. Monnereau, A. Casalot, F. Dahan, J. Galy, *J. Solid State Chem.* **1988**, *76*, 167.
Rep 1971   V. P. Repko, V. P. Orlovskii, G. M. Safranov, Kh. M. Kurbanov, M. N. Tseitlin, V. I. Pakhomov, I. V. Tananaev, A. N. Volodina, *Izv. Akad. Nauk. SSSR, Neorg. Mater* **1971**, *7*, 251.
Rin 1967   K. Rinke, M. Klein, H. Schäfer, *J. Less-Common Met.* **1967**, *12*, 497.
Rit 1977   M. Ritschel, N. Mattern, W. Brückner, H. Oppermann, G. Stöver, W. Moldenhauer, J. Henke, E. Wolf, *Krist. Tech.* **1977**, *12*, 1221.
Rit 1978   M. Ritschel, H. Oppermann, *Krist. Tech.* **1978**, *13*, 1035.
Rit 1978a   M. Ritschel, H. Oppermann, N. Mattern, *Krist. Tech.* **1978**, *13*, 1421.
Rit 1980   M. Ritschel, H. Oppermann, *Krist. Tech.* **1980**, *15*, 395.
Rog 1969   D. B. Rogers, R. D. Shannon, A. W. Sleight, J. L. Gillson, *Inorg. Chem.* **1969**, *8*, 841.
Rog 1971   D. B. Rogers, R. D. Shannon, J. L. Gillson, *J. Solid State Chem.* **1971**, *3*, 314.
Roh 1994   G. S. Rohrer, W. Lu, R. L. Smith, *Mat. Res. Soc. Symp. Proc.* **1994**, *332*, 507.
Ros 1987 A. Rossberg, H. Oppermann, R. Starke, *Z. Anorg. Allg. Chem.* **1987**, *554*, 151.
Ros 1988 A. Rossberg, H. Oppermann, *Z. Anorg. Allg. Chem.* **1988**, *556*, 109.
Ros 1990   R. Roß, R. Gruehn, *Z. Anorg. Allg. Chem.* **1990**, *591*, 95.
Ros 1990a   R. Roß, *Dissertation,* University of Gießen, **1990**.
Ros 1991a   R. Ross, R. Gruehn, *Z. Anorg. Allg. Chem.* **1991**, *605*, 75.
Ros 1992   R. Roß, R. Gruehn, *Z. Anorg. Allg. Chem.* **1992**, *612*, 63.
Ros 1992a   R. Roß, R. Gruehn, *Z. Anorg. Allg. Chem.* **1992**, *614*, 47.
Rou 1997   P. Roussel, D. Groult, C. Hess, Ph. Labbé, C. Schlenker, *J. Phys. Condens. Matter* **1997**, *9*, 7081.
Roy 1962   V. Royen, W. Forweg, *Naturwissenschaften* **1962**, *49*, 85.
Roy 1963   P. Royen, W. Forweg, *Z. Anorg. Allg. Chem.* **1963**, *326*, 113.
Roy 1963a   P. Royen, W. Forweg, *Naturwissenschaften* **1963**, *50*, 41.

Rüs 2008   C. H. Rüscher, K. R. Dey, T. Debnath, I. Horn, R. Glaum, A. Hussain, *J. Solid State Chem.* **2008**, *181*, 90.
Sae 1973   M. Saeki, N. Kimizuka, M. Ishii, I. Kawada, M. Nakano, A. Ichinose, M. Nakahira, *J. Cryst. Growth* **1973**, *18*, 101.
Sah 1983   W. Sahle, M. Nygren, *J. Solid State Chem.* **1983**, *48*, 154.
Sak 1972   T. Sakata, K. Sakata, G. Höfer, T. Horiuchi, *J. Cryst. Growth* **1972**, *12*, 88.
San 1974   N. Sano, G. R. Belton, *Metallurg. Trans.* **1974**, *5*, 2151.
Sav 2007   A. I. Savchuk, V. I. Fediv, G. I. Kleto, S. V. Krychun, S. A. Savchuk, *Phys. Stat. Sol.* **2007**, *1*, 106.
Sch 1956   H. Schäfer, H. Jacob, K. Etzel, *Z. Anorg. Allg. Chem.* **1956**, *286*, 27.
Sch 1957   H. Schäfer, K. Etzel, *Z. Anorg. Allg. Chem.* **1957**, *291*, 294.
Sch 1957a   H. Schäfer, B. Morcher, *Z. Anorg. Allg. Chem.* **1957**, *291*, 221.
Sch 1960   H. Schäfer, H. J. Heitland, *Z. Anorg. Allg. Chem.* **1960**, *304*, 249.
Sch 1960a   H. Schäfer, E. Sibbing, *Z. Anorg. Allg. Chem.* **1960**, *305*, 341.
Sch 1962   H. Schäfer, W. Huesker, *Z. Anorg. Allg. Chem.* **1962**, *317*, 321.
Sch 1962a   H. Schäfer, *Chemische Transportreaktionen*, Verlag Chemie, Weinheim, **1962**.
Sch 1963   H. Schäfer, G. Schneidereit, W. Gerhardt, *Z. Anorg. Allg. Chem.* **1963**, *319*, 327.
Sch 1963a   H. Schäfer, A. Tebben, W. Gerhardt, *Z. Anorg. Allg. Chem.* **1963**, *321*, 41.
Sch 1964   H. Schäfer, *Chem. Transport Reactions*, Academic Press, N.Y., London, **1964**.
Sch 1964a   H. Schäfer, F. Schulte, R. Gruehn, *Angew. Chemie* **1964**, *76*, 536.
Sch 1966   H. Schäfer, R. Gruehn, F. Schulte, *Angew. Chemie* **1966**, *78*, 28.
Sch 1971   H. Schäfer, *J. Cryst. Growth* **1971**, *9*, 17.
Sch 1972   H. Schäfer, V. P. Orlovskii, *Z. Anorg. Allg. Chem.* **1972**, *390*, 13.
Sch 1972a   H. Schäfer, *Nat. Bur. Standards Special Publ.* **1972**, *364*, 413.
Sch 1973   H. Schäfer, *Z. Anorg. Allg. Chem.* **1973**, *400*, 242.
Sch 1973a   H. Schäfer, T. Grofe, M. Trenkel, *J. Solid State Chem.* **1973**, *8*, 14.
Sch 1973b   H. Schäfer, M. Bode, M. Trenkel, *Z. Anorg. Allg. Chem.* **1973**, *400*, 253.
Sch 1976   H. Schäfer, *Festschrift für Leo Brandt*, Opladen, Köln, **1976**, 91.
Sch 1977   H. Schäfer, M. Binnewies, H. Rabeneck, C. Brendel, M. Trenkel, *Z. Anorg. Allg. Chem.* **1977**, *435*, 5.
Sch 1978   H. Schäfer, M. Trenkel, *Z. Naturforsch.* **1978**, *33b*, 1318.
Sch 1980   H. Schäfer, *Z. Anorg. Allg. Chem.* **1980**, *471*, 35.
Sch 1985   H. Schäfer, W. Jagusch, H. Wenderdel, U. Griesel, *Z. Anorg. Allg. Chem.* **1985**, *529*, 189.
Sch 1988   H. Schäfer, *Z. Anorg. Allg. Chem.* **1988**, *564*, 127.
Scha 1974   E. Schaum, *Diplomarbeit*, University of Gießen, **1974**.
Scha 1986   U. Schaffrath, *Diplomarbeit*, University of Gießen, **1986**.
Scha 1988   U. Schaffrath, R. Gruehn, *Z. Anorg. Allg. Chem.* **1988**, *565*, 67.
Scha 1988a   U. Schaffrath, G. Steinmann, *Z. Anorg. Allg. Chem.* **1988**, *565*, 54.
Scha 1989   U. Schaffrath, *Dissertation*, University of Gießen, **1989**.
Scha 1989a   U. Schaffrath, R. Gruehn, *Z. Anorg. Allg. Chem.* **1989**, *573*, 107.
Scha 1990   U. Schaffrath, R. Gruehn, *Z. Anorg. Allg. Chem.* **1990**, *588*, 43.
Scha 1991   U. Schaffrath, R. Gruehn, *Synthesis of Lanthanide and Actinide Compounds*, Hrsg., G. Meyer, L. Morss, Kluver Academic, Dordrecht, **1991**.
Schl 1999   M. Schleifer, J. Busch, R. Gruehn, *Z. Anorg. Allg. Chem.* **1999**, *625*, 1985.
Schl 1999a   M. Schleifer, *Dissertation*, University of Gießen **1999**.
Schl 2000   M. Schleifer, J. Busch, B. Albert, R. Gruehn, *Z. Anorg. Allg. Chem.* **2000**, *626*, 2299.
Schm 1964   H. Schmid, *Acta Crystallogr.* **1964**, *17*, 1080.
Schm 1964a   H. Schmid, *Z. Anorg. Allg. Chem.* **1964**, *327*, 110.

Schm 1981   G. Schmidt, R. Gruehn, *Z. Anorg. Allg. Chem.* **1981**, *478*, 75.
Schm 1981a   G. Schmidt, R. Gruehn, *Z. Anorg. Allg. Chem.* **1981**, *478*, 111.
Schm 1983   G. Schmidt, R. Gruehn, *Z. Anorg. Allg. Chem.* **1983**, *502*, 89.
Schm 1983a   G. Schmidt, R. Gruehn, *Z. Anorg. Allg. Chem.* **1983**, *503*, 151.
Schm 1985   G. Schmidt, R. Gruehn, *Z. Anorg. Allg. Chem.* **1985**, *528*, 69.
Schm 1989   G. Schmidt, R. Gruehn, *J. Less-Common Met.* **1989**, *156*, 75.
Schm 1990   G. Schmidt, R. Gruehn, Posterbeitrag *XVth Congress of the International Union of Crystallography*, Bordeaux PS – 07. 06. 03 **1990**.
Schm 1990a   G. Schmidt, R. Gruehn, *J. Less-Common Met.* **1990**, *158*, 275.
Schm 1991   G. Schmidt, R. Gruehn, in G. Meyer, L. R. Morss, Eds. *Synthesis of Lanthanide and Actinide Compounds* Kluwer, Netherlands, **1991**.
Schm 1991a   G. Schmidt, *Dissertation,* University of Gießen, **1991**.
Schm 1997   P. Schmidt, O. Bosholm, H. Oppermann, *Z. Naturforsch.* **1997**, *52b*, 1461.
Schm 1999   P. Schmidt, O. Rademacher, H. Oppermann, *Z. Anorg. Allg. Chem.* **1999**, *625*, 255.
Schm 2000   P. Schmidt, O. Rademacher, H. Oppermann, S. Däbritz, *Z. Anorg. Allg. Chem.* **2000**, *626*, 1999.
Schm 2002   M. Schmidt, M. Armbrüster, U. Schwarz, H. Borrmann, R. Cardoso-Gil, *Report MPI CPfS Dresden* **2001/2002**, 229.
Schm 2003   M. Schmidt, R. Cardoso-Gil, S. Gerlach, U. Müller, U. Burkhardt, *Report MPI CPfS Dresden* **2003–2005**, 288.
Schm 2004   M. Schmidt, B. Ewald, Yu. Prots, M. Armbrüster, I. Loa, L. Zhang, Ya-Xi Huang, U. Schwarz, R. Kniep, *Z. Anorg. Allg. Chem.* **2004**, *630*, 655.
Schm 2005   M. Schmidt, U. Müller, R. Cardoso Gil, E. Milke, M. Binnewies, *Z. Anorg. Allg. Chem.* **2005**, *631*, 1154.
Schm 2005a   M. Schmidt, R. Ramlau, W. Schnelle, H. Borrmann, E. Milke, M. Binnewies, *Z. Anorg. Allg. Chem.* **2005**, *631*, 284.
Schm 2008   R. Schmidt, J. Feller, U. Steiner, *Z. Anorg. Allg. Chem.* **2008**, *634*, 2076.
Scho 1988   H. Schornstein, R. Gruehn, *Z. Anorg. Allg. Chem.* **1988**, *561*, 103.
Scho 1989   H. Schornstein, R. Gruehn, *Z. Anorg. Allg. Chem.* **1989**, *579*, 173.
Scho 1990   H. Schornstein, R. Gruehn, *Z. Anorg. Allg. Chem.* **1990**, *587*, 129.
Scho 1990a   H. Schornstein, R. Gruehn, *Z. Anorg. Allg. Chem.* **1990**, *582*, 51.
Scho 1992   H. Schornstein, *Dissertation,* University of Gießen, **1992**.
Schu 1971   A. N. Schukov, R. K. Nikolaev, V. T. Uschakovski, V. Sch. Schektman, *Krist. Techn.* **1971**, *5*, 16.
Schw 1982   H. J. Schweizer, R. Gruehn, *Z. Naturforsch.* **1982**, *37b*, 1361.
Sea 1960   G. W. Sears, R. C. DeVries, *J. Chem. Phys.* **1960**, *32*, 93.
Sea 1963   G. W. Sears, R. C. DeVries, *J. Chem. Phys.* **1963**, *39*, 2837.
Sei 1983   F.-J. Seiwert, R. Gruehn, *Z. Anorg. Allg. Chem.* **1983**, *503*, 151.
Sei 1984   F.-J. Seiwert, R. Gruehn, *Z. Anorg. Allg. Chem.* **1984**, *510*, 93.
Sek 2000   T. Sekiya, K. Ichimura, M. Igarashi, S. Kurita, *J. Phys. Chem. Solids* **2000**, *61*, 1273.
Sha 1979   M. W. Shafer, R. A. Figat, B. Olson, S. J. La Placa, J. Angslello, *J. Electrochem. Soc.* **1979**, *126*, 1625.
Shi 1971   M. Shiloh, J. Gutman, *J. Cryst. Growth* **1971**, *11*, 105.
Shi 1973   M. Shiloh, J. Gutman, *J. Electrochem. Soc.* **1973**, *120*, 438.
Shi 1998   I. Shiozaki, *J. Magn. Mater.* **1998**, *177*, 261.
Shl 1995   L. Shlyk, J. Stępién-Damm, R. Troć, *J. Cryst. Growth* **1995**, *154*, 418.
Sie 1982   K. Sieber, K. Kourtakis, R. Kershaw, K. Dwight, A. Wold, *Mater. Res. Bull* **1982**, *17*, 721.
Sie 1983   K. Sieber, H. Leiva, K. Kourtakis, R. Kershaw, K. Dwight, A.Wold, *J. Solid State Chem.* **1983**, *47*, 361.

Sin 1974	R. N. Singh, R. L. Coble, *J. Cryst. Growth* **1974**, *21*, 261.
Skv 2000	V. Skvortsova, N. Mironova-Ulmane, in *Proceedings of the International Conference on Mass and Charge Transport in Inorganic Materials- Fundamentals to Devices, Faenza, Techna* **2000**, 815.
Sle 1970	A. Sleight, J. L. Gillson, B. L. Chamberland, *Mater. Res. Bull.* **1970**, *5*, 807.
Sle 1971	A. Sleight, J. L. Gillson, *Mater. Res. Bull.* **1971**, *6*, 781.
Smi 1979	A. R. R. Smith, A. K. Cheetham, *J. Solid State Chem.* **1979**, *30*, 345.
Sob 1960	B. P. Sobolev, J. P. Klyagina, *Russ. J. Inorg. Chem.* **1960**, *5*, 1112.
Söh 1997	T. Söhnel, W. Reichelt, H. Oppermann, *Z. Anorg. Allg. Chem.* **1997**, *623*, 1190.
Spi 1930	V. Spitzin, *Z. Anorg. Allg. Chem.* **1930**, *189*, 337.
Spi 1979	J. C. Spirlet, *J. Phys.* **1979**, *4*, 87.
Spi 1980	J. C. Spirlet, E. Bednarczyk, J. Rebizant, C. T. Walker, *J. Cryst. Growth* **1980**, *49*, 171.
Ste 1996	U. Steiner, W. Reichelt, H. Oppermann, *Z. Anorg. Allg. Chem.* **1996**, *622*, 1428.
Ste 2000	U. Steiner, W. Reichelt, *Z. Anorg. Allg. Chem.* **2000**, *626*, 2525.
Ste 2003	U. Steiner, W. Reichelt, S. Däbritz, *Z. Anorg. Allg. Chem.* **2003**, *629*, 116.
Ste 2003a	U. Steiner, W. Reichelt, *Z. Anorg. Allg. Chem.* **2003**, *629*, 1632.
Ste 2004a	U. Steiner, S. Daminova, W. Reichelt, *Z. Anorg. Allg. Chem.* **2004**, *630*, 2541.
Ste 2005	U. Steiner, W. Reichelt, S. Daminova, E. Langer, *Z. Anorg. Allg. Chem.* **2005**, *631*, 364.
Ste 2005a	U. Steiner, *Z. Anorg. Allg. Chem.* **2005**, *631*, 1706.
Ste 2005b	U. Steiner, W. Reichelt, *Z. Anorg. Allg. Chem.* **2005**, *631*, 1877.
Ste 2006	U. Steiner, W. Reichelt, *Z. Anorg. Allg. Chem.* **2006**, *632*, 1257.
Ste 2006a	U. Steiner, W. Reichelt, *Z. Anorg. Allg. Chem.* **2006**, *632*, 1781.
Ste 2008	U. Steiner, *Z. Anorg. Allg. Chem.* **2008**, *634*, 2083.
Stei 1987	G. Steinmann, *Diplomarbeit,* University of Gießen, **1987**.
Stei 1990	G. Steinmann-Möller, *Dissertation,* University of Gießen, **1990**.
Sto 1966	C. van de Stolpe, *Phys. Chem. Solids* **1996**, *27*, 1952.
Str 1980	P. Strobel, F. P. Koffyberg, A. Wold, *J. Solid State Chem.* **1980**, *31*, 209.
Str 1982	P. Strobel, Y. Le Page, *J. Cryst. Growth* **1982**, *56*, 723.
Str 1982a	P. Strobel, Y. Le Page, S. P. McAlister, *J. Solid State Chem.* **1982**, *42*, 242.
Str 1982b	P. Strobel, Y. Le Page, *J. Mater. Sci.* **1982**, *17*, 2424.
Str 1983	P. Strobel, Y. Le Page, *J. Cryst. Growth* **1983**, *61*, 329.
Str 1983a	P. Strobel, S. P. McAlister, Y. Le Page, *Stud. in Inorg. Chem.* **1983**, *3*, 307.
Stu 1964	W. I. Stuart, G. H. Price, *J. Nucl. Mater.* **1964**, *14*, 417.
Stu 1975	J. Sturm, R. Gruehn, *Naturwissenschaften* **1975**, *62*, 296.
Stu 1976	J. Sturm, *Dissertation,* University of Gießen, **1976**.
Sun 2002	Sun Yan-Hui, LiQuing Zhang, Peng-Xiang Lei, Zhi-Chang Wang, Lei Guo, *J. Alloys Compd.* **2002**, *335*, 196.
Sun 2004	Sun Yan-hui, Chen Zhen-fei, Wang Zhi-chang, *Trans. Nonferrous Met. Soc. China* **2004**, *14*, 412.
Szy 1970	H. Szydłowski, H. Lübbe, P. Kleinert, *Phys. Status Solidi* **1970**, *3*, 769.
Tan 1968	I. V. Tananaev, G. M. Safronov, V. P. Orlovskii, V. P. Repko, V. I. Pakhomov, A. N. Volodina, E. A. Ionkina, *Dokl. Akad. Nauk SSSR* **1968**, *1836*, 1357.
Tar 1984	J. A. K. Tareen, A. Małecki, J. P. Doumerc, J. C. Launay, P. Dordor, M. Pouchard, P. Hagenmuller, *Mater. Res. Bull.* **1984**, *19*, 989.
Ter 1976	E. I. Terukov, F. A. Chudnovskii, W. Reichelt, H. Oppermann, W. Brückner, H.- P. Brückner, W. Moldenhauer, *Phys. Status Solidi* **1976**, *37*, 541.
Tim 1964	V. A. Timofeeva, *Kristallografja* **1964**, *9*, 642.

Tor 1976   D. K. Toropov, M. G. Degen, V. P. Bolgartseva, *Izv. Akad. Nauk. SSSR, Neorg. Mater.* **1976**, *12*, 241.
Tos 1988   A. Toshev, P. Peshev, *Mater. Res. Bull.* **1988**, *23*, 1045.
Tra 1994   O. Trappe, *Diplomarbeit*, University of Gießen, **1994**.
Tri 1982   P. Triggs, C. A. Georg, F. Lévy, *Mater. Res. Bull.* **1982**, *17*, 671.
Tri 1983   P. Triggs, H. Berger, C. A. Georg, F. Lévy, *Mater. Res. Bull.* **1983**, *18*, 677.
Tri 1985   P. Triggs, *Helv. Phys. Acta* **1985**, *58*, 657.
Trö 1972   M. Trömel, *Z. Anorg. Allg. Chem.* **1972**, *387*, 346.
Tsu 1966   K. Tsushima, *J. Appl. Phys.* **1966**, *37*, 443.
Udo 2008   H. Udono, Y. Sumi, S. Yamada, I. Kukuma, *J. Cryst. Growth* **2008**, *310*, 1827.
Uek 1965   T. Ueki, A. Zalkin, D. H. Templeton, *Acta Crystallogr.* **1965**, *19*, 157.
Vei 1976   A. Veispals, E. Latsis, *Izv. Akad. Nauk. SSSR, Neorg. Mater.* **1976**, *12*, 1318.
Vei 1979   A. Veispals, A. Patmalnieks, *Fiz. Teh. Zinat.* **1979**, 99.
Vla 1976   M. Vlasse, J.-P. Doumerc, P. Peshev, J.-P. Chaminade, M. Pouchard, *Rev. Chim. minér.* **1976**, *13*, 451.
Vol 1976   D. S. Volzhenskii, V. A. Grin, V. G. Savitskii, *Kristallografiya* **1976**, *21*, 1238.
Wan 1996   Zh. Ch. Wang, L. Ch. Wang, R. J. Gao, Y. Su, *Faraday Trans.* **1996**, *92*, 1887.
Wan 1997   Zh. Ch. Wang, L. Ch. Wang, *Inorg. Chem.* **1997**, *36*, 1536.
Wan 1998   Zh. Ch. Wang, L. Ch. Wang, *J. Alloys Comp.* **1998**, *265*, 153.
Wan 1998a   Zh. Ch. Wang, J. Yu, Y. Li Yu, Y. H. Sun, *J. Alloys Comp.* **1998**, *264*, 147.
Wäs 1972   E. Wäsch, *Krist. Tech.* **1972**, 7, 187.
Weh 1970   F. H. Wehmeier, *J. Cryst. Growth* **1970**, *6*, 341.
Wei 2000   M. Weil, *Z. Naturforsch.* **2000**, *55b*, 699.
Wei 2001   M. Weil, *Acta Cryst.* **2001**, *E57*, i22.
Wei 2001a   M. Weil, *Acta Cryst.* **2001**, *E57*, i28.
Wei 2004   M. Weil, C. Lengauer, E. Füglein, E. J. Baran, *Cryst. Growth Des.* **2004**, *4*, 1229.
Wei 2004b   M. Weil, *Acta Cryst.* **2004**, *E60*, i139.
Wei 2004c   M. Weil, E. Füglein, C. Lengauer, *Z. Anorg. Allg. Chem.* **2004**, *630*, 1768.
Wei 2005   M. Weil, U. Kolitsch, *Z. Krist. Suppl.* **2005**, *22*, 183.
Wei 2008   X. Wei, Y. Zhao, Z. Dong, J. Li, *J. Cryst. Growth* **2008**, *310*, 639.
Wen 1989   M. Wenzel, R. Gruehn, *Z. Anorg. Allg. Chem.* **1989**, *568*, 95.
Wen 1990   M. Wenzel, R. Gruehn, *Z. Anorg. Allg. Chem.* **1990**, *582*, 75.
Wen 1991   M. Wenzel, R. Gruehn, *Z. Anorg. Allg. Chem.* **1991**, *594*, 139.
Wer 1996   J. Werner, G. Behr, W. Bieger, G. Krabbes, *J. Cryst. Growth* **1996**, *165*, 258.
Wes 1980   G. H. Westphal, F. Rosenberger, *J. Cryst. Growth* **1980**, *49*, 607.
Whi 1974   E. A. D. White, J. D. C. Wood, *J. Mater. Sci.* **1974**, *9*, 1999.
Wid 1971   R. Widmer, *J. Cryst. Growth* **1971**, *8*, 216.
Wit 1971   J. H. W. De Wit, *J. Cryst. Growth* **1971**, *12*, 183.
Woe 1997   F. v. Woedtke, L. Kirsten, H. Oppermann, *Z. Naturforsch.* **1997**, *52b*, 1155.
Wol 1965   E. G. Wolf, T. D. Coshren, *J. Amer. Ceramic Soc.* **1965**, *48*, 279.
Wol 1978   E. Wolf, H. Oppermann, G. Krabbes, W. Reichelt, *Current Topics in Material Science*, Vol.1, Ed. E. Kaldis, North- Holland, Amsterdam, **1978**, 697.
Yam 1972   N. Yamamoto, K. Nagasawa, Y. Bando, T. Takada, *Jpn. J. Appl. Phys.* **1972**, *11*, 754.
Yam 1973   N. Yamamoto, Y. Bando, T. Takada, *Jpn. J. Appl. Phys.* **1973**, *12*, 1115.
Yan 2009   C. Yan, P. S. Lee, *J. Phys. Chem. C* **2009**, *113*, 14135.
Yen 2004   P. C. Yen, R. S. Chen, Y. S. Huang, K. K. Tong, P. C. Liao, *J. Alloys Compd.* **2004**, *383*, 277.
Yen 2004a   P. C. Yen, R. S. Chen, Y. S. Huang, K. K. Tiong, *J. Cryst. Growth* **2004**, *262*, 271.

You 1960   W. A. Young, *J. Phys. Chem.* **1960**, *64*, 1003.
Yu 1993   F. Yu, U. Schanz, E. Schmidbauer, *J. Cryst. Growth* **1993**, *132*, 606.
Zen 1999   L.-P. Zenser, M. Weil, R. Gruehn, *Z. Anorg. Allg. Chem.* **1999**, *625*, 423.
Zen 1999a   L.-P. Zenser, *Dissertation*, University of Gießen, **1999**.
Zhe 1998   X. G. Zheng, M. Suzuki, C. N. Xu, *Mater. Res. Bull.* **1998**, *33*, 605.

6 Chemical Vapor Transport of Oxido Compounds with Complex Anions

In the following, the term "complex oxides" is understood as an expression for multinary oxide compounds that contain one or more metal cation and one or more complex anion with typical non-metal atoms as the central atom. In thermodynamic terms, these complex oxides differ from other multinary oxides ("double oxides") by their comparatively high heats of reaction for the formation from binary oxides (see section 12.2.1). In terms of their chemical structure, they differ by the low co-ordination number of the non-metal. Due to the fact that most of the non-metal oxides are readily volatile at high temperatures without the addition of a transport agent, clear differences in transport behavior result compared to the "double oxides". The following deals with the CVT of representatives of the following classes of compound.

- sulfates, selenates, and tellurates
- phosphates, arsenates, and antimonates
- silicates
- borates

6.1 Transport of Sulfates

CuSO$_4$

Chlorine as transport agent

$$ZnSO_4(s) + Cl_2(g) \rightleftharpoons ZnCl_2(g) + SO_3(g) + \frac{1}{2}O_2(g)$$

Thionyl chloride as transport agent

$$Al_2(SO_4)_3(s) + 3\,SOCl_2(g) \rightleftharpoons 2\,AlCl_3(g) + 3\,SO_2(g) + 3\,SO_3(g)$$

The crystallization of anhydrous *sulfates* used to be a challenge. Most of the representatives of this compound class show a comparatively low thermal stability (decomposition to SO$_3$ and SO$_2$/O$_2$); under normal laboratory conditions, only the sulfates of the alkali metals melt without decomposing. Due to the fact that crystallization from a solution (concentrated H$_2$SO$_4$) can only be applied for a few anhydrous sulfates, there was a lack of a synthetic method for the crystallization of these compounds. Here the application of CVT experiments on crystal growth by the gas phase led to remarkable progress. As can be seen in Table 6.1, the series of sulfates that have been transported so far extends from Ag_2SO_4 (Spi 1978a) to $VOSO_4$ (Dah 1994) and includes many sulfates MSO_4 and $M_2(SO_4)_3$. Apart from the great preparative use, the in-depth investigation of the transport behavior of the sulfates provided detailed information on the heterogeneous and homogeneous gas-phase equilibria that participate in the migration. It shows that *several transport reactions* must be considered for a quantitative description. This necessity is a special demand that must be considered during thermodynamic descriptions of the experimental observations. The according insights are most important as they serve as a model for the CVT of other complex oxides that contain a highly volatile component (phosphates, arsenates, and borates).

Often, chlorine or hydrogen chloride can be used as *transport agents* for sulfates as can be seen in Table 6.1. In individual cases, migration in a temperature gradient could be observed when I_2, NH_4Cl, $HgCl_2$, $PbCl_2$, $PbBr_2$, or $SOCl_2$ were added. The migration of sulfates is always due to an endothermic reaction ($T_2 \to T_1$). Equations 6.1.1 to 6.1.3 completely describe the experimental conditions for the transport of $ZnSO_4$ with chlorine.

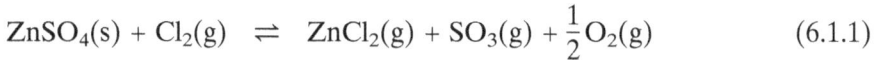

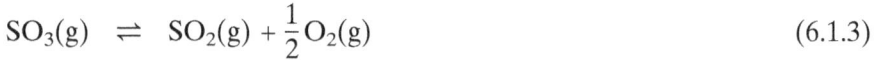

$$ZnSO_4(s) + Cl_2(g) \rightleftharpoons ZnCl_2(g) + SO_3(g) + \frac{1}{2}O_2(g) \tag{6.1.1}$$

$$3\,ZnSO_4(s) \rightleftharpoons Zn_3O(SO_4)_2(s) + SO_3(g) \tag{6.1.2}$$

$$SO_3(g) \rightleftharpoons SO_2(g) + \frac{1}{2}O_2(g) \tag{6.1.3}$$

It is noteworthy that $ZnSO_4$ shows increased thermal stability when chlorine is added, which is a consequence of the interaction of reactions 6.1.1 and 6.1.2 (Spi 1978b, Bal 1984). This effect is less distinct if hydrogen chloride is used.

According to detailed thermodynamic model calculations, equations 6.1.4 and 6.1.5 are transport determining for the transport of $Fe_2(SO_4)_3$ with chlorine at low temperatures (500 to 600 °C) (Dah 1992, Dah 1993).

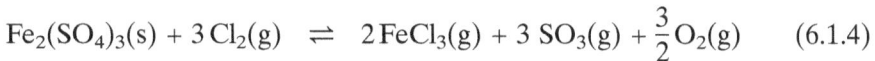

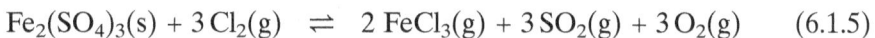

$$Fe_2(SO_4)_3(s) + 3\,Cl_2(g) \rightleftharpoons 2\,FeCl_3(g) + 3\,SO_3(g) + \frac{3}{2}O_2(g) \tag{6.1.4}$$

$$Fe_2(SO_4)_3(s) + 3\,Cl_2(g) \rightleftharpoons 2\,FeCl_3(g) + 3\,SO_2(g) + 3\,O_2(g) \tag{6.1.5}$$

Equation 6.1.4 becomes less important at higher temperatures (600 to 750 °C). The calculations show that under these conditions, $p(SO_3)$ is determined as a function of the temperature via the (endothermic) homogeneous equilibrium 6.1.3. The *oxidizing character of the equilibrium gas phase* during the transport of sulfates, even if hydrogen chloride is used as the transport agent, can show in different ways. Hence, during the CVT of $FeSO_4$ with HCl an additional solid always appears:

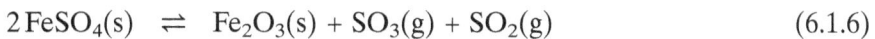

$$2\,FeSO_4(s) \rightleftharpoons Fe_2O_3(s) + SO_3(g) + SO_2(g) \tag{6.1.6}$$

$$2\,FeSO_4(s) + H_2(g) \rightleftharpoons Fe_2O_3(s) + 2\,SO_2(g) + H_2O(g) \tag{6.1.7}$$

If instead of HCl a respective amount of NH_4Cl is added, which decomposes to HCl, N_2, and H_2, the amount of Fe_2O_3 even increases (equation 6.1.7). The formation of Fe_2O_3 from $FeSO_4$ in the presence of the reduction agent H_2 is surprising at first. It becomes clear with the help of equation 6.1.7. Apparently, the oxygen partial pressure, which is defined by the ratio $p(H_2O)/p(H_2)$, leads to the oxidation of iron as well as to the reduction of SO_3. See the literature for a detailed thermodynamic discussion of the transport behavior of iron sulfates (Dah 1992, Dah 1993).

An oxidizing equilibrium gas phase is also the requirement for the use of $PbCl_2$ as transport additive for some anhydrous sulfates, such as $NiSO_4$ or $CuSO_4$ (Pli 1989). In the process, chlorine is released in a pre-reaction, e.g., 6.1.8, which contains a metathesis and a redox reaction. The chlorine functions as the actual transport agent for $NiSO_4$ according to equation 6.1.9.

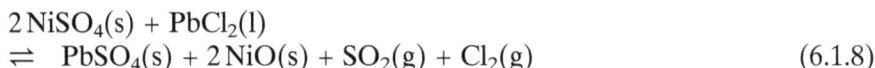

$$2\,NiSO_4(s) + PbCl_2(l)$$
$$\rightleftharpoons PbSO_4(s) + 2\,NiO(s) + SO_2(g) + Cl_2(g) \qquad (6.1.8)$$

$$NiSO_4(s) + Cl_2(g) \rightleftharpoons NiCl_2(g) + SO_3(g) + \frac{1}{2}O_2(g) \qquad (6.1.9)$$

With the help of the transport balance (see section 14.4), it was possible to detect the sequential transport behavior of the three equilibrium solids $PbSO_4$, $NiSO_4$, and NiO. It was also possible to show that $PbSO_4$ migrates before $NiSO_4$ in the temperature gradient (850 → 750 °C), while NiO remained in the source until the end of the experiments (Pli 1989).

$Cr_2(SO_4)_3$, $Ga_2(SO_4)_3$, and $In_2(SO_4)_3$ can be transported with chlorine. In contrast, a volatilization of aluminum oxide with chlorine (as with other transport agents) in a temperature gradient is impossible due to the unfavorable equilibrium position of reaction 6.1.10. Therefore, it is surprising that experiments aiming at the crystallization of aluminum sulfate by CVT are successful (625 → 525 °C; transport agent $SOCl_2$). Apparently, using $SOCl_2$ as the transport agent avoids the formation of free oxygen. In doing so, it causes a favorable position of the heterogeneous transport equilibrium 6.1.11. It is not yet clear whether the transport of rare earth metal sulfates is possible this way, too.

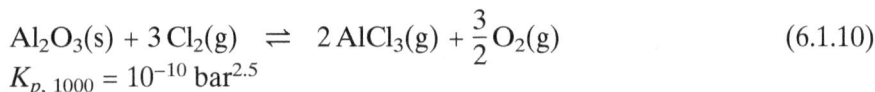

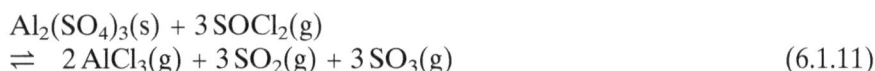

$$Al_2O_3(s) + 3\,Cl_2(g) \rightleftharpoons 2\,AlCl_3(g) + \frac{3}{2}O_2(g) \qquad (6.1.10)$$
$$K_{p,\,1000} = 10^{-10}\,bar^{2.5}$$

$$Al_2(SO_4)_3(s) + 3\,SOCl_2(g)$$
$$\rightleftharpoons 2\,AlCl_3(g) + 3\,SO_2(g) + 3\,SO_3(g) \qquad (6.1.11)$$

At the end of this section, we will refer to the possibility of transporting $VOSO_4$ with chlorine (Dah 1994a).

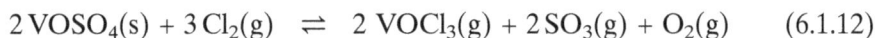

$$2\,VOSO_4(s) + 3\,Cl_2(g) \rightleftharpoons 2\,VOCl_3(g) + 2\,SO_3(g) + O_2(g) \qquad (6.1.12)$$

$VOSO_4$ appears as a solid in the source and sink in the temperature gradient from 525 to 425 °C; the transport rates are approximately 5 mg · h^{-1} and increase considerably with rising temperatures (700 → 600 °C). However, noticeable amounts of V_2O_5 do appear beside the oxide sulfate at these higher temperatures (Dah 1994a).

6.2 Transport of Phosphates, Arsenates, Antimonates, and Vanadates

GdPO$_4$

Chlorine as transport agent

$$Co_2P_2O_7(s) + 2\,Cl_2(g) \;\rightleftharpoons\; 2\,CoCl_2(g) + \frac{1}{2}P_4O_{10}(g) + O_2(g)$$

Transport agent P + I

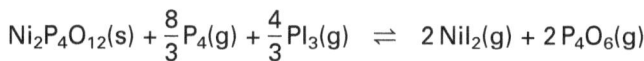

$$Ni_2P_4O_{12}(s) + \frac{8}{3}P_4(g) + \frac{4}{3}PI_3(g) \;\rightleftharpoons\; 2\,NiI_2(g) + 2\,P_4O_6(g)$$

Coupled transport of phosphate and phosphide

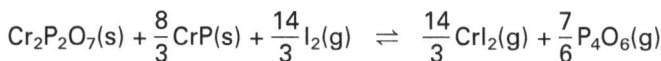

$$Cr_2P_2O_7(s) + \frac{8}{3}CrP(s) + \frac{14}{3}I_2(g) \;\rightleftharpoons\; \frac{14}{3}CrI_2(g) + \frac{7}{6}P_4O_6(g)$$

The overview in Table 6.1 shows the wide application of CVT reactions for the synthesis, crystallization, and purification of compounds with complex oxido-anions that contain a central atom with the oxidation number V. The preparative possibilities are illustrated by the crystallization of the thermally delicate phosphates $Re_2O_3(PO_4)_2$ (Isl 2009) and CuP_4O_{11} (Gla 1996), the mixed-valent iron(II, III) orthoarsenate $Fe_7(AsO_4)_6$ (Wei 2004c), and different vanadates(V) of transition metals. Phosphates of transition metals with oxidation states that are not easily accessible in another ways (low numbers) can be synthesized in sealed silica ampoules and crystallized in "one-pot reactions" by CVT (e. g., $TiPO_4$, V_2OPO_4, $Cr_3(PO_4)_2$, and $Cr_2P_2O_7$). Apart from the elemental halogens Cl_2, Br_2, and I_2, halogen compounds (NH_4X and HgX_2; $X = Cl$, Br, I) as well as mixtures P + X_2 ($X = Cl$, Br, I) are used. In some cases, such as $Fe_3O_3PO_4$ or UP_2O_7 and chlorinating compounds, such as VCl_4, $ZrCl_4$, $HfCl_4$, and $NbCl_5$, are suitable transport agents (Kos 1997). The best results, as far as transport rates and crystal growth of anhydrous phosphates are concerned, were achieved with chlorine or mixtures of phosphorus + iodine as transport agents (Gla 1999).

6.2.1 Chlorine as Transport Agent for Anhydrous Phosphates

Only those phosphates that are stable in a chlorine atmosphere and that are not oxidized by completely consuming the transport agent can be transported.

If chlorine is added, P_4O_{10} is considered as phosphorus-transferring gas species similar to the CVT of sulfates (see reactions 6.2.1.1 and 6.2.1.4). The determining equilibrium 6.2.1.1 for $RhPO_4$ is exemplary of the CVT of phosphates with chlorine.

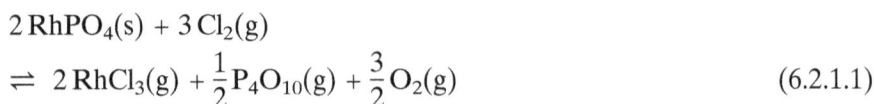

$$2\,RhPO_4(s) + 3\,Cl_2(g)$$
$$\rightleftharpoons 2\,RhCl_3(g) + \frac{1}{2}P_4O_{10}(g) + \frac{3}{2}O_2(g) \tag{6.2.1.1}$$

Detailed experiments show that depending on the concentration of the transport agent and the temperature, $RhCl_3(s)$, $Rh(PO_3)_3(s)$, and elemental rhodium appear in the equilibrium solid apart from the orthophosphate (Gör 1997). The calculated composition of solid and gas phase in the source chamber at the beginning of a representative transport experiment is graphically illustrated as a function of the temperature Figure 6.2.1.1. The calculation's results are in agreement with the experimental observations.

At temperatures up to 825 °C, equation 6.2.1.2 determines the composition of the solid and gas phases. At 825 ≤ ϑ ≤ 950 °C, equation 6.2.1.3 dominates. Above ϑ = 950 °C, the thermal decomposition of $RhPO_4$, equation 6.2.1.4 gains importance apart from the transport reaction 6.2.1.1.

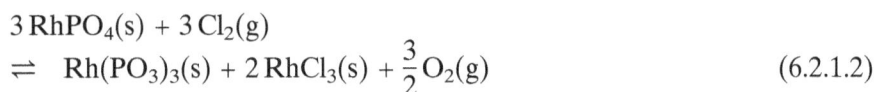

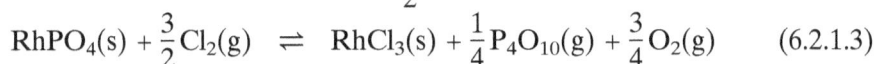

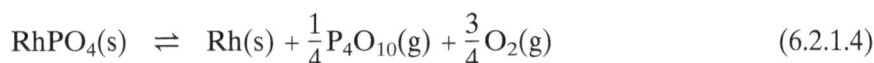

$$3\,RhPO_4(s) + 3\,Cl_2(g)$$
$$\rightleftharpoons Rh(PO_3)_3(s) + 2\,RhCl_3(s) + \frac{3}{2}O_2(g) \tag{6.2.1.2}$$

$$RhPO_4(s) + \frac{3}{2}Cl_2(g) \rightleftharpoons RhCl_3(s) + \frac{1}{4}P_4O_{10}(g) + \frac{3}{4}O_2(g) \tag{6.2.1.3}$$

$$RhPO_4(s) \rightleftharpoons Rh(s) + \frac{1}{4}P_4O_{10}(g) + \frac{3}{4}O_2(g) \tag{6.2.1.4}$$

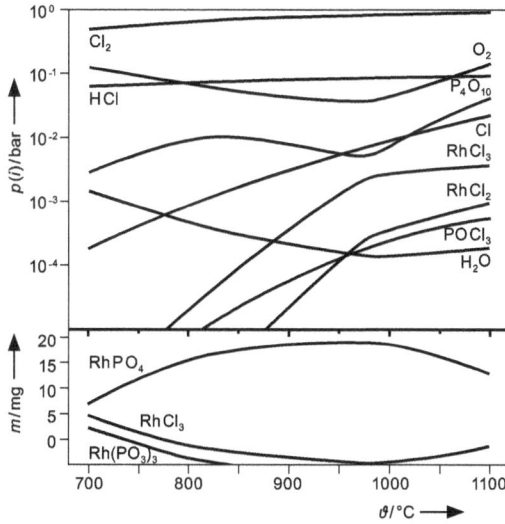

Figure 6.2.1.1 Transport of $RhPO_4$ with chlorine. Calculated equilibrium of solid- and gas-phase composition in the source as a function of temperature at the beginning of an experiment.[1]

Significant participation of phosphorus oxide chlorides, such as $POCl_3$ or PO_2Cl (Bin 1983), and their polymers $(PO_2Cl)_x$ (Ban 1990), in the transport of the phosphates with chlorine as the transport agent is rendered unlikely by the thermodynamic assessment of the homogeneous gas-phase equilibrium 6.2.1.5. The ratio $p(P_4O_{10}) : p(POCl_3) = 278$ (1273 K, $p(P_4O_{10}) = p(Cl_2) = 1$ bar) results from the thermodynamic data of reaction 6.2.1.5. Obviously, the strong positive heat of reaction, which is caused by the exchange of very stable P–O bonds by less stable P–Cl bonds, cannot be compensated for by the entropy gain at the normal temperatures of CVT experiments. It has to be taken into account that the equilibrium partial pressure of gaseous P_4O_{10} over a phosphate solid is mainly limited by the possible formation of adjacent phases with a higher content of P_4O_{10} under the mentioned transport conditions.

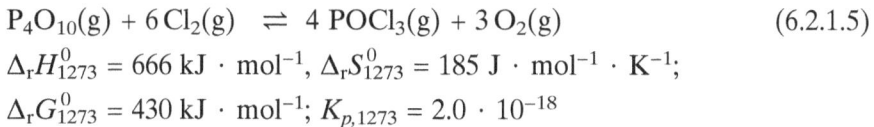

$$P_4O_{10}(g) + 6\,Cl_2(g) \rightleftharpoons 4\,POCl_3(g) + 3\,O_2(g) \tag{6.2.1.5}$$
$$\Delta_r H^0_{1273} = 666\;kJ \cdot mol^{-1},\; \Delta_r S^0_{1273} = 185\;J \cdot mol^{-1} \cdot K^{-1};$$
$$\Delta_r G^0_{1273} = 430\;kJ \cdot mol^{-1};\; K_{p,1273} = 2.0 \cdot 10^{-18}$$

6.2.2 Halogens Combined with Reducing Additives as Transport Agents for Phosphates

Because of the reduced activity of the metal oxide component in a phosphate, compared with pure oxides, a lower solubility of the phosphates in a chlorine

---

[1] For the thermodynamic data used see *Görzel* (Gör 1997). (Initial weight: 25 mg $RhPO_4$, 16.23 mg Cl_2; $V_{amp} = 22.4$ cm^3).

atmosphere, in comparison with metal oxides, is observed. Heterogeneous equilibria such as reaction 6.2.1 are generally far on the side of the source solid and often lead to low transport rates (< 1 mg $\cdot$ h^{-1}).

Because of this, *Schäfer* and *Orlovskii* used mixtures of phosphorus and chlorine or phosphorus and bromine as the transport agent for the orthophosphates *Ln*PO$_4$ (*Ln* = La, Ce, Pr, Nd) (Orl 1971, Schä 1972). In this way, the release of oxygen during reactions that are comparable to equilibrium 6.2.1.1 is avoided, and a more favorable equilibrium position results. A comparable effect is used during the CVT of Al$_2$(SO$_4$)$_3$ with SOCl$_2$ according to equation 6.1.11. Investigations of different phosphates of transition metals with mixtures consisting of phosphorus and chlorine or phosphorus and bromine as transport agent, did show a migration of the phosphates via the gas phase (Gla 1990b). At the same time, however, the wall of the silica ampoule was heavily attacked with the formation of different *silicophosphates* (see below). All in all, these experiments cannot be reproduced appropriately and are not suited for systematic investigations. Apparently, the halogenating power of the gas mixtures of chlorine, bromine, and phosphorus are so powerful that, apart from the phosphate solids, SiO$_2$ is also well dissolved in the gas phase. A peculiarity in this context is the CVT of V$_2$OPO$_4$ and VPO$_4$ with the additive "VCl" (formed from the in situ reaction of vanadium with PtCl$_2$) in the temperature gradient 900 $\rightarrow$ 800 °C. From the experimental observations one expects a significant content of PCl$_3$ and phosphorus vapor in the gas phase, which together could function as transport agent (equations 6.2.2.1 and 6.2.2.2) (Dro 2004). The CVT of phosphates with PCl$_3$ as transport agent has not yet been systematically examined.

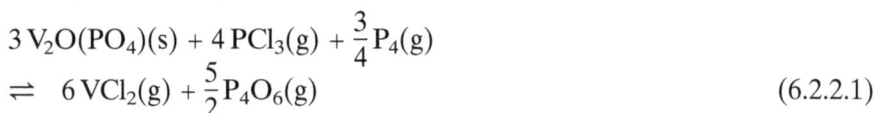

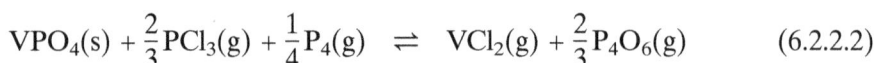

$$3\,V_2O(PO_4)(s) + 4\,PCl_3(g) + \frac{3}{4}P_4(g)$$
$$\rightleftharpoons\ 6\,VCl_2(g) + \frac{5}{2}P_4O_6(g) \tag{6.2.2.1}$$

$$VPO_4(s) + \frac{2}{3}PCl_3(g) + \frac{1}{4}P_4(g)\ \rightleftharpoons\ VCl_2(g) + \frac{2}{3}P_4O_6(g) \tag{6.2.2.2}$$

The search for "selectively" halogenating transport agents applicable to phosphates eventually led to combinations of iodine with the addition of small amounts of metal, phosphorus, or phosphide (Gla 1990b, Gla 1999). The use of iodine without further transport agents for anhydrous phosphates only rarely results in a transport effect (TiPO$_4$ (Gla 1990), Cr$_2$P$_2$O$_7$ (Gla 1991)). Equilibria such as 6.2.2.3 for the transport of orthophosphates contain a chemically less reasonable oxidation of O^{2-} by I$_2$ and are thermodynamically very unfavorable.

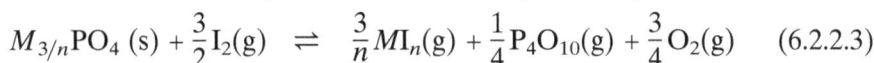

$$M_{3/n}PO_4\,(s) + \frac{3}{2}I_2(g)\ \rightleftharpoons\ \frac{3}{n}MI_n(g) + \frac{1}{4}P_4O_{10}(g) + \frac{3}{4}O_2(g) \tag{6.2.2.3}$$

The great preparative use of iodine in combination with reducing additives as transport agent (see Table 6.1) raises the question of the determining heterogeneous and homogeneous equilibria under these experimental circumstances. In particular, it is at first unclear which phosphorus-containing gas species participate in the transport.

Besides P$_4$O$_{10}$ and P$_4$O$_6$, there exist PO, PO$_2$, and P$_4$O$_n$ ($7 \le n \le 9$) known as gaseous phosphorus oxides (Mue 1970). Apart from P$_4$O$_{10}$ (Cha 1998, 1971, Glu

1989), the thermal behavior of the phosphorus oxides is insufficiently characterized; information about the heat of formation of $P_4O_6(g)$ range from -1590 to
-2142 kJ $\cdot$ mol^{-1} (Bar 1973, Glu 1989).

The thermal stability of gaseous phosphorus iodides PI_3 and P_2I_4 is dealt within
the context of the transport of phosphides (section 9.1).

If PI_3 is used as the transport agent and if it is assumed that the migration of
phosphorus, contained in the solid, takes place via P_4O_{10} or P_4O_6, transport reactions like 6.2.2.4 and 6.2.2.5 result. These heterogeneous equilibria that contribute
to chemical transport of $Mn_2P_2O_7$ are shown as examples. According equilibria
with further lower phosphorus oxides might also be possible. Finally, transport
according to equation 6.2.2.6 with the transport agent HI, which could develop
from phosphorus vapor, iodine, and moisture traces, also appears to be reasonable.

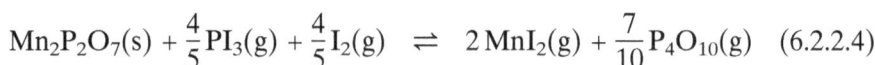

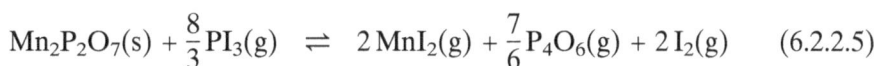

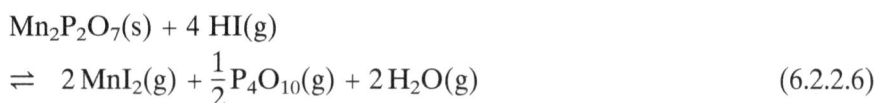

$$Mn_2P_2O_7(s) + \frac{4}{5}PI_3(g) + \frac{4}{5}I_2(g) \rightleftharpoons 2\,MnI_2(g) + \frac{7}{10}P_4O_{10}(g) \quad (6.2.2.4)$$

$$Mn_2P_2O_7(s) + \frac{8}{3}PI_3(g) \rightleftharpoons 2\,MnI_2(g) + \frac{7}{6}P_4O_6(g) + 2\,I_2(g) \quad (6.2.2.5)$$

$$Mn_2P_2O_7(s) + 4\,HI(g)$$
$$\rightleftharpoons 2\,MnI_2(g) + \frac{1}{2}P_4O_{10}(g) + 2\,H_2O(g) \quad (6.2.2.6)$$

A direct *in situ* examination of the gas phase (MS, IR, Raman) is hardly possible
due to comparatively high pressures and temperatures. Neither does thermodynamic modeling of the behavior of phosphates in CVT experiments provide con-
clusive evidence. (Gla 1990, Gla 1990b, Ger 1996). That is why only some observations can be summarized that provide indications on the most important phosphorus-containing gas species.

Apparently, transport of anhydrous phosphates with iodine is possible only if
the oxygen partial pressure is "low". In experimental terms, this can be achieved
by adding small amounts of metal phosphide, metal, or phosphorus. As far as the
mentioned exceptions $TiPO_4$ and $Cr_2P_2O_7$ are concerned, the reducing effect of
titanium(III) and chromium(II) is sufficient to cause a transport effect with iodine. In this context, it is remarkable that the related phosphates VPO_4 and
$Fe_2P_2O_7$ are not transportable with iodine alone, but only if the respective monophosphide is added (Gla 1990b).

It has not yet been proven whether, during the transport of phosphates with
iodine and reducing additives, the oxygen partial pressure is always low enough
so that, in the gas phase only, lower phosphorus oxides instead of P_4O_{10} do occur.
At least in some cases, however, the migration of phosphates is possible under
conditions that do not allow a significant P_4O_{10} pressure. Hence, $Mn_3(PO_4)_2$ migrates in a temperature gradient from 850 to 800 °C with phosphorus + iodine
as the transport additive. $Mn_2P_2O_7$, which is richer in phosphoric oxide, whose
decomposition to orthophosphate determines the P_4O_{10} co-existence pressure, is
thermally stable up to a melting temperature of approximately 1200 °C (Kon
1977). In these circumstances, it seems most unlikely that P_4O_{10} can appear as an
essential gas species with a partial pressure $p(P_4O_{10}) \geq 10^{-5}$ bar under the condi-

tions of the transport of $Mn_3(PO_4)_2$. Similar principles apply for the transport of TiO_2 adjacent to $TiPO_4$ where lower phosphorus oxides can be regarded as carrier of oxygen and phosphorus (Gla 1990) ($1000 \rightarrow 900$ °C; phosphorus + iodine as the transport additive). In addition to the two mentioned examples, it can be shown with the help of model calculations for an entire series of further phosphates (Table 6.1) that no heterogeneous equilibria can be formulated with P_4O_{10} as transport determining, phosphorus- and oxygen-transferring gas species. Thus, P_4O_{10} can not account for many transport effects observed under reducing conditions (Gla 1999).

If the observed transport of anhydrous phosphates with iodine and reducing additives does not take place via P_4O_{10}, the questions arises which phosphorus oxide plays the dominant role. Observations during the transport of $Cr_2P_2O_7$ with iodine ($1050 \rightarrow 950$ °C) in the presence of a surplus of CrP are as remarkable in this context as the transport of $WOPO_4$ and WP_2O_7 adjacent to WP (Lit 2003). In all three cases, a *simultaneous transport* of phosphides and phosphates due to an endothermic reaction is found experimentally. Experiments with the transport balance (see section 14.4) show that phosphide and phosphate migrate from the source to the sink in a single stationary state if the two condensed phases are provided in a certain ratio with respect to their amounts of substance. If there is more phosphide in the source solid, the surplus migrates to the sink in a second stationary state. This behavior indicates a *coupled vapor transport reaction* of the two phases. Considering the complexity of the gas-phase composition, it is surprising that in the three cases, the experimentally determined ratio is in good agreement with the calculated ratio that is determined for the assumption of P_4O_6 as the transport determining species in the heterogeneous equilibria 6.2.2.7 to 6.2.2.9. The formulation of $WO_2I_2(g)$ and $CrI_2(g)$ as essential metal-transferring gas species is in accordance with many other studies (Dit 1983, Gla 1989, Scho 1991).

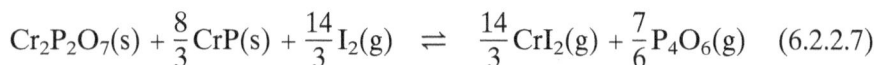

$$Cr_2P_2O_7(s) + \frac{8}{3}CrP(s) + \frac{14}{3}I_2(g) \rightleftharpoons \frac{14}{3}CrI_2(g) + \frac{7}{6}P_4O_6(g) \quad (6.2.2.7)$$

$$WOPO_4(s) + \frac{3}{7}WP(s) + \frac{10}{7}I_2(g)$$
$$\rightleftharpoons \frac{10}{7}WO_2I_2(g) + \frac{5}{14}P_4O_6(g) \quad (6.2.2.8)$$

$$WP_2O_7(s) + \frac{4}{7}WP(s) + \frac{11}{7}I_2(g)$$
$$\rightleftharpoons \frac{11}{7}WO_2I_2(g) + \frac{9}{14}P_4O_6(g) \quad (6.2.2.9)$$

Model calculations of the CVT of $Mg_2P_2O_7$ and $Mn_2P_2O_7$ also suggest crucial participation of lower gaseous phosphorus oxides, P_4O_6, in the transport of anhydrous phosphates with iodine under reducing conditions (Ger 1996). If gaseous P_4O_6 with a heat of formation of $\Delta_f H^0_{298}(P_4O_6) = -1841$ kJ · mol^{-1}, a value that was used as a suitable parameter, is considered, a quantitative description of the experiments in the model calculations is possible.

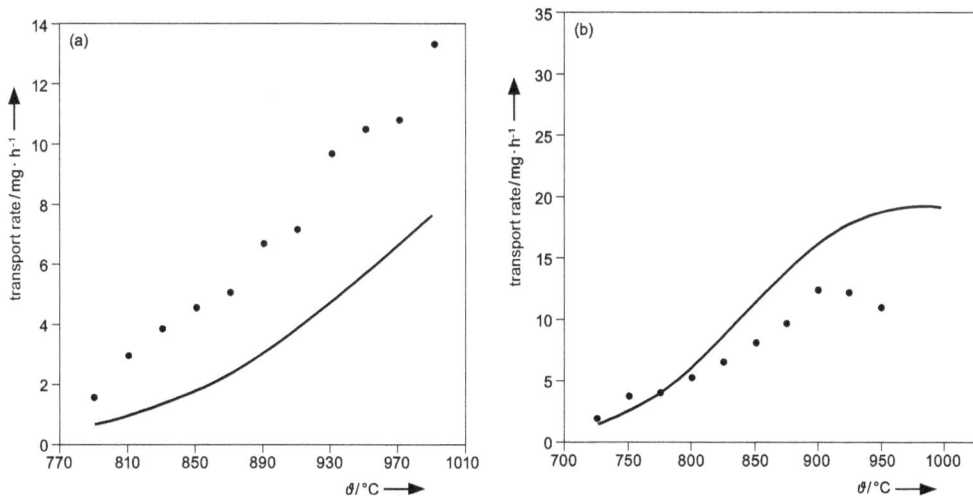

Figure 6.2.2.1 Transport rate of $Mg_2P_2O_7$ (a) and $Mn_2P_2O_7$ (b) with P/I mixtures as a function of the temperature ϑ according to *Gerk* (Ger 1996)[2].

The dependency of the transport rates of $Mg_2P_2O_7$ and of $Mn_2P_2O_7$ on the medium temperature, as well as its dependency on the amount of the transport agent, is well reflected in these calculations. According to the calculations, the transport of $Mg_2P_2O_7$ and $Mn_2P_2O_7$ can be described by the equation 6.2.2.5 with PI_3 as transport agent. In the process, P_2I_4 is also effective as a transport agent to a small extent.

When transport agents from mixtures of iodine and reducing additives (metal, metal phosphide, or phosphorus) are used, one has to consider that the latter could lead to the formation of multi-phase solids during the transport experiments. Hence, during the transport of copper(II)-pyrophosphate, $Cu_2P_2O_7$, with phosphorus + iodine, the metaphosphate $Cu_2P_4O_{12}$ develops as another phase (see Figure 6.2.2.2a). The CVT of chromium(II)-pyrophosphate can be achieved with iodine and the additives chromium, CrP, or phosphorus. In the first case, small amounts of Cr_2O_3 and CrP develop in the solid adjacent to $Cr_2P_2O_7$. In the second case, some CrP and, in the third, $Cr(PO_3)_3$, appear as adjacent phases (Gla 1993, Gla 1999). These observations are understandable with the help of the Cr/P/O phase diagram (Figure 6.2.2.2b). The (undesired) appearance of multi-phase solids during CVT reactions of anhydrous phosphates can be understood when the equilibrium relations for a given Gibbs metal/phosphorus/oxygen triangle are known (Gla 1999).

---

[2] (a) $Mg_2P_2O_7$: $\Delta T = 100$ K, 131.0 mg iodine + 10.5 mg P as transport agent, $s = 8$ cm, $q = 2$ cm^2, experiments with transport balance, Δ observed transport rates, calculates transport rates (——) with $\Delta_r H^0_{298}(P_4O_6(g)) = -1841$ kJ $\cdot$ mol^{-1} and consideration of 0.1 mmol H_2O. (b) $Mn_2P_2O_7$: $\Delta T = 50$ K, 146.5 mg iodine + 8.5 mg P as transport agent, $s = 8$ cm, $q = 2$ cm^2, experiments with transport balance, Δ observed transport rates, calculated transport rates (——) with $\Delta_r H^0_{298}$ $(P_4O_6(g)) = -1841$ kJ $\cdot$ mol^{-1} and consideration of 0.1 mmol H_2O.

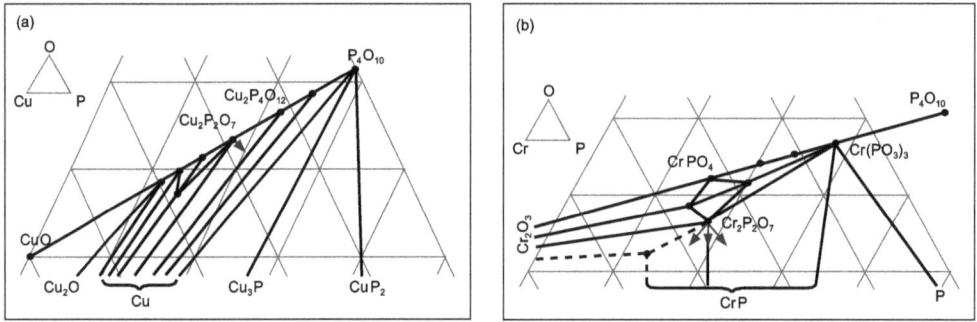

Figure 6.2.2.2 Sections from the phase diagrams Cu/P/O (Öza 1993, Gla 1999) (a) and Cr/P/O (Gla 1999) (b) to illustrate the formation of neighbouring phases during the chemical vapor transport with iodine and different reducing additives.

In unfavorable cases, the phosphates which are to be transported, react quantitatively with the reducing additives without forming volatile reaction products. A CVT is no longer observed in the process. $CrPO_4$, which is transported with chlorine with low transport rates, is reduced by phosphorus vapor with the formation of mixed-valent chromium(II, III)-phosphates and chromium(II)-phosphates without achieving a transport effect. In general, only those phosphates that co-exist in the thermal equilibrium with the respective metal, phosphorus, or the phosphides, are transportable with mixtures of iodine and a reductant.

Similar to the CVT of the oxides M_2O_5 with pentahalides MCl_5 (M = Nb, Ta; see section 5.2.5), transport experiments with uranium pyrophosphate UP_2O_7 have been carried out with metal halides ($NbCl_5$; MCl_4 where M = Ti, V, Zr, Hf) (Kos 1997) as transport additives. In the process, crystals of the pyrophosphate were obtained for single-crystal structure investigations. The observed transport rates are approximately 0.5 mg $\cdot$ h^{-1}. If the tetrahalides MCl_4 (M = Ti, Zr, Hf) were used, the deposition of mixed-crystals $U_{1-x}M_xP_2O_7$ took place at the sink. In the source, the pseudomorphic precipitation of amorphous SiO_2 in place of the source solid was observed. There are no thermodynamic investigations of the transport determining heterogeneous equilibria yet. However, equilibria such as 6.2.2.10 and 6.2.2.11 seem plausible.

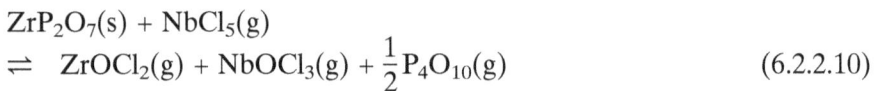

$$ZrP_2O_7(s) + NbCl_5(g)$$
$$\rightleftharpoons \ ZrOCl_2(g) + NbOCl_3(g) + \frac{1}{2}P_4O_{10}(g) \tag{6.2.2.10}$$

$$ZrP_2O_7(s) + ZrCl_4(g) \ \rightleftharpoons \ 2\,ZrOCl_2(g) + \frac{1}{2}P_4O_{10}(g) \tag{6.2.2.11}$$

The mineralizing effect of the additive $FeCl_2$ is described for the crystallization of an entire series of mixed-valent iron(II, III)-phosphates (e. g., Fe_2OPO_4 (Mod 1981), $Fe_9O_8PO_4$ (Ven 1984), $Fe_4O(PO_4)_2$ (Bou 1982), $Fe_3(P_2O_7)_2$ (Ijj 1991), and $Fe_7(P_2O_7)_4$ (Mal 1992)). If these are indeed reversible vapor transport reactions, and which compounds function as transport agent, has not yet been examined.

6.2.3 Transport of Multinary Phosphates

Apart from the CVT of ternary phosphates (containing only one metal), experiments aiming at the crystallization of multinary phosphates (containing several metals) have been successful in recent years. A detailed thermodynamic examination of these complex transport systems is not yet possible due to a lack of data. However, the following rule for further transport experiments can be laid down:

> Multinary phosphates are transportable if the ternary components (phosphates) can be transported under similar conditions (temperature and kind of transport agent).

The following examples shall clarify this rule. $Cr_3^{II}Ti_4^{III}(PO_4)_6$ (Lit 2009) and $Co_3^{II}Cr_4^{III}(PO_4)_6$ (Gru 1996).

$Cr_3^{II}Ti_4^{III}(PO_4)_6$ migrates in the temperature gradient from $1000 \rightarrow 900\,°C$ when iodine or a mixture of TiP + I_2 is added, due to the endothermic reactions, to the cooler side of the ampoule. Transport rates of several mg · h^{-1} are observed in the process. The same applies for the pyrophosphate $Cr^{II}Ti_2^{III}(P_2O_7)_2$ (Lit 2009). The ternary phosphates $Cr_3(PO_4)_2$, $Cr_2P_2O_7$, and $TiPO_4$ can be transported with similar rates under similar conditions (see Table 6.1).

Chlorine as transport agent is suited for the CVT of $Co_2P_2O_7$ (Gla 1991) as well as for $CrPO_4$ (Gla 1986) in a temperature gradient from $1050 \rightarrow 950\,°C$. As expected, $Co_3^{II}Cr_4^{III}(PO_4)_6$ can be crystallized under the same conditions (Gru 1996, Lit 2009).

The oxide phosphates $MTi_2O_2(PO_4)_2$ (M = Fe, Co, Ni) (Schö 2007) and the orthophosphates $Mn_{1.65}Ti_4^{III,\,IV}(PO_4)_6$, $Fe^{II}Ti_4^{IV}(PO_4)_6$, and $CoTi_4(PO_4)_6$, which belong to the NASICON-structure family, also belong to the group of other multinary phosphates that can be crystallized in a temperature gradient by CVT (Schö 2008c). Analogous to the problems during the CVT of vanadium(IV)-phosphates (Dro 2004), the transport of oxide phosphates $MV_2O_2(PO_4)_2$ (M = Co, Ni, Cu) does not take place, either. At least, adding small amounts of chlorine as mineralizer leads to a better re-crystallization of the solids (Ben 2008).

The migration and crystallization of multinary phosphates in the temperature gradient led to a series of interesting observations as far as thermodynamics and reaction kinetics are concerned. Starting from $Fe_2P_2O_7$ and $V_4(P_2O_7)_3$, the pyrophosphates $Fe_5^{II}V_2^{III}(P_2O_7)_4$, $Fe_3^{II}V_2^{III}(P_2O_7)_4$, and $Fe^{II}V_2^{III}(P_2O_7)_4$ can be obtained by isothermal tempering (in closed silica ampoules). The equilibration between ternary educts in the presence of iodine as mineralizer is massively accelerated and achieved in less than 24 hours. Crystallization of multinary pyrophosphates is achieved by CVT experiments when iodine is added ($1050 \rightarrow 950\,°C$) (Lit 2009). This observation is surprising because iron(II)-pyrophosphate is only transportable with iodine when Fe, FeP, or P is added (Gla 1990b, Gla 1991). Despite some experimental efforts, vanadium(III)-pyrophosphate has not yet been transported (Kai 1996, Dro 2004).

After some days during a transport experiment $Fe_2P_2O_7$, $Fe_5^{II}V_2^{III}(P_2O_7)_4$, $Fe_3^{II}V_2^{III}(P_2O_7)_4$, and $Fe^{II}V_2^{III}(P_2O_7)_4$ were detected in the sink when pure

$Fe_5^{II}V_2^{III}(P_2O_7)_4$ previously prepared in a separate experiment, had been used as starting material in the source. In the source solid, $V_4(P_2O_7)_3$ and $Fe^{II}V_2^{III}(P_2O_7)_4$ were also present after the same time. Under these experimental conditions, a partial segregation of the multinary pyrophosphate takes place in the source accompanied by an enrichment of $Fe_2P_2O_7$ in the sink. The reason why $V_4(P_2O_7)_3$ (in the form of the multinary pyrophosphates) is deposited in the sink under these circumstances is not yet understood in thermodynamic terms. Eventually, based on the observation of a four-phase sink solid, one might conclude that deposition of the phosphates at the sink did not occur under equilibrium conditions. Apparently, the non-stationary transport behavior leads to the deposition of pyrophosphates with an increasing content of vanadium in the sink. The equilibration (isothermal, by heterogeneous equilibria with participation of the gas phase or by solid state reactions) between phosphates with different ratios $n(Fe) : n(V)$ takes place slowly. The observations made here, which led to a favorable formation of iron(II)/vanadium(III)-pyrophosphate as well as to their tendency to segregation when a temperature gradient is applied, is an example of the transport behavior of multinary phosphates. The reason for the observed segregation tendency might be the comparatively low exothermic heats of reaction for the formation of the multinary phosphate from the ternary components. Similar principles apply for the isothermal formation of $In_2P_2O_7$ and its decomposition to $InPO_4$ and InP when exposed to a temperature gradient in the presence of a transport agent (iodine) (Tha 2003, Tha 2006).

6.2.4 Deposition of Thermodynamically Metastable Phosphates from the Gas Phase

The explanations in Chapter 2 about the understanding of CVT experiments are based on the assumption of a migration of solids between equilibrium spaces. Even though this assumption idealizes the actual situation, most of the quantitative studies of CVT experiments suggest at least the establishment of states close to equilibrium between solid and the respective gas phase in the source and sink of a transport ampoule. Under these conditions, there should not be a noteworthy super-saturation of the gas phase in the sink. Therefore, the *deposition of thermodynamically metastable solids from the gas phase* seems unlikely. The higher temperatures during CVT, compared to typical solvothermal syntheses, should additionally favor the deposition of thermodynamically stable solids. Considering this, the crystallization of metastable solids during CVT experiments is particularly remarkable. As far as the transport of anhydrous phosphates is concerned, two examples are well investigated: the deposition of β-$Ni_2P_2O_7$ instead of σ-$Ni_2P_2O_7$ (Gla 1991, Fun 2004) and the deposition of $(Mo^VO)_4(P_2O_7)_3$ instead of a two-phase mixture from Mo^VOPO_4 and $(Mo^VO)_2P_4O_{13}$ (Len 1995, Isl 2009b).

The CVT of nickel(II)-pyrophosphate can take place with different transport additives (Cl_2, $HCl + H_2$ from the *in situ* decomposition of NH_4Cl, mixture P + I) over a wide temperature range ($\vartheta_{source, max} = 1100$ °C, $\vartheta_{sink, min} = 600$ °C) due to endothermic reactions (see Table 6.1). In doing so, σ-$Ni_2P_2O_7$ (own structure

Figure 6.2.4.1 Deposition of metastable $(MoO)_4(P_2O_7)_3$. (a) Crystals from a transport experiment (initial weight: 200 mg $(Mo^{VI}O_2)_2(P_2O_7)$ + 5.8 mg P, 30 mg iodine, 800 → 650 °C; (b) ampoule wall with blue, amorphous film and nucleation sites of crystalline $(MoO)_4(P_2O_7)_3$ (top ampoule), deposition of $MoOPO_4$ (bottom ampoule).

type (Mas 1979)) is obtained in most cases. Only sometimes, there is a deposition of $Ni_2P_2O_7$ with thortveitite-type structure ($α$-, $β$-modification, ϑ_{tr} = 577 °C (Pie 1968)). A reversible conversion between the σ-modification and the thortveitite-type structure is not observed. The higher density of σ-$Ni_2P_2O_7$ suggests a slightly higher thermodynamic stability. Indeed, experimental conditions, which favor a higher super-saturation of the gas phase in the deposition region (ΔT = 200 K; P + I_2 as transport additive), led to a quantitative deposition of β-$Ni_2P_2O_7$ (Fun 2004).

With iodine as the transport agent, $MoOPO_4$ as well as $(MoO)_2P_4O_{13}$ migrate in the temperature gradient due to endothermic reactions (Len 1995). Isothermal temper experiments in evacuated silica ampoules, with iodine as mineralizer, show that the two phosphates co-exist with each other over a wide temperature range. The reduction of $(Mo^{VI}O_2)_2P_2O_7$ with phosphorus (equation 6.2.4.1) and a subsequent CVT in a "one-pot reaction", however, often results in product mixtures of Mo^VOPO_4, $(Mo^VO)_2P_4O_{13}$, and $(Mo^VO)_4(P_2O_7)_3$ in varying amounts. Experiments on the directed synthesis of $(Mo^VO)_4(P_2O_7)_3$ lead to different results, which are dependent on the experimental conditions. Small temperature gradients ($\Delta T \leq$ 100 K) lead to the deposition of two-phase mixtures of Mo^VOPO_4 and $(Mo^VO)_2P_4O_{13}$. If $\Delta T \geq$ 150 K is applied, only the metastable compound $(Mo^VO)_4(P_2O_7)_3$ is obtained at T_1 apart from the other reaction product $MoOPO_4$.nb Sometimes the deposition of a blue, amorphous film of the composition $Mo_4P_6O_{25}$ in the sink, has been observed during such experiments (see Figure 6.2.4.1). These observations show that product formation is influenced by the super-saturation of the gas phase as well as by heterogeneous nucleation on the silica surface. Tempering of $(Mo^VO)_4(P_2O_7)_3$ at 650 °C with iodine as mineralizer leads to the decomposition of the pyrophosphate with the formation of orthophosphate and tetraphosphate.

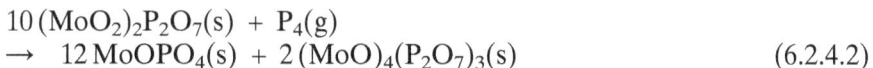

$$10\,(MoO_2)_2P_2O_7(s)\ +\ P_4(g)$$
$$\rightarrow\ 16\ MoOPO_4(s)\ +\ 2\,(MoO)_2P_4O_{13}(s) \tag{6.2.4.1}$$

$$10\,(MoO_2)_2P_2O_7(s)\ +\ P_4(g)$$
$$\rightarrow\ 12\,MoOPO_4(s)\ +\ 2\,(MoO)_4(P_2O_7)_3(s) \tag{6.2.4.2}$$

6.2.5 Formation of Silicophosphates during the Transport of Phosphates

The synthesis and crystallization of anhydrous phosphates succeeds in silica ampoules in many cases without problems. However, if the conditions are unfavorable, reactions with the ampoule wall with the formation of silicophosphates can occur (Gla 1990b, Schö 2008b). Three cases can be distinguished.

- P_4O_{10} migrates to the wall
- SiO_2 migrates to the phosphate solid (vapor transport)
- Dissolution of SiO_2 and phosphate in the gas phase, and deposition as silicophosphate in the sink

In the most simple case, $P_4O_{10}(g)$, which is used as the starting material during phosphate synthesis, reacts directly with the ampoule wall with the formation of SiP_2O_7 (Til 1973) and $Si_3^o[Si_2^iO(PO_4)_6]$ (May 1974). Sufficiently high P_4O_{10}-decomposition pressures of a phosphate solid function in the same way. The CVT of SiO_2 in a temperature gradient is possible with HCl, HCl + H_2 (from NH_4Cl) or with PCl_3 + Cl_2 (see section 5.2.14). Isothermal tempering of phosphate solids in silica ampoules in presence of these transport agents often leads to the formation of silicophosphates at the site of the initial solid. Apparently, an isothermal transport of SiO_2 takes place in the gradient of its chemical potential ($a(SiO_2$, wall) ≈ 1, $a(SiO_2$, silicophosphate) << 1). This reaction, which is undesired during preparation and crystallization of phosphates, occurs faster than the solid state reaction between phosphate and SiO_2 and can be used for a direct synthesis of silicophosphates (e. g., $M_2Si(P_2O_7)_2$ with M^{2+} = Mn, Fe, Co (Gla 1995), Ni, Cu (Schm 2002)). The gross reactions (6.2.5.1 to 6.2.5.3) provide examples of the formation of other silicophosphates that are easily and reproducible accessed as single-phase products. Silicophosphates can often be crystallized well, which suggests an additional mobilization via the gas phase in such experiments.

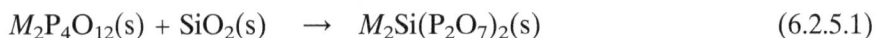

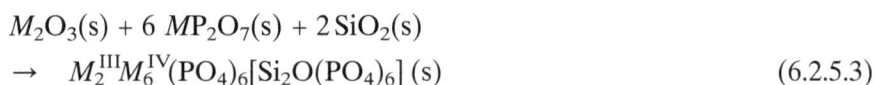

$$M_2P_4O_{12}(s) + SiO_2(s) \quad \rightarrow \quad M_2Si(P_2O_7)_2(s) \qquad (6.2.5.1)$$

$$M_4(P_2O_7)_3(s) + 2\,SiO_2(s) \quad \rightarrow \quad M_4[Si_2O(PO_4)_6](s) \qquad (6.2.5.2)$$

$$M_2O_3(s) + 6\,MP_2O_7(s) + 2\,SiO_2(s)$$
$$\rightarrow \quad M_2^{III}M_6^{IV}(PO_4)_6[Si_2O(PO_4)_6]\,(s) \qquad (6.2.5.3)$$

If the transport of phosphates takes place under conditions that also allow dissolution of SiO_2 in the gas phase, the deposition of one or more silicophosphates apart from phosphate will be observed in the sink. These are simultaneous and independent transport reactions of the phosphate and SiO_2, which can lead to different products according the respective molar ratio of the reactants in the sink region. Respective experiments allow access to well crystallized samples; however, they are often difficult to reproduce. Examples can be found in different works (see crystallization of $M_4P_6Si_2O_{25}$ with M^{3+} = Ti, V, Cr, Mo, Fe (Gla 1990b, Rei 1998, Schö 2008b), MP_3SiO_{11} with M^{3+} = Ti, Cr, Mo (Rei 1998), and $MP_3Si_2O_{13}$ with M^{3+} = Ti, Cr, Rh).

6.2.6 Transport of Arsenates(V), Antimonates(V), and Vanadates(V)

In contrast to anhydrous phosphates, metal arsenates(V), antimonates(V), and vanadates(V) show a clearly lower thermal stability. The compounds tend, more easily than phosphates, toward the formation of oxygen and gaseous As_4O_6 and Sb_4O_6 respectively. As a result from this behavior, it can be concluded that arsenates(V) and vanadates(V) with reducing cations (e. g., Cr^{2+}, Ti^{3+}) do not exist. The same applies for antimonates(V) to an even greater extent.

The limited stability of arsenates and antimonates, combined with the volatility of As_4O_6 and Sb_4O_6, seems favorable for CVT of these compounds. Hence, *Weil* describes the successful CVT experiments aiming at the crystallization of different anhydrous arsenates, such as $Mn_3(AsO_4)_2$, $M_2As_2O_7$, and MAs_2O_6 (M = Mn, Co, Ni, Cu, Zn, Cd). These migrate in the temperature gradient from 880 to 800 °C with chlorine as the transport agent (Wei 2004, Wei 2005). Experiments on preparation and transport of the so-far unknown "$Fe_2^{II}As_2O_7$" led to the formation and crystallization of $Fe_3^{II}Fe_4^{III}(AsO_4)_6$ besides $FeAsO_4$ despite using a mixture of $HCl + H_2$ (addition of NH_4Cl) as transport agent (Wei 2004c). The oxidizing effect of arsenic(V) is emphasized by this result (see transport behavior of $FeSO_4$, section 6.1). $HgCl_2$ proves valuable as a transport agent to crystallize different mercury arsenates (650 $\rightarrow$ 550 °C; see Table 6.1) (Wei 2000).

The transport behavior of the thermally stable rare earth metal arsenates(V) and rare earth metal antimonates(V) has been examined in detail. Due to the fact that it is influenced predominantly by the transport behavior of the rare earth metal oxide, it is dealt with in section 5.2.3.

Prokofiev describes the CVT of $CuSb_2O_6$ with tellurium(IV)-chloride and hydrogen chloride (Pro 2003). With $TeCl_4$, the transport takes place via the gas species Cu_3Cl_3, $SbCl_3$, O_2, and Cl_2. The gas species CuCl and Sb_4O_6 join at temperatures above 1000 °C. The transfer of oxygen occurs by $TeOCl_2$ at 870 °C, and higher by TeO_2. The transport with chlorine can be described by the same gas species, whereby, however, oxygen is transported as O_2 only. If hydrogen chloride functions as transport agent, H_2O is formed in the transport equilibrium. The following transport equations can be derived from this:

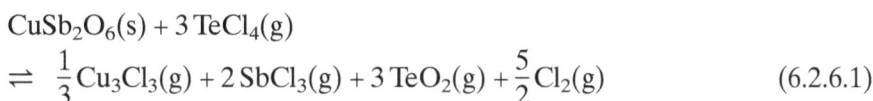

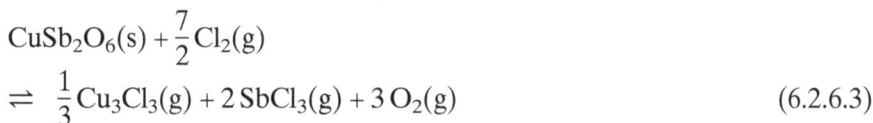

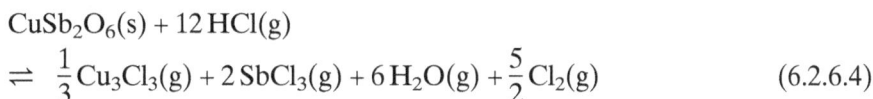

$$CuSb_2O_6(s) + 3\,TeCl_4(g)$$
$$\rightleftharpoons \frac{1}{3}Cu_3Cl_3(g) + 2\,SbCl_3(g) + 3\,TeO_2(g) + \frac{5}{2}Cl_2(g) \qquad (6.2.6.1)$$

$$CuSb_2O_6(s) + 6\,TeCl_4(g)$$
$$\rightleftharpoons \frac{1}{3}Cu_3Cl_3(g) + 2\,SbCl_3(g) + 6\,TeOCl_2(g) + \frac{5}{2}Cl_2(g) \qquad (6.2.6.2)$$

$$CuSb_2O_6(s) + \frac{7}{2}Cl_2(g)$$
$$\rightleftharpoons \frac{1}{3}Cu_3Cl_3(g) + 2\,SbCl_3(g) + 3\,O_2(g) \qquad (6.2.6.3)$$

$$CuSb_2O_6(s) + 12\,HCl(g)$$
$$\rightleftharpoons \frac{1}{3}Cu_3Cl_3(g) + 2\,SbCl_3(g) + 6\,H_2O(g) + \frac{5}{2}Cl_2(g) \qquad (6.2.6.4)$$

Thermodynamic calculations suggest that transport with chlorine should result in maximum transport rates in the temperature gradient from 900 to 800 °C. If tellurium(IV)-chloride is used as the transport agent, the range of 1000 to 870 °C was calculated. In experiments, 920 → 800 °C (chlorine) and 920 → 880 °C (TeCl$_4$) proved particularly favorable temperature ranges.

Transport of vanadates(V) Despite the limited thermal stability of the vanadates(V) of transition metals, CVT experiments are well suited for the crystallization of this class of compound. Chlorine (addition of PtCl$_2$) and tellurium(IV)-chloride have been used as transport agents so far. Orienting model calculations, which were based on estimated thermodynamic data for the vanadates FeVO$_4$, CrVO$_4$ (Gro 2009), and Co$_2$V$_2$O$_7$ (Bro 2010), suggest a gas-phase composition that can also be found during the transport of the binary oxides with chlorine (equations 6.2.6.5 to 6.2.6.7).

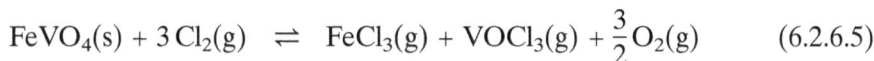

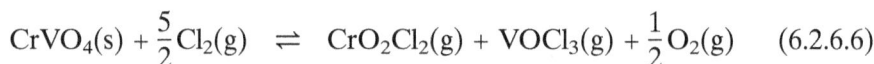

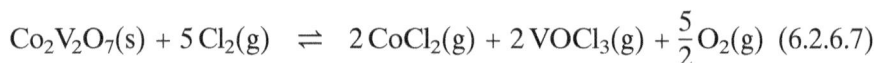

$$FeVO_4(s) + 3\,Cl_2(g) \;\rightleftharpoons\; FeCl_3(g) + VOCl_3(g) + \frac{3}{2}O_2(g) \tag{6.2.6.5}$$

$$CrVO_4(s) + \frac{5}{2}Cl_2(g) \;\rightleftharpoons\; CrO_2Cl_2(g) + VOCl_3(g) + \frac{1}{2}O_2(g) \tag{6.2.6.6}$$

$$Co_2V_2O_7(s) + 5\,Cl_2(g) \;\rightleftharpoons\; 2\,CoCl_2(g) + 2\,VOCl_3(g) + \frac{5}{2}O_2(g) \tag{6.2.6.7}$$

As far as CVT of different copper vanadates (Cu$_2$V$_2$O$_7$, CuV$_2$O$_6$ (Bec 2008)) as well as of RhVO$_4$ (Jen 2009) is concerned, tellurium(IV)-chloride is a suitable transport additive as well. According to thermodynamic model calculations, the heterogeneous equilibrium 6.2.6.8 determines the transport behavior of rhodium(III)-vanadate(V).

$$RhVO_4(s) + 3\,TeCl_4(g) \\ \rightleftharpoons RhCl_3(g) + VOCl_3(g) + 3\,TeOCl_2(g) \tag{6.2.6.8}$$

6.3 Transport of Carbonates, Silicates, and Borates

There are no indications on *chemical vapor transport of carbonates* in the literature. The following equilibrium of a hypothetical transport of carbonates with hydrogen chloride can be formulated.

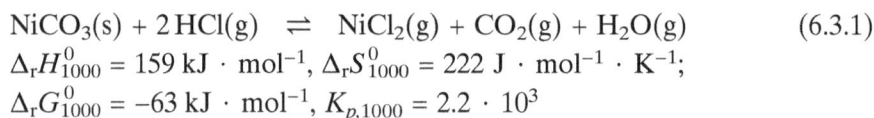

$$NiCO_3(s) + 2\,HCl(g) \;\rightleftharpoons\; NiCl_2(g) + CO_2(g) + H_2O(g) \tag{6.3.1}$$
$\Delta_rH^0_{1000} = 159\ kJ \cdot mol^{-1}$, $\Delta_rS^0_{1000} = 222\ J \cdot mol^{-1} \cdot K^{-1}$;
$\Delta_rG^0_{1000} = -63\ kJ \cdot mol^{-1}$, $K_{p,1000} = 2.2 \cdot 10^3$

Equation 6.3.2 could offer a possible transport for the special case of NiCO$_3$.

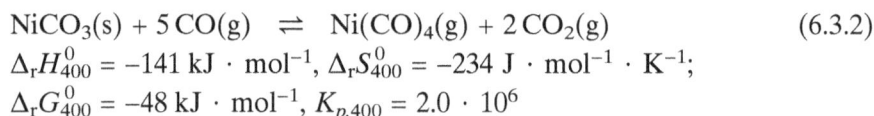

$$NiCO_3(s) + 5\,CO(g) \;\rightleftharpoons\; Ni(CO)_4(g) + 2\,CO_2(g) \tag{6.3.2}$$
$\Delta_rH^0_{400} = -141\ kJ \cdot mol^{-1}$, $\Delta_rS^0_{400} = -234\ J \cdot mol^{-1} \cdot K^{-1}$;
$\Delta_rG^0_{400} = -48\ kJ \cdot mol^{-1}$, $K_{p,400} = 2.0 \cdot 10^6$

The thermodynamic data of the given equilibria suggest a transport of nickel carbonate. However, the kinetic stability of carbon dioxide during the deposition reaction could be a problem.

There are a number of reports on the *chemical vapor transport of silicates*. Early on, the assumption was made that transport reactions with participation of the gas phase are involved in mineral-forming processes in nature (Dau 1841, Dau 1849). If one takes a deeper look into the literature, it is striking that, in most cases, the authors did not describe reversible transport. Only the migration of europium(II)-silicates (Eu_2SiO_4, $EuSi_2O_5$) at high temperatures (1980 $\rightarrow$ 1920 °C) with HCl as transport agent (Kal 1970, Kal 1971) and the crystallization of Be_2SiO_4 with SiF_4 as the transport agent are based on transport reactions of the silicates (equations 6.3.3 to 6.3.5)

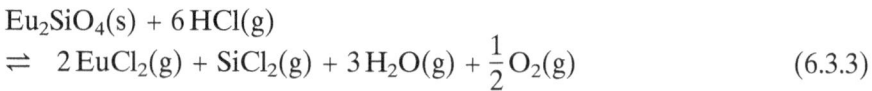

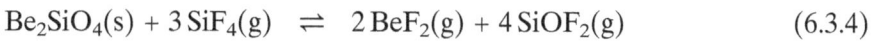

$$Eu_2SiO_4(s) + 6\,HCl(g)$$
$$\rightleftharpoons\ 2\,EuCl_2(g) + SiCl_2(g) + 3\,H_2O(g) + \frac{1}{2}O_2(g) \qquad (6.3.3)$$

$$Be_2SiO_4(s) + 3\,SiF_4(g)\ \rightleftharpoons\ 2\,BeF_2(g) + 4\,SiOF_2(g) \qquad (6.3.4)$$

$$Be_2SiO_4(s) + 3\,SiF_4(g) + 2\,NaF(g)$$
$$\rightleftharpoons\ 2\,NaBeF_3(g) + 4\,SiOF_2(g) \qquad (6.3.5)$$

The gas species $SiOF_2$, which appears during the transport of Be_2SiO_4, was detected by mass spectroscopy. Transport effects can also be observed when Na_2BeF_4 and BeF_2 are added. Hydrolysis reactions leading to the formation of hydrogen fluoride as a transport agent cannot completely be excluded for these experiments.

In contrast to reversible CVT reactions in the direct sense, the crystallization of silicates with participation of the gas phase, which is described in the literature, can be traced back to partial transport reactions. The formation of zircon $ZrSiO_4$ from zirconium dioxide in silica ampoules when silicon(IV) fluoride is added, has been discussed by *Schäfer* (Sch 1962). The reactions 6.3.6 (over the solid ZrO_2/ $ZrSiO_4$) and 6.3.7 (over the solid SiO_2/$ZrSiO_4$) describe the process completely.

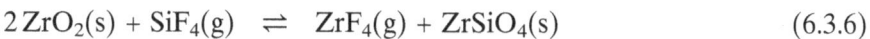

$$2\,ZrO_2(s) + SiF_4(g)\ \rightleftharpoons\ ZrF_4(g) + ZrSiO_4(s) \qquad (6.3.6)$$
$$\Delta_rH^0_{1273} = 115\ kJ \cdot mol^{-1},\ \Delta_rS^0_{1273} = 14.8\ J \cdot mol^{-1} \cdot K^{-1}$$
$$K_{p,1273} = 1.2 \cdot 10^{-4}$$

$$2\,SiO_2(s) + ZrF_4(g)\ \rightleftharpoons\ SiF_4(g) + ZrSiO_4(s) \qquad (6.3.7)$$
$$\Delta_rH^0_{1273} = -164\ kJ \cdot mol^{-1},\ \Delta_rS^0_{1273} = -38.9\ J \cdot mol^{-1} \cdot K^{-1}$$
$$K_{p,1273} = 5.6 \cdot 10^4$$

The equilibria 6.3.6 and 6.3.7 all allow sufficiently high pressures for SiF_4 and ZrF_4, so that the interdependent transport of silicon and zirconium becomes possible. Here, the migration takes place via the fluorides under isothermal (!) conditions in the gradient of the respective chemical potentials. These are, caused by the formation of $ZrSiO_4$, clearly decreased compared to those of the pure oxides. At 1000 °C and $\Sigma p = 1$ bar, a partial pressure difference of $\Delta p = 10^{-4}$ bar is calculated for SiF_4 and ZrF_4. The formation of Ni_2SiO_4 from NiO and SiO_2 in the presence of gaseous SiF_4 (Hof 1977), topaz ($Al_2SiO_4F_2$) from AlF_3 and SiO_2

(Scho 1940), and Co_2SiO_4 from CoF_2 and SiO_2 (Schm 1964), take place in a similar way.

The reversible CVT of Co_2SiO_4 with hydrogen fluoride gas should be possible despite the unfavorable thermodynamic data of the reaction 6.3.8, as detailed model calculations lead us to expect. However, experimental investigations are yet to come.

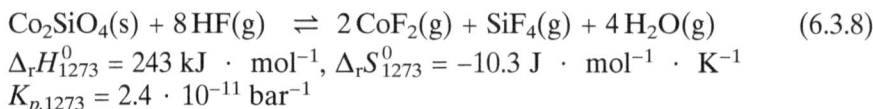

$$Co_2SiO_4(s) + 8\,HF(g) \; \rightleftharpoons \; 2\,CoF_2(g) + SiF_4(g) + 4\,H_2O(g) \qquad (6.3.8)$$
$$\Delta_r H^0_{1273} = 243\,kJ \cdot mol^{-1}, \; \Delta_r S^0_{1273} = -10.3\,J \cdot mol^{-1} \cdot K^{-1}$$
$$K_{p,1273} = 2.4 \cdot 10^{-11}\,bar^{-1}$$

The formation of Co_2SiO_4 from CoO in silica ampoules under the influence of chlorine or $TeCl_4$ has been thermodynamically analyzed in detail (Str 1981). Accordingly, cobalt(II)-oxide is dissolved in the gas phase in form of the gas species $CoCl_2$ and O_2 (equation 6.3.9). In the same way, the reaction of other oxides with SiO_2 from the wall of the silica ampoule is possible under the conditions of CVT experiments, and favorable with respect to the thermodynamic aspects. The formation of $HfSiO_4$ (Fuh 1986), $Ce_2Ti_2SiO_9$ (Zen 1999), and Ni_2SiO_4 (Fun 2004) are understood in the same way.

$$2\,CoCl_2(g) + O_2(g) + SiO_2(s) \; \rightleftharpoons \; Co_2SiO_4(s) + 2\,Cl_2(g) \qquad (6.3.9)$$

The detailed examinations of the deposition of nickel(II)-orthosilicate (Fun 2004) indicates that in the presence of HCl the deposition of the silicate is favored in comparison to the transport and deposition of pure metal oxide. This can be due to thermodynamic and kinetic reasons. SiO_2 is more soluble in a hydrogen chloride atmosphere than in chlorine. It also seems possible that the cracking of Si–O bonds in silica glass is easier if hydrogen atoms are present.

Boron(III)-oxide forms numerous ternary and multinary oxido compounds. There are hardly any indications on the *chemical vapor transport of borates*. Only the migration of $CrBO_3$, $FeBO_3$, BPO_4, and $Cr_2BP_3O_{12}$ in the temperature gradient is detected for sure. The transport of $FeBO_3$ takes place from 760 to 670 °C via the chlorinating equilibria with the formation of BCl_3 and $FeCl_3$ as well as O_2 and H_2O as gas species when chlorine or hydrogen chloride are used as the transport agent (Die 1975). The source solid $FeBO_3$ was provided in a platinum crucible to minimize reaction with silica. The deposition of crystals occurs endothermically. The temperature of the crystallization zone is in the range of 800 to 905 °C. The following transport equation is given:

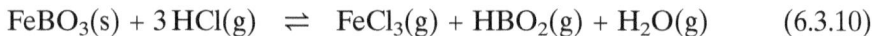

$$FeBO_3(s) + 3\,HCl(g) \; \rightleftharpoons \; FeCl_3(g) + HBO_2(g) + H_2O(g) \qquad (6.3.10)$$

Boron phosphate BPO_4 could be deposited in crystalline form via the following transport equilibrium when phosphorus(V)-chloride is added (Schm 2004):

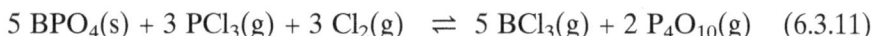

$$5\,BPO_4(s) + 3\,PCl_3(g) + 3\,Cl_2(g) \; \rightleftharpoons \; 5\,BCl_3(g) + 2\,P_4O_{10}(g) \qquad (6.3.11)$$

The deposition partly took place on a glassy carbon target, which was introduced into the ampoule in order to avoid the growing of the boron phosphate on the silica glass wall. This way, particularly pure single-crystals were obtained for further examination. Several other borates were obtained as a by-product during the synthesis of boracites $M_3B_7O_{13}X$ (M = metal atom with the oxidation number

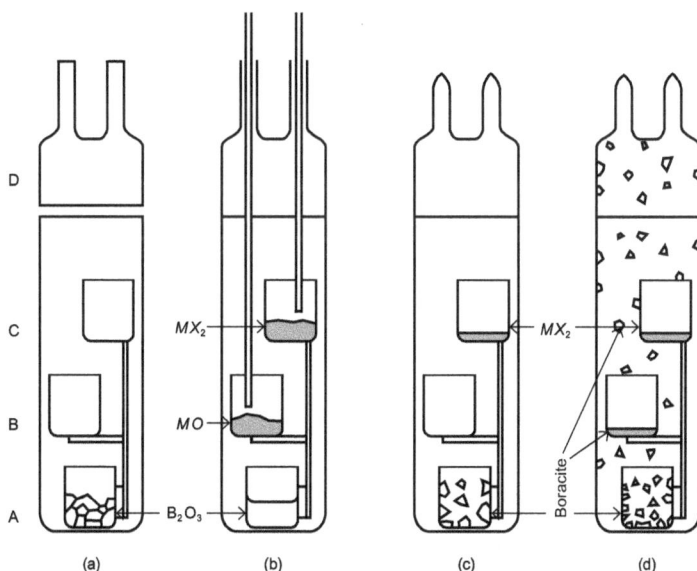

Figure 6.3.1 Steps of the reactions during the isothermal three-crucible process for preparing and crystallizing boracites $M^{II}_3B_7O_{13}X$ (M = metal; X = Cl, Br, I) according to *Schmid* (Schm 1965).

II, X = Cl, Br, I) (e. g., MgB_2O_4, ZnB_2O_4, ZnB_4O_7, CuB_2O_4) (Schm 1965). $TiBO_3$, VBO_3, and $CrBO_3$ also crystallize under participation of the gas phase with the help of titanium(II)-iodide, vanadium(II)-chloride, or chromium(II)-chloride in the presence of water (Schm 1964).

Boracites can be crystallized well with the help of CVT reactions, in contrast to the halogen-free borates. *Schmid* suggested a process of preparing and crystallizing boracites $M_3B_7O_{13}X$ (M = metal atom with the oxidation number II, X = Cl, Br, I) with the help of isothermal CVT (Figure 6.3.1) (Schm 1965). In order to do so, carefully dried B_2O_3 is filled in crucible A (a); after the welding on of the lid D with vacuum joint, the B_2O_3 is dehydrated at 1100 °C. After cooling in a vacuum, the dry oxide is put into crucible B under a dry $N_2(g)$ atmosphere. Dry halide in crucible C and an H_2O/HX-dosing agent (e. g., H_3BO_3) are filled in crucible B or C (b); after welding the loading tubes under vacuum at 20 °C, an isothermal reaction (at 600 to 1100 °C) follows. If the B_2O_3 transport via the gas phase can be neglected completely, crucible B is MO-free after the end of the reaction and boracite only grows in crucible A (c); if there is noteworthy B_2O_3 transport, a few boracite crystals develop in crucible A, in crucible B a boracite cake and sometimes big and small crystals which are spread in the ampoule (d). More recent examinations show that boracites are accessible with only little effort in simple silica ampoules in a temperature gradient by reversible transport reactions (Schm 1995).

Schmid concluded from thermodynamic considerations and the results of mass spectrometric studies (Mes 1960) that the migration of B_2O_3 via the gas phase at temperatures around 1000 °C takes place via $HBO_2(g)$ and to a small extent $H_3BO_3(g)$ and $(HBO_2)_3(g)$ (Schm 1965).

Table 6.1 Examples of the chemical vapor transport of complex oxides.

Sink solid	Transport additive	Temperature / °C	Reference
Ag_2SO_4	Cl_2	$800 \rightarrow T_1$	Spi 1978a
$Al_2(SO_4)_3$	$SOCl_2$	$625 \rightarrow 525$	Dah 1995
$Al_2(SiO_4)F_2$ (topas, from SiO_2)	AlF_3	800	Scho 1940
BPO_4	PCl_5	$800 \rightarrow 700$	Schm 2004
$BeSO_4$	Cl_2	$700 \rightarrow 600$	Koh 1988
$Cd_2As_2O_7$	$PtCl_2$	$650 \rightarrow 600$	Wei 2001b
$CdAs_2O_6$	$PtCl_2$	$720 \rightarrow 680$	Wei 2001a
CdP_4O_{11}	$P + I_2$	$510 \rightarrow 480$	Wei 1998
$CdSO_4$	Cl_2	$840 \rightarrow 740$	Spi 1978a
	HCl	$840 \rightarrow 740$	Spi 1978a
$Ce_2Ti_2SiO_9$ (from $Ce_2Ti_2O_7$)	NH_4Cl	$1050 \rightarrow 900$	Zen 1999
$Co_2As_2O_7$	Cl_2	$880 \rightarrow 800$	Wei 2005
$Co_2P_2O_7$	$CoP + I_2, P + I_2$	$1000 \rightarrow 900$	Schm 2002a
	Cl_2	$1100 \rightarrow 1000$	Gla 1991
$Co_2P_4O_{12}$	$P + I_2$	$850 \rightarrow 750$	Schm 2002a
$CoTi_2O_2(PO_4)_2$	$NH_4Cl + Cl_2,$ $P + I_2, TiP + Cl_2$	$1000 \rightarrow 900$	Schö 2007
$CoTi_4(PO_4)_6$	$NH_4Cl + Cl_2,$ $P + I_2, TiP + Cl_2$	$1000 \rightarrow 900$	Schö 2008c
$Co_3In_4(PO_4)_6$	$NH_4Cl + Cl_2$	$1000 \rightarrow 900$	Lit 2009
$CoSO_4$	Cl_2, HCl	$650 \rightarrow 550$	Spi 1978a
Co_2SiO_4	SiF_4	$1000 \rightarrow T_1$	Schm 1964
(from CoO)	$TeCl_4$	$930 \rightarrow 710$	Str 1981
Co_3TeO_6	HCl	$700 \rightarrow 600$	Bec 2006
$Co_2V_2O_7$	Cl_2	$700 \rightarrow 600$	Bro 2010
$CrBO_3$	$Cr + I_2$	$1000 \rightarrow 900$	Schm 1995
	$CrCl_2 + H_2O$	$1000 \rightarrow 900$	Schm 1964
$Cr_2BP_3O_{12}$	$I_2, P + I_2$	$1100 \rightarrow 1000$	Schm 2002b
$CrPO_4$	Cl_2	$1000 \rightarrow 900$	Gla 1986
$Cr_4(P_2O_7)_3$	$PtCl_2$	$1050 \rightarrow 950$	Tha 2006
$Cr(PO_3)_3$ (from $CrPO_4 + P$)	$P + I_2$	$1050 \rightarrow 950$	Gru 1996
$Cr_7(PO_4)_6$	I_2	$1050 \rightarrow 950$	Gla 1993
$Cr_3(P_2O_7)_2$	$P + I_2$	$1050 \rightarrow 950$	Gla 1992
$Cr_2P_2O_7$	I_2	$1050 \rightarrow 950$	Gla 1991
	$CrP + I_2$	$1050 \rightarrow 950$	Gla 1999
$Cr_3Ti_4^{III}(PO_4)_6$	$I_2, TiP + I_2$	$1000 \rightarrow 900$	Lit 2009

Table 6.1 (continued)

Sink solid	Transport additive	Temperature / °C	Reference
$CrTi_2^{III}(P_2O_7)_2$	I_2	$900 \rightarrow 850$	Lit 2009
$Cr_3V_4^{III}(PO_4)_6$	I_2	$1000 \rightarrow 900$	Lit 2009
$Cr_2(SO_4)_3$	Cl_2	$690 \rightarrow 630$	Dah 1994b
	HCl	$690 \rightarrow 630$	Dah 1993b
$Cr_2Te_4O_{11}$	$TeCl_4$	$700 \rightarrow T_1$	Meu 1976
$CrVO_4$	Cl_2	$750 \rightarrow 650$	Gro 2009
$Cr_2V_4O_{13}$	Cl_2	$750 \rightarrow 650$	Gro 2009
$Cu_2As_2O_7$	Cl_2	$880 \rightarrow 800$	Wei 2004a
$Cu_2P_2O_7$	$CuP_2 + I_2$	$900 \rightarrow 800$	Öza 1993
$Cu_2P_4O_{12}$	$P + I_2$	$850 \rightarrow 750$	Öza 1993
	$CuP_2 + I_2$	$850 \rightarrow 750$	Gla 1996
CuP_4O_{11}	$P + I_2$	$600 \rightarrow 500$	Öza 1993
	$CuP_2 + I_2$	$600 \rightarrow 500$	Gla 1996
$Cu_2O(SO_4)$	$HgCl_2$	$750 \rightarrow 650$	Bal 1983
$CuSO_4$	Cl_2, HCl, NH_4Cl, I_2	$700 \rightarrow 600$	Spi 1978a
	$HgCl_2$	$700 \rightarrow 600$	Bal 1983
$Cu_2V_2O_7$	$TeCl_4$	$600 \rightarrow 500$	Bec 2008
CuV_2O_6	$TeCl_4$	$600 \rightarrow 500$	Bec 2008
$Fe_7(AsO_4)_6$	NH_4Cl	$900 \rightarrow 800$	Wei 2004c
$FeBO_3$	HCl	$760 \rightarrow 670$	Die 1975
$Fe_3O_3(PO_4)$	$ZrCl_4$	$1000 \rightarrow 900$	Dro 1997
$(FePO_4 + ZrCl_4)$			
$FePO_4$	Cl_2	$1100 \rightarrow 1000$	Gla 1990b
$Fe_2P_2O_7$	$FeP + I_2$	$850 \rightarrow 750$	Gla 1991
$Fe_2P_4O_{12}$	$P + I_2$	$850 \rightarrow 750$	Wei 1998
$FeTi_2O_2(PO_4)_2$	$TiP + Cl_2$	$1000 \rightarrow 900$	Schö 2007
$FeTi_4(PO_4)_6$	$TiP + Cl_2$	$1000 \rightarrow 900$	Schö 2008c
$Fe_3Ti_4(PO_4)_6$	$NH_4Cl + Cl_2$	$1000 \rightarrow 900$	Lit 2009
$Fe_3V_4(PO_4)_6$	$NH_4Cl + Cl_2$	$1000 \rightarrow 900$	Lit 2009
$Fe_3Cr_4(PO_4)_6$	$NH_4Cl + Cl_2$	$1000 \rightarrow 900$	Lit 2009
$FeSO_4$	NH_4Cl	$650 \rightarrow 550$	Dah 1992
	Cl_2	$775 \rightarrow 675$	Dah 1992
	Cl_2	$775 \rightarrow 675$	Dah 1993
$FeVO_4$	Cl_2	$700 \rightarrow 600$	Gro 2009
$Ga_2(SO_4)_3$	Cl_2	$775 \rightarrow 675$	Kra 1995a
	Cl_2	$775 \rightarrow 675$	Kra 1995b
GeP_2O_7	$PtCl_2$	$1150 \rightarrow 1050$	Kai 1996
(cubic)			
GeP_2O_7	$PtCl_2$	$950 \rightarrow 850$	Kai 1996
(triclinic)			
HfP_2O_7	$P + I_2$	$1150 \rightarrow 1050$	Kos 1997

Table 6.1 (continued)

Sink solid	Transport additive	Temperature / $^\circ C$	Reference
$HgAs_2O_6$	$HgCl_2$	$650 \rightarrow 550$	Wei 2000
$(Hg_2)_2As_2O_7$	$HgCl_2$	$550 \rightarrow 500$	Wei 2004b
$(Hg_2)_2P_2O_7$	Hg_2Cl_2	$500 \rightarrow 450$	Wei 1999
$Hg_2P_2O_7$	$PCl_3 + Cl_2$	$550 \rightarrow 500$	Wei 1997
$HgSO_4$	HCl, Cl_2	$550 \rightarrow 470$	Spi 1978a
In_2OPO_4	$P + I_2$	$800 \rightarrow 700$	Tha 2004
(from In_2O_3)			
$InPO_4$	$InP + I_2$	$800 \rightarrow 700$	Tha 2003
$In_4(P_2O_7)_3$	$P + I_2$	$1000 \rightarrow 900$	Tha 2003
(from $InPO_4$)			
$In_2(SO_4)_3$	Cl_2	$625 \rightarrow 575$	Kra 1995b
	HCl	$625 \rightarrow 575$	Kra 1994
$Ir(PO_3)_3$	$IrCl_3 \cdot xH_2O$	$900 \rightarrow 800$	Pan 2008
$LaPO_4$	$PBr_3 + Br_2,$	$1120 \rightarrow 920$	Sch 1972
	$CO + Br_2$		
$LuPO_4$	$PBr_5 + Br_2$	$1000 \rightarrow 900$	Orl 1978
$MgSO_4$	Cl_2	$800 \rightarrow 700$	Koh 1988
$Mg_2P_2O_7$	$P + I_2$	$1000 \rightarrow 900$	Gla 1991,
			Ger 1996
$Mn_3(AsO_4)_2$	$PtCl_2$	$900 \rightarrow 820$	Wei 2008
$(Mn_7(AsO_4)_4Cl_2,$			
$Mn_{11}(AsO_4)_7Cl)$			
$Mn_2As_2O_7$	Cl_2	$880 \rightarrow 800$	Wei 2005
$Mn_3(PO_4)_2$	$P + I_2$	$850 \rightarrow 800$	Ger 1996
$Mn_2P_2O_7$	$P + I_2$	$1000 \rightarrow 900$	Ger 1996
	$P + I_2$	$1000 \rightarrow 900$	Gla 2002
$Mn_2P_4O_{12}$	$P + I_2$	$850 \rightarrow 750$	Gla 2002
$Mn_{1.65}Ti_4(PO_4)_6$	$P + I_2$	$1000 \rightarrow 900$	Schö 2008c
$Mn_3In_4(PO_4)_6$	$NH_4Cl + Cl_2$	$1000 \rightarrow 900$	Lit 2009
$MnSO_4$	Cl_2, HCl	$840 \rightarrow 740$	Spi 1978a
Mn_3TeO_6	Cl_2	$830 \rightarrow 750$	Wei 2006a
$Mn_2V_2O_7$	Cl_2	$700 \rightarrow 600$	Gro 2009
$MoOPO_4$	$MoP + I_2$	$800 \rightarrow 700$	Len 1995
$(MoO)_4(P_2O_7)_3$	I_2	$800 \rightarrow 650$	Len 1995
	I_2	$800 \rightarrow 650$	Isl 2009b
$(MoO)_2P_4O_{13}$	I_2, HgX_2	$900 \rightarrow 800$	Len 1995
	$(X = Cl, Br)$		
MoP_2O_7	$MoP + I_2,$	$1000 \rightarrow 900$	Len 1995
	$MoP + HgBr_2$		
$Mo(PO_3)_3$	I_2	$900 \rightarrow 800$	Wat 1994

Table 6.1 (continued)

Sink solid	Transport additive	Temperature / °C	Reference
$Mo(PO_3)_3$	I_2	$900 \rightarrow 800$	Len 1995
$Na_{2+x}Nb_6O_{10}(PO_4)_4$	NH_4Cl, $NaCl$	$1150 \rightarrow 1120$	Xu 1996
$K_3Nb_6O_{10}(PO_4)_4$	KCl	$1150 \rightarrow 1120$	Xu 1993
Nb_9PO_{25}	$NbP + I_2$	$900 \rightarrow 800$	Kai 1990
$NbOPO_4$,	$NbP + NH_4Cl$,	$1000 \rightarrow 900$	Kai 1992
$NbO_{1-x}PO_4$	$I_2 + NbP$		
$Nb_2(PO_4)_3$	$P + I_2$	$700 \rightarrow 600$	Sie 1989
	$P + I_2$	$700 \rightarrow 600$	Kai 1990
β-$Ni_2As_2O_7$	Cl_2	$880 \rightarrow 800$	Wei 2005
$Ni_3(PO_4)_2$	$PtCl_2$	$1100 \rightarrow 1000$	Fun 2004
β-$Ni_2P_2O_7$	$P + I_2$	$800 \rightarrow 600$	Fun 2004
σ-$Ni_2P_2O_7$	$PtCl_2$	$1100 \ldots 700 \rightarrow 1000 \ldots 600$	Pet 1987,
	NH_4Cl	$900 \rightarrow 750$	Gla 1991,
	$P + I_2$	$800 \rightarrow 700$	Blu 1997,
			Fun 2004
$Ni_2P_4O_{12}$	$P + I_2$	$850 \rightarrow 750$	Blu 1997,
			Fun 2004
$NiTi_2O_2(PO_4)_2$	$TiP + Cl_2$	$1000 \rightarrow 900$	Schö 2007
Ni_2SiO_4 (from NiO)	SiF_4	$1190 \rightarrow 1040$	Hof 1977
	NH_4Cl	$1000 \rightarrow 900$	Fun 2004
$NiSO_4$	Cl_2, HCl	$675 \rightarrow 560$	Spi 1978a
	$PbCl_2$	$675 \rightarrow 560$	Pli 1989
$PbSO_4$	Cl_2	$830 \rightarrow 730$	Pli 1989
	HCl, $HgCl_2$, I_2	$830 \rightarrow 730$	Spi 1978a
$PdAs_2O_6$	$PdCl_2$	$700 \rightarrow 600$	Pan 2009
$Pd_2P_2O_7$	$PdCl_2$	$850 \rightarrow 750$	Pan 2005
$Pd(PO_3)_2$	$PdCl_2$	$950 \rightarrow 850$	Gör 1997
$REAsO_4$	$TeCl_4$	$1100 \rightarrow 950$	Schm 2005a
$REPO_4$	PX_5,($X = Cl$, Br)	$1100 \rightarrow 1000$	Orl 1971,
	HBr		Sch 1972,
			Rep 1971,
			Tan 1968,
			Orl 1974
$Re_2O_3(PO_4)_2$	I_2	$600 \rightarrow 500$	Isl 2009a
$Re_2O_3(P_2O_7)$	$O_2 + H_2O$,	$700 \rightarrow 650$	Isl 2009b
	$H_2O + I_2$		
ReP_2O_7	$O_2 + H_2O$,	$800 \rightarrow 700$	Isl 2009b
	$H_2O + I_2$, $HgBr_2$		
$RhAsO_4$	$RhCl_3$	$820 \rightarrow 760$	Gör 1997
$RhPO_4$	Cl_2	$1000 \rightarrow 900$	Rit 1994
$Rh(PO_3)_3$	Cl_2	$950 \rightarrow 850$	Rit 1994

Table 6.1 (continued)

Sink solid	Transport additive	Temperature / °C	Reference
$RhVO_4$	$TeCl_4$	$1000 \rightarrow 900$	Jen 2009
SiP_2O_7	$P + I_2$	$1030 \rightarrow 900$	Kos 1996
SnP_2O_7	$PtCl_2$	$980 \rightarrow 840$	Kai 1996
$Ti_5O_4(PO_4)_4$	$TiP + Cl_2$	$1000 \rightarrow 900$	Rei 1994
TiP_2O_7	$TiP + I_2, P + I_2$	$1000 \rightarrow 900$	Gla 1990a
$Ti_{31}O_{24}(PO_4)_{24}$	$TiP + Cl_2$	$1000 \rightarrow 900$	Rei 1994
	$TiP + Cl_2$	$1000 \rightarrow 900$	Schö 2008
$Ti_4O_3(PO_4)_3$	$Ti + I_2$	$1050 \rightarrow 950$	Rei 1994
	$Ti + I_2$	$1050 \rightarrow 950$	Schö 2008
$Ti_9O_4(PO_4)_7$	I_2	$1000 \rightarrow 900$	Rei 1994
$TiPO_4$	$I_2, P + I_2$	$1000 \rightarrow 900$	Gla 1990a
$Ti(PO_3)_3$	$P + I_2$	$1000 \rightarrow 900$	Gla 1990a
$U_2O(PO_4)_2$	$UP_2 + I_2$	$800 \rightarrow 900$	Alb 1995
UP_2O_7	$NbCl_5, VCl_4$	$1000 \rightarrow 900$	Kos 1997
UPO_4X	ZrX_4	$1000 \rightarrow 900$	Dro 2004
$(X = Cl, Br)$ from			
$UP_2O_7 +$			
ZrX_4			
$VOPO_4$	NH_4Cl, Cl_2	$700 \rightarrow 600$	Vog 2006
$V_2O(PO_4)$	I_2, VCl_3	$1000 \rightarrow 900$	Gla 1989, Dro 2004
VPO_4	$VP + I_2, VCl_3$	$1000 \rightarrow 900$	Gla 1990b, Gla 1992a, Dro 2004
$V(PO_3)_3$	$P + I_2$	$1000 \rightarrow 900$	Dro 2004
$VOSO_4$	Cl_2, NH_4Cl	$525 \rightarrow 425, 550 \rightarrow 450$	Dah 1994a
$(WO_3)_n(PO_2)_2$ "MPTB"	I_2, NH_4Cl, KI	$900 \rightarrow 800$	Mat 1991, Mat 1994, Rou 1996, Tew 1992
$WOPO_4$ (next WP)	I_2	$1000 \rightarrow 900$	Mat 1991
WP_2O_7 (next WP)	I_2	$1000 \rightarrow 900$	Mat 1991
$Zn_2As_2O_7$	Cl_2	$880 \rightarrow 800$	Wei 2005
$Zn_2P_2O_7$	NH_4Cl, H_2	$920 \rightarrow 820$	Rüh 1994
ZnP_4O_{11}	$ZnP_2 + I_2$	$1000 \rightarrow 900$	Wei 1998
	$P + I_2$	$1000 \rightarrow 900$	Wei 1998
$Zn_3Ti_4(PO_4)_6$	$NH_4Cl + Cl_2$	$1000 \rightarrow 900$	Lit 2009
$Zn_3In_4(PO_4)_6$	$NH_4Cl + Cl_2$	$1000 \rightarrow 900$	Lit 2009

Table 6.1 (continued)

Sink solid	Transport additive	Temperature / °C	Reference
$Zn_3O(SO_4)_2$	$PbCl_2$	$710 \rightarrow 620$	Bal 1984
N-$ZnSO_4$	Cl_2	$T_2 \rightarrow T_1, T_1 < 690$	Spi 1978a
H-$ZnSO_4$	Cl_2	$T_2 \rightarrow T_1, T_1 > 690$	Spi 1978b
Zn_3TeO_6	Cl_2	$830 \rightarrow 750$	Wei 2006b
ZrP_2O_7	$P + I_2$	$1100 \rightarrow 1000$	Kos 1997

Bibliography

Alb 1995   J. H. Albering, W. Jeitschko, *Z. Kristallogr.* **1995**, *210*, 878.
Bal 1983   L. Bald, M. Spiess, R. Gruehn, T. Kohlmann, *Z. Anorg. Allg. Chem.* **1983**, *498*, 153.
Bal 1984   L. Bald, R. Gruehn, *Z. Anorg. Allg. Chem.* **1984**, *509*, 23.
Ban 1990   H. W. Bange, *Diplomarbeit*, University of Freiburg, **1990**.
Bec 2006   R. Becker, M. Johnsson, H. Berger, *Acta Crystallogr.* **2006**, *C62*, i67.
Bec 2008   P. Becker-Bohatý, *personal communication*, University of Köln, **2008**.
Ben 2008   E. Benser, M. Schöneborn, R. Glaum, *Z. Anorg. Allg. Chem.* **2008**, *634*, 1677.
Bin 1983   M. Binnewies, *Z. Anorg. Allg. Chem.* **1983**, *507*, 77.
Bin 2002   M. Binnewies, E. Milke, *Thermodynamic Data of Elements and Compounds*, 2. Aufl., Wiley-VCH, **2002**.
Blu 1997   M. Blum, *Diplomarbeit*, University of Gießen, **1997**.
Bou 1982   M. Bouchdoug, A. Courtois, R. Gerardin, J. Steinmetz, C. Gleitzer, *J. Solid State Chem.* **1982**, *42*, 149.
Bro 2010   A. Bronova, R. Glaum, *personal communication*, University of Bonn, **2010**.
Dah 1992   T. Dahmen, R. Gruehn, *Z. Anorg. Allg. Chem.* **1992**, *609*, 139.
Cha 1998   M. W. Chase, *NIST-JANAF Thermochemical Tables*, ACS, **1992**.
Dah 1993a   T. Dahmen, R. Gruehn, *J. Cryst. Growth* **1993**, 130, 636.
Dah 1993b   T. Dahmen, R. Gruehn, *Z. Kristallogr.* **1993**, 204, 57.
Dah 1994a   T. Dahmen, R. Gruehn, *personal communication*, University of Gießen, **1994**.
Dah 1994b   T. Dahmen, R. Gruehn, *Z. Anorg. Allg. Chem.* **1994**, *620*, 1569.
Dah 1995   T. Dahmen, R. Gruehn, *Z. Anorg. Allg. Chem.* **1995**, *621*, 417.
Dau 1841   M. A. Daubrée, *Ann. Mines* **1841**, *20*, 65. Cited in W. Schrön, *Eur. J. Mineral.* **1989**, *1*, 739.
Dau 1849   M. A. Daubrée, *Comptes Rendus* **1849**, *29*, 227. Cited in W. Schrön, *Eur. J. Mineral.* **1989**, *1*, 739.
Die 1975   R. Diehl, A. Räuber, F. Friedrich, *J. Cryst. Growth* **1975**, *29*, 225.
Die 1976   R. Diehl, F. Friedrich, *J. Cryst. Growth* **1976**, *36*, 263.
Dit 1983   G. Dittmer, U. Niemann, *Mater. Res. Bull.* **1983**, *18*, 355.
Dro 1997   T. Droß, *Diplomarbeit*, University of Gießen, **1997**.
Dro 2004   T. Droß, *Dissertation*, University of Bonn, **2004**.
Fuh 1986   J. Fuhrmann, J. Pickardt, *Z. Anorg. Allg. Chem.* **1986**, *532*, 171.
Fun 2004   M. Funke, R. Glaum, *personal communication*, University of Bonn, **2004**.

Ger 1996   M. Gerk, *Dissertation*, University of Gießen, **1996**.
Ger 2007   S. Gerlach, R. Cardoso-Gil, E. Milke, M. Schmidt, *Z. Anorg. Allg. Chem.* **2007**, *633*, 83.
Gla 1986   R. Glaum, R. Gruehn, M. Möller, *Z. Anorg. Allg. Chem.* **1986**, *543*, 111.
Gla 1989   R. Glaum, R. Gruehn, *Z. Kristallogr.* **1989**, *186*, 91.
Gla 1990a   R. Glaum, R. Gruehn, *Z. Anorg. Allg. Chem.* **1990**, *580*, 78.
Gla 1990b   R. Glaum, *Dissertation*, University of Gießen, **1990**.
Gla 1991   R. Glaum, M. Walter-Peter, D. Özalp, R. Gruehn, *Z. Anorg. Allg. Chem.* **1991**, *601*, 145.
Gla 1992a   R. Glaum, R. Gruehn, *Z. Kristallogr.* **1992**, 198, 41.
Gla 1992b   R. Glaum, *Z. Anorg. Allg. Chem.* **1992**, *616*, 46.
Gla 1993   R. Glaum, *Z. Kristallogr.* **1993**, 205, 69.
Gla 1995   R. Glaum, A. Schmidt, *Acta Crystallogr.* **1995**, *C52*, 762.
Gla 1996   R. Glaum, M. Weil, D. Özalp, *Z. Anorg. Allg. Chem.* **1996**, *622*, 1839.
Gla 1997   R. Glaum, A. Schmidt, *Z. Anorg. Allg. Chem.* **1997**, *623*, 1672.
Gla 1999   R. Glaum, *Neue Untersuchungen an wasserfreien Phosphaten der Übergangsmetalle*, Habilitationsschrift, University of Gießen, **1999**. URL: http://geb.uni-giessen.de/
Gla 2002   R. Glaum, H. Thauern, A. Schmidt, M. Gerk, *Z. Anorg. Allg. Chem.* **2002**, *628*, 2800.
Gla 2009   R. Glaum, *personal communication*, University of Bonn, **2009**.
Glu 1989   V. P. Glushko et al., *Thermodynamic Properties of Individual Substances*, Vol. 1/1, Hemisphere Publishing Corporation, **1989**.
Gör 1997   H. Görzel, *Dissertation*, University of Gießen, **1997**.
Gro 2009   R. Groher, *Diplomarbeit*, University of Bonn, **2010**.
Gru 1996   M. Gruß, R. Glaum, *Acta Crystallogr.* **1996**, *C52*, 2647.
Hof 1977   J. Hofmann, R. Gruehn, *J. Cryst. Growth* **1977**, *37*, 155.
Ijj 1991   M. Ijjaali, G. Venturini, R. Gerardin, B. Malaman, C. Gleitzer, *Eur. J. Solid State Inorg. Chem.* **1991**, *28*, 983.
Isl 2009a   M. S. Islam, R. Glaum, *Z. Anorg. Allg. Chem.* **2009**, *635*, 1008.
Isl 2010   M. S. Islam, *part of a planned Dissertation*, University of Bonn.
Jen 2009   J. Jentsch, *Diplomarbeit*, HTW Dresden, **2009**.
Kai 1990   U. Kaiser, *Diplomarbeit*, University of Gießen, **1990**.
Kai 1992   U. Kaiser, G. Schmidt, R. Glaum, R. Gruehn, *Z. Anorg. Allg. Chem.* **1992**, *607*, 113.
Kai 1994   U. Kaiser, R. Glaum, *Z. Anorg. Allg. Chem.* **1994**, *620*, 1755.
Kai 1996   U. Kaiser, *Dissertation*, University of Gießen, **1996**.
Koh 1988   T. Kohlmann, R. Gruehn, *personal communication*, University of Gießen, **1988**.
Kon 1977   Z. A. Konstant, A. I. Dimante, *Inorg. Mater.* [USSR] **1977**, *13*, 83.
Kos 1997   A. Kostencki, *Dissertation*, University of Gießen, **1997**.
Kra 1994   M. Krause, T. Dahmen, R. Gruehn, *Z. Anorg. Allg. Chem.* **1994**, *620*, 672.
Kra 1995a   M. Krause, R. Gruehn, *Z. Anorg. Allg. Chem.* **1995**, *621*, 1007.
Kra 1995b   M. Krause, R. Gruehn, *Z. Kristallogr.* **1995**, *210*, 427.
Len 1995   M. Lenz, *Dissertation*, University of Gießen, **1995**.
Lit 2003   C. Litterscheid, *Diplomarbeit*, University of Bonn, **2003**.
Lit 2009   C. Litterscheid, *Dissertation*, University of Bonn, **2009**. URL: http://hss.ulb.uni-bonn.de/2009/1928/1928.htm
Mal 1992   B. Malaman, M. Ijjaali, R. Gerardin, G. Venturini, C. Gleitzer, *Eur. J. Solid State Inorg. Chem.* **1992**, *29*, 1269.
Mas 1979   R. Masse, J. C. Guitel, A. Durif, *Mater. Res. Bull.* **1979**, *14*, 327.

Mat 1981   K. Matsumoto, T. Kawanishi, K. Takagi, S. Kaneko, *J. Cryst. Growth* **1981**, *55*, 376.
Mat 1991   H. Mathis, R. Glaum, R. Gruehn, *Acta Chem. Scand.* **1991**, *45*, 781.
Mat 1994   H. Mathis, R. Glaum, R. Gruehn, *7. Vortragstagung der GDCH-Fachgruppe „Festkörperchemie"*, Bonn, **1994**.
May 1974   H. Mayer, *Monatsh. Chem.* **1974**, *105*, 46.
Mes 1960   D. J. Meschi, W. A. Chupka, J. Berkowitz, *J. Chem. Phys.* **1960**, *33*, 530.
Meu 1976   G. Meunier, B. Frit, J. Galy, *Acta Crystallogr.* **1976**, *B32*, 175.
Mod 1981   A. Modaressi, A. Courtois, R. Gerardin, B. Malaman, C. Gleitzer, *J. Solid State Chem.* **1981**, *40*, 301.
Mue 1970   D. W. Muenow, O. M. Uy, J. L. Margrave, *J. Inorg. Nucl. Chem.* **1970**, *32*, 3459.
Öza 1993   D. Özalp, *Dissertation*, University of Gießen, **1993**.
Opp 1974   H. Oppermann, G. Stöver, E. Wolf, *Z. Anorg. Allg. Chem.* **1974**, *410*, 179.
Opp 1977a   H. Oppermann, E. Wolf, *Z. Anorg. Allg. Chem.* **1977**, *437*, 33.
Opp 1977b   H. Oppermann, *Z. Anorg. Allg. Chem.* **1977**, *434*, 239.
Orl 1971   V. P. Orlovskii et al., *Izv. Akad. Nauk SSSR, Neorg. Mater.* **1971**, *7*, 251.
Orl 1974   V. P. Orlovskii, Kh. M. Kurbanov, B. S. Khalikov, V. I. Bugakov, I. V. Tananaev, *Izv. Akad. Nauk SSSR, Neorg. Mater.* **1974**, *10*, 670.
Orl 1978   V. P. Orlovskii, B. Khalikov, Kh. M. Kurbanov, V. I. Bugakov, L. N. Kargareteli, *Zh. Neorg. Khim.* **1978**, *23*, 316.
Pan 2005   K. Panagiotidis, W. Hoffbauer, J. Schmedt auf der Günne, R. Glaum, H. Görzel, *Z. Anorg. Allg. Chem.* **2005**, *631*, 2371.
Pan 2008   K. Panagiotidis, R. Glaum, W. Hoffbauer, J. Weber, J. Schmedt auf der Günne, *Z. Anorg. Allg. Chem.* **2008**, *634*, 2922.
Pan 2009   K. Panagiotidis, *Dissertation*, University of Bonn, **2009**.
Pie 1968   A. Pietraszko, K. Lukaszewicz, *Bull. Acad. Polon. Sci., Ser. Sci. Chim.* **1968**, *16*, 183.
Pli 1989   V. Plies, T. Kohlmann, R. Gruehn, *Z. Anorg. Allg. Chem.* **1989**, *568*, 62.
Pro 2004   A. V. Prokofiev, W. Assmus, R. K. Kremer, *J. Cryst. Growth* **2004**, *271*, 113.
Rei 1994   F. Reinauer, R. Glaum, R. Gruehn, *Eur. J. Solid State Inorg. Chem.* **1994**, *31*, 779.
Rei 1998   F. Reinauer, *Dissertation*, University of Gießen, **1998**.
Rep 1971   V. P. Repko, V. P. Orlovskii, G. M. Safronov, Kh. M. Kurbanov, M. N. Tseitlin, V. I. Pakhomov, I. V. Tananaev, A. N. Volodina, *Izv. Akad. Nauk SSSR, Neorg. Mater.* **1971**, *7*, 251.
Rit 1994   P. Rittner, R. Glaum, *Z. Kristallogr.* **1994**, *209*, 162.
Rou 1996   P. Roussel, Ph. Labbe, D. Groult, B. Domenges, H. Leligny, D. Grebille, *J. Solid State Chem.* **1996**, *122*, 281.
Rüh 1994   H. Rühl, R. Glaum, *personal communication*, University of Gießen, **1994**.
Ruž 1997   M. Ružička, *Cryst. Res. Tech.* **1997**, *32*, 743.
Sch 1962   H. Schäfer, *Chemische Transportreaktionen*, Verlag Chemie, Weinheim, **1962**.
Sch 1972   H. Schäfer, V. P. Orlovskii, *Z. Anorg. Allg. Chem.* **1972**, *390*, 13.
Schm 1964   H. Schmid, *Z. Anorg. Allg. Chem.* **1964**, *327*, 110.
Schm 1995   A. Schmidt, *Diplomarbeit*, University of Gießen, **1995**.
Schm 2002a A. Schmidt, *Dissertation*, University of Gießen, **2002**. URL: http://geb.uni-giessen.de/geb/volltexte/2002/805/
Schm 2002b M. Schmidt, M. Armbrüster, U. Schwarz, H. Borrmann, R. Cardoso-Gil, *Report MPI CPfS Dresden* **2001/2002**, 229.
Schm 2004   M. Schmidt, B. Ewald, Yu. Prots, R. Cardoso Gil, M. Armbrüster, I. Loa, L. Zhang, Ya-Xi Huang, U. Schwarz, R. Kniep, *Z. Anorg. Allg. Chem.* **2004**, *630*, 655.

Schm 2005a   M. Schmidt, U. Müller, R. Cardoso Gil, E. Milke, M. Binnewies *Z. Anorg. Allg. Chem.* **2005**, *631*, 1154.

Schm 2005b   M. Schmidt, R. Ramlau, W. Schnelle, H. Borrmann, E. Milke, M. Binnewies, *Z. Anorg. Allg. Chem.* **2005**, *631*, 284.

Scho 1940   R. Schober, E. Thilo, *Chem. Ber.* **1940**, *73*, 1219.

Scho 1991   H. Schornstein, *Dissertation*, University of Gießen, **1991**.

Schö 2007   M. Schöneborn, R. Glaum, *Z. Anorg. Allg. Chem.* **2007**, *633*, 2568.

Schö 2008   M. Schöneborn, R. Glaum, F. Reinauer, *J. Solid State Chem.* **2008**, *181*, 1367.

Schö 2008b   M. Schöneborn, *Dissertation*, University of Bonn, **2008**. URL: http://hss.ulb.uni-bonn.de/diss_online/math_nat_fak/2008/schoeneborn_marcos/

Schö 2008c   M. Schöneborn, R. Glaum, *Z. Anorg. Allg. Chem.* **2008**, *634*, 1843.

Sie 1989   S. Sieg, *Staatsexamensarbeit*, University of Gießen, **1989**.

Spi 1978a   M. Spieß, *Dissertation*, University of Gießen, **1978**.

Spi 1978b   M. Spieß, R. Gruehn, *Naturwissenschaften* **1978**, *65*, 594.

Str 1981   P. Strobel, Y. Le Page, *Mater. Res. Bull.* **1981**, *16*, 223.

Tan 1968   I. V. Tananaev, G. M. Safronov, V. P. Orlovskii, V. P. Repko, V. I. Pakhomov, A. N. Volodina, E. A. Ionkina, *Dokl. Akad. Nauk SSSR* **1968**, *183*, 1357.

Tew 1992   Z. S. Teweldemedhin, K. V. Ramanujachary, M. Greenblatt, *Phys. Rev., Cond. Matt. Mater.* **1992**, *46*, 7897.

Tha 2003   H. Thauern, R. Glaum, *Z. Anorg. Allg. Chem.* **2003**, *629*, 479.

Tha 2004   H. Thauern, R. Glaum, *Z. Anorg. Allg. Chem.* **2004**, *630*, 2463.

Tha 2006   H. Thauern, *Dissertation*, University of Bonn, **2006**. URL: http://hss.ulb.uni-bonn.de/2006/0790/0790.htm

Til 1973   E. Tillmanns, W. Gebert, W. H. Baur, *J. Solid State Chem.* **1973**, 7, 69.

Ven 1984   G. Venturini, A. Courtois, J. Steinmetz, R. Gerardin, C. Gleitzer, *J. Solid State Chem.* **1984**, *53*, 1.

Vog 2006   D. Vogt, *Staatsexamensarbeit*, University of Bonn, **2006**.

Wal 1987   M. Walter-Peter, *Staatsexamensarbeit*, University of Gießen, **1987**.

Wat 1994   I. M. Watson, M. M. Borel, J. Chardon, A. Leclaire, *J. Solid State Chem.* **1994**, *111*, 253.

Wei 1997   M. Weil, R. Glaum, *Acta Crystallogr.* **1997**, *C53*, 1000.

Wei 1998   M. Weil, R. Glaum, *Eur. J. Inorg. Solid State Chem.* **1998**, *35*, 495.

Wei 1999   M. Weil, R. Glaum, *Z. Anorg. Allg. Chem.* **1999**, *625*, 1752.

Wei 2000   M. Weil, *Z. Naturforsch.* **2000**, *55b*, 699.

Wei 2001a   M. Weil, *Acta Crystallogr.* **2001**, *E57*, i22.

Wei 2001b   M. Weil, *Acta Crystallogr.* **2001**, *E57*, i28.

Wei 2004a   M. Weil, C. Lengauer, E. Fueglein, E. J. Baran, *Cryst. Growth Des.* **2004**, *4*, 1229.

Wei 2004b   M. Weil, *Z. Anorg. Allg. Chem.* **2004**, *630*, 213.

Wei 2004c   M. Weil, *Acta Crystallogr.* **2004**, *E60*, i139.

Wei 2005   M. Weil, U. Kolitsch, *Z. Kristallogr.* **2005**, *22 Suppl.*, 183.

Wei 2006a   M. Weil, *Acta Crystallogr.* **2006**, *E62*, i244.

Wei 2006b   M. Weil, *Acta Crystallogr.* **2006**, *E62*, i246.

Wei 2008   M. Weil, *personal communication* **2008**.

Wic 1998   M. Wickleder, *Z. Anorg. Allg. Chem.* **1998**, *624*, 1347.

Wic 2000a   M. Wickleder, *Z. Anorg. Allg. Chem.* **2000**, *626*, 1468.

Wic 2000b   M. Wickleder, *Z. Anorg. Allg. Chem.* **2000**, *626*, 547.

Won 2002   J. Wontcheu, T. Schleid, *Z. Anorg. Allg. Chem.* **2002**, *628*, 1941.

Xu 1993   J. Xu, K. V. Ramanujachary, M. Greenblatt, *Mater. Res. Bull.* **1993**, *28*, 1153.

Xu 1996   J. Xu, M. Greenblatt, *J. Solid State Chem.* **1996**, *121*, 273.

Zen 1999   L.-P. Zenser, M. Weil, R. Gruehn, *Z. Anorg. Allg. Chem.* **1999**, *625*, 423.

7 Chemical Vapor Transport of Sulfides, Selenides, and Tellurides

The CVT of metal sulfides, metal selenides, and metal tellurides has been examined in detail. The number of examples that are known from the literature, is only exceeded by those of the oxides. The first examinations were made in the 1960s are especially connected to the name *Nitsche*, who was the first to describe the transport of a large number of these compounds. The CVT of these compounds clearly differs from that of the oxides. Often other transport agents are used. This is due to the higher thermodynamic stability of the metal oxides compared to the sulfides, selenides, and tellurides. Because of this, most often iodine or iodine compounds, which are not ideally suited for oxides, are used as transport agents. Another difference to the transport of the oxides results from the fact that the gaseous oxide halides appear more often than sulfide halides and selenide halides according to the current level of knowledge. Gaseous telluride halides are not yet known. In contrast, the stability of the halogen compounds increases clearly from oxygen to tellurium. Tellurium halides play an important part, and selenide halides a certain part, during transport reactions. Binary sulfur/halogen compounds and oxygen/halogen compounds are not involved in transport reactions. Sulfides and selenides behave in very similar ways. This is because of the similar ionic radii of the sulfide and selenide ions and also because sulfur and selenium show the same electronegativity. Both properties together cause similar chemical behavior, similar thermodynamic stabilities, and often the metal sulfides and metal selenides have the same structure types. Metal sulfides and metal selenides are often mixable completely in the solid state. The essential aspects that apply for the transport of sulfides, also apply for the selenides.

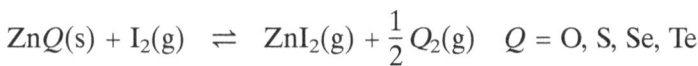

$$ZnQ(s) + I_2(g) \rightleftharpoons ZnI_2(g) + \frac{1}{2}Q_2(g) \quad Q = O, S, Se, Te$$

ZnQ	$\Delta_r G^0_{1000}/kJ \cdot mol^{-1}$	$K_{p,\,1000}/bar^{\frac{1}{2}}$
ZnO	205	$2 \cdot 10^{-11}$
ZnS	52	$2 \cdot 10^{-3}$
ZnSe	13	$2 \cdot 10^{-1}$
ZnTe	−82	$2 \cdot 10^{4}$

7.1 Transport of Sulfides

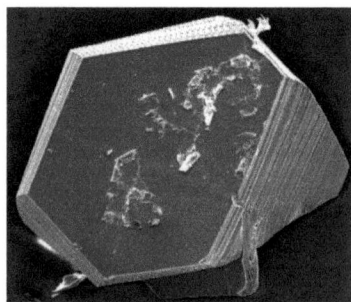

$Nb_{0.6}Ta_{0.4}S_2$

Iodine as transport agent

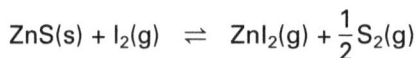

$$ZnS(s) + I_2(g) \;\rightleftharpoons\; ZnI_2(g) + \frac{1}{2}S_2(g)$$

Hydrogen halides as transport agents

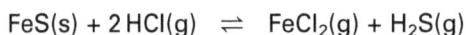

$$FeS(s) + 2\,HCl(g) \;\rightleftharpoons\; FeCl_2(g) + H_2S(g)$$

Hydrogen as transport agent

$$CdS(s) + H_2(g) \;\rightleftharpoons\; Cd(g) + H_2S(g)$$

The sulfides of most metals can be prepared by CVT reactions. Hence a large number of examples of the transport of binary and ternary sulfides are known. Furthermore, some quaternary and even multinary sulfides, such as

$FeSn_4Pb_3Sb_2S_{14}$ (Men 2006) are available by CVT reactions (Table 7.1.1). This is indeed noteworthy, because in these cases the transport agent is apparently able to transfer *all* cations that are present in the compound to the gas phase and deposit them at another temperature. Sulfides with a phase range, such as FeS_x, can be transported systematically as well (Kra 1975, Wol 1978). Mixed-crystals with substitution in the cationic sublattice, such as $Co_{1-x}Fe_xS$ (Kra 1979), in the anionic sublattice, such as $TiS_{2-x}Se_x$ (Nit 1967a, Rim 1974, Hot 2004b), or in the cationic *and* anionic sublattice, such as $Ge_xPb_{1-x}S_{1-y}Se_y$ (Kim 1993) are accessible with defined compositions and in crystalline form. The transport of doped sulfides, such as $CdAl_2S_4:Er^{3+}$ allows the study of the interesting optical characteristics of single-crystals (Oh 1997b).

Although many sulfides can be prepared by precipitation from aqueous solution, the preparation of larger crystals succeeds only in exceptional cases. The composition of so formed sulfides sometimes deviates from the ideal composition. If metal sulfides are prepared from the elements at higher temperatures, one has to expect the formation of covering layers. This leads to a reduced reaction velocity or even to a full standstill. If such a reaction is conducted in a closed ampoule, a high pressure can build due to sulfur that was not brought to reaction. The addition of small amounts of a suitable *mineralizer*, such as iodine or ammonium chloride, can remedy this. This way the chemical conversion takes place with the participation of the gas phase, and the reaction is clearly accelerated. In the process, a rougher crystalline product is obtained.

In many cases, synthesis and CVT of the sulfides can happen in one work process: the respective elements are put into a transport ampoule in a certain composition and a suitable transport agent is added. The ampoule is sealed and heated to a temperature around 450 °C for approximately one day (boiling temperature of sulfur 444 °C). During this time the reaction of the starting materials with each other takes place, and metal sulfides are formed. The transport agent accelerates the reaction; it functions as a mineralizer. Afterwards the transport ampoule is exposed to a temperature gradient where the CVT takes place. This is usually done at 700 to 1000 °C.

A pre-reaction at approximately 450 °C is highly recommended, because otherwise the sulfur, which has not been brought to a reaction yet, would evaporate partly or completely during heating to the actual reaction temperature. The pressure that develops during this process can lead to the bursting of the ampoule.

We discourage the use of commercially available sulfides as source solids in many cases. These substances, which are often prepared by precipitation reactions and consist of very fine species, can have quite large surface areas. They adsorb considerable amounts of foreign matter, for example water. These preparations tend to spray and contaminate the vacuum apparatus. Furthermore, it is time consuming and tedious to remove the absorbed foreign matter completely.

Thermal behavior When heated up, most of the metal sulfides decompose completely or partly to the elements. The heavy volatile metal generally remains as a solid. If the metal has a sufficiently high vapor pressure at the decomposition temperature, one can observe in some cases a *decomposition sublimation*. Examples of this are the sulfides of zinc and cadmium. If such a vapor is condensed, the metal sulfide will reform.

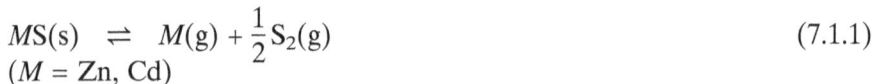

$$MS(s) \ \rightleftharpoons \ M(g) + \frac{1}{2}S_2(g)$$ (7.1.1)
$$(M = Zn, Cd)$$

Only a few metal sulfides can be sublimed undecomposed. Examples are gallium(I)-sulfide, germanium(II)-sulfide, tin(II)-sulfide, lead(II)-sulfide: Ga_2S (ϑ_m = 960 °C; K_p = 3 · 10^{-3} bar); GeS (ϑ_m = 655 °C; K_p = 10^{-1} bar); SnS (ϑ_m = 880 °C; K_p = 2 · 10^{-2} bar); PbS (ϑ_m = 1114 °C; K_p = 10^{-1} bar).

$$Ga_2S(s) \ \rightleftharpoons \ Ga_2S(g)$$ (7.1.2)

$$MS(s) \ \rightleftharpoons \ MS(g)$$ (7.1.3)
$$(M = Ge, Sn, Pb)$$

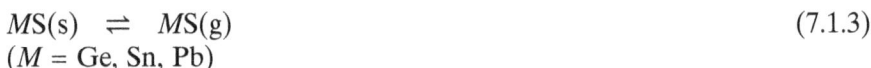

Some sulfides decompose to a metal-rich solid and gaseous sulfur, for example pyrite, which forms FeS(s) and S_2(g) (7.1.4) at high temperatures. In some cases, the metal-rich sulfides, which were formed by thermal decomposition, can appear in the gas phase as well. These compounds show noticeable effects of the gas-phase transport by a decomposition sublimation (7.1.5).

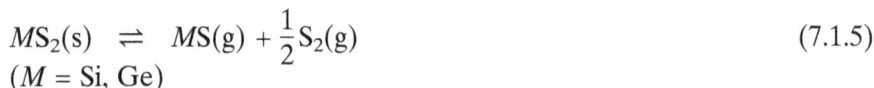

$$FeS_2(s) \ \rightleftharpoons \ FeS(s) + \frac{1}{2}S_2(g)$$ (7.1.4)

$$MS_2(s) \ \rightleftharpoons \ MS(g) + \frac{1}{2}S_2(g)$$ (7.1.5)
$$(M = Si, Ge)$$

Transport agent Mainly iodine is used as the transport agent for sulfides (as is the case for selenides and tellurides) – in contrast to the transport of oxides. This is due to the different stability of the solid chalcogenides.

The example of the zinc compounds shows that the metal oxide is generally more stable than the analogous sulfide, selenide, or telluride (ΔG_{298}^0/ kJ · mol^{-1}: ZnO (−363); ZnS (−212); ZnSe (−180); ZnTe (−142)). The way in which this affects the transport is dependent on the transport agent in use. Let us consider the CVT of zinc oxide and zinc sulfide with elemental halogen (1000 → 900 °C) to find out about it. Equations 7.1.6 and 7.1.7 describe the transport in very good approximation.

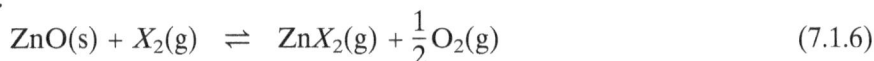

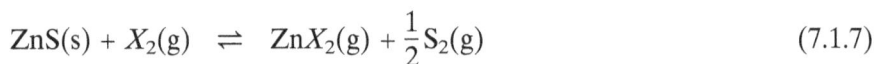

$$ZnO(s) + X_2(g) \ \rightleftharpoons \ ZnX_2(g) + \frac{1}{2}O_2(g)$$ (7.1.6)

$$ZnS(s) + X_2(g) \ \rightleftharpoons \ ZnX_2(g) + \frac{1}{2}S_2(g)$$ (7.1.7)

The equilibrium during the reaction of zinc sulfide with chlorine is so far on the side of the reaction products (K_p > 10^4) that the reversal, which would lead to the deposition of the solid, is hardly possible. This reaction is not well suited for

Table 7.1.1 Thermodynamic data of the reactions of ZnO and ZnS with halogens.

	ZnO		ZnS	
	$\Delta_r G^0_{1223}/\text{kJ} \cdot \text{mol}^{-1}$	$K_{p,\,1223\,\text{K}}/\text{bar}^{\frac{1}{2}}$	$\Delta_r G^0_{1223}/\text{kJ} \cdot \text{mol}^{-1}$	$K_{p,\,1223}/\text{bar}^{\frac{1}{2}}$
Cl_2	−59	$3.3 \cdot 10^3$	128	$3 \cdot 10^5$
Br_2	−15	4.3	−84	$3 \cdot 10^3$
I_2	69	$1 \cdot 10^{-3}$	1	0.9

CVT. The equilibrium positions during the reaction with bromine and iodine, on the other hand, are in a range that allows the dissolution of the solid and its deposition within the equilibrium range. Iodine should be the best transport agent among the halogens because the calculated equilibrium constant comes as close as possible to the ideal value of 1 bar$^{\frac{1}{2}}$. An equilibrium constant of 10^{-3} bar$^{\frac{1}{2}}$ is calculated for the reaction of zinc oxide with iodine for 1223 K. Thus the equilibrium is clearly on the side of the initial solids, and there is no noticeable dissolution in the gas phase. During transport equilibria of zinc sulfide with chlorine and bromine as transport agents, on the other hand, the numerical values of the equilibrium constants are closer to one. Here, transport effects are expected. This comparative consideration applies for most of the sulfides in a similar way. Chlorine has been successfully used as a transport agent only in exceptional cases. In some cases, hydrogen chloride or ammonium chloride have been used as a transport additive.

A series of studies report the CVT reactions of sulfides with the additives $CrCl_3$, $AlCl_3$, $CdCl_2$, or $TeCl_4$. The reactions that occur here are mostly unexplained. In cases of the transport agents $AlCl_3$ and $TeCl_4$, which are prone to hydrolysis, one assumes that hydrogen chloride is formed, which can be effective as a transport medium.

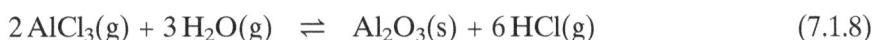

$$2\,AlCl_3(g) + 3\,H_2O(g) \quad \rightleftharpoons \quad Al_2O_3(s) + 6\,HCl(g) \tag{7.1.8}$$

During the CVT of sulfides with halogens or halogen compounds, the corresponding metal halides and sulfur are generally formed as transport effective species. Sulfur halides hardly play a role: sulfur iodides are completely unknown and sulfur bromides exist only at low temperatures. Only gaseous S_2Cl_2 shows sufficient stability to be present during transport reactions. The formation of gaseous metal sulfide halides also plays a subordinate role according to today's level of knowledge. Hence sulfur virtually always appears in elemental form in the gas phase during transport with the elemental halogen. In the temperature range that is often used for transport reactions (around 800 to 1000 °C), sulfur is mostly present as the S_2-molecule. At lower temperatures, the formation of larger sulfur molecules (S_2, S_3 ... S_8) is additionally expected.

During a few transport reactions, the transport agent does not react with the metal atoms of the solid but with the sulfur atoms instead. In particular, hydrogen is one of these transport agents, which can be used successfully for CdS and $ZnS_{1-x}Se_x$. The transport is made possible by the fact that zinc and cadmium, respectively, can be formed elementally in gaseous form during these reactions.

At transport conditions, these metals show a vapor pressure, which is so high that there is no formation of condensed metal during these reactions. The transport can be described by the following transport equation:

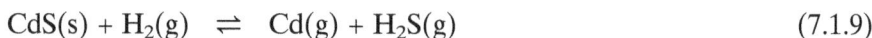

$$CdS(s) + H_2(g) \rightleftharpoons Cd(g) + H_2S(g) \tag{7.1.9}$$

Phosphorus has to be mentioned as a transport agent in this context too. During the reaction with zinc sulfide and cadmium sulfide, phosphorus forms the respective metal vapor and gaseous PS (Loc 2005c).

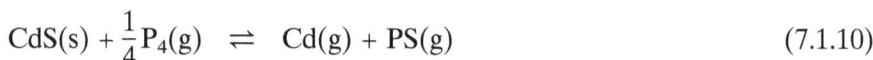

$$CdS(s) + \frac{1}{4}P_4(g) \rightleftharpoons Cd(g) + PS(g) \tag{7.1.10}$$

In some cases (SiS_2, TiS_2, TaS_2), CVT with sulfur as transport additive was successful. The transport effect was ascribed to the formation of gaseous polysulfides (Sch 1982).

Selected examples In the following, the CVT of sulfides shall be explained in more detail with the help of selected examples.

The transport of SnS$_2$ with iodine

The endothermic CVT of SnS_2 with iodine (950 → 600 °C) was first described by *Nitsche* in 1960 (Nit 1960). In the following years, two further works of this group were published (Nit 1961, Gre 1965). They aimed to grow SnS_2 crystals. *Greenway* and *Nitsche* assumed that the transport is caused by the reaction 7.1.11 (Gre 1965).

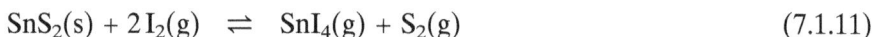

$$SnS_2(s) + 2I_2(g) \rightleftharpoons SnI_4(g) + S_2(g) \tag{7.1.11}$$

This assumption was carried over to the text book *Inorganic Syntheses* (Con 1970). *Wiedemeier et al.* used elemental iodine and also tin(IV)-iodide as transport agents (Wie 1979). If SnI_4 is used as the transport agent, the following transport equation is given:

$$SnS_2(s) + SnI_4(g) \rightleftharpoons 2SnI_2(g) + S_2(g) \tag{7.1.12}$$

Here the transport effective species is tin(II)-iodide. It was shown by experimental examination and thermodynamic considerations that the transport of SnS_2 can be both exothermic and endothermic, dependent on the pressure of the transport agent and the temperature conditions (Wie 1979). *Schäfer* (Sch 1981) arrived at a similar result although he apparently did not know *Wiedemeier*'s work (Wie 1979). The calculation of the partial pressures of the gas species that appear in the system SnS_2/I_2 (Figure 7.1.1), led to the conclusion that during the endothermic transport of SnS_2 with iodine under the conditions described by *Nitsche* SnI_2, not SnI_4, is formed as the dominant transport effective species. SnI_4 is the dominant gas species only at a temperature below approximately 700 °C, as *Greenway* assumed. The temperature dependency of the partial pressure offers an unclear picture, which does not allow conclusions on the transport effect. The introduction of the gas-phase solubility λ(Sn) makes the picture clearer (Figure 7.1.2).

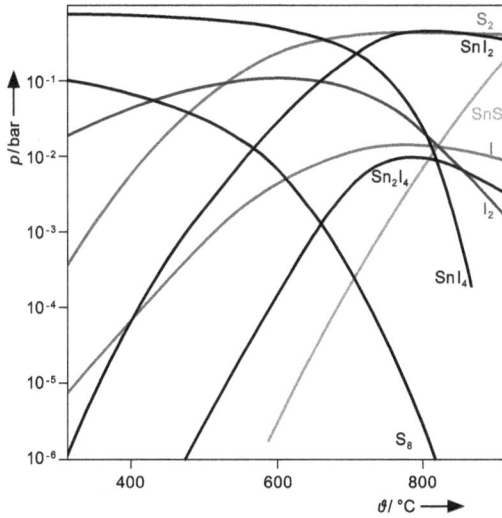

Figure 7.1.1 Calculated temperature dependency of gas species in the SnS_2/I_2 system according to *Schäfer* (Sch 1981).

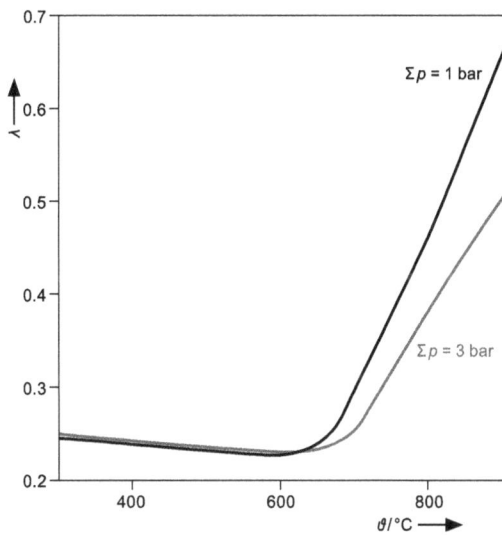

Figure 7.1.2 Calculated temperature dependency of the gas-phase solubility λ of SnS_2 according to *Schäfer* (Sch 1981).

Due to the fact that a transport always runs from the higher to the lower solubility, an exothermic transport $(T_1 \rightarrow T_2)$ with low transport rates is expected below 600 °C. At temperatures above 600 °C, there is an endothermic transport $(T_2 \rightarrow T_1)$ with higher transport rates. The experiments confirm the calculations. *Rao* and *Raeder* (Rao 1995) discussed the substance flows during the transport of SnS_2 with SnI_4 in detail, apparently not knowing *Schäfer*'s work (Sch 1981). The exothermic transport of tin(IV)-sulfide with iodine at temperatures below the temperature of the solubility minimum can be described by equation 7.1.13. Iodine functions as the transport agent; the transport effective gas species is SnI_4. Hence a transport to the hotter zone is expected. Above the temperature of solubility minimum up to around 900 °C, iodine functions as transport agent; SnI_2 is the transport effective gas species (7.1.14). The partial pressure of iodine decreases with rising temperature; that of SnI_2 increases according to the endothermic reaction 7.1.14. At even higher temperatures (above approximately 900 °C), SnS_2 is transferred to the less hot zone mainly by decomposition sublimation (7.1.15).

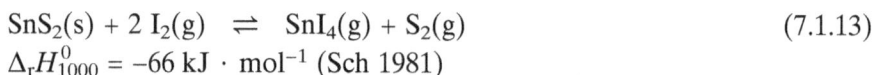

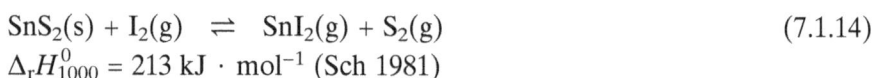

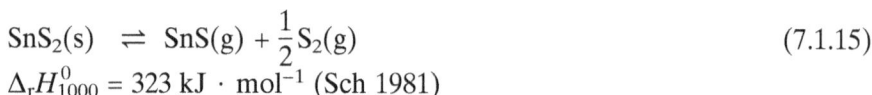

$$SnS_2(s) + 2\ I_2(g) \ \rightleftharpoons \ SnI_4(g) + S_2(g) \qquad (7.1.13)$$
$$\Delta_r H^0_{1000} = -66\ kJ \cdot mol^{-1}\ (Sch\ 1981)$$

$$SnS_2(s) + I_2(g) \ \rightleftharpoons \ SnI_2(g) + S_2(g) \qquad (7.1.14)$$
$$\Delta_r H^0_{1000} = 213\ kJ \cdot mol^{-1}\ (Sch\ 1981)$$

$$SnS_2(s) \ \rightleftharpoons \ SnS(g) + \frac{1}{2}S_2(g) \qquad (7.1.15)$$
$$\Delta_r H^0_{1000} = 323\ kJ \cdot mol^{-1}\ (Sch\ 1981)$$

The early examinations of this system show that it is possible to grow single-crystals of SnS_2 without detailed knowledge of the appearing gas species and the reactions occurring. *Wiedemeier, Schäfer,* and *Rao*'s works allow a deeper understanding of the ongoing reactions; in particular, the helpfulness of the term *gas-phase solubility* becomes clear. By using it, transport reactions that involve numerous gas species, can be well and clearly described.

The transport of FeS$_x$

"FeS" is a compound with a considerable homogeneity range. In the literature, it is often described as FeS_{1+x}, FeS_x, or $Fe_{1-x}S$. The phase range of "FeS" is dependent on the temperature and the sulfur partial pressure over an "FeS"-solid. The temperature dependency of the homogeneity range ($p(S_x)$ = const.) is described by the phase diagram (Figure 7.1.3) (Mas 1990). Figure 7.1.4 clarifies the dependency of the homogeneity range on the temperature and the sulfur pressure.

Schäfer (Sch 1962, Sch 1971), *van den Berg* (Van 1969) *et al.*, and *Gibart et al.* (Gib 1969) reported on the transport of "FeS" with iodine in early times. Only in *Gibart*'s work, the phase range of "FeS" is discussed in the context of its transport. *Krabbes, Oppermann,* and *Wolf* would eventually show that the transport does not always succeed under the conditions given in the literature (Kra 1975). On this occasion, they dealt with the problem of the transport of a compound with a homogeneity range on the basis of thermodynamic criteria. The composition of the gas phase over the solid with a composition between $FeS_{1.0}$ and $FeS_{1.17}$ (Figure 7.1.5) follows the thermodynamic functions, which are dependent on the composition within the phase range (see section 2.4).

When iodine is added, the gas phase over FeS_x basically contains FeI_2, Fe_2I_4, FeI_3, I, I_2, and S_2. Their partial pressures are dependent on the temperature and the composition of the solid. At 1000 °C, sulfur gets to the gas phase in noteworthy scale only in the case of solids with $x > 0.05$; the sulfur pressure is very

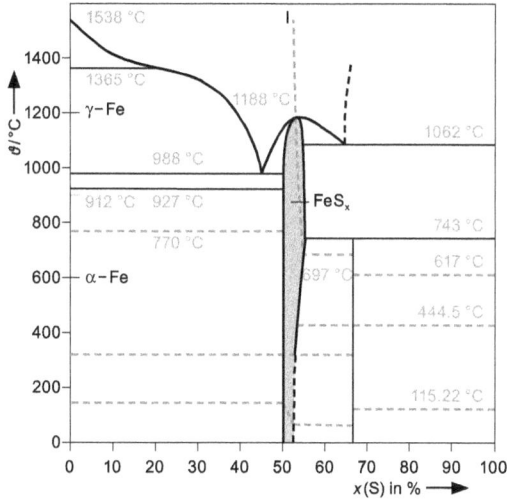

Figure 7.1.3 Phase diagram of the iron/sulfur system according to *Massalski* (Mas 1990).

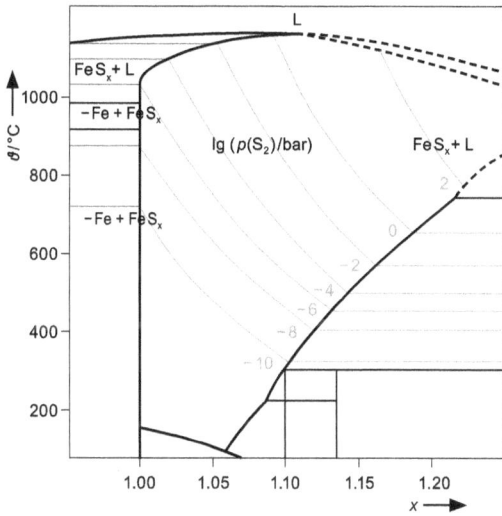

Figure 7.1.4 Section of the phase diagram of the iron/sulfur system for different temperatures and sulfur pressures according to *Krabbes et al.* and *Wolf et al.* (Kra 1975, Wol 1978).

low for stochiometrically composed $FeS_{1.0}$. Iron is transferred to the gas phase in the form of its iodides (FeI_3, FeI_2, Fe_2I_4; Figure 7.1.5) during a reaction of the source solid $FeS_{1.0}$ with iodine ($\vartheta \approx 1000\,°C$); however, sulfur is released in the process. Instead a sulfur-rich solid $FeS_{x+\delta}$ is formed. The transport of FeS is not possible because the gas phase does not contain transport effective amounts of sulfur. This is in agreement with the experimental findings – $FeS_{1.0}$ cannot be transported with iodine (Kra 1976a). The calculated transport rates of $FeS_{1.0}$ and $FeS_{1.1}$ are illustrated in Figure 7.1.6. One can see that the composition of the solid massively influences the transport rate.

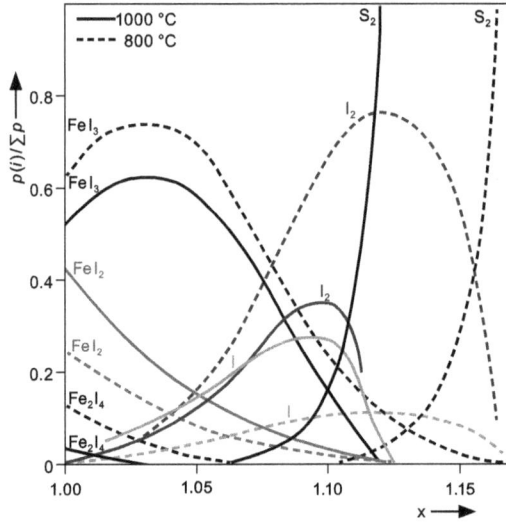

Figure 7.1.5 Normalized partial pressure in the FeS_x/I_2 system as a function of x at 800 and 1000 °C according to *Krabbes et al.* and *Wolf et al.* (Kra 1975, Wol 1978).

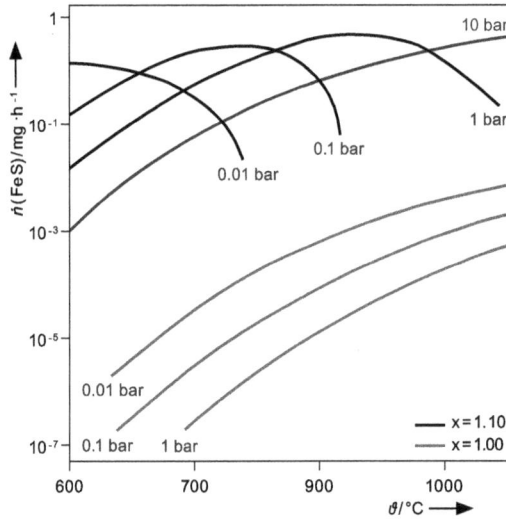

Figure 7.1.6 Calculated transport rates of solids of the composition $FeS_{1.0}$ and $FeS_{1.1}$ at different total pressures according to *Krabbes et al.* (Kra 1975).

However, *Krabbes, Oppermann,* and *Wolf* showed that the transport of $FeS_{1.0}$ succeeds with other transport agents (HCl, HBr, HI, GeI_2, GeI_4, $GeCl_2$) (Kra 1976b). In these cases, sulfur is not present in the gas phase in elemental form but as H_2S or GeS respectively. Thus the solubility of sulfur increases by some orders of magnitude, and the transport succeeds with transport rates of some milligrams per hour. The transport of hydrogen halide can be advantageous for sulfur-poor compounds when there is no transport effective solution due to the low partial pressure of sulfur.

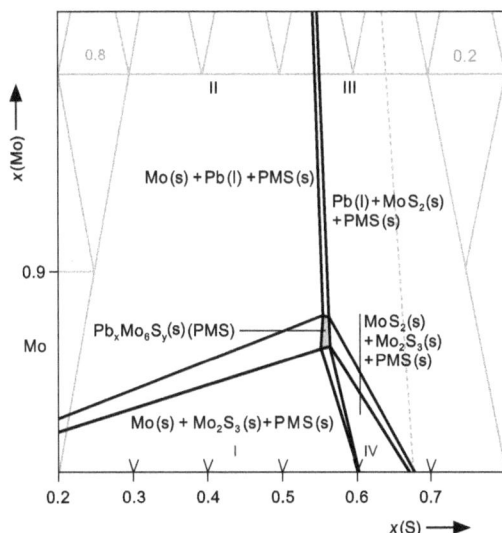

Figure 7.1.7 Section from the Pb/Mo/S phase diagram according to *Krabbes* and *Oppermann* (Kra 1981).

Transport of $Pb_xMo_6S_y$

Chevrel phases of the type $M_nMo_6Q_8$ (M = Pb, Ca, Sr, Ba, Sn, Ni; Q = S, Se, Te) are of particular interest due to their superconductive character and particularly high critical magnetic field strength. The examinations of these electric characteristics were conducted and published, apart from a few exceptions, on polycrystalline materials due to the lack of suitable single-crystals. Hence the interest in single-crystals of representatives of this compound class was huge. *Krabbes* and *Oppermann* succeeded in growing single-crystals of one representative of this compound class, "$PbMo_6S_8$" (PMS), with the help of directed transport reactions (Kra 1981). Br_2 and/or $PbBr_2$ were used as transport agents, the transport was endothermic at temperatures around 1000 °C. "$PbMo_6S_8$" is a compound with a homogeneity range of $Pb_xMo_6S_y$ ($0.9 \leq x \leq 1.1$; $7.6 \leq y \leq 7.9$). The existence area of this compound is small; it is surrounded by four three-phase areas (I to IV). The phase relations are illustrated in the form of an isothermal section (T = 1250 K) in Figure 7.1.7.

By estimating the thermodynamic data of PMS and taking into account known data of the remaining condensed and gaseous substances involved in the transport reaction, the transport behavior in the system Pb/Mo/S could be calculated. In doing so, the deposition is possible only within a certain area of the composition in the homogeneity area of PMS. If the composition of the source solid is within the light grey marked part of the homogeneity range of PMS in the two-phase area of PMS + Mo, phase-pure PMS can be obtained in the sink (Figure 7.1.8). If the composition of the source solid is in one of the areas A, B, C, or D, there will a deposition of two phases adjacent to each other in the sink (A: PMS + Mo; B: PMS + Pb; C: PMS + MoS_2; D: PMS + Mo_2S_3). If the source solid is set within the dark grey marked areas, the deposition of a three-phase solid will be expected.

This example impressively shows that, even with complex phase relations, defined substances can be obtained free from phase impurities in crystalline, often

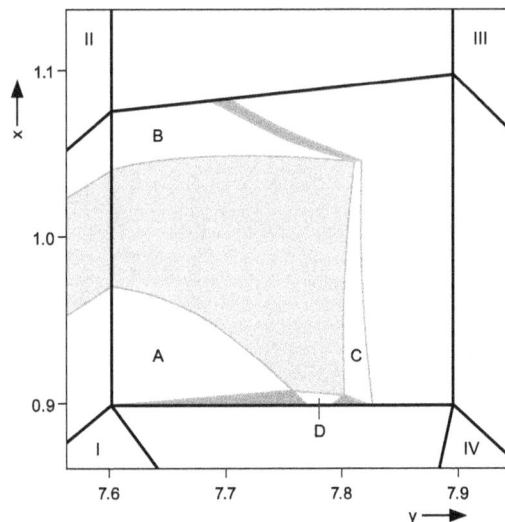

Figure 7.1.8 Calculated transport behavior in the PMS/Br$_2$ system (1 bar Br$_2$ 1100 → 1000 °C) according to *Krabbes* and *Oppermann* (Kra 1981).

even in single-crystalline, form with the help of CVT reactions. Nevertheless, a substantiated thermodynamic dealing with the problem is generally necessary to succeed in complex cases such as this one.

Nitsche made some general, qualitative considerations of CVT of ternary sulfides and described them illustratively (Nit 1971).

Transport of ZnS$_{1-x}$Se$_x$

Solid solutions are of interest because the physical characteristics of these materials can be influenced in a directed way by a defined change of composition. Especially semi-conductor materials have been examined in detail in this context. Chemical vapor transport reactions are well suited for preparing these substances in the form of homogeneously composed, large crystals (Bin 2009). The cubic mixed-phases in the system ZnS/SnSe were examined in particularly detail. Zinc sulfide and zinc selenide are completely mixable with each other in the solid state. The CVT of the binary compounds is well examined. Both can be transported under virtually the same conditions. Hence it is expected that the mixed-phases can be obtained this way, too.

Catano and *Kun* reported on the growing of single-crystals with edge lengths of more than one centimeter (transport agent iodine), as well as on the content of foreign atoms before and after the transport process (Cat 1976). *De Murcia et al.* used transport with hydrogen in order to grow epitactic layers of ZnS$_{1-x}$Se$_x$ on a calcium fluoride substrate (DeM 1979). The study of the characteristics of photoluminescence of ZnS$_{1-x}$Se$_x$ is the emphasis of another work (Hig 1980). Furthermore, it reported on the formation of epitactic layers of ZnS$_{1-x}$Se$_x$ on a gallium arsenide substrate. Different transport agents were used to optimize the transport (Har 1987). *Matsumo et al.* transported mixed-phases in the system ZnS/ZnSe with the help of ammonium chloride (HCl) as transport additive, and

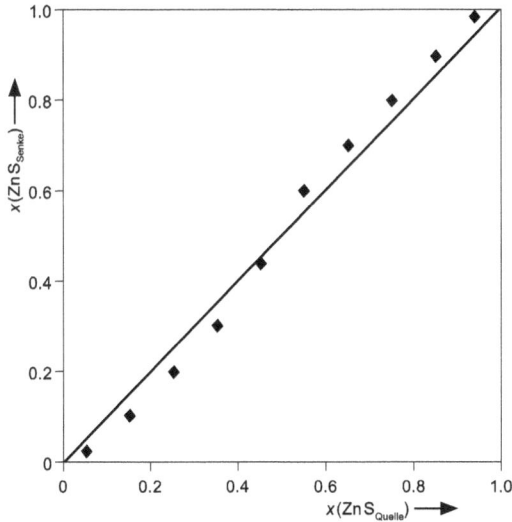

Figure 7.1.9 Relation between the composition of the solid in the source and sink during the transport of $ZnS_{1-x}Se_x$ mixed phases with iodine according to *Hotje et al.* (Hot 2005e).

reported, among other things, that their photoluminescence was dependent on the composition (Mat 1988). The same objective was pursued in work in which hydrogen was used as transport agent (Kor 2002). The nucleation kinetics of $ZnS_{1-x}Se_x$ were examined during transport with iodine (Sen 2004). *Hotje* and *Binnewies* introduced a thermodynamic model for the stability of $ZnS_{1-x}Se_x$ mixed-phases and described expected enrichment effects by the transport process (Hot 2005e). The experimentally determined composition of the solid in the source and sink are illustrated in Figure 7.1.9. It showed that there are no significant changes of the composition during the transport process in this case.

Transport of CdS with hydrogen

Transport reactions in which hydrogen is used as transport agent are unusual. The transport agent reacts with sulfur atoms of the solid under formation of hydrogen sulfide. A substance transport becomes possible by the fact that elemental cadmium, which is formed in the process, is present in the gaseous state at transport temperatures. Transport reactions in which the transport agent reacts solely with the non-metal of the solid are exceptions. They can only be successful if the respective metal has a transport effective vapor pressure. The transport of zinc sulfide with phosphorus, which was described by *Locmelis* and *Binnewies*, has to be mentioned in this context. Here the transport agent likewise reacts with sulfur atoms and gaseous PS is formed (Loc 2005c).

Ito and *Matsuura* use the transport of cadmium sulfide with hydrogen to deposit epitactic layers on a gallium phosphide substrate (Ito 1979). *Attolini et al.* (Att 1982, Att 1983) show that transport with hydrogen is also suited to growing larger single-crystals; in the transport with hydrogen, compared to iodine, they see the advantage that contamination by the transport agent is excluded.

Table 7.1.2 Examples of the chemical vapor transport of sulfides.

Sink solid	Transport additive	Temperature / °C	Reference
Ag_2S	I_2	$825 \rightarrow 700$	Mus 1975
$Ag_{1-x}Al_{1-x}Sn_{1+x}S_4$	I_2, $AlCl_3$	$850 \rightarrow 750$	Mäh 1984
$Ag_{1-x}Cr_{1-x}Sn_{1+x}S_4$	I_2, $AlCl_3$	$850 \rightarrow 750$	Mäh 1984
$AgGaS_2$	I_2	$840 \rightarrow 740$	Hon 1971
	I_2	$925 \ldots 975 \rightarrow 825 \ldots 900$	Nod 1990b
	I_2	$930 \rightarrow 890$	Bal 1994
	$AgCl$, $CdBr_2$	$925 \ldots 975 \rightarrow T_1$	Nod 1991
		$(\Delta T = 25 \ldots 100)$	
	I_2, Br_2, Cl_2	$975 \rightarrow 950$	Nod 1998
$AgGaSeS$	I_2	$935 \rightarrow 855$	Bal 1994
$Ag_{0.5}In_{0.5}Cr_2S_4$	not specified	$800 \rightarrow T_1$	Phi 1971
$AgIn_2S_4$	I_2	not specified	Bri 1985
	I_2	$780 \rightarrow 660$	Jos 1981
$AgIn_5S_8$	I_2	$740 \ldots 850 \rightarrow 700 \ldots 750$	Pao 1977a
	I_2	$740 \ldots 830 \rightarrow 750$	Pao 1980
Al_2S_3	I_2	not specified	Hel 1979
	I_2	$860 \rightarrow 750$	Kre 1993
$AlInS_3$	I_2	$800 \rightarrow 750$	Schu 1979
$Al_2In_4S_9$	Br_2	$800 \rightarrow 750$	Schu 1982
$BaGaS_4$	I_2	$970 \rightarrow 780$	Cho 2005
$BaGaS_4{:}Ho$	I_2	$910 \rightarrow 790$	Cho 2005
$BaIn_2S_4$	I_2	$750 \ldots 900 \rightarrow 650 \ldots 800$	Gul 1992
Bi_2S_3	I_2	$745 \rightarrow 670$	Car 1972
	I_2	$680 \rightarrow 600$	Krä 1976
$Bi_2In_4S_9$	I_2	$680 \rightarrow 600$	Cha 1972
$BiPS_4$	I_2	$660 \rightarrow 610$	Nit 1970
$(BiS)_xVS_2$	NH_4Cl	$900 \rightarrow 700 \ldots 850$	Got 2004
$(BiS)_xNbS_2$	NH_4Cl	$900 \rightarrow 700 \ldots 850$	Got 2004
$(BiS)_xTaS_2$	NH_4Cl	$900 \rightarrow 700 \ldots 850$	Got 2004
$(BiS)_{1.16}VS_2$	NH_4Cl	$850 \rightarrow 700$	Got 1995
CaS	I_2	$1200 \rightarrow 800$	Bri 1982
$CaAl_2S_4$	I_2	$960 \rightarrow 870$	Oh 1997a
$CaAl_2S_4{:}Er^{3+}$	I_2	$950 \rightarrow 870$	Oh 1999
$CaGa_2S_4$	I_2	$750 \rightarrow 580 \ldots 650$	Gul 1992
$CaIn_2S_4$	I_2	$700 \ldots 1000 \rightarrow 580 \ldots 800$	Gul 1992
$Ca_{3.1}In_{6.6}S_{13}$	I_2	$900 \rightarrow 750$	Cha 1971
CdS	H_2	$T_2 \rightarrow 350$	Ito 1979
	H_2	$T_2 \rightarrow T_1$	Att 1982
	H_2	$980 \rightarrow 910$	Att 1983
	I_2	$850 \rightarrow 650$	Nit 1960
	I_2	$850 \ldots 900 \rightarrow 650 \ldots 870$	Nit 1961
	I_2	$T_2 \rightarrow T_1$	Nit 1962

Table 7.1.2 (continued)

Sink solid	Transport additive	Temperature / $°C$	Reference
CdS	I_2	$900 \rightarrow 700$	Beu 1962
	I_2	$1000 \rightarrow 700$	Bon 1970
	I_2	$900 \rightarrow 870$	Pao 1977b
	I_2	$600 \dots 1000 \rightarrow 400 \dots 800$	Mat 1983
	I_2	$900 \rightarrow 700$	Shi 1988
	I_2	$T_2 \rightarrow T_1$	Shi 1989
	HCl	$1100 \rightarrow 700$	Uji 1976
	P_4	$1000 \rightarrow 900$	Loc 2005c
$CdS_{1-x}Te_x$	I_2	$900 \rightarrow 800$	Hot 2005c
	I_2	$780 \rightarrow 740$	Nit 1967a
$CdAl_2S_4$	I_2	$950 \rightarrow 850$	Oh 1997b
	$AlCl_3$	$830 \rightarrow 760, 675 \rightarrow 650$	Kra 1997
$CdAl_2S_4:Co^{2+}$	I_2	$950 \rightarrow 850$	Oh 1997b
$CdAl_2S_4:Er^{3+}$	I_2	$950 \rightarrow 850$	Oh 1997b
$CdCr_2S_4$	Cl_2	$775 \dots 900 \rightarrow 725 \dots 800$	Pin 1970
	Cl_2	$825 \rightarrow 775$	Phi 1971
	Cl_2	$1000 \rightarrow 960$	Wid 1971
	$AlCl_3$	$1000 \rightarrow 850$	Lut 1968
	$AlCl_3$	$1000 \rightarrow 800$	Lut 1970
	$AlCl_3$	$1000 \rightarrow 800$	Phi 1971
	$CrCl_3$	$950 \rightarrow 900$	Bar 1972
	$CrCl_3$	$1050 \rightarrow 800$	Oka 1974
	$CrCl_3$	$1000 \rightarrow 900$	Rad 1980
	$HCl + Cl_2$	$1100 \rightarrow 1030$	Gib 1974
$CdCr_2S_{4-x}Se_x$	$I_2, AlCl_3$	$900 \rightarrow 700$	Phi 1971
	$I_2, AlCl_3$	$900 \rightarrow 700$	Pic 1970
$CdCr_{2-x}In_xS_4$	$CrCl_3, CdCl_2$	$1050 \rightarrow 700$	Phi 1971
$Cd_{1-x}Fe_xCr_2S_4$	$CrCl_3$	$950 \rightarrow 900$	Bar 1974
$Cd_{1-x}Fe_xS$	I_2	$965 \rightarrow 870$	Dic 1990
$CdGa_2S_4$	I_2	$650 \rightarrow 600$	Nit 1961
	I_2	$650 \rightarrow 600$	Beu 1961
	not specified	not specified	Ant 1968
	I_2	$650 \rightarrow 600$	Cur 1970
	I_2	$880 \rightarrow 840$	Wu 1988
	I_2	$810 \dots 860 \rightarrow 730 \dots 760$	Bod 2004
$CdGaCrS_{4-x}Se_x$	$CdCl_2, CrCl_3$	$1000 \rightarrow 900$	Sag 2004
Cd_4GeS_5	I_2	$750 \rightarrow 700$	Nit 1964
Cd_4GeS_6	Cl_2	$580 \rightarrow 510$	Que 1975
	I_2	$900 \rightarrow 800$	Kal 1965
	I_2	$850 \rightarrow 750$	Nit 1967a
	I_2	$750 \dots 800 \rightarrow 700 \dots 790$	Kal 1974
	I_2	$580 \rightarrow 510$	Que 1975
$CdIn_2S_4$	I_2	$1000 \rightarrow 600$	Nit 1960
	I_2	$850 \rightarrow 750$	Nit 1961
	I_2	$850 \rightarrow 750$	Cur 1970
	I_2	$1000 \rightarrow 600$	Val 1970
	I_2	$830 \rightarrow 780$	Taf 1996b

Table 7.1.2 (continued)

Sink solid	Transport additive	Temperature / °C	Reference
$CdIn_2S_4$	I_2	$770 \ldots 830 \rightarrow T_1$	Pao 1982
	I_2	$800 \ldots 850 \rightarrow 750 \ldots 800$	Ven 1986b
	I_2	$650 \ldots 850 \rightarrow 630 \ldots 820$	Neu 1989
$CdIn_2S_4$:Co	I_2	$850 \rightarrow 600$	Lee 2003
$CdInGaS_4$:Er^{3+}	I_2	not specified	Cho 2001
$Cd_{1-x}Mn_xIn_2S_4$	I_2	$950 \ldots 1000 \rightarrow 875 \ldots 900$	Del 2006
$Cd_2P_2S_6$	I_2	$630 \rightarrow 600$	Nit 1970
$CdSc_2S_4$	I_2, HCl	not specified	Yim 1973
Cd_4SiS_6	I_2	$450 \rightarrow 380$	Que 1975
	I_2	$830 \rightarrow 800$	Nit 1967a
CoS	I_2, NH_4I, GeI_2	$800 \ldots 1100$ $\rightarrow 700 \ldots 1000$	Kra 1978
	$AlCl_3$	$880 \rightarrow 800$	Lut 1970
CoS_2	Cl_2	$730 \rightarrow 695$	Bou 1968
	$AlCl_3$	$880 \rightarrow 800$	Lut 1970
$Co_xCd_yCr_2S_4$	$AlCl_3$	$990 \ldots 1020 \rightarrow 870 \ldots 920$	Lut 1989
$CoCr_2S_4$	I_2	$1150 \rightarrow 1000$	Nit 1967a
	I_2	$1150 \rightarrow 1000$	Phi 1971
	NH_4Cl	$1150 \rightarrow 1000$	Cur 1970
	$HCl + Cl_2$	$1040 \rightarrow 950$	Gib 1974
	$CrCl_3$	$960 \ldots 1080$ $\rightarrow 930 \ldots 1035$	Wat 1972
	$CrCl_3$, $CuCl_2$	$1070 \rightarrow 1020, 1000 \rightarrow 960$	Wat 1978
$Co_{1-x}Fe_xS$	I_2, NH_4I, GeI_2	$920 \rightarrow 850$	Kra 1979
	GeI_2	$920 \rightarrow 850$	Kra 1984b
$CoIn_2S_4$	I_2	$850 \rightarrow 800$	Nit 1967a
	I_2	$850 \rightarrow 800$	Cur 1970
$Co_{0.86}In_{2.09}S_4$	$I_2 + AlCl_3$	$900 \rightarrow 780$	Lut 1989
$Co_{1-x}Ni_xS_2$	Br_2	$750 \rightarrow 700$	But 1971
$Co_{1-x}Fe_xS$	GeI_2	$950 \rightarrow 900$	Kra 1984b
	NH_4Cl	$1050 \rightarrow 1030 \ldots 1035$	Gu 1990
CrS	$AlCl_3$	$1000 \rightarrow 980$	Lut 1970
Cr_2S_3	Br_2	$920 \rightarrow 800$	Nit 1967a
	I_2	not specified	Dis 1970
	$AlCl_3$	not specified	Lut 1968
	$AlCl_3$	$1000 \rightarrow 920$	Lut 1970
	$AlCl_3$	$900 \ldots 1000 \rightarrow 800 \ldots 850$	Lut 1974
	$CrCl_3$	$1000 \rightarrow 900$	Nak 1978
$Cr_2S_{3-x}Se_x$	$CrCl_3$, $AlCl_3$	$930 \ldots 1000 \rightarrow 760 \ldots 850$	Lut 1973b
Cr_3S_4	$AlCl_3$	$1000 \rightarrow 920$	Lut 1970
	$CrCl_3$, $AlCl_3$	$1000 \rightarrow 920$	Lut 1973a
	$AlCl_3$, $CrCl_3$	$1000 \rightarrow 920$	Lut 1974
Cr_5S_6	$AlCl_3$	$1000 \rightarrow 950$	Lut 1970
Cr_7S_8	$AlCl_3$	$1000 \rightarrow 920 \ldots 980$	Lut 1970
$CrDyS_3$	I_2	$1100 \ldots 1180$ $\rightarrow 1000 \ldots 1080$	Kur 1980

Table 7.1.2 (continued)

Sink solid	Transport additive	Temperature / °C	Reference
$CrErS_3$	I_2	1100 ... 1180 $\rightarrow$ 1000 ... 1080	Kur 1980
$CrGdS_3$	I_2	1100 ... 1180 $\rightarrow$ 1000 ... 1080	Kur 1980
$CrTbS_3$	I_2	1100 ... 1180 $\rightarrow$ 1000 ... 1080	Kur 1980
$CrTmS_3$	I_2	1100 ... 1180 $\rightarrow$ 1000 ... 1080	Kur 1980
$CrYbS_3$	I_2	1100 ... 1180 $\rightarrow$ 1000 ... 1080	Kur 1980
CuS	HBr	470 $\rightarrow$ 430	Rab 1967
Cu_3PS_4	I_2	825 $\rightarrow$ 775	Mar 1983
Cu_3PS_3Se	I_2	825 $\rightarrow$ 775	Mar 1983
$CuAlS_2$	I_2	860 ... 950 $\rightarrow$ 870	Pao 1980
	I_2	1020 ... 1070 $\rightarrow$ 920 ... 1000	Bod 1984
	I_2	900	Ohg 1994
	I_2	not specified	Lip 1993
	I_2	950 $\rightarrow$ 900	Moc 2001
$CuAl_2S_4$	I_2	800 $\rightarrow$ 700	Hon 1969
	I_2	950 $\rightarrow$ 1050	Bri 1975
	I_2	not specified	Aks 1988a
$CuAl_{1-x}Ga_xS_2$	I_2	1000 $\rightarrow$ 965	He 1988
	I_2	not specified	Bod 1998
$Cu(Al_{1-x}Ga_x)S_{2-y}Se_y$	I_2	885 $\rightarrow$ 800	Chi 1994
$CuAl_{1-x}In_xS_2$	I_2	880 $\rightarrow$ 730	Aks 1988b
$CuAl_{1-x}In_xS_2$	I_2	$T_2 \rightarrow T_1$ ($\Delta T = 80 ... 100$)	Bod 1996
$Cu_{0.5}Al_{2.5}S_4$	$TeCl_4$, $AlI_3 + I_2$	850 $\rightarrow$ 750	Mäh 1982
$Cu_{1-x}Al_{1-x}Sn_{1+x}S_4$	I_2, $AlCl_3$	850 $\rightarrow$ 750	Mäh 1984
$Cu_{1-x}Al_xCr_2S_4$	$CuCl_2$	not specified	Phi 1971
Cu_3BiS_3	HI	440 $\rightarrow$ 400	Mar 1998
$Cu_4Bi_4S_9$	I_2	440 $\rightarrow$ 410	Kry 2007
Cu_2CdGeS_4	I_2	800 $\rightarrow$ 750	Nit 1967b
Cu_2CdGeS_7	I_2	750 $\rightarrow$ 700	Fil 1991
Cu_2CdSiS_4	I_2	800 $\rightarrow$ 750	Nit 1967b
Cu_2CdSnS_4	I_2	800 $\rightarrow$ 750	Nit 1967b
$Cu_2CeNb_2S_5$	I_2	930 $\rightarrow$ 880	Ohn 1996
$CuCrSnS_4$	$TeCl_4$, $AlI_3 + I_2$	900 $\rightarrow$ 800	Mäh 1982
$Cu_{1-x}Cr_{1-x}Sn_{1+x}S_4$	I_2, $AlCl_3$	850 $\rightarrow$ 750	Mäh 1984
$CuCrZrS_4$	$TeCl_4$, $AlI_3 + I_2$	850 $\rightarrow$ 750	Mäh 1982
$Cu_{1-x}Fe_xCr_2S_4$	HCl	800 $\rightarrow$ 725	Phi 1971
Cu_2FeGeS_4	I_2	800 $\rightarrow$ 750	Nit 1967b
	I_2	800 $\rightarrow$ 750	Nit 1967b
$Cu_2Fe_{1-x}Ge_xS_2$	I_2	825 ... 850 $\rightarrow$ 780 ... 790	Ack 1976
Cu_2FeSnS_4	I_2	800 $\rightarrow$ 750	Nit 1967b

Table 7.1.2 (continued)

Sink solid	Transport additive	Temperature / °C	Reference
CuGaS$_2$	I$_2$	850 → 700	Hon 1971
	I$_2$	1000 → 940	Bod 1976
	I$_2$	900 ... 950 → 910	Pao 1978
	I$_2$	1020 ... 1070 → 920 ... 1000	Bod 1984
	I$_2$	900 → 750	Tan 1989
	I$_2$	840 → 750	Bal 1994
	I$_2$	900 → 850	Pra 2007
CuGa$_2$S$_4$	I$_2$	900 → 750 ... 800	Yam 1975
	I$_2$	900 ... 950 → 910	Pao 1980
	I$_2$	800 ... 850 → 750 ... 800	Bin 1983
	I$_2$	1000 → 965	He 1988
CuGaS$_{2-x}$Se$_x$	I$_2$	980 ... 1000 → 940 ... 970	Bod 1976
	I$_2$	850 ... 900 → 650 ... 700	Kim 1993
Cu$_{1-x}$Ga$_x$Cr$_2$S$_4$	CuCl$_2$	not specified	Phi 1971
Cu$_2$GeS$_3$	I$_2$	750 → 700	Nit 1967a
CuInS$_2$	I$_2$	880 ... 940 → 890	Pao 1978
	I$_2$	820 → 740	Hwa 1978
	I$_2$	950 → 940	Pro 1979
	I$_2$	not specified	Hwa 1980
	I$_2$	800 → 600	Lah 1981
	I$_2$	1020 ... 1070 → 920 ... 1000	Bod 1984
	I$_2$	not specified	Ven 1986a
	I$_2$	810 → 790	War 1987
	I$_2$	830 ... 850 → 800	Bal 1990
	I$_2$	830 → 800	Bal 1994
	I$_2$	850 → 800	Taf 1996a
	I$_2$	800 → 750	Schö 1999
	I$_2$	850 → 750	Sav 2000
	I$_2$	850 → 830	Tsu 2006
CuInSeS	I$_2$	830 → 800	Bal 1990
	I$_2$	950 → 900	Bal 1994
CuIn$_2$S$_4$	I$_2$	900 ... 950 → 910	Pao 1980
	I$_2$	780 ... 835 → T_1	Pao 1982
	I$_2$	800 ... 850 → 750 ... 800	Bin 1983
CuInS$_{2-x}$Se$_x$	I$_2$	780 ... 920 → 700 ... 860	Bod 1980
	I$_2$	950 → 900	Bal 1989
	not specified	not specified	Bod 1997
	not specified	not specified	Lop 1998
CuIn$_5$S$_8$	I$_2$	740 ... 830 → 750	Pao 1980
Cu$_{1-x}$In$_x$Cr$_2$S$_4$	CuCl$_2$	not specified	Phi 1971
CuInSnS$_4$	TeCl$_4$	830 → 720	Mäh 1982
Cu$_{1-x}$InSnS$_4$	TeCl$_4$, AlCl$_3$ + I$_2$	850 → 750	Mäh 1982
Cu$_2$MnGeS$_4$	I$_2$	800 → 750	Nit 1967b
Cu$_x$Nb$_{1+y}$S$_2$	I$_2$	1050 → 950	Har 1989

Table 7.1.2 (continued)

Sink solid	Transport additive	Temperature / °C	Reference
Cu_3NbS_4	I_2	$780 \rightarrow 700$	Nit 1967a
$Cu_xNb_yS_z$	I_2	$800 \dots 850 \rightarrow 750 \dots 780$	Nit 1968
$Cu_xNb_yS_2$	I_2	$1050 \rightarrow 950$	Har 1989
Cu_2NiGeS_4	I_2	$800 \rightarrow 750$	Nit 1967b
Cu_3PS_4	I_2	$850 \rightarrow 800$	Nit 1970
Cu_3SmS_3	I_2	$1050 \rightarrow 950$	Ali 1972
Cu_3TaS_4	I_2	$780 \rightarrow 700$	Nit 1967a
$Cu_xTa_{1+y}S_2$	I_2	$1050 \rightarrow 950$	Har 1989
$CuTi_2S_4$	$TeCl_4$, $AlI_3 + I_2$	$900 \rightarrow 800$	Mäh 1982
$Cu_2U_3S_7$	UBr_4	$600 \rightarrow 540$	Dao 1996
CuV_2S_4	$TeCl_4$	not specified	Cra 2005
	$TeCl_4$	$830 \rightarrow 720$	Mäh 1982
	Cl_2	$760 \rightarrow 690$	LeN 1979
$CuVTiS_4$	$TeCl_4$	$900 \rightarrow 800$	Mäh 1982
$Cu_2Zn_{1-x}Mn_xGeS_4$	I_2	$800 \rightarrow 850, 800 \rightarrow 750$	Hon 1988
Cu_2ZnGeS_4	I_2	$800 \rightarrow 750$	Nit 1967b
Cu_2ZnSiS_4	I_2	$800 \rightarrow 750$	Nit 1967b
Cu_2ZnSnS_4	I_2	$800 \rightarrow 750$	Nit 1967b
Er_2CrS_4	I_2	$1000 \rightarrow 900$	Vaq 2009
Er_3CrS_6	I_2	$1000 \rightarrow 900$	Vaq 2009
Er_4CrS_7	I_2	$1000 \rightarrow 900$	Vaq 2009
$Er_6Cr_2S_{11}$	I_2	$1000 \rightarrow 900$	Vaq 2009
EuS	I_2	$T_2 \rightarrow 1700$	Kal 1969
	S	$2050 \rightarrow T_1$	Kal 1972
	S	not specified	Kal 1974
$EuSb_4S_7$	I_2	$980 \rightarrow 870$	Ali 1978
Fe_2GeS_4	I_2	$1050 \rightarrow 1000$	Nit 1967a
FeS	$TeCl_4$, I_2, $FeCl_3$	$880 \rightarrow 700$, $900 \rightarrow 700 \dots 800$, $800 \rightarrow 700$	Mer 1973
	I_2		Kra 1976a
	HCl, HBr, NH_4Cl, NH_4Br, NH_4I, GeI_2, GeI_4, $GeCl_2$	$750 \dots 1050$ $\rightarrow 700 \dots 1000$	Kra 1976b
	HCl, HBr, HI, $GeCl_2$, $GeCl_4$	$700 \dots 1100$ $\rightarrow 600 \dots 1000$	Wol 1978
	I_2, HCl, NH_4Cl, GeI_2, GeI_4	$850 \rightarrow 800, 850 \rightarrow 800$, $900 \rightarrow 800$, $900 \rightarrow 800$	Kra 1984a
FeS_x	I_2		Kra 1975
	I_2, Cl_2, HCl		Ber 1976
	I_2	$850 \dots 1100$ $\rightarrow 750 \dots 1050$	Wol 1978
$Fe_{1-x}Mn_xS$	I_2	$900 \rightarrow 800$	Kni 2000

Table 7.1.2 (continued)

Sink solid	Transport additive	Temperature / °C	Reference
$Fe_{1-x}Ni_xS$	GeI_2	$870 \rightarrow 780$	Kra 1984b
$Fe_{1-x}Zn_xS$	I_2	$900 \rightarrow 800$	Kni 2000
FeS_2	Cl_2	$715 \rightarrow 655$	Bou 1968
	Cl_2	$705 \rightarrow 665$	Yam 1974
	Cl_2	$705 \rightarrow 665$	Yam 1979
	I_2, Cl_2	$700 \rightarrow 600, 650 \rightarrow 550$	Kra 1984a
	Cl_2, $Cl_2 + H_2$, ICl_3, HCl, NH_4Cl, Br_2, NH_4Br, I_2	$630 \rightarrow 580 \dots 700$	Fie 1986
	Br_2	$T_2 \rightarrow 370 \dots 500$	Ble 1992
	Cl_2, Br_2	$630 \dots 700 \rightarrow 580 \dots 680$	Tom 1995
$FeCr_2S_4$	Cl_2	$875 \rightarrow 850$	Gib 1969
	$CrCl_3$	$905 \rightarrow 855$	Phi 1971
	Cl_2, $HCl + Cl_2$	not specified	Gol 1973
	not specified	$800 \rightarrow 750$	Ish 1978a
	not specified	$800 \rightarrow 750$	Ish 1978b
	$CrCl_3$	not specified	Ish 1978c
	I_2	$1050 \rightarrow 1000$	Cur 1970
	$CrCl_3$	$900 \dots 1045 \rightarrow 845 \dots 980$	Wat 1972
	$CrCl_3$	$1040 \rightarrow 980$	Wat 1978
	$CrCl_3$	$1120 \dots 1180 \rightarrow 1150 \dots 1190$	Vol 1993
$FeCr_2S_{4-x}Se_x$	$HCl + Cl_2$	$920 \dots 940 \rightarrow 870 \dots 880$	Gib 1974
$FeIn_2S_4$	I_2	$780 \rightarrow 680$	Lut 1989
$FeIn_{2-x}Cr_xS_4$	I_2	$900 \rightarrow 850$	Sag 1998
Fe_xNbS_2	I_2	not specified	Koy 2000
$Fe_xNb_yS_2$	I_2	$950 \rightarrow 850$	Hin 1987
$FePS_3$	Cl_2	$750 \rightarrow 690, 700 \rightarrow 640$	Tay 1973
$Fe_2P_2S_6$	I_2	$670 \rightarrow 620$	Nit 1970
$FePb_3Sb_2Sn_4S_{14}$	I_2	$660 \rightarrow 620 \dots 630$	Men 2006
FeV_2S_4	I_2	not specified	Mur 1973
$FeTi_2S_4$	I_2	not specified	Mur 1973
$Fe_{1-x}Zn_xCr_2S_4$	$CrCl_3$	$990 \rightarrow 935$	Wat 1978
GaS	Br_2	$790 \rightarrow 700$	Zul 1974
	I_2	$925 \rightarrow 850$	Nit 1961
	I_2	$930 \rightarrow 850$	Lie 1969
	I_2	$800 \rightarrow 700$	Rus 1969
	I_2	$930 \rightarrow 800 \dots 870$	Lie 1972
	I_2	$870 \rightarrow 780$	Azi 1975
	I_2	$850 \rightarrow 800$	AlA 1977b
	I_2	$850 \rightarrow 750$	Whi 1978
	I_2	$860 \rightarrow 800$	Mic 1990
Ga_2S_3	I_2	$1000 \rightarrow 800$	Nit 1961
$GaPS_4$	I_2	$550 \rightarrow 700$	Per 1978
	I_2	$450 \rightarrow 430$	Nit 1970
Ga_2S_3:Co	I_2	$800 \rightarrow 700$	Rus 1969

Table 7.1.2 (continued)

Sink solid	Transport additive	Temperature / °C	Reference
$Ga_2S_{3-x}Se_x$	I_2	$950 \rightarrow 800$	Hot 2005a
$Ga_{2-x}Er_xS_3$	I_2	$970 \rightarrow 820$	Jin 1998
$Ga_{2-x}In_xS_3$	I_2	$650 \rightarrow 150$	Ami 1990
$Ga_{0.5}Fe_{0.5}InS_3$	I_2	$660 \rightarrow 540$	Gus 2003
$Ga_2In_4S_9$	I_2	$750 \rightarrow 700$	Krä 1970
$Ga_2In_8S_{15}$	I_2	$750 \rightarrow 700$	Krä 1970
$Ga_6In_4S_{15}$	I_2	$750 \rightarrow 700$	Krä 1970
Gd_2S_3	I_2	$950 \dots 1150$ $\rightarrow 900 \dots 1000$	Pes 1971
	I_2	$950 \dots 1150$ $\rightarrow 900 \dots 1000$	Pie 1970
Gd_2S_3:Nd	I_2	$1200 \rightarrow 1150$	Lei 1980
GeS	I_2	$520 \rightarrow 450$	Nit 1967a
GeS_2	I_2	$600 \rightarrow 500$	Gol 1998
	I_2	$700 \rightarrow 600$	Nit 1967a
$Ge_xPb_{1-x}S_{1-y}Se_y$	I_2, $PbCl_2$	$885 \dots 935 \rightarrow 755 \dots 850$	Kim 1993
HfS_2	I_2	$900 \rightarrow 800$	Gre 1965
	I_2	$900 \rightarrow 800$	Nit 1967a
	I_2	$1010 \rightarrow 1000$	Rim 1972
	I_2, ICl_3 u. a.	$780 \rightarrow 740$	Lev 1983
	Cl_2, Br_2, I_2	$T_1 \rightarrow T_2$	Fie 1988
HfS_3	I_2, S_2Cl_2	$650 \rightarrow 600$	Lev 1983
HgS	$NH_4Cl + I_2$	$400 \rightarrow 285$	Fai 1978
	I_2, NH_4Cl	not specified	Sim 1980
$HgCr_2S_4$	Cl_2	$900 \rightarrow 780$	Phi 1971
$HgGa_2S_4$	I_2	$730 \rightarrow 580$	Beu 1961
	I_2	$750 \rightarrow 600$	Kra 1965
	I_2	$1100 \rightarrow 1000$	Cur 1970
Hg_4GeS_6	Cl_2	$400 \rightarrow 470$	Que 1975
$HgIn_2S_4$	I_2	$950 \rightarrow 650$	Nit 1960
	I_2	$950 \rightarrow 650$	Nit 1961
	I_2	$950 \rightarrow 650$	Cur 1970
$HgIn_2S_4$:Co	I_2	$850 \rightarrow 570$	Lee 2003
Hg_4SiS_6	I_2	$480 \rightarrow 420$	Que 1975
In_2S_3 (α)	Br_2	$300 \rightarrow 400$	Gul 1979
In_2S_3 (β)	Br_2	$680 \rightarrow 450$	Gul 1979
In_2S_3 (γ)	Br_2	$990 \rightarrow 800$	Gul 1979
In_2S_3	I_2	$950 \rightarrow 450$	Nit 1960
	I_2	$950 \rightarrow 450$	Nit 1961
	I_2	$1100 \rightarrow 730$	Hol 1965
	I_2	$850 \rightarrow 800$	Die 1973
	I_2	$850 \rightarrow 800$	Die 1975
	I_2	$680 \rightarrow 600$	Krä 1976

Table 7.1.2 (continued)

Sink solid	Transport additive	Temperature / °C	Reference
$In_2S_{3-x}Se_x$	I_2	700	And 1970
$InPS_4$	I_2	$550 \rightarrow 700$	Per 1978
	I_2	$640 \rightarrow 580$	Nit 1970
In_2S_3:Co	I_2	$1000 \rightarrow 800$	Kim 1991
$InBiS_3$	I_2	$680 \rightarrow 600$	Krä 1976
$In_4Bi_2S_9$	I_2	$680 \rightarrow 600$	Krä 1971
	I_2	$680 \rightarrow 600$	Krä 1976
$Ir_{0.667x}Ru_{1-x}S_2$	I_2	$1100 \rightarrow 1050$	Col 1994
La_2S_3:Nd	I_2	$1200 \rightarrow 1150$	Lei 1980
$LaIn_3S_6$	I_2	$830 \ldots 980 \rightarrow 740 \ldots 900$	Ali 2000
MgS	I_2	$870 \rightarrow 710$	Mar 1999
$MgAl_2S_4$	I_2	$960 \rightarrow 870$	Oh 1997a
$MgAl_{1.04}S_{1.84}$	I_2	$780 \rightarrow 660$	Kha 1984
$MgAl_2S_4$:Er^{3+}	I_2	$950 \rightarrow 870$	Oh 1999
$MgIn_2S_4$:Co	I_2	$950 \rightarrow 530$	Lee 2003
MnS	I_2	$1000 \rightarrow 550$	Nit 1960
	I_2	$1000 \rightarrow 550$	Nit 1961
	HCl	$825 \ldots 900 \rightarrow 800$	Paj 1983
	Cl_2, $AlCl_3$	$900 \rightarrow 875 \ldots 885$, $950 \rightarrow 1000$	Paj 1980
$MnS_{1-x}O_x$	Br_2	$1000 \rightarrow 900$	Loc 2005a
$MnS_{1-x}Se_x$	I_2	$945 \rightarrow 850$	Wie 1969
$Mn_{1-x}Zn_xS$	I_2	$1000 \rightarrow 980$	Gar 1997
$MnAlS_2$	I_2	$785 \rightarrow 655$	Kha 1984
$MnCr_2S_4$	$AlCl_3$	$1000 \rightarrow 900$	Lut 1968
	$AlCl_3$	$1000 \rightarrow 900$	Lut 1970
	$AlCl_3$	$1000 \rightarrow 800$	Phi 1971
Mn_2GeS_4	I_2	$1050 \rightarrow 1000$	Nit 1967a
$MnPS_3$	Cl_2	$750 \rightarrow 690$, $700 \rightarrow 640$	Tay 1973
	I_2	$850 \rightarrow 800$	Cur 1970
$Mn_2P_2S_6$	I_2	$670 \rightarrow 620$	Nit 1970
$MnIn_2S_4$	I_2	$1100 \rightarrow 1000$	Cur 1970
	I_2	$1100 \rightarrow 1000$	Nit 1967a
	$I_2 + AlCl_3$	$900 \rightarrow 780$	Lut 1989
$MnIn_{2-2x}Ga_{2x}S_4$	I_2	$900 \rightarrow 850$	Sag 1998
$MnIn_{2-2x}Cr_{2x}S_4$	I_2	$850 \rightarrow 750$	Sag 1995
Mn_2SiS_4	I_2	$830 \rightarrow 780$	Nit 1967a
MoS_2	Br_2	$900 \rightarrow 800$	Nit 1967a
	Br_2	$950 \rightarrow 890$	AlH 1972
	Br_2, $Br_2 + S$, Cl_2	$800 \ldots 1000 \rightarrow 750 \ldots 950$	Kra 1980
	I_2	$820 \ldots 900 \rightarrow 700 \ldots 750$	Sch 1973a
	I_2	$T_2 \rightarrow 1000$	Rem 1999

Table 7.1.2 (continued)

Sink solid	Transport additive	Temperature / °C	Reference
MoS_2	I_2	$T_2 \rightarrow 740$	Rem 2002
	I_2	$790 \rightarrow T_1$	Vir 2007
$MoS_{2-x}Se_x$	I_2	$1000 \rightarrow 900$	Hot 2005f
$Mo_{1-x}Nb_xS_2$	I_2	$1000 \rightarrow 900$	Hot 2005f
Mo_2S_3	Br_2, NH_4Br	$1000 \ldots 1050$	Kra 1980
		$\rightarrow 900 \ldots 950$	
NbS_2	I_2	$850 \rightarrow 800$	Nit 1967a
	I_2	$950 \rightarrow 850$	Fuj 1979b
	I_2, ICl_3 u. a.	$780 \rightarrow 730$	Lev 1983
$NbS_{2-x}Se_x$	I_2	$1000 \rightarrow 900$	Hot 2005f
NbS_3	S u. a.	$670 \rightarrow 610$	Lev 1983
Nb_3S_4	I_2	$T_2 \rightarrow 1000$	Nak 1984
Nb|Pb|Bi|S	Cl_2	$960 \rightarrow 910$	Ohn 2005
$NdIn_3S_6$	I_2	$830 \ldots 980 \rightarrow 740 \ldots 900$	Ali 2000
NiS	I_2, HCl, NH_4I, GeI_2	$900 \rightarrow 800$	Kra 1984a
	$AlCl_3$	$850 \rightarrow 820$	Lut 1970
NiS_2	Cl_2	$715 \rightarrow 655$	Bou 1968
	Cl_2	$715 \rightarrow 655$	Kra 1984a
	Cl_2, Br_2	$670 \rightarrow 500$	Yao 1994
	$AlCl_3$	$830 \rightarrow 800$	Lut 1970
$NiS_{2-x}Se_x$	Cl_2, Br_2	$670 \rightarrow 500$	Yao 1994
Ni_3S_2	NH_4I, GeI_2	$800 \rightarrow 700$	Kra 1984a
	$AlCl_3$	$860 \rightarrow 830$	Lut 1970
$NiS_{2-x}Se_x$	Cl_2	$780 \rightarrow 760$	Bou 1973
$NiPS_3$	I_2	$720 \ldots 750 \rightarrow 690 \ldots 720$	Aru 1989
	Cl_2	$750 \rightarrow 690, 700 \rightarrow 640$	Tay 1973
$NiCr_2S_4$	$AlCl_3$	$1000 \rightarrow 850$	Lut 1970
$Ni_{1-x}Cr_{2+x}S_4$	$AlCl_3$, $CrCl_3$	$1050 \rightarrow 660$	Lut 1973a
$NiIn_{2-x}Cr_xS_4$	I_2	$900 \rightarrow 850$	Sag 1998
Ni_3CrS_4	$AlCl_3$	$1020 \rightarrow 970$	Lut 1970
Ni_5CrS_6	$AlCl_3$	$1030 \rightarrow 980$	Lut 1970
$NiIn_2S_4$	$I_2 + AlCl_3$	$820 \rightarrow 730$	Lut 1989
$Ni_{0.95}In_{2.03}S_4$	$I_2 + InCl_3$	$800 \rightarrow 730$	Lut 1989
$Ni_{1-x}Co_xGa_2S_4$	I_2	$925 \rightarrow 850$	Nam 2008
$Ni_{1-x}Fe_xGa_2S_4$	I_2	$925 \rightarrow 850$	Nam 2008
$Ni_{1-x}Mn_xGa_2S_4$	I_2	$925 \rightarrow 850$	Nam 2008
Ni_6SnS_2	I_2	$600 \rightarrow 570$	Bar 2003
$Ni_9Sn_2S_2$	I_2	$600 \rightarrow 570$	Bar 2003
$Ni_{1-x}Zn_xGa_2S_4$	I_2	$925 \rightarrow 850$	Nam 2008
PbS	NH_4Cl	$745 \rightarrow 670$	Car 1972
$Pb_3(PS_4)_2$	I_2	$850 \rightarrow 800$	Pos 1984
$PbIn_2S_4$	I_2	> 700	Krä 1980
$PbMo_6S_8$	$PbBr_2$	$1320 \rightarrow 1260$	Kra 1981

Table 7.1.2 (continued)

Sink solid	Transport additive	Temperature / °C	Reference
$Pb_6In_{10}S_{21}$	I_2	> 700	Krä 1980
$(Pb_{1-y}Bi_yS)_{1+x}(NbS_2)_n$	Cl_2	960 → 910	Ohn 2005
PdPS	Cl_2	760 → 740	Fol 1987
$Pd_3(PS_4)_2$	Cl_2	760 → 740	Fol 1987
$Pt_{1-x}S_2$	Cl_2	800 → 740	Fin 1974
$PtS_{2-x}Se_x$	P_4, Cl_2	850 → 690 ... 720,	Sol 1976
		850 ... 875 → 690 ... 750	
$Pt_{1-x}Sn_xS_2$	I_2	950 → 750	Tom 1998
ReS_2	H_2O, I_2 + H_2O	900 → 800	Sch 1973b
	Br_2	1125 → 1075	Mar 1984
	I_2	1050 → 990	Lia 2009
	Br_2	1050 → 990	Ho 2005
ReS_2:Mo	Br_2	1040 → 1000	Yen 2002
$ReS_{2-x}Se_x$	Br_2	1100 → 1050	Ho 1999
RuS_2	Cl_2	1100 → 1050	Fie 1987
	ICl_3, ICl_3 + S_2Cl_2	1040 → 1020	Bic 1984
	ICl_3	T_2 → 960	Hua 1988
$RuS_{2-x}Se_x$	ICl_3	1080 → 980	Lin 1992
	ICl_3	1120 → 1090	Sti 1992
$Ru_{1-x}Fe_xS_2$	ICl_3	1000 → 960	Tsa 1994
Sb_2S_3	I_2	455 ... 490 → 395 ... 450	Sch 1978
	I_2	250 ... 500 → 150 ... 400	Bal 1986
	I_2	490 → 420	Ven 1987
	I_2	490 → 410	Ven 1988
Sc_2S_3	I_2	not specified	Dis 1970
SiS_2	S	700 → 600	Sch 1982
$(SmS)_{1.19}(TaS_2)_2$	Cl_2	960 → 910	Ohn 2005
SnS	I_2	950 → 600	Nit 1961
	I_2	950 → 850	Cru 2003
SnS_2	Cl_2	640 → 590, 730 → 680	Kou 1988
	Cl_2	420 ... 440 → 450 ... 455	Shi 1990
	I_2	950 → 600	Nit 1960
	I_2	950 → 600	Nit 1961
	I_2	800 → 700	Gre 1965
	I_2	687 → 647	Whi 1979
	I_2	680 → 640	Min 1980
	I_2	700 → 600, 400 → 500	Sch 1981

Table 7.1.2 (continued)

Sink solid	Transport additive	Temperature / °C	Reference
SnS$_2$	I$_2$	525 → T_1	Rao 2000
	I$_2$	950 → 850	Cru 2003
	SnI$_4$	580 ... 750 → 550 ... 740	Pal 1986
	SnI$_4$	650 → 550	Rao 1995
	SnI$_4$	650 → 550	Wie 1979
	SnCl$_4$ · 5 H$_2$O	420 ... 440 → 450 ... 455	Shi 1991
SnS$_{2-x}$O$_x$	I$_2$	600 → 550	Nit 1967a
SnS$_{2-x}$Se$_x$	I$_2$	620 → 578	AlA 1977a
	I$_2$	680 → 620	Pat 1997
	I$_2$	680 → 620	Rim 1972
Sn$_2$S$_3$	I$_2$	950 → 850	Cru 2003
Sn$_2$P$_2$S$_6$	I$_2$	600 → 630	Nit 1970
Sn$_{1-x}$Zr$_x$S$_2$	I$_2$	690 ... 900 → 650 ... 820	AlA 1973
	I$_2$	690 ... 900 → 650 ... 840	AlA 1977a
SrGa$_2$S$_4$	I$_2$	900 → 700	Tan 1995
SrIn$_2$S$_4$	I$_2$	700 ... 900 → 600 ... 800	Gul 1992
TaS$_2$	I$_2$	850 → 800	Nit 1967a
	I$_2$	950 → 300	Eno 2004
	I$_2$, ICl$_3$ u. a.	950 → 900	Lev 1983
	S	800 → 700	Sch 1968
	S	800 → 700, 800 → 1000	Sch 1980
TaS$_2$:Cu	I$_2$	1000 → 900	Zhu 2008
TaS$_{2-x}$Se$_x$	I$_2$	850 ... 900 → 700 ... 800	AlA 1977a
	I$_2$	1000 → 900	Hot 2005b
Ta$_{1.08}$S$_2$	NH$_4$Cl	1100 → 700 ... 800	Got 1998
Ta$_{1-x}$Mo$_x$S$_2$	I$_2$	1000 → 800	Hot 2005d
Ta$_{1-x}$Nb$_x$S$_2$	I$_2$	1000 → 900	Hot 2005g
TaS$_3$	S$_2$Cl$_2$ u. a.	550 → 500	Lev 1983
Tb$_2$S$_3$	I$_2$	not specified	Ebi 2006
TiS$_2$	I$_2$	900 → 800	Gre 1965
	I$_2$	900 → 800	Nit 1967a
	I$_2$	800 → 720	Rim 1972
	I$_2$	950 → 850	Sae 1976
	I$_2$	900 → 800	Sae 1977
	I$_2$	800 → 700	Ma 2008
	S	700 → 600	Sch 1982
	I$_2$	1000 → 770	Kus 1998
	I$_2$	not specified	Kul 1987
	I$_2$, ICl$_3$ u. a.	630 → 625	Lev 1983
	I$_2$, S	800 ... 900 → 700 ... 800	Ino 1984
TiS$_{2-x}$Se$_x$	I$_2$	850 → 800	Nit 1967a
	I$_2$	T_2 → 720 ... 820	Rim 1974
	I$_2$	1000 → 900	Hot 2005b

Table 7.1.2 (continued)

Sink solid	Transport additive	Temperature / °C	Reference
$TiS_{2-x}Te_x$	I_2	$780 \rightarrow 740$	Nit 1967a
	I_2	$750 \ldots 800 \rightarrow 690 \ldots 720$	Rim 1974
$Ti_{1-x}Mo_xS_2$	I_2	$1000 \rightarrow 800$	Hot 2005d
$Ti_{1-x}NbS_2$	I_2	$850 \rightarrow 800$	Nit 1967a
	I_2	$1000 \rightarrow 900$	Hot 2005g
$Ti_{1-x}TaS_2$	I_2	$850 \rightarrow 800$	Nit 1967a
	I_2	$1000 \rightarrow 900$	Hot 2005b
$Ti_{1-x}VS_2$	I_2	$850 \rightarrow 800$	Nit 1967a
	I_2	$900 \rightarrow 800$	Sae 1978
$Ti_{1-x}ZrS_3$	Br_2	$625 \ldots 900 \rightarrow 530 \ldots 785$	Sie 1983
Ti_2S_3	I_2	$700 \rightarrow 800$	Sae 1982
Ti_5S_8	I_2	$500 \rightarrow 600$	Sae 1982
TiS_3	ICl_3 u. a.	$500 \rightarrow 450$	Lev 1983
$TlFeS_2$	I_2	$500 \rightarrow 400$	Wan 1972
TlV_5S_8	I_2	$1100 \rightarrow 1000$	Ben 1987
$TlV_5S_{8-x}Se_x$	I_2	$1100 \rightarrow 1000$	Ben 1987
US_x ($x = 1.65 \ldots 1.99$)	Br_2	$930 \rightarrow 830$	Sev 1970
US_2	I_2	$940 \rightarrow 700$	Smi 1967
	Br_2	$970 \rightarrow 870$	Slo 1966
$U_xPd_3S_4$	I_2	$940 \rightarrow 880$	Dao 1986
VS_2	I_2	$900 \rightarrow 850$	Nit 1967a
V_2S_3	I_2	$900 \rightarrow 700$	Tan 1980
V_3S_4	Cl_2	$900 \rightarrow 750$	Sae 1974
	I_2	$900 \rightarrow 700 \ldots 800$	Wak 1982
V_5S_8	Cl_2	$900 \rightarrow 750$	Sae 1974
	I_2	$900 \rightarrow 650 \ldots 750$	Tan 1980
	I_2	$900 \rightarrow 700 \ldots 800$	Wak 1982
WS_2	Cl_2, Br_2	$1200 \rightarrow 1170$	Bag 1983
	Br_2	$900 \rightarrow 800$	Nit 1967a
	$I_2, H_2O, I_2 + H_2O$	$900 \rightarrow 700$	Sch 1973a
	I_2	$T_2 \rightarrow 790$	Rem 1998
	I_2	$T_2 \rightarrow 740$	Rem 2002
	I_2	$790 \rightarrow T_1$	Vir 2007
WS_2:Re	Br_2	$1000 \rightarrow 950$	Yen 2004a
$WS_{2-x}Se_x$	I_2	$1000 \rightarrow 960$	Jos 1993
	I_2	not specified	Jos 1994
$W_{1-x}Re_xS_2$	Br_2	$1000 \rightarrow 950$	Yen 2004a
Y_2S_3:Nd	I_2	$1200 \rightarrow 1150$	Lei 1980
Y_2HfS_5	I_2	not specified	Jei 1975
$YbAs_4S_7$	I_2	$530 \rightarrow 460$	Mam 1988
$Yb_3As_4S_9$	I_2	$880 \rightarrow 810$	Mam 1988

Table 7.1.2 (continued)

Sink solid	Transport additive	Temperature / °C	Reference
YbIn$_3$S$_6$	I$_2$	830 ... 980 → 740 ... 900	Ali 2000
ZnS	I$_2$	1000 → 750	Nit 1960
	I$_2$	1000 → 750	Nit 1961
	I$_2$	1050 → 900	Jon 1964
	I$_2$	1000 ... 1200 → 700 ... 1000	Har 1967
	I$_2$	800 → 725 ... 800	Dan 1973
	I$_2$	950 → 750	Har 1974
	I$_2$	1160 → 930	Aot 1976
	I$_2$	1050 → 850	Har 1977
	I$_2$	850 → 840	Fuj 1979a
	I$_2$	T_2 → 840	Tho 1983
	I$_2$	not specified	Pal 1983
	I$_2$	900 → 840 ... 890	Mat 1986
	I$_2$	850 → 840	Kit 1987
	I$_2$	1000 → 800	Shi 1992
	I$_2$	not specified	Zuo 2002a
	I$_2$	not specified	Zuo 2002b
	I$_2$, NH$_4$Cl	920 ... 1025 → 630 ... 1000, 1160 → 700 ... 1070	Len 1971
	I$_2$, NH$_4$Cl	1010 → 900, 920 → 700	Len 1975
	HCl	T_2 → T_1	Jon 1962
	HCl	T_2 → > 755	Sam 1962
	HCl	1050 → 940	Jon 1963
	HCl	700 → 900	Uji 1971
	HCl	1100 → 900	Uji 1976
	NH$_4$Cl	T_2 → T_1	Nod 1990a
	H$_2$S	1200 → 1000	Sam 1961a
	H$_2$S	1100 ... 1230 → 900 ... 1200	Sam 1961b
	H$_2$S, HCl	1200 → 1060, not specified	Ska 1963
	P$_4$	1000 → 900	Loc 2005c
ZnO$_{1-x}$S$_x$	Br$_2$	1000 → 900	Loc 2007
ZnS:P	P$_4$	1000 → 900	Loc 2005c
ZnS$_{1-x}$Se$_x$	H$_2$	800 → 650 ... 780	DeM 1979
	H$_2$	not specified	Kor 2002
	I$_2$	850 → 840	Cat 1976
	I$_2$	850 → 840	Fuj 1979a
	I$_2$	1000 → 850	Hig 1980
	I$_2$	900 → 840 ... 890	Mat 1986
	I$_2$	900 → 800	Sen 2004
	I$_2$	900 → 800	Hot 2005e
	I$_2$	1000 → 900	Gru 2005
	I$_2$	850 → 800	Gra 1995
	H$_2$, H$_2$ + I$_2$, H$_2$ + HCl	850 → 450 ... 700	Har 1987
	NH$_4$Cl	900 → T_1	Mat 1988

Table 7.1.2 (continued)

Sink solid	Transport additive	Temperature / °C	Reference
$ZnS_{1-x}Se_x$	I_2	$725 \rightarrow 650$	Nis 1982
$ZnS_{1-x}Se_x$:Fe	I_2	$850 \rightarrow 800$	Gra 1995
$ZnS_{1-x}Te_x$	I_2	$1000 \rightarrow 900$	Ros 2004
$Zn_{1-x}Cd_xS$	H_2	$760 \rightarrow 700$	Fra 1979
	H_2	$760 \rightarrow 700$	Ant 1980
	H_2	$760 \rightarrow 700$	Fra 1981
	I_2	$T_2 \rightarrow T_1$	Pal 1982a
	I_2	$T_2 \rightarrow T_1$	Pal 1982b
$Zn_{1-x}Co_xS$	I_2	$900 \rightarrow 875$	Pas 1998
$Zn_{1-x}Fe_xS$	I_2	$965 \rightarrow 870$	Dic 1990
$Zn_{1-x}Fe_xPS_3$	Cl_2	$500 \rightarrow 480$	Odi 1975
$Zn_{1-x}Mn_xS$	I_2	$1050 \rightarrow 700 \dots 900$	Nit 1961
	I_2	$850 \rightarrow 750$	Nit 1971
	I_2	$900 \dots 1100$ $\rightarrow 800 \dots 1030$	Kni 1999
$Zn_{1-x}Ni_xS$	I_2	$950 \rightarrow 925$	Wu 1989
$Zn_{1-x-y}Mn_xFe_yS$	I_2	$900 \rightarrow 800$	Kni 2000
Zn_2AgInS_4	I_2	$750 \rightarrow 700$	Lam 1972
Zn_3AgInS_5	I_2	$750 \rightarrow 700$	Lam 1972
$ZnAl_2S_4$	I_2	$T_2 \rightarrow 740$	Ber 1981
	I_2	$780 \rightarrow 700$	Kai 1995
$Zn_{1-x}Cd_xGa_2S_4$	I_2	$1000 \rightarrow 800$	Wu 1988
$Zn_{1-x}Cd_xIn_2S_4$	I_2	$1000 \rightarrow 800$	Cur 1987
$Zn_{1-x}Cd_xGeS_6$	I_2	not specified	Dub 1991
$ZnCr_2S_4$	Cl_2	$775 \dots 900 \rightarrow 725 \dots 800$	Pin 1970
	$AlCl_3$	$1000 \rightarrow 850$	Lut 1968
	$AlCl_3$	$1000 \rightarrow 850$	Lut 1970
	$AlCl_3$	$1000 \rightarrow 800$	Phi 1971
	$CrCl_3$	$950 \rightarrow 900$	Phi 1971
$ZnCr_2S_{4-x}Se_x$	I_2, $AlCl_3$	$950 \rightarrow 700$	Pic 1971
$ZnGa_2S_4$	I_2	$1100 \rightarrow 1000$	Nit 1961
	I_2	$1100 \rightarrow 1000$	Beu 1961
	I_2	$1100 \rightarrow 1000$	Cur 1970
	I_2	$850 \rightarrow 750$	Wu 1988
$ZnIn_2S_4$	I_2	$1000 \rightarrow 700$	Nit 1960
	I_2	$750 \rightarrow 700$	Nit 1961
	I_2	$1000 \rightarrow 700$	Lap 1962
	I_2	$750 \rightarrow 700$	Cur 1970
	I_2	$1000 \rightarrow 700$	Val 1970
	I_2	$790 \rightarrow 750$	Buc 1974
$ZnIn_2S_{4-x}Se_x$	I_2	$725 \rightarrow 675$	Loc 2005b
$Zn_3In_2S_4$	I_2	$T_2 \rightarrow >1000$	Buc 1974
$Zn_5In_2S_8$	I_2	$1000 \rightarrow 900$	Kal 1987
$ZnLu_2S_4$	I_2, HCl	not specified	Yim 1973
$ZnSc_2S_4$	I_2, HCl	not specified	Yim 1973
$ZnTm_2S_4$	I_2, HCl	not specified	Yim 1973
$(ZnS)_{1-x}(CuAlS_2)_x$	I_2	$950 \rightarrow 925$	Do 1992

Table 7.1.2 (continued)

Sink solid	Transport additive	Temperature / °C	Reference
$(ZnS)_{1-x}(CuFeS_2)_x$	I_2	$950 \rightarrow 925$	Do 1992
$(ZnS)_{1-x}(CuInS_2)_x$	I_2	$950 \rightarrow 925$	Do 1992
$(ZnS)_{1-x}(GaP)_x$	I_2	$945 \rightarrow 700$	Han 1995
	I_2	$1000 \rightarrow 900$	Loc 2004b
$(ZnS)_x(Zn_3P_2)_y$	I_2	$1000 \rightarrow 900$	Loc 2004a
ZrS_2	I_2	$900 \rightarrow 800$	Gre 1965
	I_2	$900 \rightarrow 800$	Nit 1967a
	I_2	$900 \rightarrow 820$	Rim 1972
	I_2	$900 \rightarrow 800$	Fuj 1979b
	I_2, ICl_3 u. a.	$760 \rightarrow 730$	Lev 1983
$ZrO_{2-x}S_x$	I_2	$780 \rightarrow 700$	Nit 1967a
$ZrS_{2-x}Se_x$	I_2	$850 \rightarrow 800$	Nit 1967a
	I_2	$860 \dots 900 \rightarrow 800 \dots 820$	Bar 1995
	I_2	$930 \rightarrow 900$	Pat 1998
ZrS_3	I_2	$900 \rightarrow 850$	Nit 1967a
	I_2	$1010 \rightarrow 930$	Pat 1993
	I_2	$900 \rightarrow 850$	Pat 2005
	S_2Cl_2 u. a.	$750 \rightarrow 730$	Lev 1983
$ZrP_{1.4}S_{0.6}$	I_2	$880 \rightarrow 980$	Schl 2009
$ZrSiS$	I_2	$850 \rightarrow 750$	Nit 1967a

Bibliography of Section 7.1

Ack 1976   J. Ackermann, S. Soled, A. Wold, E. Kostiner, *J. Solid State Chem.* **1976**, *19*, 75.

Aks 1988a   I. A. Aksenov, L. A. Makovetskaya, V. A. Savchuk, *Phys. Status Solidi* **1988**, *108*, K63.

Aks 1988b   I. A. Aksenov, S. A. Gruzo, L. A. Makowezkaja, G. P. Popelnjuk, W. A. Rubzov, *Izv. Akad. Nauk SSSR, Neorg. Mater.* **1988**, *24*, 560.

AlA 1973   F. A. S. Al-Alamy, A. A. Balchin, *Mater. Res. Bull.* **1973**, *8*, 245.

AlA 1977a   F. A. S. Al-Alamy, A. A. Balchin, *J. Cryst. Growth* **1977**, *38*, 221.

AlA 1977b   F. A. S. Al-Alamy, A. A. Balchin, *J. Cryst. Growth* **1977**, *39*, 275.

AlH 1972   A. A. Al-Hilli, B. L. Evans, *J. Cryst. Growth* **1972**, *15*, 93.

Ali 1972   U. M. Aliev, R. S. Gamidov, G. G. Guseinov, *Izv. Akad. Nauk. SSSR, Neorg. Mater.* **1972**, *8*, 1855.

Ali 1978   U. M. Aliev, P. G. Rustamov, G. G. Guseinov, *Izv. Akad. Nauk. SSSR, Neorg. Mater.* **1978**, *14*, 1346.

Ali 2000   V. O. Aliev, E. R. Guseinov, O. M. Aliev, R. Ya. Alieva, *Inorg. Mater.* **2000**, *36*, 753.

Ame 1976   K. Ametani, *Bull. Chem. Soc. Jpn.* **1976**, *49*, 450.

Ami 1990   I. R. Amiraslamov, F. J. Asadov, B. A. Maksimov, W. N. Moltschanov, A. A. Musaev, H. G. Furmapova, *Kristallografija* **1990**, *35*, 332.

And 1970   I. Ya. Andronik, V. P. Mushinskii, *Uchenye Zapiski – Kishinevskii Gosudarst-vennyi Universitet* **1970**, *110*, 19.

Ant 1968   V. B. Antonov, G. G. Gusejnov, D. T. Gusejnov, R. C. Nani, *Dokl. Akad. Nauk Azerbajdzh. SSSR* **1968**, *24*, 12.

Ant 1980   G. Antonioli, D. Bianchi, *Thin Solid Films* **1980**, *70*, 71.

Aot 1976   S. Aotsu, M. Takahashi, H. Fujisaki, *Tohoku Daigaku Kagaku Keisoku Ken-kyusho Hokuku* **1980**, *24*, 109.

Aru 1989   A. Aruchamy, H. Berger, F. Levy, *J. Electrochem. Soc.* **1989**, 136, 2261.

Att 1982   G. Attolini, C. Paorici, L. Zanotti. *J. Cryst. Growth* **1982**, *56*, 254.

Att 1983   G. Attolini, C. Paorici, *Mater. Chem. Phys.* **1983**, *9*, 65.

Azi 1975   T. K. Azizov, I. Y. Aliev, A. S. Abbasov, M. K. Alieva, *Azerb. Khim. Zh.* **1975**, *3*, 126.

Bag 1983   J. Baglio, E. Kamieniecki, *J. Solid State Chem.* **1983**, *49*, 166.

Bal 1986   C. Balarew, M. Ivanova, *Cryst. Res. Technol.* **1986**, *21*, K171.

Bal 1989   K. Balakrishnan, B. Vengatesan, N. Kanniah, P. Ramasamy, *Bioelectroche-mistry* **1989**, *5*, 841.

Bal 1990   K. Balakrishnan, B. Vengatesan, N. Kanniah, P. Ramasamy, *Cryst. Res. Tech-nol.* **1990**, *25*, 633.

Bal 1994   K. Balakrishnan, B. Vengatesan, P. Ramasamy, *J. Mat. Sci.* **1994**, *29*, 1879.

Bar 1972   K. G. Barraclough, A. Meyer, *J. Cryst. Growth* **1972**, *16*, 265.

Bar 1974   K. G. Barraclough, W. Lugschneider, *Phys. Status Solidi* **1974**, *22*, 401.

Bar 1995   K. S. Bartwal, O. N. Srivastava, *Mat. Sci. Eng.* **1995**, *B33*, 115.

Bar 2003   A. I. Baranov, A. A. Isaeva, L. Kloo, B. A. Popovkin, *Inorg. Chem.* **2003**, *42*, 6667.

Ben 1987   W. Bensch, R. Schlögl, *Micron Microsc. Acta* **1987**, *18*, 89.

Ber 1976   C. Bernhard, G. Fourcaudot, J. Mercier, *J. Cryst. Growth* **1976**, *35*, 192.

Ber 1981   H. J. Berthold, K. Köhler, *Z. Anorg. Allg. Chem.* **1981**, *475*, 45.

Beu 1961   J. A. Beun, R. Nitsche, M: Lichtensteiger, *Physika* **1961**, *27*, 448.

Beu 1962   J. A. Beun, R. Nitsche, H. U. Bölsterli, *Physika* **1962**, *28*, 184.

Bic 1984   R. Bichsel, F. Levy, H. Berger, *J. Phys. C: Solid State Phys.* **1984**, *17*, L19.

Bin 1983   J. J. M. Binsma, W. J. P. van Enckevort, G. W. M. Staarink, *J. Cryst. Growth* **1983**, *61*, 138.

Bin 2009   M. Binnewies, S. Locmelis, B. Meyer, A. Polity, D. M. Hofmann, H. v. Wenckstern, *Progr. Solid State Chem.* **2009**, *37*, 57.

Ble 1992   O. Blenk, E. Bucher, G. Willeke, *Appl. Phys. Lett.* **1993**, *62*, 2093.

Bod 1976   I. V. Bodnar, *Vestsi Akad. Nauk BSSSR, Ser. Khim. Navuk* **1976**, *2*, 124.

Bod 1980   I. V. Bodnar, A. P. Bologa, B. V. Korzun, *Krist. Technik,* **1980**, *15*, 1285.

Bod 1984   I. V. Bodnar, I. T. Bodnar, A. A. Vaipolin, *Cryst. Res. Technol.* **1984**, *19*, 1553.

Bod 1996   I. V. Bodnar, *Inorg. Mater.* **1996**, *32*, 936.

Bod 1997   I. V. Bodnar, *Fiz. Techn. Poluprov.* **1997**, *31*, 49.

Bod 1998   I. V. Bodnar, *Zhurnal Neorg. Khim.* **1998**, *43*, 2090.

Bod 2004   I. V. Bodnar, V. Yu. Rud, Yu. V. Rud, *Inorg. Mater.* **2004**, *40*, 102.

Bon 1970   S. Bontscheva-Mladenova, I. Dukov, *Godisnik na Vissija Chimiko-Technolo-giceski Institut Sofija* **1970**, *14*, 335.

Bou 1968   R. J. Bouchard, *J. Cryst. Growth* **1968**, *2*, 40.

Bou 1973   R. J. Bouchard, H. J. Kent, *Mater. Res. Bull.* **1973**, *8*, 489.

Bri 1975   P. Bridenbaugh, B. Tell, *Mater. Res. Bull.* **1975**, *10*, 1127.

Bri 1985   J. M. Briceno-Valero, S. A. Lopez-Rivera, L. Martinez, G. Gonzales de Ar-mengol, G. Frias, *Progr. Cryst. Growth Charact.* **1985**, *10*, 159.

Bri 1982   J. W. Brigthwell, B. Ray, C. N. Buckley, *J. Cryst. Growth,* **1982**, *59*, 210.

Buc 1974   P. Buck, *J. Cryst. Growth* **1974**, *22*, 13.

But 1971	S. R. Butler, J. Bouchard, *J. Cryst. Growth* **1971**, *10*, 163.
Car 1972	E. H. Carlson, *J. Cryst. Growth* **1972**, *12*, 162.
Cat 1976	E. Catano, Z. K. Kun, *J. Cryst. Growth* **1976**, *33*, 324.
Cha 1971	J. P. Chapius, A. Niggli, R. Nitsche, *Naturwissenschaften* **1971**, *58*, 94.
Cha 1972	J. P. Chapius, C. H. Gnehn, V. Krämer, *Acta Crystallogr.* **1972**, *B28*, 3128.
Chi 1994	S. Chichibu, S. Shirakata, A. Ogawa, R. Sudo, M. Uchida, Y. Harada, T. Wakiyama, M. Shishikura, S. Matsumoto, *J. Cryst. Growth* **1994**, *140*, 388.
Cho 2001	S.-H. Choe, H.-L. Park, W.-T. Kim, *J. Korean Phys. Soc.* **2001**, *38*, 155.
Cho 2005	S.-H. Choe, M.-S. Jin, W.-T. Kim, *J. Korean Phys. Soc.* **2005**, *47*, 866.
Col 1994	H. Colell, N. Alonso-Vante, S. Fiechter, R. Schieck, R. Diesner, W. Henrion, H. Tributsch, *Mater. Res. Bull.* **1994**, *29*, 1065.
Con 1970	L. E. Conroy, R. J. Bouchard, *Inorg. Synth.* **1979**, *12*, 163.
Cra 2005	D. A. Crandles, M. Reedyk, G. Wardlaw, F. S. Razavi, T. Hagino, S. Nagata, I. Shimono, R. K. Kremer, *J. Phys.: Condens. Matter* **2005**, *17*, 4813.
Cru 2003	M. Cruz, J. Morales, J. P. Espinos, J. Sanz, *J. Solid State. Chem.* **2003**, *175*, 359.
Cur 1970	B. J. Curtis, F. P. Emmenegger, R. Nitsche, *R. C. A. Rev.* **1970**, *31*, 647.
Cur 1987	M. Curti, P. Gastaldi, P. P. Lottici, C. Paorici, C. Razetti, S. Viticoli, L. Zanotti, *J. Solid State Chem.* **1987**, *69*, 289.
Dan 1973	P. N. Dangel, B. J. Wuensch, *J. Cryst. Growth* **1973**, *19*, 1.
Dao 1986	A. Daoudi, H. Noel, *Inorg. Chim. Acta* **1986**, *117*, 183.
Dao 1996	A. Daoudi, M. Lamire, J. C. Levet, H. Noel, *J. Solid State Chem.* **1996**, *123*, 331.
Dav 2002	G. Ye. Davyduk, O. V. Parasyuk, Ya. E. Romanyuk, S. A. Semenyk, V. I. Zaremba, L. V. Piskach, J. J. Koziol, V. O. Halka, *J. Alloys Compd.* **2002**, *339*, 40.
Del 2006	G. E. Delgado, L. Betancourt, V. Sagredo, M. N. C. Moron, *Phys. Status Solidi* **2006**, *203*, 3627.
DeM 1979	M. De Murcia, D. Etienne, J. P. Fillard, *Surf. Sci.* **1979**, *8*, 280.
Dic 1990	J. Dicarlo, K. Albert, K. Dwigth, A. Wold, *J. Solid State Chem.* **1990**, *87*, 443.
Die 1973	R. Diehl, R. Nitsche, *J. Cryst. Growth* **1973**, *20*, 38.
Die 1975	R. Diehl, R. Nitsche, *J. Cryst. Growth* **1975**, *28*, 306.
Dis 1970	J. P. Dismukes, R. T. Smith, *Z. Kristallogr.* **1970**, *132*, 272.
Do 1992	Y. R. Do, K. Dwigth, A. Wold, *Chem. Mater.* **1992**, *4*, 1014.
Dub 1991	I. V. Dubrovin, L. D. Budennaya, E. V. Sharkina, *Izv. Akad. Nauk SSSR, Neorg. Mater.* **1991**, *27*, 244.
Ebi 2006	S. Ebisu, M. Gorai, K. Maekawa, S. Nagata, CP 850, *Low Temperature Physics, 24 th. International Conference on Low Temp. Physics*, **2006**.
Eno 2004	H. Enomoto, T. Kawano, M. Kawaguchi, Y. Takano, K. Sekizawa, *Jpn. J. Appl. Phys.* **2004**, *43*, L123.
Fai 1978	S. P. Faile, *J. Cryst. Growth*, **1978**, *43*, 129.
Fie 1986	S. Fiechter, J. Mai, A. Ennaoui, *J. Cryst. Growth* **1986**, *78*, 438.
Fie 1987	S. Fiechter, H.-M. Kühne, *J. Cryst. Growth* **1987**, *83*, 517.
Fie 1988	S. Fiechter, H. Eckert, *J. Cryst. Growth* **1988**, *88*, 435.
Fil 1991	V. V. Filonenko, B. D. Nechiporuk, N. E. Novoseletskii, V. A. Yukhimchuk, Yu. F. Lavorik, *Izv. Akad. Nauk SSSR, Neorg. Mater.* **1991**, *27*, 1166.
Fin 1974	A. Finley, D. Schleich, J. Ackermann, S. Soled, A. Wold, *Mater. Res. Bull.* **1974**, *9*, 1655.
Fol 1987	J. C. W. Folmer, J. A. Turner, B. A. Parkinson, *J. Solid State Chem.* **1987**, *68*, 28.
Fra 1979	P. Franzosi, C. Ghezzi, E. Gombia, *Mater. Chem.* **1979**, *4*, 557.
Fra 1981	P. Franzosi, C. Ghezzi, E. Gombia, *J. Cryst. Growth* **1981**, *51*, 314.

Fuj 1979a   S. Fujita, H. Mimoto, H. Takebe, T. Noguchi, *J. Cryst. Growth* **1979**, *47*, 326.
Fuj 1979b   S. Fujiki, Y. Ishazawa, Z. Inoue, *Mineral. J.* **1979**, *9*, 339.
Gar 1997   V. J. Garcia, J. M. Briceno-Valero, L. Martinez, A. Mora, S. Adan Lopez-Rivera, W. Giriat, *J. Cryst. Growth* **1997**, *173*, 222.
Gib 1969   P. Gibart, A. Begouen-Demeaux, *Compt. Rend. Acad. Sci. Paris C* **1969**, *268*, 816.
Gib 1974   P. Gibart, *J. Cryst. Growth* **1974**, *24/25*, 147.
Gol 1973   L. Goldstein, J. L. Dormann, R. Druilhe, M.Guittard, P. Gibart, *J. Cryst. Growth* **1973**, *20*, 24.
Gol 1998   A. V. Golubkov, G. B. Dubrovskii, A. I. Shelyk, A. V. Golubkov, G. B. Dubrovskii, A. I. Shelykh, A. F. Loffe, *Fizika i Tekhnika Poluprovodnikov* **1998**, *32*, 827.
Got 1995   Y. Gotoh, J. Akimoto, Y. Oosawa, M. Onoda, *Jpn. J. Appl. Phys.* **1995**, Letters, *34(12B)*, L1662.
Got 1998   Y. Gotoh, J. Akimoto, Y. Oosawa, *J. Alloys Compd.* **1998**, *270*, 115.
Got 2004   Y. Gotoh, I. Yamaguchi, Y. Takahashi, J. Akimoto, M. Goto, K. Kawaguchi, N. Yamamoto, M. Onoda, *Solid State Ionics*, **2004**, *172*, 519.
Gra 1995   K. Grasza, E. Janik, A. Mycielski, J. Bak-Misiuk, *J. Cryst. Growth* **1995**, *146*, 75.
Gre 1965   D. L. Greenway, R. Nitsche, *J. Phys. Chem. Solids* **1965**, *26*, 1445.
Gru 2005   S. Gruhl, C. Vogt, J. Vogt, U. Hotje, M. Binnewies, *Microchim. Acta* **2005**, *149*, 43.
Gu 1990   X. Gu, W. Giriat, J. K. Furdyna, *Rare Metals* **1990**, *9*, 139.
Gul 1979   T. N. Guliev, D. I. Zul'fugarly, N. F. Gakhramanov, *Azerb. Khim. Zh.* **1979**, *5*, 89.
Gul 1992   T. N. Guliev, *Izv. Vyssh. Uchebn. Zaved., Khim. Khim. Tekhnol.* **1992**, *35*, 15.
Gus 2003   G. G. Guseynov, N. N. Musaeva, M. G. Kyazumov, I. B. Asadova, O. M. Aliyev, *Inorg. Mater.* **2003**, *39*, 924.
Han 1995   Y. Han, M. Acinc, *J. Amer. Ceram. Soc.* **1995**, *78*, 1834.
Har 1967   H. Hartmann, *Rost Kristallov.* **1967**, *7*, 252.
Har 1974   H. Hartmann, *Kristall Technik* **1974**, *9*, 743.
Har 1977   H. Hartmann, *J. Cryst. Growth* **1977**, *42*, 144.
Har 1987   H. Hartmann, R. Mach, N. Testova, *J. Cryst. Growth* **1987**, *84*, 199.
Har 1989   B. Harbrecht, G. Kreiner, *Z. Anorg. Allg. Chem.* **1989**, *572*, 47.
He 1988   X-C. He, H-S. Shen, P. Wu, K. Dwight, A. Wold, *Mater. Res. Bull.* **1988**, *23*, 799.
Hel 1979   E. E. Hellstrom, R. A. Huggins, *Mater. Res. Bull.* **1979**, *14*, 127.
Hig 1980   S. Higo, M. Oka, T. Numata M. Aoki, *Kagoshima Daigaku Kogakubu Kenkyu Hokoku* **1980**, *22*, 175.
Hin 1987   H. Hinode, M. Wakihara, M. Taniguchi, *J. Cryst. Growth* **1987**, *84*, 413.
Ho 1999   C. H. Ho, Y. S. Huang, P. C. Liao, K. K. Tiong, *J. Phys. Chem. Sol.* **1999**, *60*, 1797.
Ho 2005   C. H. Ho, *Optics Express*, **2005**, *13*, 8.
Hol 1965   H. Holzapfel, E. Butter, U. Stottmeister, *Z. Chem.* **1965**, *5*, 31.
Hon 1969   W. N. Honeyman, *J. Phys. Chem. Solids* **1969**, *30*, 1935.
Hon 1971   W. N. Honeyman, K. H. Wilkinson, *J. Phys. D. Appl. Phys.* **1971**, *4*, 1182.
Hon 1988   E. Honig, H-S. Shen, G-Q. Yao, K. Doverspike, R. Kershaw, K. Dwight, A. Wold, *Mater. Res. Bull.* **1988**, *23*, 307.
Hot 2005a   U. Hotje, R. Wartchow, E. Milke, M. Binnewies, *Z. Anorg. Allg. Chem.* **2005**, *631*, 1675.
Hot 2005b   U. Hotje, R. Wartchow, M. Binnewies, *Z. Anorg. Allg. Chem.* **2005**, *631*, 403.

Hot 2005c	U. Hotje, M. Binnewies, *Z. Anorg. Allg. Chem.* **2005**, *631*, 1682.
Hot 2005d	U. Hotje, R. Wartchow, M. Binnewies, *Z. Naturforsch.* **2005**, *60b*, 1235.
Hot 2005e	U. Hotje, C. Rose, M. Binnewies, *Z. Anorg. Allg. Chem.* **2005**, *631*, 2501.
Hot 2005f	U. Hotje, M. Binnewies, *Z. Anorg. Allg. Chem.* **2005**, *631*, 2467.
Hot 2005g	U. Hotje, R. Wartchow, M. Binnewies, *Z. Naturforsch.* **2005**, *60b*, 1241.
Hua 1988	Y.-S. Huang, S.-S. Lin, *Mater. Res. Bull.* **1988**, *23*, 277.
Hwa 1978	H. L. Hwang, C. Y. Sun, C. Y. Leu, C. L. Cheng, C. C. Tu, *Rev. Phys. Appl.* **1978**, *13*, 745.
Hwa 1980	H. L. Hwang, B. H. Tseng, *Solar Energy Mater.* **1980**, *4*, 67.
Ino 1984	M. Inoue, H. Negishi, *J. Phys. Soc. Jpn.* **1984**, *53*, 943.
Ish 1978a	H. Ishizuki, I. Nakada, *Jpn. J. Appl. Phys.* **1978**, *17*, 43.
Ish 1978b	H. Ishizuki, *Jpn. J. Appl. Phys.* **1978**, *17*, 1171.
Ish 1978c	H. Ishizuki, I. Nakada, *J. Cryst. Growth* **1978**, *44*, 632.
Ito 1979	K. Ito, Y. Matsuura, *Proc. Electrochem. Soc.* **1979**, *79*, 281.
Jei 1975	W. Jeitschko, P. C. Donohue, *Acta Crystallogr.* **1975**, *B31*, 1890.
Jin 1998	M.-S. Jin, Y.-G. Kim, B.-S. Park, D.-I. Yang, H.-J. Lim, H.-L. Park, W.-T. Kim, *Inst. Phys. Conf. Ser. No. 152:Section C,* Salford **1997**, IOP Publishing Ltd., **1998**.
Jon 1962	F. Jona, *J. Phys. Chem. Solids* **1962**, *23*, 1719.
Jon 1963	F. Jona, *J. Chem. Phys.* **1963**, *38*, 346.
Jon 1964	F. Jona, G. Mandel, *J. Phys. Chem. Solids* **1964**, *25*, 187.
Jos 1981	N. V. Joshi, L. Martinez, E. Echeverria, *J. Phys. Chem. Solids* **1981**, *42*, 281.
Jos 1993	S. Joshi, D. Lakshminarayana, P. K. Garg, M. K. Agarwal, *Ind. J. Pure Appl. Phys.* **1993**, *31*, 651.
Jos 1994	S. Joshi, D. Lakshminarayana, P. K. Garg, M. K. Agarwal, *Cryst. Res. Technol.* **1994**, *29*, 109.
Kai 1995	T. Kai, M. Kaifuku, I. Aksenov, K. Sato, *Jpn. J. Appl. Phys.* **1995**, *34*, 4682.
Kal 1965	E. Kaldis, *J. Phys. Chem. Solids,* **1965**, *26*, 1697.
Kal 1969	E. Kaldis, *Z. Kristallogr.* **1969**, *128*, 444.
Kal 1972	E. Kaldis, *J. Cryst. Growth* **1972**, *17*, 3.
Kal 1974	E. Kaldis, *Principles of the vapor growth of single crystals.* In: C. H. L. Goodman (Ed.) *Crystal Growth, Theory and Techniques.* Vol. 1. **1974**, 49.
Kal 1987	J. A. Kalomiros, A. N. Anagnostopoulos, J. Spyridelis, *Mater. Res. Bull.* **1987**, *22*, 1307.
Kha 1984	A. Khan, N. Abreu, L. Gonzales, O. Gomez, N. Arcia, D. Aguilera. S. Tarantino, *J. Cryst. Growth* **1984**, *69*, 241.
Kim 1991	D. T. Kim, K. S. Yu, W. T. Kim, *New Phys.* (Korean Physical Soc.) **1991**, *31*, 477.
Kim 1993	K. Kimoto, K. Masumoto, Y. Noda, K. Kumazawa, T. Kiyosawa, N. Koguchi, *Jpn. J. Appl. Phys.,* Part 1, **1993**, 32 (Suppl. 32–3, Proceedings of the 9th International Conference of Ternary and Multinary Compounds, **1993**), 187.
Kit 1987	M. Kitagawa, Y. Tomomura, S. Yamaue, S. Nakajima, *Sharp Technical Journal* **1987**, *27*, 13.
Kni 1999	S. Knitter, M. Binnewies, *Z. Anorg. Allg. Chem.* **1999**, *625*, 1582.
Kni 2000	S. Knitter, M. Binnewies, *Z. Anorg. Allg. Chem.* **2000**, *626*, 2335.
Kor 2002	Y. V. Korostelin, V. I. Kozlovsky, *Phys. Status Solidii* **2002**, *B229*, 5.
Kou 1988	K. Kourtakis, J. DiCarlo, R. Kershaw, K. Dwigth, A. Wold, *J. Solid State Chem.* **1988**, *76*, 186.
Koy 2000	M. Koyano, H. Watanabe, Y. Yamura, T. Tsuji, S. Katayama, *Mol. Cryst. Liq. Cryst.* **2000**, *341*, 33.
Kra 1965	L. Krausbauer, R. Nitsche, P. Wild, *Phys.* **1965**, *31*, 113.

Kra 1975   G. Krabbes, H. Oppermann, E. Wolf, *Z. Anorg. Allg. Chem.* **1975**, *416*, 65.
Kra 1976a   G. Krabbes, H. Oppermann, E. Wolf, *Z. Anorg. Allg. Chem.* **1976**, *421*, 111.
Kra 1976b   G. Krabbes, H. Oppermann, E. Wolf, *Z. Anorg. Allg. Chem.* **1976**, *423*, 212.
Kra 1978   G. Krabbes, H. Oppermann, J. Henke, *Z. Anorg. Allg. Chem.* **1978**, *442*, 79.
Kra 1979   G. Krabbes, H. Oppermann, *Z. Anorg. Allg. Chem.* **1979**, *450*, 27.
Kra 1980   G. Krabbes, H. Oppermann, H. Henke, *Z. Anorg. Allg. Chem.* **1980**, *470*, 7.
Kra 1981   G. Krabbes, H. Oppermann, *Z. Anorg. Allg. Chem.* **1981**, *481*, 13.
Kra 1984a   G. Krabbes, H. Oppermann, *Z. Anorg. Allg. Chem.* **1984**, *511*, 19.
Kra 1984b   G. Krabbes, J. Klosowski, H. Oppermann, H. Mai, *Cryst. Res. Technol.* **1984**, *19*, 491.
Kra 1997   G. Krauss, V. Kramer, A. Eifler, V. Riede, S. Wenger, *Cryst. Res. Technol.* **1997**, *32*, 223.
Krä 1970   V. Krämer, R. Nitsche, J. Ottemann, *J. Cryst. Growth* **1970**, *7*, 285.
Krä 1971   V. Krämer, R. Nitsche, *Z. Naturforsch.* **1971**, *26b*, 1074.
Krä 1976   V. Krämer, *Thermochim. Acta* **1976**, *15*, 205.
Krä 1980   V. Krämer, K. Berroth, *Mater. Res. Bull.* **1980**, *15*, 299.
Kre 1993   B. Krebs, A. Schiemann, M. Läge, *Z. Anorg. Allg. Chem.* **1993**, *619*, 983.
Kry 2007   G. Kryukova, M. Heuer, Th. Doering, K. Bente, *J. Cryst. Growth* **2007**, *306*, 212.
Kul 1987   L. M. Kulikov, *Izv. Akad. Nauk SSSR, Neorg. Mater.* **1987**, *23*, 681.
Kur 1980   T. K. Kurbanov, P. G. Rustamov, *Izv. Akad. Nauk SSSR, Neorg. Mater.* **1980**, *16*, 611.
Kus 1998   T. Kusawake, Y. Takahashi, K. Oshima, *Mater. Res. Bull.* **1998**, *33*, 1009.
Lah 1981   N. Lahlou, G. Masse, *J. Appl. Phys.* **1981**, *52*, 978.
Lam 1972   V. G. Lamprecht, *Mater. Res. Bull.* **1972**, *7*, 1411.
Lap 1962   F. Lappe, A. Niggli, R. Nitsche, J. G. White, *Z. Kristallogr.* **1962**, *117*, 146.
Lau 2010   R. Lauck *J. Cryst. Growth* **2010**, *312*, 3642.
Lee 2003   S.-J. Lee, J.-E. Kim, H. Y. Park, *J. Mater. Res.* **2003**, *18*, 733.
Lei 1980   M. Leiss, *J. Phys. Solid State Phys.* **1980**, *13*, 151.
Len 1971   E. Lendvay, *J. Cryst. Growth* **1971**, *10*, 77.
Len 1975   E. Lendvay, *Acta Technica Scientiarium Hungaria* **1975**, *80*, 151.
LeN 1979   N. Le Nagard, A. Katty, G. Collin, O. Gorochov, A. Willig, *J. Solid State Chem.* **1979**, *27*, 267.
Lev 1983   F. Levy, H. Berger, *J. Cryst. Growth* **1983**, *61*, 61.
Lia 2009   C. H. Lia Y. H. Chan, K. K. Tiog, Y. S. Huang, Y. M. Chen, D. O. Dumenco, C. H. Ho, *J. Alloys and Compd.*, **2009**, *480*, 94.
Lie 1969   R. M. A. Lieth, C. W. M. van der Heijden, J. W. M. van Kessel, *J. Cryst. Growth* **1969**, *5*, 251.
Lie 1972   R. M. A. Lieth, *Phys. Status Solidi* **1972**, *12*, 399.
Lin 1992   S.-S. Lin, J.-K. Huang, Y.-S. Huang, *Mater. Res. Bull.* **1992**, *27*, 177.
Lip 1993   V. I. Lipnitskii, V. A. Savchuk, B. V. Korzun, G. I. Makovetskii, G. P. Popelnjuk, *Jpn. J. Appl. Phys.* **1993**, *32*, Suppl. 32–3, 635.
Loc 2004a   S. Locmelis, M. Binnewies, *Z. Anorg. Allg. Chem.* **2005**, *630*, 1301.
Loc 2004b   S. Locmelis, M. Binnewies, *Z. Anorg. Allg. Chem.* **2005**, *630*, 1308.
Loc 2005a   S. Locmelis, U. Hotje, M. Binnewies, *Z. Anorg. Allg. Chem.* **2005**, *631*, 3080.
Loc 2005b   S. Locmelis, E. Milke, M. Binnewies, S. Gruhl, C. Vogt, *Z. Anorg. Allg. Chem.* **2005**, *631*, 1667.
Loc 2005c   S. Locmelis, U. Hotje, M. Binnewies, *Z. Anorg. Allg. Chem.* **2005**, *631*, 672.
Loc 2007   S. Locmelis, C. Brünig, M. Binnewies, A. Börger, K.-D. Becker, T. Homann, T. Bredow, *J. Mater. Sci.* **2007**, *42*, 1995.

Lop 1998   S. A. Lopez-Rivera, B. Fontal, J. A. Henao, E. Mora, W. Giriat, R. Vargas, *Institute of Physics Conference Series* **1998**, *152* (Ternary and Multinary Compounds), 175.

Lut 1968   H. D. Lutz, Cs. Lovasz, *Angew. Chem.* **1968**, *80,* 562.

Lut 1970   H. D. Lutz, Cs. Lovasz, K. H. Bertram, M. Sreckovic , U. Brinker, *Monatsh. Chemie* **1970**, *101*, 519.

Lut 1973a   H. D. Lutz, K. H. Bertram, *Z. Anorg. Allg. Chem.* **1973**, *401*, 185.

Lut 1973b   H. D. Lutz, K. H. Bertram, M. Sreckovic, W. Molls, *Z. Naturforsch.* **1973**, *28b,* 685.

Lut 1974   H. D. Lutz, K. H. Bertram, G. Wribel, M. Ridder, *Monatsh. Chemie* **1974**, *105*, 849.

Lut 1989   H. D. Lutz, W. Becker, B. Mueller, M. Jung, *J. Raman Spectrosc.* **1989**, *20*, 99.

Mäh 1982   D. Mähl, J. Pickardt, B. Reuter, *Z. Anorg. Allg. Chem.* **1982**, *491,* 203.

Mäh 1984   D. Mähl, J. Pickardt, B. Reuter, *Z. Anorg. Allg. Chem.* **1984**, *516,* 102.

Ma 2008   J. Ma, H. Jin, X. Liu, M. E. Fleet, J. Li, X. Cao, S. Feng, *Cryst. Growth Des.* **2008**, *8*, 4460.

Mam 1988   A. I. Mamedov, T. M. Iljasov, P. G. Rustamov, F. G. Akperov, *Zh. Neorg. Khim.* **1988**, *33*, 1103.

Mar 1983   J. V. Marzik, A. K. Hsieh, K. Dwight, A. Wold, *J. Solid State Chem.* **1983**, *49*, 43.

Mar 1984   J. V. Marzik, R. Kershaw, K. Dwight, A. Wold, *J. Solid State Chem.* **1984**, *51*, 170.

Mar 1998   K. Mariolacos, *N. Jahrb. Mineral. Monatsh.* **1998**, 164. K. Mariolacos, *N. Jb. Miner. Mh.* **1999**, 415.

Mar 1999   K. Mariolacos, *N. Jahrb. Mineral. Monatsh.* **1999**, 415.

Mas 1990   H. Massalski, *Binary Alloy Phase Diagrams*, 2nd Ed. ASN International, **1990.**

Mat 1983   K. Matsumoto, K. Takagi, *J. Cryst. Growth* **1983**, *62*, 389.

Mat 1986   K. Matsumoto, G. Shimaoka, *J. Cryst. Growth* **1986**, *79*, 723.

Mat 1988   K. Matsumoto, *Shizuoka Daigaku Denshi Kogaku Kenkyusho Kenkyu Hokoku* **1988**, *23*, 127.

Men 2006   F. Menzel, R. Kaden, D. Spemann, K. Bente, T. Butz, *Nucl. Instr. Meth. Phys. Res. B* **2006**, *249*, 478.

Mer 1973   J. Mercier, J. C. Bruyere, *Bull. Soc. Scie. De Bretagne* **1973**, *48*, 135.

Mic 1990   G. Micocci, R. Rella, P. Siciliano, A. Tepore, *J. Appl. Phys.* **1990**, *68*, 138.

Min 1980   T. Minagawa, *J. Phys. Soc. Jpn.* **1980**, *49,* 2317.

Moc 2001   K. Mochizuki, N. Kuroishi, K. Kimoto, *Ishinomaki Senshu Daigaku Kenkyu Kiyo* **2001**, *12*, 37.

Mur 1973   S. Muranka, T. Takada, *Bull Inst. Chem. Res. Kyoto Univ.* **1973,** *51*, 287.

Mus 1975   F. M. Mustafaev, F. I. Ismailiv, *Izv. Akad. Nauk SSSR, Neorg. Mater.* **1975**, *11*, 1552.

Nak 1978   I. Nakada, M. Kubota, *J. Cryst. Growth* **1978**, *43*, 711.

Nak 1984   I. Nakada, Y. Ishihara, *Jpn. J. Appl. Phys.* **1984**, *23*, 677.

Nam 2008   Y. Nambu, M. Ichihara, Y. Kiuchi, S. Nakatsuji, Y. Maeno, *J. Cryst. Growth* **2008**, *310*, 1881.

Neu 1989   H. Neumann, W. Kissinger, F. Levy, H. Sobotta, V. Riede, *Cryst. Res. Technol.* **1989**, *24*, 1165.

Nis 1978   T. Nishinaga, R. M. A. Lieth, *J. Cryst. Growth* **1973**, *20*, 109.

Nit 1960   R. Nitsche, *J. Phys. Chem. Solids* **1960**, *17,* 163.

Nit 1961   R. Nitsche, H. U. Bölsterli, M. Lichtensteiger, *J. Phys. Chem. Solids* **1961**, *21*, 199.

Nit 1962   R. Nitsche, D. D. Richman, *Z. Elektrochem.* **1962**, *66*, 709.
Nit 1964   R. Nitsche, *Z. Kristallogr.* **1964**, *120*, 229.
Nit 1967a   R. Nitsche, *J. Phys. Chem. Solids, Suppl.* **1967**, *1*, 215.
Nit 1967b   R. Nitsche, D. Sargant, P. Wild, *J. Cryst. Growth* **1967**, *1*, 52.
Nit 1968   R. Nitsche, P. Wild, *J. Cryst. Growth* **1968**, 3/4, 153.
Nit 1970 R. Nitsche, P. Wild, *Mater. Res. Bull.* **1970**, *5*, 419.
Nit 1971 R. Nitsche, *J. Cryst. Growth* **1971**, *9*, 238.
Nod 1990a K. Noda, N. Matsumura, S. Otsuka, K. Matsumoto, *J. Electrochem. Soc.* **1990**, *137*, 1281.
Nod 1990b   K. Noda, T. Kurasawa, N. Sugai, Y. Furukawa, *J. Cryst. Growth* **1990**, *99*, 757.
Nod 1991   K. Noda, T. Kurasawa, Y. Furukawa, *J. Cryst. Growth* **1991**, *115*, 802.
Nod 1998 K. Noda, Y. Furukawa, S. Nakazawa, *Mem. Fac. Sci. Eng. Stimane Univ. Ser.* **1998**, A 32, 19.
Odi 1975   J. P. Odile, J. J. Steger, A. Wold, *Inorg. Chem.* **1975**, *14*, 2400.
Oh 1997a   S.-K. Oh, W.-T. Kim, E.-J. Cho, M.-S. Jin, H.-G. Kim, C.-D. Kim, *J. Kor. Phys. Soc.* **1997**, *31*, 677.
Oh 1997b   S.-K. Oh, W.-T. Kim, M.-S. Jin, C.-S. Yun, S.-H. Choe, C.-D. Kim, *J. Kor. Phys. Soc.* **1997**, *31*, 681.
Oh 1999   S.-K. Oh, H.-J. Song, T.-Y. Park, H.-G. Kim, S.-H. Choe, *Semicond. Sci. Technol.* **1999**, *14*, 848.
Ohg 1994   T. Ohgoh, I. Aksenov, Y. Kudo, K. Sato, *Jpn. J. Appl. Phys.* **1994**, *33*, 962.
Ohn 1996   Y. Ohno, *Phys. Rev. B: Condens. Matter Mater. Phys.* **1996**, *54*, 11693.
Ohn 2005   Y. Ohno, *J. Solid State Chem.* **2005**, *178*, 1539.
Oka 1974   F. Okamoto, K. Ametani, T. Oka, *Jpn. J. Appl. Phys.* **1974**, *13*, 187.
Paj 1980 A. Pajaczkowska, *J. Cryst. Growth* **1980**, *49*, 563.
Paj 1983 A. Pajaczkowska, *Mater. Res. Bull.* **1983**, *18*, 397.
Pal 1982a   W. Palosz, M. J. Kozielsky, B. Palosz, *J. Cryst. Growth* **1982**, *58*, 185.
Pal 1982b W. Palosz, *J. Cryst. Growth* **1982**, *60*, 57.
Pal 1983   W. Palosz, *J. Cryst. Growth* **1983**, *61*, 412.
Pal 1986   B. Palosz, W. Palosz, S. Gierlotka, *Bull. Mineral.* **1986**, *109*, 143.
Pao 1977a   C. Paorici, L. Zanotti, *Mater. Res. Bull.* **1977**, *12*, 1207.
Pao 1977b   C. Paorici, C. Pelosi, G. Attolini, *J. Cryst. Growth* **1977**, *37*, 9.
Pao 1978   C. Paorici, L. Zanotti, G. Zucalli, *J. Cryst. Growth* **1978**, *43*, 705.
Pao 1980   C. Paorici, L. Zanotti, *Mater. Chem.* **1980**, *5*, 337.
Pao 1982   C. Paorici, L. Zanotti, M. Curti, *Cryst. Es. Techn.* **1982**, *17*, 917.
Pas 1998   W. Paskowicz, J. Domagala, Z. Golacki, *J. Alloys Compd.* **1998**, *274*, 128.
Pat 1993   S. G. Patel, S. H. Chaki, A. Agarwal, *Phys. Status Solidi A* **1993**, *140*, 207.
Pat 1997   D. H. Patel, S. G. Patel, S. K. Arora, M. K. Agarwal, *Cryst. Res. Technol.* **1997**, *32*, 701.
Pat 1998   D. H. Patel, S. K. Arora, M. K. Agarwal, *Bull. Mater. Sci.* **1998**, *21*, 297.
Pat 2005   K. Patel, J. Prajapati, R. Vaidya, S. G. Patel, *Ind. J. Phys.* **2005**, *79*, 373.
Per 1973   A. Perrin, C. Perrin, R. Chevrel, M. Sergent, R. Brochu, J. Padiou, *Bull. Soc. Sci. Bretagne* **1973**, *48*, 141.
Per 1978   E. Yu. Peresch, W. W. Zigika, N. P. Stasjuk, J. W. Galagowez, A. W. Gapak, *Izv. Wyssch. Utsch. Saw.* **1978**, *21*, 1070.
Pes 1971   P. Peshev, W. Piekarczyk, S. Gazda, *Mater. Res Bull.* **1971**, *6*, 479.
Phi 1971   H. von Philipsborn, *J. Cryst. Growth* **1971**, *9*, 296.
Pic 1970   J. Pickardt, E. Riedel, B. Reuter, *Z. Anorg. Allg. Chem.* **1970**, *373*, 15.
Pic 1971   J. Pickardt, E. Riedel, *J. Solid State Chem.* **1971**, *3*, 67.
Pie 1970   W. Piekarczyk, P. Peshev, *J. Cryst. Growth* **1970**, *6*, 357.
Pin 1970   H. L. Pinch, L. Ekstrom, *R. C. A. Rev.* **1970**, *31*, 692.

Pos 1984   E. Post, V. Krämer, *Mater. Res. Bull.* **1984**, *19*, 1607.
Pra 2007   P. Prabukanthan, R. Dhansekaran, *Cryst. Growth Des.* **2007**, *7*, 618.
Pro 1979   V. A. Prokhorov, E. N. Kholina, A. V. Voronin, *Izv. Akad. Nauk SSSR, Neorg. Mater.* **1979**, *15*, 1923.
Que 1975   P. Quenez, A. Maurer, O. Gorochov, *J. Phys.* **1975**, *36*, 83.
Rab 1967   A. Rabenau, H. Rau, *Z. Phys. Chem.* **1967**, *53*, 155.
Rad 1980   S. I. Radautsan, V. I. Tezlevan, K. G. Nikiforov, *J. Cryst. Growth* **1980**, *49*, 67.
Rao 1995   Y. K. Rao, C. H. Raeder, *J. Mater. Synth. Process.* **1995**, *3*, 49.
Rao 2000   Y. K. Rao, C. H. Raeder, *Schriften des Forschungszentrums Jülich, Reihe Energietechnik, High Temperature Materials Chemistry* **2000**, *15*, 189.
Rem 1998   M. Remskar, Z. Skraba, C. Ballif, R. Sanjines, F. Levy, *Adv. Mater.* **1998**, *10*, 246.
Rem 1999   M. Remskar, Z. Skraba, C. Ballif, R. Sanjines, F. Levy, *Surface Sci.* **1999**, *433–435*, 637.
Rem 2002   M. Remskar, A. Mrzel, F. Levy, in: *Perspectives of Fullerene Nanotechnology*, 113 Kluwer Academic Publ. Dordrecht, **2002**.
Rim 1972   H. P. B. Rimmington, A. Balchin, B. K. Tanner, *J. Cryst. Growth* **1972**, *15*, 51.
Rim 1974   H. P. B. Rimmington, A. Balchin, B. K. Tanner, *J. Cryst. Growth* **1972**, *21*, 171.
Ros 2004   C. Rose, M. Binnewies, *Z. Anorg. Allg. Chem.* **2004**, *630*, 1296.
Rus 1969   P. G. Rustamov, B. A. Geidarov, *Azerb. Khim. Zh.* **1969**, *2*, 143.
Sae 1974   M. Saeki, M. Nakano, M. Nakahira, *J. Cryst. Growth* **1974**, *24/25*, 154.
Sae 1976   M. Saeki, *J. Cryst. Growth* **1976**, *36*, 77.
Sae 1977   M. Saeki, *Mater. Res. Bull.* **1977**, *12*, 773.
Sae 1978   M. Saeki, *J. Cryst. Growth* **1978**, *45*, 25.
Sae 1982   M. Saeki, M. Onoda, *Bull. Chem. Soc. Jpn.* **1982**, *55*, 113.
Sag 1995   V. Sagredo, H. Romero, L. Betancourt, J. Alvarez, G. Attolini, C. Pelosi, *Mater. Sci. Forum* **1995**, *182–184*, 467.
Sag 1998   V. Sagredo, L. Nieves, G. Attolini, *Institute of Physics Conference Series* **1998**, *152*, 677.
Sag 2004   V. Sagedo, L. Betancourt, L. M. de Chalbaud, G. E. Delgado, *Cryst. Res. Technol.* **2004**, *39*, 873.
Sam 1961a   H. Samelson, *J. Appl. Phys.* **1961**, *32*, 309.
Sam 1961b   H. Samelson, V. A. Brophy, *J. Electrochem Soc.* **1961**, *108*, 150.
Sam 1962   H. Samelson, *J. Appl. Phys.* **1962**, *33*, 1779.
Sav 2000   *Proc. 12th Int. Conf. Ternary and Multinary Compounds, Jpn. J. Appl. Phys.* **2000**, *39*, Suppl. 39–1, 69.
Sch 1962   H. Schäfer, *Chemische Transportreaktionen*, Verlag Chemie, Weinheim, **1962**.
Sch 1968   H. Schäfer, F. Wehmeier, M. Trenkel, *J. Less-Common Met.* **1968**, *16*, 290.
Sch 1971   H. Schäfer, *J. Cryst. Growth* **1971**, *9*, 17.
Sch 1973a   H. Schäfer, T. Grofe, M. Trenkel, *J. Solid State Chem.* **1971**, *8*, 14.
Sch 1973b   H. Schäfer, *Z. Anorg. Allg. Chem.* **1973**, *400*, 253.
Sch 1978   H. Schäfer, H, Plautz, C. Balarew, J. Bazelkov, *Z. Anorg. Allg. Chem.* **1978**, *440*, 130.
Sch 1980   H. Schäfer, *Z. Anorg. Allg. Chem.* **1980**, *471*, 21.
Sch 1981   H. Schäfer, *Z. Anorg. Allg. Chem.* **1981**, *475*, 201.
Sch 1982   H. Schäfer, *Z. Anorg. Allg. Chem.* **1982**, *486*, 33.
Schl 2009   A. Schlechte, K. Meier, R. Niewa, Y. Prots, M. Schmidt, R. Kniep, *Z. Kristallogr. NCS* **2009**, *224*, 375.
Schö 1999   J. H. Schön, E. Bucher, *Phys. Status Solidi* **1999**, *171*, 511.
Schu 1979   M. Schulte-Kellinghaus, V. Krämer, *Acta Crystallogr.* **1979**, *B35*, 3016.

Schu 1982   M. Schulte-Kellinghaus, V. Krämer, *Z. Naturforsch.* **1982**, *37B*, 390.
Sen 2004   O. Senthil Kumar, S. Soundeswaran, R. Dhanasekaran, *Mater. Chem. Phys.* **2004**, *87*, 75.
Sev 1970   V. G. Sevast`yanov, G. V. Ellert, V. K. Slovyanskikh, *Zh. Neorg. Khim.* **1972**, *17*, 16.
Shi 1988   Y. J. Shin, B. H. Park, H. K. Min, T. S. Jeong, Y. W. Seo, P. W. Rho, P. Y. Yu, *New Physics (Kor. Phys. Soc.)* **1988**, *28*, 616.
Shi 1989   Y. J. Shin, B. H. Park, T. S. Jeong, Y. W. Seo, K. S. Rheu, T. S. Kim, S. Kang, P. Y. Yu, *New Physics (Kor. Phys. Soc.)* **1989**, *29*, 275.
Shi 1990   T. Shibata, N. Kambe, Y. Muranushi, T. Miura, T. Kishi, *J. Phys. D: Appl. Phys.* **1990**, *23*, 719.
Shi 1991   T. Shibata, N. Kambe, Y. Muranushi, T. Miura, T. Kishi, *J. Phys. Chem. Solids* **1991**, *52*, 551.
Shi 1992   Y. J. Shin, T. S. Jeong, H. K. Shin, K. S. Rheu, T. S. Kim, S. Kang, J. H. Song, *New Physics (Kor. Phys. Soc.)* **1992**, *32, 91*.
Sie 1983   K. Sieber, B. Fotouhi, O. Gorochov, *Mater. Res. Bull.* **1983**, *18*, 1477.
Sim 1980   C. T. Simpson, W. Imaino, W. M. Becker, *Phys. Rev.* **1980**, *B 22*, 911.
Ska 1963   M. Skala, K. Hauptmann, *Z. Naturforsch.* **1963**, *18a*, 368.
Slo 1966   V. K. Slovyanskikh, G. V. Ellert, E. I. Yarembash, M. D. Korsakova, *Izv. Akad. Nauk SSR, Neorg. Mater.* **1966**, *6*, 973.
Smi 1967   P. K. Smith, L. Cathey, *J. Electrochem. Soc.* **1967**, *114*, 973.
Sol 1976   S. Soled, A. Wold, O. Gorochov, *Mater. Res. Bull* **1976**, *11*, 927.
Sti 1992   T. Stingl, B. Mueller, H. D. Lutz, *J. Alloys Compd.* **1992**, *184*, 275.
Taf 1996a   M. J. Tafreshi, K. Balakrishnan, R. Dhanasekaran, *Il Nuovo Cimento* **1996**, *18D*, 471.
Taf 1996b   M. J. Tafreshi, K. Balakrishnan, J. Kumar, R. Dhanasekaran, *Ind. J. Pure Appl. Phys.* **1996**, *34*, 18.
Tan 1995   K. Tanaka, T. Ohgoh, K. Kimura, H. Yamamoto, *Jpn. J. Appl. Phys. Lett.* **1995**, 34, L1651.
Tan 1997   S. Tanaka, S. Kawami, H. Kobayashi, H. Sasakura, *J. Phys. Chem. Solids* **1977**, *38*, 680.
Tan 1980   M. Taniguchi, M: Wakihara, Y. Shirai, *Z. Anorg. Allg. Chem.* **1980**, *461*, 234.
Tan 1989   K. Tanaka, K. Ishii, S. Matsuda, Y. Hasegawa, K. Sato, *Jpn. J. Appl. Phys. Lett.* **1989**, *28*, 12.
Tay 1973   B. E. Taylor, J. Steger, A. Wold, *J. Solid State Chem.* **1973**, *7*, 461.
Tho 1983   A. E. Thomas, G. J. Russel, J. Woods, *J. Cryst. Growth* **1983**, *63*, 265.
Tom 1995   Y. Tomm, R. Schiek, K. Ellmer, S. Fiechter, *J. Cryst. Growth* **1995**, *146*, 271.
Tom 1998   Y. Tomm, S. Fiechter, K. Diener, H. Tributsch, *Int. Phys. Conf. Ser.* No 152: *Section A, Int. Conf. on Ternary and Multinary Compounds*, IMTMC-11, Salford **1997**, p. 167.
Tsa 1994   M.-Y. Tsay, S.-H. Chen, C.-S. Chen, Y.-S. Huang, *J. Cryst. Growth* **1994**, *144*, 91.
Tsu 2006   N. Tsujii, Y. Imanaka, T. Takamasu, H. Kitazawa, G. Kido, *J. Alloys Compd.* **2006**, *408–412*, 791.
Uji 1971   S. Ujiie, Y. Kotera, *J. Cryst. Growth* **1971**, *10*, 320.
Uji 1976   S. Ujiie, *Denki Kagaku* **1976**, *44*, 22.
Val 1970   Y. Valov, A. Paionchkovska, *Izv. Akad. Nauk SSSR, Neorg. Mater.* **1970**, *6*, 241.
Van 1969   C. B. Van den Berg, J. E. van Delden, J. Boumann, *Phys. Status Solidi* **1969**, *36*, K 89.
Vaq 2009   P. Vaqueiro, I. Szkoda, D. Sanchez, A. V. Powell, *Inorg. Chem.* **2009**, *48*, 1284.

Ven 1986a	B. Vengatesan, N. Kanniah, P. Ramasamy, *J. Mater. Sci. Lett.* **1986**, *5*, 984.
Ven 1986b	B. Vengatesan, N. Kanniah, P. Ramasamy, *J. Mater. Sci. Lett.* **1986**, *5*, 595.
Ven 1987	B. Vengatesan, N. Kanniah, P. Ramasamy, *Mater. Chem. Phys.* **1987**, *17,* 311.
Ven 1988	B. Vengatesan, N. Kanniah, P. Ramasamy, *Mater. Sci. Eng.* **1988**, *A104*, 245.
Vir 2007	M. Virsek, A. Jesih, I. Milosevic, M. Damnjanovic, M. Remskar, *Surface Science* **2007**, *601,* 2868.
Vol 1993	V. V. Volkov, C. van Heurck, J. van Landuyt, S. Amelinckx, E. G. Zhukov, E. S. Polulyak, V. M. Novotortsev, *Cryst. Res. Technol.* **1993**, *28*, 1051.
Wak 1982	M. Wakihara, K. Kinoshita, H. Hinode, M. Taniguchi, *J. Cryst. Growth* **1982**, *56*, 157.
Wan 1972	R. Wandji, J. K. Kom, *Compt. Rend. Acad. Sci.* **1972**, *C 275*, 813.
War 1987	T. Warminski, M. Kwietniak, W. Giriat, L. L. Kazmerski, J. J. Loferski, *Ternary Multinary Comp. Proc. Int. Conf* 7th **1987**, 127.
Wat 1972	T. Watanabe, *J. Phys. Soc. Jpn.* **1972**, *32*, 1443.
Wat 1978	T. Watanabe, I. Nakada, *Jpn. J. Appl. Phys.* **1978**, *17,* 1745.
Whi 1978	C. R. Whitehouse, A. A. Balchin, *J. Cryst. Growth* **1978**, *43*, 727.
Whi 1979	C. R. Whitehouse, A. A. Balchin, *J. Cryst. Growth* **1979**, *47*, 203.
Wid 1971	R. Widmer, *J. Cryst. Growth* **1971**, *8*, 216.
Wie 1969	H. Wiedemeier, A. G. Sigai, *J. Cryst. Growth* **1969**, *6*, 67.
Wie 1979	H. Wiedemeier, F. J. Csillag, *J. Cryst. Growth* **1979**, *46*, 189.
Wol 1978	E. Wolf, H. Oppermann, G. Krabbes, W. Reichelt, *Current Topics in Materials Science* **1978**, *1*, 697.
Wu 1988	P. Wu, X-C. He, K. Dwight, A. Wold, *Mater. Res. Bull.* **1988**, *23,* 1605.
Wu 1989	P. Wu, R. Kershaw, K. Dwight, A. Wold, *Mater. Res. Bull.* **1989**, *24*, 49.
Yam 1974	S. Yamada, Y. Matsuno, J. Nanjo, S. Nomura, S. Hara, *Muroran Kogyo Daigaku Kenkyu Hokoku* **1974**, *8*, 451.
Yam 1975	Y. Yamamoto, N. Toghe, T. Miyauchi, *Jpn. J. Appl. Phys.* **1975**, *14*, 192.
Yam 1979	S. Yamada, J. Nanjo, S. Nomura, S. Hara, *J. Cryst. Growth* **1979**, *46*, 10.
Yao 1994	X. Yao, J. M. Honig, *Mater. Res. Bull.* **1994**, *29*, 709.
Yen 2002	P. C. Yen, M. J. Chen, Y. S. Huang, C. H. Ho, K. K. Tiong, *J. Phys.: Condens. Matt.* **2002**, *14*, 4737.
Yen 2004	P. C. Yen, Y. S. Huang, K. K. Tiong, *J. Phys.: Condens. Matt.* **2004**, *16*, 2171.
Yim 1973	W. M. Yim, A. K. Fan, E. J. Stofko, *J. Electrochem. Soc.* **1973**, *120*, 441.
Zhu 2008	X. D. Zhu, Y. P. Sun, X. B. Zhu, X. Luo, B. S. Wang, G. Li, Z. R. Yang, W. H. Song, J. M. Dai, *J. Cryst. Growth* **2008**, *311*, 218.
Zul 1974	D. I. Zul'fugarly, B. A. Geidarov, *Azerb. Khim. Zh.* **1974**, *2*, 110.
Zuo 2002a	R. Zuo, W. Wang, *J. Cryst. Growth* **2002**, *236*, 687.
Zuo 2002b	R. Zuo, W. Wang, *J. Cryst. Growth* **2002**, *236*, 695.

7.2 Transport of Selenides

Sb_2Se_3

H																	He
Li	Be											B	C	N	O	F	Ne
Na	Mg											Al	Si	P	S	Cl	Ar
K	Ca	Sc	Ti	V	Cr	Mn	Fe	Co	Ni	Cu	Zn	Ga	Ge	As	Se	Br	Xe
Rb	Sr	Y	Zr	Nb	Mo	Tc	Ru	Rh	Pd	Ag	Cd	In	Sn	Sb	Te	I	Kr
Cs	Ba	La	Hf	Ta	W	Re	Os	Ir	Pt	Au	Hg	Tl	Pb	Bi	Po	At	Rn

Ce	Pr	Nd	Pm	Sm	Eu	Gd	Tb	Dy	Ho	Er	Tm	Yb	Lu
Th	Pa	U	Np	Pu	Am	Cm	Bk	Cf	Es	Fm	Md	No	Lr

Iodine as transport agent

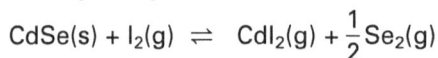

$$CdSe(s) + I_2(g) \rightleftharpoons CdI_2(g) + \frac{1}{2}Se_2(g)$$

Hydrogen halide as transport agent

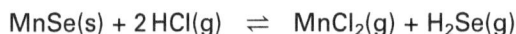

$$MnSe(s) + 2\,HCl(g) \rightleftharpoons MnCl_2(g) + H_2Se(g)$$

Hydrogen as transport agent

$$ZnSe(s) + H_2(g) \rightleftharpoons Zn(g) + H_2Se(g)$$

Today, many examples of CVT reactions of selenides of the main group elements (groups 2, 13, 14, and 15) are known. Almost all transition metal elements and some lanthanoids (see Table 7.2.1) are included, too. The alkali metal selenides cannot be transported with halogens or halogen compounds due to the high stability of the alkali metal halides. Chemical vapor transport reactions of selenides of the platinum metals are likewise limited due to the fact that the stability of the metal halides is too low. A general overview of the preparation of crystalline alkali metal selenides is provided by *Böttcher* and *Doert*, in particular by ammonothermal syntheses (Böt 1998).

The selenides are generally less stable than the analogous sulfides. Thus the transport reactions are less endothermic and the positions of transport equilibria shift to the side of the reaction products. Consequences are higher partial pressures of the transport effective species and lower dissolution temperatures. The first reports of *Nitsche* and his co-workers on the preparation and purification of selenides by means of heterogeneous gas-phase reactions coincide with the methodological development of CVT reactions (Nit 1957, Nit 1960, Nit 1962). Based on the definition of CVT reactions according to *Schäfer*, the basic equilibria of binary selenides were mainly established with the help of the examples of ZnSe and CdSe. For ternary compounds the examples of the selenide spinels ZnM_2Se_4 and CdM_2Se_4 were significant.

Thermal behavior   Some selenides sublime undecomposed. This applies for the compounds of groups 13 and 14, MSe (M = Ge, Sn, Pb) and M_2Se (M = Ga, In, Tl), respectively. In the process, the sublimation pressures decrease towards the heavy homologues of the respective group: GeSe ($K_{p, 700} = 3 \cdot 10^{-5}$ bar), SnSe ($K_{p, 700} = 2 \cdot 10^{-8}$ bar), PbSe ($K_{p, 700} = 2 \cdot 10^{-9}$ bar). However, PbSe melts at only 1077 °C so that sublimation is possible up to this temperature. Thus, high sublimation rates result: GeSe ($\vartheta_m = 675$ °C; $K_p > 10^{-2}$ bar), SnSe ($\vartheta_m = 540$ °C; $K_p > 10^{-6}$ bar), PbSe ($\vartheta_m = 1077$ °C; $K_p > 10^{-1}$ bar).

Experiments of sublimation of GeSe in transport ampoules filled with xenon ($p_{tot} = 2$ bar) under microgravity (experiments in the space laboratory) and intensified gravity up to 10 g (centrifuge), respectively, were used for verifying the respective contribution of diffusion and convection for the substance transport (Wie 1992) (see section 2.8).

A vast number of selenides show noticeable effects of dissolution by decomposition reactions in the gas phase (see Figure 7.2.1). In the process, high volatile, selenium-poor selenides of elements of groups 13, 14 and, 15, as well as selenium, are formed.

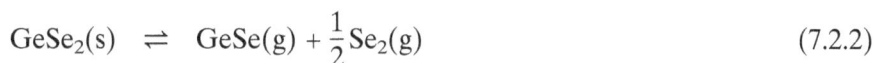

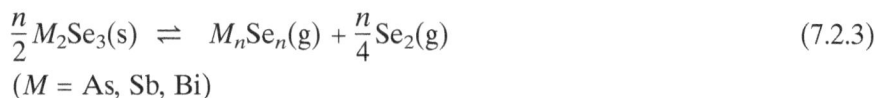

$$M_2Se_3(s) \rightleftharpoons M_2Se(g) + Se_2(g) \tag{7.2.1}$$
$$(M = \text{Al, Ga, In})$$

$$GeSe_2(s) \rightleftharpoons GeSe(g) + \frac{1}{2}Se_2(g) \tag{7.2.2}$$

$$\frac{n}{2}M_2Se_3(s) \rightleftharpoons M_nSe_n(g) + \frac{n}{4}Se_2(g) \tag{7.2.3}$$
$$(M = \text{As, Sb, Bi})$$

The thermal decomposition of ZnSe and CdSe to the elements is of importance, too.

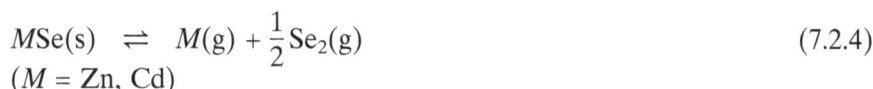

$$MSe(s) \rightleftharpoons M(g) + \frac{1}{2}Se_2(g) \tag{7.2.4}$$
$$(M = \text{Zn, Cd})$$

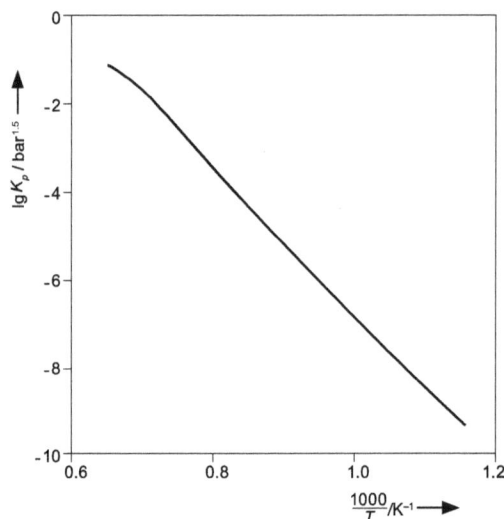

Figure 7.2.1 Decomposition pressure of CdSe as a function of the temperature according to *Sigai* and *Wiedemeier* (Sig1972).

In Figure 7.2.1, the decomposition pressure over cadmium selenide is illustrated as a function of temperature. Applying the decomposition equlibria, the deposition of crystalline ZnSe and CdSe over the gas phase is possible at temperatures above 1000 °C ($K_{p,\ 1200}$(ZnSe) = 1.5 · 10^{-4} bar; $K_{p,\ 1200}$(CdSe) = 3 · 10^{-4} bar) without adding a transport agent. Thus the thermodynamic basis of physical vapor deposition (PVD) processes for deposition of layers of these two compounds (Kle 1967, Sig 1972, Sha 1995) is provided.

Iodine as transport agent   More than three quarters of all known CVT reactions of selenides take place with the addition of iodine. The other halogens only play a marginal role. The stability of the gaseous metal halides, which decrease from chloride to iodide, compensates for the stability of the solids, which decreases from the metal oxides to the selenides. This way, balanced equilibrium positions and thus good results of vapor transport reactions of selenides with iodine are obtained.

If we take a look at the example that was already dealt with – the vapor transport of CdSe – it becomes clear that the temperature of the source can be decreased by up to 200 K by adding iodine compared to decomposition sublimation. Transport of CdSe is possible in the temperature range from 1000 to 700 °C ($K_{p,\ 1000}$ ≈ 1 bar) at the source and 700 to 500 °C at the sink (Kle1967); transport proceeds according to the following equilibrium:

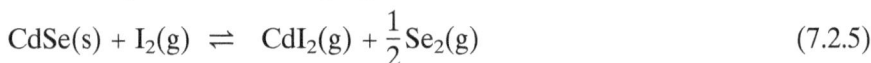

$$CdSe(s) + I_2(g) \;\rightleftharpoons\; CdI_2(g) + \frac{1}{2}Se_2(g) \tag{7.2.5}$$

The transport of most of the selenides takes place in this way. At temperatures above 600 °C, Se_2 dominates in the gas phase. Below this temperature, the higher condensed molecules Se_n (n = 3 … 8) have to be considered.

The CVT of a number of ternary selenides follows this mechanism. Thus the transport of mixed-crystals, such as $Cd_{1-x}Mn_xSe$ (Wie 1970, Sig1971), $Cd_{1-x}Fe_xSe$ (Smi 1988), or $CdS_{1-x}Se_x$ is described (Moc1978). If the ternary compound contains two different metal atoms, each of them will be transferred to the iodide while selenium predominates in the gas phase in elemental form.

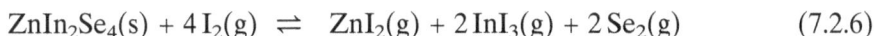

$$ZnIn_2Se_4(s) + 4\,I_2(g) \;\rightleftharpoons\; ZnI_2(g) + 2\,InI_3(g) + 2\,Se_2(g) \tag{7.2.6}$$

With increasing numbers of different kinds of atoms, however, the relations of all heterogeneous and homogeneous equilibria involved in the transport process become more complex. Often only *one* independent reaction is no longer sufficient for the description of the CVT. For reasons of clarity, a *dominating transport reaction* (7.2.7) is given.

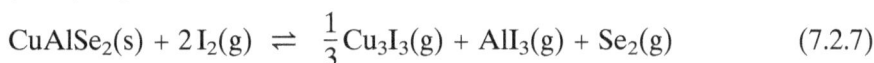

$$CuAlSe_2(s) + 2\,I_2(g) \;\rightleftharpoons\; \frac{1}{3}Cu_3I_3(g) + AlI_3(g) + Se_2(g) \tag{7.2.7}$$

Nevertheless, further independent equilibria ($r_u = s - k + 1 = 4$) influence the composition of the gas phase for the transport of $CuAlSe_2$:

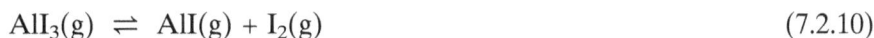

$$I_2(g) \;\rightleftharpoons\; 2\,I(g) \tag{7.2.8}$$

$$Cu_3I_3(g) \;\rightleftharpoons\; 3\,CuI(g) \tag{7.2.9}$$

$$AlI_3(g) \;\rightleftharpoons\; AlI(g) + I_2(g) \tag{7.2.10}$$

Which equilibrium dominates depends on the temperature and the pressure. This can be shown with the example of the transport of GaSe with iodine (Figure 7.2.2). At low iodine pressure, the sublimation of GaSe dominates; while the transport rate at a medium partial pressure is minimal due to the unfavorable

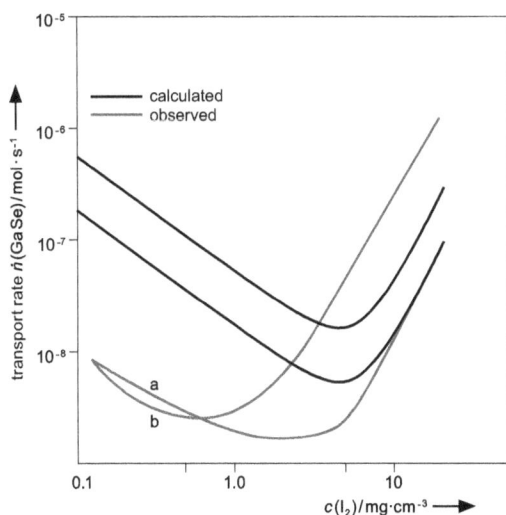

Figure 7.2.2 Calculated and experimentally determined transport rates for the transport of GaSe. The calculated curves represent the mean diffusion coefficients $D^0 = 0.035$ (top) and $0.012\ cm^2 \cdot s^{-2}$ (bottom) according to N*ishinaga et al.* (Nis 1975) (a: ampoule diameter 11 mm; b: ampoule diameter 20 mm).

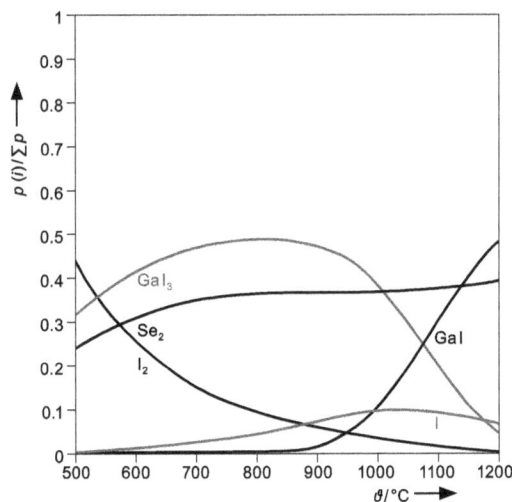

Figure 7.2.3 Normalized partial pressure in the Ga_2Se_3/I_2 system at $p^0(I_2) = 1$ bar according to *Hotje et al.* (Hot 2005b).

equilibrium position for the formation of $GaI(g)$. When the iodine pressure is increased, the formation of $GaI_3(g)$ becomes predominant and CVT takes place with high transport rates (Nis 1975). If the ampoule diameter is larger than 20 mm, the additional convective contribution to the gas movement will lead to even higher transport rates.

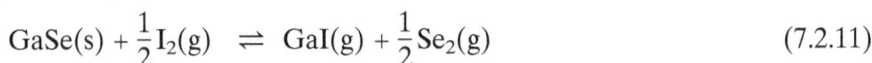

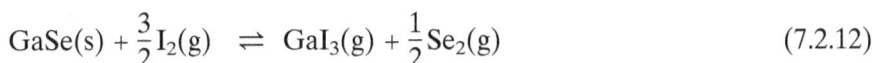

$$GaSe(s) + \frac{1}{2}I_2(g) \rightleftharpoons GaI(g) + \frac{1}{2}Se_2(g) \qquad (7.2.11)$$

$$GaSe(s) + \frac{3}{2}I_2(g) \rightleftharpoons GaI_3(g) + \frac{1}{2}Se_2(g) \qquad (7.2.12)$$

The transport effects of different equilibria for the transport of Ga_2Se_3 with iodine are dependent on the temperature. At low temperatures ($\vartheta < 800\ °C$) the formation of GaI_3 dominates (Figure 7.2.3); above approx. 900 °C, the formation of $GaI(g)$ becomes transport effective (Hot 2005b). The transport direction does not change because both reactions are endothermic.

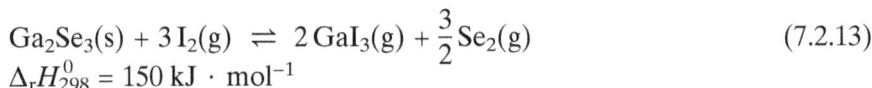

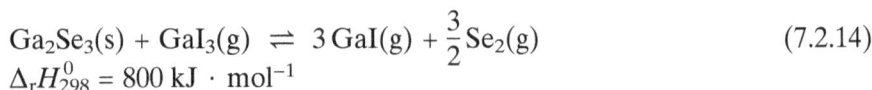

$$Ga_2Se_3(s) + 3I_2(g) \rightleftharpoons 2GaI_3(g) + \frac{3}{2}Se_2(g) \qquad (7.2.13)$$
$$\Delta_r H^0_{298} = 150\ kJ \cdot mol^{-1}$$

$$Ga_2Se_3(s) + GaI_3(g) \rightleftharpoons 3GaI(g) + \frac{3}{2}Se_2(g) \qquad (7.2.14)$$
$$\Delta_r H^0_{298} = 800\ kJ \cdot mol^{-1}$$

In contrast, the transport direction of $GeSe_2$ can be reversed by the directed choice of concentration of the transport agent. In any case, GeI_4 is formed. At a high initial pressure of iodine, it functions as transport agent and an endothermic transport ($520 \rightarrow 420\ °C$) is observed according to equation 7.2.15 (Buc 1987). At low initial pressure, iodine is present in atomic form and becomes effective as transport agent in the equilibrium 7.2.16; there is a reversal of the transport direction and an exothermic transport from 400 to 550 °C (Wie 1972) results:

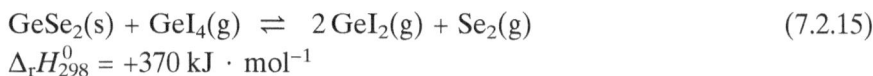

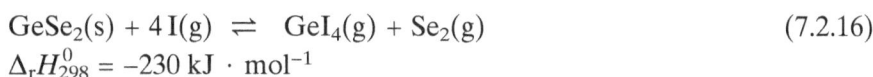

$$GeSe_2(s) + GeI_4(g) \rightleftharpoons 2\,GeI_2(g) + Se_2(g) \tag{7.2.15}$$
$$\Delta_r H_{298}^0 = +370 \text{ kJ} \cdot \text{mol}^{-1}$$

$$GeSe_2(s) + 4\,I(g) \rightleftharpoons GeI_4(g) + Se_2(g) \tag{7.2.16}$$
$$\Delta_r H_{298}^0 = -230 \text{ kJ} \cdot \text{mol}^{-1}$$

Schäfer and co-workers reported on the reversal of the transport direction, which is dependent on the composition of the solid MQ_n within a chemical system M/Q (Sch 1965). The examples of some niobium compounds (Nb/Q; Q = P, As, Sb, S, Se, Te) show that metal-poor compounds are transported from the zone of higher temperature to the cooler in an endothermic reaction, while metal-rich compounds migrate from the cooler side to the hotter in an exothermic equilibrium. In this manner, the selenium-rich compounds $NbSe_2$ (Bri 1962) and Nb_3Se_4 (Nak 1985) are transportable in the temperature gradient T_2 to T_1. The selenide $Nb_{1+x}Se$, on the other hand, follows the gradient from T_1 to T_2 by vapor transport with iodine (Sch 1965). An estimation of the transport direction is possible with the help of the following partial reactions:

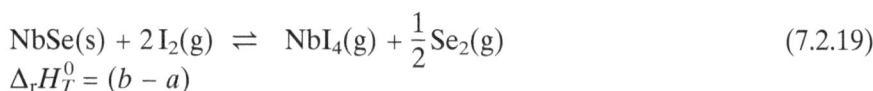

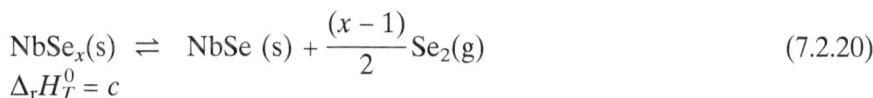

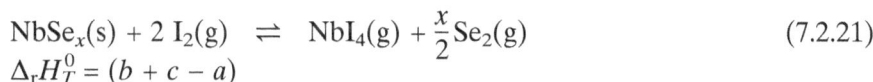

$$Nb(s) + 2\,I_2(g) \rightleftharpoons NbI_4(g) \tag{7.2.17}$$
$$\Delta_r H_T^0 = -a$$

$$NbSe(s) \rightleftharpoons Nb(s) + \frac{1}{2}Se_2(g) \tag{7.2.18}$$
$$\Delta_r H_T^0 = b$$

$$NbSe(s) + 2\,I_2(g) \rightleftharpoons NbI_4(g) + \frac{1}{2}Se_2(g) \tag{7.2.19}$$
$$\Delta_r H_T^0 = (b - a)$$

$$NbSe_x(s) \rightleftharpoons NbSe\,(s) + \frac{(x-1)}{2}Se_2(g) \tag{7.2.20}$$
$$\Delta_r H_T^0 = c$$

$$NbSe_x(s) + 2\,I_2(g) \rightleftharpoons NbI_4(g) + \frac{x}{2}Se_2(g) \tag{7.2.21}$$
$$\Delta_r H_T^0 = (b + c - a)$$

Accordingly, the dissolution of niobium in the form of the iodide takes place in the exothermic reaction 7.2.17 ($\Delta_r H_T^0 = -a$). This is in contrast to the endothermic decomposition 7.2.18, which does not compensate for the exothermic part $((b - a) < 0)$ in the balance of the total reaction (7.2.19) for weakly endothermic compounds NbSe. If the amount c of the heat of reaction of the metal-poorer compounds $NbSe_x$ is included in the balance of the total reaction, the characteristics of the resulting transport equilibrium (7.2.21) will change $((b + c - a) > 0)$ and an endothermic transport results $(T_2 \rightarrow T_1)$. Such behavior can be used for the separation of the compounds because the deposition occurs at different sites of the ampoule.

The example of niobium(IV)-selenide, $NbSe_2$, shows another advantage of CVT reactions. By choosing a suitable temperature gradient, different modifications of the compounds can be crystallized.

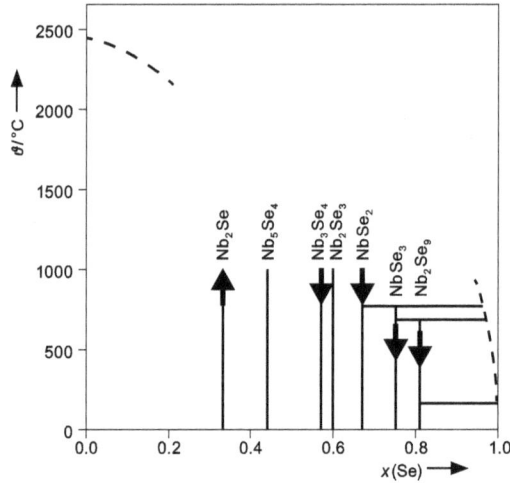

Figure 7.2.4 Schematic phase diagram of the Nb/Se binary system and the temperature ranges of the transport of niobium selenides.

The result of the CVT of selenides is not only caused by the choice of the dissolution temperature and the amount of transport agents. If the transport equilibrium involves the formation of $Se_2(g)$, the temperature at the deposition side must not fall below approx. 400 °C, because otherwise selenium will condense at the sink. In this case, the composition of the source solid will change and the transport does not lead to the desired product or is disrupted completely.

Some systems require special attention with regard to the choice of the transport conditions because of the formation of condensed selenide halides (see section 8.2). Too high pressure of the transport agent as well as too large temperature gradients can lead to the deposition of the respective solid selenide halide. The transport of selenide halides takes place first. The condensation of the halide-containing solid in the sink causes a decrease of the vapor pressure of the transport agent. If the equilibrium pressure of the selenide halide is no longer reached on the source side, the transport of pure selenide occurs.

The transport of Bi_2Se_3 and Sb_2Se_3 (Schö 2010) takes place under such limited conditions of the co-existence of the selenide iodides BiSeI and SbSeI, respectively. Characteristic courses of transports are illustrated in Figure 7.2.5 for the transport of Bi_2Se_3:

Case I: The stationary transport of Bi_2Se_3 is possible at a dissolution temperature $\vartheta_{source} = 550$ °C and a temperature gradient up to $\Delta T = 100$ K if the equilibrium pressure of $BiI_3(g)$ does not exceed a value of 10^{-3} bar. The concentration of iodine, which is used as transport additive, is limited to $\beta^0 \leq 0.5$ mg/cm^{-3} ampoule.

Case II: The transport will change if the amount of iodine is increased. If the equilibrium pressure of $p(BiI_3(g)) = 10^{-3}$ bar is exceeded, the equilibrium condition for the formation of BiSeI is fulfilled at the deposition side and the compound will be transported to the sink. During the deposition of BiSeI the partial

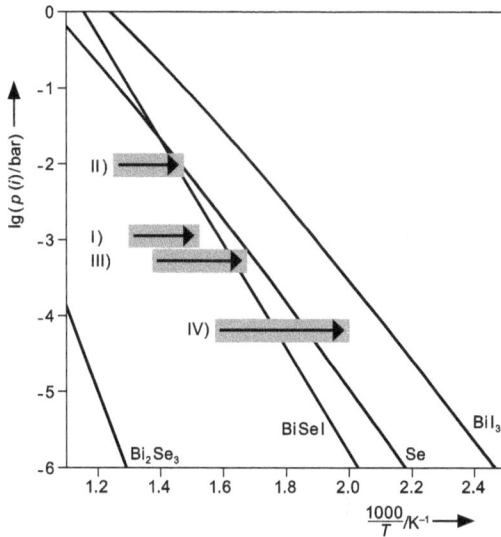

Figure 7.2.5 Phase barogram of the Bi_2Se_3/BiI_3 system and illustration of phase relations during the CVT of bismuth(III)-selenide. I: stationary transport of Bi_2Se_3; II and III: sequential transport of BiSeI followed by Bi_2Se_3; IV: simultaneous transport of BiSeI and Se according to *Schöneich* et al. (Schö 2010).

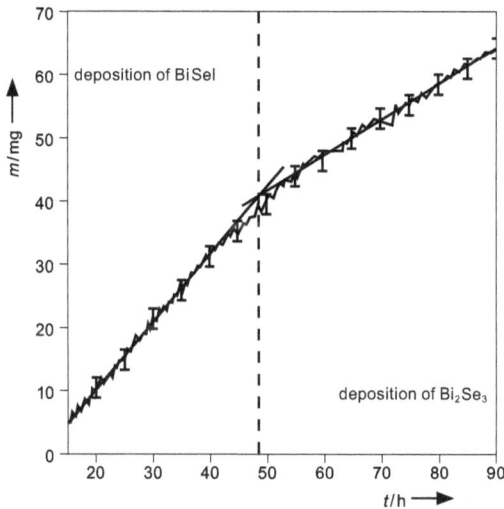

Figure 7.2.6 Diagram of time-dependent mass transport of Bi_2Se_3. Sequential transport of BiSeI followed by Bi_2Se_3 during a transport from 500 to 425 °C as described in case II, according to *Schöneich et al.* (Schö 2010).

pressure of BiI_3 decreases. In the process, the transport behavior changes to the initially described case I and, consequently, Bi_2Se_3 is transported (Figure 7.2.6).

Case III: If the equilibrium pressure $p(BiI_3(g)) < 10^{-3}$ bar (starting concentration of iodine: $\beta^0 \leq 0.5$ mg $\cdot$ cm^{-3}) as well as the temperature gradient are kept constant at $\Delta T = 100$ K, the transport behavior will be dependent on the dissolution temperature: at $\vartheta_{source} \leq 450$ °C, BiSeI is deposited in the sink.

Case IV: If the temperature gradient is increased to $\Delta T > 100$ K, the saturation pressure of selenium is achieved at the cooler zone of the ampoule and a simultaneous transport of BiSeI and selenium is observed at the sink.

As the largest number of known CVT reactions of selenides take place with the addition of iodine, the ternary compounds Cd_4MSe_6 (M = Si, Ge) are unusual insofar as they are not only transportable with iodine but also when chlorine and bromine are added. The transports are realized in the temperature range from 680 to 520 °C ($\Delta T = 100$ K) (Kal 1967, Que 1974).

Hydrogen halide and hydrogen as transport agents If hydrogen or hydrogen halides are used as transport agents, the solubility of selenium in the gas phase can be supported by the formation of hydrogen selenide, H_2Se. In the process, the use of a hydrogen halide, in particular hydrogen chloride, leads to the formation of the respective metal chloride (7.2.22). Often ammonium chloride, which is easy to handle in experiments, is used as a source for hydrogen chloride (see Chapter 14).

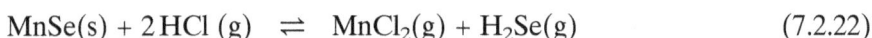
$$MnSe(s) + 2\,HCl\ (g) \quad \rightleftharpoons \quad MnCl_2(g) + H_2Se(g) \tag{7.2.22}$$

Hydrogen is suited as a transport agent if the metal that is to be transported has a sufficient vapor pressure in the temperature range of the transport. Hence this method is limited to compounds of group 12. This transport is of importance as far as the deposition of thin layers is concerned (see (Har 1987)).

$$ZnSe(s) + H_2(g) \quad \rightleftharpoons \quad Zn(g) + H_2Se(g) \tag{7.2.23}$$

The transport with hydrogen is also successful with the mixed-crystals $ZnS_{1-x}Se_x$ (Eti 1980b) because H_2Se and H_2S are likewise transport effective.

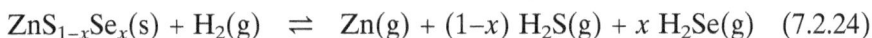
$$ZnS_{1-x}Se_x(s) + H_2(g) \quad \rightleftharpoons \quad Zn(g) + (1-x)\,H_2S(g) + x\,H_2Se(g) \tag{7.2.24}$$

Ga_2Se_3 transports in equilibrium via GaCl and H_2Se in a flow of H_2 and HCl (Rus 2003).

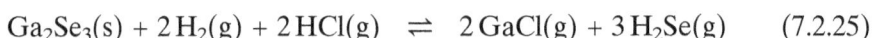
$$Ga_2Se_3(s) + 2\,H_2(g) + 2\,HCl(g) \quad \rightleftharpoons \quad 2\,GaCl(g) + 3\,H_2Se(g) \tag{7.2.25}$$

Ternary selenides can be transported this way, too (Iga 1993). It is required that the gaseous halides of both metal components show transport effective species. Also the difference of the partial pressure (the "flow" of the components in the temperature gradient) must reflect exactly the composition of the compound.

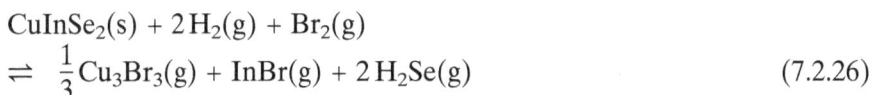
$$CuInSe_2(s) + 2\,H_2(g) + Br_2(g)$$
$$\rightleftharpoons \quad \frac{1}{3}Cu_3Br_3(g) + InBr(g) + 2\,H_2Se(g) \tag{7.2.26}$$

Metal halides as transport agent Apart from the use of iodine, a considerable number of CVT reactions of selenides with added $AlCl_3$, $CrCl_3$ or $CdCl_2$, respectively, are known. Hence the CVT of $CoSe_2$ with $AlCl_3$ is described with a mechanism that generally applies for transition metals, such as Cr_2Se_3, MnSe, FeSe, $FeSe_2$, and NiSe (Lut 1974). See section 14.2 about the problem of transports with added $AlCl_3$.

Metal halides are especially suitable for the transport of ternary transition metal compounds. While, for example, the selenido spinels $CdGa_2Se_4$ and $CdIn_2Se_4$ are accessible by transport reaction when iodine is added, transports with chlorine-containing transport agents are described for the analogous chromium spinel $CdCr_2Se_4$ and its substitution variants. In doing so, $CdCl_2$ (Weh 1969, Lya 1982, Sag 2004), $CrCl_3$ (Oka 1974, Oko 1999, Jen 2004), as well as $AlCl_3$ (Lut 1970) are used as additives. The formation of Cd(g) and $CrCl_3$(g) are described during dominating transport reactions.

However, calculations of transport reactions of the transition metal selenides with $AlCl_3$ show that $CrCl_2$ is present in the equilibrium in the gas phase adjacent to AlCl, but not the higher chlorides $CrCl_3$ and $CrCl_4$ (Lut 1974). Thus equilibrium calculations of the transport with $CrCl_3$ show the formation of the dichlorides $CdCl_2$ and $CrCl_2$. Accordingly, the mentioned equilibrium of Cd adjacent to $CrCl_3$ is not possible and should be formulated according to equations 7.2.27 and 7.2.28, respectively.

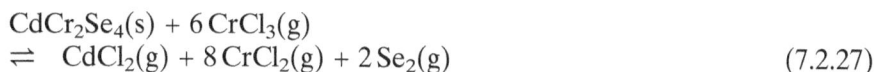

$$CdCr_2Se_4(s) + 6\,CrCl_3(g)$$
$$\rightleftharpoons\ CdCl_2(g) + 8\,CrCl_2(g) + 2\,Se_2(g) \qquad (7.2.27)$$

$$CdCr_2Se_4(s) + 2\,CdCl_2(g)\ \rightleftharpoons\ 3\,Cd(g) + 2\,CrCl_2(g) + 2\,Se_2(g) \quad (7.2.28)$$

Table 7.2.1 Examples of the chemical vapor transport of selenides.

Sink solid	Transport additive	Temperature / °C	Reference
AgGaSSe	I_2	935 → 855	Bal 1994
$Ag_{0.5}Cu_{0.5}InSSe$	I_2	940 → 850	Taf 1995
Al_2Se_3	I_2	not specified	Sch 1963
BaSe	I_2	1050 ... 950 → 950 ... 850	Jin 2001
$BaSe:Co^{2+}$	I_2	1050 ... 950 → 950 ... 850	Jin 2001
Bi_2Se_3	I_2	500 → 450	Schö 2010
CaSe	I_2	1050 ... 950 → 950 ... 850	Jin 2001
$CaSe:Co^{2+}$	I_2	1050 ... 950 → 950 ... 850	Jin 2001
CdSe	decomposition sublimation	1000 → 800 ... 500	Kle 1967
	decomposition sublimation	950 → 875	Sig 1972
	I_2	1000 → 500	Nit 1960
	I_2	1000 ... 700 → 800 ... 500	Kle 1967
	I_2	875 → 840	Wie 1970, Sig 1971
	I_2	950 → 875	Sig 1972
	H_2	1050 → 950	Var 1973
$CdSe_{1-x}S_x$	decomposition sublimation	1000 → 950	Moc 1978
$Cd_{1-x}Mn_xSe$	I_2	875 → 840	Wie 1970, Sig 1971
$Cd_{1-x}Fe_xSe$	I_2	705 → 725	Smi 1988
$CdGa_2Se_4$	I_2	720 → 700	Gas 1985
$CdAl_2Se_4$	$AlCl_3$	675 → 650	Kra 1997
$Cd_{1-x}Mn_xGa_2Se_4$	I_2	720 → 700	Sim 1987
	I_2	720 → 700	Sim 1988
$CdCr_2Se_4$	Cl_2	800 → 750	Mer 1981
	$AlCl_3$	850 → 700	Lut 1970
	$CdCl_2$	800 → 700	Weh 1969
	$CdCl_2$	not specified	Lya 1982
	$CrCl_3$	800 → 750	Oka 1974
$Cd_{1-x}Ga_{\frac{2}{3}x}Cr_2Se_4$	$CrCl_3$	800 ... 750 → 700 ... 675	Oko 1999
$CdCr_2Se_4:Cu^{2+}$	$CrCl_3$	800 → 750	Oka 1974
$Cd_{1-x}Ni_xCr_2Se_4$	$CrCl_3$	800 ... 700 → 760 ... 680	Jen 2004
$CdCr_2Se_4:Ag$	$CrCl_3$	800 → 750	Oka 1974
$CdCr_{2-x}Ga_xSe_4$	$CdCl_2$	800 → 700	Sag 2004
$CdCr_2Se_4:In^{3+}$	$CrCl_3$	800 → 750	Oka 1974
$CdIn_2Se_2S_2$	I_2	950 → 900	Ven 1988
$CdIn_2Se_4$	I_2	1000 → 700	Nit 1960
	I_2	not specified	Beu 1962
	I_2	950 → 900	Ven 1987

Table 7.2.1 (continued)

Sink solid	Transport additive	Temperature / °C	Reference
Cd_4GeSe_6	I_2	not specified	Nit 1964
	Cl_2, Br_2, I_2	$660 \rightarrow 560, 680 \rightarrow 580,$ $520 \dots 400 \rightarrow 460 \dots 390$	Que 1974
	I_2	$560 \rightarrow 480$	Kov 2003
$Cd_4GeSe_6:Cu^{2+}$	I_2	$520 \rightarrow 440$	Que 1974
	Br_2	$680 \rightarrow 580$	Que 1974
	Cl_2	$660 \rightarrow 560$	Que 1974
Cd_4SiSe_6	I_2	800 (mineralization)	Kal 1967
$CoSe_2$	$AlCl_3$	$650 \rightarrow 500$	Lut 1974
Cr_2Se_3	I_2, Se_2	$1010 \rightarrow 920,$ $1100 \rightarrow 1030$	Weh 1970
	I_2	not specified	Sat 1990
	$AlCl_3$	$950 \rightarrow 900$	Lut 1970
$Cr_2Se_xS_{3-x}$	$AlCl_3$	$1000 \rightarrow 800$	Lut 1974
Cr_7Se_8	$AlCl_3$	$980 \rightarrow 920$	Lut 1970
$Cu_{0.5}Ag_{0.5}InSSe$	I_2	$940 \rightarrow 850$	Taf 1995
$CuAlSe_2$	I_2	$820 \rightarrow 750$	Geb 1990
	I_2	$800 \rightarrow 700$	Chi 1991
	I_2	not specified	Moc 1993
	I_2	$700 \rightarrow 800$	Bod 2002
$CuAlSe_{2-x}S_x$	I_2	$700 \rightarrow 800$	Bod 2002
$CuAlSe_2:Cd^{2+}$	I_2	$800 \rightarrow 700$	Chi 1991
$CuAlSe_2:Zn^{2+}$	I_2	$800 \rightarrow 700$	Chi 1991
$CuAl_{1-x}Ga_xSSe$	I_2	$800 \rightarrow 700$	Chi 1994
$CuAl_{1-x}Ga_xSe_2$	I_2	$800 \rightarrow 700$	Dev 1988
	I_2	not specified	Shi 1997
$CuAl_{1-x}In_xSe_2$	I_2	$820 \rightarrow 750$	Geb 1990
$CuCr_2Se_4$	I_2	not specified	Lot 1964
	$AlCl_3$	$900 \rightarrow 700$	Lut 1970
$Cu_{1-x}Ga_xCr_2Se_4$	$CrCl_3$	$900 \rightarrow 700$	Oko 1992
$Cu_{1-x}In_xCr_2Se_4$	$CrCl_3$	$900 \rightarrow 700$	Oko 1995
$Cu|In|Cr|Se$	$CrCl_3$	$900 \rightarrow 700$	Oko 1995
$CuCrSnSe_4$	$I_2 + AlCl_3$	$850 \rightarrow 750$	Mae 1984
$CuCrZrSe_4$	$I_2 + AlCl_3$	$850 \rightarrow 750$	Mae 1984
$CuGaSe_2$	I_2	$900 \rightarrow 700$	Tan 1977
	I_2	$900 \rightarrow 600$	Sus 1978
	I_2	$570 \rightarrow 550$	Mas 1993
	I_2	$810 \rightarrow 750$	Tom 1998
$CuGa_3Se_5$	I_2	not specified	Lev 2006
$CuGaSe_2:Sn$	I_2	$830 \rightarrow 780$	Schö 1996
$CuGa_{1-x}In_xSe_2$	I_2	$570 \rightarrow 550$	Mas 1993
	I_2	$600 \rightarrow 550$	Dje 1993
	I_2	$600 \dots 700$	Schö 2000

Table 7.2.1 (continued)

Sink solid	Transport additive	Temperature / °C	Reference
$CuGaS_{2-x}Se_x$	I_2	$850 \ldots 900 \rightarrow 650 \ldots 700$	Tan 1977
$CuGa_3Se_5$	I_2	not specified	Aru 2006
$CuInSe_2$	I_2	$810 \rightarrow 780 \ldots 730$	Cis 1984
	I_2	$820 \rightarrow 770$	Bal 1990a
	I_2	$810 \rightarrow 770$	Bal 1990b
	I_2	$570 \rightarrow 550$	Mas 1993
	I_2	$600 \ldots 400 \rightarrow 575 \ldots 350$	Dje 1994
	$H_2 + Br_2$	$620 \rightarrow 540$	Iga 1993
$CuInSe_xS_{2-x}$	I_2	$780 \ldots 900 \rightarrow 700 \ldots 860$	Bod 1980
$CuInSSe$	I_2	$820 \rightarrow 770$	Bal 1989
	I_2	$950 \rightarrow 900$	Bal 1990b
$CuIn_5Se_8$	I_2	not specified	Aru 2006
$Cu_2ZnGeSe_4$	I_2	$800 \rightarrow 750$	Nit 1967a
$Cu_2ZnSiSe_4$	I_2	$800 \rightarrow 750$	Nit 1967a
$Cu_2U_3Se_7$	I_2	$600 \rightarrow 540$	Dao 1996
Cu_3NbSe_4	I_2	$800 \rightarrow 750$	Nit 1967b
Cu_3TaSe_4	I_2	$800 \rightarrow 750$	Nit 1967b
Dy_8Se_{15}	I_2	$850 \rightarrow 700$	Doe 2007
$Dy_4U_5Se_{16}$	I_2	not specified	Pak 1981
Er_8Se_{15}	I_2	$850 \rightarrow 700$	Doe 2007
$EuSe$	I_2	$1700 \rightarrow T_1$	Kal 1968
$FeSe$	$AlCl_3$	$830 \rightarrow 500$	Lut 1974
$FeSe_2$	$AlCl_3$	$650 \rightarrow 540$	Lut 1974
$FeAs_{2-x}Se_x$	Cl_2	$800 \rightarrow 780$	Bag 1974
$FePSe_3$	Cl_2	$650 \rightarrow 610$	Tay 1974
$Fe_{1-x}Mn_xIn_2Se_4$	I_2	$800 \rightarrow 750$	Att 2005
$Ga_{1-x}Zn_xAs_{1-x}Se_x$	I_2	$800 \rightarrow 780$	Bru 2006
$GaSe$	I_2	$870 \rightarrow 750$	Nit 1961
	I_2	$870 \ldots 805 \rightarrow 750 \ldots 700$	Kuh 1972
	I_2	$920 \rightarrow 880 \ldots 800$	Egm 1974
	I_2	$920 \rightarrow 880 \ldots 800$	Nis 1975
	I_2	$950 \ldots 750 \rightarrow 850 \ldots 650$	Whi 1978a
	I_2	$870 \rightarrow 820$	Ish 1986
	$SnCl_2$	$790 \rightarrow 740$	Zul 1984
$GaSe_{2-x}S_x$	I_2	$910 \ldots 850 \rightarrow 850 \ldots 750$	Whi 1978b
Ga_2Se_3	$H_2 + HCl$	$580 \ldots 520 \rightarrow 480 \ldots 420$	Rus 2003
$Ga_2Se_{3-x}S_3$	I_2	$950 \rightarrow 800$	Hot 2005b
Gd_8Se_{15}	I_2	$850 \rightarrow 700$	Doe 2007

Table 7.2.1 (continued)

Sink solid	Transport additive	Temperature / °C	Reference
GeSe	I_2	$570 \rightarrow 450$	Wie 1972
	I_2	$520 \rightarrow 420$	Rao 1984
	I_2, NH_4Cl	$600 \rightarrow 530$, $570 \rightarrow 490$	Sol 2003
	GeI_4	$520 \rightarrow 420$	Wie 1981
	GeI_4	$520 \rightarrow 420$	Cha 1982
	GeI_4	$520 \rightarrow 420$	Buc 1987
$GeSe_{1-x}Te_x$	Sublimation	$600 \rightarrow 590$	Wie 1991a, Wie 1991b
	Sublimation	$625 \rightarrow 325$	Lia 1992
$GeSe_2$	I_2	$400 \rightarrow 550$	Wie 1972
	GeI_4	$520 \rightarrow 420$	Buc 1987
$HfSe_2$	I_2	$900 \rightarrow 800$	Gre 1965
	I_2	$900 \rightarrow 800$	Nit 1967c
	I_2	$900 \rightarrow 860$	Rim 1972
	I_2	$900 \rightarrow 800$	Rad 2008
$HfS_{2-x}Se_x$	I_2, Se	$900 \ldots 950 \rightarrow 750 \ldots 850$	Gai 2004
$HfSe_3$	I_2	$650 \rightarrow 600$	Lev 1983
$Hg_{1-x}Cd_xSe$		not specified	Wan 1986
$HgCr_2Se_4$	$AlCl_3$	$750 \rightarrow 600$	Lut 1970
	$CrCl_3$	$1000 \rightarrow 900$	Bel 1989
$HgGa_2Se_4$	I_2		Beu 1962
	I_2	$720 \rightarrow 700$	Gas 1984b
Ho_8Se_{15}	I_2	$850 \rightarrow 700$	Doe 2007
InSe	I_2	$600 \rightarrow 560$	Med 1965
	NH_4Cl	$450 \rightarrow 500$, $600 \rightarrow 400$	Che 1981
In_2Se	I_2	480	Med 1965
In_2Se_3	I_2	$500 \rightarrow 460$	Med 1965
	I_2	not specified	Zor 1965
	I_2	$850 \rightarrow 400$	Gri 1975
	NH_4Cl	$600 \rightarrow 400$	Che 1981
$In_{1.9}As_{0.1}Se_3$	I_2	$650 \rightarrow 600$	Kat 1978
In_5Se_6	I_2	$600 \rightarrow 560$	Med 1965
$In_{0.667}PSe_3$	Cl_2	$630 \rightarrow 560$	Kat 1977
$IrSe_2$	ICl_3	$1080 \rightarrow 930$	Lia 2009
$LaSe_{1.9}$	I_2	not specified	Gru 1991
$MgIn_2Se_4$	I_2	$950 \rightarrow 900$	Gas 1984a
$MgIn_2Se_4{:}Co^{2+}$	I_2	$950 \rightarrow 900$	Gas 1984a
MnSe	decomposition sublimation	$1600 \rightarrow 1400$	Wie 1968
	I_2	$875 \rightarrow 840$	Wie 1970, Sig 1971

Table 7.2.1 (continued)

Sink solid	Transport additive	Temperature / °C	Reference
MnSe	Cl_2	830 → 780	Paj 1983
	NH_4Cl	800 → 775 … 700	Paj 1983
	$AlCl_3$	830 → 500	Lut 1974
	$AlCl_3$	750 … 650 → 740 … 620	Paj 1980
$Mn_{1.04}AlSe_{1.77}$	I_2	720 → 620	Kha 1984
$MnGa_{1-x}In_xSe_2$	I_2	800 → 750	Sag 2005
$MnIn_2Se_4$	I_2	not specified	Neu 1986
	$AlCl_3$	840 → 820	Doe 1990, Doe 1991
$Mn_{1-x}Ho_xInSe_4$	I_2	705 … 670 → 675 … 640	Kha 1997
$Mn_{1-x}Zn_xIn_2Se_4$	I_2, $AlCl_3$	850 → 800, 900 → 950	Man 2004
$MoSe_2$	I_2	900 → 700	Bri 1962
	Cl_2	1000 … 650 → 990 … 640	Ols 1983
$MoSe_{2-x}S_x$	I_2	1000 (mineralization)	Aga 1986a
	I_2	1000 → 900	Hot 2005c
$MoSe_{2-x}Te_x$	Br_2	900 … 800 → 800 … 650	Aga 1986b
$MoSe_2$:Nb^{4+}	I_2	880 → 850	Leg 1991
$Mo_{1-x}Nb_xSe_2$	I_2	1000 → 900	Hot 2005c
$MoSe_2$:Re^{4+}	I_2	880 → 850	Leg 1991
$Mo_{1-x}Ta_xSe_2$	I_2	1000 → 800	Hot 2005e
$Mo_{1-x}W_xSe_2$			Hof 1988
	Br_2	1020 … 1000 → T_1	You 1990
NbSe	I_2	880 → 1050	Sch 1965
	I_2	900 → 700	Bri 1962
	I_2	800 → 730	Bay 1976
	I_2	825 → 725	Vac 1977
	I_2	825 → 725	Vac 1993
Nb_3Se_4	I_2	1000 → 950	Nak 1985
$NbSe_xS_{2-x}$	I_2	1000 → 900	Hot 2005c
$Nb_{1-x}Ta_xSe_2$	I_2	780 → 700	Dal 1986
	I_2	1000 → 900	Hot 2005f
$Nb_{1-x}Ti_xSe_2$	I_2	1000 → 900	Hot 2005f
$Nb_{1-x}V_xSe_2$	I_2	800 (mineralization)	Bay 1976
$NbSe_3$	I_2	700 (mineralization)	Mee 1975, Hae 1978
$NbSe_4$:I	I_2	730 → 670	Nak 1986
Nb_2Se_9	I_2	600 → 500	Mee 1979
	$SeCl_4$, ICl_3	700 → 680	Lev 1983
$NdSe_2$	I_2	800 → 600	Doe 2005
$NbSe_3$	I_2, S_2Cl_2, Se, ICl_3	650 … 750 → 600 … 700	Lev 1983
PbSe	sublimation	800 → 795	Sto 1992
	AgI	700 → 695	Sto 1992
PbSe:Sn^{2+}	sublimation	not specified	Zlo 1990

Table 7.2.1 (continued)

Sink solid	Transport additive	Temperature /°C	Reference
$PrSe_2$	I_2	800 → 600	Doe 2005
NiSe	$AlCl_3$	830 → 500	Lut 1974
$ReSe_2$	Br_2	1075 → 1025	Lut 1974
$ReSe_2$:Mo	Br_2	1050 → 1000	Hu 2004
	Br_2	1060 → 1000	Hu 2007
$ReSe_2$:W	Br_2	1050 → 1000	Hu 2004
	Br_2	1050 → 1000	Hu 2006
$RuSe_xS_{2-x}$	ICl_3	not specified	Mar 1984
Sb_2Se_3	I_2	500 → 450	Schö 2010
$SiSe_2$	I_2	810 → 730	Hau 1969
Sc_2Se_3	I_2	not specified	Dis 1964
$SnSe_2$	I_2	560 → 410	Mct 1958
	I_2	500 → 400	Nit 1961
	I_2	560 → 500	Lee 1968
	I_2	650 → 610	Rim 1972
	I_2	600 → 400	Aga 1989a
$SnSe_xS_{2-x}$	I_2	650 → 610	Rim 1972
	I_2	690 ... 550 → 640 ... 510	Ala 1977
	I_2	650 → 600	Har 1978
$Sn_{1-x}Zr_xSe_2$	I_2	550 ... 850 → 510 ... 800	Ala 1977
$TaSe_2$	I_2	900 → 700	Bri 1962
	I_2	not specified	Asl 1963
	I_2	not specified	Bro 1965
$TaSe_{2-x}S_x$	I_2	850 ... 900 → 800 ... 700	Ala 1977
$TaSe_3$	I_2, ICl_3, Se	650 ... 750 → 600 ... 710	Lev 1983
	I_2	1000 → 900	Hot 2005a
$Ta_{1-x}Ti_xSe_2$	I_2	1000 → 900	Hot 2005a
$TaSe_3$	I_2	900 (mineralization)	Bje 1964
	I_2	700 (mineralization)	Hae 1978
Tb_8Se_{15}	I_2	850 → 700	Doe 2007
$TiSe_2$	I_2	900 → 800	Gre 1965
	I_2	900 → 800	Nit 1967c
	I_2	780 → 740	Rim 1972
	Se_2	880 → 790	Weh 1970

Table 7.2.1 (continued)

Sink solid	Transport additive	Temperature /°C	Reference
$TiSe_{2-x}S_x$	I_2	$850 \rightarrow 800$	Nit 1967c
	I_2	$800 \ldots 780 \rightarrow 720 \ldots 740$	Rim 1974
	I_2	$1000 \rightarrow 900$	Hot 2005a
$TiSe_{2-x}Te_x$	I_2	$780 \ldots 750 \rightarrow 740 \ldots 690$	Rim 1974
USe_2	Se_2Br_2	$970 \rightarrow 870$	Slo 1966
USe_3	Se_2Br_2	$770 \rightarrow 610$	Slo 1966
VSe_2	I_2	$850 \rightarrow 800$	Nit 1967c
	I_2	$800 \rightarrow 730$	Bay 1976
	Se_2	$870 \rightarrow 780$	Weh 1970
$V_{1+x}Se_2$	I_2	$820 \rightarrow 720$	Hay 1983
$(x = 0 \ldots 0.25)$			
$V_{1+x}Se_2$	I_2	$820 \rightarrow 720$	Oht 1987
V_2Se_9	I_2	$325 \rightarrow 300$	Fur 1984
WSe_2	I_2	$900 \rightarrow 700$	Bri 1962
	Cl_2	$1000 \ldots 650 \rightarrow 990 \ldots 640$	Ols 1983
	$Br_2, SeCl_4,$	$970 \rightarrow 950, 990 \rightarrow 960$	Aga 1989b
	$TeCl_4$	$985 \rightarrow 945$	
	$SeCl_4$	$980 \rightarrow 950$	Aga 1989c
	$TeCl_4$	$1100 \rightarrow 1050$	Pra 1986
	$TeBr_4$	$1100 \rightarrow 1050$	Pra 1986
WSe_2:Nb	$Se, Br_2, I_2, SeCl_4,$	$880 \rightarrow 850$	Leg 1991
	$TeCl_4$		
WSe_2:Re	$Se, Br_2, I_2, SeCl_4,$	$880 \rightarrow 850$	Leg 1991
	$TeCl_4$		
WSe_{2-x}	I_2	$950 \ldots 915 \rightarrow 700$	Aga 1982
$(x = 0 \ldots 0.1)$			
Y_8Se_{15}	I_2	$850 \rightarrow 700$	Doe 2007
$Yb_2U_{0.87}Se_4$	I_2	not specified	Slo 1982
$ZnSe$	I_2	$1050 \rightarrow 800$	Nit 1960
	I_2	$900 \rightarrow 800$	Kal 1965b
	I_2	$1050 \rightarrow 800$	Ari 1966
	I_2	$855 \rightarrow 800$	Sch 1966
	I_2	$800 \rightarrow 700$	Sim 1967
	I_2	$900 \rightarrow 800$	Poi 1979
	I_2	$810 \rightarrow 775$	Tri 1982
	I_2	$900 \ldots 800 \rightarrow 850 \ldots 700$	Böt 1995
	GeI_4	$1050 \rightarrow 800$	Ari 1966
	HCl	$760 \ldots 620 \rightarrow 660 \ldots 570$	Hov 1969
	HCl	$950 \ldots 750 \rightarrow 750 \ldots 450$	Har 1987
	$HCl,$	$910 \ldots 1030 \rightarrow T_1$	Liu 2010
	$ZnCl_2 \cdot 2\,NH_4Cl$		
	$ZnCl_2 \cdot 3\,NH_4Cl$	$915 \rightarrow 900$	Li 2003
	H_2	$950 \rightarrow 450$	Voh 1971
	H_2	$930 \rightarrow 880$	Che 1972

Table 7.2.1 (continued)

Sink solid	Transport additive	Temperature /°C	Reference
ZnSe	H_2	850 → 800 ... 650	Eti 1980a
	H_2	1000 → 750	Bes 1981
	H_2	950 ... 850 → 800 ... 650	Har 1987
	H_2	1200 → 1150	Kor 1996
	H_2O	1030 → 880	Mim 1995
ZnSe:Cu^{2+}	H_2	800 → 500	Fal 1984
ZnSe:Fe^{2+}	I_2	not specified	Jan 1993
ZnSe:Fe	H_2	850 → 800	Gra 1995
ZnSe:Ga^{2+}	H_2	800 → 500	Fal 1984
ZnSe:Mn^{2+}	I_2	not specified	Jan 1993
ZnSe:Mn	H_2	850 → 800	Gra 1995
ZnSe:Ni	I_2	not specified	Rab 1990
	I_2	not specified	Jan 1993
	H_2	850 → 800	Gra 1995
ZnSe:Ti	H_2	1190 → 1180	Kli 1994
$ZnSe_{1-x}S_x$	I_2	850 → 840	Cat 1976
	I_2	1000 → 900	Hot 2005d
	H_2	850 → 780 ... 630	Eti 1980b
$ZnSe_{1-x}Te_x$	decomposition sublimation	1250 → 1200	Tsu 1967
$ZnCr_2Se_4$	$AlCl_3$	850 → 650	Lut 1970
	$CrCl_3$	850 → 700	Oko 1989
Zn|In|Cr|Se	$CrCl_3$	850 → 700	Oko 1989
$ZnGa_2Se_4$	I_2	not specified	Beu 1962
$ZnGa_{1.02}Se_{1.89}$	I_2	680 → 580	Kha 1984
Zn|Ga|Cr|Se	$CrCl_3$	850 ... 700 → 775 ... 600	Oko 1999
$ZnIn_2Se_4$	I_2	1000 → 700	Nit 1960
	I_2	not specified	Beu 1962
$ZnIn_2S_{4-x}Se_x$	I_2	725 → 675	Loc 2005
$ZrAs_{1.4}Se_{0.5}$	I_2	750 → 850	Schm 2005
$ZrSe_2$	I_2	900 → 800	Gre 1965
	I_2	900 → 800	Nit 1967c
	I_2	850 → 800	Rim 1972
	I_2	850 → 800	Whi 1973
	I_2	850 → 800	Pat 1998
	I_2	800 → 700	Czu 2010
$ZrSe_3$	I_2	900 → 850	Nit 1967c
	I_2	400 → 350	Pro 2001
	I_2	600 → 750	Pat 2009
	I_2	650 → 600	Pro 2001
$Zr_{0.9}Ti_{0.1}Se_3$	$I_2 + Se_2Cl_2$	700 → 690	Lev 1983
Zr_3Se_4	I_2	870 → 770	Wie 1986
Zr_4Se_3	I_2	870 → 770	Wie 1986

Table 7.2.1 (continued)

Sink solid	Transport additive	Temperature / °C	Reference
ZrSe$_{2-x}$S$_x$	I$_2$	850 → 800	Nit 1967c
	I$_2$	870 → 770	Wie 1986
	I$_2$	900 ... 850 → 820 ... 800	Bar 1995
ZrSe$_{3-x}$S$_x$	I$_2$	780 → 830	Pat 2009
ZrSiSe	I$_2$	900 → 850	Nit 1967c

Bibliography of Section 7.2

Aga 1982   M. K. Agarwal, J. D. Kshatriya, P. D. Patel, P. K. Garg, *J. Cryst. Growth* **1982**, *60*, 9.

Aga 1986a   M. K. Agarwal, L. T. Talele, *Solid State Comm.* **1986**, *59*, 549.

Aga 1986b   M. K. Agarwal, P. D. Patel, R. M. Joshi, *J. Mater. Science Lett.* **1986**, *5*, 66.

Aga 1989a   M. K. Agarwal, P. D. Patel, S. S. Patel, *J. Mater. Science Lett.* **1989**, *8*, 660.

Aga 1989b   M. K. Agarwal, V. V. Rao, *Cryst. Res. Technol.* **1989**, *24*, 1215.

Aga 1989c   M. K. Agarwal, V. V. Rao, V. M. Pathak, *J. Cryst. Growth* **1989**, *97*, 675.

Aga 1994   M. K. Agarwal, P. D. Patel, D. Lakshminarayana, *J. Cryst. Growth* **1994**, *142*, 344.

Ala 1977   F. A. S. Al-Alamy, A. A. Balchin, *J. Cryst. Growth* **1977**, *38*, 221.

Ari 1966   T. Arizumi, T. Nishinaga, M. Kakehi, *Jpn. J. Appl. Phys.* **1966**, *5*, 588.

Aru 2006   E. Arushanov, S. Levcenko, N. N. Syrbu, A. Nateprov, V. Tezlevan, J. M. Merino, *Phys, Stat. Sol.* **2006**, *203*, 2909.

Asl 1963   L. A. Aslanov, Y. M. Ukrainskii, Y. P. Simanov, *Russ. J. Inorg. Chem.* **1963**, *8*, 937.

Att 2005   G. Attolini, V. Sagredo, L. Mogollon, T. Torres, C. Frigeri, *Cryst. Res. Technol.* **2005**, *40*, 1064.

Bag 1974   A. Baghdadi, A. Wold, *J. Phys. Chem. Solids* **1974**, *35*, 811.

Bal 1989   K. Balakrishnan, B. Vengatesan, N. Kanniah, P. Ramasamy, *Bull. Electrochem.* **1989**, *5*, 841.

Bal 1990a   K. Balakrishnan, B. Vengatesan, N. Kanniah, P. Ramasamy, *J. Mater. Science Lett.* **1990**, *9*, 785.

Bal 1990b   K. Balakrishnan, B. Vengatesan, N. Kanniah, P. Ramasamy, *Cryst. Res. Technol.* **1990**, *25*, 633.

Bal 1994   K. Balakrishnan, B. Vengatesan, P. Ramasamy, *J. Mater. Science* **1994**, *29*, 1879.

Bar 1973   K. G. Barraclough, A. Meyer, *J. Cryst. Growth* **1973**, *20*, 212.

Bar 1995   K.S. Bartwal, O. N. Srivastava, *Mater. Science Engin. B* **1995**, *B33*, 115.

Bay 1976   M. Bayard, B.F. Mentzen, M. J. Sienko, *Inorg. Chem.* **1976**, *15*, 1763.

Bel 1989   V. K. Belyaev, K. G. Nikiforov, S. I. Radautsan, V. A. Bazakutsa, *Cryst. Res. Technol.* **1989**, *24*, 371.

Bes 1981   P. Besomi, B. W. Wessels, *J. Cryst. Growth* **1981**, *55*, 477.

Beu 1962   J. A. Beun, R. Nitsche, M. Lichtensteiger, *Phys.* **1961**, *27*, 448.

Bje 1964   E. Bjerkelund, A. Kjekshus, *Z. Anorg. Allg. Chem.* **1964**, *328*, 235.

Bod 1980   I. V. Bodnar, A. P. Bologa, B.V. Korzun, *Kristall Technik* **1980**, *15*, 1285.

Bod 2002   I. V. Bodnar, *Inorg. Mater.* **2002**, *38*, 647.

Boe 1962   H. U. Boelsterli, E. Mooser, *Helv. Phys. Acta* **1962**, *35*, 538.

Böt 1995	K. Böttcher, H. Hartmann, *J. Cryst. Growth* **1995**, *146*, 53.
Böt 1998	P. Böttcher, Th. Doert, *Phosphorus, Sulfur and Silicon and the Related Elements* **1998**, *136*, 255.
Bri 1962	L. H. Brixner, *J. Inorg. Nucl. Chem.* **1962**, *24*, 257.
Bro 1965	B. Brown, D. Beerntsen, *Acta Crystallogr.* **1965**, 18, 31.
Bru 2006	C. Bruenig, S. Locmelis, E. Milke, M. Binnewies, *Z. Anorg. Allg. Chem.* **2006**, *632*, 1067.
Buc 1987	N. Buchan, F. Rosenberger, *J. Cryst. Growth* **1987**, 84, 359.
Cat 1976	A. Catano, Z. Kun, *J. Cryst. Growth* **1976**, *33*, 324.
Cha 1982	D. Chandra, H. Wiedemeier, *J. Cryst. Growth* **1982**, *57*, 159.
Che 1972	J. Chevrier, D. Etienne, J. Camassel, D. Auvergne, J. C. Pons, H. Mathieu, G. Bougnot, *Mater. Res. Bull.* **1972**, *7*, 1485.
Che 1981	A. Chevy, *J. Cryst. Growth* **1981**, *51*, 157.
Chi 1991	S. Chichibu, M. Shishikura, J. Ino, S. Matsumoto, *J. Appl. Phys.* **1991**, *70*, 1648.
Chi 1994	S. Chichibu, M. Shirakata, A.Ogawa, R. Sudo, M. Uchida, Y. Harada, T. Wakiyama, M. Shishikura, S. Matsumoto, *J. Cryst. Growth* **1994**, *140*, 388.
Cis 1984	T. F. Ciszek, *J. Cryst. Growth* **1984**, *70*, 405.
Czu 2010	A. Czulucki, *Dissertation*, University of Dresden, **2010**.
Dal 1986	B. J. Dalrymple, S. Mroczkowski, D. E. Prober, *J. Cryst. Growth* **1986**, *74*, 575.
Dao 1996	A. Daoudi, M. Lamire, J. C. Levet, H. Noeel, *J. Solid State Chem.* **1996**, *123*, 331.
Dev 1988	W. E. Devaney, R. A. Mickelsen, *Solar Cells* **1988**, *24*, 19.
Dis 1964	J. P. Dismukes, J. G. White, *Inorg. Chem.* **1964**, *3*, 1220.
Dje 1993	K. Djessas, G. Masse, *Thin Solid Films* **1993**, *232*, 194.
Dje 1994	G. Masse, K. Djessas, *Thin Solid Films* **1994**, *237*, 129.
Doe 1990	G. Doell, M. C. Lux-Steiner, C. Kloc, J. R. Baumann, E. Bucher, *J. Cryst. Growth* **1990**, *104*, 593.
Doe 1991	G. Doell, M. C. Lux-Steiner, C. Kloc, J. R. Baumann, E. Bucher, *Cryst. Prop. Prepar.* **1991**, *36–38*, 152.
Doe 2005	T. Doert, C. Graf, *Z. Anorg. Allg. Chem.* **2005**, *631*, 1101.
Doe 2007	T. Doert, E. Dashjav, B. P. T. Fokwa, *Z. Anorg. Allg. Chem.* **2007**, *633*, 261.
Egm 1974	G. E. van Egmond, R. M. Lieth, *Mater. Res. Bull.* **1974**, *9*, 763.
Eti 1980a	D. Etienne, G. Bougnot, *Thin Solid Films* **1980**, *66*, 325.
Eti 1980b	D. Etienne, L. Soonckindt, G. Bougnot, *J. Electrochem. Soc.* **1980**, *127*, 1800.
Fal 1984	C. Falcony, F. Sanchez-Sinencio, J. S. Helman, O. Zelaya, C. Menezes, *J. Appl. Phys.* **1984**, *56*, 1752.
Fur 1984	S. Furuseth, B. Klewe, *Acta Chem. Scand.* **1984**, *A38*, 467.
Gai 2004	C. Gaiser, T. Zandt, A. Krapf, R. Severin, C. Janowitz, R. Manzke, *Phys. Rev. B* **2005**, *69*, 075205.
Gas 1984a	L. Gastaldi, A. Maltese, S. Viticoli, *J. Cryst. Growth* **1984**, *66*, 673.
Gas 1984b	L. Gastaldi, M. P. Leonardo, *Compt. Rend. Acad. Sci. II* **1984**, *298*, 37.
Gas 1985	L. Gastaldi, M. G. Simeone, S. Viticoli, *Solid State Comm.* **1985**, *55*, 605.
Geb 1990	W. Gebicki, M. Igalson, W. Zajac, R. Trykozko, *J. Phys. D: Appl. Phys.* **1990**, *23*, 964.
Gra 1995	K. Grasza, E. Janik, A. Mycielski, J. Bak-Misiuk, *J. Cryst. Growth* **1995**, *146*, 75.
Gre 1965	D. L. Greenaway, R. Nitsche, *Phys. Chem. Solids* **1965**, *26*, 1445.
Gri 1975	Y. K. Grinberg, R. Hillel, V. A. Boryakova, V. F. Shevelkov, *Izv. Akad. Nauk SSSR, Neorg. Mater.* **1975**, *11*, 1945.
Gru 1991	M. Grupe, W. Urland, *J. Less-Common Met.* **1991**, *170*, 271.

Hae 1978   P. Haen, F. Lapierre, P. Monceau, M. Nunez Regueiro, J. Richard, *Solid State Comm.* **1978**, *26*, 725.

Har 1978   J. Y. Harbec, Y. Paquet, S. Jandl, *Canad. J. Phys.* **1978**, *56*, 1136.

Har 1987   H. Hartmann, R. Mach, N. Testova, *J. Cryst. Growth* **1987**, *84*, 199.

Hau 1969   E. A. Hauschild, C. R. Kannewurf, *J. Phys. Chem. Solids* **1969**, *30*, 353.

Hay 1983   K. Hayashi, T. Kobashi, M. Nakahira, *J. Cryst. Growth* **1983**, *63*, 185.

Hof 1988   W. K. Hofmann, H. J. Lewerenz, C. Pettenkofer, *Solar Energy Mater.* **1988**, *17*, 165.

Hot 2005a   U. Hotje, R. Wartchow, M. Binnewies, *Z. Anorg. Allg. Chem.* **2005**, *631*, 403.

Hot 2005b   U. Hotje, R. Wartchow, E. Milke, M. Binnewies, *Z. Anorg. Allg. Chem.* **2005**, *631*, 1675.

Hot 2005c   U. Hotje, M. Binnewies, *Z. Anorg. Allg. Chem.* **2005**, *631*, 2467.

Hot 2005d   U. Hotje, C. Rose, M. Binnewies, *Z. Anorg. Allg. Chem.* **2005**, *631*, 2501.

Hot 2005e   U. Hotje, R. Wartchow, M. Binnewies, *Z. Naturforsch. B* **2005**, *60*, 1235.

Hot 2005f   U. Hotje, R. Wartchow, M. Binnewies, *Z. Naturforsch. B* **2005**, *60*, 1241.

Hov 1969   H. J. Hovel, A. G. Milnes, *J. Electrochem. Soc.* **1969**, *116*, 843.

Hu 2004   S. Y. Hu, S. C. Lin, K. K. Tiong, P. C. Yen, Y. S. Huang, C. H. Ho, P. C. Liao, *J. Alloys Compd.* **2004**, *383*, 63.

Hu 2006   S. Y. Hu, C. H. Liang, K. K. Tiong, Y. S. Huang, Y. C. Lee, *J. Electrochem. Soc.* **2006**, *153*, J100.

Hu 2007   S. Y. Hu, Y. Z. Chen, K. K. Tiong, Y. S. Huang, *Mater. Chem. Phys.* **2007**, *104*, 105.

Ish 1986   T. Ishii, N. Kambe, *J. Cryst. Growth* **1986**, *76*, 489.

Iga 1993   O. Igarashi, *J. Cryst. Growth* **1993**, *130*, 343.

Jan 1993   E. Janik, K. Grasza, A. Mycielski, J. Bak-Misiuk, J. Kachniarz, *Acta Phys. Polon. A* **1993**, *84*, 785.

Jen 2004   I. Jendrzejewska, M. Zelechower, K. Szamocka, T. Mydlarz, A. Waskowska, I. Okonska-Kozlowska, *J. Cryst. Growth* **2004**, *270*, 30.

Jin 2001   M. S. Jin, N. O. Kim, H. G. Kim, C. S. Yoon, C. I. Lee, M. Y. Kim, *J. Korean Phys. Soc.* **2001**, *39*, 692.

Kal 1965a   E. Kaldis, R. Widmer, *J. Phys. Chem. Solids* **1965**, *26*, 1697.

Kal 1965b   E. Kaldis, *J. Phys. Chem. Solids* **1965**, *26*, 1701.

Kal 1967   E. Kaldis, L. Krausbauer, R. Widmer, *J. Electrochem. Soc.* **1967**, *114*, 1074.

Kal 1968   E. Kaldis, *J. Cryst. Growth* **1968**, *3–4*, 146.

Kal 1984   A. Kallel, H. Boller, *J. Less-Common Met.* **1984**, *102*, 213.

Kat 1978   A. Katty, C.A. Castro, J. P. Odile, S. Soled, A. Wold, *J. Solid State Chem.* **1978**, *24*, 107.

Kat 1977   A. Katty, S. Soled, A. Wold, *Mater. Res. Bull.* **1977**, *12*, 663.

Kha 1984   A. Khan, N. Abreu, L. Gonzales, O. Gomez, N. Arcia, D. Aguilera. S. Tarantino, *J. Cryst. Growth* **1984**, *69*, 241.

Kha 1997   A. Khan, J. Diaz, V. Sagredo, R. Vargas, *J. Cryst. Growth* **1997**, *174*, 783.

Kle 1967   W. Kleber, I. Mietz, U. Elsasser, *Kristall Technik* **1967**, *2*, 327.

Kli 1994   A. Klimakow, J. Dziesiaty, J. Korostelin, M. U. Lehr, P. Peka, H. J. Schulz, *Adv. Mater. Optics Electron.* **1994**, *3*, 253.

Kor 1996   Yu. V. Korostelin, V. I. Kozlowsky, A. S. Nasibov, P. V. Shapki, *J. Cryst. Growth* **1996**, *161*, 51.

Kov 2003   S. Kovach, A. Nemcsics, Z. Labadi, S. Motrya, *Inorg. Mater.* **2003**, *39*, 108.

Kuh 1972   A. Kuhn, A. Chevy, E. Lendvay, *J. Cryst. Growth* **1972**, *13–14*, 380.

Kra 1997   G. Krauss, V. Kramer, A. Eifler, V. Riede, S. Wenger, *Cryst. Res. Technol.* **1997**, *32*, 223.

Kyr 1976   D. S. Kyriakos, T. K. Karakostas, N. A. Economou, *J. Cryst. Growth* **1976**, *35*, 223.

Lee 1968   P. A. Lee, G. Said, *Brit. J. Appl. Phys.* **1968**, *2*, 837.
Lee 1994   Y. E. Lee, H. J. Kim, Y. J. Kim, K. L. Lee, B. H. Choi, K. H. Yoon, J. S. Song, *J. Electrochem. Soc.* **1994**, *141*, 558.
Leg 1991   J. B. Legma, G. Vacquier, H. Traore, A. Casalot, *Mater. Science Engin. B* **1991**, *B8*, 167.
Lev 1983   F. Levy, H. Berger, *J. Cryst. Growth* **1983**, *61*, 61.
Lev 2006   S. Levcenko, N. N. Syrbau, A. Nateprov, E. Arushanov, J. M. Merino, M. Leon, *J. Phys. D: Appl. Phys.* **2006**, *39*, 1515.
Lew 1975   N. E. Lewis, T. E. Leinhardt, J. G. Dillard, *Mater. Res. Bull.* **1975**, *10*, 967.
Lot 1964   F. K. Lotgering, *Proc. Int. Conf Magnetism, Nottingham* **1964**, 533.
Li 2003   H. Li, W. Jie, *J. Cryst. Growth* **2003**, *257*, 110.
Lia 1992   B. Liautard, M. Muller, S. Dal Corso, G. Brun, J. C. Tedenac, A. Obadi, C. Fau, S.Charar, F. Gisbert, M. Averous, *Phys. Status Solidi A* **1992**, *133*, 411.
Lin 1992   S. S. Lin, J. K. Huang, Y. S. Huang, *Mater. Res. Bull.* **1992**, *27*, 177.
Liu 2010   C. Liu, T. Hu, W. Jie, *J. Cryst. Growth* **2010**, *312*, 933.
Loc 2005   S. Locmelis, E. Milke, M. Binnewies, S. Gruhl, C. Vogt, *Z. Anorg. Allg. Chem.* **2005**, *631*, 1667.
Lut 1970   H. D. Lutz, C. Lovasz, K. H. Bertram, M. Sreckovic, U. Brinker, *Monatsh. Chem.* **1970**, *101*, 519.
Lut 1974   H. D. Lutz, K. H. Bertram, G. Wrobel, M. Ridder, *Monatsh. Chem.* **1974**, *105*, 849.
Lya 1982   R. Y. Lyalikova, A. I. Merkulov, S. I. Radautsan, V. E. Tezlevan, *Izv. Akad. Nauk SSSR, Neorg. Mater.* **1982**, *18*, 1968.
Mae 1984   D. Maehl, J. Pickardt, B. Reuter, *Z. Anorg. Allg. Chem.* **1984**, *508*, 197.
Man 2004   J. Mantilla, G. E. S. Brito, E. Ter Haar, V. Sagredo, V. Bindilatti, *J. Phys.: Cond. Matter* **2004**, *16*, 3555.
Mas 1993   G. Masse, K. Djessas, *Thin Solid Films* **1993**, *226*, 254.
Mct 1958   F. McTaggert, *Austr. J. Chem.* **1958**, *13*, 458.
Mee 1975   A. Meerschaut, J. Rouxel, *J. Less-Common Met.* **1975**, *39*, 197.
Mee 1979   A. Meerschaut, L. Guemas, R. Berger, J. Rouxel, *Acta Cryst.logr.* **1979**, *B35*, 1747.
Med 1965   Z. S. Medvedeva, T. N. Guliev, *Izv. Akad. Nauk SSSR, Neorg. Mater.* **1965**, *1*, 848.
Mer 1981   A. I. Merkulov, R. Y. Lyalikova, S. I. Radautsan, V. E. Tezlevan, Y. M. Ya-kovlev, *Izv. Akad. Nauk SSSR, Neorg. Mater.* **1981**, *17*, 926.
Mim 1995   J. Mimila, R. Triboulet, *Mater. Lett.* **1995**, *24*, 221.
Moc 1978   K. Mochizuki, K. Igaki, *J. Cryst. Growth* **1978**, *45*, 218.
Moc 1993   K. Mochizuki, E. Niwa, K. Kimoto, *Japan. Jpn. J. Appl. Phys.* **1993**, *Suppl. 32–3*, 168.
Nak 1985   I. Nakada, Y. Ishihara, *Jpn. J. Appl. Phys.* **1985**, *24*, 31.
Nak 1986   I. Nakada, E. Bauser, *J. Cryst. Growth* **1986**, *79*, 837.
Neu 1986   H. Neumann, C. Bellabarba, A. Khan, V. Riede, *Cryst. Res. Technol.* **1986**, *21*, 21.
Nis 1975   T. Nishinaga, R.M. Lieth, G. E. van Egmond, *Jpn. J. Appl. Phys.* **1975**, *14*, 1659.
Nit 1957   R. Nitsche, *Angew. Chem.* **1957**, *69*, 333.
Nit 1960   R. Nitsche, *Phys. Chem. Solids* **1960**, *17*, 163.
Nit 1961   R. Nitsche, H. Boelsterli, M. Lichtensteiger, *Phys. Chem. Solids* **1961**, *21*, 199.
Nit 1964   R. Nitsche, *Z. Kristallogr.* **1964**, *120*, 1.
Nit 1967a   R. Nitsche, D. F. Sargent, P. Wild, *J. Cryst. Growth* **1967**, *1*, 52.
Nit 1967b   R. Nitsche, P. Wild, *J. Appl. Physics* **1967**, *38*, 5413.

Nit 1967c   R. Nitsche, *J. Phys. Chem. Solids* **1967**, *Suppl. 1*, 215.

Oht 1987   T. Ohtani, T. Kohashi, *Jpn. Chem. Lett.* **1987**, *7*, 1413.

Oka 1974   F. Okamoto, K. Ametani, T. Oka, *Jpn. J. Appl. Phys.* **1974**, *13*, 187.

Oko 1989   I. Okonska-Kozlowska, J. Kopyczok, M. Jung, *Z. Anorg. Allg. Chem.* **1989**, *571*, 157.

Oko 1992   I. Okonska-Kozlowska, J. Kopyczok, K.Wokulska, J. Kammel, *J. Alloys Compds.* **1992**, *189*, 1.

Oko 1995   I. Okonska-Kozlowska, E. Maciazek, K. Wokulska, J. Heimann, *J. Alloys Compds.* **1995**, *219*, 97.

Oko 1999   I. Okonska-Kozlowska, E. Malicka, R. Nagel, H.D. Lutz, *J. Alloys Compds.* **1999**, *292*, 90.

Ols 1983   J. M. Olson, R. Powell, *J. Cryst. Growth* **1983**, *63*, 1.

Pak 1981   V. I. Pakhomov, G. M. Lobanova, V. K. Slovyanskikh, N. T. Kuznetsov, N. V. Gracheva, V. I. Chechernikov, P. V. Nutsubidze, *Zh. Neorg. Khim.* **1981**, *26*, 1961.

Paj 1980   A. Pajaczkowska, *J. Cryst. Growth* **1980**, *49*, 563.

Paj 1983   A. Pajaczkowska, *Mater. Res. Bull.* **1983**, *18*, 397.

Pat 1998   S. G. Patel, M. K. Agarwal, N. M. Batra, D. Lakshminarayana, *Bull. Mater. Sci* **1998**, *21*, 213.

Pat 2009   K. R. Patel, R. D. Vaidya, M. S. Dave, S. G. Patel, *Pramana-J. Phys.* **2009**, *73*, 945.

Poi 1979   R. Poindessault, *J. Electronic Mater.* **1979**, *8*, 619.

Pra 1986   G. Prasad, N. N. Rao, O. N. Srivastava, *Cryst. Res. Technol.* **1986**, *21*, 1303.

Pro 2001   A. Prodan,V. Marinkovic, N. Jug, H. J. P. van Midden, H. Bohm, F. W. Boswell, J. Bennett, *Surface Science* **2001**, *482–485*, 1368.

Que 1974   P. Quenez, Y. Gorochov, *J. Cryst. Growth* **1974**, *26*, 55.

Rab 1990   F. Rabago, A. B. Vincent, N. V. Joshi, *Mater. Lett.* **1990**, *9*, 480.

Rad 2008   K. Radhakrishnan, K. M. Pilla, *Asian J. Chem.* **2008**, *20*, 3774.

Rao 1984   Y. K. Rao, M. Donley, H. G. Lee, *J. Electr. Mater.* **1984**, *13*, 523.

Rim 1972   H. P. B. Rimmington, A. A. Balchin, B. K. Tanner, *J. Cryst. Growth* **1972**, *13*, 51.

Rim 1974a   H .P. B. Rimmington, A. A. Balchin, B. K. Tanner, *J. Cryst. Growth* **1974**, *21*, 171.

Rim 1974b   H. P. B. Rimmington, A. A. Balchin, *J. Mater. Sci.* **1974**, *9*, 343.

Rus 2003   M. Rusu, S. Wiesner, S. Lindner, E. Strub, J. Roehrich, R. Wuerz, W. Fritsch, W. Bohne, Th. Schedelniedrig, M.Ch. Luxsteiner, *J. Phys. Cond. Matter* **2003**, *15*, 8185.

Sag 2004   V. Sagredo, L. M. de Chalbaud, *Cryst. Res. Technol.* **2004**, *39*, 877.

Sag 2005   V. Sagredo, G. E. Delgado, E. ter Haar, G. Attolini, *J. Cryst. Growth* **2005**, *275*, 521.

Sat 1990   K. Sato, Y. Aman, M. Hirai, M. Fujisawa, *J. Phys. Soc. Jpn.* **1990**, *59*, 435.

Sch 1963   H. Schäfer, *Naturwissenschaften* **1963**, *50*, 53.

Sch 1965   H. Schäfer, W. Fuhr, Werner *J. Less-Common Met.* **1965**, *8*, 375.

Sch 1966   H. Schäfer, H. Odenbach, *Z. Anorg. Allg. Chem.* **1966**, *346*, 127.

Schö 1996   J. H. Schön, F. P. Baumgaertner, E. Arushanov, H. Riazi-Nejad, Ch. Kloc, *J. Appl. Phys:* 1996, *79*, 6961.

Schö 2000   J. H. Schön, Ch. Kloc, E. Bucher, *Thin Solid Films* **2000**, *361/362*, 411.

Schö 2010   M. Schöneich, M. Schmidt, P. Schmidt, *Z. Anorg. Allg. Chem.* **2010**, *636*, 1810.

Schm 2005   M. Schmidt, T. Cichorek, R. Niewa, A. Schlechte, Y. Prots, F. Steglich, R. Kniep, *J. Phys.: Cond. Matter* **2005**, *17*, 5481.

Shi 1997   S. Shirataka, S. Shigefusa, S. Isomura, *Jpn. J. Appl. Phys.* **1997**, *37*, 7160.
Sig 1971   A. G. Sigai, H. Wiedemeier, *J. Cryst. Growth* **1971**, *9*, 244.
Sig 1972   A.G. Sigai, H. Wiedemeier, *J. Electrochem. Soc.* **1972**, *119*, 910.
Sim 1967   A. A. Simanovskii, *Rost Kristallov* **1967**, *7*, 258.
Sim 1987   M. G. Simeone, S. Viticoli, *J. Cryst. Growth* **1987**, *80*, 447.
Sim 1988   M. G. Simeone, S. Viticoli, *Mater. Res.* **1988**, *23*, 1219.
Sha 1995   Y. G. Sha, C. H. Su, W. Palosz, M P. Volz, D. C. Gillies, F. R. Szofran, S. L. Lehoczky, H. C. Liu, R. F. Brebrick, *J. Cryst. Growth* **1995**, *146*, 42.
Slo 1966   V. K. Slovyanskikh, G. V. Ellert, E. I. Yarembash, M. D. Korsakova, *Izv. Akad. Nauk SSR, Neorg. Mater.* **1966**, *6*, 973.
Slo 1982   V. K. Slovyanskikh, N. T. Kuznetsov, N. V. Gracheva, *Zh. Neorg. Khim.* **1982**, *27*, 1327.
Smi 1988   K. Smith, J. Marsella, R. Kershaw, K. Dwight, A. Wold, *Mater. Res. Bull.* **1988**, *23*, 1423.
Sol 2003   G. K. Solamnki, M. P. Deshpande, M. K. Agarwal, P. D. Patel, *J. Mater. Sci. Lett.* **2003**, *22*, 985.
Sto 1992   D. Stöber, B. O. Hildmann, H. Böttner, S. Schelb, K. H. Bachem, M. Binnewies, *J. Cryst. Growth* **1992**, *121*, 656.
Sus 1978   M. Susaki, T. Miyauchi, H. Horinaka, N. Yamamoto, *Jpn. J. Appl. Phys.* **1978**, *17*, 1555.
Taf 1995   M. J. Tafreshi, K. Balakrishnan, R. Dhanasekaran, *Mater. Res. Bull.* **1995**, *30*, 1371.
Tan 1977   S. Tanaka, S. Kawami, H. Kobayashi, H. Sasakura, *J. Phys. Chem. Solids* **1977**, *38*, 680.
Tay 1974   B. Taylor, J. Steger, A. Wold, E. Kostiner, *Inorg. Chem.* **1974**, *13*, 2719.
Tri 1982   R. Triboulet, F. Rabago, R. Legros, H. Lozykowski, G. Didier, *J. Cryst. Growth* **1982**, *59*, 172.
Tom 1998   Y. Tomm, S. Fiechter, C. Fischer, *Institute of Physics Conference Series* **1998**, *152*, 181.
Tsu 1967   Y. Tsujimoto, T. Nakajima, Y. Onodera, F. Masakazu, *Jpn. J. Appl. Phys.* **1967**, *6*, 1014.
Vac 1977   G. Vachier, O. Cerclier, A. Casalot, *J. Cryst. Growth* **1977**, *41*, 157.
Vac 1992   G. Vacquier, A. Casalot, *J. Cryst. Growth* **1993**, *130*, 259.
Var 1973   J. Varvas, T. Nirk, *Zh. Khim.* **1973**, *8B*, 562.
Ven 1987   B. Vengatesan, N. Kanniah, R. Gobinathan, P. Ramasamy, *Indian J. Phys.* **1987**, *61A*, 393.
Ven 1988   B. Vengatesan, K. Chinnakali, N. Kanniah, P. Ramasamy, *J. Mater. Science Lett.* **1988**, *7*, 654.
Voh 1971   P. Vohl, W. R. Buchan, J. E. Genthe, *J. Electrochem. Soc.* **1971**, *118*, 1842.
Wan 1986   D. Wang, W. Shi, B. Wang, *Jilin Daxue Ziran Kexue Xuebao* **1986**, *3*, 45.
Whi 1973   C. R. Whitehouse, H. P. B. Rimmington, A. A. Balchin, *Phys. Status Solidi* **1973**, *A18*, 623.
Whi 1978a   C. R. Whitehouse, A. A. Balchin, *J. Cryst. Growth* **1978**, *43*, 727.
Whi 1978b   C. R. Whitehouse, A. A. Balchin, *J. Mater. Sci.* **1978**, *13*, 2394.
Weh 1969   F. H. Wehmeier, *J. Cryst. Growth* **1969**, *5*, 26.
Weh 1970   F. H. Wehmeier, E. T. Keve, S. C. Abrahams, *Inorg. Chem.* **1970**, *9*, 2125.
Wie 1968   H. Wiedemeier, W. J. Goyette, *J. Chem. Phys.* **1968**, *48*, 2936.
Wie 1970   H. Wiedemeier, A. G. Sigai, *J. Solid State Chem.* **1970**, *2*, 404.
Wie 1972   H. Wiedemeier, E. A. Irene, A. K. Chaudhuri, *J. Cryst. Growth* **1972**, *13–14*, 393.
Wie 1981   H. Wiedemeier, D. Chandra, F. Klaessig, *J. Cryst. Growth* **1981**, *51*, 345.

Wie 1986   H. Wiedemeier, H. Goldman, *J. Less-Common Met.* **1986**, *116*, 389.
Wie 1991a   H. Wiedemeier, Y. R. Ge, *Z. Anorg. Allg. Chem.* **1991**, *598–599*, 339.
Wie 1991b   H. Wiedemeier, Y. R. Ge, *Z. Anorg. Allg. Chem.* **1991**, *602*, 129.
Wie 1992   H. Wiedemeier, L. L. Regel, W. Palosz, *J. Cryst. Growth* **1992**, *119*, 79.
You 1990   G. H. Yousefi, *J. Mater. Science Lett.* **1990**, *9*, 1216.
Zlo 1990   V. P. Zlomanov, A. M. Gaskov, I. M. Malinskii, *Izv. Akad. Nauk SSSR, Ne-org. Mater.* **1990** 26, 744.
Zor 1965   E. L. Zorina, V. B. Velichkova, T. N. Guliev, *Inorg. Mater. (USSR)* **1965**, *1*, 633.
Zul 1984   D. I. Zulfugarly, B. A. Geidarov, T. A. Nadzhafova, *Azerbaidzh. Khim. Zh.* **1984**, *3*, 136.

7.3 Transport of Tellurides

$HfTe_2$

H																	He
Li	Be											B	C	N	O	F	Ne
Na	Mg											Al	Si	P	S	Cl	Ar
K	Ca	Sc	Ti	V	Cr	Mn	Fe	Co	Ni	Cu	Zn	Ga	Ge	As	Se	Br	Xe
Rb	Sr	Y	Zr	Nb	Mo	Tc	Ru	Rh	Pd	Ag	Cd	In	Sn	Sb	Te	I	Kr
Cs	Ba	La	Hf	Ta	W	Re	Os	Ir	Pt	Au	Hg	Tl	Pb	Bi	Po	At	Rn

Ce	Pr	Nd	Pm	Sm	Eu	Gd	Tb	Dy	Ho	Er	Tm	Yb	Lu
Th	Pa	U	Np	Pu	Am	Cm	Bk	Cf	Es	Fm	Md	No	Lr

Iodine as transport agent

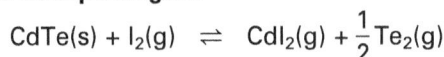

$$CdTe(s) + I_2(g) \;\rightleftharpoons\; CdI_2(g) + \frac{1}{2}Te_2(g)$$

Hydrogen halide as transport agent

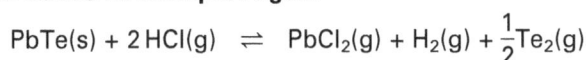

$$PbTe(s) + 2\,HCl(g) \;\rightleftharpoons\; PbCl_2(g) + H_2(g) + \frac{1}{2}Te_2(g)$$

Examples of CVT reactions of tellurides (see Table 7.3.1) are almost as numerous as those of sulfides and selenides. References can be found for the main group elements (groups 2, 13, 14 and 15) as well as for almost all transition metal elements and some lanthanoids. The alkali metal tellurides cannot be transported with halogens or halogen compounds due to the high stability of alkali metal halides.

The tellurides have a lower thermal stability than the analogous selenides and sulfides. For this reason, the transport reactions are less endothermic, thus the transport equilibrium shifts to the side of the reaction products. As a consequence, the transport conditions change to higher partial pressures of the transport effective species and to lower dissolution temperatures. In contrast to sulfur and selenium, tellurium forms the stable diiodide, TeI_2, which is also stable in the gas phase. This was not considered in older literature but it is discussed in a more recent work (Czu 2010)(see section 9.2).

Due to the methodological developments of CVT in the 1960s, gas-phase reactions for preparation and purification of tellurides also came into focus (Bro 1962, Pia 1966, Mei 1967, Gib 1969, Wie 1969). Due to their potential for technical applications, the compounds CdTe, ZnTe, and $Hg_{1-x}Cd_xTe$ and their basic transport equilibria were mainly investigated.

Thermal behavior   Some tellurides sublime undecomposed. Above all, this applies for compounds of group 14 MTe (M = Ge, Sn, Pb). The sublimation pressures decrease toward the heavy homologues of the respective group: GeTe ($K_{p,\,800} = 5 \cdot 10^{-5}$ bar); SnTe ($K_{p,\,800} = 1 \cdot 10^{-6}$ bar); and PbTe ($K_{p,\,700} = 3 \cdot 10^{-7}$ bar). Within the line of the chalcogenides of an element, the sublimation pressures increase toward the tellurides. In doing so, the equilibrium pressures of selenides and tellurides hardly differ (Figure 7.3.1).

Like the selenides, the tellurides of the elements of groups 13 and 15 show the effect of a decomposition sublimation. In the process, high volatile metal-rich tellurides form apart from tellurium. The total pressures, which result from the decomposition sublimation, have a maximum at the tellurides in the series of the chalcogenides.

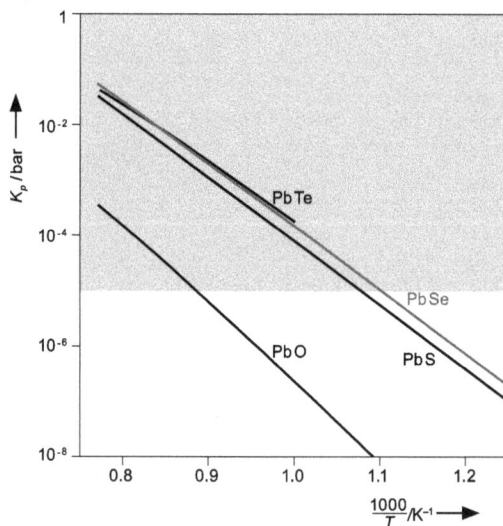

$$M_2Te_3(s) \;\rightleftharpoons\; M_2Te(g) + Te_2(g) \tag{7.3.1}$$
$$(M = Al,\ Ga,\ In)$$

Figure 7.3.1   Equilibrium constants for the sublimation of the lead chalcogenides as a function of the temperature in logarithmic scale.

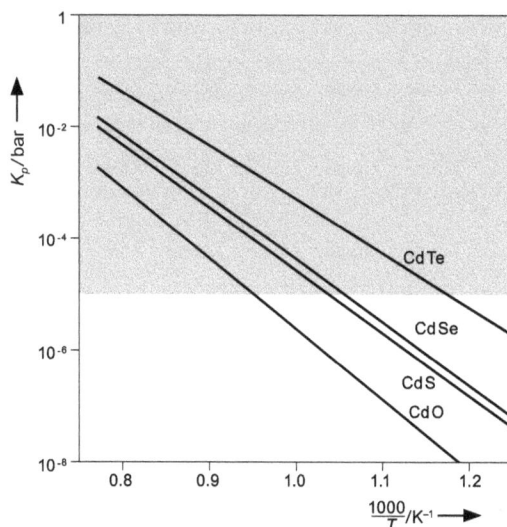

Figure 7.3.2 Equilibrium constants for the decomposition sublimation of the cadmium chalcogenides as a function of the temperature in logarithmic scale.

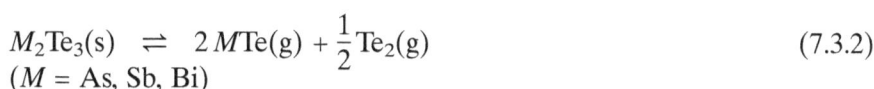

$$M_2Te_3(s) \;\rightleftharpoons\; 2\,MTe(g) + \frac{1}{2}Te_2(g) \tag{7.3.2}$$
$$(M = \text{As, Sb, Bi})$$

The congruent thermal decomposition of ZnTe and CdTe to the elements is important, too.

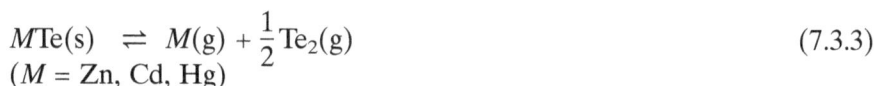

$$MTe(s) \;\rightleftharpoons\; M(g) + \frac{1}{2}Te_2(g) \tag{7.3.3}$$
$$(M = \text{Zn, Cd, Hg})$$

The partial pressure of the decomposition sublimation clearly increases from the oxides to the tellurides (Figure 7.3.2). This way, the deposition of crystalline CdTe and ZnTe succeeds over the gas phase already at temperatures of approx. 900 °C ($K_{p,\,1100}(\text{CdTe}) = 4 \cdot 10^{-3}$ bar; $K_{p,\,1100}(\text{ZnTe}) = 5 \cdot 10^{-4}$ bar) without adding a transport agent. The equilibria and their thermodynamic description provide the basis of PVD processes for depositing thin layers of both compounds (Aku 1971, Iga 1976, Ido 1968). In this context, the term *closed spaced vapor transport* – CSVT (Cas 1993) – has to be mentioned: during a PVD-process, good deposition rates at low temperatures are achieved due to short transport distances (a few millimeters).

Iodine as transport agent   Although the CVT of tellurides is described with chlorine, bromine, as well as with iodine, the majority of the known transport reactions utilize iodine as the transport agent. Due to the similar thermal stabilities of tellurides and selenides, a similar transport behavior results. Hence what applies for selenides, often applies for tellurides, too. A distinctive feature, however, is the possibility of the formation of tellurium iodides with transport effective partial pressures.

Although the decomposition sublimation of CdTe shows sufficient deposition rates above 900 °C, the CVT of the compound has also been examined in detail. The temperature of the source side can be decreased by about 100 K compared to the decomposition sublimation. Transport of CdTe is possible in the temperature range of 1000 to 800 °C ($K_{p,\ 1000} \approx 1$ bar) at the source side and 750 to 600 °C at the sink (Pia 1966, Pao 1974); the transport follows equilibrium 7.3.4.

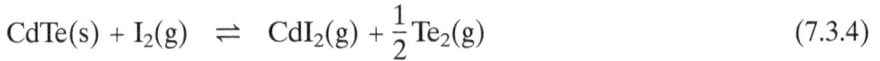

$$CdTe(s) + I_2(g) \ \rightleftharpoons \ CdI_2(g) + \frac{1}{2}Te_2(g) \tag{7.3.4}$$

The transport of most of the binary tellurides takes place accordingly. The formation of $Te_2(g)$ dominates in the gas phase; at higher temperatures $Te(g)$ is also transport relevant. Higher condensed molecules, which appear with sulfur and selenium, do not have to be considered. Furthermore, the transport of a number of ternary tellurides follows this transport mechanism. Detailed examinations describe the transport of mixed-crystals, such as $Cd_{1-x}Co_xTe$ (Red 2008), and $Cd_{1-x}Mn_xTe$ (Mel 1990), or $CdTe_{1-x}S_x$ (Hot 2005); $CdTe_{1-x}Se_x$ (Hot 2005), $ZnTe_{1-x}S_x$ (Ros 2004), and $ZnTe_{1-x}Se_x$ (Tsu 1967, Su 2000).

In this context, the system CdTe/HgTe gained special attention. Its importance results from the possibility of changing the band gap within the mixed-crystal series $Hg_{1-x}Cd_xTe$ almost linearly with the composition. Although both binary boundary phases can be obtained well by a decomposition sublimation (7.3.3), problems with the directed deposition of mixed-crystals of defined composition arise due to different partial pressures ($K_{p,\ 900}(CdTe) = 5 \cdot 10^{-5}$ bar; $K_{p,\ 900}(HgTe) = 6 \cdot 10^{-1}$ bar). These problems can be solved by evaporating the components from spatially separated sources of different temperatures. Homogeneous materials with defined compositions are particularly obtained by CVT reactions with iodine (Wie 1982, Ire 1983, Wie 1987, Shi 1987). In the process, cadmium is transferred to the high volatile iodine while mercury becomes transport effective as an elemental gas species.

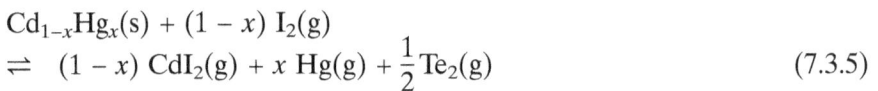

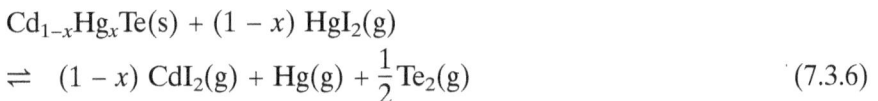

$$Cd_{1-x}Hg_x(s) + (1 - x)\,I_2(g)$$
$$\rightleftharpoons \ (1 - x)\,CdI_2(g) + x\,Hg(g) + \frac{1}{2}Te_2(g) \tag{7.3.5}$$

$$Cd_{1-x}Hg_xTe(s) + (1 - x)\,HgI_2(g)$$
$$\rightleftharpoons \ (1 - x)\,CdI_2(g) + Hg(g) + \frac{1}{2}Te_2(g) \tag{7.3.6}$$

It is not known in how far iodine is really effective as a transport agent, or if gaseous HgI_2 is formed immediately with iodine as additive. The transport effects during transport with iodine are identical to those when HgI_2 is directly added (7.3.6) (Wie 1983a, Wie 1983b, Ire 1983, Wie 1989, Hut 2002). The resulting equilibria of formation of the transport effective species are analogous. The partial pressures of the individual species that are involved in the transport depend on the temperature and the pressure of the transport agent (Figure 7.3.3).

The problem of a presumable contamination of the transported crystals with the transport agent becomes clear with the help of the example of the transport of $Cd_{1-x}Hg_xTe$, especially in the characteristic composition $Hg_{0.8}Cd_{0.2}Te$, with

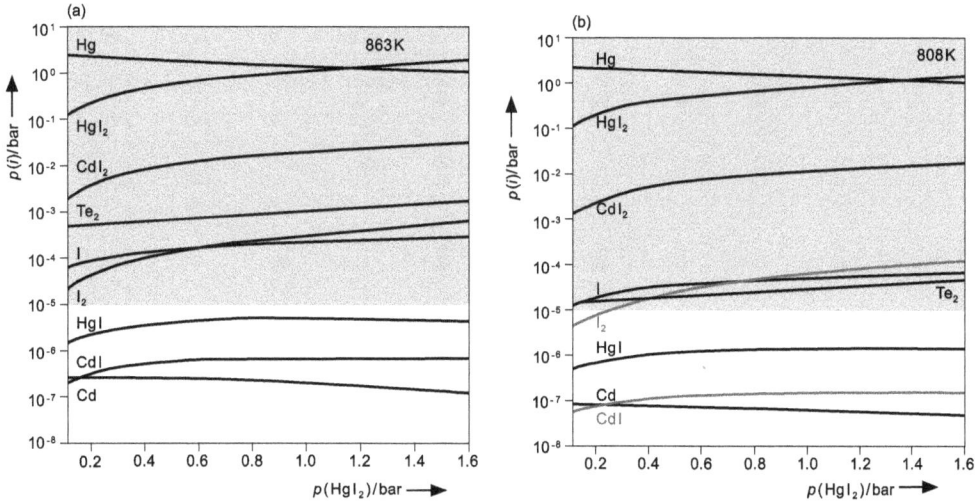

Figure 7.3.3 Equilibrium partial pressure in the $Hg_{0.8}Cd_{0.2}Te/I_2$ system as a function of the HgI_2 partial pressure at 863 K (a) and 808 K (b) according to *Wiedemeier* and *Chandra* (Wie 1987).

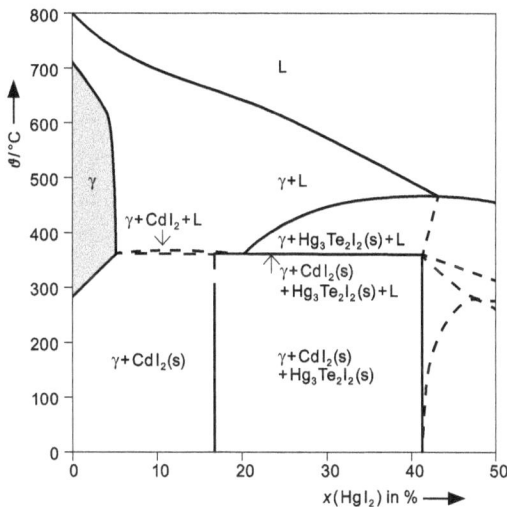

Figure 7.3.4 Phase diagram of $Hg_{0.8}Cd_{0.2}Te/HgI_2$ according to *Hutchins* and *Wiedemeier* (Hut 2002).

HgI_2. The phase diagram (Hut 2002) shows that the γ-phase, $Hg_{0.8}Cd_{0.2}Te$, has a solubility of up to 5 % for HgI_2 in the temperature range between 300 and 700 °C. Transport of phase-pure $Hg_{0.8}Cd_{0.2}Te$ must be realized with small amounts of transport agent and deposition temperatures below 290 °C. Furthermore, low concentrations of the transport agent are appropriate in order to maintain the composition x of the initial solid $Cd_{1-x}Hg_xTe$ during transport: Higher concentrations of HgI_2 lead to the deposition of CdI_2 and $Hg_3Te_2I_2$ (Hut 2002), respectively, at the sink and thus to changes in composition of the source.

The CVT of other ternary tellurides, such as $CuInTe_2$ and $CuGaTe_2$ (Pao 1980), can be described – as for the binary compounds – as dissolution reactions of the metallic components in the respective iodide while tellurium is getting to the gas phase in elemental form:

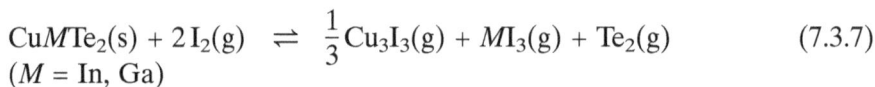

$$CuMTe_2(s) + 2\,I_2(g) \;\rightleftharpoons\; \frac{1}{3}Cu_3I_3(g) + MI_3(g) + Te_2(g) \tag{7.3.7}$$
$$(M = In, Ga)$$

The dominating gas-phase equilibrium is given with this reaction equation. The total reaction must be described by participation of gas species I, Te_n, CuI, MI, M_2Te, and Cu_2Te (M = In, Ga). The principles of the transport of phase-pure tellurides with iodine in co-existence with the respective telluride iodides and elemental tellurium apply in the same way as described for the selenides (Schö 2010).

Hydrogen halides and hydrogen as transport agent   The use of hydrogen or hydrogen halides as transport agent is important for the transport of oxides and sulfide because the solubility of oxygen and sulfur, respectively, in the gas phase is supported by the formation of water and hydrogen sulfide, respectively. However, the stability of hydrogen compounds H_2Q (Q = O, S, Se, Te) constantly decreases, the participation of H_2Se and, in particular, of H_2Te in CVT reactions must be discussed critically.

H_2O as well as H_2S is still stable above 1000 °C (equilibrium constants of decomposition (7.3.8): $K_{p,\ 1300}(H_2O) = 2 \cdot 10^{-5}\ bar^{1/3}$, $K_{p,\ 1300}(H_2S) = 2 \cdot 10^{-1}\ bar^{1/3}$). H_2Se, however, decomposes already between 700 and 800° ($K_{p,\ 1000}(H_2Se) = 1\ bar^{1/3}$). If the hydrogen pressures (p approx. 1 bar) are sufficiently high, transport relevant partial pressures of H_2Se are possible up to 1000 °C.

$$\frac{2}{3}H_2Q(g) \;\rightleftharpoons\; \frac{2}{3}H_2(g) + \frac{1}{3}Q_2(g) \tag{7.3.8}$$

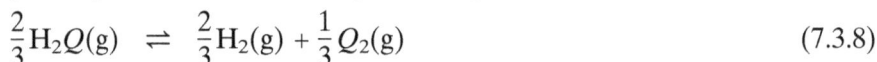

$H_2Te(g)$ is unstable in the entire temperature range ($K_{p,\ 1000}(H_2Te) = 10^2\ bar^{1/3}$). The partial pressure of $Te_2(g)$ resulting from the equilibrium is higher by orders of magnitude. It determines the transport equilibrium.

Accordingly, the described transport of cadmium telluride in the presence of hydrogen (Pia 1966, Pao 1972, Akh 1981b, Ant 1984) should rather be seen as a decomposition sublimation. The given temperature ranges of the vapor deposition (1090 ... 800 → 700 ... 500 °C) support this statement. Transport reactions that take place with added halogen/hydrogen mixtures occur under formation of the respective metal halides and Te_2, but not H_2Te (Tok 1979, Akh 1981a, Pao 1974).

Often ammonium halides are described as transport additives for the transport of tellurides. The ammonium halides, which are easy to handle in experiments, serve as a hydrogen halide source (see Chapter 14).

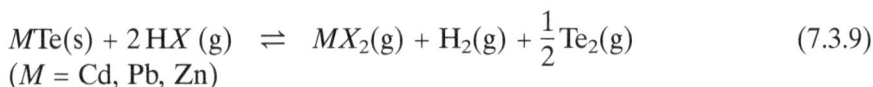

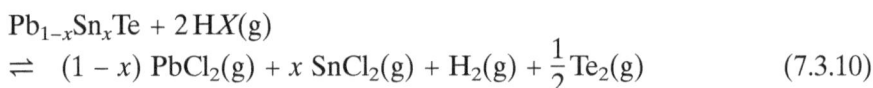

$$MTe(s) + 2\,HX\,(g) \;\rightleftharpoons\; MX_2(g) + H_2(g) + \frac{1}{2}Te_2(g) \tag{7.3.9}$$
$$(M = Cd, Pb, Zn)$$

$$Pb_{1-x}Sn_xTe + 2\,HX(g)$$
$$\rightleftharpoons\; (1-x)\,PbCl_2(g) + x\,SnCl_2(g) + H_2(g) + \frac{1}{2}Te_2(g) \tag{7.3.10}$$

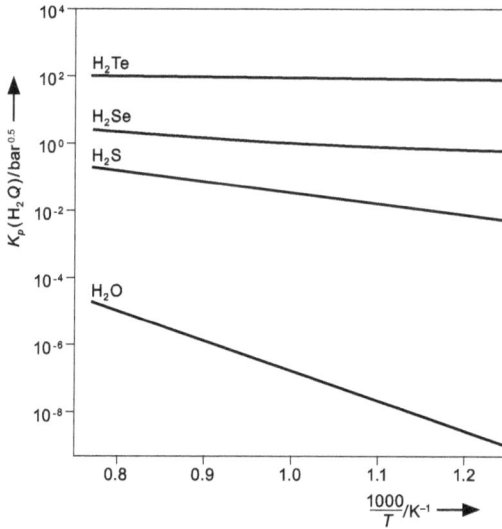

Figure 7.3.5 Equilibrium constants of the composition of the hydrogen chalcogenides as a function of the temperature in logarithmic scale.

As mentioned above, in contrast to the transport equilibria of the sulfides and selenides, the participation of H_2Te must not be considered and the respective reactions with participation of $H_2(g)$ and $Te_2(g)$ dominate. The only function of the hydrogen halide is the halogenation the metallic components.

Metal halides as transport agents The CVT of tellurides with added metal halides is described by only a few examples. Often it is about additives that contain a component of the solid ($CdTe/CdCl_2$ (Vac 1991); $Cu_xTe/CuBr$ (Abb 1987); and $Gd_4NiTe_2/GdBr_3$ (Mag 2004)).

The efficiency of the additive $TeCl_4$ can be seen in a similar way. In reactions with base metals, $TeCl_4$ transfers chlorine while the actual transport agent MCl_x and Te_2 are being formed (Phi 2008b). Hence the metal halide MCl_x, here $TiCl_4$, becomes active as transport agent, equation 7.3.12.

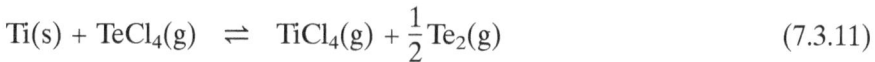

$$Ti(s) + TeCl_4(g) \; \rightleftharpoons \; TiCl_4(g) + \frac{1}{2}Te_2(g) \tag{7.3.11}$$

$$2\,Ti_2PTe_2(s) + 12\,TiCl_4(g) \; \rightleftharpoons \; 16\,TiCl_3(g) + P_2(g) + 2\,Te_2(g) \tag{7.3.12}$$

In the same way, the calculations of the transport equilibria of the cationic clathrate $Si_{46-2x}P_{2x}Te_x$ show the formation of $SiCl_4$ as dominant species, as a product of the reaction of the starting solid with $TeCl_4$ (Phi 2008a). Thus endothermic transport in the temperature gradient from 900 to 800 °C can be described as follows:

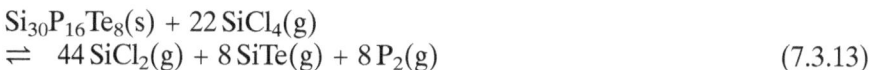

$$Si_{30}P_{16}Te_8(s) + 22\,SiCl_4(g) \; \rightleftharpoons \; 44\,SiCl_2(g) + 8\,SiTe(g) + 8\,P_2(g) \tag{7.3.13}$$

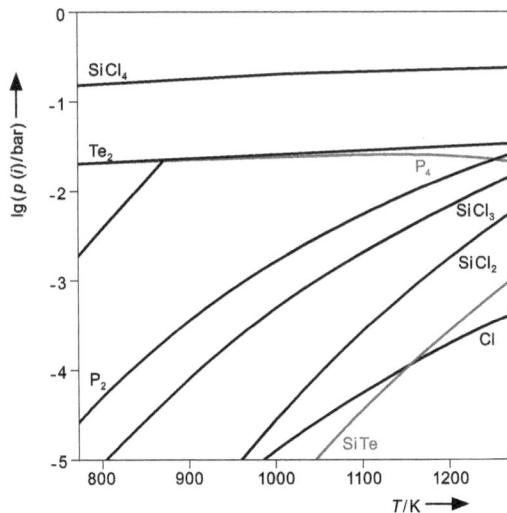

Figure 7.3.6 Gas-phase composition in the $\mathrm{Si/P/Te/Cl}$ system as a function of the temperature according to *Philipp* and *Schmidt* (Phi 2008a).

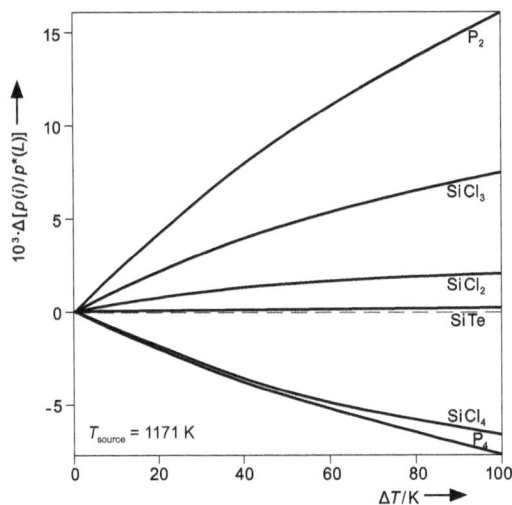

Figure 7.3.7 Transport efficiency of the gas species in the transport of $\mathrm{Si_{30}P_{16}Te_8}$ according to *Philipp* and *Schmidt* (Phi 2008a).

A reverse of the transport was observed at lower temperatures, which is not comprehensible with the equilibrium calculations at hand. Only when hydrogen-containing species are considered, an exothermic transport at lower temperatures ($650 \rightarrow 730\,°\mathrm{C}$) is possible. In this process, HCl becomes effective as the transport agent.

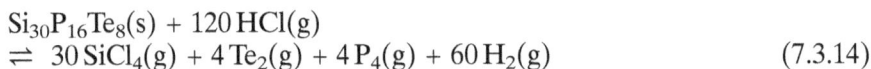

$$\mathrm{Si_{30}P_{16}Te_8(s) + 120\,HCl(g)}$$
$$\rightleftharpoons \mathrm{30\,SiCl_4(g) + 4\,Te_2(g) + 4\,P_4(g) + 60\,H_2(g)} \qquad (7.3.14)$$

It is apparent that the formation of the hydrogen halide through the interaction with moisture adhering to the ampoule walls cannot be excluded. However, such

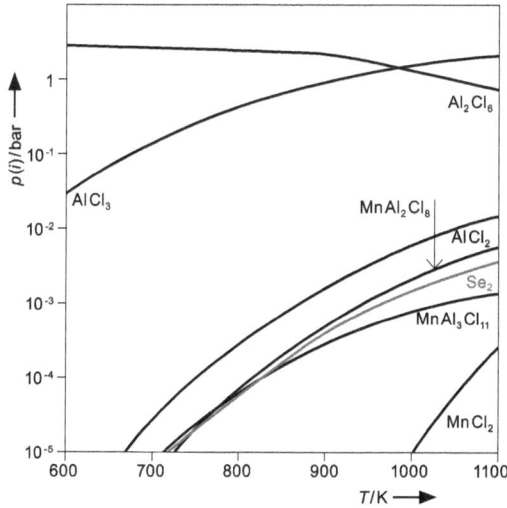

Figure 7.3.8 Gas-phase composition in the $MnTe/AlCl_3$ system as a function of the temperature.

a serious effect of a transport reversal, as observed with the clathrate $Si_{46-2x}P_{2x}Te_x$, is rare. Similar effects were found for the transport of SiAs (Bol 1994) (see section 9.1).

When $AlCl_3$ is added, the CVT of MnTe takes place under the formation of the ternary gas-phase complexes $MnAl_2Cl_8$ and $MnAl_3Cl_{11}$ (Paj 1980). In the author's opinion, the "trick" is that, compared to the transport with Cl_2, the partial pressures are decreased. This way, the absolute solubilities of the components in the gas phase are reduced but due to this, larger differences in the gas-phase solubilities and thus better transports are achieved. The appearance of hydrogen chloride is not considered in this work.

Table 7.3.1 Examples of the chemical vapor transport reactions of tellurides.

Sink solid	Transport additive	Temperature / °C	Reference
Ag_2Te	Br_2, I_2	950 ... 900 → 650 ... 450	Bon 1969
	Br_2	1140 ... 725	Bon 1972
		→ 940 ... 540	
	I_2	930 → 830	Pav 1978
	not specified	not specified	Chu 2001
	not specified	not specified	Mus 2005
Bi_2Te_3	I_2	500 → 450	Sch 2010
CdTe	$Cl_2 + H_2$, $Br_2 + H_2$ $I_2 + H_2$	not specified	Tok 1979
	$Br_2 + H_2$	825 → 425	Akh 1981a
	I_2	1100 ... 850 → 600	Pia 1966
	I_2	not specified	Pao 1972a
	I_2	not specified	Pao 1977
	$I_2 + H_2$	820 → 750	Pao 1974
	$CdCl_2$	620 → T_1	Vac 1991
	NH_4Cl, NH_4Br, NH_4I	not specified	Ilc 2002
	NH_4Cl	800 ... 700 → 730 ... 500	Pao 1972b
	NH_4Cl	800 ... 700 → 730 ... 500	Pao 1973
	NH_4Cl	800 → 600	Ghe 1974
	NH_4Cl	800 → 650	Pao 1975
	NH_4Cl	750 ... 650 → 650 ... 550	Man 1983
	NH_4Cl	800 → 660	Pao 1986
	NH_4I	850 ... 800 → 630 ... 510	Zha 1983
$Cd_{1-x}Co_xTe$	I_2	1100 → T_1	Red 2008
$Cd_{1-x}Mn_xTe$	I_2	725 ... 715 → 685 ... 675	Mel 1990
$Cd_{1-x}Zn_xTe$	NH_4Cl, NH_4Br, NH_4I	not specified	Ilc 2002
$CdTe_{1-x}S_x$	I_2	900 → 800	Hot 2005
$CdTe_{1-x}Se_x$	I_2	900 → 800	Hot 2005
Ce_2Te_3	I_2	900 → 700	Bro 1962
$CeTe_2$	I_2	900 → 700	Bro 1962
	I_2	950 → 850	Sto 2000
CoTe	I_2	870 ... 675 → 845 ... 640	Gib 1969a
$Co_{1-x}Te$	I_2	870 ... 650 → 845 ... 615	Gib 1969b
Co_2Te_3	I_2	675 → 650	Boc 1981
$CoCr_2Te_4$	I_2	870 ... 650 → 845 ... 615	Gib 1969b
$Cr_{1-x}Te$	I_2	1010 → 930	Str 1973
	I_2	800 → T_1	Shi 1985
Cr_3Te_4	I_2	not specified	Sat 1990
Cr_2Te_3	I_2	not specified	Sat 1990

Table 7.3.1 (continued)

Sink solid	Transport additive	Temperature / °C	Reference
$Cr_{1-x}Fe_{1-y}Te$	I_2	$800 \rightarrow T_1$	Shi 1985
Cr_2FeTe_4	Cl_2	$840 \rightarrow 780$	Beg 1975
Cu_2Te	I_2	not specified	Mus 2005
	CuBr	$900 \ldots 600 \rightarrow 750 \ldots 350$	Abb 1987
$Cu_{2-x}Te$	CuBr	$900 \ldots 600 \rightarrow 750 \ldots 350$	Abb 1987
CuTe	CuBr	$900 \ldots 600 \rightarrow 750 \ldots 350$	Abb 1987
$CuGaTe_2$	I_2	not specified	Pao 1980
	I_2	not specified	Lec 1983
	I_2	not specified	Mas 1993
	I_2	$700 \rightarrow 650$	Gom 1983
$CuInTe_2$	I_2	not specified	Pao 1980
	I_2	not specified	Lec 1983
	I_2	$720 \rightarrow 680$	Bal 1990
	I_2	not specified	Mas1993
	I_2, $TeCl_4$	$710 \rightarrow 670$, $710 \rightarrow 670$	Bal 1994
	I_2	$700 \rightarrow 650$	Gom 1983
$DyTe_2$	Br_2, I_2	not specified	Slo 1985
$Dy_{1.5}U_{1.5}Te_5$	Br_2, I_2	not specified	Slo 1985
$Dy_{0.5}U_{0.5}Te_2$	Br_2, I_2	not specified	Slo 1985
$Dy_{0.5}U_{0.5}Te_3$	Br_2, I_2	not specified	Slo 1985
Er_2Te_3	$ErCl_3$	$950 \rightarrow 800$	Sto 1998
EuTe	I_2	$1700 \rightarrow T_1$	Kal 1968
$Fe_{1-x}Te$	I_2	$800 \rightarrow T_1$	Shi 1985
GaTe	Br_2	$780 \rightarrow 690$	Zul 1983
	$SnCl_2$	$780 \rightarrow 690$	Zul 1983
Gd_4NiTe_2	$GdBr_3$	$1000 \rightarrow T_1$	Mag 2004
GeTe	I_2	$590 \rightarrow 440$	Wie 1972
$GeTe_{1-x}Se_x$		$600 \rightarrow 590$	Wie 1991a
	I_2	$600 \rightarrow 590$	Wie 1991b
$HfTe_2$	I_2	not specified	Bra 1973
$HfTe_2$	I_2	$600 \rightarrow 700$	Lev 1983
$HfTe_5$	I_2	$450 \rightarrow 400$	Lev 1983
HgTe	H_2	$T_2 \rightarrow 420 \ldots 320$	Iga 1976b
$Hg_{1-x}Cd_xTe$	I_2	$590 \rightarrow 535$	Wie 1982, Ire 1983, Wie 1987
	I_2	not specified	Shi 1987

Table 7.3.1 (continued)

Sink solid	Transport additive	Temperature / °C	Reference
$Hg_{1-x}Cd_xTe$	HgI_2	590 → 535	Wie 1983a, Wie 1983b, Ire 1983
	HgI_2	590 → 585 … 520	Wie 1989, Hut 2002
	HgI_2	590 → 540	Wie 1991c, Wie 1992
	HgI_2	600 → 570 … 500	Sha 1993, Sha 1994
	HgI_2	595 → 545	Ge 1996, Ge 1999
	NH_4Br, NH_4Cl NH_4I	580 → 300, 590 → 290 590 → 290	Akh 1983
	NH_4I	670 … 520 (HgTe), 850 … 800 (CdTe) → 630 … 510	Gol 1979
$Hg_{1-x}Cd_xTe:In^{3+}$	NH_4I	not specified	Tom 1980
$Hg_{1-x}Mn_xTe$	HgI_2	590 → 535	Pal 1989
La_2Te_3	I_2	900 → 700	Bro 1962
$LaTe_2$	I_2	900 → 700	Bro 1962
$MgTe$		960 → 760	Kuh 1971
$MnTe$	I_2	800 → 750 … 600	Wie 1969
	I_2	800 … 670 → 770 … 640	Mel 1991
	$AlCl_3$	730 → 630	Paj 1980
$MnAl_{1.04}Te_{2.18}$	I_2	740 → 630	Kha 1984
$MoTe_2$	Br_2	900 → 700	Bri 1962
	Br_2	not specified	Bro 1966
	Br_2	800 → 750	Hil 1972
	Br_2	895 → 845	Alb 1992
	$TeCl_4$	1000 → 900	Fou 1979
	$TeCl_4$	825 → 730	Bal 1994a
$MoTe_{2-x}Se_x$	Br_2	900 … 700 → 800 … 650	Aga 1986
Nb_3Te_4	not specified	1160 → T_1	Edw 2005
$NbTe_2$	I_2	1000 → T_1	All 1969
	I_2	not specified	Bha 2004
$NbTe_4$	I_2, $TeCl_4$	420 → 360	Lev 1983
	Cl_2, $TeCl_4$	550 → 530	Lev 1991
$Nb_2FeCu_{0.35}Te_4$	$TeCl_4$	875 → 810	Li 1994
Nd_2Te_3	I_2	900 → 700	Bro 1962
$NdTe_2$	I_2	900 → 700	Bro 1962
	I_2	950 → 850	Sto 2001

Table 7.3.1 (continued)

Sink solid	Transport additive	Temperature / °C	Reference
PbTe	Br$_2$	not specified	Bez 1972
	I$_2$	700 → T_1	Sto 1992
	I$_2$	540 → 490	Ker 1998
	NH$_4$Cl	830 → 370	Akh 1986
Pb$_{1-x}$Sn$_x$Te	Br$_2$	not specified	Bez 1972
	NH$_4$Cl, NH$_4$Br, NH$_4$I	680 → 380	Akh 1987b
Pd$_{13}$Te$_3$	PdBr$_2$, PdCl$_2$	600 → 650	Jan 2006
Nd$_2$Te$_3$	I$_2$	900 → 700	Bro 1962
PrTe$_2$	I$_2$	900 → 700	Bro 1962
	I$_2$	950 → 850	Sto 2000
Pr$_2$Te$_3$	I$_2$	900 → 700	Bro 1962
RuTe$_2$	ICl$_3$	1060 → 960	Hua 1994
Sb$_2$Te$_3$	I$_2$	500 → 450	Sch 2010
Si$_2$Te$_3$	I$_2$	750 → T_1	Bai 1966
	I$_2$	750 → T_1	Pet 1973
Si$_{46-2x}$P$_{2x}$Te$_x$	TeCl$_4$	650 → 730, 900 → 800	Phi 2008a
SnTe	I$_2$	not specified	Vor 1973
TaTe$_4$	I$_2$, TeCl$_4$	520 → 460	Lev 1983
	Cl$_2$, TeCl$_4$	550 → 530	Lev 1991
Ti$_3$Te$_4$	I$_2$	not specified	Pan 1994
TiTe$_2$	I$_2$	900 → 800	Gre 1965
	I$_2$	750 → 690	Rim 1974
TiTe$_{2-x}$S$_x$	I$_2$	800 ... 750 → 720 ... 690	Rim 1974
TiTe$_{2-x}$Se$_x$	I$_2$	780 ... 750 → 740 ... 690	Rim 1974
Ti$_2$PTe$_2$	TeCl$_4$	800 → 700	Phi 2008b
UTe$_2$	AgBr	970 → 870	Slo 1966
	Br$_2$, I$_2$	not specified	Slo 1985
U$_7$Te$_{12}$	I$_2$	1030 → 1000	Tou 1998
UTe$_3$	AgBr	770 → 610	Slo 1966
U$_3$Ge$_{0.7}$Te$_5$	I$_2$	870 → 840	Tou 2002
U$_3$Sn$_{0.5}$Te$_5$	I$_2$	840 → 800	Tou 2002
WTe$_2$	Br$_2$	not specified	Bro 1966
ZnTe	Cl$_2$, Br$_2$, I$_2$	not specified	Mei 1967
	I$_2$	T_2 → 650 ... 550	Nis 1979, Nis 1980, Nis 1982
	I$_2$	675 ... 600 → 650 ... 550	Kak 1981

Table 7.3.1 (continued)

Sink solid	Transport additive	Temperature / °C	Reference
ZnTe	I_2	$725 \rightarrow 650 \dots 550$	Oga 1981
	H_2	$930 \rightarrow 610$	Ido 1968
	HCl	$725 \rightarrow 650$	Nis 1986,
			Nis 1988
	NH_4I	$1100 \rightarrow 700$	Ilc 1999
$ZnS_{1-x}Te_x$	I_2	$1000 \rightarrow 900$	Ros 2004
$ZnGa_{1.01}Te_{2.13}$	I_2	$650 \rightarrow 550$	Kha 1984
$ZrTe_2$	I_2	not specified	Bra 1973
	I_2	$700 \rightarrow 800$	Czu 2010
$ZrTe_3$	I_2	$400 \rightarrow 350$	Pro 2001
	I_2, $TeCl_4$	$650 \dots 800 \rightarrow 600 \dots 700$	Lev 1983
$ZrTe_5$	I_2	$530 \rightarrow 480$	Lev 1983
Zr_2PTe_2	I_2	$800 \rightarrow T_2$	Tsc 2009
$ZrSb_{0.85}Te_{1.15}$	I_2	$600 \rightarrow 700$	Czu 2010
$ZrAs_{0.6}Te_{1.4}$	I_2	$900 \rightarrow 950$	Czu 2010
$ZrAs_{1.6}Te_{0.4}$	I_2	$900 \rightarrow 950$	Czu 2010

Bibliography of Section 7.3

Abb 1987   A. S. Abbasov, T. Kh. Azizov, N. A. Alieva, U. Ya. Aliev, F. M. Mustafaev, *Dokl. Akad. Nauk Azerbaidzhanskoi SSR* **1987**, *42*, 41.

Aga 1986   M. K. Agarwal, P. D. Patel, R. M. Joshi, *J. Mater. Sci. Lett.* **1986**, *5*, 66.

Akh 1981a   Y. G. Akhromenko, G. A. Il'chuk, I. E. Lopatinskii, S. P. Pavlishin, *Izv. Akad. Nauk SSSR, Neorg. Mater.* **1981**, *17*, 2016.

Akh 1983   Y. G. Akhromenko, G. A. Il'chuk, S. P. Pavlishin, V. I. Ivanov-Omskii, *Pisma Zh. Tekhn. Fiz.* **1983**, *9*, 564.

Akh 1986   Y. G. Akhromenko, Y. G. Belashov, G. A. Il'chuk, S. P. Pavlishin, S. I. Petrenko, *Izv. Akad. Nauk SSSR, Neorg. Mater.* **1986**, *22*, 1275.

Akh 1987a   Y. G. Akhromenko, G. A. Il'chuk, S. P. Pavlishin, S . I. Petrenko, O. I. Gorbova, *Izv. Akad. Nauk SSSR, Neorg. Mater.* **1987**, *23*, 762.

Akh 1987b   Y. G. Akhromenko, G. A. Il'chuk, S. P. Pavlishin, S. I. Petrenko, *Vestnik L'vovskogo Politekhnich. Inst.* **1987**, *215*, 122.

Akh 1990   Y. Akhromenko, G. Il'chuk, S. Pavlishin, I. Lopatinskii, V. Ukrainets, *Izv. Akad. Nauk SSSR, Neorg. Mater.* **1990**, *26*, 739.

Alb 1992   M. Albert, R. Kershaw, K. Dwight, A. Wold, *Solid State Commun.* **1992**, *81*, 649.

All 1969   K. R. Allakhverdiev, E. A. Antonova, G. A. Kalyuzhnaya, *Izv. Akad. Nauk SSSR, Neorg. Mater.* **1969**, *5*, 1653.

Bai 1966   L. G. Bailey, *Phys. Chem. Solids* **1966**, *27*, 1593.

Bal 1990   K. Balakrishnan, B. Vengatesan, N. Kanniah, P. Ramasamy, *Crystal Res. Technol.* **1990**, *25*, 633.

Bal 1994   K. Balakrishnan, B. Vengatesan, P. Ramasamy, *J. Mater. Sci.* **1994**, *29*, 1879.

Bal 1994a   K. Balakrishnan, P. Ramasamy, *J. Cryst. Growth* **1994**, *137*, 309.

Beg 1975   A. Begouen-Demeaux, G. Villers, P. Gibart, *J. Solid State Chem.* **1975**, *15*, 178.

Bez 1972   L. I. Bezrodnaya, N. I. Makarova, E. P. Strukova, Y. S. Kharionovskii, S. G. Yudin, *Rost Kristallov* **1972**, *9*, 231.

Bha 2004   N. Bhatt, R. Vaidya, S. G. Patel, A. R. Jani, *Bull. Mater. Sci.* **2004**, *27*, 23.

Boc 1981   C. Bocchl, P. Franzosi, F. Leccabue, R. Panizzieri, *J. Cryst. Growth* **1981**, *54*, 335.

Bol 1994   P. Bolte, R. Gruehn, *Z. Anorg. Allg. Chem.* **1994**, *620*, 2077.

Bon 1969   Z. Boncheva-Mladenova, G. Bachvarov, *Dokl. Bolg. Akad. Nauk* **1969**, *22*, 125.

Bon 1972   Z. Boncheva-Mladenova, St. Karbanov, M. Mitkova, *Dokl. Bolg. Akad. Nauk* **1972**, *25*, 225.

Bra 1973   L. Brattas, A. Kjekshus, *Acta Chem. Scand.* **1973**, *27*, 1290.

Bri 1962   L. H. Brixner, *J. Inorg. Nucl. Chem.* **1962**, *24*, 257.

Bro 1962   P. Bro, *J. Electrochem. Soc.* **1962**, *109*, 1110.

Bro 1966   B. E. Brown, *Acta Cryst.logr.* **1966**, *20*, 268.

Chu 2001   I. S. Chuprakov, V. B. Lyalikov, K. H. Dahmen, P. Xiong, *Mater. Res. Soc. Symp. Proc.* **2001**, *602 (Magnetoresistive Oxides and Related Materials)*, 472.

Czu 2010   A. Czulucki, *Dissertation*, University of Dresden, **2010**.

Edw 2005   H. K. Edwards, P. A. Salyer, M. J. Roe, G. S. Walker, P. D. Brown, D. H. Gregory, *Angew. Chem.* **2005**, *117*, 3621; *Angew. Chem., Intern. Edt.* **2005**, *44*, 3555.

Fou 1979   G. Fourcaudot, M. Gourmala, J. Mercier, *J. Cryst. Growth* **1979**, *46*, 132.

Ge 1996   Y. R. Ge, H. Wiedemeier, *J. Electron. Mater.* **1996**, *25*, 1067.

Ge 1999   Y. R. Ge, H. Wiedemeier, *J. Electron. Mater.* **1999**, *28*, 91.

Ghe 1974   C. Ghezzi, C. Paorici, *J. Cryst. Growth* **1974**, *21*, 58.

Gib 1969a   P. Gibart, C. Vacherand, *J. Cryst. Growth* **1969**, *5*, 111.

Gib 1969b   P. Gibart, G. Collin, *Croissance Composes Miner. Monocrist.* **1969**, *2*, 127.

Gol 1979   Z. Golacki, J. Makowski, *J. Cryst. Growth* **1979**, *47*, 749.

Gom 1983   E. Gombia, F. Leccabue, C. Pelosi, D. Seuret, *J. Cryst. Growth* **1983**, *65*, 391.

Gre 1965   D. L. Greenaway, R. Nitsche, *Phys. Chem. Solids* **1965**, *26*, 1445.

Hil 1972   A. A. Al Hilli, B. L. Evans, *J. Cryst. Growth* **1972**, *15*, 193.

Hot 2005   U. Hotje, M. Binnewies, *Z. Anorg. Allg. Chem.* **2005**, *631*, 1682.

Hua 1994   J. K. Huang, Y. S. Huang, T. R. Yang, *J. Cryst. Growth* **1994**, *135*, 224.

Hut 2002   M. A. Hutchins, H. Wiedemeier, *Z. Anorg. Allg. Chem.* **2002**, *628*, 1489.

Ido 1968   T. Ido, S. Oshima, M. Saji, *Jpn. J. Appl. Phys.* **1968**, *7*, 1141.

Iga 1976b   O. Igarashi, *Oyo Butsuri* **1976**, *45*, 864.

Ilc 1999   G. A. Ilchuk, *Inorg. Mater. (Transl. Neorg. Mater.)* **1999**, *35*, 682.

Ilc 2002   G. A. Ilchuk, V. Ukrainetz, A. Danylov, V. Masluk, J. Parlag, O. Yaskov, *J. Cryst. Growth* **2002**, *242*, 41.

Ire 1983   E. A. Irene, E. Tierney, H. Wiedemeier, D. Chandra, *Appl. Phys. Lett.* **1983**, *42*, 710.

Jan 2006   M. Janetzky, B. Harbrecht, *Z. Anorg. Allg. Chem.* **2006**, *632*, 837.

Kak 1981   M. Kakehi, T. Wada, *Jpn. J. Appl. Phys.* **1981**, *20*, 429.

Kal 1968   E. Kaldis, *J. Cryst. Growth* **1968**, *3–4*, 146.

Ker 1998   S. Kertoatmodjo, G. Nugraha; *Thin Solid Films* **1998**, *324*, 25.

Kha 1984   A. Khan, N. Abreu, L. Gonzales, O. Gomez, N. Arcia, D. Aguilera. S. Tarantino, *J. Cryst. Growth* **1984**, *69*, 241.

Kuh 1971   A. Kuhn, A. Chevy, M. Naud, *J. Cryst. Growth* **1971**, *9*, 263.

Lec 1983   F. Leccabue, C. Pelosi, *Mater. Lett.* **1983**, *2*, 42.

Lev 1983   F. Levy, H. Berger, *J. Cryst. Growth* **1983**, *61*, 61.

Lev 1991   F. Levy, H. Berger, *J. Chim. Phys. Phys.-Chim. Biolog.* **1991**, *88*, 1985.
Li 1994   J. Li, F. McCulley, M J. Dioszeghy, S. C. Chen, K. V. Ramanujachary, M. Greenblatt, *Inorg. Chem.* **1994**, *33*, 2109.
Mag 2004   C. Magliocchi, F. Meng, T. Hughbanks, *J. Solid State Chem.* **2004**, *177*, 3896.
Man 1983   A. M. Mancini, P. Pierini, A. Quirini, A. Rizzo, L. Vasanelli, *J. Cryst. Growth* **1983**, *62*, 34.
Mas 1993   G. Masse, L. Yarzhou, K. Djessas, *Fr. J. Phys. III* **1993**, *3*, 2087.
Mei 1967   W. M. DeMeis, A. G. Fischer, *Mater. Res. Bull.* **1967**, *2*, 465.
Mel 1990   O. De Melo, F. Leccabue, R. Panizzieri, C. Pelosi, G. Bocelli, G. Calestani, V. Sagredo, M. Chourio, E. Paparazzo, *J. Cryst. Growth* **1990**, *104*, 780.
Mel 1991   O. De Melo, F. Leccabue, C. Pelosi, V. Sagredo, M. Chourio, J. Martin, G. Bocelli, G. Calestani, *J. Cryst. Growth* **1991**, *110*, 445.
Moc 1981   K. Mochizuki, *J. Cryst. Growth* **1981**, *51*, 453.
Mus 2005   F. M. Mustafaev, *Kim. Problemlari J.* **2005**, *2*, 113.
Nem 1984   Y. Nemirovsky, A. Kepten, *J. Electronic Mater.* **1984**, *13*, 867.
Nis 1979   M. Nishio, K. Tsuru, H. Ogawa, *Jpn. J. Appl. Phys.* **1979**, *18*, 1909.
Nis 1980   M. Nishio, K. Tsuru, H. Ogawa, *Rikogakubu Shuho* **1980**, *8*, 59.
Nis 1982   M. Nishio, H. Ogawa, *Jpn. J. Appl. Phys. Part 1: Regular Papers, Short Notes & Review Papers* **1982**, *21*, 90.
Nis 1986   M. Nishio, H. Ogawa, *J. Cryst. Growth* **1986**, *78*, 218.
Nis 1988   M. Nishio, H. Ogawa, K. Komorita, *Rikogakubu Shuho (Saga Daigaku)* **1988**, *17*, 19.
Oga 1981   H. Ogawa, M. Nishio, T. Arizumi, *J. Cryst. Growth* **1981**, *52*, 263.
Paj 1980   A. Pajaczkowska, *J. Cryst. Growth* **1980**, *49*, 563.
Pal 1989   W. Palosz, H. Wiedemeier, *J. Less-Common Met.* **1989**, *156*, 299.
Pan 1994   O. Y. Pankratova, S. A. Novozhilova, L. I. Grigor'eva, R. A. Zvinchuk, *Zh. Neorg. Khim.* **1994**, *39*, 1609.
Pao 1972a   C. Paorici, *Proc. Int. Symp. Cadmium Telluride, Mater. Gamma-Ray Detectors* **1972**, VII-1–VII-4.
Pao 1972b   C. Paorici, C. Pelosi, G. Zuccalli, *Phys. Status Solidi A* **1972**, *13*, 95.
Pao 1973   C. Paorici, G. Attolini, C. Pelosi, G. Zuccalli, *J. Cryst. Growth* **1973**, *18*, 289.
Pao 1974   C. Paorici, G. Attolini, C. Pelosi, G. Zuccalli, *J. Cryst. Growth* **1974**, *21*, 227.
Pao 1975   C. Paorici, C. Pelosi, G. Attolini, G. Zuccalli, *J. Cryst. Growth* **1975**, *28*, 358.
Pao 1977   C. Paorici, C. Pelosi, *Rev. Phys. Appl.* **1977**, *12*, 155.
Pao 1980   C. Paorici, L. Zanotti, *Mater. Chem.* **1980**, *5*, 337.
Pav 1978   O. Pavlov, Z. Boncheva-Mladenova, T. Dzhubrailov, *Izv. Khim. (Bolg.)* **1978**, *11*, 124.
Pet 1973   K. E. Peterson, U. Birkholz, D. Adler, *Phys. Rev. B* **1973**, *8*, 1453.
Pro 2001   A. Prodan, V. Marinkovic, N. Jug, H. J. P. van Midden, H. Bohm, F. W. Boswell, J. Bennett, *Surface Science* **2001**, *482–485*, 1368.
Phi 2008   F. Philipp, P. Schmidt, *J. Cryst. Growth* **2008**, *310*, 5402.
Red 2008   Y. D. Reddy, B. K. Reddy, D. S. Reddy, D. R. Reddy, *Spectrochim. Acta A* **2008**, *70A*, 934.
Rim 1974   H. P. B. Rimmington, A. A. Balchin, *J. Cryst. Growth* **1974**, *21*, 171.
Ros 2004   C. Rose, M. Binnewies, *Z. Anorg. Allg. Chem.* **2004**, *630*, 1296.
Sat 1990   K. Sato, Y. Aman, M. Hirai, M. Fujisawa, *J. Phys. Soc. Jpn.* **1990**, *59*, 435.
Schö 2010   M. Schöneich, M. Schmidt, P. Schmidt, *Z. Anorg. Allg. Chem.* **2010**, 1810.
Sha 1993   Y. G. Sha, C. H. Su, F. R. Szofran, *J. Cryst. Growth* **1993**, *131*, 574.
Sha 1994   Y. G. Sha, M. P. Volz, S. L. Lehoczky, *J. Electron. Mater.* **1994**, *23*, 25.
Shi 1985   T. Shingyoji, T. Nakamura, *Report Res. Lab. Engin. Mater., Tokyo Institute of Technology* **1985**, *10*, 79.

Shi 1987   W. Shi, H. Kong, Z. Li, B. Wang, *Yibiao Cailiao* **1987**, *18*, 16.
Slo 1966   V. K. Slovyanskikh, G. V. Ellert, E. I. Yarembash, M. D. Korsakova, *Izv. Akad. Nauk SSR, Neorg. Mater.* **1966**, *6*, 973.
Slo 1985   V. K. Slovyanskikh, N. T. Kuznetsov, N. V. Gracheva, *Zh. Neorg. Khim.* **1985**, *30*, 558.
Sto 1992   D. Stoeber, B. O. Hildmann, H. Boettner, S. Schelb, K. H. Bachem, M. Binnewies, *J. Cryst. Growth* **1992**, *121*, 656.
Sto 1998   K. Stöwe, *Z. Anorg. Allg. Chem.* **1998**, *624*, 872.
Sto 2000   K. Stöwe, *Z. Anorg. Allg. Chem.* **2000**, *626*, 803.
Sto 2000a   K. Stöwe, *J. Alloys Compd.* **2000**, *307*, 101.
Sto 2001   K. Stöwe, *Z. Kristallogr.* **2001**, *216*, 215.
Str 1973   G. B. Street, E. Sawatzky, K. Lee, *J. Phys. Chem. Solids* **1973**, *34*, 1453.
Tok 1979   V. V. Tokmakov, I. I. Krotov, A. V. Vanyukov, *Izv. Akad. Nauk SSSR, Neorg. Mater.* **1979**, *15*, 1546.
Tom 1980   A. S. Tomson, N. V. Baranova, *Izv. Akad. Nauk SSSR, Neorg. Mater.* **1980**, *16*, 2059.
Tou 1998   O. Touhait, M. Potel, H. Noel, *Inorg. Chem.* **1998**, *37*, 5088.
Tou 2002   O. Tougait, M. Potel, H. Noël, *J. Solid State Chem.* **2002**, *168*, 217.
Tsc 2009   K. Tschulik, M. Ruck, M. Binnewies, E. Milke, S. Hoffmann, W. Schnelle, B. P. T. Fokwa, M. Gilleßen, P. Schmidt, *Eur. J. Inorg. Chem.* **2009**, 3102.
Vac 1991   P. O. Vaccaro, G. Meyer, J. Saura, *J. Phys. D: Appl. Phys.* **1991**, *24*, 1886.
Vig 2008   O. Vigil-Galan, E. Sanchez-Meza, J. Sastre-Hernandez, F. Cruz-Gandarilla, E. Marin, G. Contreras-Puente, E. Saucedo, C. M. Ruiz, M. Tufino-Velazquez, A. Calderon, *Thin Solid Films* **2008**, *516*, 3818.
Vig 2009a   O. Vigil-Galan, F. Cruz-Gandarilla, J. Sastre-Hernandez, F. Roy, E. Sanchez-Meza, G. Contreras-Puente, *Mex. J. Phys. Chem. Solids* **2009**, *70*, 365.
Vor 1973   L. P. Voropaeva, L. A. Firsanov, V. V. Nechaev, Strukt. *Svoistva Termoelektr. Mater.* **1973**, 79.
Wie 1969   H. Wiedemeier, A. G. Sigai, *J. Cryst. Growth* **1969**, *6*, 67.
Wie 1972   H. Wiedemeier, E. A. Irene, A. K. Chaudhuri, *J. Cryst. Growth* **1972**, *13–14*, 393.
Wie 1982   H. Wiedemeier, D. Chandra, *Z. Anorg. Allg. Chem.* **1982**, *488*, 137.
Wie 1983a   H. Wiedemeier, A. E. Uzpurvis, *J. Electrochem. Soc.* **1983**, *130*, 252.
Wie 1983b   H. Wiedemeier, A. E. Uzpurvis, D. Wang, *J. Cryst. Growth* **1983**, *65*, 474.
Wie 1987   H. Wiedemeier, D. Chandra, *Z. Anorg. Allg. Chem.* **1987**, *545*, 109.
Wie 1989   H. Wiedemeier, W. Palosz, *J. Cryst. Growth* **1989**, *96*, 933.
Wie 1991a   H. Wiedemeier, Y. R. Ge, *Z. Anorg. Allg. Chem.* **1991**, *598–599*, 339.
Wie 1991b   H. Wiedemeier, Y. R. Ge, *Z. Anorg. Allg. Chem.* **1991**, *602*, 129.
Wie 1991c   H. Wiedemeier, G. Wu, *J. Electron. Mater.* **1991**, *20*, 891.
Wie 1992   H. Wiedemeier, Y. G. Sha, *J. Electron. Mater.* **1992**, *21*, 563.
Zha 1983   S. N. Zhao, C. Y. Yang, C. Huang, A. S. Yue, *J. Cryst. Growth* **1983**, *65*, 370.
Zul 1983   D. I. Zulfugarly, B. A. Geidarov, Kh. S. Khalilov, T. A. Nadzhafova, *Azerbaidzh. Khim. Zh.* **1983**, *1*, 111.

8 Chemical Vapor Transport of Chalcogenide Halides

BiOBr

Halogens as transport agents

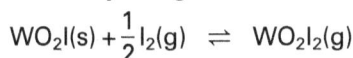

$$WO_2I(s) + \frac{1}{2}I_2(g) \; \rightleftharpoons \; WO_2I_2(g)$$

Halides as transport agents

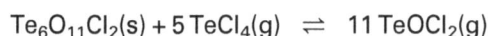

$$Te_6O_{11}Cl_2(s) + 5\,TeCl_4(g) \; \rightleftharpoons \; 11\,TeOCl_2(g)$$

Water as transport agent

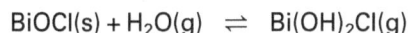

$$BiOCl(s) + H_2O(g) \; \rightleftharpoons \; Bi(OH)_2Cl(g)$$

The CVT of chalcogenide halides is probably described less detailed than the transports of the corresponding binary chalcogenides and halides, respectively. If there are sufficiently stable solid compounds MQ_yX_z in a system MQ_q/MX_x, the knowledge of the gas-phase composition over this solid is important for the understanding of the transport characteristics of the total system. Especially during transport reactions of binary oxides, sulfides, selenides, and tellurides, the co-existing chalcogenide halides can condense by adding larger amounts of halogens or halogen compounds as trans-

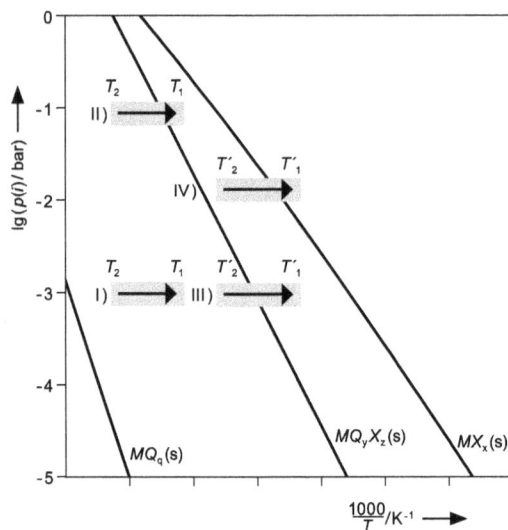

Figure 8.1 Schematic phase barogram of a MQ_n/MX_m system with the illustration of the phase relations during transport experiments dependent on the pressure of the transport agent, the dissolution temperature, and the temperature gradient.

I = Transport of MQ_n at low pressure of the transport agent and high dissolution temperature;

II = Deposition of MQ_yX_z at high pressure of the transport agent and high dissolution temperature;

III = Deposition of MQ_yX_z at low dissolution temperature and great temperature gradient, respectively;

IV = Deposition of MX_m at high pressure of the transport agent, very low dissolution temperature and great temperature gradient, respectively.

port agents. This way, simultaneous transport or transport with a phase sequence can be observed (Opp 1990). During the transport of oxides, very stable solid oxide halides form in some cases. For instance, the formation of BiOX and REOX occur under condensation of the halogen or halogen compound. Thus the transport agent is removed from the system and the transport of the binary oxide is inhibited as is the case in the Bi_2O_3/X_2 system or made difficult as in the RE_2O_3/X_2 systems (see sections 5.2.15 and 5.2.3).

The respective behavior results from the decomposition pressure of the co-existing phases. These are clearly illustrated in the phase barograms (Figure 8.1; see also section 15.6). As a general rule, the temperature dependent course of the evaporation and decomposition pressure over a solid chalcogenide halide MQ_yX_z is between those of the halide and those of the chalcogenide (Figure 8.1).

- Case I: If the average temperature in a temperature gradient ($T_2 \rightarrow T_1$) is high ($T > T'$; Figure 8.1) and the partial pressure of the halides and the halogen are low during the transport, the chalcogenide halide inclines to thermal decomposition and the co-existing binary chalcogenide MQ_n is transported free from phase impurities.

- Case II: At constant temperature and increased partial pressure, the equilibrium conditions for the deposition of the chalcogenide halide MQ_yX_z can be fulfilled. In doing so, MQ_yX_z transports in the first step of a sequential transport. If the partial pressure of the halide is lowered sufficiently by condensation of MQ_yX_z, the transport of the binary chalcogenide takes place in a second step.
- Case III: The transport of MQ_yX_z can be carried out at a relatively low partial pressure if the temperature of deposition is decreased sufficiently $(T' < T)$.
- Case IV: In the combination of low deposition temperature and high partial pressure of the halide, the binary halide MX_m itself can be deposited in sublimation. The behavior, which is shown exemplarily here, has already been discussed in detail with the focus on the transport of Bi_2Se_3 in co-existence with BiSeI (section 7.3).

The transport of a chalcogenide halide MQ_yX_z always takes place at conditions that are between those of the deposition of the binary chalcogenide MQ_n and the halide MX_m of the respective element; hence at medium temperatures and medium pressure of the transport agent as a rule. The concrete conditions of the examined system have to be determined by comparison of the transport experiments of the binary chalcogenide and the halide.

This rule can be applied with good reliability to estimate the transport conditions of chalcogenide halides. If a cation appears in several oxidation numbers, the transport conditions of the compounds with the same oxidation numbers have to be considered for the described comparison.

Thus one observes a transport of CrOCl with the additive Cl_2 or $CrCl_4$ in a temperature gradient from 1000 to 840 °C (Sch 1961b), while chromium(III)-oxide can be transported with chlorine from 1050 °C on the dissolution side to 950 °C on the deposition side (see section 5.2.6) and the corresponding chloride $CrCl_3$ can be transported from 500 to 400 °C (see section 4.1).

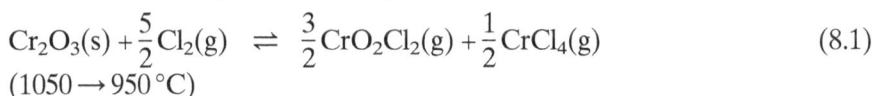

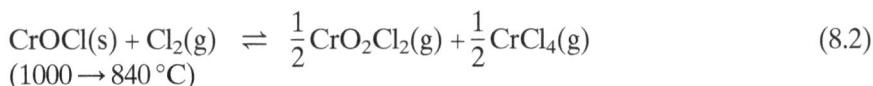

$$Cr_2O_3(s) + \frac{5}{2}Cl_2(g) \rightleftharpoons \frac{3}{2}CrO_2Cl_2(g) + \frac{1}{2}CrCl_4(g) \tag{8.1}$$
$$(1050 \rightarrow 950\,°C)$$

$$CrOCl(s) + Cl_2(g) \rightleftharpoons \frac{1}{2}CrO_2Cl_2(g) + \frac{1}{2}CrCl_4(g) \tag{8.2}$$
$$(1000 \rightarrow 840\,°C)$$

$$CrCl_3(s) + \frac{1}{2}Cl_2(g) \rightleftharpoons CrCl_4(g) \tag{8.3}$$
$$(500 \rightarrow 400\,°C)$$

In the same way, the conditions for the transport of the vanadium(IV)-compounds VO_2, $VOCl_2$, and VCl_4 can be related to one another. The temperature ranges for the transport of compounds that are free from phase impurities differ in parts clearly from those of the vanadium(III)-compounds (V_2O_3: 1050 $\rightarrow$ 950 °C; VOCl: 800 $\rightarrow$ 700 °C (Sch 1961a)):

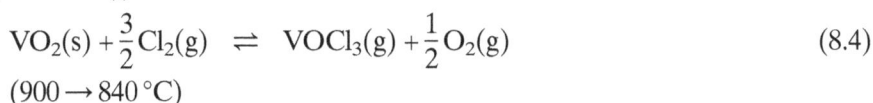

$$VO_2(s) + \frac{3}{2}Cl_2(g) \rightleftharpoons VOCl_3(g) + \frac{1}{2}O_2(g) \tag{8.4}$$
$$(900 \rightarrow 840\,°C)$$

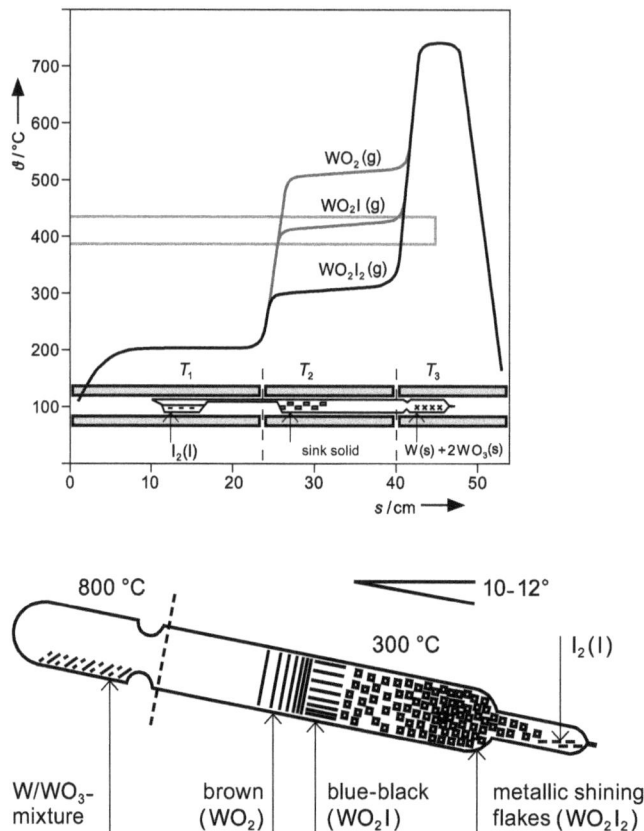

Figure 8.2 Illustration of the phase relations during transport experiments of WO_2 with iodine $(7 \text{ mg} \cdot \text{cm}^{-3})$ from 800 °C to T_1 according to *Tillack* (Til 1968).

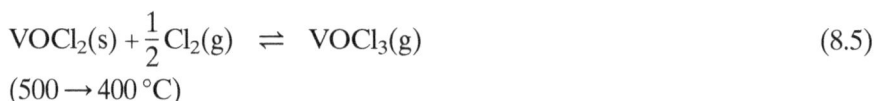

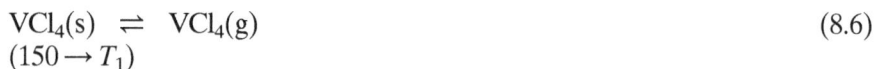

$$VOCl_2(s) + \frac{1}{2}Cl_2(g) \; \rightleftharpoons \; VOCl_3(g) \qquad (8.5)$$
$$(500 \rightarrow 400 \, ^{\circ}\text{C})$$

$$VCl_4(s) \; \rightleftharpoons \; VCl_4(g) \qquad (8.6)$$
$$(150 \rightarrow T_1)$$

The medium transport temperatures decrease with increasing oxidation number of the metal atom in chalcogenide halides. Alternatively, at constant temperatures higher pressures of the transport agent lead to phases with higher oxidation state of the metal atom.

The temperature dependency of transport effects at changing oxidation states can be demonstrated with the help of the conditions of deposition of WO_2, WO_2I, and WO_2I_2 (Til 1968). At a temperature of 800 °C on the source and a pressure of the transport agent of iodine of approx. 1 bar, one observes a phase sequence in the ampoule along the temperature gradient. The gradient can be realized using a three-

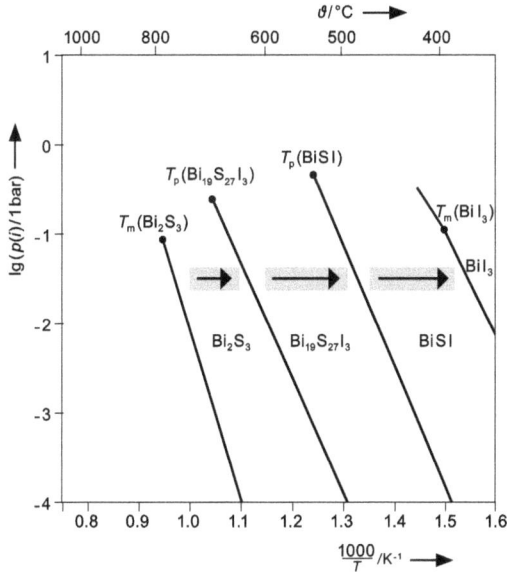

Figure 8.3 Phase barogram of the Bi_2S_3/BiI_3 system and the illustration of the phase relations during transport experiments of: Bi_2S_3 (750 → 650 °C), $Bi_{19}S_{27}I_3$ (600 → 500 °C), and BiSI (500 → 400 °C) according to *Oppermann et al.* (Opp 2003).

zone furnace with a minimum temperature at the sink of 200 °C. Tungsten(IV)-oxide can be deposited free from phase impurities in the sink up to 500 °C; the oxide iodides of tungsten remain completely dissolved in the gas phase. Tungsten(V)-oxide iodide WO_2I is obtained between 500 and 400 °C while tungsten(VI)-oxide iodide WO_2I_2 is transported at a temperature of approx. 300 °C in the sink. If the end of the ampoule is cooled below 200 °C, iodine will condense out eventually.

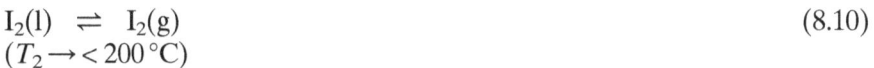

$$WO_2(s) + I_2(g) \ \rightleftharpoons \ WO_2I_2(g) \tag{8.7}$$
$$(800 \rightarrow 500\,°C)$$

$$WO_2I(s) + \frac{1}{2}I_2(g) \ \rightleftharpoons \ WO_2I_2(g) \tag{8.8}$$
$$(T_2 \rightarrow 400\,°C)$$

$$WO_2I_2(s) \ \rightleftharpoons \ WO_2I_2(g) \tag{8.9}$$
$$(T_2 \rightarrow 300\,°C)$$

$$I_2(l) \ \rightleftharpoons \ I_2(g) \tag{8.10}$$
$$(T_2 \rightarrow < 200\,°C)$$

One can also observe a temperature-dependent phase sequence during the transport if there are several chalcogenide halides with metal atoms of the *same oxidation state*. Thus the transport of the compounds $Bi_{19}S_{27}I_3$ (8.12) and BiSI (8.13) can be observed sequentially in the Bi_2S_3/BiI_3 system (Opp 2003, Opp 2005) (Figure 8.3).

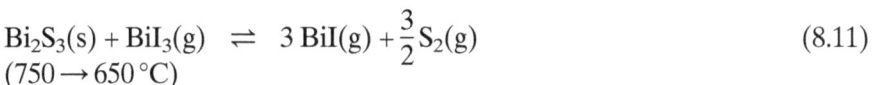

$$Bi_2S_3(s) + BiI_3(g) \ \rightleftharpoons \ 3\,BiI(g) + \frac{3}{2}S_2(g) \tag{8.11}$$
$$(750 \rightarrow 650\,°C)$$

$$Bi_{19}S_{27}I_3(s) + 8\,BiI_3(g) \;\rightleftharpoons\; 27\,BiI(g) + \frac{27}{2}S_2(g) \qquad (8.12)$$
$$(600 \rightarrow 500\,°C)$$

$$BiSI(s) \;\rightleftharpoons\; BiI(g) + \frac{1}{2}S_2(g) \qquad (8.13)$$
$$(500 \rightarrow 400\,°C)$$

$$BiI_3(s) \;\rightleftharpoons\; BiI_3(g) \qquad (8.14)$$
$$(400 \rightarrow T_1)$$

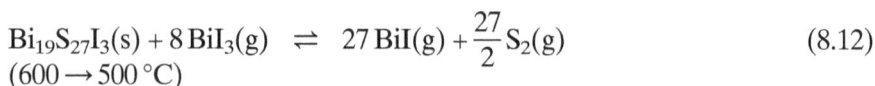

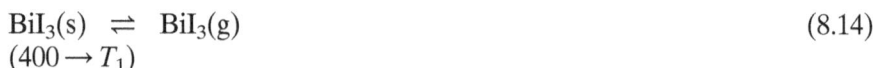

If there are several chalcogenide halides in a system, the chalcogen-richer and halogen-poorer compound is transported at higher temperature or lower pressure of the transport agent. The halogen-richer compound is transported at lower temperature or higher pressure of the halide.

However, the described cascade of the transport with phase sequence is not extendable at will. At condensation of the chalcogenide halide, the partial pressure of the halide can fall below the transport-effective range. In order to counteract this, the medium transport temperature must be increased; even though the increase of the temperature is limited by the melting temperatures and the peritectic transition temperatures of the compounds. This way, a deposition is only possible in a short distance transport or the transport will disrupt completely. The compounds of bismuth show this kind of transport behavior: Bi_2O_3 (equilibrium 8.15) as well as the oxygen-rich halides oxides $Bi_{12}O_{17}Cl_2(s)$ and $Bi_3O_4Cl(s)$ as well as $Bi_{12}O_{17}Br_2$ and Bi_5O_7I (equilibrium 8.16) are not transportable (Schm 1997, Schm 1999, Opp 2005). The compound $Bi_7O_9I_3$, which is in equilibrium with Bi_5O_7I, can be deposited at high temperatures, below the melting temperature ($\vartheta_m = 960\,°C$) in crystalline form (Schm 1997, Opp 2005) (equilibrium 8.17; Figure 8.4).

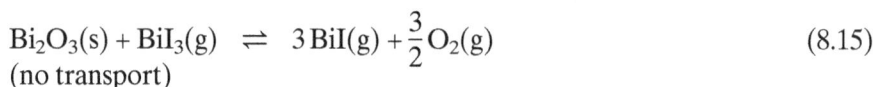

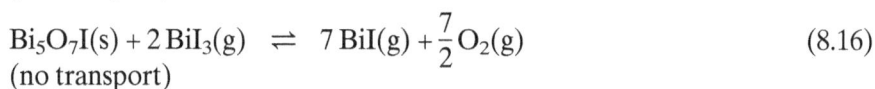

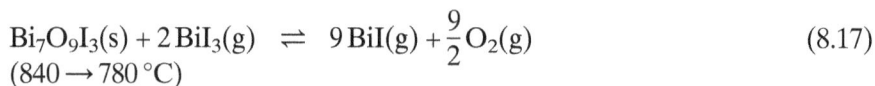

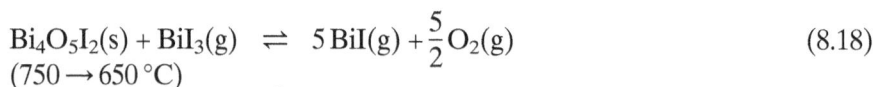

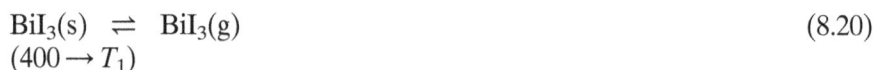

$$Bi_2O_3(s) + BiI_3(g) \;\rightleftharpoons\; 3\,BiI(g) + \frac{3}{2}O_2(g) \qquad (8.15)$$
$$(\text{no transport})$$

$$Bi_5O_7I(s) + 2\,BiI_3(g) \;\rightleftharpoons\; 7\,BiI(g) + \frac{7}{2}O_2(g) \qquad (8.16)$$
$$(\text{no transport})$$

$$Bi_7O_9I_3(s) + 2\,BiI_3(g) \;\rightleftharpoons\; 9\,BiI(g) + \frac{9}{2}O_2(g) \qquad (8.17)$$
$$(840 \rightarrow 780\,°C)$$

$$Bi_4O_5I_2(s) + BiI_3(g) \;\rightleftharpoons\; 5\,BiI(g) + \frac{5}{2}O_2(g) \qquad (8.18)$$
$$(750 \rightarrow 650\,°C)$$

$$BiOI(s) \;\rightleftharpoons\; BiI(g) + \frac{1}{2}O_2(g) \qquad (8.19)$$
$$(500 \rightarrow 400\,°C)$$

$$BiI_3(s) \;\rightleftharpoons\; BiI_3(g) \qquad (8.20)$$
$$(400 \rightarrow T_1)$$

The example of WO_2I_2 shows that the gas species of the chalcogenide halides are sometimes stable enough to become transport effective themselves (see Chapter 11). In such a case, the gas-phase deposition of the condensed compounds is understood

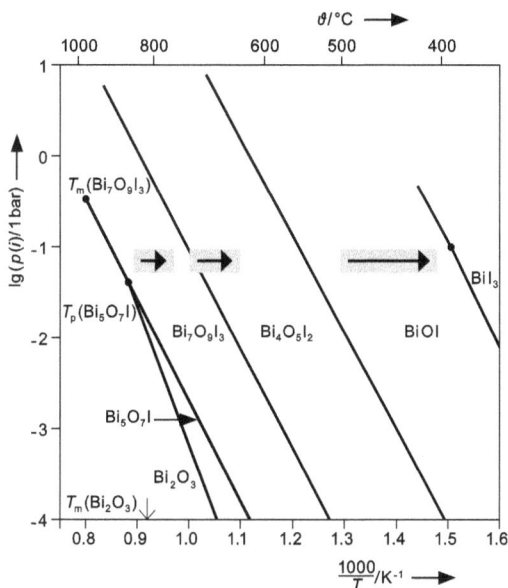

Figure 8.4 Phase barogram of the Bi$_2$O$_3$/BiI$_3$ system with characteristic transport behavior of the oxide iodides of bismuth: Bi$_7$O$_9$I$_3$ (840 → 780 °C), Bi$_4$O$_5$I$_2$ (750 → 650 °C), and BiOI (500 → 400 °C); no transport of Bi$_2$O$_3$ and Bi$_5$O$_7$I.

in the simplest terms as sublimation. Vanadium(V)-oxide chloride, VOCl$_3$, shows such behavior already at room temperature. The tungsten species are dissolved at the source above 100 °C (WOCl$_4$), above 200 °C (WOCl$_3$), and above 300 °C (WO$_2$Cl$_2$). WOBr$_3$ sublimes above 300 °C as do WO$_2$Br$_2$ and WO$_2$I$_2$. Due to their volatility these species are important for the CVT of oxides by adding halogens and halogen compounds, respectively (see Chapter 5 and section 15.1).

The behavior during gas-phase depositions in the sense of sublimation is not always clear. Bismuth chalcogenide halides and antimony chalcogenide halides MQX (M = Sb, Bi; Q = Se, Te; X = Cl, Br, I) show gas-phase depositions that indicate a sublimation. That means the transport takes place without the addition of a transport agent. Despite the existence of ternary gas species MQX (with Q = Se; and X = Cl, Br, I) (see (Schm 2000)), which were detected by mass spectrometric analyses, the gas phase is dissolved in the sense of a decomposition sublimation (8.21). In the process, the gas particles MX and Q_2 become transport effective, as is also the case during the transport of the chalcogenides (see section 7.2). The homogenous equilibrium 8.22 is on the product side.

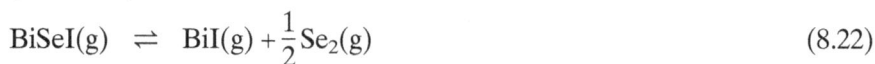

$$BiSeI(s) \;\rightleftharpoons\; BiI(g) + \frac{1}{2}Se_2(g) \tag{8.21}$$
$$(500 \to T_1)$$

$$BiSeI(g) \;\rightleftharpoons\; BiI(g) + \frac{1}{2}Se_2(g) \tag{8.22}$$

Furthermore, the thermal decomposition of a chalcogenide halide does not always have to be congruent, which applies for the decomposition sublimations. The incongruent decomposition in the equilibrium 8.23 leads to the formation of a halide-poorer solid and to the gaseous metal halide. In a second heterogeneous reaction 8.24, the halides formed can become effective as transport agent.

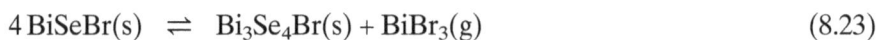

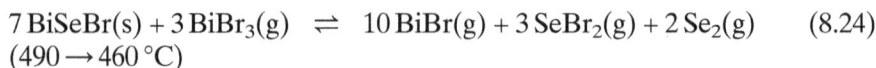

$$4\,BiSeBr(s) \;\rightleftharpoons\; Bi_3Se_4Br(s) + BiBr_3(g) \tag{8.23}$$

$$7\,BiSeBr(s) + 3\,BiBr_3(g) \;\rightleftharpoons\; 10\,BiBr(g) + 3\,SeBr_2(g) + 2\,Se_2(g) \tag{8.24}$$
$$(490 \rightarrow 460\,°C)$$

Here, an auto transport is observed, a CVT reaction without external addition of a transport agent (see section 1.6). Gas-phase depositions according to the principle of auto transport can be used for a great number of chalcogenide halides; a thorough overview is provided by *Oppermann* and co-workers (Opp 2005). Generally, hetero-geneous equilibria, which occur during auto transports, can also be conducted as regular CVT reactions when small amounts of the transport agent are added.

8.1 Transport of Oxide Halides

Transition metal oxide halides   The first reports on CVT reactions of oxide halides of the transition metals go back to the beginning of the systematic examinations of the method at the end of the 1950s and beginning of the 1960s. *Oppermann* provided the first overview of the phase relations during transport experiments of oxide halides (Opp 1990). Chemical vapor transport reactions of the compounds of the composition *M*OCl are described for titanium (Sch 1958), vanadium (Sch 1961a), chromium (Sch 1961b), and iron (Sch 1962). Chlorine (8.1.1, 8.1.2) and hydrogen chloride (8.1.3, 8.1.4) as well as the metal chlorides (8.1.5) can be used as transport agents.

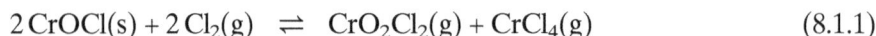

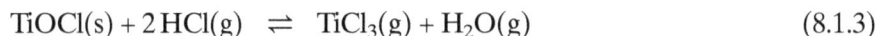

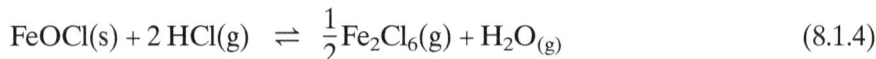

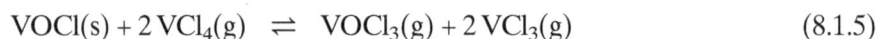

$$2\,CrOCl(s) + 2\,Cl_2(g) \;\rightleftharpoons\; CrO_2Cl_2(g) + CrCl_4(g) \tag{8.1.1}$$

$$VOCl(s) + 2\,Cl(g) \;\rightleftharpoons\; VOCl_3(g) \tag{8.1.2}$$

$$TiOCl(s) + 2\,HCl(g) \;\rightleftharpoons\; TiCl_3(g) + H_2O(g) \tag{8.1.3}$$

$$FeOCl(s) + 2\,HCl(g) \;\rightleftharpoons\; \tfrac{1}{2}Fe_2Cl_6(g) + H_2O_{(g)} \tag{8.1.4}$$

$$VOCl(s) + 2\,VCl_4(g) \;\rightleftharpoons\; VOCl_3(g) + 2\,VCl_3(g) \tag{8.1.5}$$

Apart from the CVT of VOCl with $VCl_4(g)$ via the heterogeneous reaction 8.1.5, *Schäfer* (Sch 1961) detected an auto transport of this compound for the first time. The auto transport is describable quantitatively due to the knowledge of the thermal behavior of VOCl (Opp 1967). First, the transport-effective gas phase is generated in the incongruent decomposition reaction 8.1.6 and 8.1.7 in the temperature range from 700 to 850 °C. Additionally, other independent gas-phase equilibria 8.1.8 and 8.1.9 are involved in the formation of the dominating gas-phase species, Figure 8.1.1.

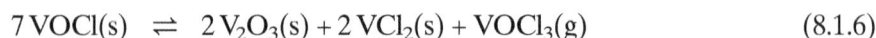

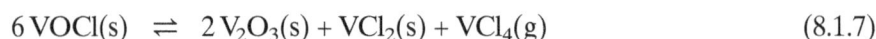

$$7\,VOCl(s) \;\rightleftharpoons\; 2\,V_2O_3(s) + 2\,VCl_2(s) + VOCl_3(g) \tag{8.1.6}$$

$$6\,VOCl(s) \;\rightleftharpoons\; 2\,V_2O_3(s) + VCl_2(s) + VCl_4(g) \tag{8.1.7}$$

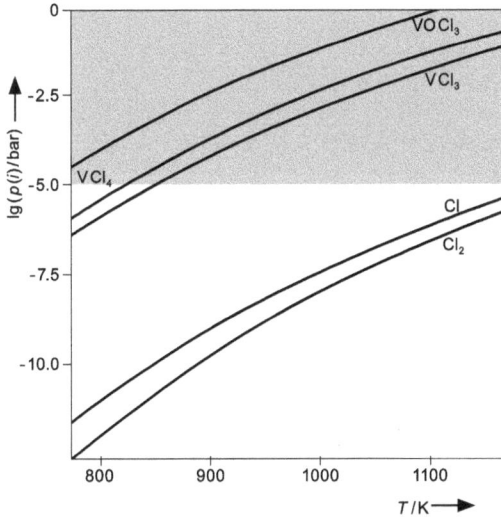

Figure 8.1.1 Composition of the gas phase during the incongruent thermal decomposition of VOCl according to *Oppermann et al.* (Opp 2005).

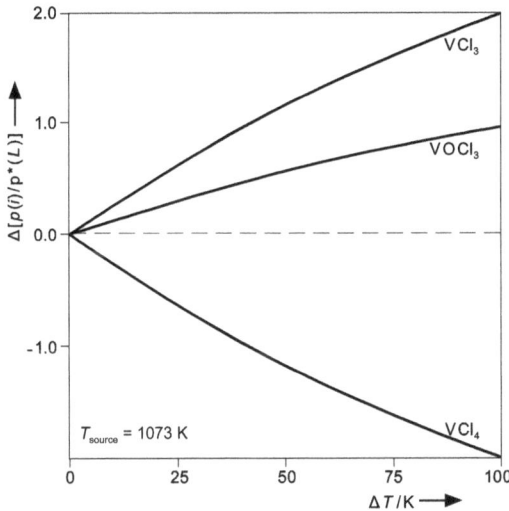

Figure 8.1.2 Transport efficiencies of the gas species during the auto transport of VOCl from 800 °C on the source side to 700 °C on the sink side according to *Oppermann et al.* (Opp 2005).

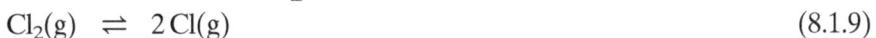

$$VCl_4(g) \;\rightleftharpoons\; VCl_3(g) + \frac{1}{2}Cl_2(g) \tag{8.1.8}$$

$$Cl_2(g) \;\rightleftharpoons\; 2\,Cl(g) \tag{8.1.9}$$

Thus an auto transport becomes principally possible with Cl_2 or Cl as transport agents according to equation 8.1.2, or with VCl_4 as transport agent according to 8.1.5 (Opp 1967). The role of VCl_4 as transport agent ($\Delta[p(i)/p^*(L)] < 0$) for the *auto transport* results from calculation of the transport efficiency of the gas species (Figure 8.1.2). Cl_2 and Cl do not become transport effective with $\Delta[p(i)/p^*(L)] \approx 0$. The spe-

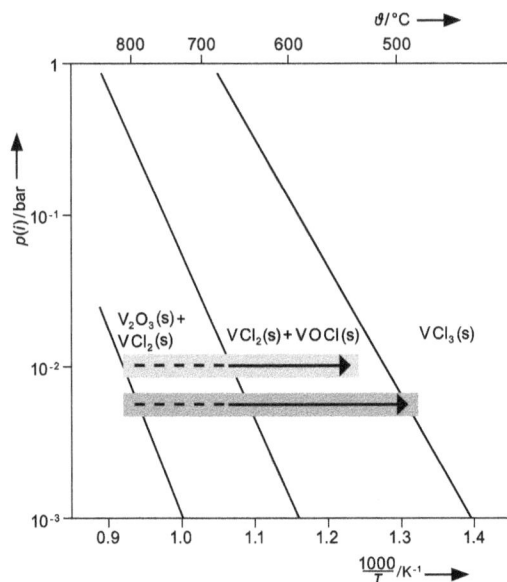

Figure 8.1.3 Phase barogram of the $V_2O_3/VOCl/VCl_3$ system and illustration of the phase relations for the auto transport of VOCl (800 → 550 °C) and the sublimation of VCl_3 (800 → 500 °C) according to *Oppermann*. (Opp 1967).

cies VCl_3 and $VOCl_3$ with $[p(i)/p^*(L)] > 0$ are responsible for the transport of the components in the temperature gradient from T_2 to T_1. Due to the ratio of the transport efficiencies of VCl_4, $VOCl_3$, and VCl_3, the *auto transport* has to be formulated as the regular transport 8.1.5 (Opp 2005).

The range (p, T) of the auto transport of VOCl results from decomposition pressures of VOCl and VCl_3, Figure 8.1.3. The source temperature can be chosen from 700 to 800 °C ($p_{tot} = 10^{-2} \ldots 1$ bar), the deposition temperature between 550 °C and 650 °C. The gradient of the auto transport is not determined by the total pressure over the solid VOCl but by the partial pressure $p(VCl_3)$ that is contained in the total pressure (Figure 8.1.8). If on the deposition side the saturation pressure of VCl_3 is exceeded, VCl_3 will condense; the transport of VOCl is disrupted this way (Opp 1967).

Additionally, the elements of group 5 and 6 form oxide halides of the composition MOX_2. Their transport is based on the formation of a respective transport-effective gas species MOX_3 during the dissolution of the solid with the metal halides of the elements concerned. The transport reactions of $NbOBr_2$, $NbOI_2$ (Sch 1962b), $TaOX_2$ (X = Cl, Br, I) (Sch 1961c), and $MoOCl_2$ (Sch 1964) follow the exemplary shown equilibrium 8.1.10 with $NbOCl_2$ as solid (Sch 1961c).

$$NbOCl_2(s) + NbCl_5(g) \rightleftharpoons NbOCl_3(g) + NbCl_4(g) \qquad (8.1.10)$$

The transport of $VOCl_2$ occurs with chlorine as the transport agent in the following equilibrium:

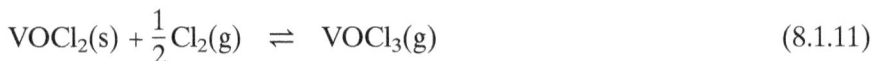

$$VOCl_2(s) + \frac{1}{2}Cl_2(g) \rightleftharpoons VOCl_3(g) \qquad (8.1.11)$$

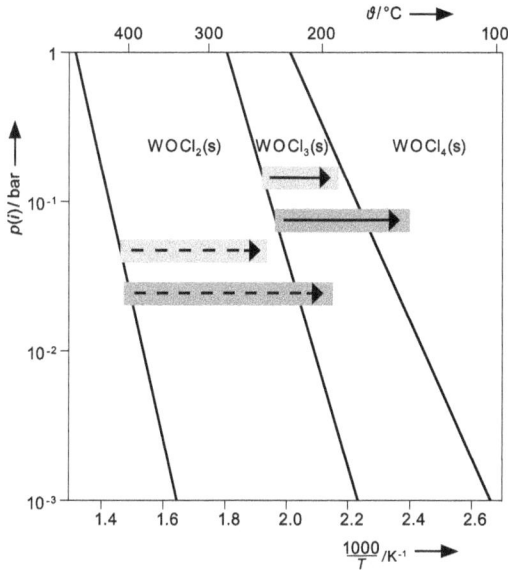

Figure 8.1.4 Phase barogram of the $WOCl_2/WOCl_3/WOCl_4$ system and illustration of the phase relations for the auto transport of $WOCl_2$ (400 → 300 °C), $WOCl_3$ (400 → 200 °C), $WOCl_3$ (250 → 200 °C), and $WOCl_4$ (250 → 150 °C) according to *Oppermann et al.* (Opp 1986).

The assumption of the existence of a gas species $VOCl_2$ (Ros 1990, Hac 1996), which contributes significantly to a sublimation during the gas-phase deposition of solid $VOCl_2$, results from endothermic transport experiments of V_2O_3 and V_3O_5 with the transport agent Cl_2 (Opp 2005).

In contrast, the according tungsten oxide chloride $WOCl_2$ is, like the tungsten(V)-oxide chloride, subjected to an incongruent decomposition 8.1.12 to 8.1.14.

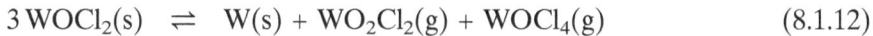

$$3\,WOCl_2(s) \;\rightleftharpoons\; W(s) + WO_2Cl_2(g) + WOCl_4(g) \tag{8.1.12}$$

$$2\,WOCl_3(s) \;\rightleftharpoons\; WOCl_2(s) + WOCl_4(g) \tag{8.1.13}$$

$$2\,WOCl_4(g) \;\rightleftharpoons\; WO_2Cl_2(g) + WCl_6(g) \tag{8.1.14}$$

Thus the compounds do not sublime; accordingly, the gas species $WOCl_2$ and $WOCl_3$, respectively, do not play a role for the transport of oxides and oxide chlorides. $WOCl_2$ and $WOCl_3$ have to be crystallized by CVT with WCl_6 (Til 1970, Opp 1972).

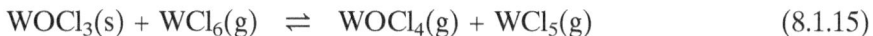

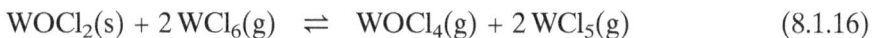

$$WOCl_3(s) + WCl_6(g) \;\rightleftharpoons\; WOCl_4(g) + WCl_5(g) \tag{8.1.15}$$

$$WOCl_2(s) + 2\,WCl_6(g) \;\rightleftharpoons\; WOCl_4(g) + 2\,WCl_5(g) \tag{8.1.16}$$

The presence of WCl_6 as transport agent can result from the homogeneous equilibrium 8.1.14. Thus the transport agent is formed internally and an auto transport of $WOCl_2$ and $WOCl_3$ results. The auto transport of the compounds follows the position of the decomposition pressure curves in the phase barogram (Figure

8.1.4). Hence the transport of phase-pure $WOCl_2$ takes place from 350 to 450 °C at the source and from 250 to 350 °C at the sink. Based on an initial solid consisting of $WOCl_2$, $WOCl_3$ is transported at a sufficiently high temperature gradient of more than 150 K. $WOCl_3$ is deposited directly from approx. 250 to 200 °C. If $WOCl_3$ is used, $WOCl_4$ condenses at gradients ΔT higher than 50 K (Figure 8.1.4). This behavior follows the rule that was laid down at the beginning of this chapter. If there are several chalcogenide halides in a system, the halogen-poorer compounds are transported at higher, and the halogen-richer compounds at lower, temperatures.

The CVT reactions that form rare earth metal and transition metal oxide halides deserve special attention. While the transport of binary rare earth metal oxides with halogens or halogen compounds are only partly possible, there is already a great number of gas-phase transports of the rare earth metal oxidometallates (see section 5.2.3). It is from their transport behavior that the behavior of the oxide halide of the respective system is generally derived: The compounds should be transportable at lower temperatures than the oxides (8.1.17 and 8.1.18). Thus the transport of $Sm_2Ti_2O_7$ with chlorine from 1050 °C to 950 °C is observed (Hüb 1992); the oxide chloride $SmTiO_3Cl$ is transportable from 950 to 850 °C. On the other hand there is, at temperatures between 1000 and 900 °C, using the same solid $SmTiO_3Cl$, a simultaneous transport of TiO_2 (rutile) and $SmTiO_3Cl$ (Hüb 1991).

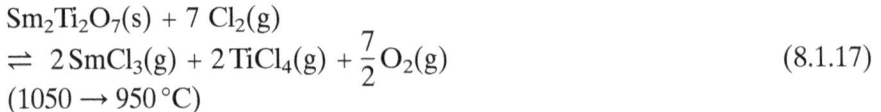

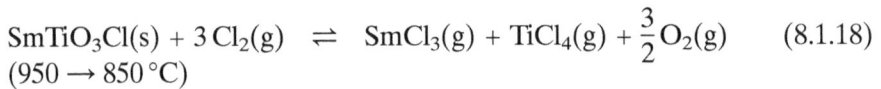

$$Sm_2Ti_2O_7(s) + 7\,Cl_2(g)$$
$$\rightleftharpoons\ 2\,SmCl_3(g) + 2\,TiCl_4(g) + \frac{7}{2}O_2(g) \tag{8.1.17}$$
$$(1050 \rightarrow 950\,°C)$$

$$SmTiO_3Cl(s) + 3\,Cl_2(g)\ \rightleftharpoons\ SmCl_3(g) + TiCl_4(g) + \frac{3}{2}O_2(g) \tag{8.1.18}$$
$$(950 \rightarrow 850\,°C)$$

The transportability of the oxides and oxidometallates is indeed limited by the existence of stable oxide halides. An example of this is the transport of $NdTaO_4$ with the additives of Cl_2 and $TaCl_5$ from 1000 to 900 °C. Under chlorinating conditions of the transport additive (approx. 1 bar) $Nd_2Ta_2O_7Cl_2$ adjacent to $NdTa_7O_{19}$ is formed at the source. At the sink, $NdTaO_4$, $NdTa_7O_{19}$, $Nd_{7.33}Ta_8O_{28}Cl_6$, and $Nd_2Ta_2O_7Cl_2$ are then deposited, dependent on the local temperature gradient (Scha 1988). The transport of phase-pure $NdTaO_4$ only succeeds at temperatures above 1000 °C and at low pressure of the transport agent (Sch 1989).

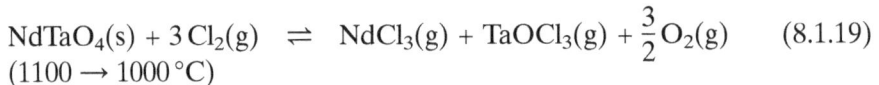

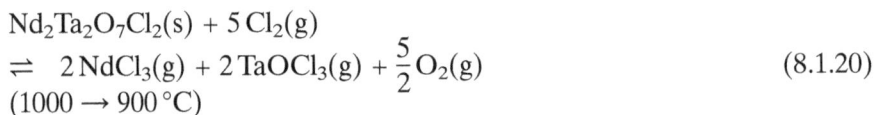

$$NdTaO_4(s) + 3\,Cl_2(g)\ \rightleftharpoons\ NdCl_3(g) + TaOCl_3(g) + \frac{3}{2}O_2(g) \tag{8.1.19}$$
$$(1100 \rightarrow 1000\,°C)$$

$$Nd_2Ta_2O_7Cl_2(s) + 5\,Cl_2(g)$$
$$\rightleftharpoons\ 2\,NdCl_3(g) + 2\,TaOCl_3(g) + \frac{5}{2}O_2(g) \tag{8.1.20}$$
$$(1000 \rightarrow 900\,°C)$$

Knowledge of the very complex temperature dependency and pressure dependency allows the successful preparation of a series of oxidotantalate chlorides

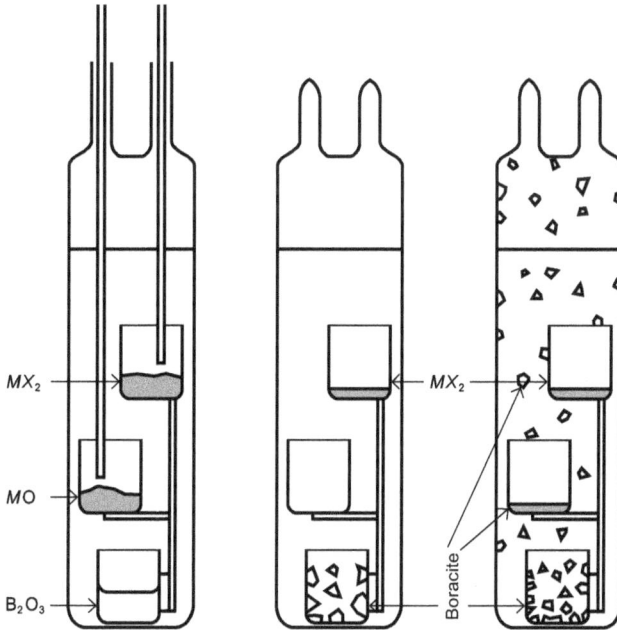

Figure 8.1.5 Schematic arrangement of the crucibles for the transport of the boracites according to *Schmid* (Schm 1965).
a) Reservoir of the binary educts
b) Transport of MO/MX_2 toward B_2O_3 at low pressure of the transport additive H_2O/HX
c) Transport of B_2O_3 at high pressures H_2O/HX; partly joined transport of $B_2O_3/MO/MX_2$.

and oxidoniobate chlorides of the rare earth metals of the composition $RE_2CeMO_6Cl_3$ and $RE_{3.25}MO_6Cl_{3.5-x}$ (RE = La ... Sm; M = Nb, Ta) (Wei 1999).

Boracites $M_3B_7O_{13}X$ (M = Mg, Cr, Mn, Fe, Co, Ni, Cu, Zn, Cd; X = Cl, Br, I) are constituted in a similar complex way but accessible by CVT reactions. Their representatives are examined because of their physical characteristics in particular. Some of them show *ferroelectric* and *ferromagnetic* behavior at the same time; some compounds show *thermochroism* ($Ni_3B_7O_{13}X$ and $Ni_3B_7O_{13}X_{1-x}X'_x$; X, X' = Cl, Br, I) and some the *Alexandrit* effect ($Mg_3B_7O_{13}Cl$, $Cu_3B_7O_{13}I$). The mentioned characteristics of the boracites are connected to a ferroic phase transition from the cubic high temperature modification (T_d^5-symmetry, space group $F\bar{4}3m$) to the orthorhombic low temperature phase (C_{2v}^5-symmetry, space group $Pca2_1$). The temperature of the phase transition of the mineral boracite $MgB_7O_{13}Cl$ is 538 K. The transition temperatures of other representatives vary in a wide range from approx. 10 K ($Cr_3B_7O_{13}I$) up to 800 K ($Cd_3B_7O_{13}Cl$) (Schm 1965) whereby the value rises with decreasing radius of the halide anion and increasing radius of metal ions.

Boracites are transported with water and the corresponding hydrogen halide. Above all the metal dihalides MX_2 as well as BX_3, $B_3O_3X_3$, and HBO_2 are considered as active transport species (Tak 1976).

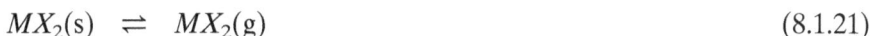

$$MX_2(s) \rightleftharpoons MX_2(g) \tag{8.1.21}$$

$$MO(s) + 2\,HX(g) \;\rightleftharpoons\; MX_2(g) + H_2O(g) \tag{8.1.22}$$

$$B_2O_3(s) + 6\,HX(g) \;\rightleftharpoons\; 2\,BX_3(g) + 3\,H_2O(g) \tag{8.1.23}$$

$$3\,B_2O_3(s) + 6\,HX(g) \;\rightleftharpoons\; 2\,(BOX)_3(g) + 3\,H_2O(g) \tag{8.1.24}$$

$$B_2O_3(s) + H_2O(g) \;\rightleftharpoons\; 2\,HBO_2(g) \tag{8.1.25}$$

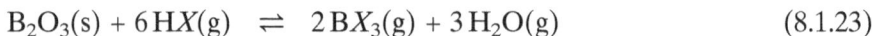

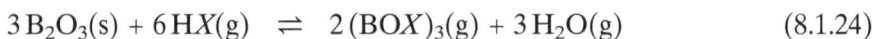

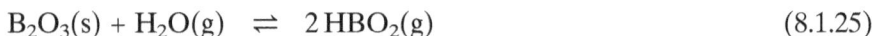

The specific feature of the transport of boracites is the experimental arrangement that was first described by *Schmid*. The constituent compounds MO, MX_2 and, B_2O_3 are separately put in a *two-crucible technique* apparatus (Nas 1972) or a *three-crucible technique* apparatus (Schm 1965, Cas 2005, Cam 2006) (Figure 8.1.5). The transport takes place isothermally at approx. 800 to 900 °C along an *activity gradient* of the components. The transport direction is determined by the transport additives water and hydrogen halide. At low total pressure H_2O/HX a transport takes place from MO/MX_2 toward the liquid boron oxide; while at higher pressure, B_2O_3 migrates toward the metal oxide (Figure 8.1.5) (Schm 1965). The concrete conditions are specific for the transition metal M and its transport behavior in the compounds MO/MX_2. If additionally a small temperature gradient is applied (ΔT = 5 to 20 K), a simultaneous transport of B_2O_3 and MO/MX_2 can take place toward the cooler part of the ampoule (Schm 1965, Nas 1972). If separate solids are used in a closed quartz tube, B_2O_3 even migrates from the cooler side (ϑ_1 = 800 ... 770 °C) contrary to the temperature gradient toward the hotter side of the ampoule (ϑ_2 = 850 ... 820 °C) (Tak 1976). Thereby the deposition always starts from the binary educts along the activity gradient that is given by phase formation.

Predictions of the according substance flows were done with the help of *Richardson* and *Noläng*'s flux functional diagram (see section 2.4) using the example of the formation of $Ni_3B_7O_{13}Cl$ (Dep 1979). The description is done according to the five degrees of freedom: p_{tot}, T, $a(Ni)$, $a(O)$, $a(Cl)$. The direct transport of already pre-reacted boracites has not been described yet. Apparently the compounds formed are so stable that the equilibrium of their dissolution in a transport reaction is too extreme.

Oxide halides of the main group elements In a way that is similar to the transport of boracites, the formation of topaz – $Al_2SiO_4(OH, F)_2$ – can be described. Crystals form when AlF_3 and SiO_2 are provided with SiF_4 and HF, respectively, as transport agent (Scho 1940). This is one of the few examples of the transport of *oxide fluorides*. Already metal fluorides (see section 4.4) are hardly suitable for transport reactions due to their high stability. This effect is even amplified during the formation of oxide fluorides. Hence, the transport of an oxide fluoride can only be expected if a binary fluoride can be deposited via the gas phase.

The transport of AlOCl runs according to the equilibrium 8.1.26 when $NbCl_5$ is added whereby Al_2Cl_6 alongside $NbOCl_3$ become transport effective (see section 11.1). Compared to other synthesis methods, this way one obtains well-formed crystals of AlOCl that are suited for structure analysis (Sch 1962, Sch 1972).

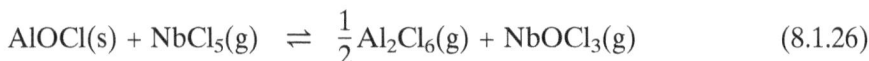

$$AlOCl(s) + NbCl_5(g) \;\rightleftharpoons\; \frac{1}{2}Al_2Cl_6(g) + NbOCl_3(g) \tag{8.1.26}$$

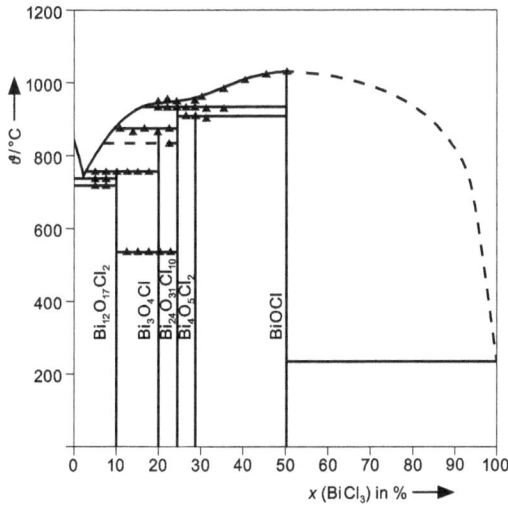

Figure 8.1.6 Phase diagram of the $Bi_2O_3/BiCl_3$ system according to *Schmidt* (Schm 1999).

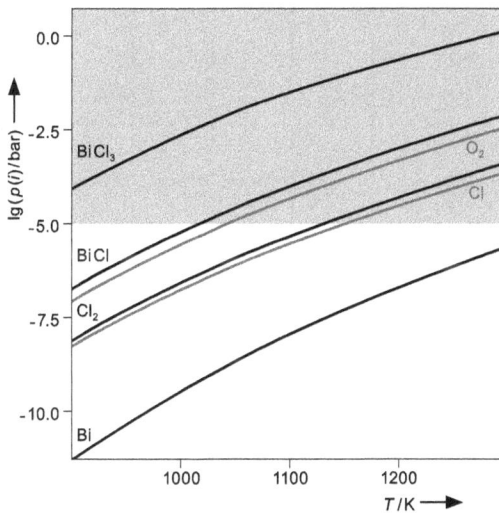

Figure 8.1.7 Composition of the gas phase over solid BiOCl according to *Schmidt* (Schm 1999).

The compounds of antimony and bismuth have been the most extensively examined as far as CVT reactions of oxide halides of the main group elements are concerned. Several phases exist on the quasi-binary sections (see section 8.1.6) in each M_2O_3/MX_3 system (M = Sb, Bi; X = Cl, Br, I). Using the example of bismuth oxide chlorides, the compounds show an incongruent decomposition behavior whereby the respective oxygen-richer oxide chloride and gaseous chloride are formed (Figure 8.1.7).

$$5\,BiOCl(s) \;\rightleftharpoons\; Bi_4O_5Cl_2(s) + BiCl_3(g) \tag{8.1.27}$$

$$\frac{31}{4}\,\mathrm{Bi_4O_5Cl_2(s)} \;\rightleftharpoons\; \frac{5}{4}\,\mathrm{Bi_{24}O_{31}Cl_{10}(s)} + \mathrm{BiCl_3(g)} \tag{8.1.28}$$

$$\frac{4}{3}\,\mathrm{Bi_{24}O_{31}Cl_{10}(s)} \;\rightleftharpoons\; \frac{31}{3}\,\mathrm{Bi_3O_4Cl(s)} + \mathrm{BiCl_3(g)} \tag{8.1.29}$$

$$\frac{17}{3}\,\mathrm{Bi_3O_4Cl(s)} \;\rightleftharpoons\; \frac{4}{3}\,\mathrm{Bi_{12}O_{17}Cl_2(s)} + \mathrm{BiCl_3(g)} \tag{8.1.30}$$

$$\frac{3}{2}\,\mathrm{Bi_{12}O_{17}Cl_2(s)} \;\rightleftharpoons\; \frac{17}{2}\,\mathrm{Bi_2O_3(s)} + \mathrm{BiCl_3(g)} \tag{8.1.31}$$

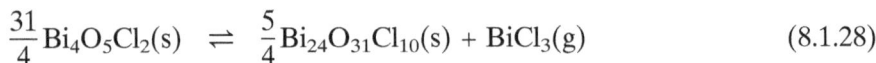

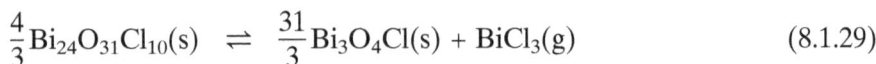

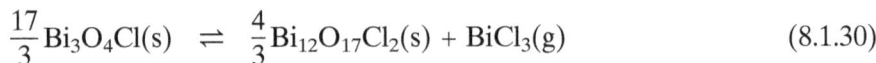

Dependent on the decomposition pressures, auto transport of the compounds can take place with $\mathrm{BiCl_3}$ as transport agent (8.1.32 and 8.1.33).

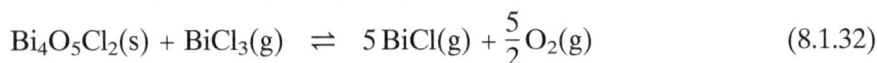

$$\mathrm{Bi_4O_5Cl_2(s)} + \mathrm{BiCl_3(g)} \;\rightleftharpoons\; 5\,\mathrm{BiCl(g)} + \frac{5}{2}\,\mathrm{O_2(g)} \tag{8.1.32}$$

$$\mathrm{Bi_{24}O_{31}Cl_{10}(s)} + 7\,\mathrm{BiCl_3(g)} \;\rightleftharpoons\; 31\,\mathrm{BiCl(g)} + \frac{31}{2}\,\mathrm{O_2(g)} \tag{8.1.33}$$

The more oxide-rich phases $\mathrm{Bi_3O_4Cl(s)}$ and $\mathrm{Bi_{12}O_{17}Cl_2(s)}$ cannot be obtained by auto transport. The total pressures at 880 °C (peritectic decomposition of $\mathrm{Bi_3O_4Cl}$) and 760 °C (peritectic decomposition of $\mathrm{Bi_{12}O_{17}Cl_2}$) are too low so that transport of these phases is not possible in finite time (Schm 1999, Opp 2005).

The deposition of BiOCl cannot be described in the sense of an auto transport: a suitable gas species does not result from the reaction with $\mathrm{BiCl_3(g)}$. The transport can rather take place with the transport relevant gas particles BiCl and $\mathrm{O_2}$ in a decomposition sublimation (8.1.34) ($p(i) > 10^{-5}$ bar; Figure 8.1.7).

$$\mathrm{BiOCl(s)} \;\rightleftharpoons\; \mathrm{BiCl(g)} + \frac{1}{2}\,\mathrm{O_2(g)} \tag{8.1.34}$$

The analogous compounds BiOBr and BiOI can only be deposited in a short distance transport because the oxygen partial pressure of the decomposition reaction falls below the limit of 10^{-5} bar (BiOBr) and 10^{-7} bar (BiOI), respectively (Opp 2005).The regular CVT of the bismuth oxide halides BiOX is discussed for a series of transport agents that appear suitable. Transport with the halogens and hydrogen halide compounds, respectively, should run via the equilibria 8.1.35 and 8.1.36. In fact, sufficiently high partial pressures are achieved only for BiOCl and BiOBr, the transport rates decrease from BiOCl to BiOBr. Transport of BiOI with the transport agents $\mathrm{I_2}$ and HI, respectively, cannot be expected (Opp 2000).

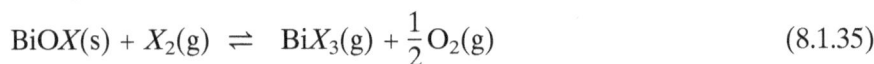

$$\mathrm{BiO}X\mathrm{(s)} + X_2\mathrm{(g)} \;\rightleftharpoons\; \mathrm{Bi}X_3\mathrm{(g)} + \frac{1}{2}\,\mathrm{O_2(g)} \tag{8.1.35}$$

$$\mathrm{BiO}X\mathrm{(s)} + 2\,\mathrm{H}X\mathrm{(g)} \;\rightleftharpoons\; \mathrm{Bi}X_3\mathrm{(g)} + \mathrm{H_2O(g)} \tag{8.1.36}$$

In contrast, transport with water shows high transport rates of up to $15\ \mathrm{mg} \cdot \mathrm{h}^{-1}$ for all compounds. The formation of gaseous $\mathrm{Bi(OH)_2}X$ species in all three systems, which become transport effective according 8.1.37, is the reason for this (Figure 8.1.8) (Opp 2000). The gas-phase depositions of BiOBr and BiOI, which are sometimes observed as sublimations or decomposition sublimations, can rather be attributed to traces of moisture in the ampoule.

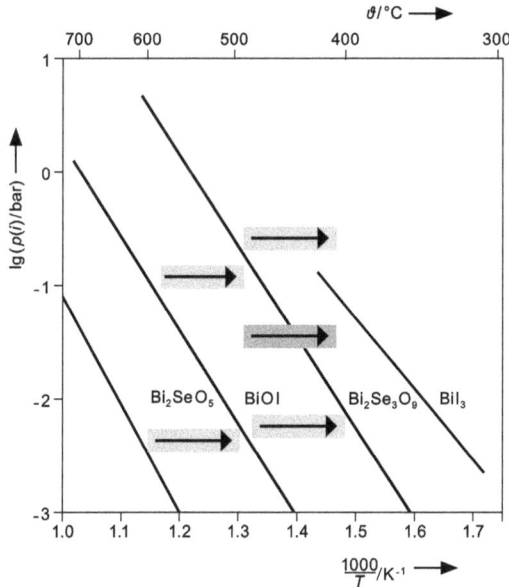

Figure 8.1.8 Phase barogram of the $Bi_2O_3/SeO_2/BiI_3$ system and illustration of the phase relations during transport experiments according to *Oppermann et al.* (Opp 2002b).
Bi_2SeO_5, $600 \rightarrow 500$ °C, $p(BiI_3) \approx 10^{-2}$ bar (see section 5.2)
BiOI, $600 \rightarrow 500$ °C, $p(SeO_2) \approx 10^{-1}$ bar
BiOI, $500 \rightarrow 400$ °C, $p(SeO_2) < 10^{-2}$ bar
$Bi_2Se_3O_9$, $500 \rightarrow 400$ °C, $p(SeO_2) > 10^{-2}$ bar (see section 5.2)
$Bi_2Se_3O_9$, $500 \rightarrow 400$ °C, $p(BiI_3) > 10^{-1}$ bar (see section 5.2).

$$BiOX(s) + H_2O(g) \rightleftharpoons Bi(OH)_2X(g) \tag{8.1.37}$$

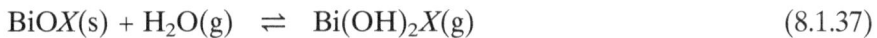

Furthermore, the transport of bismuth oxide halides with selenium(IV)-oxide via the common gas species $BiSeO_3X$ (X = Cl, Br, I) is described (8.1.39) (Opp 2001, Opp 2002a, Opp 2002b). These species are especially notable due to their transport efficiency for four components at the same time. The knowledge about them allows an understanding of the transport of oxides as well as of the oxide halides of bismuth. Thus the transport of Bi_2SeO_5 and $Bi_2Se_3O_9$ with BiI_3 takes place via the gas species $BiSeO_3I$ (see section 5.2). The phase relations during the transport of phase-pure BiOI result accordingly from the equilibrium pressures of BiOI and the co-existing phase $Bi_2Se_3O_9$. If the partial pressure $p(SeO_2)$ is too high or the deposition temperature is too low, the ternary bismuth selenate(IV) instead of BiOI can be transported (Figure 8.1.8).

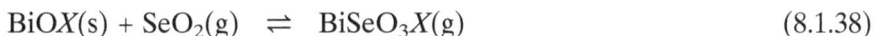

$$BiOX(s) + SeO_2(g) \rightleftharpoons BiSeO_3X(g) \tag{8.1.38}$$

$$Bi_2Se_3O_9(s) + BiI_3(g) \rightleftharpoons 3\,BiSeO_3X(g) \tag{8.1.39}$$

During CVT experiments of TeO_2 with $TeCl_4$, the behavior of a system with condensed oxide halides shows decidedly. The formation of tellurium(IV)-oxide chloride $Te_6O_{11}Cl_2$ could be observed, dependent on the partial pressure, the dissolution temperature, and the temperature gradient, in more than 20 variations

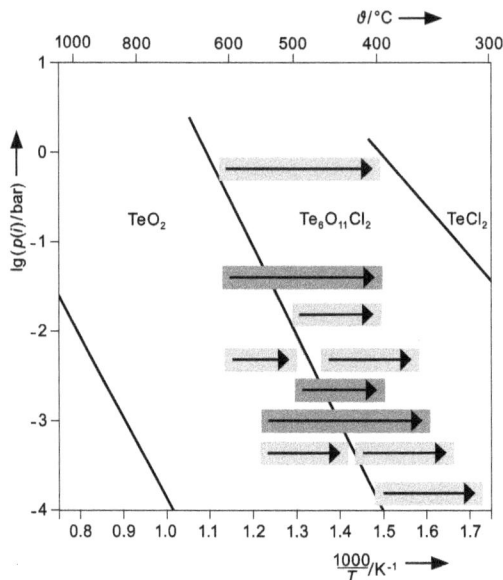

Figure 8.1.9 Phase barogram of the $TeO_2/TeCl_4$ system and illustration of the phase relations during transport experiments according to *Oppermann et al.* (Opp 1977b).

of transport conditions, via a wide temperature range from 600 to 400 °C on the dissolution side and 500 to 300 °C on the deposition side. Following the general trend that the oxide halides are obtained at high partial pressure and low average transport temperatures, respectively, numerous variations of pressure and temperature dependent transport experiments are possible, as shown in Figure 8.1.9 by selected examples in the system $TeO_2/TeCl_4$. Thus $Te_6O_{11}Cl_2$ is transportable from 600 to 400 °C as well as in the temperature gradient from 400 to 300 °C (8.1.41). On the other hand, TeO_2 is deposited from 550 to 450 °C free from phase impurities (8.1.40) – hence in a medium temperature range. Here a detailed consideration of the equilibrium pressure is necessary; the transport of TeO_2 only takes place under these conditions at very low transport agent pressures $p(TeCl_4)$ $\leq 10^{-3}$ bar.

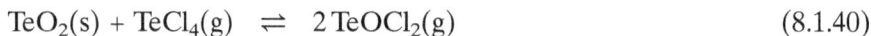

$$TeO_2(s) + TeCl_4(g) \rightleftharpoons 2\,TeOCl_2(g) \tag{8.1.40}$$

$$Te_6O_{11}Cl_2(s) + 5\,TeCl_4(g) \rightleftharpoons 11\,TeOCl_2(g) \tag{8.1.41}$$

Beyond the deposition of tellurium-containing compounds (see section 2.4, Chapter 5, and Chapter 11), the transport-effective species $TeOCl_2(g)$ is of essential importance for the transport of metal oxides with the additive $TeCl_4$. If the homologues $TeBr_4$ or TeI_4 are used as transport additives, the corresponding species $TeOBr_2$ and $TeOI_2$, respectively, can appear. The compound $Te_6O_{11}Br_2$ can additionally be formed in the $TeO_2/TeBr_4$ system (Opp 1978). Their deposition follows the same pattern as in the case of $Te_6O_{11}Cl_2$.

8.2 Transport of Sulfide Halides, Selenide Halides, and Telluride Halides

In many cases, the crystalline preparation of sulfide halides, selenide halides, and telluride halides is possible by gas-phase deposition. In the process, there is a general trend that corresponds to the transport behavior of binary compounds. In the same way as the oxides are rather transportable with chlorine or chlorine compounds (see section 5.1), oxide chlorides can be deposited via the gas phase better than the respective oxide bromides or oxide iodides. As for the other chalcogenide halides, as well as for the binary chalcogenides, especially bromides and above all iodides are suitable for CVT. The stability of the solid phases, which decreases from O to Te, is balanced out in the heterogeneous equilibrium by the likewise decreasing stability of the gaseous metal halides (Cl ... I) (see Chapter 7).

The gas-phase depositions of chalcogenide halides frequently take place as decomposition sublimations or in the sense of an auto transport, far more often than the corresponding reaction of the oxide halides (Opp 2005). Due to the decreasing stability of the chalcogenides, the rising partial pressures of the transport relevant species (S_2, Se_2, Te_2) reach the range of the partial pressures of the transport-effective metal halides ($p(i) > 10^{-5}$ bar). Adding an external transport agent is not necessary in this case.

Hence many experimental works mention the formation of crystals in the temperature gradient without discussion of the vapor transport behavior of the respective compounds. *Fenner* provided an extensive overview on the synthesis and characteristics of chalcogenide halides, which is at the level of knowledge in 1980 (Fen 1980). Detailed, thermodynamically motivated examinations of the CVT behavior of the chalcogenide halides are rather rare.

Transition metal chalcogenide halides   As an example, let us consider the existence and the deposition behavior of the chalcogenide halides of chromium, $CrQX$. The transport of the oxide chloride CrOCl with chlorine can be derived from the transport conditions of chromium(III)-oxide with Cl_2 (980 → 860 °C; see Figure 5.2.6); the oxide chloride transports at lower temperatures in the gradient from 940 to 840 °C. Chromium(III)-sulfide is transported endothermically with bromine and the selenide as well as telluride being transported with iodine in the range of 1000 °C according to the general trend of the suitability of transport agents for chalcogenides. This tendency can also be found in the transport behavior of the ternary compounds.

The chromium sulfide halides are known with chlorine (Meh 1980) as well as with bromine and iodine (Kat 1966). The transport of CrSCl is not known but the deposition of a solid solution phase $CrSCl_{0.33}Br_{0.67}$ is possible at temperatures from 950 to 880 °C (Sas 2000). CrSBr shows the best transport results of the chromium(III)-sulfide halides. The deposition of CrSBr is described over a wide temperature range (870 °C → T_1 (Kat 1966) to 950 → 880 °C (Bec 1990c)) as auto transport, while Cr_2S_3 is transported with bromine from 920 to 800 °C. The transport of phase-pure CrSBr does not work at high dissolution temperatures; Cr_2S_3, CrSBr, and $CrBr_3$ are deposited spatially divided (Bec 1990c). In relation

Table 8.2.1 Transport conditions of binary compounds Cr_2Q_3 and behavior of the ternary phases $CrQX$ of the $Cr/Q/X$ systems (Q = O, S, Se, Te; X = Cl, Br, I) during auto transport.

	O	S	Se	Te
Cr_2Q_3	$Cr_2O_3 + Cl_2$ (Emm 1968)	$Cr_2S_3 + Br_2$ (Nit 1967)	$Cr_2Se_3 + I_2$ (Weh 1970)	$Cr_{1-x}Te + I_2$ (Str 1973)
	1000 → 900 °C	920 → 800 °C	1000 → 900 °C	1000 → 900 °C
$CrQCl$	CrOCl (Sch 1961b)	$CrSCl_{0.33}Br_{0.67}$ (Sas 2000)		
	940 → 840 °C	950 → 880 °C		
$CrQBr$		CrSBr (Kat 1966)		
		870 °C → T_1		
$CrQI$		$CrSI_{0.83}$ (Kat 1966)	CrSeI (Kat 1966)	$CrTe_{0.73}I$ (Kat 1966)
		420 °C → T_1	400 °C → T_1	315 °C → T_1
				CrTeI (Bat1966)
				190 °C → T_1

to the transport temperature of Cr_2S_3, the source temperature of 950 °C seems too high to dissolve CrSBr. Under these circumstances, a noticeable decomposition of CrSBr forming solid Cr_2S_3 and gaseous chromium bromide takes place. Chromium(III)-bromide migrates to the coolest spot in the ampoule while the solid phases, which remained in the source, are transported in succession to the sink. A dissolution temperature of 900 °C (Kat 1966) seems more appropriate for the directed transport of CrSBr.

The preparation of the iodides $CrQI$ occurs at clearly lower temperatures, apparently by mineralization when the reaction takes place isothermally. Already at low temperatures, the compounds show sufficiently high decomposition pressures so that an ideal composition $CrQI$ cannot always be achieved (see $CrSI_{0.83}$ and $CrTe_{0.73}I$ (Kat 1966)). Only at temperatures below 200 °C almost defect-free CrTeI can be obtained (Bat 1966).

This example can provide important indications of the realization of transport experiments for the experimenter. Most of the gas-phase transport reactions of the chalcogenide halides are possible without adding a transport additive. Transport always takes place endothermically. The transport agent and the transport-effective species, respectively, are set free by thermal decomposition of the compounds. If transport agents are added, this should be done in small amounts (a few milligrams) in order to avoid condensation of halogen-rich phases in the system.

If there is already information about the transport of binary chalcogenides and ternary chalcogenide halides of one element, the conditions of deposition of further chalcogenide halides can be estimated. The temperature of the source of the chalcogenide halide should be below the dissolution temperature of the binary chalcogenide. The temperature on the dissolution side decreases through the series Cl ... I as well as for the chalcogenides S ... Te. Accordingly, oxide chlorides generally require the highest dissolution temperatures; telluride iodides the lowest.

These rules are meant to be reference points for planning experiments. The concrete conditions are always dependent on the complexity of the system. If there appear to be several compounds in a system, the suitable conditions of transports of phase-pure compounds have to be found in further experiments as far as dissolution temperature and temperature gradient are concerned.

Representatives of the phases NbQ_2X_2 (Q = S, Se; X = Cl, Br, I) (Sch 1964b, Rij 1979b) are characterized by the appearance of dichalcogenide anions $[Q_2]^{2-}$ (Schn 1966). They can be prepared by auto transport at source temperatures of 500 °C (Sch 1964b, Fen 1980). After the transport agent is set free in an incongruent decomposition reaction, transport can take place according to the equilibrium 8.2.1. The transport of the analogous compounds NbS_2Br_2, $NbSe_2Cl_2$, and $NbSe_2Br_2$ should proceed in the same way. Transport according to the reaction 8.2.2 is expected for the transport of the iodides NbS_2I_2 and $NbSe_2I_2$ (Fen 1980).

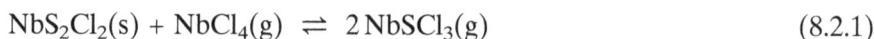

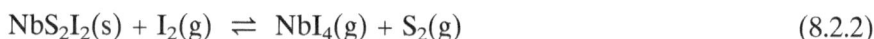

$$NbS_2Cl_2(s) + NbCl_4(g) \rightleftharpoons 2\,NbSCl_3(g) \tag{8.2.1}$$

$$NbS_2I_2(s) + I_2(g) \rightleftharpoons NbI_4(g) + S_2(g) \tag{8.2.2}$$

Small temperature gradients of about 5 to 20 K have to be applied for the deposition of phase-pure compounds, because otherwise a series of ternary phases (e.g., $Nb_3Se_5Cl_7$, $Nb_3Q_{12}X$) would appear adjacent to the binary phases (Rij 1979a, Rij 1979b). If higher temperature gradients are used, more halogen-rich compounds are deposited in the sink, possibly even the halide NbX_3 itself. The source solid becomes poorer if the transported halide and a new heterogeneous equilibrium appears. If the compounds formed are transportable themselves in the given temperature range (see NbS_2 (Nit 1967), Nb_2Se_9 (Mee 1979)), sequential transport takes place. Contrary to the phase rule, there can be several compounds in the sink. The equilibrium between different crystalline products is established only slowly through solid-state reactions. The simultaneous appearance of halogen-rich *and* chalcogen-rich compounds in *one* experiment suggests very small differences in thermodynamic stabilities of the phases involved – so the transport can occur under constant conditions. The principle of the temperature dependent transport of chalcogenide halides hardly applies.

In contrast to the respective oxides, the chalcogen-rich sulfides, selenides, and tellurides of rare earth metals of the composition between RE_2Q_3 and REQ_2 are generally easily transportable (see Chapter 7). Under this pre-condition, transport of the chalcogenide halides of the rare earth metals is expected.

The synthesis of the sulfide halides of the rare earth metals $RESX$ (RE = La, Ce, Pr, Nd, Sm, Gd, Tb, Y, Tb, Dy, Ho, Er, Tm, Yb, Lu; X = Cl, Br, I) (Fen 1980) can take place variably through the reaction of the sulfide RE_2S_3 with the rare

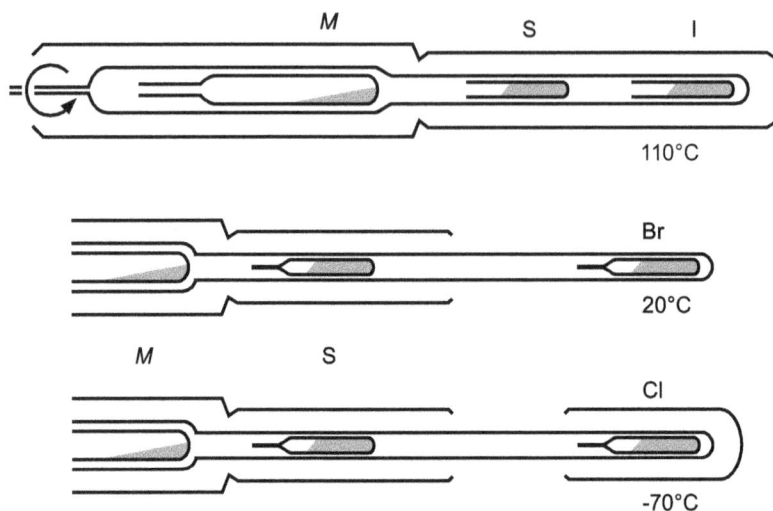

Figure 8.2.1 Crystallization of the sulfide halides of the rare earth metals by isopiestic transport of sulfur and halogen toward the rare earth metal in the sink (M: metal; S: sulfur) according to *Fenner* (Fen 1980).

earth metal and the respective halide. The direct synthesis from the sulfide and the halide does not usually lead to phase-pure products (Fen 1980, Bec 1986, Kle 1995).

The formation of the ternary phases can also take place by migration of sulfur and halides along the activity gradient toward the rare earth metal (Figure 8.2.1). Basically this is an *isopiestic method* of phase formation. Equal or at least similar equilibrium partial pressures of the volatile elements are realized for the migration to the sink (to the metal) by different source temperatures (here, approx. 10^{-1} bar for sulfur at approx. 400 °C as well as for chlorine/bromine/iodine/ at $-70/20/110$ °C).

The sulfides RE_2S_3 are dissolved at source-side temperatures from 950 to 1150 °C (see section 7.1), which is why the transport of sulfide halides can be expected at lower temperatures. Hence crystals of DySBr and DySI can be formed at 750 and 900 °C, for example (Kle 1995). They form during short distance transport along the natural gradient of the furnace.

The preparation of the selenide bromides RESeBr (RE = Dy ... Lu) (Pro 1985) and the selenide iodides RESeI (RE = Gd ... Lu) (Pro 1984) succeeds at approx. 500 °C. The gas-phase transport is explained in detail with the help of the example of the compounds of erbium ErSeX (X = Br, I) (Stö 1997). Depending on the temperature gradient along the ampoule and on the amount of the trihalide ErX_3 used in mixture with Er$_2$Se$_3$, a phase sequence can be observed. ErX_3 sublimates at the coolest spot in the ampoule followed by the selenide bromide and selenide iodide, respectively. At low partial pressure, caused by adding only a small amount of ErX_3, or after condensation of the ternary compounds ErSeX, Er$_2$Se$_3$ is the eventual transported. The temperature at the source is equal to that for the transport of Er$_2$Se$_3$ (850 $\rightarrow$ 700 °C) (Doe 2007). The temperature of transport of selenide bromide is slightly higher (850 °C $\rightarrow$ T_1) than for the selenide iodide (800 °C $\rightarrow$ T_1) (Sto 1997). It is not clear which gas species are effective for the transport of Er_2Se_3 and $ErSeX$. The transport of Er_2Se_3 with the halide as well as the auto transport of $ErSeX$ can succeed only if the species ErX or ErX_3 are present with sufficient partial pressures in the gas phase.

The transport of chalcogenide halides with more than one metallic component is only possible if *all* components form gas species with transport-effective partial pressure in *one* temperature range. This condition is ideally fulfilled if the binary chalcogenides can be transported under similar conditions or if a ternary or multinary chalcogenide, respectively, can be obtained by transport reactions. This can be proven with the help of the example of the chalcogenide halide spinels $CuCr_2Q_3X$ (Q = S, Se, Te; X = Cl, Br, I) (Miy 1968).

The ternary spinels $CuCr_2Q_4$ (Q = S, Se) have to be transported by adding $AlCl_3$ or iodine from 850 to 750 °C ($CuCr_2S_4$) (Mäh 1984) and from 900 to 700 °C (Lut 1970), respectively. The transport of the two binary tellurides CuTe (with CuBr, 900 → 750 °C) (Abb 1987) and $Cr_{1-x}Te$ (with I_2, 1000 → 900 °C) (Str 1973) succeeds under similar conditions. The migration of the ternary compound $CuCr_2Te_4$ by CVT has not yet been documented.

All of these transports indicate that the transport-effective species of copper, chromium, and the chalcogens exist in a temperature range of approx. 900 °C. Consistently the transport of all compounds $CuCr_2Q_3X$ (Q = S, Se, Te; X = Cl, Br, I) is successful in the gradient from 900 to 850 °C (Miy 1968). Copper(II)-halides can be used as transport additives. The halogens are eventually released as active transport agents by the decomposition of copper halides at the source temperature (Miy 1968). The variations of the ratio of halide components and chalcogen components lead to mixed-crystals of the composition $CuCr_2S_{4-x}Cl_x$, $CuCr_2Se_{4-x}Br_x$, and $CuCr_2Te_{4-x}I_x$ (Sle 1968).

If the transport conditions of the binary chalcogenides differ clearly, the transport of the common chalcogenide should succeed in a medium temperature range. There are phases of the general composition $MM'Q_2X$: $MnSbS_2Cl$ and $MnBiS_2Cl$ (Dou 2006), $MnBiS_2Br$ (Pfi 2005a), $MnBiSe_2I$ (Pfi 2005b) as well as $CdSbS_2X$ (X = Cl, Br), $CdBiS_2X$ (X = Cl, Br) and $CdBiSe_2X$ (X = Br, I) (Wan 2006) of the transition metals manganese and cadmium with antimony and bismuth. The comparison of the transport conditions of the binary chalcogenides shows clear differences of CdQ and MnQ with $\vartheta_2 > 850$ °C on the one hand and Sb_2Q_3 and Bi_2Q_3 (Q = S, Se) with $\vartheta_2 \approx 500$ °C on the other. As a consequence, the quaternary compounds of manganese are obtained by auto transport from 600 °C → T_1 (Dou 2006, Pfi 2005a, Pfi 2005b). The sulfide halides and the selenide halides of cadmium are dissolved at 550 to 600 °C (Wan 2006).

Chalcogenide halides of the main group elements The CVT and auto transport of the chalcogenide halides of the main group elements are particularly well examined for compounds of antimony and bismuth. The interest in compounds MQX (M = Sb, Bi; Q = S, Se, Te; X = Cl, Br, I) results above all from their physical characteristics as ferroelectric semi-conductor and piezoelectric material. The need of larger crystals for physical measurements inspired manifold examinations of CVT of these compounds.

As has been shown already, the transport of the chalcogenides M_2Q_3 with iodine is possible at temperatures of the source of approx. 500 °C. Thus good transport effects are expected for the chalcogenide halides antimony and bismuth at dissolution temperatures below 500 °C.

The transport of the chalcogenide iodides MQI (M = Sb, Bi; Q = S (Nee 1971, Ale 1981a; Ale 1981b, Ale 1990), Se (Ale 1981a), Te (Tur 1973)) succeeds by adding iodine according to equilibrium 8.2.3 in the temperature range from 440 to 380 °C on the dissolution side and 400 to 340 °C in the sink. The also described transport by addition of sulfur and selenium (Ale 1981a), respectively, requires further heterogeneous equilibria, which precede the actual transport. The partial pressure of the pure chalcogen species Q_2 is fixed in the equilibrium 8.2.4. This is due to the fact that, apart from the transport-effective species MI_3, MI, and MQ, there are no chalcogen-rich gas species in the system that can be formed by oxidation in transport equilibrium. Thus a surplus of the chalcogen leads to a shift in the composition of the source solid toward the chalcogenide while releasing I_2. Iodine is formed as transport agent in a heterogeneous equilibrium, which precedes the actual transport. The transport then takes place as described in equilibrium 8.2.3. The transport rate, which has been found experimentally, for the transport of SbSI with the additive sulfur, accordingly corresponds to that of transport with iodine (Ale 1981a). *Neels* pointed out that sulfur can sublime at higher temperature gradients (400 % 250 … 350 °C) and the transport effect does not take place because of the absence of formation of transport agent in equilibrium 8.2.4. SbSI, however, can also be deposited in the temperature gradient from 400 °C → T_1 without adding a transport agent (equilibrium 8.2.5) (Nee 1971).

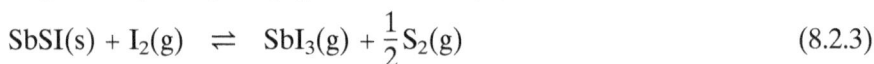

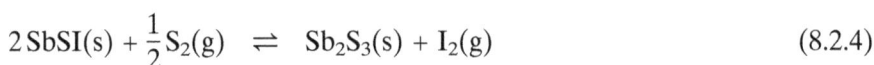

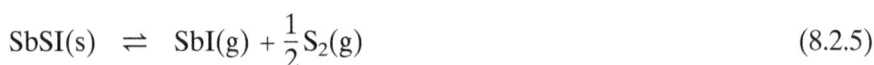

$$SbSI(s) + I_2(g) \; \rightleftharpoons \; SbI_3(g) + \frac{1}{2}S_2(g) \tag{8.2.3}$$

$$2\,SbSI(s) + \frac{1}{2}S_2(g) \; \rightleftharpoons \; Sb_2S_3(s) + I_2(g) \tag{8.2.4}$$

$$SbSI(s) \; \rightleftharpoons \; SbI(g) + \frac{1}{2}S_2(g) \tag{8.2.5}$$

In order to obtain larger cross sections of the thin, needle-shaped crystals, *Neels* suggests the use of a *pendulum process* of auto transport (see (Sch 1962a)). By varying the temperature on the deposition side, nucleation probability is supposed to be reduced and crystal growth is enhanced.

Basically, the pendulum process can take place by two versions, which differ in their ratios of the maximum temperatures of the sink and the source. Thus the temperature of the sink T_1 can temporarily exceed the temperature of the source T_2 during the pendulum process. The periodic increase of the crystallization temperature above the temperature of the source causes smaller, unstable nuclei to be transported back, which results in the fact that only a few nuclei grow.

A second possibility for the pendulum process results by variation of the deposition temperature T_1 below the dissolution temperature T_2. In the process, the minimum of T_1 drops slightly below the temperature difference of the beginning of nucleation (*Ostwald–Miers* range). By oscillation of the temperature, the range of nucleation is left again immediately. This way, only a few nuclei form, which then grow further in the range of low supersaturation. Above a certain quantity, the crystals incorporate so much substance from the gas phase that no new nucleation sets in when the concentration falls below the *Ostwald–Miers* range. This way SbSI crystals up to 3 mm thick are formed at a source temperature of 390 °C and an oscillation of the sink temperature between 320 and 285 °C (two cycles/ hour) (Nee 1971).

The preparation of the other phases SbQX (Q = S, Se, Te; X = Cl, Br, I) (Dön 1950a, Dön 1950b, Nit 1960) generally only succeeds via gas-phase equilibria. A local temperature gradient does not always have to be used for crystallization. Often crystals are obtained in a temporal temperature gradient by cooling down the ampoule. Furthermore, the mixed-crystal systems SbS$_{1-x}$Se$_x$I (Nit 1964, Cha 1982, Kal 1983a), SbSI$_{1-x}$Br$_x$ (Bar 1976, Kve 1996, Aud 2009), and Sb$_{1-x}$Bi$_x$SI (Ish 1974, Ten 1981, Kal 1983b) are suited for gas-phase transport reactions due to similar transport conditions with dissolution temperatures of the ternary phases of 450 to 350 °C. The examinations aim above all to a decided variation of ferroelectric characteristics. The *Curie* temperature for crystals of SbSI$_{1-x}$Br$_x$ changes from 22 to 293 K for x = 1 ... 0 (Aud 2009).

The reaction of bismuth chalcogenides Bi$_2Q_3$ (Q = S, Se, Te) with bismuth halides BiX_3 (X = Cl, Br, I) form ternary compounds with varying compositions. The phases that exist on pseudo-binary sections Bi$_3Q_3$/BiX_3, have been descri-bed in detail as far as their synthesis, structure, and thermodynamic behavior are concerned, by different work groups (Krä 1972, Krä 1973, Krä 1974a, Krä 1978, Krä 1979, Rya 1970, Vor 1979, Tri 1997, Opp 1996, Pet 1997, Pet 1998, Pet 1999a, Pet 1999b). *Oppermann* provided a summary of the chemistry of sulfide halides, selenide halides, and telluride halides of bismuth (Opp 2003, Opp2004). Many compounds of the bismuth chalcogenide halides obtainable by auto transports are mono-crystalline (Opp 2005). In doing so, the "volatility", that is the transpor-tability, increases under the decomposition pressure of the bismuth sulfide halides via the bismuth selenide halides to the bismuth telluride halides. As for the chal-cogen, the iodides are transportable best, and the bromides better than the chlorides. *Bismuth sulfide halides:* Bismuth sulfide Bi$_2$S$_3$ forms the ternary phases BiSCl, Bi$_4$S$_5$Cl$_2$, and Bi$_{19}$S$_{27}$Cl$_3$ in solid-state reactions with BiCl$_3$ (Dön 1950, Vor 1979, Krä 1974a, Krä 1976a, Krä 1979, Opp 2003), the compounds BiSBr and Bi$_4$S$_5$Br$_2$ with BiBr$_3$ (Dön 1950, Krä 1972, Krä 1973, Vor 1990, Opp 2003), as well as the phases BiSI and Bi$_{19}$S$_{27}$I$_3$ with BiI$_3$ (Dön 1950, Rya 1970, Krä 1979, Opp 2003). The sulfide chlorides are not transportable due to the low partial pressures $p(i)$ < 10^{-5} bar of the sulfur-containing species (see also BiSeCl; Figure 8.2.2) (Opp 2003).

In contrast, the auto transport of the sulfide bromides with bismuth(III)-bro-mide, which was formed in a preceding heterogeneous decomposition equilib-rium, is according to the equilibria 8.2.6 and 8.2.7 (Opp 2003). The deposition of the bismuth sulfide iodides is similar. The conditions of a transport of phase-pure products concerning the dissolution temperature and the temperature gradient have been discussed at the beginning of the chapter (see Figure 8.3).

$$Bi_{19}S_{27}Br_3(s) + 8\,BiBr_3(g) \; \rightleftharpoons \; 27\,BiBr(g) + \frac{27}{2}S_2(g) \qquad (8.2.6)$$

$$7\,BiSBr(s) + 3\,BiBr_3(g) \; \rightleftharpoons \; 10\,BiBr(g) + 3\,SBr_2(g) + 2\,S_2(g) \qquad (8.2.7)$$

Bismuth selenide halides: Apart from the compound BiSeCl (Pet 1997), there exists a ternary phase with a homogeneity area between Bi$_8$Se$_9$Cl$_6$ (Pet 1997) and

Table 8.2.2 Transport conditions of the binary compounds Bi_2Q_3 and behavior of the ternary phases $BiQX$ of the $Bi/Q/X$ systems (Q = O, S, Se, Te; X = Cl, Br, I) during auto transport.

	O	S	Se	Te
Bi_2Q_3	$Bi_2O_3 + X_2$	$Bi_2S_3 + I_2$ (Krä 1976a)	$Bi_2Se_3 + I_2$ (Schö 2010)	$Bi_2Te_3 + I_2$ (Schö 2010)
	–	$680 \rightarrow 600\,°C$	$500 \rightarrow 450\,°C$	$500 \rightarrow 450\,°C$
$BiQCl$	BiOCl (Opp 2005)	BiSCl (Opp 2003, Opp 2005)	$Bi_{11}Se_{12}Cl_9$ (Opp 2004, Opp 2005)	BiTeCl (Opp 2004, Opp 2005)
	$850 \rightarrow 800\,°C$	–	$500 \rightarrow 480\,°C$	$410 \rightarrow 390\,°C$
$BiQBr$	BiOBr (Opp 2005)	BiSBr (Opp 2003, Opp 2005)	BiSeBr (Opp 2004, Opp 2005)	BiTeBr (Opp 2004, Opp 2005)
	$750 \rightarrow 700\,°C$	$500 \rightarrow T_1$	$490 \rightarrow 460\,°C$	$450 \rightarrow 400\,°C$
$BiQI$	BiOI (Schm 1997, Opp 2005)	BiSI (Opp 2003, Opp 2005)	BiSeI (Opp 2004, Opp 2005)	BiTeI (Opp 2004, Opp 2005)
	–	$500 \rightarrow 450\,°C$	$520 \ldots 400 \rightarrow$ $470 \ldots 350\,°C$	$500 \ldots 450 \rightarrow$ $450 \ldots 400\,°C$

$Bi_{11}Se_{12}Cl_9$ (Pet 1997, Tri 1997, Egg 1999) in the $Bi_2Se_3/BiCl_3$ system. The gas-phase deposition of BiSeCl is not possible by auto transport. While the partial pressure of the transport-effective species $BiCl_3$ and BiCl are sufficiently high for the transport of bismuth, the pressures of the selenium-containing gas species BiSe and Se_2 remain below 10^{-8} bar (Figure 8.2.2).

If the phase barogram of the system (Figure 8.2.3) is known, a gas-phase deposition based on the Bi_2Se_3-richer solid ($Bi_8Se_9Cl_6 \ldots Bi_{11}Se_{12}Cl_9$) is realizable. So a "false" solid is consciously provided. In doing so, the partial pressure of the transport-effective species for the transport of selenium is increased. The conditions for the condensation of the required compound are achieved in a sufficiently high temperature gradient. This behavior, concerning the different compositions of the starting mixture and the crystals obtained, can be compared to crystallization from peritectic melts. The estimation of the transport range of the ternary phases $Bi_{11}Se_{12}Cl_9$ and BiSeCl is possible according to the phase barogram of the $Bi_2Se_3/BiCl_3$ system (Figure 8.2.3).

The incongruent decomposition equilibrium 8.2.8, where the transport agent $BiCl_3(g)$ is formed, precedes the transport of $Bi_{11}Se_{12}Cl_9$ (equilibrium 8.2.9).

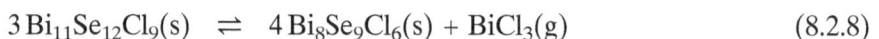

$$3\,Bi_{11}Se_{12}Cl_9(s) \rightleftharpoons 4\,Bi_8Se_9Cl_6(s) + BiCl_3(g) \tag{8.2.8}$$

$$Bi_{11}Se_{12}Cl_9(s) + BiCl_3(g) \rightleftharpoons 12\,BiCl(g) + 6\,Se_2(g) \tag{8.2.9}$$

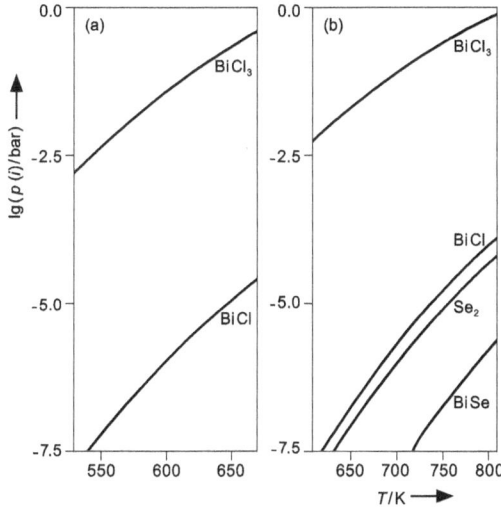

Figure 8.2.2 Composition of the gas phase above the source solids BiSeCl (a) and $Bi_8Se_9Cl_6$ (b) according to *Oppermann et al.* (Opp 2005).

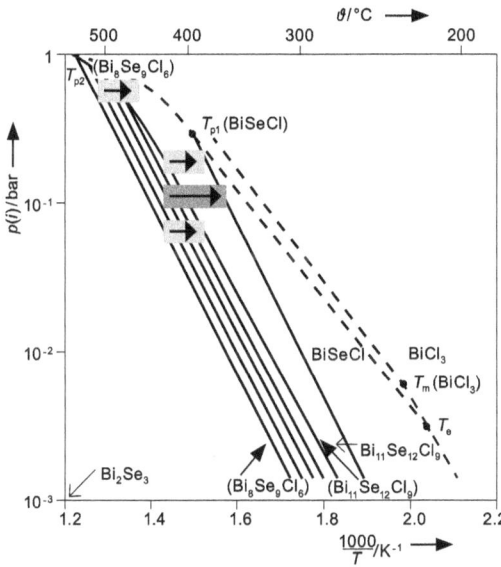

Figure 8.2.3 Phase barogram of the $Bi_2Se_3/BiCl_3$ system and illustration of the phase relations during transport experiments according to *Oppermann et al.* (Opp 2005).
Gray: transport of phase-pure $Bi_{11}Se_{12}Cl_9$
Dark gray: transport of BiSeCl (starting material $Bi_{11}Se_{12}Cl_9$).

During the auto transport, based on a solid of the lower, Bi_2Se_3-rich phase boundary, $Bi_8Se_9Cl_6$, a solid of the composition close to the upper phase boundary $Bi_{11}Se_{12}Cl_9$ is always deposited in the sink if a minimal gradient of 20 K is applied. This behavior proves typical for the auto transport of compounds with a *homogeneity range*.

During the auto transport of compounds with a homogeneity range along the temperature gradient $T_2 \rightarrow T_1$, the component that is more volatile is enriched at T_1. One always obtains a composition close to the upper phase boundary in the sink, if a phase of the lower phase boundary is provided. The extent of change of the composition is dependent on the width of the homogeneity range and on the temperature gradient provided.

During auto transport, the selenide bromides BiSeBr (Hor 1968, Vor 1987) and Bi_3Se_4Br (Vor 1987) have to be deposited with the transport agent $BiBr_3(g)$, which was formed internally. The transports are according to equilibria 8.2.10 and 8.2.11 (Opp 2004, Opp 2005). The bismuth selenide iodide BiSeI, on the other hand, is transferred to the sink without a transport additive in a decomposition sublimation 8.2.12.

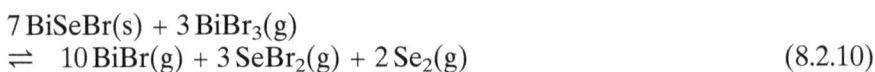

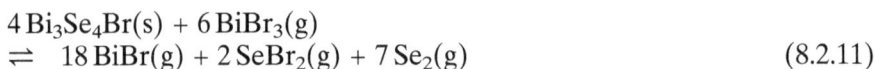

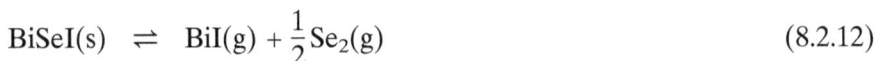

$$7\,BiSeBr(s) + 3\,BiBr_3(g)$$
$$\rightleftharpoons\ 10\,BiBr(g) + 3\,SeBr_2(g) + 2\,Se_2(g) \tag{8.2.10}$$

$$4\,Bi_3Se_4Br(s) + 6\,BiBr_3(g)$$
$$\rightleftharpoons\ 18\,BiBr(g) + 2\,SeBr_2(g) + 7\,Se_2(g) \tag{8.2.11}$$

$$BiSeI(s)\ \rightleftharpoons\ BiI(g) + \frac{1}{2}Se_2(g) \tag{8.2.12}$$

Bismuth telluride halide   The transport of the bismuth telluride halides can be described in a similar way to the gas-phase deposition of the respective selenide halides. The derivation of the existence conditions and transport ranges is clear because the only ternary compounds are $BiTeX$ (X = Cl, Br, I) (Opp 2004, Opp 2005).

The phase barogram of the $Bi_2Te_3/BiCl_3$ system (Pet 1999b) suggests that the transport of BiTeCl is only possible with small temperature gradients because the decomposition pressure of BiTeCl is only slight below the equilibrium pressure of the $BiCl_3$-rich melt. Hence the transport succeeds in a temperature gradient of $\Delta T = 15 - 20$ K under the conditions of the *short distance transport* (see section 14.1 (Krä 1974b).

BiTeBr is deposited in crystalline form in a temperature gradient from 450 ... 500 °C to 400 ... 450 °C. Several equilibria have to be formulated for the migration. Accordingly, BiTeBr is deposited significantly in an incongruent decomposition sublimation 8.2.13; in contrast, during a simultaneous auto transport 8.2.14, it is deposited in small amounts.

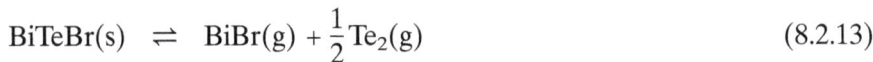

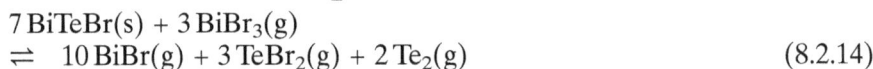

$$BiTeBr(s)\ \rightleftharpoons\ BiBr(g) + \frac{1}{2}Te_2(g) \tag{8.2.13}$$

$$7\,BiTeBr(s) + 3\,BiBr_3(g)$$
$$\rightleftharpoons\ 10\,BiBr(g) + 3\,TeBr_2(g) + 2\,Te_2(g) \tag{8.2.14}$$

The deposition of BiTeI basically follows the heterogeneous equilibrium of the decomposition sublimation of BiSeI (8.1.12). *Valitova*, however, obtained single-crystals during the transport of the compound from 520 to 455 °C whose analytical composition corresponds to the upper phase boundary composition $BiTe_{0.973}I_{1.054}$ (Val 1976). Here the large gradient of more than 50 K leads to the deposition of the phase at the upper phase boundary.

Polynary chalcogenide halides We have already seen several times that the transport of multinary compounds is expected, especially if transport of the binary or ternary phases takes place under similar conditions. This rule is not bound to certain substances, so there are examples of chalcogenide halides with several main group elements as well. This way, the auto transport of $InBi_2S_4Br$ can be attributed to the CVT of In_2S_3 as well as Bi_2S_3, $InBiS_3$ and $In_4Bi_2S_9$ (each $680 \rightarrow 600$) (Krä 1976a) as well as BiSBr ($500 \rightarrow T_1$) (Opp 2003, Opp 2005). $InBi_2S_4Br$ migrates in a medium temperature gradient from 600 to 550 °C (Krä 1976b).

8.3 Transport of Compounds with Chalcogen Poly-cations and Chalcogenate(IV)-halides

Up to now, only chalcogen- and halogen-containing compounds with chalcogenide anions have been the subject of our study. Furthermore, the chemistry of the chalcogens, in particular selenium and tellurium, offers a wide spectrum of existence of these elements in positive oxidation states as well. Thus compounds with chalcogen poly-cations right up to oxide chalcogenate(IV)-halides are known. Just because there are hardly any methodological examinations of CVT reactions of these compounds, the following section shall stimulate an intensive dealing with the reactions during gas-phase depositions.

Compounds with chalcogen poly-cations A number of compounds with homo-atomic and hetero-atomic poly-cations of sulfur, selenium, and tellurium can be obtained in crystalline form by gas-phase reactions (Bec 1994, Bec 2002, Bau 2004). Often the migration of the phases along the temperature gradient is observed in the ampoule; the transport equilibria, which occur in the course of the process, generally remain unexplained. The theoretical background and experimental conditions of the CVT of poly-cationic compounds shall be explained with the help of a clear example.

The compounds $Te_4[WCl_6]_2$ and $Te_8[WCl_6]_2$ are obtained by reaction of elemental tellurium and WCl_6 at approx. 200 °C; the crystalline products are transportable in pure phase in the range between 250 and 180 °C (Bec 1990a, Bec 1990b). If the temperature at the source is increased to 280 ... 300 °C, both phases decompose forming elemental tellurium (8.3.1, 8.3.2); whereas, at lower temperatures, $WCl_x(l)$ condenses at the sink (Bec 1990a, Bec 1990b). However, thermal decomposition does not take place, as would be expected superficially, via $WCl_6(g)$ but with the dominant species $TeCl_2$ and WCl_4 (Figure 8.3.2). The formation of the species $WCl_5(g)$ in the course of the decomposition equilibrium is of minor importance; nevertheless, the partial pressure is in a relevant scale $p(WCl_5) > 10^{-5}$ bar for gas-phase deposition (Figure 8.3.1) (Schm 2007). A phase barogram with the corresponding co-existence decomposition pressures (Figure 8.3.2) can be derived from further information on the thermodynamic behavior, such as the peritectioid transformation of $Te_8[WCl_6]_2$.

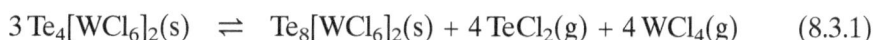

$$3\,Te_4[WCl_6]_2(s) \;\rightleftharpoons\; Te_8[WCl_6]_2(s) + 4\,TeCl_2(g) + 4\,WCl_4(g) \qquad (8.3.1)$$

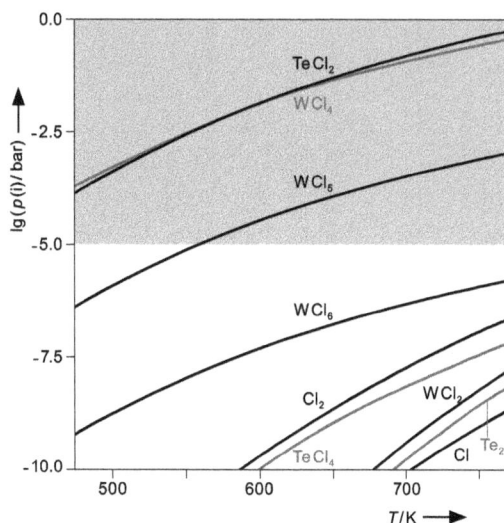

Figure 8.3.1 Composition of the gas phase over the source material $Te_4[WCl_6]_2$ according to *Schmidt* (Schm 2007).

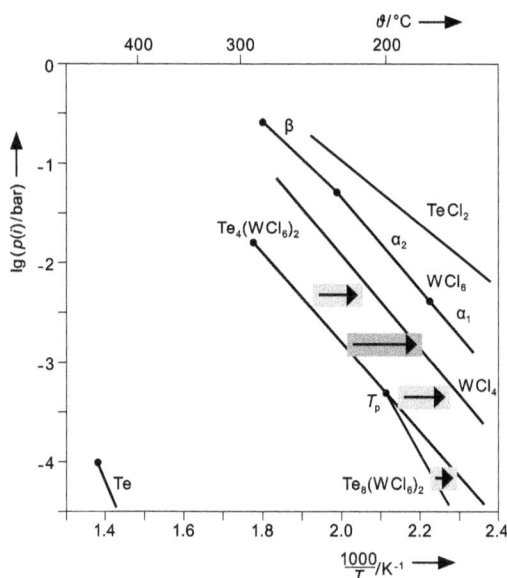

Figure 8.3.2 Schematic presentation of the phase barogram of the Te/WCl_6 system and illustration of the phase relations during transport experiments according to *Schmidt* (Schm 2007).
Light gray: transport of $Te_8(WCl_6)_2$ and $Te_4(WCl_6)$
Dark gray: condensation of WCl_4.

$$Te_8[WCl_6]_2(s) \; \rightleftharpoons \; 6\,Te(s,\,l) + 2\,TeCl_2(g) + 2\,WCl_4(g) \qquad (8.3.2)$$

The deposition can take place by auto transport because all components of the system are sufficiently dissolved in the gas phase (Figure 8.3.1). The temperatures of the source and sink, which are suitable for the phase-pure transport of the compounds, result from the position of the equilibrium curves in the phase barogram (Figure 8.3.2). $Te_4[WCl_6]_2$ can be transported in closed ampoules at temperatures from 180 to 250 °C at the source and deposited free from phase impurities at small temperature gradients (below 30 K). At higher gradients, the equilibrium condition for condensation of WCl_4 is reached at T_1 and WCl_4 is then transported away from the source. If a temperature gradient of more than 70 K is chosen for the arrangement of the experiment, WCl_6 is additionally deposited in the sink.

The tellurium-rich compound $Te_8[WCl_6]_2$ can be deposited phase-pure if dissolved just under the peritectioid decomposition temperature of 200 °C in a gradient below 15 K. If there are both ternary phases present at T_2, $Te_4[WCl_6]_2$ should be transported first in a chronological sequence. In a mixture of the phases $Te_8[WCl_6]_2$ and tellurium, the deposition of $Te_8[WCl_6]_2$ is expected first in the sink then, afterwards, the tellurium sublimes.

The heterogeneous equilibria, which occur during auto transport, can be characterized with the help of further calculations. The species $TeCl_2$, WCl_4, and WCl_5 are relevant for transport at partial pressures of $p(i) > 10^{-5}$ bar. WCl_5 functions as transport agent during the auto transport of the ternary compounds with a transport efficiency $\Delta[p(i)/p^*(L)] < 0$ while $TeCl_2$ and WCl_4 with $\Delta[p(i)/p^*(L)] > 0$ are the effective species for the migration to the sink (Figure 8.3.3). The transport equilibria (8.3.3 and 8.3.4) follow from the relation of the efficiency of the gas species.

$$Te_4[WCl_6]_2(s) + 4\,WCl_5(g) \quad \rightleftharpoons \quad 4\,TeCl_2(g) + 6\,WCl_4(g) \tag{8.3.3}$$

$$Te_8[WCl_6]_2(s) + 12\,WCl_5(g) \quad \rightleftharpoons \quad 8\,TeCl_2(g) + 14\,WCl_4(g) \tag{8.3.4}$$

The reaction equations 8.3.3 and 8.3.4 allegorize a simplification insofar as the gas-phase composition is in fact more complex. Apart from $TeCl_2$, WCl_4 and WCl_5, WCl_6, $TeCl_4$ and Cl_2 can be involved in the transport to a small degree (Schl 1991) (see Figure 8.3.1).

The claim, which was made for the transport behavior of the tellurium-chloride tungstates, also applies for further systems of poly-cationic compounds of selenium and tellurium $Q_x[MX_y]_z$. Basically it should be possible to deposit them by auto transport if the constituent metal halides MX_y show sufficient vapor pressures; selenium and tellurium sublime at temperatures above 400 °C. A suitable transport temperature can be derived from the sublimation behavior of the metal halides MX_y. The temperature of the dissolution of the ternary compound should be around 100 K above the temperature of marked sublimation of the metal halide; the gradient to the sink should not exceed 20 to 30 K. In a system with several ternary phases, chalcogen-rich compounds are transportable at higher temperatures than halide-rich ones. The dissolution temperature should not exceed 300 to 350 °C because, otherwise, the incongruent decomposition of the phases toward the chalcogens will take place.

Chalcogenate(IV)-halides $BiSeO_3Cl$ is the only compound on the quasi-binary section $BiOCl/SeO_2$ (Opp 2001). The incongruent equilibrium 8.3.5 under formation of the dominant gas species SeO_2 can be observed in the temperature range

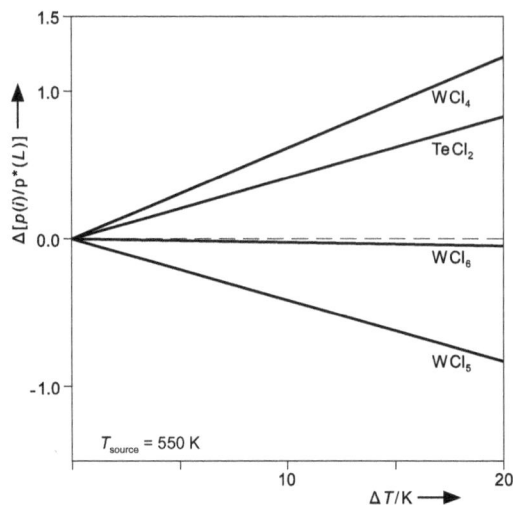

Figure 8.3.3 Transport efficiency ($\Delta[p(i)/p^*(L)]$) during the auto transport of $Te_4[WCl_6]_2$ according to *Schmidt* (Schm 2007).

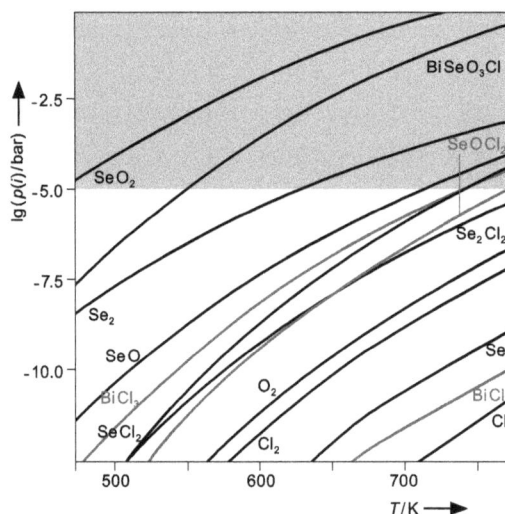

Figure 8.3.4 Composition of the gas phase over a starting solid $BiSeO_3Cl$ according to *Oppermann et al.* (Opp 2001).

from 250 to 400 °C. Adjacent to it, the quaternary species $BiSeO_3Cl$ forms, with partial pressure in the transport-effective range (Figure 8.3.4).

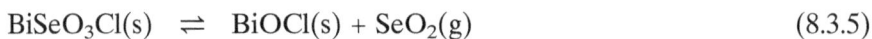

$$BiSeO_3Cl(s) \;\rightleftharpoons\; BiOCl(s) + SeO_2(g) \tag{8.3.5}$$

The conditions of the CVT of solid $BiSeO_3Cl$ result from the position of the equilibrium pressure in the phase barogram. The solid can be dissolved at a source temperature between 300 and 400 °C along the decomposition equilibrium

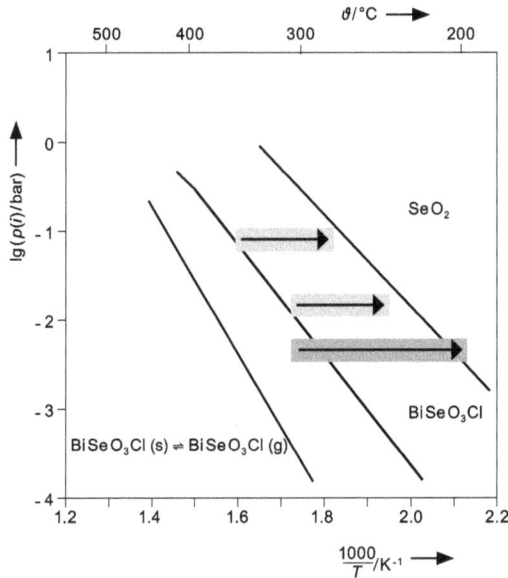

Figure 8.3.5 Schematic presentation of the phase barogram of the BiOCl/SeO$_2$ system and illustration of the phase relations during transport experiments according to *Oppermann et al.* (Opp 2001, Opp 2005).
Light gray: phase-pure transport of BiSeO$_3$Cl
Dark gray: condensation of SeO$_2$.

curve. The deposition temperature T_1 is above the sublimation pressure of SeO$_2$ at a temperature gradient of up to 50 K and the deposition of BiSeO$_3$Cl can be observed in the sink (Figure 8.3.5).

Despite the incongruent decomposition, the gas-phase deposition of BiSeO$_3$Cl can be described with a dominating equilibrium 8.3.6 as sublimation. Principally, it is a special case of the incongruent decomposition sublimation. The dominating gas species SeO$_2$ does not contribute to the flow of the components in the temperature gradient $(\Delta[p(SeO_2)]/p^*(L)] = 0)$ while the gas species, which is subordinated in the partial pressure, is solely responsible for the transport: $(\Delta[p(BiSeO_3Cl)]/p^*(L)] > 0$; Figure 8.3.6).

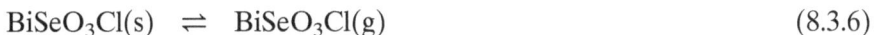

$$BiSeO_3Cl(s) \quad \rightleftharpoons \quad BiSeO_3Cl(g) \tag{8.3.6}$$

For the homologue compound BiSeO$_3$Br, the gas species SeOBr$_2$ has a transport-effective partial pressure $(p(SeOBr_2) > 10^{-5}$ bar) at temperatures above 400 °C. If the dissolution temperatures are too high, the source solid BiSeO$_3$Br decomposes towards BiBr$_3$-poorer compositions; Bi$_8$(SeO$_3$)$_9$Br$_6$ is formed. The gas-phase deposition of Bi$_8$(SeO$_3$)$_9$Br$_6$ can be described in the sense of an auto transport by the following equilibrium (Ruc 2003):

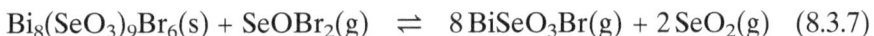

$$Bi_8(SeO_3)_9Br_6(s) + SeOBr_2(g) \quad \rightleftharpoons \quad 8\,BiSeO_3Br(g) + 2\,SeO_2(g) \tag{8.3.7}$$

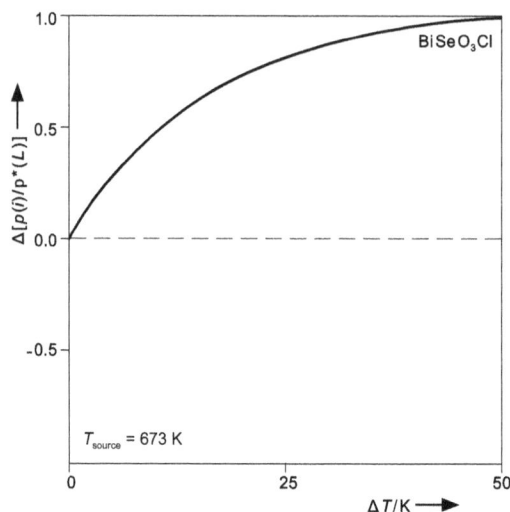

Figure 8.3.6 Transport efficiency ($\Delta[p(i)/p^*(L)]$) during the auto transport of BiSeO₃Cl according to *Oppermann et al.* (Opp 2005).

Among the substance classes that were introduced in Chapters 3 to 10, there are many examples of auto transport with chalcogenide halides. During auto transport, a solid phase is decomposed to a second, co-existing solid phase and a gaseous reaction product. The gas species formed function as transport agents. The auto transport is always endothermic due to the preceding decomposition reaction. The transport can also be realized by adding the transport agent externally. Due to its appearance, auto transport is often described as sublimation in the literature. As far as the substance class of the chalcogenide halides is concerned, there are many examples of isothermal crystallization along the activity gradient. This method is often used in the process without being mentioned as a transport reaction in the literature.

Tabelle 8.1 Examples of the chemical vapor transport of chalcogenide halides.

Sink solid	Transport additive	Temperature / °C	Reference
AlOCl	NbCl$_5$	400 → 380	Sch 1962, Sch 1972
BiOBr	auto transport	710 → 670	Ket 1985
	auto transport	750 → 700	Schm 1999b, Opp 2005
	Br$_2$	800 → 700	Opp 2000
	HBr	700 → 600	Sht 1973, Sht 1983
	HBr	800 → 600	Opp 2000
	BiBr$_3$	700 → 600	Sht 1972, Sht 1983
	H$_2$O	700 → 600	Sht 1972, Sht 1983
	H$_2$O	550 ... 800 → 450	Opp 2000
	SeO$_2$	500 ... 700 → 450	Opp 2002
BiOCl	auto transport	$T_2 \to T_1$	Ket 1985
	auto transport	720 → 600	Gan 1993
	auto transport	850 → 800	Schm 1999a, Opp 2005
	Cl$_2$	800 → 700	Opp 2000
	HCl	720 → 600	Sht 1972, Sht 1983
	HCl	800 → 600	Opp 2000
	BiCl$_3$	720 → 600	Sht 1972, Sht 1983
	H$_2$O	720 → 600	Sht 1972, Sht 1983
	H$_2$O	550 ... 800 → 450	Opp 2000
	SeO$_2$	550 ... 700 → 500	Opp 2001
BiOI	auto transport	580 → 450	Gan 1993
	HI	580 → 450	Sht 1972, Sht 1983
	BiI$_3$	580 → 450	Sht 1972, Sht 1983
	H$_2$O	580 → 450	Sht 1972, Sht 1983
	H$_2$O	550 ... 800 → 450	Opp 2000
	SeO$_2$	500 ... 700 → 400 ... 600	Opp 2002b
BiPbO$_2$Br	auto transport	670 → 640	Ket 1985
BiPbO$_2$Cl	auto transport	670 → 640	Ket 1985
BiPbO$_2$I	auto transport	670 → 640	Ket 1985
BiSI	auto transport	500 → 400	Opp 2003, Opp 2005
BiSb$_9$O$_{14}$I$_2$	auto transport	$T_2 \to T_1$	Ket 1985

Table 8.1 (continued)

Sink solid	Transport additive	Temperature / °C	Reference
BiSeBr	auto transport	$490 \to 460$	Opp 2004, Opp 2005
BiSeI	auto transport	$560 \to T_1$	Gan 1993
	auto transport	$600 \to T_1$	Bra 2000
	auto transport	$500 \to T_1$	Opp 2004, Opp 2005
$BiSeO_3Br$	auto transport	$400 \ldots 490 \to 300 \ldots 390$	Opp 2002, Opp 2005
$BiSeO_3Cl$	auto transport	$300 \ldots 400 \to 250 \ldots 350$	Opp 2001, Opp 2005
BiTeBr	auto transport	$450 \ldots 500 \to 400 \ldots 450$	Opp 2004, Opp 2005
BiTeCl	auto transport	$410 \to 395$	Opp 2004, Opp 2005
α-BiTeI	auto transport	$530 \to 490$	Opp 2004, Opp 2005
β-BiTeI	auto transport	$450 \to 400$	Opp 2004, Opp 2005
Bi_3O_4Br	auto transport	$650 \to 620$	Ket 1985
	auto transport	$835 \to 815$	Schm 1999b, Opp 2005
$Bi_3Sb_7O_{14}I_2$	auto transport	$600 \to 540$	Ket 1985
Bi_3Se_4Br	auto transport	$550 \to 510$	Opp 2004, Opp 2005
$Bi_4O_4SeCl_2$	auto transport	$700 \ldots 800 \to 650 \ldots 750$	Schm 2000, Opp 2005
$Bi_4O_5Br_2$	auto transport	$T_2 \to T_1$	Ket 1985
	auto transport	$660 \to 650$	Schm 1999b, Opp 2005
$Bi_4O_5Cl_2$	auto transport	$T_2 \to T_1$	Ket 1985
	auto transport	$650 \to 645$	Schm 1999a, Opp 2005
$Bi_4O_5I_2$	auto transport	$825 \to 795$	Ket 1985
	auto transport	$750 \to 650$	Schm 1997, Opp 2005
$(Bi, Sb)_4O_5I_2$	auto transport	$630 \to 600$	Ket 1985
$Bi_7O_9I_3$	auto transport	$750 \to 680$	Ket 1985
	auto transport	$840 \to 780$	Schm 1997, Opp 2005
$Bi_7Sb_3O_{14}I_2$	auto transport	$720 \to 680$	Ket 1985
$Bi_8(SeO_3)_9Br_6$	auto transport	$400 \to T_1$	Ruc 2003, Opp 2005
$Bi_{10}O_{12}SeCl_4$	auto transport	$750 \ldots 800 \to 700 \ldots 750$	Schm 2000, Opp 2005

Table 8.1 (continued)

Sink solid	Transport additive	Temperature / °C	Reference
$Bi_{11}Se_{12}Cl_9$	auto transport	$500 \rightarrow 480$	Opp 2004, Opp 2005
$Bi_{19}S_{27}I_3$	auto transport, BiI_3	$500 \dots 700$ $\rightarrow 400 \dots 600$	Opp 2003, Opp 2005
$Bi_{24}O_{31}Br_{10}$	auto transport	$T_2 \rightarrow T_1$	Ket 1985
	auto transport	$900 \rightarrow 840$	Schm 1999b, Opp 2005
$Bi_{24}O_{31}Cl_{10}$	auto transport	$750 \rightarrow 700$	Ket 1985
	auto transport	$850 \rightarrow 825$	Schm 1999a, Opp 2005
$Cd_3B_7O_{13}Br$	$H_2O + HBr$	$800 \dots 900 \rightarrow T_1$	Schm 1965
	$H_2O + HBr$	$820 \rightarrow 770$	Tak 1976
$Cd_3B_7O_{13}Cl$	$H_2O + HCl$	$800 \dots 900 \rightarrow T_1$	Schm 1965
	$H_2O + HCl$	$850 \rightarrow 800$	Tak 1976
$Cd_3B_7O_{13}I$	$H_2O + HI$	$800 \dots 900 \rightarrow T_1$	Schm 1965
$CeNb_7O_{19}$	Cl_2	$850 \rightarrow 800$	Hof 1991
$Ce^{III}_{2+x}Ce^{IV}_{1-x}TaO_6Cl_{3-x}$	$Cl_2 + TaCl_5$	$1100 \rightarrow 1000$	Scha 1990
$Ce_3TaO_6Cl_3$	Cl_2	$1000 \rightarrow 900$	Scha 1988
$Ce_{3.25}NbO_6Cl_{3.5-x}$	Cl_2	$950 \dots 1050$ $\rightarrow 900 \dots 1000$	Wei 1999
$Ce_{3.25}TaO_6Cl_{3.5-x}$	Cl_2	$950 \dots 1050$ $\rightarrow 900 \dots 1000$	Wei 1999
$Ce_{3.5}NbO_6Cl_{4-x}$	HCl	$900 \dots 1050$ $\rightarrow 850 \dots 1000$	Wei 1999
$Ce_{3.5}TaO_6Cl_{4-x}$	HCl	$900 \dots 1050$ $\rightarrow 850 \dots 1000$	Wei 1999
$Ce_{12.33}V_6O_{23}(OH)Cl_2$	Cl_2	$900 \rightarrow 800$	Käm 1998
$Co_3B_7O_{13}Br$	$H_2O + HBr$	$800 \dots 900 \rightarrow T_1$	Schm 1965
$Co_3B_7O_{13}Cl$	$H_2O + HCl$	$800 \dots 900 \rightarrow T_1$	Schm 1965, Nas 1972
$Co_3B_7O_{13}I$	$H_2O + HI$	$800 \dots 900 \rightarrow T_1$	Schm 1965
CrOCl	Cl_2	$940 \rightarrow 840$	Sch 1961b
	$CrCl_4$	$1000 \rightarrow 840$	Sch 1961b
	$CrCl_3$	$1000 \rightarrow 840$	Sch 1961b
CrSBr	auto transport	$870 \rightarrow T_1$	Kat 1966, Bec 1990
$CrSCl_{0.33}Br_{0.67}$	auto transport	$950 \rightarrow 880$	Sas 2000
CrSI	auto transport	$420 \rightarrow T_1$	Kat 1966
CrSeI	auto transport	$400 \rightarrow T_1$	Kat 1966
CrTeI	auto transport	$190 \rightarrow T_1$	Bat 1966
$CrTe_{0.73}I$	auto transport	$315 \rightarrow T_1$	Kat 1966

Table 8.1 (continued)

Sink solid	Transport additive	Temperature / °C	Reference
$Cr_3B_7O_{13}Br$	$H_2O + HBr$	$800 \ldots 900 \to T_1$	Schm 1965
$Cr_3B_7O_{13}Cl$	$H_2O + HCl$	$800 \ldots 900 \to T_1$	Schm 1965
$Cr_3B_7O_{13}I$	$H_2O + HI$	$800 \ldots 900 \to T_1$	Schm 1965
$CuCr_2S_3Br$	auto transport	$900 \to 850$	Miy 1968
$CuCr_2S_{4-x}Cl_x$	auto transport	$800 \to T_1$	Sle 1968
$CuCr_2S_3Cl$	auto transport	$900 \to 850$	Miy 1968
$CuCr_2S_3I$	auto transport	$900 \to 850$	Miy 1968
$CuCr_2Se_3Br$	auto transport	$800 \to T_1$	Rob 1968
	auto transport	$900 \to 850$	Miy 1968
$CuCr_2Se_{4-x}Br_x$	auto transport	$800 \to T_1$	Sle 1968
$CuCr_2Se_3Cl$	auto transport	$800 \to T_1$	Rob 1968
	auto transport	$900 \to 850$	Miy 1968
$CuCr_2Se_3I$	auto transport	$900 \to 850$	Miy 1968
$CuCr_2Te_3Br$	auto transport	$900 \to 850$	Miy 1968
$CuCr_2Te_3Cl$	auto transport	$900 \to 850$	Miy 1968
$CuCr_2Te_3I$	auto transport	$800 \to T_1$	Rob 1968
	auto transport	$900 \to 850$	Miy 1968
$CuSe_2Cl$	auto transport	$300 \to 280$	Car 1976
$CuSe_3Br$	auto transport	$340 \to 290$	Car 1976
$CuTeBr$	auto transport	$420 \to 200$	Car 1976
$CuTeCl$	auto transport	$390 \to 200$	Car 1976
$CuTeI$	auto transport	$500 \to 300$	Car 1976
Cu_2OCl_2	auto transport	$470 \to 370$	Kri 2002
$Cu_3B_7O_{13}Br$	$H_2O + HBr$	$800 \ldots 900 \to T_1$	Schm 1965
$Cu_3B_7O_{13}Cl$	$H_2O + HCl$	$800 \ldots 900 \to T_1$	Schm 1965
Cu_6PS_5Br	auto transport	$T_2 \to T_1$	Fie 1983
	CuBr	$765 \to 740$	Kuh 1976
Cu_6PS_5Cl	auto transport	$625 \to 600$	Fie 1983
	CuCl	$765 \to 740$	Kuh 1976
Cu_6PS_5I	auto transport	$T_2 \to T_1$	Fie 1983
	I_2	$800 \to 700$	Kuh 1976
	CuI	$765 \to 740$	Kuh 1976
ErSeBr	$ErBr_3$	$850 \to T_1$	Sto 1997
ErSeI	ErI_3	$800 \to T_1$	Sto 1997
$FeMoO_4Cl$	auto transport	$400 \ldots 450 \to T_1$	Cho 1989
FeOCl	HCl	$350 \to T_1$	Sch 1962a
$Fe_3B_7O_{13}Br$	$H_2O + HBr$	$800 \ldots 900 \to T_1$	Schm 1965, Nas 1972
$Fe_3B_7O_{13}Cl$	$H_2O + HCl$	$800 \ldots 900 \to T_1$	Schm 1965
$Fe_3B_7O_{13}I$	$H_2O + HI$	$800 \ldots 900 \to T_1$	Schm 1965
$Hg_3S_2Cl_2$	HCl	$400 \to 295$	Car 1967

Table 8.1 (continued)

Sink solid	Transport additive	Temperature / °C	Reference
$LaNb_7O_{19}$	Cl_2	$800 \ldots 820 \rightarrow 780 \ldots 800$	Hof 1991
$LaTiO_4Cl_5$	$S + Cl_2$	$1050 \rightarrow 950$	Hüb 1990
$(La, Ce)_{3.25}NbO_6Cl_{3.5-x}$	Cl_2	$950 \ldots 1050$ $\rightarrow 900 \ldots 1000$	Wei 1999
$(La, Ce)_{3.25}TaO_6Cl_{3.5-x}$	Cl_2	$950 \ldots 1050$ $\rightarrow 900 \ldots 1000$	Wei 1999
$(La, Ce)_{3.5}NbO_6Cl_{4-x}$	HCl	$900 \ldots 1050$ $\rightarrow 850 \ldots 1000$	Wei 1999
$(La, Ce)_{3.5}TaO_6Cl_{4-x}$	HCl	$900 \ldots 1050$ $\rightarrow 850 \ldots 1000$	Wei 1999
$(La, Tb)_{3.5}TaO_6Cl_{4-x}$	HCl	$900 \ldots 1050$ $\rightarrow 850 \ldots 1000$	Wei 1999
$La_xCe_yTaO_6Cl_z$	Cl_2	$1100 \rightarrow 1000$	Scha 1990
$La_2TaO_4Cl_3$	Cl_2	$1100 \rightarrow 1000$	Scha 1988
La_2TeI_2	auto transport	$900 \rightarrow T_1$	Rya 2006
$La_2ThTaO_6Cl_3$	Cl_2	$1080 \rightarrow 940$	Scha 1988
$La_3TaO_5(OH)Cl_3$	Cl_2	$1100 \rightarrow 1000$	Scha 1988
$La_3UO_6Cl_3$	Cl_2	$1000 \rightarrow 900$	Hen 1993
$La_{12.33}V_6O_{23}(OH)Cl_2$	Cl_2	$900 \rightarrow 800$	Käm 1998
$LuTiO_3Cl$	Cl_2	$950 \rightarrow 850$	Hüb 1993
$Mg_3B_7O_{13}Br$	$H_2O + HBr$	$800 \ldots 900 \rightarrow T_1$	Schm 1965
$Mg_3B_7O_{13}Cl$	$H_2O + HCl$	$800 \ldots 900 \rightarrow T_1$	Schm 1965, Nas 1972
$Mg_3B_7O_{13}I$	$H_2O + HI$	$800 \ldots 900 \rightarrow T_1$	Schm 1965
$Mn_3B_7O_{13}Br$	$H_2O + HBr$	$800 \ldots 900 \rightarrow T_1$	Schm 1965
$Mn_3B_7O_{13}Cl$	$H_2O + HCl$	$800 \ldots 900 \rightarrow T_1$	Schm 1965, Nas 1972
$Mn_3B_7O_{13}I$	$H_2O + HI$	$800 \ldots 900 \rightarrow T_1$	Schm 1965
$MoOBr_3$	auto transport	$350 \rightarrow 270$	Opp 1972b
$MoOCl_2$	$MoCl_5$	$350 \rightarrow 300$	Sch 1964a
$MoOCl_3$	auto transport	$300 \rightarrow 250$	Opp 1972c
$MoOCl_4$	auto transport	$120 \rightarrow 80$	Opp 1972c
MoO_2Br_2	auto transport	$180 \rightarrow 130$	Opp 1970
MoO_2Cl_2	auto transport	$250 \rightarrow T_1$	Opp 1970
MoS_2Cl_2	auto transport	$515 \rightarrow 510$	Rij 1979b
$NbOBr_2$	$NbBr_5$	$450 \rightarrow 400$	Sch 1986
	auto transport	$500 \rightarrow T_1$	Bec 2006
$NbOBr_3$	auto transport, Br_2	$T_2 \rightarrow T_1$	Fai 1959
$NbOCl_2$	$NbCl_5$	$370 \rightarrow 350$	Sch 1961c
	$NbCl_5$	$420 \rightarrow 375$	Sch 1974
	$NbCl_5$	$400 \rightarrow 360$	Sch 1986
$NbOCl_3$	$NbCl_5$	$350 \rightarrow 210$	Sch 1960

Table 8.1 (continued)

Sink solid	Transport additive	Temperature / °C	Reference
$NbOI_2$	I_2	$500 \rightarrow 450$	Sch 1962b
$NbOI_3$	auto transport, I_2	$400 \rightarrow 275$	Sch 1962b
NbO_2Br	Br_2	$450 \rightarrow 400$	Sch 1986
NbO_2I	NbI_5	$500 \rightarrow 475$	Sch 1965
	I_2	$500 \rightarrow 475$	Har 2007
NbS_2Br_2	auto transport	$500 \rightarrow T_1$	Sch 1964b, Fen 1980
	auto transport	$505 \rightarrow 500$	Rij 1979b
NbS_2Cl_2	auto transport	$500 \rightarrow T_1$	Sch 1964b, Schn 1966, Fen 1980
	auto transport	$480 \rightarrow 475$	Rij 1979b
NbS_2I_2	auto transport	$500 \rightarrow T_1$	Sch 1964b, Fen 1980
	auto transport	$400 \rightarrow 380$	Rij 1979b
$NbSe_2Br_2$	auto transport	$500 \rightarrow T_1$	Sch 1964b, Fen 1980
	auto transport	$480 \rightarrow 475$	Rij 1979b
$NbSe_2Cl_2$	auto transport	$500 \rightarrow T_1$	Sch 1964b, Fen 1980
	auto transport	$410 \rightarrow 405$	Rij 1979b
$NbSe_2I_2$	auto transport	$500 \rightarrow T_1$	Sch 1964b, Fen 1980
	auto transport	$470 \rightarrow 460$	Rij 1979b
$NbSe_4I_{0.33}$	I_2	$730 \rightarrow 670$	Nak 1985
Nb_3O_7Cl	$NbCl_5$	$600 \rightarrow 550$	Sch 1961c, Sch 1962
	$NbCl_5$	$610 \rightarrow 580$	Sch 1986
$Nd_2Ta_2O_7Cl_2$	$Cl_2 + TaCl_5$	$1000 \rightarrow 900$	Scha 1988a
$Nd_2Ti_3O_8Cl_2$	Cl_2	$950 \rightarrow 850$	Hüb 1991
$Nd_3NbO_4Cl_6$	NH_4Cl	$900 \rightarrow 800$	Tho 1992
$Nd_3UO_6Cl_3$	Cl_2	$840 \rightarrow 780$	Hen 1993
$Nd_{7.33}Ta_8O_{28}Cl_6$	$Cl_2 + TaCl_5$	$1000 \rightarrow 900$	Scha 1988a
$Ni_3B_7O_{13}Br$	$H_2O + HBr$	$800 \dots 900 \rightarrow T_1$	Schm 1965
	$H_2O + HBr$	$860 \rightarrow T_1$	Cas 2005
$Ni_3B_7O_{13}Cl$	$H_2O + HCl$	$800 \dots 900 \rightarrow T_1$	Schm 1965, Dep 1979
	$H_2O + HCl$	$920 \rightarrow T_1$	Cas 1998
$Ni_3B_7O_{13}I$	$H_2O + HI$	$800 \dots 900 \rightarrow T_1$	Schm 1965, Nas 1972
$OsO_{0.5}Cl_3$	Cl_2	$505 \rightarrow 480$	Hun 1986
	Cl_2	$500 \rightarrow 400$	Sch 1967
$OsOCl_2$	Cl_2	$500 \rightarrow 100$	Hun 1986
	Cl_2	$500 \rightarrow 470$	Sch 1967

Table 8.1 (continued)

Sink solid	Transport additive	Temperature / °C	Reference
$PbBiO_2Br$	auto transport	$670 \rightarrow 640$	Ket 1985
$PbBiO_2Cl$	auto transport	$670 \rightarrow 640$	Ket 1985
$PbBiO_2I$	auto transport	$670 \rightarrow 640$	Ket 1985
$Pr_3NbO_4Cl_6$	NH_4Cl	$900 \rightarrow 800$	Tho 1992
$Pr_3NbO_5(OH)Cl_3$	Cl_2	$900 \rightarrow 800$	Tho 1992
$Pr_3UO_6Cl_3$	Cl_2	$840 \rightarrow 780$	Hen 1993
$(Pr, Ce)_{3.25}NbO_6Cl_{3.5-x}$	Cl_2	$950 \ldots 1050$ $\rightarrow 900 \ldots 1000$	Wei 1999
$(Pr, Ce)_{3.25}TaO_6Cl_{3.5-x}$	Cl_2	$950 \ldots 1050$ $\rightarrow 900 \ldots 1000$	Wei 1999
$(Pr, Ce)_{3.5}NbO_6Cl_{4-x}$	HCl	$900 \ldots 1050$ $\rightarrow 850 \ldots 1000$	Wei 1999
$(Pr, Ce)_{3.5}TaO_6Cl_{4-x}$	HCl	$900 \ldots 1050$ $\rightarrow 850 \ldots 1000$	Wei 1999
$Re_6S_8Br_2$	Br_2	$1160 \rightarrow 1120$	Fis 1992
$Re_6S_8Cl_2$	Cl_2	$1100 \rightarrow 1060$	Fis 1992a
$Re_6Se_7Br_4$	Br_2	$1080 \rightarrow 1050$	Aru 1994
$Re_6Se_8Br_2$	Br_2	$1120 \rightarrow 1080$	Spe 1988
	Br_2	$1120 \rightarrow 1080$	Fis 1992
RhTeCl	$Cl_2 + AlCl_3$	$900 \rightarrow 700$	Köh 1997
SbSBr	auto transport	$500 \ldots 600 \rightarrow T_1$	Nit 1960
$SbSBr_{1-x}I_x$	auto transport	$370 \ldots 460 \rightarrow 300 \ldots 395$	Aud 2009
SbSI	auto transport	$500 \ldots 600 \rightarrow T_1$	Nit 1960
	auto transport	$390 \rightarrow 320 \ldots 285$	Nee 1971
	I_2	$250 \ldots 500 \rightarrow 150 \ldots 400$	Bal 1986
SbSeBr	auto transport	$360 \rightarrow 300 \ldots 320$	Ari 1987
	auto transport	$500 \ldots 600 \rightarrow T_1$	Nit 1960
SbSeI	auto transport	$500 \ldots 600 \rightarrow T_1$	Nit 1960
SbTeI	auto transport	$500 \ldots 600 \rightarrow T_1$	Nit 1960
Sb_3O_4I	auto transport	$400 \rightarrow 350$	Krä 1973a
$(Sb, Bi)_4O_5I_2$	auto transport	$630 \rightarrow 600$	Ket 1985
Sb_5O_7I	auto transport	$550 \rightarrow 530$	Krä 1973a
	auto transport	$540 \ldots 580 \rightarrow 470 \ldots 540$	Krä 1974b
	auto transport	$580 \rightarrow 550$	Nit 1977
$Sb_8O_{11}I_2$	auto transport	$500 \rightarrow 450$	Krä 1973a
$Sb_3Bi_7O_{14}I_2$	auto transport	$720 \rightarrow 680$	Ket 1985
$Sb_7Bi_3O_{14}I_2$	auto transport	$600 \rightarrow 540$	Ket 1985
$Sb_9BiO_{14}I_2$	auto transport	$T_2 \rightarrow T_1$	Ket 1985
$SmTiO_3Cl$	Cl_2	$1000 \rightarrow 900$	Hüb 1991a
$Sm_2Ta_2O_7Cl_2$	NH_4Cl	$1000 \rightarrow 960$	Guo 1994

Table 8.1 (continued)

Sink solid	Transport additive	Temperature / °C	Reference
$TaOBr_2$	$TaBr_5$	$650 \rightarrow 500$	Sch 1986
$TaOBr_3$	auto transport, Br_2	$550 \rightarrow T_1$	Fai 1959
$TaOCl_2$	$TaCl_5$	$500 \rightarrow 400$	Sch 1961c
$TaOI_2$	TaI_5	$650 \rightarrow 550$	Sch 1986
TaO_2Br	Br_2	$500 \rightarrow 400$	Sch 1986
TaO_2Cl	$TaCl_5$	$500 \rightarrow 400$	Sch 1986
TaO_2I	I_2	$500 \rightarrow 450$	Sch 1965
Ta_3O_7Cl	$TaCl_5$	$500 \rightarrow 400$	Sch 1986
$Te_4[HfCl_6]$	auto transport	$220 \rightarrow 200$	Bau 2004
$(Te_4)(Te_{10})[Bi_4Cl_{16}]$	auto transport	$160 \rightarrow 90$	Bec 2002
$Te_6[HfCl_6]$	auto transport	$220 \rightarrow 200$	Bau 2004
$Te_6[ZrCl_6]$	auto transport	$220 \rightarrow 200$	Bau 2004
$Te_6O_{11}Br_2$	auto transport, $TeBr_4$	$500 \dots 550 \rightarrow 450 \dots 500$	Opp 1978
$Te_6O_{11}Cl_2$	auto transport, $TeCl_4$	$400 \dots 600 \rightarrow 350 \dots 550$	Opp 1977b, Sch 1977
$Te_8[HfCl_6]$	auto transport	$220 \rightarrow 200$	Bau 2004
$ThOI_2$	HI	$530 \dots 780 \rightarrow 600 \dots 800$	Cor 1969
TiOCl	HCl, $TiCl_3$	$650 \dots 800 \rightarrow 520 \dots 600$	Sch 1957, Sch 1958
$Tl[NbOBr_4]$	auto transport	$350 \rightarrow 300$	Bec 2005
$Tl[NbOCl_4]$	auto transport	$350 \rightarrow 300$	Bec 2005
$TmTiO_3Cl$	Cl_2	$950 \rightarrow 850$	Hüb 1993
VOCl	auto transport	$700 \dots 850 \rightarrow 550 \dots 700$	Opp 2005
	Cl_2, VCl_4	$800 \rightarrow 700$	Sch 1961a
	Cl_2, VCl_4	$800 \rightarrow 700$	Opp 1967, Opp 1990
$VOCl_2$	Cl_2	$400 \rightarrow 300$	Opp 1967, Opp 1990
	Cl_2	$500 \rightarrow 400$	Hac 1996
	auto transport	$450 \dots 500 \rightarrow T_1$	Opp 2005
$WOBr_2$	auto transport	$550 \rightarrow 470$	Til 1969
	auto transport	$580 \rightarrow 450$	Opp 1972e
	auto transport	$400 \dots 500 \rightarrow T_1$	Opp 2005
$WOBr_3$	auto transport	$400 \rightarrow 350$	Opp 2005
	WBr_6	$400 \rightarrow 350$	Opp 1972e
$WOBr_4$	auto transport	$300 \rightarrow 230$	Opp 1971a
$WOCl_2$	auto transport	$400 \dots 500 \rightarrow 250 \dots 400$	Opp 1972a, Opp 2005
$WOCl_3$	auto transport	$250 \dots 350 \rightarrow T_1$	Opp 1972a, Opp 2005

Table 8.1 (continued)

Sink solid	Transport additive	Temperature / °C	Reference
$WOCl_4$	auto transport	$220 \rightarrow 150$	Opp 1971b
WO_2Br_2	auto transport	$400 \rightarrow 320$	Opp 1971a
WO_2Cl_2	auto transport	$320 \rightarrow 260$	Opp 1971b
WO_2I	I_2	$T_2 \rightarrow 400$	Til 1968
WO_2I_2	I_2	$T_2 \rightarrow 300$	Til 1968
$YbTiO_3Cl$	Cl_2	$950 \rightarrow 850$	Hüb 1993
$Zn_3B_7O_{13}Br$	$H_2O + HBr$	$800 \dots 900 \rightarrow T_1$	Schm 1965
	$H_2O + HBr$	$920 \rightarrow T_1$	Cam 2006
$Zn_3B_7O_{13}Cl$	$H_2O + HCl$	$800 \dots 900 \rightarrow T_1$	Schm 1965
$Zn_3B_7O_{13}I$	$H_2O + HI$	$800 \dots 900 \rightarrow T_1$	Schm 1965

Bibliography

Abb 1987   A. S. Abbasov, T. Kh. Azizov, N. A. Alieva, U. Ya. Aliev, F. M. Mustafaev, *Dokl. Akad. Nauk Azerbaidzhanskoi SSR* **1987**, *42*, 41.
Ale 1981a   V. A. Aleshin, V. I. Dernovskii, B. A. Popovkin, A. V. Novoselova, *Izv. Akad. Nauk SSSR, Neorg. Mater.* **1981**, *17*, 618.
Ale 1981b   V. A. Aleshin, B. A. Popovkin, A. V. Novoselova, *Izv. Akad. Nauk SSSR, Neorg. Mater.* **1981**, *17*, 1398.
Ale 1990   V. A. Aleshin, B. A. Popovkin, *Izv. Akad. Nauk SSSR, Neorg. Mater.* **1990**, *26*, 1391.
Ari 1987   D. Arivuoli, F. D. Gnanam, P. Ramasamy, *J. Mater. Sci. Lett.* **1987**, *6*, 249.
Aru 1994   A. Aruchamy, H. Tamaoki, A. Fujishima, H. Berger, N. L. Speziali, F. Lévy, *Mater. Res. Bull.* **1994**, *29*, 359.
Aud 2009   A. Audzijonis,L. Zigas, A. Kvedaravicius, R. Zaltauskas, *Phys. B* **2009**, *404*, 3941.
Bal 1986   C. Balarew, M. Ivanova, *Cryst. Res. Technol.* **1986**, *21*, K171.
Bar 1976   A. Bartzokas, D. Siapkas, *Ferroelectrics* **1976**, *127*, 12.
Bat 1966   S. S. Batsanov, L. M. Doronina, *Inorg. Mater.* **1966**, *2*, 423.
Bau 2004   A. Baumann, J. Beck, *Z. Anorg. Allg. Chem.* **2004**, *630*, 2078.
Bec 1986   H. P. Beck, C. Strobel, *Z. Anorg. Allg. Chem.* **1986**, *535*, 229.
Bec 1990a   J. Beck, *Angew. Chem.* **1990**, *102*, 301.
Bec 1990b   J. Beck, *Z. Naturforsch.* **1990**, *B45*, 413.
Bec 1990c   J. Beck, *Z. Anorg. Allg. Chem* **1990**, *585*, 157.
Bec 2002   J. Beck, A. Fischer, A. Stankowski, *Z. Anorg. Allg. Chem.* **2002**, *628*, 2542.
Bec 1994   J. Beck, *Angew. Chem.* **2004**, *106*, 172, *Angew. Chem. Int. Ed.* **1994**, *33*, 163.
Bec 2005   J. Beck, J. Bordinhão, *Z. Anorg. Allg. Chem.* **2005**, *631*, 1261.
Bec 2006   J. Beck, C. Kusterer, *Z. Anorg. Allg. Chem.* **2006**, *632*, 2193.
Bra 2000   T. P. Braun, F. J. DiSalvo, *Acta Cryst.* **2000**, *C56*, e1.
Cam 2006   J. Campa-Molina, S. Ulloa-Godínez, A. Barrera, L. Bucio, J. Mata, *J. Phys.: Condens. Matter* **2006**, *18*, 4827.

Cha 1982   R. Chaves, H. Amaral, A. Levelur, S. Ziolkiewicz, M. Balkanski, M. K. Teng,
             J. F. Vittori, H. Stone, *Phys. Stat. Sol.* **1982**, *73*, 367.
Cho 1989   J. H. Choy, S. H. Chang, D. Y. Noh, K. A. Son, *Bull. Korean Chem. Soc.*
             1989, *10*, 27.
Car 1967   E. H. Carlson, *J. Crystal Growth* **1967**, *1*, 271.
Car 1976   P. M. Carkner, H. M. Haendler, *J. Cryst. Growth* **1976**, *33*, 196.
Cas 1998   A. G. Castellanos-Guzman, J. Reyes-Gomez, H. H. Eulert, J. Campa-Molina,
             W. Depmeier, *J. Korean Phys. Soc.* **1998**, *32*, 208.
Cas 2005   A. G. Castellanos-Guzman, M. Trujillo-Torrez, M. Czank, *Mater. Science En-*
             gin. **2005**, *B 120*, 59.
Cor 1969   J. D. Corbett, R. A. Guidotti, D. G. Adolphson, *Inorg. Chem.* **1969**, *8*, 163.
Dag 1969   C. Dagron, E Thevet, a) *Compt. Rend. Acad. Sci.*, **1969**, *C268*, 1867; b) *Ann.*
             Chim. **1971**, *6*, 67.
Dep 1979   W. Depmeier, H. Schmid, B. I. Noläng, M. W. Richardson, *J. Cryst. Growth*
             1979, *46*, 718.
Doe 2007   T. Doert, E. Dashjav, B. P. T. Fokwa, *Z. Anorg. Allg. Chem.* **2007**, *633*, 261.
Dön 1950a   E. Dönges, *Z. Anorg. Allg. Chem.* **1950**, *263*, 112.
Dön 1950b   E. Dönges, *Z. Anorg. Allg. Chem.* **1950**, *263*, 280.
Dou 2006   C. Doussier, G. Andre, P. Leone, E. Janod, Y. Moelo, *J. Solid State Chem.*
             2006, *179*, 486.
Ebi 2009   S. Ebisu, K. Koyama, H. Omote, S. Nagata, *J. Phys. Conf. Ser.* **2009**, *150*,
             042027.
Egg 1999   U. Eggenweiler, E. Keller, V. Krämer, U. Petasch, H. Oppermann, *Z. Kristal-*
             logr. **1999**, *214*, 264.
Emm 1968   F. Emmenegger, A. Petermann, *J. Cryst. Growth* **1968**, *2*, 33.
Fai 1959   F. Fairbrother, A. H. Cowley, N. Scott, *J. Less Comm. Met.* **1959**, *1*, 206
Fen 1980   J. Fenner, A. Rabenau, G. Trageser, *Adv. Inorg. Chem. Radiochem.* **1980**,
             23, 329.
Fie 1983   S. Fiechter, J. Eckstein, R. Nitsche, *J. Cryst. Growth* **1983**, *61*, 275.
Fis 1992   C. Fischer, N. Alonso-Vante, S. Fiechter, H. Tributsch, *J. Alloys. Comp.* **1992**,
             178, 305.
Fis 1992a   C. Fischer, S. Fiechter, H. Tributsch, G. Reck, B. Schultz, *Ber. Bunsenges.*
             Phys. Chem. **1992**, *11*, 1652.
Gan 1993   R. Ganesha, D. Arivuoli, P. Ramasamy, *J. Cryst. Growth* **1993**, *128*, 1081.
Guo 1994   G. Guo, M. Wang, J. Chen, J. Huang, Q. Zhang, *J. Solid State Chem.* **1994**,
             113, 434.
Hac 1996   A. Hackert, V. Plies, R. Gruehn, *Z. Anorg. Allg. Chem* **1996**, *622*, 1651.
Har 2007   S. Hartwig, H. Hillebrecht, *Z. Naturforsch.* **2007**, *62b*, 1543.
Hen 1993   G. Henche, K. Fiedler, R. Gruehn, *Z. Anorg. Allg. Chem.* **1993**, *619*, 77.
Hof 1991   R. Hofmann, R. Gruehn, *Z. Anorg. Allg. Chem.* **1991**, *602*, 105.
Hor 1968   J. Horak, J. D. Turjanica, J. Klazar, H. Kozakova, *Krist. Tech.* **1968**, *3*, 231
             und 241.
Hüb 1990   N. Hübner, U. Schaffrath, G. Gruehn, *Z. Anorg. Allg. Chem.* **1990**, *591*, 107.
Hüb 1991   N. Hübner, R. Gruehn, *Z. Anorg. Allg. Chem.* **1991**, *597*, 87.
Hüb 1991a   N. Hübner, R. Gruehn, *Z. Anorg. Allg. Chem.* **1991**, *602*, 119. *Z. Anorg. Allg.*
             Chem. **2000**, *626*, 2515. *J. Less Comm. Met.* **1961**, *3*, 29. *J. Cryst. Growth* **1972**,
             16, 59. *Inorg. Chem.* **2006**, *45*, 10728. *Mater. Res. Bull.* **1976**, *11*, 183. *J. Solid*
             State Chem. **1979**, *30*, 365.
Hüb 1992   N. Hübner, *Dissertation*, University of Gießen, **1992**.
Hüb 1993   N. Hübner, K. Fiedler, A. Preuß, R. Gruehn, *Z. Anorg. Allg. Chem.* **1993**,
             619, 1214.

Hun 1986   K.-H. Huneke, H. Schäfer, Z. Anorg. Allg. Chem. **1986**, *534*, 216.
Ish 1974   K. Ishikawa, Y. Shikatawa, A. Toyoda, Phys. Stat. Sol. **1974**, *25*, K187.
Käm 1998   H. Kämmerer, R. Gruehn, Z. Anorg. Allg. Chem. **1998**, *624*, 1526.
Kal 1983a   V. Kalesinskas, J. Grigas, A. Audzijonis, K. Zickus, Phase Trans. **1983**, *3*, 217.
Kal 1983b   V. Kalesinskas, J. Grigas, R. Jankevicius, A. Audzijonis, Phys. Stat. Sol. **1983**, *115*, K11.
Kat 1966   H. Katscher, H. Hahn, Naturwiss. **1966**, *53*, 361.
Ket 1985   J. Ketterer, Dissertation, University of Freiburg, **1985**
Kle 1995   G. Kleeff, H. Schilder, H. Lueken, Z. Anorg. Allg. Chem. **1995**, *621*, 963.
Köh 1997   J. Köhler, W. Urland, Z. Anorg. Allg. Chem. **1997**, *623*, 583.
Krä 1972   V. Krämer, R. Nitsche, J. Cryst. Growth **1972**, *15*, 309.
Krä 1973   V. Krämer, J. Appl. Cryst. **1973**, *6*, 499.
Krä 1973   V. Krämer, M. Schumacher, R. Nitsche, Mater. Res. Bull. **1973**, *8*, 65.
Krä 1974a   V. Krämer, Z. Naturforsch. **1974**, *29b*, 688 ibid. *31b*, 1582.
Krä 1974b   V. Krämer, R. Nitsche, M. Schuhmacher, J. Cryst. Growth **1974**, *24/25*, 179.
Krä 1976a   V. Krämer, Thermochim. Acta **1976**, *15*, 205.
Krä 1976b   V. Krämer, Mater. Res. Bull. **1976**, *11*, 183.
Krä 1978   V. Krämer, J. Thermal Anal. **1978**, *16*, 303.
Krä 1979   V. Krämer, Acta Crystallogr. **1979**, *B35*, 139.
Kri 2002   S. V. Krivovichev, S. K. Filatov, P. C. Burns, Can. Mineral. **2002**, *40*, 1185.
Kuh 1976   W. F. Kuhs, R. Nitsche, K. Scheunemann, Mat. Res. Bull. **1976**, *11*, 1115.
Kve 1996   S. Kvedaravicius, A. Audzijonis, N. Mykolaitien, A. Kanceravicius, Ferroelectrics **1996**, *58*, 235.
Lut 1970   H. D. Lutz, C. Lovasz, K. H. Bertram, M. Sreckovic, U. Brinker, Monatsh. Chem. **1970**, *101*, 519.
Mäh 1984   D. Mähl, J. Pickardt, B. Reuter, Z. Anorg. Allg. Chem. **1984**, *516*, 102.
Mee 1979   A. Meerschaut, L. Guemas, R. Berger, J. Rouxel, Acta Crystallogr. B **1979**, *B35*, 1747.
Meh 1980   A. Le Mehaute, J. Rouxel, M. Spiesser, GB 79-23572 19790705, DE 79-2929778 19790723, **1980**.
Miy 1968   K. Miyatani, Y. Wada, F. Okamoto, J. Phys. Soc. Jap. **1968**, *25*, 369.
Nak 1985   I. Nakada, E. Bauser, J. Cryst. Growth **1985**, *73*, 410.
Nas 1972   K. Nassau, J. W. Shiever, J. Cryst. Growth **1972**, *16*, 59.
Nee 1971   H. Neels, W. Schmitz, H. Hottmann, N. Rössner, W. Topp, Krist. Techn. **1971**, *6*, 225.
Nit 1960   R. Nitsche, W. J. Merz, J. Phys. Chem. Solids, **1960**, *13*, 154.
Nit 1964   R. Nitsche, H. Roetschi, P. Wild, Appl. Phys. Lett. **1964**, *4*, 210.
Nit 1967   R. Nitsche, J. Phys. Chem. Solids, Suppl. **1967**, *1*, 215.
Nit 1977   R. Nitsche, V. Krämer, M. Schuhmacher, A. Bussmann, J. Cryst. Growth **1977**, *42*, 549.
Noc 1993   K. Nocker, R. Gruehn, Z. Anorg. Allg. Chem. **1993**, *619*, 1530.
Opp 1967   H. Oppermann, Z. Anorg. Allg. Chem. **1967**, *351*, 127.
Opp 1970   H. Oppermann, Z. Anorg. Allg. Chem. **1970**, *379*, 262.
Opp 1971a   H. Oppermann, G. Stöver, Z. Anorg. Allg. Chem. **1971**, *383* 14.
Opp 1971b   H. Oppermann, Z. Anorg. Allg. Chem. **1971**, *383* 1.
Opp 1972a   H. Oppermann, G. Stöver, G. Kunze, Z. Anorg. Allg. Chem. **1972**, *387*, 317.
Opp 1972b   H. Oppermann, G. Kunze, G. Stöver, Z. Anorg. Allg. Chem. **1972**, *387*, 339.
Opp 1972c   H. Oppermann, G. Stöver, G. Kunze, Z. Anorg. Allg. Chem. **1972**, *387*, 201.
Opp 1972d   H. Oppermann, G. Stöver, Z. Anorg. Allg. Chem. **1972**, *387*, 218.
Opp 1972e   H. Oppermann, G. Stöver, G. Kunze, Z. Anorg. Allg. Chem. **1972**, *387*, 329
Opp 1977a   H. Oppermann, Z. Anorg. Allg. Chem. **1977**, *434*, 239.

Opp 1977b	H. Oppermann, E. Wolf, *Z. Anorg. Allg. Chem.* **1977**, *437*, 33.
Opp 1978	H. Oppermann, V. A. Titov, G. Kunze, G. A. Kollovin, E. Wolf, *Z. Anorg. Allg. Chem.* **1978**, *439*, 13.
Opp 1986a	H. Oppermmann, G. Stöver, G. Kunze, *Z. Anorg. Allg. Chem.* **1986**, *387*, 317.
Opp 1986b	H. Oppermmann, G. Stöver, G. Kunze, *Z. Anorg. Allg. Chem.* **1986,** *387*, 317 ibid. **1986**, *387*, 329.
Opp 1990	H. Oppermann, *Solid State Ionics* **1990**, *39*, 17.
Opp 1997	H. Oppermann, H. Göbel, U. Petasch, *J. Thermal Anal.* **1996**, *47*, 595.
Opp 2000	H. Oppermann, M. Schmidt, H. Brückner, W. Schnelle, E. Gmelin, *Z. Anorg. Allg. Chem.* **2000**, *626*, 937.
Opp 2001	H. Oppermann, H. Dao Quoc, M. Zhang, P. Schmidt, B. A. Popovkin, S. A. Ibragimov, P. S. Berdonosov, V. A. Dolgikh, *Z. Anorg. Allg. Chem.* **2001**, *627*, 1347.
Opp 2002a	H. Oppermann, P. Schmidt, M. Zhang-Preße, H. Dao Quoc, R. Kucharkowski, B. A. Popovkin, S. A. Ibragimov, V. A. Dolgikh, *Z. Anorg. Allg. Chem.* **2002**, *628*, 91.
Opp 2002b	H. Oppermann, H. Dao Quoc, P. Schmidt, *Z. Anorg. Allg. Chem.* **2002**, *628*, 2509.
Opp 2003	H. Oppermann, U. Petasch, *Z. Naturforsch.* **2003**, *58b*, 725.
Opp 2004	H. Oppermann, U. Petasch, P. Schmidt, E. Keller, V. Krämer, *Z. Naturforsch.* **2004**, *59b*, 727.
Opp 2005	H. Oppermann, M. Schmidt, P. Schmidt, *Z. Anorg. Allg. Chem.* **2005**, *631*, 197.
Pes 1971	P. Peshev, W. Piekarczyk, S. Gazda, *Mater. Res Bull.* **1971**, *6*, 479.
Pet 1997	U. Petasch, H. Oppermann, *Z. Anorg. Allg. Chem.* **1997**, *623*, 169.
Pet 1998	U. Petasch, H. Göbel, H. Oppermann, *Z. Anorg. Allg. Chem.* **1998,** *624*, 1767.
Pet 1999a	U. Petasch, H. Oppermann, *Z. Naturforsch.* **1999**, *54b*, 487.
Pet 1999b	U. Petasch, C. Hennig, H. Oppermann, *Z. Naturforsch.* **1999**, *54b*, 234.
Pfi 2005a	A. Pfitzner, M. Zabel, F. Rau, *Monatsh. Chem.* **2005**, *136*, 1977.
Pfi 2005b	A. Pfitzner, M. Zabel, F. Rau, *Z. Anorg. Allg. Chem.* **2005**, *631*, 1439.
Pie 1970	W. Piekarczyk, P. Peshev, *J. Cryst. Growth* **1970**, *6*, 357.
Pro 1984	I. V. Protskaya, V. A. Trifonov, B. A. Popovkin, A. V. Novoselova, S. I. Troyanov, A.V. Astaf'ev, *Zh. Neorg. Khim.* **1985**, *29*, 1128.
Pro 1985	I. V. Protskaya, V. A. Trifonov, B. A. Popovkin, A. V. Novoselova, S. I. Troyanov, A. V. Astaf'ev, *Zh. Neorg. Khim.* **1985**, *30*, 3029.
Rij 1979a	J. Rijnsdorp, F. Jellinek, *J. Solid State Chem.* **1979**, *28*, 149.
Rij 1979b	J. Rijnsdorp, G. J. De Lange, G. A. Wiegers, *J. Solid State Chem.* **1979**, *30*, 365.
Rij 1980	J. Rijnsdorp, C. Haas, *J. Phys. Chem. Solids* **1980**, *41*, 375.
Rob 1968	M. Robbins, M. K. Baltzer, E. Lopatin, *J. Appl. Phys.* **1968**, *39*, 662.
Ros 1990	R. Ross, *Dissertation*, University of Gießen, **1990**.
Ruc 2003	M. Ruck, P. Schmidt, *Z. Anorg. Allg. Chem.* **2003**, *629*, 2133.
Rya 1970	A. A. Ryazantsev, L. M. Varekha, B. A. Popovkin, A. V. Novoselova, *Izv. Akad. Nauk, Neorg. Mater.* **1970**, *6(6)*, 1175.
Rya 2006	M. Ryazanov, A. Simon, H. J. Mattausch, *Inorg. Chem.* **2006**, *45*, 10728.
Sas 2000	M. Saßmannshausen, H. D. Lutz, *Mater. Res. Bull.* **2000**, *35*, 2431.
Sch 1958	H. Schäfer, F. Wartenpfuhl, E. Weise, *Z. Anorg. Allg. Chem.* **1958**, *295*, 268.
Sch 1960	H. Schäfer, F. Kahlenberg, *Z. Anorg. Allg. Chem.* **1960**, *305*, 327.
Sch 1961a	H. Schäfer, F. Wartenpfuhl, *J. Less Comm. Met.* **1961**, *3*, 29.
Sch 1961b	H. Schäfer, F. Wartenpfuhl, *Z. Anorg. Allg. Chem.* **1961**, *308*, 282.
Sch 1961c	H. Schäfer, E. Sibbing, R. Gerken, *Z. Anorg. Allg. Chem.* **1961**, *307*, 163.

Sch 1962a   H. Schäfer, *Chemische Transportreaktion*, Verlag Chemie, Weinheim **1962**.
Sch 1962b   H. Schäfer, R. Gerken, *Z. Anorg. Allg. Chem.* **1962**, *317*, 105.
Sch 1964a   H. Schäfer, J. Tillack, *J. Less Comm. Met.* **1964**, *6*, 152.
Sch 1964b   H. Schäfer, D. Bauer, W. Beckmann, R. Gerken, H. G. Nieder-Vahrenholz, K.-J. Niehues, H. Scholz, *Naturwiss.* **1964**, *51*, 241.
Sch 1965   H. Schäfer, L. Zylka, *Z. Anorg. Allg. Chem.* **1965**, *338*, 309.
Sch 1967   H. Schäfer, K.-H. Huneke, *J. Less Comm. Met.* **1967**, *12*, 331.
Sch 1972   H. Schäfer, P. Hagenmüller, *Prep. Methods in Solid State Chem.*, New York, **1972**.
Sch 1974   H. Schäfer, F. Schulte, *Z. Anorg. Allg. Chem.* **1974**, *405*, 307.
Sch 1977   H. Schäfer, M. Binnewies, H. Rabeneck, C. Brendel, M. Trenkel, *Z. Anorg. Allg. Chem.* **1977**, *435*, 5.
Sch 1986   H. Schäfer, R. Gerken, L. Zylka, *Z. Anorg. Allg. Chem.* **1986**, *534*, 209.
Schö 2010   M. Schöneich, M. Schmidt, P. Schmidt, *Z. Anorg. Allg. Chem.* **2010**, *636*, 1810.
Scha 1988   U. Schaffrath, R. Gruehn, *J. Less Comm. Met.* **1988**, *137*, 61.
Scha 1988a   U. Schaffrath, R. Gruehn, *Z. Naturforsch. B* **1988**, *43*, 1567.
Scha 1989   U. Schaffrath, *Dissertation*, University of Gießen, **1989**.
Scha 1990   U. Schaffrath, R. Gruehn, *Z. Anorg. Allg. Chem.* **1990**, *589*, 139.
Schl 1991   Th. Schlörb, *Diplomarbeit*, University of Gießen, **1991**.
Schm 1965 H. Schmid, *J. Phys. Chem. Solids* **1965**, *27*, 973.
Schm 1997   M. Schmidt, H. Oppermann, H. Brückner, M. Binnewies, *Z. Anorg. Allg. Chem.* **1997**, *623*, 1945.
Schm 1999 M. Schmidt, H. Oppermann, M. Binnewies, *Z. Anorg. Allg. Chem.* **1999**, *625*, 1001.
Schm 1999b M. Schmidt, *Dissertation*, TU Dresden **1999**.
Schm 2000   P. Schmidt, H. Oppermann, N. Söger, M. Binnewies, A. N. Rykov, K. O. Znamenkov, A. N. Kuznetsov, B. A. Popovkin, *Z. Anorg. Allg. Chem.* **2000**, *626*, 2515.
Schm 2007   P. Schmidt, *Habilitationsschrift, TU Dresden,* **2007**; http://nbn-resolving.de/urn:nbn:de:bsz:14-ds-1200397971615-40549
Schn 1966   H. G. von Schnering, W. Beckmann, *Z. Anorg. Allg. Chem.* **1966**, *347*, 231.
Scho 1940   R. Schober, E. Thilo, *Ber. Dt. Chem. Ges.* **1940**, *73*, 1219.
Sht 1972   M. V. Shtilikha, D. V. Chepur, I. I. Yatskovich, *Sov. Phys. Cryst.* **1972**, *16*, 732.
Sht 1983   M. V. Shtilikha, *Russ. J. Inorg. Chem.* **1983**, *28*, 154.
Sle 1968   A. W. Sleight, H. S. Harret, *J. Phys. Chem. Solids* **1968**, *29*, 868.
Som 1972   I. Sommer, *J. Cryst. Growth* **1972**, *16*, 259.
Spe 1988   N. L. Speziali, H. Berger, G. Leicht, R. Sanjinés, G. Chapuis, F. Lévy, *Mater. Res. Bull.* **1988**, *23*, 1597.
Str 1973   G. B. Street, E. Sawatzky, K. Lee, *J. Phys. Chem. Solids* **1973**, *34*, 1453.
Sto 1997   K. Stöwe, *Z. Anorg. Allg. Chem.* **1997**, *623*, 1639.
Tak 1976   T. Takahashi, O. Yamada, *J. Cryst. Growth* **1976**, *33*, 361.
Ten 1981   M. K. Teng, M. Massot, M. R. Chaves, M. H. Amoral, S. Ziolkievicz, W. Young, *Phys. Stat. Sol.* **1981**, *63*, 605.
Tho 1992   M. H. Thomas, R. Gruehn, *J. Solid State Chem.* **1992**, *99*, 219.
Til 1968   J. Tillack, *Z. Anorg. Allg. Chem.* **1968**, *357*, 11.
Til 1969   J. Tillack, R. Kaiser, *Angew. Chem.* **1969**, *8*, 142.
Til 1970   J. Tillack, *J. Less Comm. Met.* **1970**, *20*, 171.
Tri 1997   V. A. Trifonov, A. V. Shevelkov, E. V. Dikarev, B. A. Popovkin, *Zh. Neorg. Khim.* **1997**, *42(8)*, 1237.
Tur 1973   I. D. Turynitsa, I. D. Olekseyuk, I. I. Kozmanko, *Izv. Akad. Nauk SSSR, Neorg. Mater.* **1973**, *9*, 1433.

Val 1976   N. R. Valitova, V. A. Aleshin, B. A. Popovkin, A. V. Novoselova, *Izv. Akad. Nauk SSSR, Neorg. Mater.* **1976**, *12*, 225.

Vor 1979   T. A. Vorobeva, A. M. Pantschenkov, V. A. Trifonov, B. A. Popovkin, A. V. Novoselova, *Zh. Neorg. Khim.* **1979**, *24(3)*, 767.

Vor 1987   T. A. Vorobeva, E. V. Kolomnina, V. A. Trifonov, B. A. Popovkin, A. V. Novoselova, *Izv. Akad. Nauk SSSR, Neorg. Mater.* **1987**, *23*, 1843.

Vor 1990   T. A. Vorobeva, V. A. Trifonov, B. A. Popovkin, *Neorg. Mater. 1990*, *26*, 51.

Wan 2006   L. Wang, Y.-C. Hung, S.-J. Hwu, H.-J. Koo, M.-H. Whangbo, *Chem. Mater.* **2006**, *18*, 1219.

Weh 1970   F. H. Wehmeier, E. T. Keve, S. C. Abrahams, *Inorg. Chem.* **1970**, *9*, 2125.

Wei 1999   H. Weitzel, B. Behler, R. Gruehn, *Z. Anorg. Allg. Chem.* **1999**, *625*, 221.

9 Chemical Vapor Transport of Pnictides

The bond character of metal pnictides is very variable and ranges from the metallic, ionic, and covalent nitrides and phosphides through the rather covalent or metallic arsenides and antimonides to the typical metallic bismutides.

In the literature, there is only one example of the CVT of a binary nitride (Mün 1956): TiN can be transported to the hotter zone with hydrogen chloride at temperatures around $1000\,°C$; at temperatures around $1500\,°C$ it can be transported to the lower temperature zone with considerable transport rates. The transport of phosphides and arsenides is documented by numerous examples. There are only a few examples of the transport of antimonides; only one example of the transport of a bismuth-containing intermetallic phase, NiBi, is known today (Ruc 1999).

Elemental halogens, in particular iodine, and in some cases halogen compounds are the preferred transport additives. The tendency of pnictogens to form halogen compounds increases from nitrogen to bismuth. The formation of phosphorus halides is important during the transport of phosphides. The stability of phosphorus halides, arsenic halides, and antimony halides is so strong that in some cases metal halides with transport-effective partial pressures are not formed.

The boiling temperatures of the pnictogens increase from nitrogen to bismuth. While nitrogen, phosphorus, and arsenic have sufficiently high saturation pressures to be transport effective in elemental form, it is necessary to generate transport-effective compounds for the antimonides and bismuthides.

Concerning the transport of phosphides, in the gas phase mostly phosphorus(III)-halides occur. During the transport of arsenides and antimonides, one has to expect, at rising temperatures, the formation of monohalides. This applies in particular for the heavy halogens. One has also to expect the formation of pnictogen chalcogenide halides, such as AsSeI (Bru 2006).

9.1 Transport of Phosphides

TaP

Iodine as transport agent

$$VP(s) + \frac{7}{2}I_2(g) \rightleftharpoons VI_4(g) + PI_3(g)$$

Phosphorus(III)-iodide as transport agent

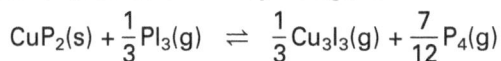

$$CuP_2(s) + \frac{1}{3}PI_3(g) \rightleftharpoons \frac{1}{3}Cu_3I_3(g) + \frac{7}{12}P_4(g)$$

Mercury(II)-bromide as transport additive

$$Fe_2P(s) + \frac{7}{2}HgBr_2(g) \rightleftharpoons 2FeBr_2(g) + PBr_3(g) + \frac{7}{2}Hg(g)$$

As Table 9.1.1 shows, CVT provides very good access to well crystallized samples of phosphides of the transition metals. Apart from the preparative use, transport experiments allowed the narrowing down of thermodynamic data of some phosphides. The indications, which were obtained from the critical comparison of all experimental observations (see Chapter 14) with the results of thermodynamic model calculations (see Chapters 2 and 13), allow a detailed chemical understanding of the transport determining equilibria. This way, a more specific selection of favorable experimental conditions for the transport of new compounds becomes possible.

In most cases, the *transport of phosphides of the transition metals with iodine* is possible. Depending on the thermodynamic stability of the phosphide and the volatile metal iodide, transport via exothermic (e. g., VP/I_2, MnP/I_2, Cu_3P/I_2) or endothermic (e. g., CrP/I_2, CoP/I_2, CuP_2/I_2) reactions can occur in a temperature gradient. The transport determining equilibria 9.1.1 to 9.1.6 result from thermodynamic model calculations of the mentioned examples.

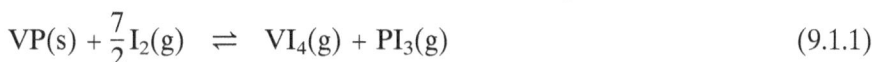

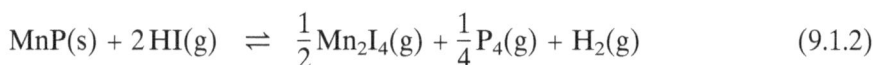

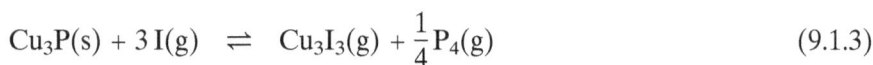

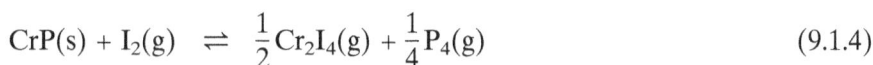

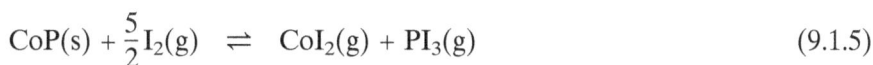

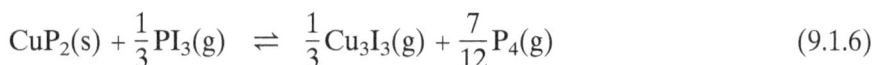

$$VP(s) + \frac{7}{2}I_2(g) \rightleftharpoons VI_4(g) + PI_3(g) \tag{9.1.1}$$

$$MnP(s) + 2\,HI(g) \rightleftharpoons \frac{1}{2}Mn_2I_4(g) + \frac{1}{4}P_4(g) + H_2(g) \tag{9.1.2}$$

$$Cu_3P(s) + 3\,I(g) \rightleftharpoons Cu_3I_3(g) + \frac{1}{4}P_4(g) \tag{9.1.3}$$

$$CrP(s) + I_2(g) \rightleftharpoons \frac{1}{2}Cr_2I_4(g) + \frac{1}{4}P_4(g) \tag{9.1.4}$$

$$CoP(s) + \frac{5}{2}I_2(g) \rightleftharpoons CoI_2(g) + PI_3(g) \tag{9.1.5}$$

$$CuP_2(s) + \frac{1}{3}PI_3(g) \rightleftharpoons \frac{1}{3}Cu_3I_3(g) + \frac{7}{12}P_4(g) \tag{9.1.6}$$

In general, a comparatively high transport agent density of approximately $5\ mg \cdot cm^{-3}$ is necessary for efficient transport. The absence of a transport effect for NbP in earlier experiments by *Schäfer* (Sch 1965) can be traced back to the fact that the amount of added iodine was not sufficiently high (see Table 9.1.1).

Qualitatively, the experimental results suggest that phosphides show the best results (high transport rates; large crystals) at a ratio of $n(M) : n(P)$ close to $1 : 1$. Chemical vapor transport of metal-rich and phosphorus-rich phosphides in a temperature gradient can only be conducted with lower efficiency. Furthermore, experiments aiming at the transport of metal-rich phosphides of the early transition metals ($M : P \geq 1$; Ti, Zr, V, Cr) led to the migration of the respective monophosphide and strongly attacked the walls of the silica ampoules. Several considerations and further experimental results clarify the circumstances. If the activity of the metal component in a phosphide is high but that of phosphorus very low (metal-rich phosphide), the only reaction that will occur is that of the transport agent iodine with the metal under formation of the volatile metal iodide. In some cases, even its saturation pressure is exceeded so that condensed metal halides appear as well. Phosphorus is kept and enriched in the solid; the simultaneous volatilization of both components is impossible. The reactions of $Cr_{12}P_7$ ($\rightarrow$ CrP), Fe_2P ($\rightarrow$ FeP), and Co_2P ($\rightarrow$ CoP) with iodine are examples of this behavior (Gla 1999). In an alternative but equivalent consideration, the absence of transport can be traced back to the limited *phosphorus co-existence pressure*, which is set by neighboring phases of higher phosphorus content. Under these conditions, the partial pressures $p(P_2)$ and $p(P_4)$, respectively, cannot take on values that are sufficiently high ($\geq 10^{-5}$ bar) to allow a traceable transport effect. The thermodynamic stability of the phosphorus iodides P_2I_4 and PI_3 does (under these conditions) not suffice to keep phosphorus in the gas phase (Fin 1965, Fin 1969, Fin 1970, Hil 1973). These considerations led to the successful application of $HgBr_2$ as transport agent for Mo_3P, Mo_4P_3 (Len 1995), and Fe_2P (Cze 1999). In these cases, the transfer of phosphorus through the gas phase takes place via the more stable phosphorus bromide (9.1.7 to 9.1.9) (Cze 1999).

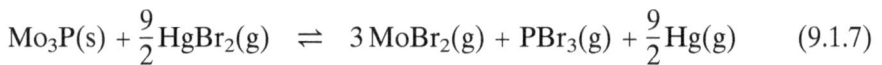

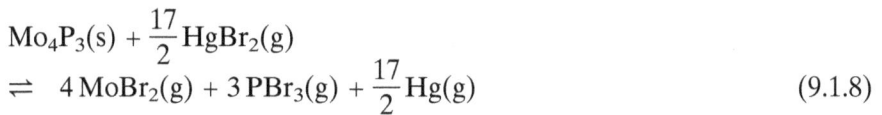

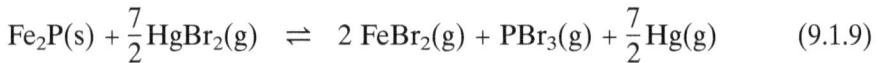

$$Mo_3P(s) + \frac{9}{2}HgBr_2(g) \;\rightleftharpoons\; 3\,MoBr_2(g) + PBr_3(g) + \frac{9}{2}Hg(g) \qquad (9.1.7)$$

$$Mo_4P_3(s) + \frac{17}{2}HgBr_2(g)$$
$$\rightleftharpoons\; 4\,MoBr_2(g) + 3\,PBr_3(g) + \frac{17}{2}Hg(g) \qquad (9.1.8)$$

$$Fe_2P(s) + \frac{7}{2}HgBr_2(g) \;\rightleftharpoons\; 2\,FeBr_2(g) + PBr_3(g) + \frac{7}{2}Hg(g) \qquad (9.1.9)$$

In reversal of these considerations, the phosphorus co-existence pressure over a phosphide and its phosphorus-rich neighboring phase must be higher than approximately 10^{-5} bar if the metal phosphide is to be transportable with iodine.

The phosphorus pressure in an ampoule during the *transport of phosphorus-rich phosphides* has a different consequence. At temperatures that allow sufficient volatility of the metal iodides, many phosphides with a ratio of $M : P < 1 : 1$ (polyphosphides) are only stable at $p(P_2/P_4) > 1$. Despite the comparatively low thermodynamic stability of gaseous P_2I_4 and PI_3, iodine can be bound in the gas phase by the phosphorus that was added as transport agent; thus it is not available to form an iodide with the metal. For this reason, the reversible CVT of many polyphosphides, which are specially known from the investigations of the groups of *Jeitschko* (Bra 1978, Jei 1984) and *von Schnering* (vSc 1988), is hardly possible. Exceptions are VP_4, FeP_4, and Cu_2P_7, whose metals form particularly stable and/or highly volatile iodides. This way, transport becomes possible at comparatively low temperatures between 500 and 600 °C with the in situ developing transport agents PI_3 and/or P_2I_4 (equations 9.1.10 to 9.1.12).

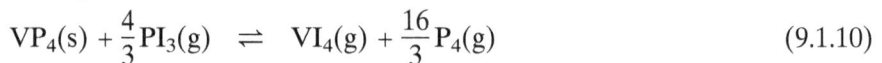

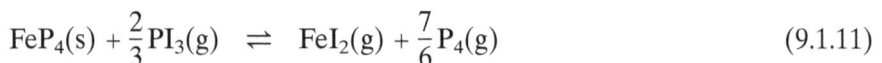

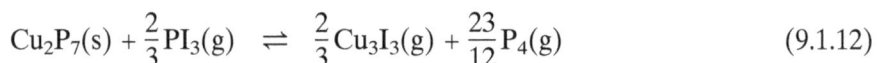

$$VP_4(s) + \frac{4}{3}PI_3(g) \;\rightleftharpoons\; VI_4(g) + \frac{16}{3}P_4(g) \qquad (9.1.10)$$

$$FeP_4(s) + \frac{2}{3}PI_3(g) \;\rightleftharpoons\; FeI_2(g) + \frac{7}{6}P_4(g) \qquad (9.1.11)$$

$$Cu_2P_7(s) + \frac{2}{3}PI_3(g) \;\rightleftharpoons\; \frac{2}{3}Cu_3I_3(g) + \frac{23}{12}P_4(g) \qquad (9.1.12)$$

Transport systems with phosphides as solids that have been described so far, show that the simple case of a single-phase equilibrium solid, identical for source and sink, which is present during the entire course of a transport experiment, is not always realized. As has been pointed out already, the reaction of a phosphide with iodine can lead to the formation of condensed metal iodides. The appearance of $VI_2(s)$ adjacent to VP (Gla 1989a); $CrI_2(l)$ adjacent to CrP (Gla 1989b); $MnI_2(l)$ adjacent to MnP (Gla 1989b); $CoI_2(l)$ adjacent to CoP (Schm 1995); and CuI(l) adjacent to Cu_3P and/or CuP_2 (Öza 1992) were examined in detail in this context. The heat of formation of the phosphide could be determined approximately from the amount of $VI_2(s)$ that appears adjacent to VP(s). In other cases, the verification of the relative thermodynamic stability of metal phosphides and iodides was possible with the help of the ratio of the amounts of substance in the solid. The transport balance, which is described in section 14.4, has proven of crucial value for these examinations.

The formation of very stable metal iodides, as described above, can lead to the development of phosphorus-rich phosphides (incongruent volatilization of phosphides) even without high metal activity in a phosphide. Thus in experiments with sufficiently high initial amounts of iodine TiP_2 (Mar 1988) adjacent to TiP (Uga 1978); ZrP_2 adjacent to ZrP; as well as CuP_2 adjacent to Cu_3P and CuI(l) appeared (equations 9.1.13 and 9.1.14).

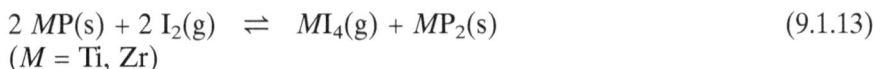

$$2\,MP(s) + 2\,I_2(g) \rightleftharpoons MI_4(g) + MP_2(s) \tag{9.1.13}$$
$$(M = Ti,\,Zr)$$

$$2\,Cu_3P(s) + \frac{5}{2}I_2(g) \rightleftharpoons 5\,CuI(l) + CuP_2(s) \tag{9.1.14}$$

The transport behavior of VP is enlightening for the understanding of the homogeneous gas-phase equilibrium between phosphorus and iodine, which is dependent on pressure and temperature. VP migrates due to the exothermic reaction 9.1.1 with iodine as transport agent. During experiments aiming at synthesis of VP_2 (800 °C, from the elements with iodine as mineralizer) vanadium monophosphide and diphosphide were present next to each other and a migration of both phosphides to T_1 could be observed. Under the experimental conditions PI_3 does not function as phosphorus-transferring species as in equation 9.1.1 or 9.1.5 instead it acts as the transport agent.

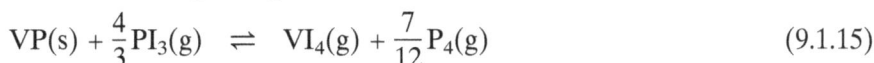

$$VP(s) + \frac{4}{3}PI_3(g) \rightleftharpoons VI_4(g) + \frac{7}{12}P_4(g) \tag{9.1.15}$$

The relation to the transport behavior of silicides (Chapter 10) and of elemental silicon is obvious (see section 3.3). CuP_2 can migrate in a temperature gradient via both exothermic and endothermic reactions (Öza 199). In the latter case, copper(I)-iodide and phosphorus vapor develop besides PI_3 and P_2I_4 during setting of the equilibrium between the starting materials CuP_2 and iodine. Consequently, evaporation of the remaining CuP_2 occurs with PI_3 as transport agent by the following endothermic reaction.

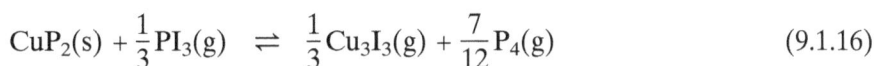

$$CuP_2(s) + \frac{1}{3}PI_3(g) \rightleftharpoons \frac{1}{3}Cu_3I_3(g) + \frac{7}{12}P_4(g) \tag{9.1.16}$$

Finally, we want to give some experimental advice. Often a comparatively small transport effect is observed, sometimes none at all, if a temperature gradient is applied immediately after the *in situ* formation of phosphides from metal and phosphorus in a silica ampoule with iodine as mineralizer and transport agent. If, in contrast, a previously synthesized phosphide is used as the source solid, the transport rates will be higher and well reproducible. This is due to two reasons. Traces of moisture, which were brought in by phosphorus and iodine, react with the phosphide to form phosphates. This reaction is accompanied by the formation of hydrogen. This way, the iodine, which is used as transport agent, is bound in the form of HI(g), which is less favorable for the transport of phosphides. This effect occurs at low to medium concentrations of the transport agent iodine of ≤ 5 mg $\cdot$ cm^{-3}. By this, the experimental transport rates are drastically decreased in contrast to what is expected in a completely dry system.

It can also not be excluded that oxygen, which can get into the ampoule in different ways (traces of moisture, partial oxidation of the metals used), influences the transport of phosphides in a negative way. This is of particular importance if for a phosphide, by means of iodine, two different transport reactions become possible: an exothermic transport for the pure phosphide and an endothermic transport in the presence of a phosphate. The latter makes it migrate simultaneously with the phosphate to the less hot side of the ampoule in a coupled transport reaction. In section 6.2, examples of this behavior are provided.

The synthesis of phosphides starting from metal oxides is very convenient. In a "one-pot reaction", aluminum (Mar 1990) or an excess of phosphorus (Mat 1991) can serve as a reducing agent.

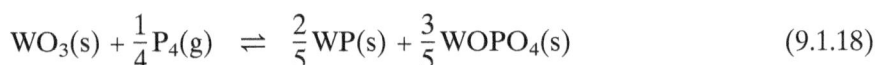

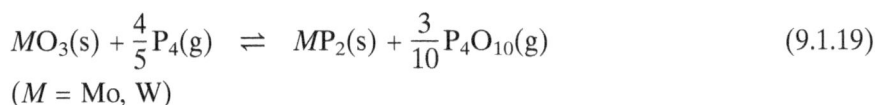

$$WO_3(s) + \frac{1}{4}P_4(g) + 2\,Al(l) \;\rightleftharpoons\; WP(s) + Al_2O_3(s) \tag{9.1.17}$$

$$WO_3(s) + \frac{1}{4}P_4(g) \;\rightleftharpoons\; \frac{2}{5}WP(s) + \frac{3}{5}WOPO_4(s) \tag{9.1.18}$$

$$MO_3(s) + \frac{4}{5}P_4(g) \;\rightleftharpoons\; MP_2(s) + \frac{3}{10}P_4O_{10}(g) \tag{9.1.19}$$
$$(M = Mo,\ W)$$

The separation of the phosphides from the other reaction products can be achieved by CVT.

The crystallization of CoP_2 is an example of the deposition of a *metastable* compound from the gas phase (see section 2.9). Isothermal heating of equimolar amounts of CoP and CoP_3 leads only in exceptional cases to the formation of some CoP_2. In contrast, higher amounts of CoP_2 are obtained in transport experiments when CoP and CoP_3 are used as starting materials (Jei 1984, Schm 1992). The diphosphide can be deposited as a single phase in the sink if an unusually high temperature gradient $\Delta T = 200$ K is applied (Schm 2002, Sel 1972, Jei 1984, Flö 1983).

In contrast to the above discussed transport reactions of phosphides with halogens or halogen compounds, the transport of InP and GaP succeeds by adding an excess of phosphorus. *Ab initio* calculation of the stability of different gas species in the system are indicating the formation of $MP_5(g)$ (*M*: In, Ga) (Köp 2003). Thus the following transport reactions are likely:

$$MP(s) + P_4(g) \rightleftharpoons MP_5(g) \qquad (9.1.20)$$
$$(M = In, Ga)$$

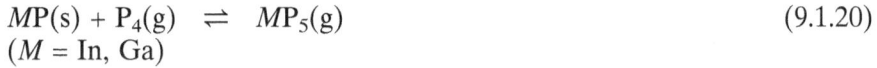

On the one hand, the thermal stability of this unusual gas species is surprising; on the other hand, this result corresponds to many observations from coordination chemistry where the P_5-fragment, which is isoelectronic to the cyclopentadienid anion, often appears as a ligand. $[P_5]^-$ and $[As_5]^-$ ligands, respectively, in complex compounds are also described in the literature (Sche 1999, Kra 2010).

During CVT experiments aiming at the crystallization of binary gold phosphides, a compound of the composition $Au_7P_{10}I$ was obtained in the sink. This is an example of a phosphide iodide. The transport agent iodine is incorporated in the sink solid during the reaction, and this way its partial pressure is gradually reduced with time (Bin 1978).

Table 9.1.1 Examples of the chemical vapor transport of phosphides

Sink solid	Transport additive	Temperature / °C	Reference
AlP	I_2	CVD	Sei 1976
AgZnREP$_2$ (RE = La, Sm)	I_2	$1000 \rightarrow T_1$	Tej 1990
Au$_2$P$_3$	Br$_2$	$930 \rightarrow 830$	Jei 1979
Au$_7$P$_{10}$I	I_2	$700 \rightarrow 650$	Bin 1978, Jei 1979
BP	I_2	$1100 \rightarrow 900$	Bou 1976, Nis 1972
	S, Se	$1145 \rightarrow 1065 \dots 880$	Arm 1967, Med 1969, Kic 1967
	BI$_3$	$1150 \rightarrow 1050$	Med 1967
	PCl$_3$	$1150 \rightarrow T_1$	Chu 1972
Cd$_3$P$_2$	decomposition sublimation	$700 \rightarrow 450$	Laz 1974, Klo 1984
CdP$_2$	decomposition sublimation	$800 \rightarrow T_1$	Laz 1977
Cd$_4$P$_2$$X_3$ (X = Cl, Br, I)	X_2	$550 \rightarrow 530$	Suc 1963
CdSiP$_2$	CdCl$_2$, I$_2$, SiCl$_4$	$1200 \dots 1100 \rightarrow$ $1000 \dots 950$	Val 1967, Val 1968, Bue 1971
CdGeP$_2$	CdCl$_2$, I$_2$, PCl$_3$	$780 \dots 750 \rightarrow 755 \dots 720$	Mio 1980, Süs 1982
CeSiP$_3$	I_2	$800 \rightarrow 1000$	Hay 1975
CeSi$_2$P$_6$	I_2	$800 \rightarrow 1000$	Hay 1975
Co$_2$P	I_2	$1050 \rightarrow 950$	Schm 2002
CoP	I_2	$850 \rightarrow 750$	Schm 1995, Schm 2002
CoP$_2$	I_2	$850 \rightarrow 650$	Schm 1992
CoP$_3$	I_2	$1000 \rightarrow 900$	Schm 2002
	Br$_2$	$900 \rightarrow 840$	Ric 1977
	Cl$_2$	$900 \rightarrow 840$	Ack 1977
CrP	I_2	$1050 \rightarrow 950$	Gla 1989b
Cu$_3$P	I_2	$800 \rightarrow 900$	Öza 1992
CuP$_2$	I_2	$750 \rightarrow 650$	Guk 1972, Öza 1992
	CuI	$730 \rightarrow 630$	Öza 1992
	Cl$_2$	$810 \rightarrow 760$	Öza 1992 Odi 1978

Table 9.1.1 (continued)

Sink solid	Transport additive	Temperature / °C	Reference
Cu_2P_7	I_2	$700 \dots 550 \rightarrow T_1$	Möl 1982
	$P + I_2$	$580 \rightarrow 530$	Öza 1992
$CuZnSmP_2$	I_2	$1000 \rightarrow T_1$	Tej 1990
Fe_2P	$HgBr_2$	$1000 \rightarrow 900$	Cze 1999
FeP	I_2	$800 \rightarrow 550$	Sel 1972,
			Ric 1977,
			Nol 1980,
			Bel 1973,
			Gla 1990
FeP_2	I_2	$800 \rightarrow 700$	Bod 1971,
			Flö 1983
FeP_4	I_2	$700 \rightarrow 600$	Flö 1983
DyP	I_2	$800 \rightarrow 1000$	Kal 1971
GaP	P_4	$900 \rightarrow 800$	Köp 2003
	H_2O	$1100 \rightarrow T_1$	Dja 1965,
			Nic 1963,
			Fro 1964
	I_2	$1050 \rightarrow 950$	Wid 1971,
			Fai 1978
	sublimation	$1060 \rightarrow 975$	Ger 1961
GdP	I_2	$800 \rightarrow 1000$	Kal 1971
InP	P_4	$950 \rightarrow 850$	Köp 2003
	I_2	$800 \rightarrow 700$	Nic 1974,
			Tha 2006
	PCl_3	$730 \rightarrow 630$	Nic 1974
HoP	I_2	$800 \rightarrow 1000$	Kal 1971
$LaSiP_3$	I_2	$800 \rightarrow 1000$	Hay 1975
$LaSi_2P_6$	I_2	$800 \rightarrow 1000$	Hay 1975
MnP	I_2	$1000 \rightarrow 1100$	Gla 1989b,
			Gla 1990
MnP_4	I_2	$700 \dots 500 \rightarrow 600 \dots 450$	Rüh 1981
Mo_3P	$HgBr_2$	$1100 \dots 900 \rightarrow 1000 \dots 800$	Len 1995,
			Len 1997
Mo_4P_3	$HgBr_2$	$1050 \dots 900 \rightarrow 950 \dots 800$	Len 1995,
			Len 1997
MoP	$I_2 + O_2$	$1000 \rightarrow 900$	Mar 1988,
			Len 1995,
			Len 1997

Table 9.1.1 (continued)

Sink solid	Transport additive	Temperature / °C	Reference
MoP	$HgBr_2$	$1000 \rightarrow 900$	Len 1995, Len 1997
MoP_2	$I_2 + O_2$, $HgCl_2 + O_2$	$1000 \rightarrow 900$	Mat 1990
NbP	I_2	$850 \rightarrow 950$	Mar 1988a
Ni_5P_4	I_2	$900 \rightarrow 800$	Blu 1997
PrP	I_2	$800 \rightarrow 1000$	Mir 1968, Hay 1975
PrP_2	I_2	$800 \rightarrow 1000$	Hay 1975
PrP_5	$P + I_2$	$800 \rightarrow 1000$	Hay 1975
$PrSiP_3$	I_2	$800 \rightarrow 1000$	Hay 1975
Re_6P_{13}	I_2	$930 \rightarrow T_1$	Rüh 1980
SiP	I_2	not specified	Wad 1969
SiP_2	I_2, Br_2, Cl_2	600 ... 1200 crystallization at 800	Don 1968
TaP	I_2	$850 \rightarrow 950$	Mar 1988a
Th_3P_4	Br_2	$\sim 500 \rightarrow 1250$	Hen 1977, Mar 1990
TiP	I_2	$800 \rightarrow 900$	Gla 1990, Uga 1978
TiP_2	$P + I_2$	$650 \rightarrow 700$	Mar 1988a
UP_2	I_2, Br_2	$T_1 \rightarrow T_2$	Hen 1968
U_3P_4	I_2, Br_2	$990 \rightarrow 1040$	Hen 1968, Buh 1969, Mar 1990
VP	I_2	$810 \rightarrow 930$	Gla 1989a
	PI_3	$800 \rightarrow 700$	Gla 2000
VP_2	I_2	$800 \rightarrow 700$	Mar 1988a
VP_4	I_2	$700 \rightarrow 600$	Mar 1988b
WP	$I_2 + O_2$	$1000 \rightarrow 900$	Mar 1988a
WP_2	$I_2 + O_2$	$1000 \rightarrow 900$	Mat 1990
Zn_3P_2	decomposition sublimation	$900 \rightarrow 450$	Klo 1984
	I_2	860 ... 830 $\rightarrow$ 720 ... 700	Wan 1981

Table 9.1.1 (continued)

Sink solid	Transport additive	Temperature / °C	Reference
ZnP_2	decomposition sublimation	$1050 \rightarrow T_1$	Laz 1977
$ZnGeP_2$	$GeCl_4$, $ZnCl_2$	$1200 \rightarrow 800$	Bue 1971, Bau 1978, Win 1977
$ZnSiP_2$	$SiCl_4$, $ZnCl_2$	$1200 \rightarrow 800$	Bue 1971, Bau 1978, Win 1977
Zn_3SmP_3	I_2	$1000 \rightarrow T_1$	Tej 1995
ZrP	I_2	$950 \rightarrow 1050$	Mar 1988a
ZrP_2	I_2	$850 \rightarrow 900$	Mar 1988a

Bibliography of Section 9.1

Ack 1977   J. Ackermann, A. Wold, *J. Phys. Chem. Solids* **1977**, *38*, 1013.
Arm 1967   A. F. Armington, *J. Cryst. Growth* **1967**, *1*, 47.
Bel 1973   D. W. Bellavance, A. Wold, *Inorg. Synth.* **1973**, *14*, 176.
Bin 1978   M. Binnewies, *Z. Naturforsch.* **1978**, *33b*, 570.
Blu 1997   M. Blum, *Diplomarbeit*, University of Gießen, **1997**.
Bod 1971   G. Boda, B. Strenström, V. Sagredo, D. Beckman, *Phys. Scr.* **1971**, *4*, 132.
Bou 1976   J. Bouix, R. Hillel, *J. Less-Common Met.* **1976**, *47*, 67.
Bau 1978   H. Baum, K. Winkler, *Krist. Tech.* **1978**, *13*, 645.
Bra 1978   D. J. Braun, W. Jeitschko, *Z. Anorg. Allg. Chem.* **1978**, *445*, 157.
Bue 1971   E. Buehler, J. H. Wernick, *J. Cryst. Growth* **1971**, *8*, 324.
Buh 1969   C. F. Buhrer, *J. Phys. Chem. Solids* **1969**, *30*, 1273.
Chu 1972   T. L. Chu, J. M. Jackson, R. K. Smeltzer, *J. Cryst. Growth* **1972**, *15*, 254.
Cze 1999   K. Czekay, R. Glaum, *unpublished results*, University of Gießen, **1999.**
Dja 1965   L. I. Djakonov, A. V. Lisina, V. N. Maslov, A. J. Naselskij, B. A. Sacharov, *Izv. Akad. Nauk. SSSR, Neorg. Mater.* **1965**, *1*, 2154.
Don 1968   P. C. Donohue, W. J. Siemons, J. L. Gillson, *J. Phys. Chem. Solids* **1968**, *29*, 807.
Fai 1978   P. S. Faile, *J. Cryst. Growth* **1978**, *43*, 129.
Fin 1965   A. Finch, P. J. Gardner, I. H. Wood, *J. Chem. Soc.* **1965**, 746.
Fin 1969   A. Finch, P. J. Gardner, K. K. SenGupta, *J. Chem. Soc.* (A) **1969**, 2958.
Fin 1970   A. Finch, P. J. Gardner, A. Hameed, *J. Inorg. Nucl. Chem.* **1970**, *32*, 2869.
Flö 1983   U. Flörke, *Z. Anorg. Allg. Chem.* **1983**, *502*, 218.
Fro 1964   C. J. Frosch, *J. Electrochem. Soc.* **1964**, *111*, 180.
Ger 1961   M. Gershenzon, R. M. Mikulyak, *J. Electrochem. Soc.* **1961**, *108*, 548.
Gla 1989a   R. Glaum, R. Gruehn, *Z. Anorg. Allg. Chem.* **1989**, *568*, 73.
Gla 1989b   R. Glaum, R. Gruehn, *Z. Anorg. Allg. Chem.* **1989**, *573*, 24.
Gla 1990   R. Glaum, *Dissertation*, University of Gießen **1990**.
Gla 1999   R. Glaum, *unpublished results*, University of Gießen, **1999**
Gla 2000   R. Glaum, R. Gruehn, *Angew. Chem.* **2000**, *112*, 706, *Angew. Chem. Int. Ed.* **2000**, *39*, 692.

Guk 1972   O. Ya. Gukov, Y. A. Ugai, W. R. Rshestanchik, V. Anokhin, *Izv. Akad. Nauk SSSR Neorg. Mater.* **1972**, *8*, 167.
Hay 1975   H. Hayakawa, T. Sekine, S. Ono, *J. Less-Common Met.* **1975**, *41*, 197.
Hen 1968   Z. Henkie, *Rocznicki Chemii* **1968**, *42*, 363.
Hen 1981   Z. Henkie, P. Markowski, *J. Cryst. Growth* **1977**, *41*, 303.
Hil 1973   R. Hillel, J.-M. Letoffe, J. Bouix, *J. Chim. Phys.* **1973**, *73*, 845.
Jei 1979   W. Jeitschko, M. Möller, *Acta Crystallogr.* **1979**, *B35*, 573.
Jei 1984   W. Jeitschko, U. Flörke, U. D. Scholz, *J. Solid State Chem.* **1984**, *52*, 320.
Kal 1971   E. Kaldis, *J. Cryst. Growth* **1971**, *9*, 281.
Klo 1984   K. Kloc, W. Zdanowicz, *J. Cryst. Growth* **1984**, *66*, 451.
Köp 2003   *Z. Anorg. Allg. Chem.* **2003**, *629*, 2168.
Kra 2010   H. Krauss , G. Balázs , M. Bodensteiner, M. Scheer, *Chem. Sci.*, **2010,** *1*, 337.
Laz 1974   V. B. Lazarev, V. J. Shevchenko, *J. Cryst. Growth* **1974**, *23*, 237.
Laz 1977   V. B. Lazarev, V. J. Shevchenko, S. F. Marenkin, G. Magomedgadghiev, *J. Cryst. Growth* **1977**, *38*, 275.
Len 1995   M. Lenz, *Dissertation*, University of Gießen, **1995**.
Len 1997   M. Lenz, R. Gruehn, *Chem. Rev.* **1997**, *97*, 2967.
Mar 1988a   J. Martin, R. Gruehn, *Z. Kristallogr.* **1988**, *182*, 180.
Mar 1988b   J. Martin, Diplomarbeit, University of Gießen, **1988**.
Mar 1990   J. Martin, R. Gruehn, *Solid State Ionics* **1990**, *43*, 19.
Mat 1990   H. Mathis, Diplomarbeit, University of Gießen, **1990**.
Mat 1991   H. Mathis, R. Glaum, R. Gruehn, *Acta Chem. Scand.* **1991**, *45*, 781.
Med 1969   Z. S. Medvedeva, J. H. Greenberg, E. G. Zhukov, *Krist. Tech.* **1969**, *4*, 487.
Mir 1968   K. E. Mironov, *J. Cryst. Growth* **1968**, *3,4*, 150.
Möl 1982   M. Möller, W. Jeitschko, *Z. Anorg. Allg. Chem.* **1982**, *491*, 225.
Nic 1963   F. H. Nicoll, *J. Electrochem. Soc.* **1963**, *110*, 1165.
Nic 1974   I. F. Nicolau, *Z. Anorg. Allg. Chem.* **1974**, *407*, 83.
Nis 1972   T. Nishinaga, H. Ogawa, H. Watanabe, T. Arizumi, *J. Cryst. Growth* **1972**, *13/14*, 346.
Nol 1980   B. I. Noläng, M. W. Richardson, *J. Chem. Soc. Spec. Publ.* **1980**, *3*, 75.
Odi 1978   J. P. Odile, S. Soled, C. A. Castro, A. Wold, *Inorg. Chem.* **1978**, *17*, 283.
Öza 1992   D. Özalp, *Dissertation*, University of Gießen, **1992**.
Ric 1977   M. W. Richardson, B. I. Noläng, *J. Cryst. Growth* **1977**, *42*, 90.
Rüh 1980   R. Rühl, W. Jeitschko, *Z. Anorg. Allg. Chem.* **1980**, *466*, 171.
Sch 1965   H. Schäfer, W. Fuhr, *J. Less-Common Met.* **1965**, *8*, 375.
Sch 1971   H. Schäfer, *J. Cryst. Growth* **1971**, *9*, 17.
Sche 1999   O. J. Scherer, *Acc. Chem. Res.* **1999,** *32*, 751.
Schm 1992   A. Schmidt, *Diplomarbeit*, University of Gießen, **1992**.
Schm 2002   A. Schmidt, *Dissertation*, University of Gießen, **1999**.
Schn 1988   H.-G. von Schnering, W. Hönle, *Chem. Rev.* **1988**, *88*, 243.
Sei 1976   E. Seidowski, S. O. Newiak, *Krist. Tech.* **1976**, *11*, 329.
Sel 1972   K. Selte, A. Kjekshus, *Acta Chem. Scand.* **1972**, *26*, 1276.
Süs 1982   B. Süss, K. Hein, E. Buhrig, H. Oettel, *Cryst. Res. Tech.* **1982**, *17*, 137.
Suc 1963   L. Suchow, N. R. Stemple, *J. Electrochem. Soc.* **1963**, *110*, 766.
Tej 1990   P. Tejedor, A. M. Stacy, *J. Cryst. Growth* **1990**, *89*, 227.
Tej 1995   P. Tejedor, F. J. Hollander, J. Fayos, A. M. Stacy, *J. Cryst. Growth* **1995**, *155*, 223.
Uga 1978   Y. A. Ugai, O. Y. Gukov, A. A. Illarionov, *Izv. Akad. Nauk SSSR Neorg. Mater.* **1978**, *14*, 1012.
Val 1967   Yu. A. Valov, R. L. Plečko, *Krist. Tech.* **1967**, *2*, 535.
Val 1968a   Yu. A. Valov, R. L. Plečko, *Izv. Akad. Nauk Neorg. Mater.* **1968**, *4*, 993.

Val 1968b   Yu. A. Valov, T. N. Ushakova, *Izv. Akad. Nauk Neorg. Mater.* **1968**, *4*, 1054.
Wad 1969a   T. Wadsten, *Acta chem. Scand.* **1969**, *23*, 331.
Wad 1969b   T. Wadsten, *Acta chem. Scand.* **1969**, *23*, 2532.
Wan 1981   F.-C. Wang, R. H. Bube, R. S. Feigelson, R. K. Route, *J. Cryst. Growth* **1981**, *55*, 268.
Wid 1971   R. Widmer, *J. Cryst. Growth* **1971**, *8*, 216.
Win 1977   K. Winkler, K. Hein, *Krist. Tech.* **1977**, *12*, 211.
Win 1978   K. Winkler, U. Schulz, K. Hein, *Krist. Tech.* **1978**, *13*, 137.

9.2 Transport of Arsenides

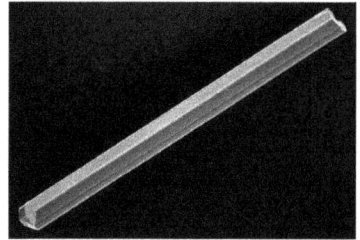

ZrAs$_2$

H																	He
Li	Be											B	C	N	O	F	Ne
Na	Mg											Al	Si	P	S	Cl	Ar
K	Ca	Sc	Ti	V	Cr	Mn	Fe	Co	Ni	Cu	Zn	Ga	Ge	As	Se	Br	Xe
Rb	Sr	Y	Zr	Nb	Mo	Tc	Ru	Rh	Pd	Ag	Cd	In	Sn	Sb	Te	I	Kr
Cs	Ba	La	Hf	Ta	W	Re	Os	Ir	Pt	Au	Hg	Tl	Pb	Bi	Po	At	Rn

Ce	Pr	Nd	Pm	Sm	Eu	Gd	Tb	Dy	Ho	Er	Tm	Yb	Lu
Th	Pa	U	Np	Pu									

Iodine as transport agent

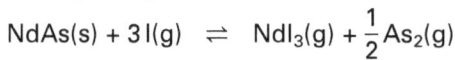

$$NdAs(s) + 3\,I(g) \;\rightleftharpoons\; NdI_3(g) + \frac{1}{2}As_2(g)$$

Hydrogen halides as transport agent

$$SiAs(s) + 4\,HI(g) \;\rightleftharpoons\; SiI_4(g) + \frac{1}{4}As_4(g) + 2\,H_2(g)$$

Water as transport agent

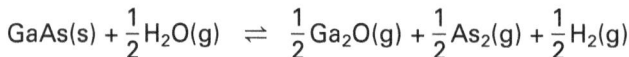

$$GaAs(s) + \frac{1}{2}H_2O(g) \;\rightleftharpoons\; \frac{1}{2}Ga_2O(g) + \frac{1}{2}As_2(g) + \frac{1}{2}H_2(g)$$

The CVT of arsenides is documented by many examples. The arsenides of group 13 are experimentally examined in a comprehensive manner and described by thermodynamic calculations because of the technical applications of gallium arsenide. Most of the other works deal with preparative aspects.

Almost all metal and semi-metal arsenides decompose below their melting temperature. In doing so, they release arsenic under formation of a metal-rich arsenide or the respective metal or semi-metal. The decomposition leads in most cases to gaseous arsenic.

A comparison of the *transport behavior of the pnictides* shows that transport of arsenides is similar that of phosphides but markedly different to those of the antimonides and bismutides. The reason for this is the relatively high saturation vapor pressure of phosphorus and arsenic at rather low temperatures: 1 bar at 277 °C and 602 °C, respectively. Hence phosphorus as well as arsenic can be transferred to the gas phase in considerable amounts at relatively low temperatures without exceeding the saturation vapor pressure and thus condensing again. The saturation vapor pressure of antimony, in contrast, reaches the value of 1 bar at 1585 °C. Because of the low saturation pressure, condensation occurs already at low concentrations of elemental antimony so that antimony in elemental form cannot be transferred with transport-effective partial pressure via the gas phase.

As far as the thermodynamic stability of the pnictides is concerned, phosphides and arsenides are similar as well. This is shown by a comparison of the standard heat of formation of the uranium pnictides. The differences between the standard heats of formation of phosphides and arsenides are lower than those between arsenides and antimonides.

Table 9.2.1 Comparison of the standard heats of formation of uranium pnictides $(kJ \cdot mol^{-1})$.

UN	−294.6	UN_2	−431.0
UP	−262.3	UP_2	−292.9
UAs	−234.3	UAs_2	−251.0
USb	−138.5	USb_2	−176.0
UBi	−73.6	UBi_2	−102.5

Transport agent   A number of transport additives have proven of value for the transport of arsenides. The most important are halogens, in particular iodine. Halogens do not always work directly as transport agents, but the halides formed from them can also be used as transport additives. Hydrogen halides are of minor importance. Water and elemental arsenic can cause a transport effect, too.

Halogens as transport additives: The transport of arsenides occurs most often when halogens or halides, respectively, are added. About 50 % of the transport experiments that are described in the literature take place by adding iodine or iodides, respectively. This way, balanced equilibrium positions are achieved with the solid arsenide and the transport agent on one side and their gaseous reaction products on the other. If halides are used as transport additives, it will be arsenic halides AsX_3 (X = Cl, Br, I) or those that contain a component of the solid as a central atom. There are only two exceptions: the transport of ZnSiAs$_2$ and that of CdSiAs$_2$ with tellurium(IV)-chloride.

Volatile metal halides as well as volatile arsenic halides – especially the trihalides – are formed during the transport of arsenides with halides. The trihalides of arsenic are less stable than those of phosphorus, which is proven by the comparison of the standard enthalpies of formation. If the transport of an arsenide is compared to the corresponding phosphide, the part of the transport agent that is present in the form of pnictogen halide, is clearly lower for the arsenide than for the phosphide. As far as transport experiments including phosphides are concerned, it can happen that the transport agent is consumed completely for the formation of phosphorus halides so that gaseous metal halides cannot be formed. In this case, transport is impossible. Such a situation is not observed for arsenides.

Table 9.2.2 Comparison of the standard enthalpies of formation of gaseous pnictogen(III)-halides (kJ · mol^{-1}); data according to *Binnewies* and *Milke* (Bin 2002).

PCl$_3$	−288.7	PI$_3$	−18.0
AsCl$_3$	−261.5	AsI$_3$	8.9
SbCl$_3$	−313.1	SbI$_3$	6.7
BiCl$_3$	−265.3	BiI$_3$	−16.3

Thermodynamic model calculations showed that when halogens are used as *transport additives* they are not necessarily the effective *transport agents*. Often the arsenic trihalides or the metal or semi-metal halides, respectively, which are formed from the halogens and the solids, function as such. For example, the transport of neodymium arsenide, NdAs, when iodine is added (950 → 1080 °C), can be described with the help of the following transport equation:

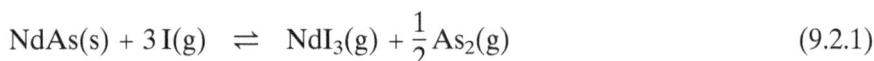

$$NdAs(s) + 3\,I(g) \; \rightleftharpoons \; NdI_3(g) + \frac{1}{2}As_2(g) \tag{9.2.1}$$

Contrary to this, model calculations prove that the transport of iron arsenide with added iodine (800 → 765 °C) takes place via the following equilibrium:

$$FeAs_2(s) + AsI_3(g) \; \rightleftharpoons \; FeI_2(g) + \frac{3}{4}As_4(g) + \frac{1}{2}I_2(g) \tag{9.2.2}$$

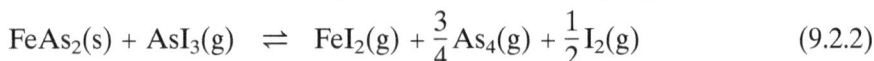

In this case, iodine is added but arsenic(III)-iodide is the effective transport agent.

The transport of gallium arsenide, GaAs, with iodine is endothermic (e.g., from 900 to 850 °C) and is based on the following equilibrium:

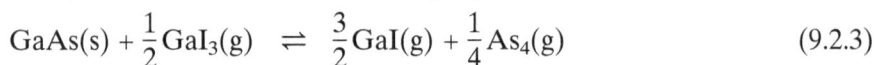

$$GaAs(s) + \frac{1}{2}GaI_3(g) \; \rightleftharpoons \; \frac{3}{2}GaI(g) + \frac{1}{4}As_4(g) \tag{9.2.3}$$

Arsenic is transferred into the gas phase mainly in elemental form due to the high saturation pressure and the comparatively low stability of gaseous arsenic iodides. Up to approximately 900 to 1000 °C the gas phase is mostly dominated by As$_4$, above that temperature, by As$_2$. Apart from these two species, As$_3$ and As appear in the gas phase. However, they are of minor importance to the CVT.

Apart from the mentioned arsenic species, the arsenic monohalides are to be mentioned, which can be formed by the following decomposition equilibrium:

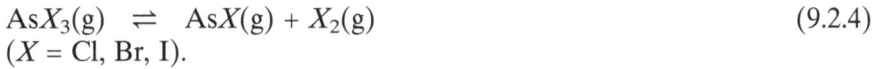

$$AsX_3(g) \rightleftharpoons AsX(g) + X_2(g) \qquad (9.2.4)$$
$$(X = Cl, Br, I).$$

The tendency toward the formation of arsenic monohalides increases at rising temperatures and from chloride to iodide.

During the transport of arsenides, mostly $As_4(g)$ and $As_2(g)$, respectively, are the transport-effective, arsenic-transferring species. In order to describe the transport processes completely, the arsenic halides, especially the arsenic trihalides, have to be considered additionally.

Hydrogen halides as transport agents: Hydrogen halides, hydrogen chloride in particular, are only important for the transport of arsenides of group 13 (BAs, GaAs, and InAs). Only one example is known in which the role of the hydrogen halides for CVT is discussed. *Bolte* and *Gruehn* described the CVT of SiAs with iodine (Bolt 1994). With the help of thermodynamic calculations and experiments, this work shows that, at low temperatures (750 → 850 °C), exothermic transport can be traced back to HI as transport agent, which is formed by traces of water desorbed off the ampoule walls. The following transport equilibrium can be formulated:

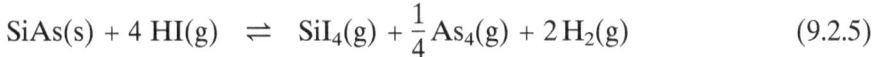

$$SiAs(s) + 4\,HI(g) \rightleftharpoons SiI_4(g) + \frac{1}{4}As_4(g) + 2\,H_2(g) \qquad (9.2.5)$$

Contrary to this, SiAs transport occurs endothermically from 1050 to 950 °C by rigorously *excluding* moisture and using iodine as the transport agent. The migration can be described by the following equilibrium:

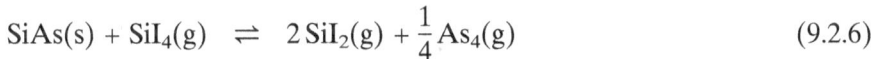

$$SiAs(s) + SiI_4(g) \rightleftharpoons 2\,SiI_2(g) + \frac{1}{4}As_4(g) \qquad (9.2.6)$$

SiI_4 functions as the transport agent. The experiment shows that traces of residual moisture can influence the transport action decisively: in this case, a reverse of the transport direction can occur.

The transport of gallium arsenide with hydrogen chloride and hydrogen bromide, respectively, is well investigated experimentally and by thermodynamic calculations. The transport is described in good approximation by the following equation:

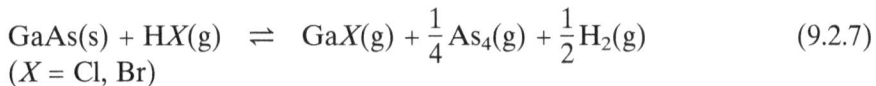

$$GaAs(s) + HX(g) \rightleftharpoons GaX(g) + \frac{1}{4}As_4(g) + \frac{1}{2}H_2(g) \qquad (9.2.7)$$
$$(X = Cl, Br)$$

During the transport of arsenides, a further arsenic-containing gas species, AsH_3, may have to be considered in the presence of hydrogen or hydrogen compounds.

Water as transport agent: GaAs, InAs, $Ga_{1-x}In_xAs$ and $InAs_{1-x}P_x$ can be transported with water. The transport occurs via the following equilibrium:

$$GaAs(s) + \frac{1}{2}H_2O(g) \rightleftharpoons \frac{1}{2}Ga_2O(g) + \frac{1}{2}As_2(g) + \frac{1}{2}H_2(g) \qquad (9.2.8)$$

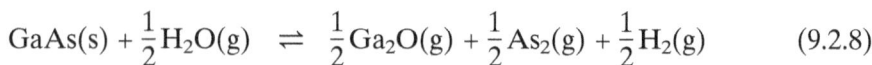

Transport of the other mentioned compounds can be described in a similar way. The transport reaction is always coupled with a redox equilibrium in which a gaseous suboxide (Coc 1962), arsenic, and hydrogen are formed. In addition, GaAs can be transported with a mixture of water and hydrogen. In the process, exothermic transport behavior (e.g., $800 \rightarrow 1070\,°C$) is observed. The mentioned transport agents are used especially in open systems with flowing gases (Fro 1964).

Arsenic as transport agent: The CVT of gallium arsenide GaAs and indium arsenide InAs, respectively, with arsenic as transport agent is unusual and therefore noteworthy. The transport effect, which has been proven by *Köppe* and *Schnöckel* experimentally, is, according to their quantum chemical considerations and thermodynamic calculations, based on the formation of gaseous $GaAs_5$ and $InAs_5$ molecules, respectively (Köp 2004). The transport occurs from 940 to 840 °C at arsenic pressures between 0.88 and 5.0 bar via the following equilibrium:

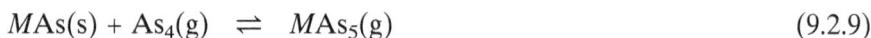

$$MAs(s) + As_4(g) \rightleftharpoons MAs_5(g) \qquad (9.2.9)$$

As calculations show, there is a further gallium- and indium-transferring gas species, respectively, of the composition MAs_3. However, it reaches transport-effective partial pressures only above 1000 °C.

Selected examples In the following, the CVT of arsenides shall be visualized with the help of selected examples.

Binary arsenides of the main group elements: Due to the technical importance, one finds in the literature many studies on the transport of GaAs. This can be transported endothermically in a wide temperature range with halogens or hydrogen halides (Fak 1973, Fak 1979, Gar 1978, Pas 1983). Arsenic(III)-chloride or gallium(III)-iodide are other halogenating additives that allow transport. Most authors assume that the transport occurs via one of the two following equilibria when halogens or the trihalides of gallium or arsenic, respectively, are added:

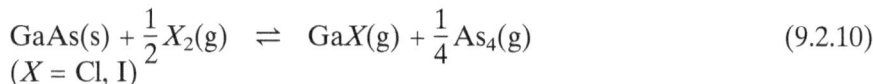

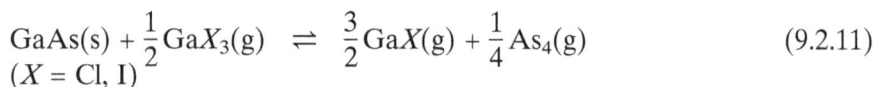

$$GaAs(s) + \frac{1}{2}X_2(g) \rightleftharpoons GaX(g) + \frac{1}{4}As_4(g) \qquad (9.2.10)$$
$$(X = Cl, I)$$

$$GaAs(s) + \frac{1}{2}GaX_3(g) \rightleftharpoons \frac{3}{2}GaX(g) + \frac{1}{4}As_4(g) \qquad (9.2.11)$$
$$(X = Cl, I)$$

Because in both equilibria GaX as well as As_4 are formed, the arsenic pressure in the system can be reduced according to *Le Chatelier*'s principle by adding the respective gallium halide GaX in exchange for a part of the halogen. This allows the transport of GaAs under defined arsenic partial pressures. In practice, the substance amount ratio is varied (Fak 1971, Hit 1994).

The transport of indium arsenide, InAs, can be described via analogous equilibria, although the equilibrium position slightly changes due to the decreasing stability of the trihalides from gallium to indium (see Chapter 11) (Cah 1971, Sah 1982, Kao 2004, Bol 2008). In some publications, the transport of other monoarsenides of group 13 elements is described: BAs (Arm 1967, Bou 1976, Bou 1977), AlAs (Vor 1972, Hil 1977), $Ga_xIn_{1-x}As$ mixed-crystals (Cah 1982).

BAs can be transported due to exothermic reactions by adding bromine, iodine, and arsenic(III) iodide. *Bouix* and *Hillel* (Bou 1976, Bou 1977) investigated and described the transport of compounds of the composition $A^{III}B^V$ (A = B, Al, Ga, In; B = P, As). Here the emphasis is put on the analysis of the gas-phase composition during the transport of these compounds with iodine and iodine-containing transport agents, respectively, by Raman spectroscopy. Based on these examinations, the transport of BAs with iodine or arsenic(III) iodide, respectively, the following transport equation was derived:

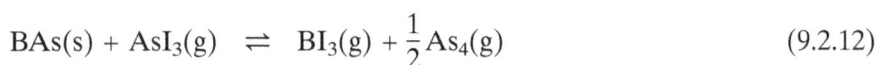

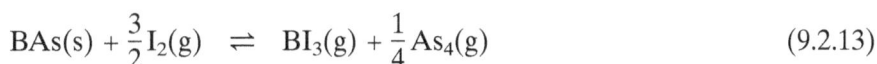

$$BAs(s) + AsI_3(g) \rightleftharpoons BI_3(g) + \frac{1}{2}As_4(g) \tag{9.2.12}$$

$$BAs(s) + \frac{3}{2}I_2(g) \rightleftharpoons BI_3(g) + \frac{1}{4}As_4(g) \tag{9.2.13}$$

Apart from BAs, B_6As (Rad 1978) and $B_{12}As_2$ (Bec 1980] can be transported with iodine as well. The transport of B_6As from 1050 to 950 °C when tellurium is added is particularly noteworthy. Here a volatile boron telluride is formed, comparable to the transport of elemental boron with selenium.

Binary arsenides of the transition elements: In the literature, the transport of a series of transition metal arsenides with added halogens, especially iodine, is described. Examples are arsenides of zirconium, niobium, tantalum, molybdenum, tungsten, cobalt, nickel, and palladium. A thermodynamic analysis of the transport behavior of the systems niobium/arsenic and molybdenum/arsenic was conducted: *Schäfer* describes and explains the transport behavior of niobium monoarsenide, NbAs, and niobium diarsenide, $NbAs_2$, (Sch 1965). In this context, comparisons of the CVT of sulfides, selenides, and tellurides as well as arsenides and antimonides are made and generalized on the basis of the observed transport behavior of the respective niobium compounds. NbAs can be transported by exothermic reaction from 940 to 1065 °C by adding iodine. In contrast, the transport of $NbAs_2$ with iodine occurs due to endothermic reaction from 1045 to 855 °C. This reversal of the transport direction with decreasing metal portion applies for the niobium antimonides, niobium sulfides, niobium selenides, and niobium tellurides as well. It is explained in section 7.2 with the example of niobium selenides.

In contrast to the transport of chalcogenides, the formation of pnictogen halides plays an important role in the case of the pnictides. The transport of $FeAs_2$ is an example of this. It can be transported from 800 to 765 °C by adding chlorine or iodine (Fan 1972, Ros 1972, Lut 1987). The values of the transport efficiency of the individual gas species in the iron/arsenic/chlorine system and the iron/arsenic/iodine system, respectively, which were shown by the thermodynamic calculations, demonstrate that not the added halogens but the arsenic trihalides formed from them function as transport agent:

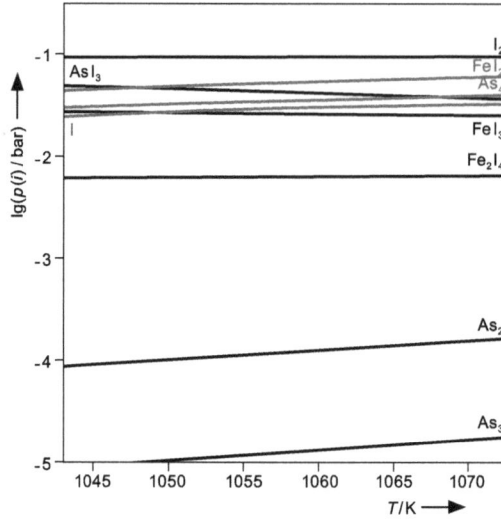

Figure 9.2.1 Gas-phase composition in the $FeAs_2/I_2$ system.

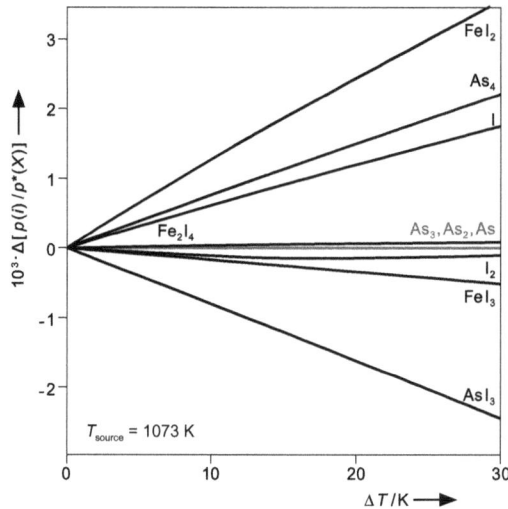

Figure 9.2.2 Transport efficiency of the essential gas species in the $FeAs_2/I_2$ system.

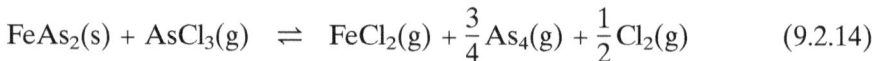

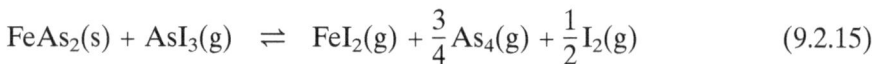

$$FeAs_2(s) + AsCl_3(g) \rightleftharpoons FeCl_2(g) + \frac{3}{4}As_4(g) + \frac{1}{2}Cl_2(g) \qquad (9.2.14)$$

$$FeAs_2(s) + AsI_3(g) \rightleftharpoons FeI_2(g) + \frac{3}{4}As_4(g) + \frac{1}{2}I_2(g) \qquad (9.2.15)$$

The CVT of molybdenum arsenides is also particularly well examined. *Murray et al.* published data of the transport of $MoAs_2$, $MoAs_3$, and Mo_5As_4 with chlorine and bromine in particular (Mur 1972). The gas phases were analyzed in detail by total pressure measurements and UV spectroscopy. At temperatures around 725 °C, the gas phase is dominated by $AsBr_3$. Molybdenum bromides were detec-

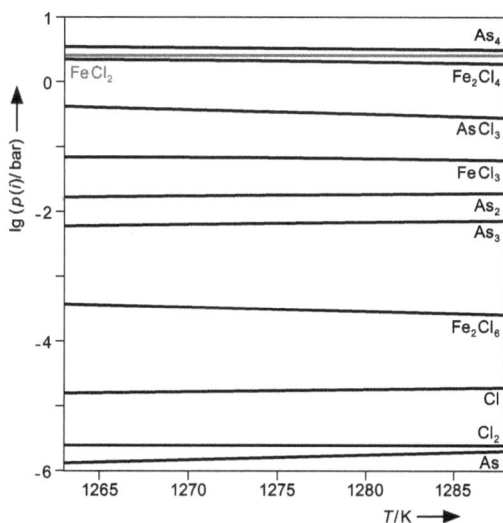

Figure 9.2.3 Gas-phase composition in the $FeAs_2/Cl_2$ system.

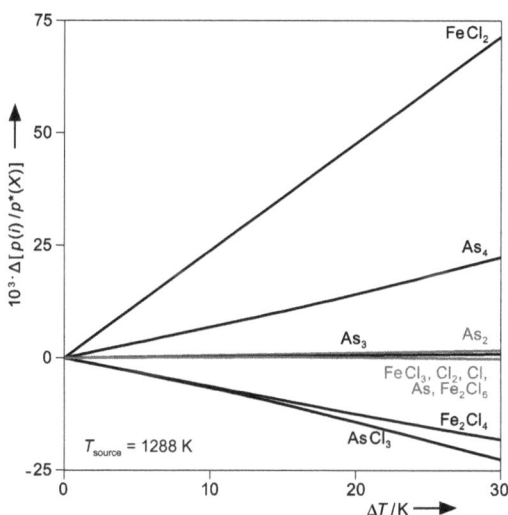

Figure 9.2.4 Transport efficiency of the essential gas species in the $FeAs_2/Cl_2$ system.

ted in small amounts. Furthermore, the examination showed that the molybdenum-containing species increase with an increasing amount of molybdenum in the solid. $MoBr_4$ was identified as the essential molybdenum-transferring species. In the Mo_5As_4/Br_2 system, gaseous $MoBr_2$ was described as well.

Rare earth metal arsenides: Apart from the transition metal arsenides, a series of rare earth metal arsenides can be transported as well. Transport always occurs via halogenating transport equilibria – usually with iodine as the transport agent. The transport of ytterbium arsenide, YbAs, with iodine from 1025 to 900 °C is an example of this (Kha 1974). The following transport equation is assumed:

$$YbAs(s) + I_2(g) \; \rightleftharpoons \; YbI_2(g) + \frac{1}{2}As_2(g) \qquad (9.2.16)$$

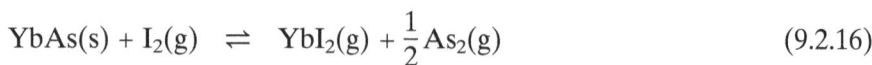

YbI_3 instead of YbI_2 is assumed to be the ytterbium-transferring species. The relation between the iodine pressure and the amount of substance transported is noteworthy: particularly high transport rates for iodine pressures of 0.5 and 3 bar are observed. The transport rate at 1.5 bar is notably low.

The exothermic CVT of the neodymium arsenide NdAs (e.g., 985 → 1080 °C) is exemplary for the transport of the other rare earth metal monoarsenides. It shows that here the added iodine functions as the transport agent. The gas phase over a NdAs solid in equilibrium with iodine is dominated by the gas species NdI_3, I, I_2, and As_2 in the given temperature range. The partial pressures of As_4, As_3, As, and AsI_3 are below 10^{-5} bar. The transport can be described in good approximation with the help of the following equation as the values of the calculated transport efficiency of the individual gas species show:

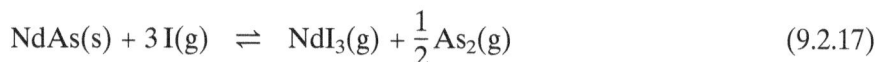

$$NdAs(s) + 3\,I(g) \; \rightleftharpoons \; NdI_3(g) + \frac{1}{2}As_2(g) \qquad (9.2.17)$$

$\Delta_r H^0_{1300} = -272.1 \text{ kJ} \cdot \text{mol}^{-1}$, $\Delta_r S^0_{1300} = -114.1 \text{ J} \cdot \text{mol}^{-1} \cdot \text{K}^{-1}$
$\Delta_r G^0_{1300} = -123.0 \text{ kJ} \cdot \text{mol}^{-1}$

The sign of the heat of reaction is in accordance with the observed transport direction. The Gibbs energy is in a range where low transport rates are expected.

Ternary compounds: Apart from a number of binary arsenides (see Table 9.2.1), the transport behavior of some ternary compounds were investigated and described as well. Among others, these are rare earth metal/coinage metal diarsenides, e.g., $CeAgAs_2$ (Dem 2004); semi-conducting cadmium arsenide halides of the composition $Cd_4As_2X_3$ (X = Cl, Br, I) (Suc 1963); clathrate compound

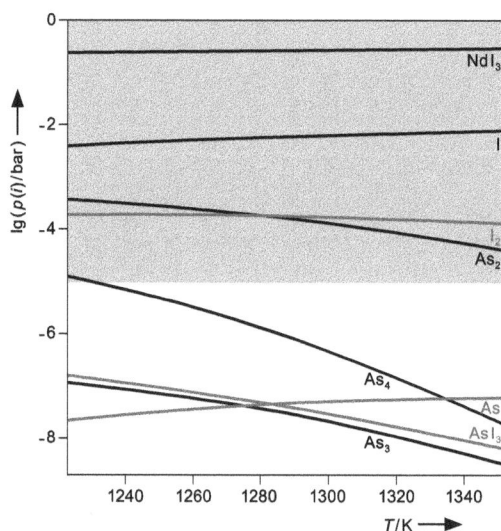

Figure 9.2.5 Gas-phase composition in the $NdAs/I_2$ system.

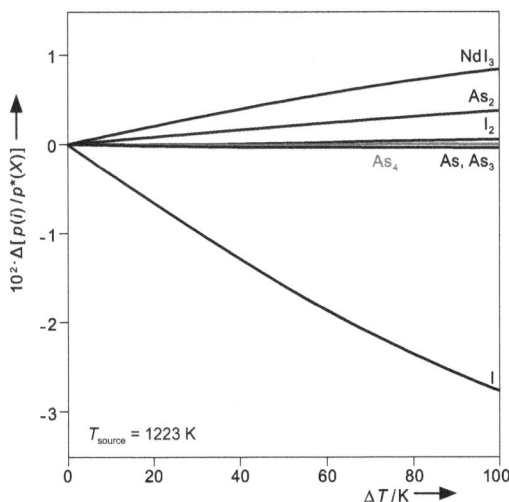

Figure 9.2.6 Transport efficiency of the essential gas species in the NdAs/I_2 system.

$Ge_{38}As_8I_8$ (vSc 1976); arsenide/phosphide mixed-crystals, e.g., $GaAs_{1-x}P_x$; rare earth metal monoarsenide mixed-crystals $RE_{1-x}RE'_xAs$ (Gol 2011); as well as a number of differently composed pnictide-chalcogenides (Hul 1968, Mam 1988, Hen 2001, Bru 2006, Czu 2010) and chalcopyrites of the composition $A^{II}B^{IV}As_2$ (A = Cd, Zn; B = Si, Ge) (Spr 1968, Win 1974, Avi 1984, Kim 1989, Wen 1997).

Due to the thermochemical characteristics of the *chalcopyrites* – they decompose below melting temperature – CVT as a preparative method is of special importance for the synthesis of single-crystals. The compound class of the chalcopyrites is of interest due to its semiconducting properties. *Kimmel* described and analyzed the thermochemical behavior and the transport of $CdSiAs_2$ with bromine or cadmium chloride (Kim 1998). Accordingly, the thermal decomposition of $CdSiAs_2$ occurs between 570 and 710 °C to solid SiAs, while cadmium and arsenic are released to the gas phase according to the following reaction equation:

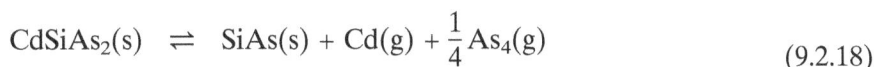

$$CdSiAs_2(s) \;\rightleftharpoons\; SiAs(s) + Cd(g) + \frac{1}{4}As_4(g) \tag{9.2.18}$$

Above 710 °C SiAs decomposes to solid silicon and gaseous arsenic:

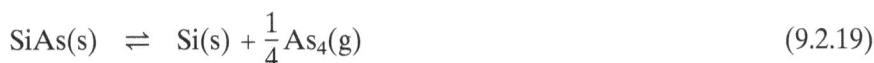

$$SiAs(s) \;\rightleftharpoons\; Si(s) + \frac{1}{4}As_4(g) \tag{9.2.19}$$

Because the endothermic transport reactions are conducted in the temperature range around 800 °C, one can assume that cadmium and arsenic are already to some extent gaseous. Due to its low volatility silicon, on the other hand, must be chemically dissolved into the gas phase by a suitable transport agent. Cadmium(II)-chloride, tellurium(IV)-chloride, as well as iodine are suitable transport additives apart from those already mentioned, especially bromine.

The following transport equation for $CdSiAs_2$ is derived from thermodynamic considerations:

$$CdSiAs_2(s) + SiX_4(g) \rightleftharpoons Cd(g) + 2SiX_2(g) + \frac{1}{2}As_4(g) \qquad (9.2.20)$$
$$(X = Cl, Br, I)$$

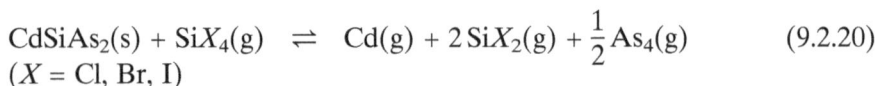

The considerations can be transferred to other compounds of the composition $A^{II}B^{IV}As_2$ (A = Cd, Zn; B = Si, Ge) as far as the transport with halogens and halides, respectively, is concerned. This is because zinc as well as cadmium have a sufficiently high transport-effective vapor pressure and germanium and silicon get to the gas phase only by adding a corresponding transport agent. The works of *Winkler*, who describes the transport behavior of chalcopyrites in detail, are particular important (Win 1974, Win 1977). The emphasis of these works is put on the optimization of crystal growth concerning the size, morphology, and quality of the crystals. In doing so, the influence of the transport additives $ZnCl_2$, $SiCl_4$, $PbCl_2$, and $TeCl_4$ as well as the concentration of the transport agent and the transport temperatures on crystal growth is examined. Optimum crystallization conditions for $ZnSiAs_2$ are achieved with the transport agents $PbCl_2$ and $ZnCl_2$, respectively, at temperatures around 1000 °C.

Finally, we will deal with the transport behavior of *pnictide chalcogenides* more closely because a number of representatives of this compound class can be crystallized by CVT reactions. These are the uranium arsenide chalcogenides (UAsS, UAsSe, UAsTe), which adopt the PbFCl structure type and the thorium arsenide chalcogenides (ThAsS, ThAsSe, ThAsTe) whose endothermic transport (e.g., 1000 → 900 °C) with bromine or iodine as transport agent is described by *Hulliger* (Hul 1968, Kac 1994, Nar 1998, Waw 2005, Kac 2005) and *Henkie et al.* (Hen 1968, Hen 2001). On the other hand, there are zirconium arsenide chalcogenides and hafnium arsenide chalcogenides and others of the composition $HfAs_{1.7}Se_{0.2}$, $ZrAs_{0.7}Se_{1.3}$, $ZrAs_{0.7}Te_{1.3}$, $ZrAs_{1.4}Se_{0.5}$, and $ZrAs_{1.6}Te_{0.4}$. The two latter phases show a wide homogeneity range. Transport takes place by adding iodine and is

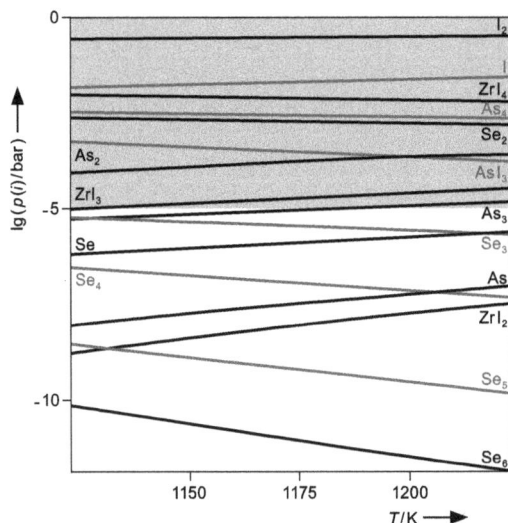

Figure 9.2.7 Gas-phase composition in the $ZrAs_{1.5}Se_{0.5}/I_2$ system according to *Czulucki* (Czu 2010).

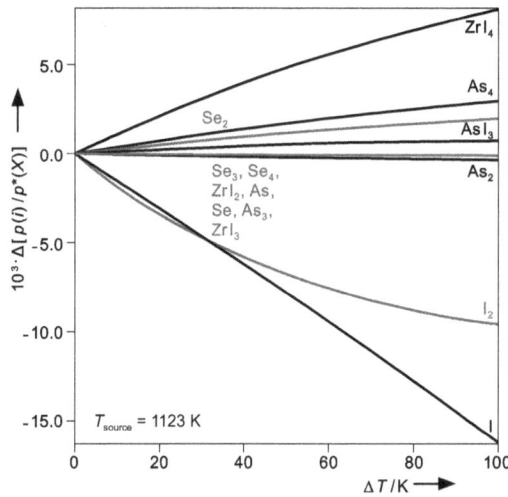

Figure 9.2.8 Transport efficiency of the essential gas species in the $ZrAs_{1.5}Se_{0.5}/I_2$ system according to *Czulucki* (Czu 2010).

always exothermic (Czu 2010). The transport behavior of the mentioned compounds was described by thermodynamic calculation with the help of the examples $ZrAs_{1.5}Se_{0.5}$ and $ZrAs_{1.5}Te_{0.5}$. The gas-phase composition and transport efficiency, which were calculated in the $ZrAs_{1.5}Se_{0.5}/I_2$ system show that ZrI_4, As_4, and Se_2, appear as essential transport-effective gas species (Figures 9.2.7 and 9.2.8). The gas species AsI_3, As_2, ZrI_3, and As_3 achieve transport-effective partial pressure as well; however, they are of minor importance. The following transport equation can be derived from the calculated transport efficiency of the individual gas species for transport from 850 to 950 °C. Due to the complex composition of the gas phase, the transport behavior is given only by approximation. It depicts the transport process adequately:

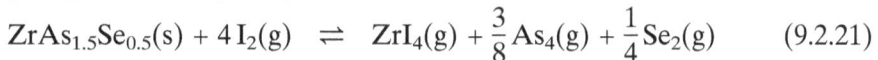

$$ZrAs_{1.5}Se_{0.5}(s) + 4\,I_2(g) \;\rightleftharpoons\; ZrI_4(g) + \frac{3}{8}As_4(g) + \frac{1}{4}Se_2(g) \qquad (9.2.21)$$

The calculated gas-phase composition and the transport efficiency of the involved gas species show that there are no significant changes for the exothermic transport of $ZrAs_{1.5}Te_{0.5}$ compared to the transport of $ZrAs_{1.5}Se_{0.5}$ as far as zirconium- and arsenic-transferring species are concerned. There is only a deviation with respect to the chalcogen-transferring species. TeI_2 is the essential tellurium-transferring gas species for the transport of $ZrAs_{1.5}Te_{0.5}$ although Te_2 and Te also have transport-effective partial pressures, which, however, are clearly below those of TeI_2 (Figures 9.2.9 and 9.2.10). Due to the increasing stability of the chalcogen halides from selenium to tellurium, TeI_2 is formed with transport-effective partial pressures in contrast to SeI_2 (see Chapter 7). In order to illustrate the transport behavior, the following essential equilibrium can be derived:

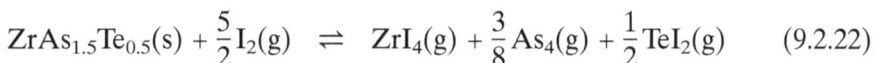

$$ZrAs_{1.5}Te_{0.5}(s) + \frac{5}{2}I_2(g) \;\rightleftharpoons\; ZrI_4(g) + \frac{3}{8}As_4(g) + \frac{1}{2}TeI_2(g) \qquad (9.2.22)$$

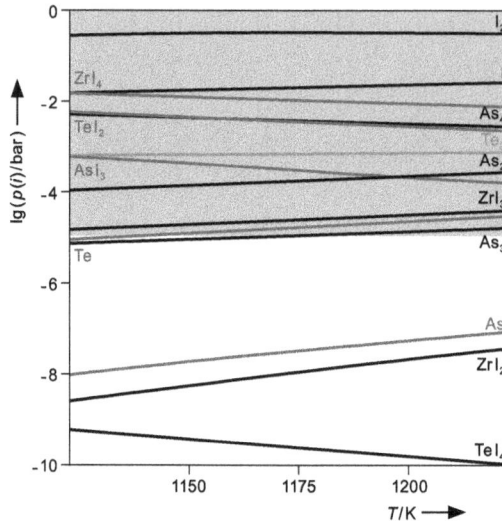

Figure 9.2.9 Gas-phase composition in the $ZrAs_{1.5}Te_{0.5}/I_2$ system according to *Czulucki* (Czu 2010).

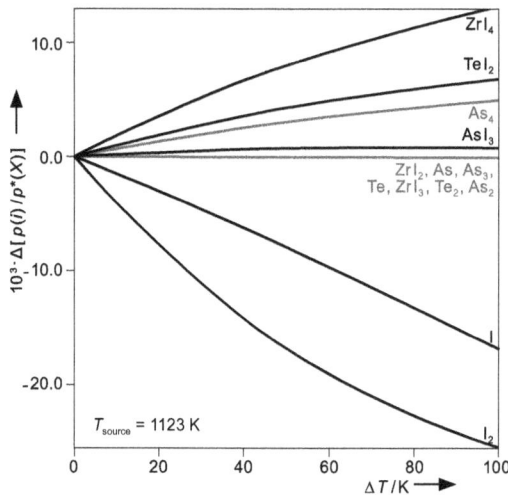

Figure 9.2.10 Transport efficiency of the essential gas species in the $ZrAs_{1.5}Te_{0.5}/I_2$ system according to *Czulucki* (Czu 2010).

Mass spectrometric investigations show that the experimentally determined gas-phase composition is in good agreement with that simulated with the help of thermodynamic data.

Other examples of the CVT of arsenide chalcogenides are the *mixed phases* in the GaAs/ZnSe system. GaAs/ZnSe mixed-crystals can be transported in a temperature gradient from 800 to 700 °C with iodine. *Brünig* published thermodynamic calculations and mass spectrometric examinations of the transport behavior of mixed-crystals (Bru 2006). Using mass spectrometry, a gaseous arsenic selenide halide (AsSeI) was detected for the first time. Only the formation of this gas species allows the transport of GaAs/ZnSe mixed-crystals.

Arsenides of most metals and semi-metals can be obtained as well crystallized materials by CVT. The often-missing thermodynamic analysis of the observed transport behavior is due to the fact that there are no, or no reliable, thermodynamic data for most of the solid arsenides. In contrast to the solid phases, the gas phase during the transport of binary arsenides can be described with relative reliability. Here the most important arsenic-containing gas species are As_4, As_3, As_2, and As and the arsenic(III)-halides as well as arsenic hydride when water or hydrogen, respectively, is present. Arsenic(I)-halides are also to be considered although thermodynamic data are not known. Due to low partial pressure, they are of minor relevance for CVT. As the example of AsSeI(g) shows, the formation of pnictogen chalcogenide halides is expected in systems that contain chalcogen atoms.

Table 9.2.3 Examples of the chemical vapor transport of arsenides.

Sink solid	Transport additive	Temperature / °C	Reference
AlAs	Cl_2, I_2		Vor 1972
	AlI_3	$300 \rightarrow 900$	Hil 1977,
			Bou 1977
B/As	Br_2	$850 \rightarrow 600$	Arm 1967
	I_2	$850 \rightarrow 400$	Arm 1967
	I_2	$920 \rightarrow 900$	Chu 1972a
	$Cl_2 + H_2O$, $Br_2 +$	$800 \rightarrow 620$	Rad 1978
	H_2O, $I_2 + H_2O$		
	I_2, AsI_3	$350 \rightarrow 800$	Bou 1976
	I_2, AsI_3	900	Bou 1977
	Te	$850 \rightarrow 750 \dots 800$	Rad 1978
B_6As	I_2, Te	$1050 \rightarrow 950$	Rad 1978
$B_{12}As_2$	I_2	$620 \rightarrow 900$	Bec 1980
$CdSiAs_2$	Br_2, $CdCl_2$	$805 \rightarrow 785$	Kim 1989
	$TeCl_4$	$790 \dots 815 \rightarrow 720 \dots 760$	Avi 1984
CeAs	I_2	$800 \rightarrow 850, 985 \rightarrow 1080$	Gol 2011
$Ce_{1-x}Nd_xAs$	I_2	$950 \rightarrow 1080$	Gol 2011
$CeAgAs_2$	I_2	$600 \rightarrow 800, 800 \rightarrow 900$	Dem 2004
CeAsSe	I_2	not specified	Czu 2010
CoAs	I_2	$850 \rightarrow 550$	Sel 1971
$CoAs_3$	Cl_2	$900 \rightarrow 800$	Ack 1977
$CdGeAs_2$	Br_2	$655 \rightarrow 635$	Bau 1990
Cd_3As_2	CdX_2	$500 \rightarrow 550$	Suc 1963
$Cd_4As_2X_3$	CdX_2	$500 \rightarrow 550$	Suc 1963
$(X = Cl, Br, I)$			
DyAs	I_2	1000	Bus 1965
ErAs	I_2	1000	Bus 1965
$FeAs_2$	Cl_2	$800 \rightarrow 765$	Fan 1972,
			Ros 1972
	I_2	$800 \rightarrow 765$	Ros 1972,
			Lut 1987
GaAs	Cl_2	$780 \rightarrow 730$	Alt 1968
	Cl_2	$830 \rightarrow 680$	Fak 1971
	Cl_2	$800 \rightarrow 750$	Wat 1975
	I_2	$800 \rightarrow 775 \dots 700,$	Ari 1965
		$700 \rightarrow 675 \dots 575$	
	I_2	$1070 \rightarrow 1030$	Ant 1959
	I_2	$200 \rightarrow 1100$	Sil 1962

Table 9.2.3 (continued)

Sink solid	Transport additive	Temperature / °C	Reference
GaAs	I_2	$825 \rightarrow 725$	Fer 1964
	I_2	$800 \dots 600 \rightarrow T_1;$ $\Delta T = 40$ K	Oka 1963
	I_2	$850 \rightarrow T_1$	Sel 1972
	I_2	$700 \rightarrow 650, 800 \rightarrow 750,$ $900 \rightarrow 850$	Hit 1994
	$Cl_2 + H_2, I_2 + H_2$	$825 \rightarrow 725$	Fer 1964
	HCl	$700 \rightarrow T_1$	Moe 1962
	HCl	$830 \rightarrow 680$	Fak 1973, Fak 1979
	HCl	$725 \rightarrow 925$	Pas 1983
	$HCl + H_2$	$1000 \dots 750 \rightarrow T_1$	Ett 1965
	$HCl + H_2,$ $H_2O + H_2$	$850 \dots 950 \rightarrow T_1$	Mic 1964
	HBr	$425 \rightarrow 1125$	Fak 1977, Fak 1979
	HBr	$830 \rightarrow 630$	Gar 1978
	H_2O	$1100 \rightarrow 1100 \dots 1080$	Fro 1964
	H_2O	$800 \rightarrow 1070$	Dem 1972
	H_2O	$800 \rightarrow 740$	Bar 1975
	As	$940 \rightarrow 840$	Köp 2004
	$AsCl_3$		Hen 1971
	$AsCl_3 + H_2$	700	Kra 1970
	$AsCl_3 + H_2$	$820 \rightarrow 680 \dots 550$	Nic 1971
	GaI_3	900	Bou 1977
	GaI_3		Hil 1977
$GaAs/GaAs_{1-x}P_x$	I_2	$1000 \rightarrow 900$	Ros 2005
GaAs/ZnSe	I_2	$800 \rightarrow 700$	Bru 2006
$GaAs_xP_{1-x}$	H_2O	$800 \rightarrow 1070 \rightarrow T_1$	Dem 1972
$Ga_xIn_{1-x}As$	HCl	$800 \rightarrow 650$	Cha 1982
$Ga_{1-y}In_yAs_{1-x}P_x$	HCl	725	Esk 1982
GdAs	I_2	1000	Bus 1965
	I_2	$960 \rightarrow 1030$	Mur 1970
$GdCuAs_2$	I_2	$900 \rightarrow 800$	Moz 2000
$GdCuAs_{1.15}P_{0.85}$	I_2	$900 \rightarrow 800$	Moz 2000
GeAs	I_2		Hil 1980
$GeAs_2$	I_2	$515 \rightarrow 505, 615 \rightarrow 605$	Hil 1982
	I_2		Hil 1980
$Ge_{38}As_8I_8$	I_2		vSc 1976
$HfAs_2$	I_2	$700 \rightarrow 800, 800 \rightarrow 900$	Czu 2010
$HfAs_{1.7}Se_{0.2}$	I_2	$800 \rightarrow 900$	Schl 2007, Czu 2010

Table 9.2.3 (continued)

Sink solid	Transport additive	Temperature / °C	Reference
HoAs	I_2	1000	Bus 1965
	I_2	950 → 1080	Gol 2011
InAs	I_2	840 → 815	Nic 1972
	HCl	635 → 520	Bol 2008
	$InCl_3$	890 → 840	Ant 1959
	InI_3	875 → 830	Ant 1959
	InI_3	900	Bou 1977
	$AsCl_3 + H_2$	825 → 775	Car 1974
	As	940 → 840	Köp 2004
	H_2O		Cha 1971
LaAs	I_2	800	Mur 1970
$LaAs_2$	I_2	960 → 840	Mur 1970
	I_2	not specified	Ono 1970
$La_{1-x}Ce_xAs$	I_2	950 → 1080	Gol 2011
$La_{1-x}Nd_xAs$	I_2	950 → 1080	Gol 2011
LaAsSe	I_2	not specified	Czu 2010
Mo_4As_5	Br_2	980 → 930	Mur 1972
Mo_2As_3	Br_2	not specified	Tay 1965
	Br_2	1080 → 1030	Mur 1972
	Br_2	not specified	Jen 1965
$MoAs_2$	Cl_2	915 → 940	Tay 1965
	Cl_2	1000 → 950	Mur 1972
	Br_2	1000 → 950	Mur 1972
	I_2	1010 → 940	Mur 1972
NbAs	Br_2	1000 → 850	Sai 1964
	I_2	940 → 1065	Sch 1965
$NbAs_2$	Cl_2	1000 → 700	Sai 1964
	Br_2	1000 → 700	Sai 1964
	I_2	1045 → 855	Sch 1965
	I_2	not specified	Fur 1965
NdAs	I_2	800	Mur 1970
	I_2	800 → 850, 985 → 1080	Gol 2011
	$PtCl_2$	950 → 1080	Gol 2011
	AsI_3	900 → 1000	Gol 2011
$NdAs_2$	I_2	960 → 1030	Mur 1970
	I_2	not specified	Ono 1970
$Nd_{1-x}Ho_xAs$	I_2	950 → 1080	Gol 2011
Ni_5As_2	Br_2, I_2	750 → 650	Sai 1964
$NpAs_2$	I_2	800 → 850	Woj 1982
Np_3As_4	I_2	720 → 760	Woj 1982

Table 9.2.3 (continued)

Sink solid	Transport additive	Temperature / °C	Reference
PaAs	I_2	$450 \to 2000$	Cal 1979
Pa_3As_4	I_2	$400 \to 1500$	Cal 1979
$PaAs_2$	I_2	$400 \to 1000$	Cal 1979
Pd_5As	Cl_2	$500 \to 650$	Sai 1964
PrAsSe	I_2	$950 \to 800$	Czu 2010
SiAs	I_2	$400 \to 1180$	Bec 1966
	I_2	$1100 \to 900$	Wad 1969
	I_2	$1100 \to T_1$	Chu 1971
	I_2	not specified	Hil 1980
	I_2	$1015 \to 985, 1015 \to 790$	Uga 1989
	I_2	$750 \to 850, 1050 \to 950$	Bolt 1994
	HI	$750 \to 850$	Bolt 1994
$SiAs_2$	I_2, AsI_3	$1100 \to 900$	Ing 1967
Ta_2As	I_2	$1000 \to T_1$	Mur 1976
Ta_5As_4	Br_2, I_2	$1000 \to T_1$	Mur 1976
TaAs	I_2	$T_2 \to T_1$	Mur 1976
$TaAs_2$	Cl_2, Br_2	$1000 \to 700$	Sai 1964
	Cl_2, Br_2, I_2	$T_2 \to T_1$	Mur 1976
TbAs	I_2	1000	Bus 1965
	I_2	$800 \to 850, 900 \to 1050$	Gol 2011
ThAsS	Br_2	$1020 \to 970$	Hen 2001
ThAsSe	Br_2	$1020 \to 970$	Hen 2001
	Br_2, I_2	$1000 \to 900$	Hul 1968
ThAsTe	Br_2, I_2	$1000 \to 900$	Hul 1968
$(Th_xU_{1-x})_3As_4$	I_2	not specified	Hen 1977
UAs	I_2	not specified	Hen 1977
U_3As_4	I_2	$890 \to 940$	Buh 1969
	I_2	$800 \to 935$	Hen 1968
	I_2	not specified	Hen 1977
	I_2	$920 \to 960$	Hen 1985
	I_2	$925 \to 975$	Onu 2007
UAs_2	I_2	$710 \to 790$	Hen 1968
	I_2	$830 \to 870$	Hen 1985
	I_2	$750 \to 900$	Wis 2000, Wis 2000a
	I_2	$925 \to 975$	Onu 2007
UAsS	Br_2	$950 \to 900$	Hen 2001
	Br_2, I_2	$1000 \to 900$	Hul 1968
UAsSe	Br_2	$950 \to 900$	Hen 2001
	Br_2, I_2	$1000 \to 900$	Hul 1968

Table 9.2.3 (continued)

Sink solid	Transport additive	Temperature / °C	Reference
UAsTe	Br_2, I_2	$1000 \rightarrow 900$	Hul 1968
WAs$_2$	Br_2	$1075 \rightarrow 900$	Tay 1965
	Br_2	not specified	Jen 1965
YbAs	I_2	$1025 \rightarrow 900$	Kha 1974
YbAs$_4$S$_7$	I_2	$525 \rightarrow 455$	Mam 1988
Yb$_3$As$_4$S$_9$	I_2	$875 \rightarrow 805$	Mam 1988
ZnSiAs$_2$	I_2	$950 \rightarrow 925$	Spr 1968
	$PbCl_2$	$1040 \rightarrow 1000$	Win 1974
	$ZnCl_2$	$1050 \rightarrow 1000$	Win 1974
	$TeCl_4$	$965 \rightarrow 940$	Wen 1997
ZnSiAs$_2$		$1150 \dots 1090 \rightarrow T_1$	Nad 1973
ZnSiP$_{2-x}$As$_x$	I_2	$950 \rightarrow 925$	Spr 1968
ZrAs$_2$	I_2	$700 \rightarrow 800$	Czu 2010
ZrAs$_{0.7}$Se$_{1.3}$	I_2	$970 \rightarrow 1020$	Czu 2010
ZrAs$_{0.6}$Te$_{1.4}$	I_2	$900 \rightarrow 950$	Czu 2010
ZrAs$_{1.4}$Se$_{0.5}$	I_2	$850 \rightarrow 950$	Czu 2010
ZrAs$_{1.6}$Te$_{0.4}$	I_2	$900 \rightarrow 950$	Czu 2010
ZrAs$_{0.5}$Ga$_{0.5}$Te	$TeCl_4$	$900 \rightarrow 950$	Wan 1995
ZrAsSi$_{0.5}$Te$_{0.5}$	$TeCl_4$	$900 \rightarrow 950$	Wan 1995

Bibliography of Section 9.2

Ack 1977   J. Ackermann, A. Wold, *J. Phys. Chem. Solids* **1977**, *38*, 1013.

Akh 1977   O. S. Akhverdov, S. A. Ershova, E. M. Novikova, K. A. Sokolovskii, *Tezisy Dokl. Vses. Soveshch. Rostu Krist. 5*[th] **1977**, *2*, 100.

Alf 1968   Z. I. Alferov, M. K. Trukan, *Izv. Akad. Nauk. SSSR, Neorg. Mater.* **1968**, *4*, 331.

Ant 1959   G. R. Anteli, D. Effer, *J. Electrochem. Soc.* **1959**, 509.

Ant 1961   G. R. Antell, *Brit. J. Appl. Phys.* **1961**, *12*, 687.

Ari 1965   T. Arizumi, T. Nishinaga, *Jap. J. Appl. Phys.* **1965**, *4*, 165.

Arm 1967   A. F. Armington, *J. Cryst. Growth* **1967**, *1*, 47.

Avi 1984   M. Avirović, M. Lux- Steiner, U. Elrod, J. Hönigschmid, E. Bucher, *J. Cryst. Growth* **1984**, *67*, 185.

Bar 1975   A. A. Barybin, A. A. Zakharov, N. K. Nedev, *Izv. Akad. Nauk SSSR, Neorg. Mater.* **1975**, *11*, 1005.

Bau 1990   F. P. Baumgartner, M. Lux-Steiner, E. Bucher, *J. Electr. Mater.* **1990**, *19*, 777.

Bec 1966   C. G. Beck, R. Stickler, *J. Appl. Phys.* **1966**, *37*, 4683.

Bec 1980   H. J. Becher, F. Thévenot, C. Brodhag, *Z. Allg. Chem.* **1980**, *469*, 7.

Bin 2002   M. Binnewies, E. Milke, *Thermochemical Data of Elements and Compounds*, Wiley-VCH, Weinheim, **2002**.

Bod 1973	I. V. Bodnar, *Kinet. Mekh. Krist.* **1973**, 248.
Bol 1963	D. E. Bolger, B. E. Barry, *Nature* **1963**, *199*, 1287.
Bol 1994	P. Bolte, R. Gruehn, *Z. Anorg. Allg. Chem.* **1994**, *620*, 2077.
Bou 1976	J. Bouix, R. Hillel, *J. Less-Common Met.* **1976**, *47*, 67.
Bou 1977	J. Bouix, R. Hillel, *J. Crystal Growth* **1977**, *38*, 61.
Boz 1974	C. O. Bozler, *Solid-State Electronics* **1974**, *17*, 251.
Bru 2006	C. Brünig, S. Locmelis, E. Milke, M. Binnewies, *Z. Anorg. Allg. Chem.* **2006**, *632*, 1067.
Buh 1969	C. F. Buhrer, *J. Phys. Chem. Solids* **1969**, *30*, 1273.
Bus 1965	G. Busch, O. Vogt, F. Hulliger, *Phys. Letters* **1965**, *15*, 301.
Cal 1979	G. Calestani, J. C. Spirlet, J. Rebizant, W. Müller, *J. Less-Common Met.* **1979**, *68*, 207.
Cha 1971	G. V. Chaplygin, T. I. Shcherballova, S. A. Semiletov, *Kristallografiya* **1971**, *16*, 207.
Cha 1976	S. Chang, J.-K. Liang, *Guoli Taiwan Daxue Gongcheng Xuekan* **1976**, *19*, 64.
Cha 1982	A. K. Chatterjee, M. M. Faktor, M. H. Lyons, R. H. Moss, *J. Cryst. Growth* **1982**, *56*, 591.
Che 1985	K. J. Chen, Z. Y. Yang, R. L. Wu, *J. Non-Cryst. Solids* **1985**, *77–78*, 1281.
Chu 1971	T. L. Chu, R. W. Kelm, S. S. C. Chu, *J. Appl. Phys.* **1971**, *42*, 1169.
Chu 1972a	T. L. Chu, A. E. Hyslop, *J. Appl. Phys.* **1972**, *43*, 276.
Coc 1962	C. N. Cochran, L. M. Foster, *J. Electrochem. Soc.*, **1962**, *109*, 149.
Czu 2010	A. Czulucki, *Dissertation,* University of Dresden, **2010**.
Dem 1972	E. Deml, J. Talpova, A. S. Popova, *Krist. Tech.* **1972**, *7*, 1089.
Dem 2004	R. Demchyna, J. P. F. Jemetio, Y. Prots, T. Doert, L. G. Akselrud, W. Schnelle, Y. Kuz'ma, Y. Grin, *Z. Anorg. Allg. Chem.* **2004**, *630*, 635.
Der 1966	H. J. Dersin, E. Sirtl, *Z. Naturforsch.* **1966**, *21A*, 332.
Eff 1965	D. Effer, *J. Electrochem. Soc.* **1965**, *112*, 1020.
Esk 1982	S. M. Eskin, E. N. Vigdorovich, T. P. Shapovalova, A. S. Pashikin, *Izv. Akad. Nauk. SSSR, Neorg. Mater.* **1982**, *18*, 729.
Fak 1971	M. M. Faktor, I. Garrett, *J. Chem. Soc. A* **1971**, *8*, 934.
Fak 1973	M. M. Faktor, I. Garrett, R. H. Moss, *J. Chem. Soc. Faraday T. 1* **1973**, *69*, 1915.
Fak 1977	M. M. Faktor, I. Garrett, M. H. Lyons, R. H. Moss, *J. Chem. Soc. Faraday Trans.* **1977**, *73*, 1446.
Fak 1979	M. M. Faktor, I. Garret, M. H. Lyons, *J. Cryst. Growth* **1979**, *46*, 21.
Fan 1972	A. K. L. Fan, G. H. Rosenthal, H. L. McKinzie, A. Wold, *J. Solid State Chem.* **1972**, *5*, 136.
Fer 1964	R. R. Fergusson, T. Gabor, *J. Electrochem. Soc.* **1964**, *111,* 585.
Fra 1981	P. Franzosi, C. Ghezzi, E. Gombia, *J. Cryst. Growth* **1981**, *51*, 314.
Fro 1964	C. J. Frosch, *J. Electrochem. Soc.* **1964**, *111*, 180.
Fur 1965	S. Furuseth, A. Kjekshus, *Acta Crystallogr.* **1965**, *18*, 320.
Gar 1978	I. Garrett, M. M. Faktor, M. H. Lyons, *J. Cryst. Growth* **1978**, *45*, 150.
Gol 1977	M. I. Golovei, M. Yu. Rigan, *Khim. Fiz. Khal'kogenidov* **1977**, 38.
Gol 2011	S. Golbs, *Dissertation,* University of Dresden, **2011** (in preparation).
Hen 1968	Z. Henkie, *Roczniki Chemii, Ann. Soc. Chim. Polonorum* **1968**, *42*, 363.
Hen 1977	Z. Henkie, P. J. Markowski, *J. Cryst. Growth* **1977**, *41*, 303.
Hen 2001	Z. Henkie, A. Pietraszko, A. Wojakowski, L. Kępiński, T. Chichorek, *J. Alloys. Compd.* **2001**, *317–318*, 52.
Hil 1977	R. Hillel, J. Bouix, *J. Cryst. Growth* **1977**, *38*, 67.
Hil 1980	R. Hillel, J. Bouix, A. Michaelides, *Thermochim. Acta* **1980**, *38*, 259.
Hil 1982	R. Hillel, C. Bec, J. Bouix, A. Michaelides, Y. Monteil, A. Tranquard, *J. Electrochem. Soc.* **1982**, *129*, 1343.

Hit 1994 L. Hitova, A. Lenchev, E. P. Trifonova, M. Apostolova, *Cryst. Res. Technol.* **1994**, *29*, 957.

Hul 1968 F. Hulliger, *J. Less-Common Met.* **1968**, *16*, 113.

Ing 1967 S. W. Ing, Y. S. Chiang, W. Haas, *J. Electrochem. Soc.* **1967**, *114*, 761.

Jen 1965 P. Jensen, A. Kjekshus, T. Skansen, *Acta Chem. Scand.* **1965**, *19*, 1499.

Jen 1969 P. Jensen, A. Kjekshus, T. Skansen, *J. Less-Common Met.* **1969**, *17*, 455.

Kac 1994 D. Kaczorowski, H. Noël, M. Potel, A. Zygmunt, *J. Phys. Chem. Solids*, **1994**, *55*, 1363.

Kac 2005 D. Kaczorowski, A. P. Pikul, A. Zygmunt, *J. Alloy. Compd.* **2005**, *398*, L1.

Kha 1974 A. Khan, J. Castro, C. Vallendilla, *J. Cryst. Growth* **1974**, *23*, 221.

Kha 1976 A. Khan, J. Castro, *Proc. Rare Earth Res. Conf.* **1976**, *2*, 961.

Kim 1989 M. Kimmel, M. Lux- Steiner, A. Klein, E. Bucher, *J. Cryst. Growth* **1989**, *97*, 665.

Köp 2004 R. Köppe, H. Schnöckel, *Angew. Chem.* **2004**, *116*, 2222, *Angew. Chem., Int. Ed.* **2004**, *43*, 2170.

Kra 1970 P. Kramer, W. Schmidt, G. Knobloch, E. Butter, *Krist. Tech.* **1970**, *5*, 523.

Kun 1976 Y. Kuniya, T. Fujii, M. Yuizumi, *Denki Kagaku oyobi Kogyo Butsuri Kagaku* **1976**, *44*, 124.

Lut 1987 H. D. Lutz, M. Jung, G. Wäschenbach, *Z. Anorg. Allg. Chem.* **1987**, *554*, 87.

Mam 1988 A. I. Mamedov, T. M. Il'yasov, P. G. Rustamov, F. G. Akperov, *Zh. Neorg. Khim.* **1988**, *33*, 1103.

Mic 1964 M. Michelitsch, W. Kappallo, G. Hellbardt, *J. Electrochem. Soc.* **1964**, *111*, 1248.

Min 1971 H. T. Minden, *J. Cryst. Growth* **1971**, *8*, 37.

Moe 1962 R. R. Moest, B. R. Shupp, *J. Electrochem. Soc.* **1962**, *109*, 1061.

Moz 2000 Y. Mozharivskyj, D. Kaczorowski, H. F. Franzen, *J. Solid State Chem.* **2000**, *155*, 259.

Mur 1970 J. J. Murray, J. B. Taylor, *J. Less-Common Met.* **1970**, *21*, 159.

Mur 1972 J. J. Murray, J. B. Taylor, L. Usner, *J. Cryst. Growth* **1972**, *15*, 231.

Mur 1976 J. J. Murray, J. B. Taylor, L. D. Calvert, Yu. Wang, E. J. Gabe, J. G. Despault, *J. Less-Common Met.* **1976**, *46*, 311.

Nad 1973 H. Nadler, M. D. Lind, *Mater. Res. Bull.* **1973**, *8*, 687.

Nar 1998 A. Narducci, J. A. Ibers, *Chem. Mater.*, **1998**, *10*, 2811.

Nic 1963 F. H. Nicoll, *J. Electrochem. Soc.* **1963**, *110*, 1165.

Nic 1971 J. J. Nickl, W. Just, *J. Cryst. Growth* **1971**, *11*, 11.

Nic 1972 W. Nicolaus, E. Seidowski, V. A. Voronin, *Krist. Technik* **1972**, 7, 589.

Oka 1963 T. Okada, T. Kano, S. Kikuchi, Jap. *J. Appl. Phys.* **1963**, *2*, 780.

Ono 1970 S. Ono, J. G. Despault, L. D. Calvert, J. B. Taylor, *J. Less-Common Met.* **1970**, *22*, 51.

Onu 2007 Y. Ōnuki, R. Settai, K. Sugiyama, Y. Inada, T. Takeuchi, Y. Haga, E. Yamamoto, H. Harima, H. Yamagami, *J. Phys.: Condens. Matter* **2007**, *19*, 125203.

Pas 1983 A. S. Pashinkin, A. S. Malkova, V. A. Fedorov, A. V. Rodionov, Yu. N. Sveshnikov, *Izv. Akad. Nauk SSSR, Neorg. Mater.* **1983**, *19*, 538.

Rad 1978 A. F. Radchenko, *Neorg. Materialy* **1978**, *14*, 1051.

Rao 1978 M. Subba Raó, R. H. Moss, M. M. Faktor, *Indian J. Pur. Ap. Phys.* **1978**, *16*, 805.

Ros 1972 G. Rosenthal, R. Kershaw, A. Wold, *Mater. Res. Bull.* **1972**, *7*, 479.

Ros 2005 C. Rose, S. Locmelis, E. Milke, M. Binnewies, *Z. Anorg. Allg. Chem.* **2005**, *631*, 530.

Sai 1964 G. S. Saini, L. D. Calvert, J. B. Taylor, *Canad. J. Chem.* **1964**, *42*, 630.

Sch 1965 H. Schäfer, W. Fuhr, *J. Less-Common Met.* **1965**, *8*, 375.

Schl 2007   A. Schlechte, R. Niewa, M. Schmidt, H. Borrmann, G. Aufermann, R. Kniep, *Z. Kristallogr. NCS* **2007**, *222*, 369.

Schl 2007a   A. Schlechte, R. Niewa, M. Schmidt, G. Aufermann, Yu. Prots, W. Schnelle, D. Gnida, T. Cichorek, F. Steglich, R. Kniep, *Sci. Technol. Adv. Mat.* **2007**, *8*, 341.

Sel 1971   K. Selte, A. Kjekshus, *Acta Chem. Scand.* **1971**, *25*, 3277.

Sel 1971a   K. Selte, A. Kjekshues, W. E. Jamison, A. I. Andresen, J. E. Engebretsen, *Acta Chim. Scand.* **1971**, *25*, 1703.

Sel 1972   K. Selte, A. Kjekshus, *Acta Chem. Scand.* **1972**, *26*, 3101.

Sil 1962   V. J. Silvestri, V. J. Lyons, *J. Electrochem. Soc.* **1962**, *109*, 963.

Suc 1963   L. Suchow, N. R. Stemple, *J. Electrochem. Soc.* **1963**, *110*, 766.

Tay 1965   J. B. Taylor, L. D. Calvert, M. R. Hunt, *Can. J. Chemistry* **1965**, *43*, 3045.

Uga 1989   Y. A. Ugai, E. G. Goncharov, A. Y. Zavrazhnov, A. E. Popov, I. V. Vavresyuk, *Neorg. Materialy* **1989**, *25*, 709.

Uts 1971   Y. Utsugi, K. Yanata, H. Fujisaki, Y. Tanabe, *Tohoku Daigaku Kagaku Keisoku Kenkyusho Hokoku* **1971**, *20*, 1.

Ven 1975   K. Venttsel, V. A. Kempel, V. I. Tomilin, *Izv. Leningradskogo Elektrotekhnicheskogo Instituta* **1975**, *167*, 138.

Vor 1972   V. A. Voronin, V. A. Prokhorov, *J. Electrochem. Soc.* **1972**, *119*, C97.

vSc 1976   H. von Schnering, H. Menke, *Z. Anorg. Allg. Chem.* **1976**, *424*, 108.

Wad 1969   T. Wadsten, *Acta Chem. Scand.* **1969**, *23*, 331.

Wan 1995   C. Wang, T. Hughbanks, *Inorg. Chem.* **1995**, *34*, 5524.

Wat 1975   H. Watanabe, T. Nishinaga, T. Arizumi, *J. Cryst. Growth* **1975**, *31*, 179.

Waw 2005   R. Wawryk, A. Wojakowski, A. Pietraszko, Z. Henkie, *Solid-State Comm.* **2005**, *133*, 295.

Wen 1997   Y. Wen, B. A. Parkinson, *J. Phys. Chem. B* **1997**, *101*, 2659.

Win 1974   K. Winkler, K. Hein, K. Leipner, *Krist. Technik* **1974**, *9*, 1223.

Win 1977   K. Winkler, K. Hein, *Krist. Technik* **1977**, *12*, 211.

Wis 2000   P. Wiśniewski, D. Aoki, K. Miyake, N. Watanabe, Y. Inada, R. Settai, Y. Haga, E. Yamamoto, Y. Ōnuki, *Physica B*, **2000**, *281/282*, 769.

Wis 2000a   P. Wiśniewski, D. Aoki, N. Watanabe, K. Miyake, R. Settai, Y. Ōnuki, Y. Haga, E. Yamamoto, Z. Henkie, *J. Phys.: Condens. Matter*, **2000**, *12*, 1971.

Woj 1982   A. Wojakowski, D. Damien, *J. Less-Common Met.* **1982**, *83*, 263.

Yan 1977   K. Yanata, *Tohoku Daigaku Kagaku Keisoku Kenkyusho Hokoku* **1977**, *26*, 43.

10 Chemical Vapor Transport of Intermetallic Phases

Cu_9Ga_4

H																	He
Li	Be											B	C	N	O	F	Ne
Na	Mg											Al	Si	P	S	Cl	Ar
K	Ca	Sc	Ti	V	Cr	Mn	Fe	Co	Ni	Cu	Zn	Ga	Ge	As	Se	Br	Xe
Rb	Sr	Y	Zr	Nb	Mo	Tc	Ru	Rh	Pd	Ag	Cd	In	Sn	Sb	Te	I	Kr
Cs	Ba	La	Hf	Ta	W	Re	Os	Ir	Pt	Au	Hg	Tl	Pb	Bi	Po	At	Rn

Ce	Pr	Nd	Pm	Sm	Eu	Gd	Tb	Dy	Ho	Er	Tm	Yb	Lu
Th	Pa	U	Np	Pu									

Iodine as transport agent

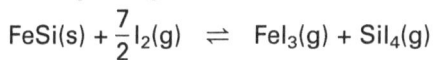

$$FeSi(s) + \frac{7}{2}I_2(g) \ \rightleftharpoons \ FeI_3(g) + SiI_4(g)$$

Iodine as transport additive – SiI_4 as transport agent

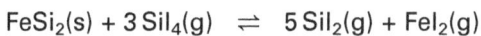

$$FeSi_2(s) + 3\,SiI_4(g) \ \rightleftharpoons \ 5\,SiI_2(g) + FeI_2(g)$$

Intermetallic phases with wide phase range

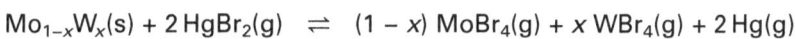

$$Mo_{1-x}W_x(s) + 2\,HgBr_2(g) \ \rightleftharpoons \ (1-x)\,MoBr_4(g) + x\,WBr_4(g) + 2\,Hg(g)$$

If one refers to **intermetallic phases**, solids are meant which are built up by two or more metals. Sometimes there is a differentiation between alloys and intermetallic compounds. In the literature, however, these terms are not used uniformly. In order to avoid misunderstanding, we solely use the term intermetallic phase. It includes metallic solids that are composed stoichiometrically as well as those with phase ranges and solid solutions, respectively.

Solid matter, which develops from metals and the semi-metals boron, silicon, germanium, and antimony, is dealt with in this chapter as well due to its behavior during CVT. In a narrow sense, intermetallic phases do not include these solids although many of them show metallic characteristics.

Usually, these phases are made with the help of common preparation methods such as metallurgical melting or powder metallurgical processes. They form during solidification of the melts or during tempering of mixtures of the individual components. Melting usually takes place in a vacuum or under an inert gas. In this process, high temperatures are partially necessary. Apart from the considerable experimental and technical complexity (electric arc-furnace, high-temperature furnace, protective gas technique, high vacuum technique), it is particularly difficult to find a suitable crucible material that does not react with its contents at high temperatures. In some cases, fluxing agents are used to crystallize intermetallic phases (Kan 2005). Melts of the metals aluminum, gallium, indium, tin, lead, antimony as well as copper, lithium, and sodium are often-used fluxing agents. There are two disadvantages of this process. First, the atoms of the fluxing agent can be integrated into the phase and, second, the crystals formed are often difficult to separate from the solidified melt.

Transport reactions are good alternatives to the synthesis or crystal growth of small amounts of intermetallic phases (Loi 1972). The CVT of intermetallic phases principally follows the one of metals. Nowadays a large number of examples of transports of intermetallic phases are known. They are basically different to those of other compounds such as oxides, sulfides, selenides, arsenides, and halides. The difference is that in intermetallic phases, there are generally no components that have a transport-effective partial pressure of their own. This means that *all* components of the solid have to be transferred to the gas phase by the transport agent under formation of volatile gas species under the same conditions. The transport equations 10.1 to 10.3 clarify the differences between the transport of a saline compound and an intermetallic phase and a metal silicide, respectively.

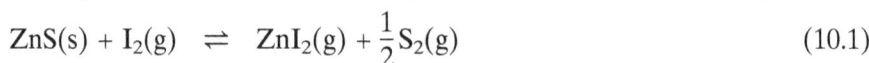

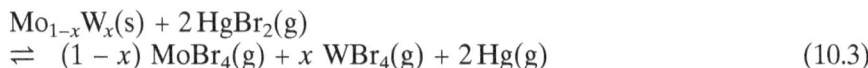

$$ZnS(s) + I_2(g) \; \rightleftharpoons \; ZnI_2(g) + \frac{1}{2}S_2(g) \tag{10.1}$$

$$CrSi_2(s) + 5\,I_2(g) \; \rightleftharpoons \; CrI_2(g) + 2\,SiI_4(g) \tag{10.2}$$

$$
\begin{aligned}
&Mo_{1-x}W_x(s) + 2\,HgBr_2(g) \\
&\rightleftharpoons \; (1-x)\,MoBr_4(g) + x\,WBr_4(g) + 2\,Hg(g)
\end{aligned}
\tag{10.3}
$$

Many intermetallic systems are labeled by the appearance of numerous solid phases with similar stabilities. An incongruent CVT, in which the compositions of the source solid and sink solid are different, is often observed. The composition of the deposited solid can be influenced by the following factors:

- The composition of the source solid.
- The kind of transport agent and its concentration.
- The temperatures of the source and sink, as well as the resulting temperature gradient.
- Chemical vapor transport reactions are not just an alternative to the synthesis and crystal growing of intermetallic phases with high melting temperatures. In particular, in the following cases, they are preferable to the just mentioned processes:

One or more components of the intermetallic phase have a high vapor pressure at melting temperature.

The intermetallic phase decomposes, e.g., peritectically before the melting temperature is reached.

The intermetallic phase shows one or more phase changes before the melting temperature is reached.

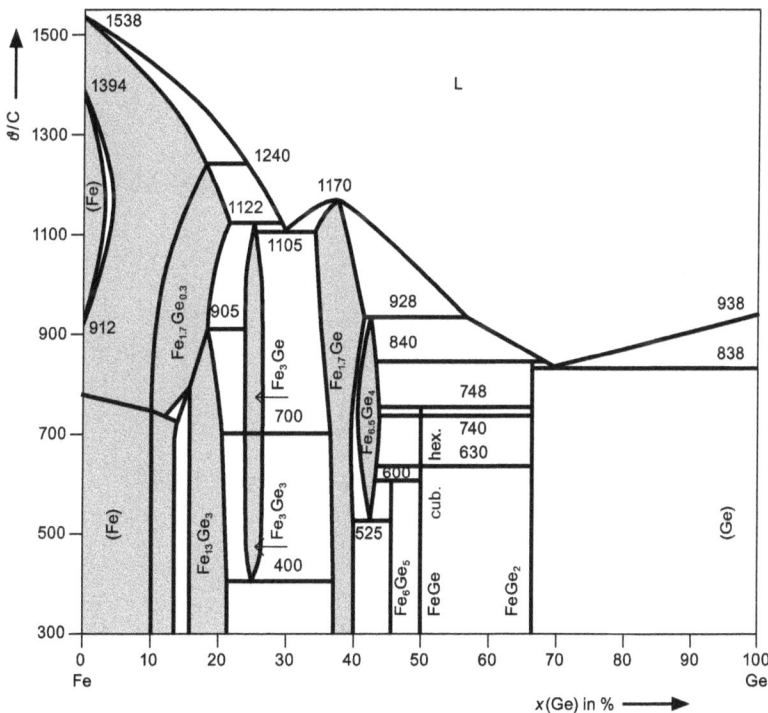

Figure 10.1 Phase diagram of the Fe/Ge system according to *Richardson* (Ric 1967).

Intermetallic phases can be obtained in particular if the elements are also obtained with the same transport agent. Examples can be found in the molybdenum/tungsten, cobalt/nickel, and copper/silver systems. Exceptions to this gen-

Figure 10.2 Scanning electron micrograph of FeGe.

eral rule can be found if the Gibbs energy of the intermetallic phases is especially high, e.g., in the chromium/germanium, cobalt/germanium, iron/germanium, nickel/tin, or copper/tin systems.

It is possible to make low temperature modifications of intermetallic phases in the form of single-crystals. Their preparation only rarely succeeds with other methods. FeGe is an example of this, in the cubic modification (Wil 2007).

Transport agent Due to the fact that the transport behavior of the intermetallic phases often follows that of the elements, iodine is the most commonly used transport agent for the transport of intermetallic phases. Apart from iodine, combinations of iodine and aluminium(III)-, gallium(III)-, or indium(III)-iodide are used as transport agents as well. Other transport agents and transport-effective additives, respectively, are the halogens chlorine and bromine as well as hydrogen chloride, copper(II)-chloride, manganese(II)-chloride, the mercury(II)-halides, tellurium(IV)-chloride, and iron(II)-bromide in individual cases.

Mineralization – short distance transport – microdiffusion Apart from transport reactions in the actual sense, where there is material transport of a distance in the centimeter range along a temperature gradient, intermetallic phases are often prepared using a mineralizer (Mer 1980, Sau 1993, Wan 2001, Wan 2002, Wan 2007). These reactions can be considered as *short distance transport* or transport with *microdiffusion*, respectively. In contrast to transport in the true sense, they are executed as isothermal reactions in most cases. Often it proves favorable for the maintenance of defined reaction products if the temperature of the mineralization reaction is chosen in a way that no liquid phase occurs. However, in some cases, this is not possible because the partial pressures of the transport-effective halides are too low at these temperatures. In intermetallic systems that contain high- and low-melting metals, educts can be provided spatially separated from each other. In general, the conditions are to be chosen in a way that the reaction takes places via the gas phase. Here the gas species, which were formed from metal components, react back under formation of an intermetallic phase and thus the mineralizer is set free again. Intermetallic phases can be made from the elements and crystallized, and polycrystalline and micro crystalline source materials, respectively, can be recrystallized with the help of a suitable mineralizer. For this kind of reaction, the partial pressure difference, which is necessary for transport reactions, is not based on a different position of the transport equilibrium on the dissolution or deposition side, which is temperature dependent. It is instead based on the difference of the Gibbs energy of the solid phases before and after mineralization, respectively, on the Gibbs energy of formation of the intermetallic phase (Nic 1973). This kind of mineralization reaction is very effective even when there are small partial pressure differences around $\Delta p = 10^{-6}$ bar at a total pressure of one bar. This method is particularly effective if all components of the solid form gaseous compounds with the mineralizer. Halogens, especially iodine, or halogen compounds of the solid's components, are often used as mineralizers. This way contamination, by the mineralizer, of the intermetallic phases formed is avoided. In some cases, the chalcogenides oxygen, sulfur, and tellurium are used as mineralizers. The mineralizer is responsible for an ongoing mass transfer between the solid and gas phases. The mineralizer concentration is to be measured in a way that at the chosen reaction temperature, the partial pressures of the gas species formed are not higher than their respective saturation pressures. Otherwise the gas species will condense when the saturating pressure is exceeded. Similar principles, as for CVT reactions, apply for the reaction of the solid with the mineralizer gas as far as the equilibrium position is concerned; however, the equilibrium position may well be more extreme, which results in smaller partial pressure differences. Additionally, for isothermal mineralizations, it is not necessary that the reaction of all components with the chosen mineralizer gas have the same transport direction. The use of a "pendulum furnace" (Sch 1962) for such mineralization reactions is advantageous in many cases. In doing so, the temperature is varied around a mean value of, e. g., an amplitude of 20 K and a frequency of 20 minutes each period. In contrast to CVT, there is not a spatial but a temporal temperature difference that leads to crystallization of the intermetallic phase. Further examples of the crystallization of compounds of other substance groups by the temperature oscillation method and the use of the "pendulum furnace", respectively, can be found in the literature (Nee 1971, Ros 1973, Scho 1974, Schi 1976, Ala 1977, Bei 1977), although the methodical aspect is given prominence.

Intermetallic phases that were synthesized or crystallized, respectively, by means of mineralization or microdiffusion, can contain almost all metals (apart from alkali metals) and the vast majority of the semi-metals, e. g., the elements beryllium, magnesium, the rare earth metals, titanium, vanadium, rhenium, iron, ruthenium, cobalt, rhodium, iridium, nickel, palladium, platinum, zinc, boron, aluminum, gallium, indium, silicon, germanium, tin, lead, antimony, and bismuth.

10.1 Selected Examples

Numerous examples of CVT reactions of intermetallic phases are known. The systems investigated cover solid solutions of two similar metals, stoichiometrically composed phases, as well as silicides, germanides, and borides. The CVT of *Zintl* phases has not yet been reported. In particular, those examples are described where one component of the solid is an element of group 14.

Intermetallic phases with wide phase range Molybdenum and tungsten are two metals with very high melting points. They are isotypic and completely mixable in the solid and liquid state (see Figure 10.1.1). It requires great experimental effort to grow molybdenum/tungsten mixed-crystals. With the help of CVT reactions, this succeeds far below the liquidus curve at temperatures around 1000 °C. Both metals can be transported with the same transport agent under the same conditions.

Molybdenum/tungsten mixed-crystals can be deposited by CVT from 1000 to 900 °C when mercury(II)-bromide is added (Ned 1996a). The following transport equation describes the process:

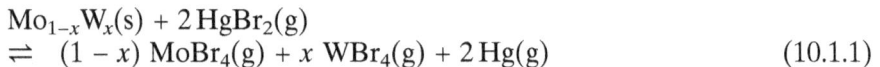

$$Mo_{1-x}W_x(s) + 2\,HgBr_2(g)$$
$$\rightleftharpoons\ (1-x)\,MoBr_4(g) + x\,WBr_4(g) + 2\,Hg(g) \tag{10.1.1}$$

In the following binary systems, mixed-crystals are transportable in an analogous way: cobalt/nickel, iron/nickel, silver/copper, gold/copper, copper/nickel, gold/nickel, and copper/gallium.

There is no miscibility gap in the cobalt/nickel system. Cobalt and nickel can be transported under similar conditions when iodine and gallium(III)-iodide are added (Sch 1977). This additive is iodizing and complex-forming too. The gas species that are essential for transferring nickel and cobalt, respectively, are: NiI_2,

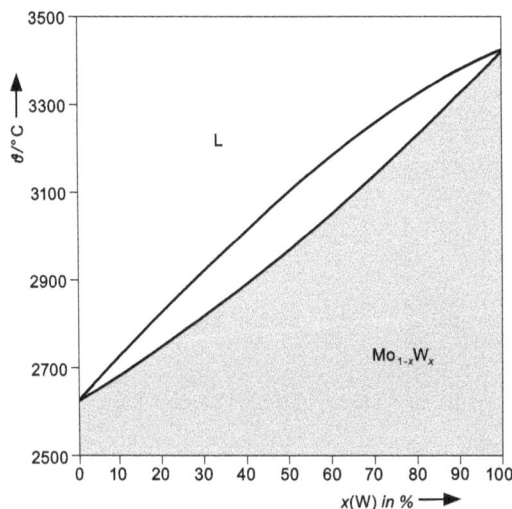

Figure 10.1.1 Phase diagram of the Mo/W system according to *Massalski* (Mas 1990).

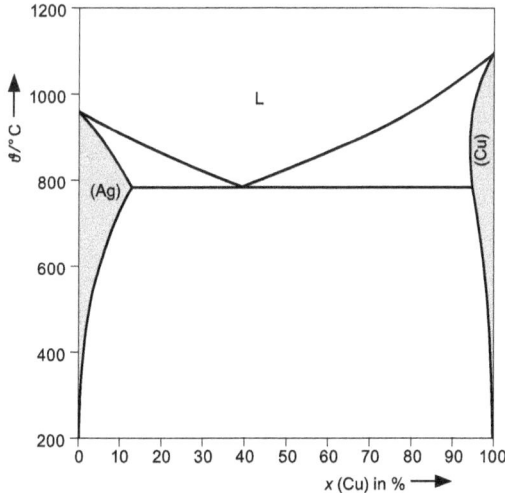

Figure 10.1.2 Phase diagram of the Cu/Ag system according to *Massalski* (Mas 1990).

Ni_2I_4, CoI_2, Co_2I_4, $NiGa_2I_8$, $CoGa_2I_8$ as well as $NiGaI_5$ and $CoGaI_5$. The gas-phase solubility of cobalt and its temperature dependency is clearly higher than that of nickel in the investigated temperature range between T_2 (1000 to 900 °C) and T_1 (800 °C). Hence, an enrichment of cobalt is expected. This enrichment effect can be experimentally compensated – if wished – by using a three-zone furnace with different diffusion distances and different temperature gradients. Based on the spatially separated elements cobalt and nickel, mixed-crystals with iodine and gallium(III)-iodide can be deposited from T_2 to T_1 with compositions between 5 and 75 % nickel (Ned 1996).

The copper/silver system is another system in which mixed-crystals have been obtained. It shows a eutectic mixture with $x(Cu) = 60.1$ % copper at 770 °C (see Figure 10.1.2). There is a wide miscibility gap below the eutectic temperature. By CVT with iodine from 600 to 700 °C mixed-crystals that are copper- or silver-rich can be deposited. It is noticeable that the individual compositions of the crystals within one experiment can be different. Mass spectrometric investigations of the gas phase above Cu/Ag in the presence of gaseous iodine prove the formation of the gas species $CuAg_2I_3$ and Cu_2AgI_3 adjacent to CuI and AgI, respectively, and the respective trimers M_3I_3 (M = Cu, Ag) (Ger 1995). This example shows that, especially during the transport of intermetallic phases, one has to expect the formation of gas complexes. This has to be considered for the description of the gas phase and above all for the thermodynamic modeling of transport experiments.

The composition of the deposited mixed-crystals can be influenced by the experimental conditions, in particular by the composition of the source solid. One cannot always assume that the composition of source and sink solid are identical. Enrichment effects by transport actions can also lead to concentration gradients within individual crystals (Ned 1996a).

Stochiometrically composed intermetallic phases   The selected examples are the intermetallic phases in the binary nickel/tantalum system (e. g., NiTa, $NiTa_2$, Ni_3Ta) and cobalt/tantalum system (e. g., $CoTa_2$, Co_2Ta, Co_7Ta_2). Further examples are Nb_3Ga, PdAl, Pd_2Al, Pd_2Ga. Pd_2In, $AlPt_3$, and V_3Ga.

All binary phases that occur in the nickel/tantalum and cobalt/tantalum systems can be deposited by CVT with iodine from 800 to 950 °C. The two mentioned systems represent binary systems in which a multiplicity of intermetallic phases do exist. In the majority of cases, one observes an incongruent transport, the compositions of source and sink solid are different (see Figures 10.1.4 and 10.1.5). Thereby the activities of the two components in the source solid are the main effects that determine the composition of the solid deposited. In particular, the composition in the source determines the composition in the sink. Additionally, the transport agent, the temperature gradient, and the total pressure have an influence as well; however, they are of minor importance.

Intermetallic phases that contain platinum or palladium atoms and an atom of group 13 (aluminum, gallium, indium) are typical intermetallic phases that can be deposited via the gas phase. Usually their transport takes place with iodine and/or the respective metal(III)-iodide (Sau 1992, Dei 1998). When the gas phase is described, the formation of gas complexes has to be taken into account (see section 11.1). Extensive investigations of CVT reactions of intermetallic phases in the palladium/aluminum, palladium/gallium, and palladium/indium systems showed that, apart from the composition of the source solid, if $MI_3 + I_2$ is used as transport agent, the ratio of metal(III)-iodide to iodine and the concentration of the transport agents is decisive for the composition of the solid deposited (Dei 1998). If at least one component is liquid at experimental conditions, it has proven of value to spatially separate the starting materials on the source side; for example in the Pd/Al (1000 → 800 °C) and Pd/Ga systems (400 → 600 °C).

Nb_3Ga and Nb_3Ga_5 can be obtained when hydrogen chloride, which is relatively rare for intermetallic phases, is used as the transport agent from 860 to 940 °C and from 910 to 970 °C, respectively (Hor 1971). The following transport equation is essential:

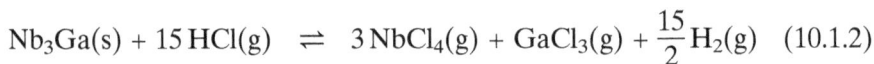

$$Nb_3Ga(s) + 15\,HCl(g) \;\rightleftharpoons\; 3\,NbCl_4(g) + GaCl_3(g) + \frac{15}{2}H_2(g) \quad (10.1.2)$$

Tetrelides   We know a lot of examples of the transport of binary and some ternary phases of which one of the components is an element of group 14. Some of them are well examined experimentally and with the help of thermodynamic model calculations. They contribute decisively to the understanding of the transport of intermetallic phases. In the following systems, the CVT of silicides, germanides, or stannides are described: **M/Si** (M = Co, Cr, Cu, Fe, Ni, Mn, Mo, Nb, Re, Ta, Ti, U, V, W), **M/Ge** (M = Co, Cr, Cu, Fe, Ni, Nb, Ti) and **M/Sn** (M = Co, Cu, Fe, Nb, Ni, Ti). Apart from the just mentioned binary systems, transport experiments are also published in the ternary systems cobalt/tantalum/germanium, cobalt/chromium/germanium as well as iron/cobalt/silicon. When iodine is added, CVT succeeds for the vast majority of these crystallized binary and ternary phases in these systems. Exceptions are the molybdenum/silicon, uranium/

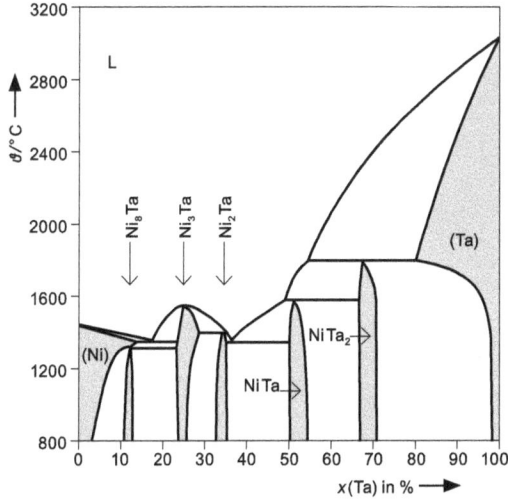

Figure 10.1.3 Phase diagram of the Ni/Ta system according to *Massalski* (Mas 1990).

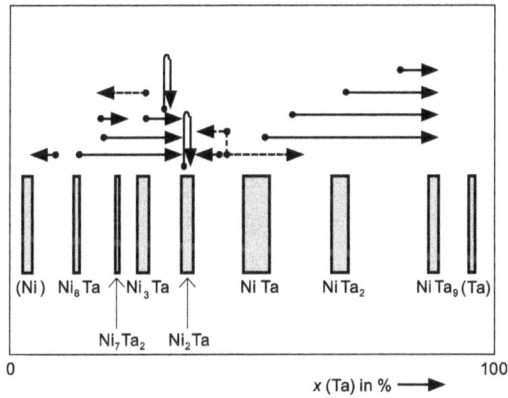

Figure 10.1.4 Transport shed in the Ni/Ta system according to *Neddermann* (Ned 1997).

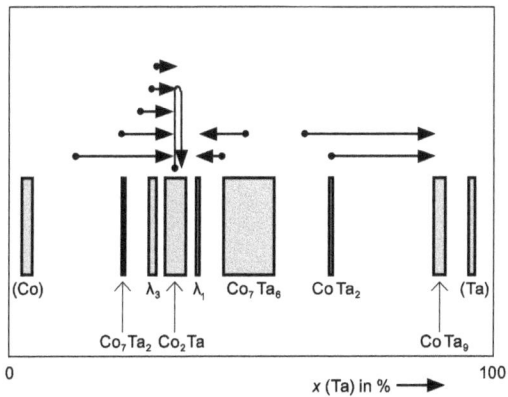

Figure 10.1.5 Transport shed in the Co/Ta system according to *Neddermann* (Ned 1997).

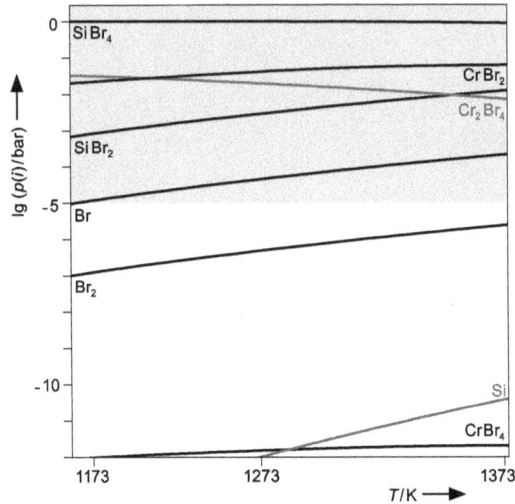

Figure 10.1.6 Gas-phase composition of the CVT of CrSi$_2$ with bromine according to *Krausze et al.* (Kra 1989).

silicon, and tungsten/silicon systems; here bromine is used as the transport agent. In the niobium/germanium and niobium/tin systems, hydrogen chloride causes a transport effect. Nickel/germanium and nickel/tin phases can be transported when gallium(III)-iodide is added.

The CVT in the Cr/Si system with the halogens chlorine, bromine, and iodine is well examined and thermodynamically understood (Kra 1989, Kra 1990). Cr$_3$Si, Cr$_5$Si$_3$ and, CrSi$_2$ can be deposited by an endothermic transport with chlorine from 1100 to 900 °C. At the same temperatures, Cr$_3$Si, Cr$_5$Si$_3$, CrSi, and CrSi$_2$ can be deposited with bromine. The transport with iodine, on the other hand, takes place exothermically from 900 to 1100 °C. In this process, Cr$_3$Si, Cr$_5$Si$_3$, CrSi, and CrSi$_2$ can be deposited. In all three cases, transport mechanisms are clearly different. If one considers the transport efficiency of the individual gas species, the following transport equations are derived:

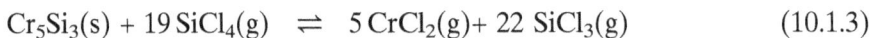

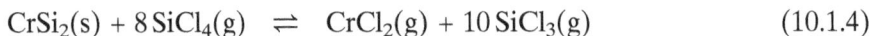

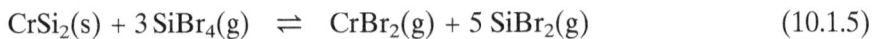

$$Cr_5Si_3(s) + 19\,SiCl_4(g) \;\rightleftharpoons\; 5\,CrCl_2(g) + 22\,SiCl_3(g) \tag{10.1.3}$$

$$CrSi_2(s) + 8\,SiCl_4(g) \;\rightleftharpoons\; CrCl_2(g) + 10\,SiCl_3(g) \tag{10.1.4}$$

$$CrSi_2(s) + 3\,SiBr_4(g) \;\rightleftharpoons\; CrBr_2(g) + 5\,SiBr_2(g) \tag{10.1.5}$$

$$CrSi_2(s) + 10\,I(g) \;\rightleftharpoons\; CrI_2(g) + 2\,SiI_4(g). \tag{10.1.6}$$

In the first three cases, the transport agent is not the added halogen chlorine or bromine, respectively, but the silicon(IV)-chloride or bromide, respectively, which was formed in a simultaneous reaction. In contrast to this, iodine functions as the transport agent directly when added.

Krauße et al. published further work comparing thermodynamic considerations of the transport behavior of silicides of the transition metals titanium, vanadium,

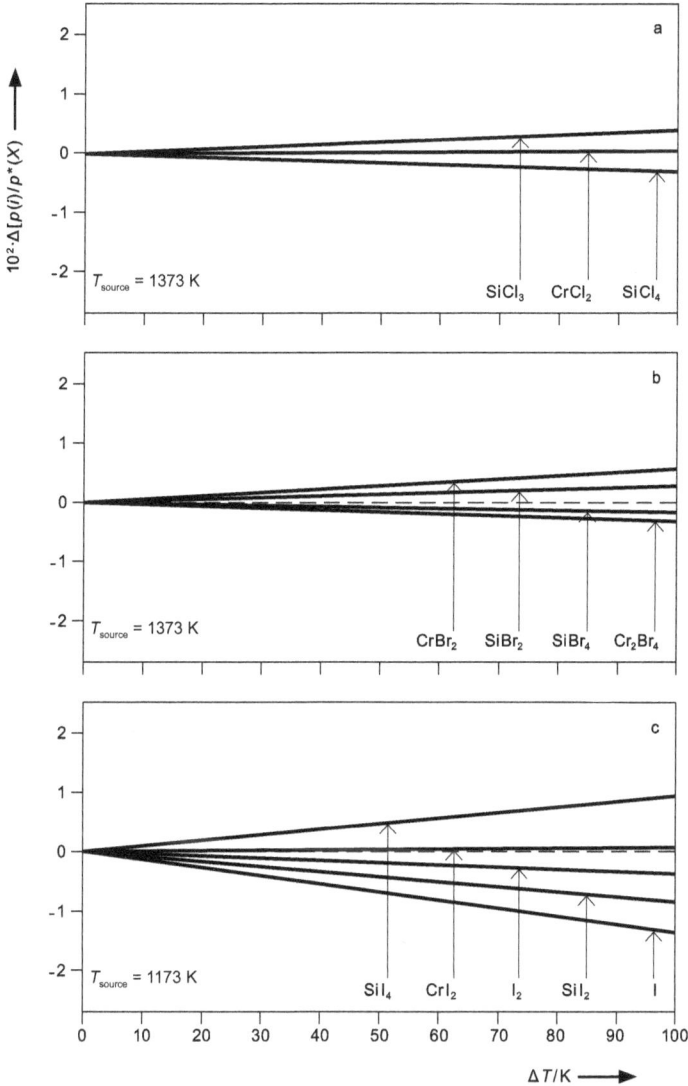

Figure 10.1.7 Transport efficiency of the different gas species during the CVT of CrSi$_2$ with a) chlorine, b) bromine, and c) iodine according to *Krausze et al.* (Kra 1989).

chromium, molybdenum, manganese, and iron with the transport agents chlorine, bromine, and iodine (Kra 1991).

All phases existing in the iron/silicon system, Fe$_2$Si, Fe$_5$Si$_3$, FeSi, and FeSi$_2$, can be crystallized by CVT reactions with iodine (Bos 2000). On the iron-rich side to FeSi, the transport takes places from cold to hot and the deposition temperatures are between 700 and 1030 °C. The transport behavior parallels that of iron. If the transport efficiency of the individual gas species is considered for the reaction of FeSi with iodine (see Figure 10.1.9), and the following transport equation can be derived:

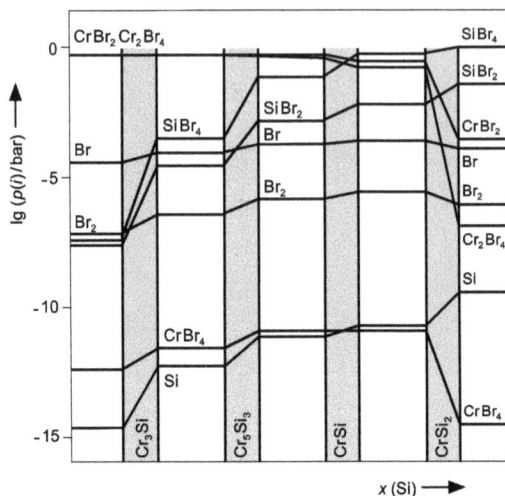

Figure 10.1.8 Gas-phase composition for the CVT of different chromium silicides with bromine according to *Krausze et al.* (Kra 1990).

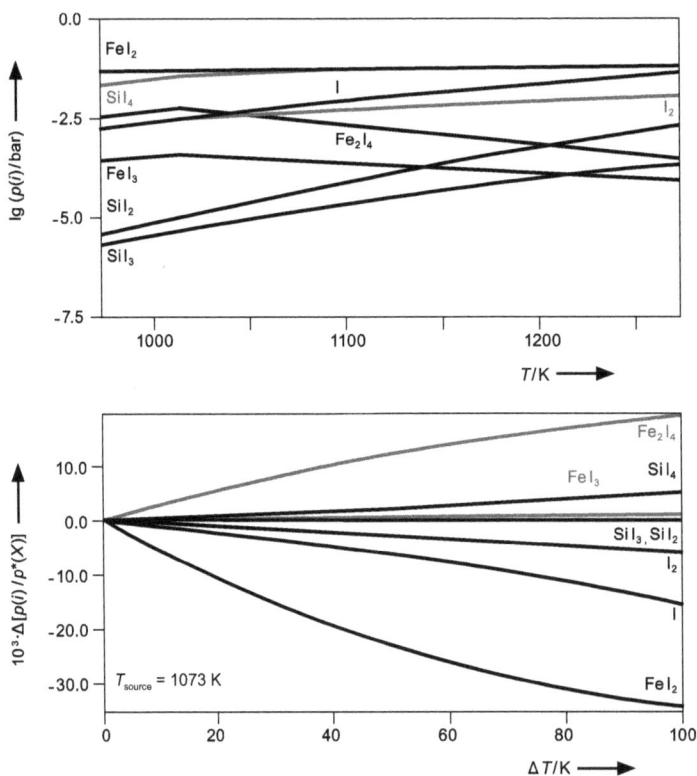

Figure 10.1.9 Gas-phase composition and transport efficiency for the CVT of FeSi with iodine according to *Bosholm et al.* (Bos 2000).

$$\text{FeSi(s)} + \frac{7}{2}\text{I}_2\text{(g)} \ \rightleftharpoons \ \text{FeI}_3\text{(g)} + \text{SiI}_4\text{(g)} \tag{10.1.7}$$

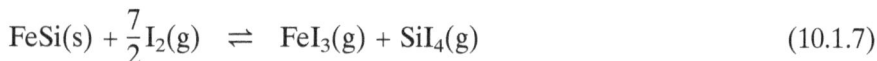

The FeI_2 and Fe_2I_4 species, which are connected by the dimerization equilibrium, contribute only little to the iron transport. The following transport equation can be formulated for the *endothermic* transport of the silicon-rich phase FeSi_2:

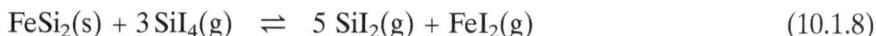

$$\text{FeSi}_2\text{(s)} + 3\,\text{SiI}_4\text{(g)} \ \rightleftharpoons \ 5\,\text{SiI}_2\text{(g)} + \text{FeI}_2\text{(g)} \tag{10.1.8}$$

In this case, the added iodine is not the transport agent but the silicon(IV)-iodide that was formed from it. Thus this transport is similar to that of silicon with SiI_4. As far as CVT reactions are concerned, FeSi_2 is one of the best examined intermetallic phases (Ouv 1972, Beh 1997, Zah 1999, Bos 2000, Beh 2001, Osa 2002, Wan 2004). Analogous investigations are described for the cobalt/silicon system (Bos 2000a). In contrast to the iron/silicon system, only the cobalt-rich phases Co_2Si, CoSi as well as a cobalt-rich mixed-crystal can be transported in the cobalt/silicon system. The deposition takes place in the hotter zone (700 → 800 °C). The following transport equation is given:

$$\text{CoSi(s)} + 6\,\text{I(g)} \ \rightleftharpoons \ \text{CoI}_2\text{(g)} + \text{SiI}_4\text{(g)} \tag{10.1.9}$$

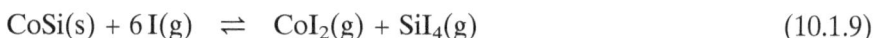

In contrast to FeSi_2, CoSi_2 cannot be obtained by CVT. Thermodynamic calculations show that the partial pressure of the only cobalt-transferring gas species CoI_2 is clearly below 10^{-5} bar in the examined temperature range, so that transport cannot take place. The added iodine transfers predominantly silicon, but virtually no cobalt, to the gas phase. The difference in the transport behavior of FeSi_2 and CoSi_2 is the low thermodynamic stability of gaseous CoI_2 compared to FeI_2 (see Figure 10.1.11).

Transport experiments in the ternary cobalt/iron/silicon system are based on these results (Bol 2003). In doing so, the formation of mixed-crystals was investigated on the quasi binary FeSi/CoSi and $\text{FeSi}_2/\text{CoSi}_2$ systems. Only mixed-crystals of the composition $\text{Fe}_{1-x}\text{Co}_x\text{Si}$ could be obtained by CVT from 700 to 900 °C with the additive iodine. Experiments as well as thermodynamic calculations show the relation between the composition of the solid deposited according to the composition of the source solid.

First transport experiments in the nickel/silicon system imply that only the metal-rich binary phases Ni_5Si_2, Ni_2Si (Bos 2000b), and Ni_3Si (Son 2007) can be prepared by exothermic CVT in the range from 500 to 900 °C with iodine.

The transport of Ti_5Si_3 (Hoa 1988), TiSi (Hoa 1988), and TiSi_2 (Pes 1986, Hoa 1988, Kra 1991) is described for the titanium/silicon system. Thereby, the transport behavior of TiSi_2 is analyzed with the help of thermodynamic model calculations for the transport agents chlorine, bromine, and iodine (Pes 1986). The transport of the three mentioned transport agents basically takes place endothermically from 1000 to 800 °C whereby the best formed crystals can be obtained at a low temperature gradient from 1000 to 950 °C. Further examples of the transport of silicides are shown in Table 10.1.

Apart from silicides, a multiplicity of transition metals are preparatively available by CVT. This way, amongst others, the titanium germanides Ti_5Ge_3 and TiGe_2 can be obtained by CVT with iodine from 900 to 700 °C (Wir 2000). The niobium germanides Nb_5Ge_3, Nb_3Ge_2, and NbGe_2 can be deposited exothermi-

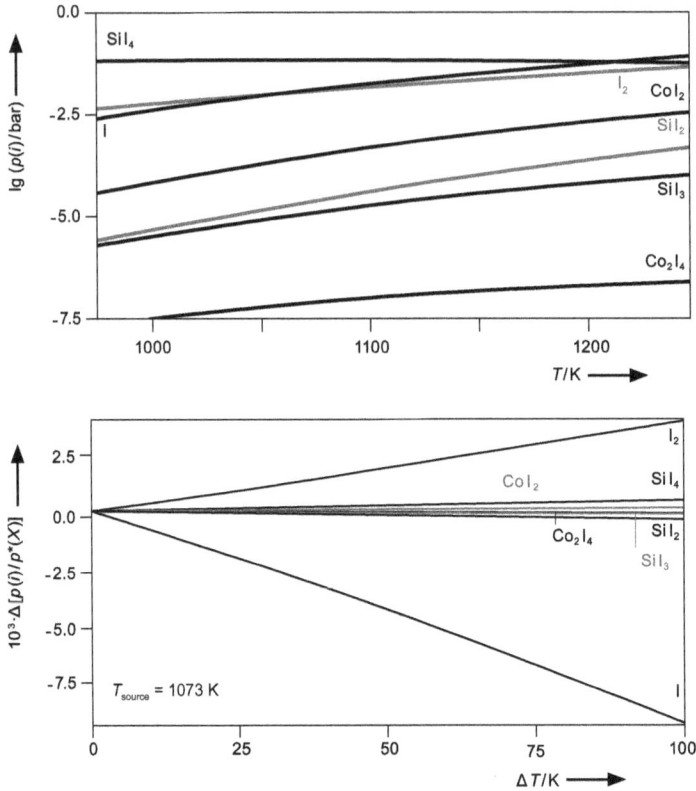

Figure 10.1.10 Gas-phase composition and transport efficiency during the CVT of CoSi with iodine according to *Bosholm et al.* (Bos 2000a).

cally with hydrogen chloride as the transport-effective additive (Hor 1971). Extensive experimental investigations and thermodynamic calculations prove that all solid phases that exist in the chromium/germanium system can be prepared by CVT with iodine from 780 to 880 °C. Only Cr_3Ge is congruently transportable. All other phases show an incongruent transport, the composition of source and sink solid differ. Source solids that are more germanium-rich than Cr_3Ge lead to further more germanium-rich sink solids. Source solids that are more chromium-rich than Cr_3Ge lead to further more chromium-rich sink solids. Thus Cr_3Ge is a so-called *transport shed* (Ned 1998).

The transport of Co_5Ge_3 with iodine from 900 to 700 °C is experimentally proven as well as explained by thermodynamic simulations (Ger 1996). Extensive experimental works and thermodynamic calculations for transport in the iron/germanium system (Ric 1967, Bos 2001) have been published. All phases that are contained in this system can be obtained by CVT reactions with iodine. The transport takes place exothermically with deposition temperatures between 800 and 950 °C on the iron-rich side (above $x(Fe) = 64$ %). The temperatures of the source solid are generally 100 K lower. Germanium-rich phases such as Fe_6Ge_5 ($560 \rightarrow 500$ °C), FeGe, and $FeGe_2$ ($700 \rightarrow 600$ °C) on the other hand are transported endothermically.

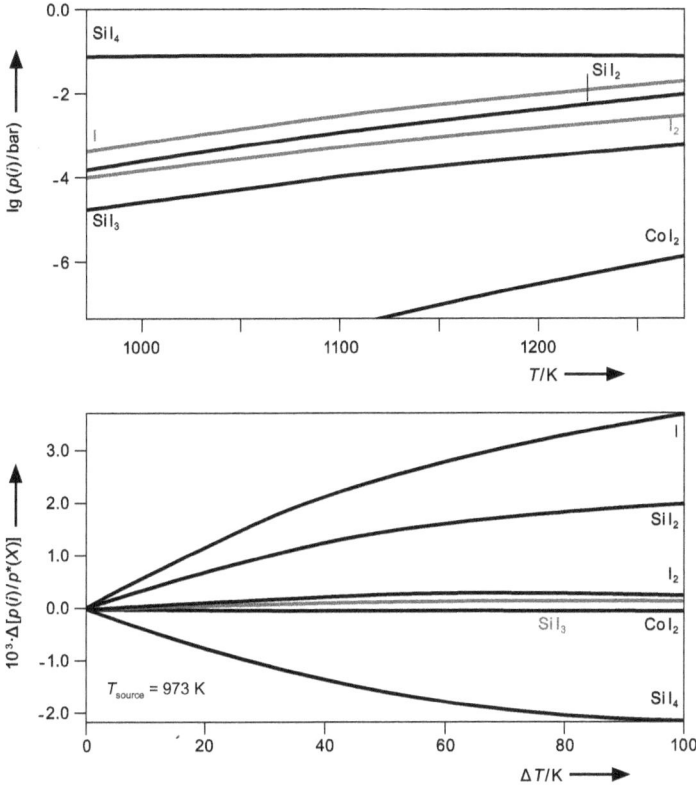

Figure 10.1.11 Gas-phase composition and transport efficiency of the individual gas species in the CoSi$_2$/iodine system according to *Bosholm et al.* (Bos 2000a).

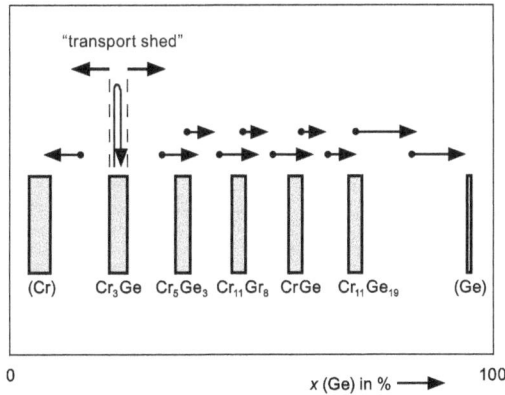

Figure 10.1.12 Transport shed in the chromium/germanium system according to *Neddermann* (Ned 1997).

The crystallization of FeGe, which forms peritectoidally and occurs in three modifications (monoclinic, hexagonal, cubic), is of particular interest for CVT. The individual modifications, especially the cubic one, which is the stable modifi-

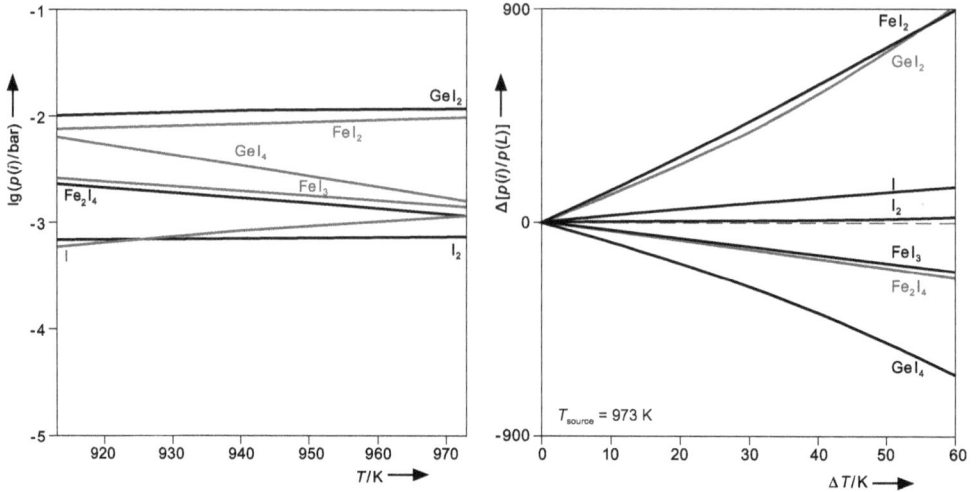

Figure 10.1.13 Gas-phase composition (a) and transport efficiency (b) of the CVT of FeGe with iodine from 700 to 640 °C under deposition of the hexagonal modification.

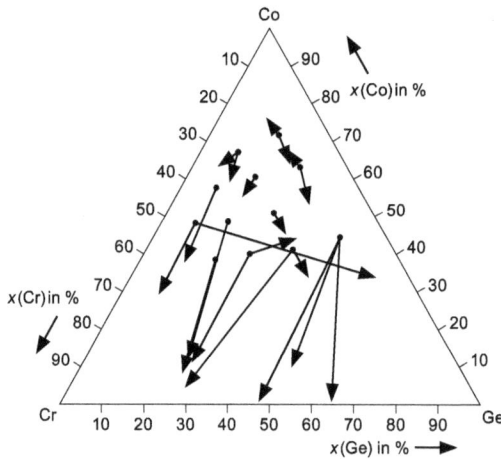

Figure 10.1.14 Relation between the composition of the solid in the source and sink in the cobalt/chromium/germanium system according to *Wirringa* and *Binnewies* (Wir 2000b). (The arrowhead indicates the composition of the sink solid.)

cation below 630 °C, is prepared by transport reactions in form of large crystals only (Wil 2007). The nickel germanides Ni_3Ge and Ni_5Ge_3 as well as a nickel-rich mixed-crystal can be transported with gallium(III)-iodide as transport-effective additive from 1000 to 800 °C (Ger 1998). Cu_3Ge and Cu_5Ge can be deposited in crystalline form as well as a copper-rich mixed-crystal by CVT with bromine and iodine from 570 to 630 °C (Wir 2000a). Apart from the multiplicity of binary transition metals, a series of ternary intermetallic phases in the chromium/cobalt/germanium and cobalt/tantalum/germanium systems can be obtained by CVT with iodine from 800 to 900 °C. These systems are characterized by relatively

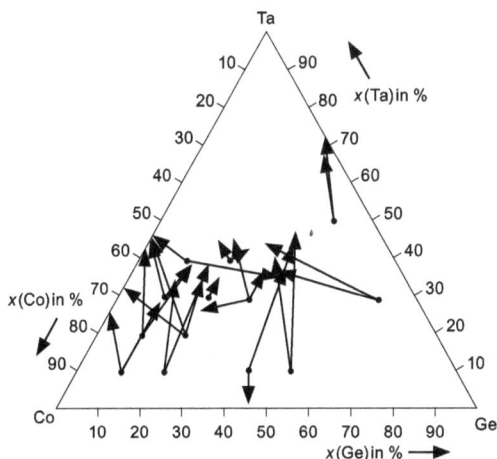

Figure 10.1.15 Relation between the composition of the solid in the source and sink in the tantalum/cobalt/germanium system according to *Wirringa* and *Binnewies* (Wir 2000b). (The arrowhead indicates to the composition of the sink solid.)

complex relations between the composition of the solid in the source and in the sink due to an incongruent transport (Wir 2000b) (see Figure 10.1.14 and 10.1.15).

In comparison with the silicides and germanides, we know only a few examples of the CVT of stannides. In most publications, the transport is only covered as a preparative method; however, the transport mechanism is not analyzed. Hence, Ti_2Sn_3 can be obtained in the presence of iodine at 500 °C. The transport of Nb_3Sn is possible when tellurium(IV)-chloride (960 → 890 °C) (Nor 1990) as well as when hydrogen chloride (850 → 900 °C) (Han 1967) is added. The transport takes place via the following equilibrium:

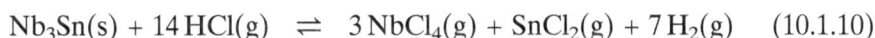

$$Nb_3Sn(s) + 14\,HCl(g) \;\rightleftharpoons\; 3\,NbCl_4(g) + SnCl_2(g) + 7\,H_2(g) \qquad (10.1.10)$$

Furthermore, Co_3Sn_2 and $CoSn$ can be transported with gallium(III)-iodide as the transport-effective additive from 900 to 700 °C. By adding gallium(III)-iodide, the solubility of cobalt in the gas phase is increased by complex formation so that cobalt-containing and tin-containing gas species SnI_2, SnI_4, Sn_2I_4, CoI_2, Co_2I_4, $CoGaI_5$ and $CoGa_2I_8$, respectively, have to be considered. Additionally, the formation of a tin-containing gas complex of the composition $SnGaI_5$ would be possible in an analogous way to $GeGaI_5$ (Ger 1998). The transport of $Cu_{41}Sn_{11}$, $Cu_{10}Sn_3$, and a copper-rich mixed-crystal with iodine from 550 to 600 °C succeeds in the copper/tin system (Ger 1998). Extensive examinations and thermodynamic model calculations have been published concerning the nickel/tin system. Accordingly, Ni_3Sn and Ni_3Sn_2 can be transported with iodine from 900 to 700 °C (Ger 1997). The crystallization of $FeSn_2$ with iodine (490 → 460 °C) took place by short distance transport, vertical to the ampoule center-line along a diffusion distance of approx. 4 cm. By this arrangement, the substance transport is intensified by convection. The convection becomes the essential substance transferring process already below 3 bar. Also, the transport rate increases due to the shorter diffusion distance and the extension of the cross section (see Chapter 2) (Arm

2007). Short distance transport is suited in particular for crystallization of phases that can principally be transported but only with very low transport rates. Due to the vast convective part of the substance transport, thermodynamic models can show if a transport is principally possible; the transport rate, however, cannot be predicted. Due to the experimental arrangement, the short distance transport is suited for endothermic transport reactions in particular.

Antimonides The CVT of only a few binary antimonides is described in the literature. These are $CrSb_2$ (Kje 1979), Mo_3Sb_7 (Jen 1966), $FeSb_2$ (Ros 1972), $CoSb_3$ (Ack 1977), $ThSb_x$ (Hen 1977), USb_2 (Hen 1979), GaSb (Ari 1971, Okh 1978), $ZrSb_2$ (Czu 2010) as well as the niobium antimonides Nb_3Sb, Nb_5Sb_4, and $NbSb_2$ (Sch 1965). Chlorine and iodine as well as hydrogen chloride, chromium(III)-chloride, and antimony(V)-chloride are used as transport agents. The CVT of antimonides is not examined systemically. The background of the observed transport actions are not described in the few published works. It is Assumed that antimony is transferred in the form of the respective halide or subhalide when a halogen is used as the transport agent. It can be transferred to the gas phase with low partial pressure in elemental form. Due to the low antimony saturation pressure, which is only $2.7 \cdot 10^{-3}$ at 827 °C, condensation sets in if elemental antimony is present as soon as the partial pressure of antimony exceeds the saturation pressure. The use of a transport agent that is able to transfer antimony is necessary because the saturation pressure is not sufficient. Antimony(III)-iodide is relatively unstable; at higher temperature, antimony(I)-iodide is increasingly formed instead; however, only to a minor degree. Chlorine should be preferred as the transport agent for the CVT of antimony and metal antimonides because the stability of the halides increases from antimony(III)-iodide through antimony(III)-bromide to antimony(III)-chloride. However, chlorine as a transport agent forms very stable gaseous chlorides with numerous metals. Hence the transport agent transfers the respective metal to the gas phase but not antimony. The equilibrium positions of the reactions of chlorine with antimony on this side and with typical metals on the other are so different that there is generally no deposition of an antimonide. The deposition of NiBi in the temperature gradient from 525 to 475 °C when bromine or iodine, respectively, is added provides an example of the CVT of a bismuthide (Ruc 1999).

Borides Examples of the transport of borides are also rare. Chemical vapor transport of borides are an attractive preparative alternative as borides are generally difficult to prepare because of their high melting temperatures. Information on the transport of the following borides can be found in the literature: CrB (Nic 1966), CrB_2 (Nic 1966), VB_2 (Nic 1966), NbB_2 (Arm 1978), TaB_2 (Arm 1978), TiB_2 (Feu 1979, Mon 1987), and ZrB_2 (Nic 1966) as well as the deposition of LaB_6 (Nie 1968) and NbB_2 (Mot 1975). The CVT of TiB_2 with chlorine, bromine, iodine, boron(III)-bromide, boron(III)-iodide, and tellurium(IV)-chloride are especially well examined and described with the help of thermodynamic model calculations (Feu 1979, Mon 1987, Ber 1981, Ber 1979). Raman spectroscopic examinations of the gas phase and thermodynamic calculations show that the boron(III)-halides are the basic boron-transferring species and the titanium(IV)-halides are the basic titanium-transferring species. Tellurium(IV)-chloride and boron(III)-bromides are particularly effective as the highest transport rates in the endothermic transport can be achieved with them. Furthermore, the publications of *Feurer* and *Constant* show clearly that different transport agents lead to different compositions of the deposited crystals. Hence during the CVT of TiB_2 with iodine, crystals of the composition $TiB_{1.89}$ are deposited. When tellurium(IV)-chloride is used, crystals of the composition $TiB_{1.96}$ develop. Besides works on the CVT of borides, there are some publications on the synthesis of borides via the gas phase in flowing systems, for example of TiB_2, CrB_2, NbB_2, TaB_2, LaB_6, and $Nd_2Fe_{14}B$ (Mur 1975, Mot 1975, Pes 2000).

Table 10.1 Examples of the chemical vapor transport of intermetallic phases.

Sink solid	Transport additive	Temperature / °C	Reference
Ag/Ga	I_2	$600 \to 400, 700 \to 500$	Pla 2000
AlPt$_3$	I_2	$400 \to 600$	Sau 1992
Au/Cu	I_2	850	Nic 1973
Au/Ni	Cl_2	900	Nic 1973
Co/Cr	I_2	$800 \to 900$	Wir 2000b
Co/Cr/Ge	I_2	$800 \to 900$	Wir 2000b
Co:Ge	I_2	$900 \to 700$	Ger 1996a
CoGe	I_2	$900 \to 700$	Ger 1996, Ger 1996a
Co$_5$Ge$_3$	I_2	$900 \to 700$	Ger 1996, Ger 1996a
CoGe$_2$	I_2	$900 \to 700$	Ger 1996a
Co/Ni	$GaI_3 + I_2$	$900 \dots 1000 \to 800$	Ned 1996
CoSb$_3$	Cl_2	$750 \to 725$	Ack 1977
CoSi	I_2	$700 \to 800$	Bos 2000a
Co$_{1-x}$Fe$_x$Si	I_2	$700 \to 900$	Bol 2003
Co$_2$Si	I_2	$800 \to 900$	Bos 2000a
CoSn	GaI_3	$900 \to 700$	Ger 1998, Ger 1996a
Co$_3$Sn	GaI_3	$900 \to 700$	Ger 1996a
Co$_3$Sn$_2$	GaI_3	$900 \to 700$	Ger 1996a, Ger 1998
Co:Ta	I_2	$800 \to 950$	Ned 1997
Co$_2$Ta	I_2	$800 \to 950$	Ned 1997
Co$_6$Ta$_7$	I_2	$800 \to 950$	Ned 1997
CoTa$_2$	I_2	$800 \to 950$	Ned 1997
Co$_7$Ta$_2$	I_2	$800 \to 950$	Ned 1997
CoTa$_9$	I_2	$800 \to 950$	Ned 1997
Co/Ta/Ge	I_2	$800 \to 950$	Wir 2000b
CrB	Cl_2, Br_2, I_2	$1100 \to 950$	Nic 1966
CrB$_2$	Cl_2, Br_2, I_2	$1050 \to 850$	Nic 1966
Cr:Ge	I_2	$780 \to 880$	Ned 1997
Cr$_3$Ge	I_2	$780 \to 880$	Ned 1997, Ned 1998
Cr$_{11}$Ge$_{19}$	I_2	$780 \to 880$	Ned 1997, Ned 1998
CrGe	I_2	$780 \to 880$	Ned 1997, Ned 1998
Cr$_{11}$Ge$_8$	I_2	$780 \to 880$	Ned 1997, Ned 1998
Cr$_5$Ge$_3$	I_2	$780 \to 880$	Ned 1997, Ned 1998
CrSb$_2$	$CrCl_3$	not specified	Kje 1979

Table 10.1 (continued)

Sink solid	Transport additive	Temperature / °C	Reference
Cr/Si	Br$_2$	1000	Nic 1971a
Cr$_3$Si	Cl$_2$, Br$_2$, I$_2$	1100 → 900, 1100 → 900, 900 → 1100	Kra 1990
Cr$_5$Si$_3$	Cl$_2$, Br$_2$, I$_2$	1100 → 1000, 1100 → 1000, 900 → 1100	Kra 1990
Cr$_3$Si$_2$	Cl$_2$, Br$_2$, I$_2$	1050 → 850	Nic 1966
	Br$_2$	1100 → 1000	Kra 1990
	I$_2$	900 → 1100	Kra 1990
CrSi	Cl$_2$, Br$_2$, I$_2$	1025	Kra 1991
CrSi$_2$	Cl$_2$, Br$_2$, I$_2$	1100 → 950, 1100 → 950, 1100 → 950	Nic 1966
	Cl$_2$	1100 ... 900 → 1000 ... 700	Nic 1971a
	Cl$_2$	1100 → 1000	Kra 1989
	Cl$_2$, Br$_2$, I$_2$	1100 → 1000, 1100 → 900, 900 → 1100	Kra 1990
	Cl$_2$, Br$_2$, I$_2$	1025	Kra 1991
	Br$_2$	1050 → 800	Nic 1971a
	Br$_2$	1100 → 900	Kra 1989
	I$_2$	1000 → 1100	Kra 1989
	I$_2$	900 → T_1	Szc 2007
	Cl$_2$	1025	Kra 1991
	Cl$_2$, Br$_2$, I$_2$	1025	Kra 1991
Cu/Ag	I$_2$	750	Nic 1973
	I$_2$	600 → 700	Ger 1995
Cu/Ga	I$_2$	800 → 700	Pla 2000
Cu$_{1-x}$Ga$_x$	I$_2$	800 → 650 ... 500	Kos 2007
Cu:Ge	Br$_2$, I$_2$	570 → 630	Wir 2000a
Cu$_5$Ge	Br$_2$, I$_2$	570 → 630	Wir 2000a
Cu$_3$Ge	Br$_2$, I$_2$	570 → 630	Wir 2000a
Cu:Si	I$_2$	600 → 700	Wir 2000a
Cu$_5$Si	I$_2$	600 → 700	Wir 2000a
Cu$_3$Si	I$_2$	600 → 700	Wir 2000a
	HCl	700 → 1150	Tej 1988
	HCl	550 ... 700 → 1075 ... 1300	Tej 1989
Cu:Sn	I$_2$	550 → 600	Ger 1998
Cu$_{41}$Sn$_{11}$	I$_2$	550 → 600	Ger 1996a, Ger 1998
Cu$_{10}$Sn$_3$	I$_2$	550 → 600	Ger 1996a, Ger 1998
Fe$_3$Ge	I$_2$	850 → 950	Bos 2001
	I$_2$	900 → 1000	Bos 2000
Fe$_6$Ge$_5$	I$_2$	560 → 500	Bos 2001

Table 10.1 (continued)

Sink solid	Transport additive	Temperature / °C	Reference
FeGe	I_2	$T_2 \rightarrow 560$	Ric 1967
	I_2	$700 \rightarrow 600$	Bos 2001
	I_2	$575 \rightarrow 535$	Will 2007
	$FeBr_2$	$T_2 \rightarrow 745$	Ric 1967
$FeGe_2$	I_2	$700 \rightarrow 600$	Bos 2001
Fe/Ni	Cl_2	900	Nic 1973
$FeSb_2$	Cl_2	$700 \rightarrow 650$	Fan 1972
	Cl_2	$700 \rightarrow 650$	Ros 1972
Fe/Si	Cl_2, Br_2, I_2	$750 \rightarrow 900$	Nic 1973
	I_2	$1050 \rightarrow 950 \dots 750$	Wan 2006
Fe_3Si	Cl_2	1025	Kra 1991
	I_2	$800 \rightarrow 900$	Bos 2000
Fe_2Si	I_2	three zone experiment	Ger 1996a
	I_2	$950 \rightarrow 1030$	Bos 2000
Fe_5Si_3	Cl_2	1025	Kra 1991
FeSi	Br_2, I_2	900	Kra 1991
	I_2	$1050 \rightarrow 900$	Ouv 1972
	I_2	three zone experiment	Ger 1996a
	I_2	$700 \rightarrow 800$	Bos 2000
$FeSi_2$	Br_2, I_2	900	Kra 1991
	I_2	$825 \rightarrow 750, 1030 \rightarrow 830$	Ouv 1972
	I_2	$1050 \rightarrow 750, 1050 \rightarrow 950$	Beh 1997
	I_2	$1050 \rightarrow 950 \dots 750$	Hei 1996
	I_2	$1050 \rightarrow 750$	Zha 1999
	I_2	$1000 \rightarrow 800$	Bos 2000
	I_2	$1050 \rightarrow 750$	Beh 2001
	I_2	$1050 \rightarrow 750, 1050 \rightarrow 950$	Li 2003
	I_2	$1050 \rightarrow 850 \dots 750$	Wan 2004
	I_2	$1050 \rightarrow 950 \dots 750$	Wan 2006
$FeSi_3$	Cl_2	1025	Kra 1991
$FeSn_2$	I_2	$490 \rightarrow 460$	Arm 2007
GaSb	I_2	not specified	Okh 1978
	I_2, HCl, $SbCl_5$	$650 \rightarrow 600$	Ari 1971
Ge:Cr	I_2	$780 \rightarrow 880$	Ned 1997, Ned 1998
Ge:Ta	I_2	$800 \rightarrow 950$	Ned 1997
$Ge_{1-x} Si_x$	Br_2	$1150 \rightarrow 800$	Dru 2005
LaB_6	$BCl_3 + H_2$	$1000 \rightarrow 1350 \dots 1450$	Nie 1968
$MnSi_{1.7}$	Cl_2, I_2	$900, 800 \rightarrow 900$	Kra 1991
$MnSi_{1.73}$	I_2, $CuCl_2$, $MnCl_2$	$800 \rightarrow 900, 900 \rightarrow 800, 800 \rightarrow 900$	Koj 1975
$Mn_{15}Si_{26}$	$CuCl_2$	$900 \rightarrow 800$	Koj 1979

Table 10.1 (continued)

Sink solid	Transport additive	Temperature /°C	Reference
Mo_3Sb_7	Cl_2, Br_2, I_2	not specified	Jen 1966
Mo/Si	Br_2	1000	Nic 1971a
Mo_3Si	I_2	1025	Kra 1991
Mo/W	$HgBr_2$	1000 → 900	Ned 1996a
NbB_2	I_2	1050 → 900	Arm 1978
Nb_3Ga	I_2	not specified	Web 1975
	HCl	860 → 940	Hor 1971
Nb_3Ga_5	HCl	860 → 920, 910 → 970	Hor 1971
Nb_5Ge_3	HCl	870 → 920	Hor 1971
Nb_3Ge_2	HCl	820 → 870	Hor 1971
$NbGe_2$	HCl	820 → 870	Hor 1971
Nb_3Sb	I_2	840 → 970	Sch 1965
Nb_5Sb_4	I_2	830 → 980	Sch 1965
$NbSb_2$	I_2	1010 → 830	Sch 1965
Nb_5Si_3	HCl	800 → 900, 870 → 930	Hor 1971
$NbSi_2$	Cl_2, Br_2, I_2	1050 → 850	Nic 1966
Nb_3Sn	HCl	850 → 900	Han 1967
	$TeCl_4$	890 → 960	Nor 1990
NiBi	Br_2, I_2	525 → 475	Ruc 1999
Ni/Cu	Cl_2, Br_2, I_2	700, 850, 1000	Nic 1973
$Ni_{1-x}Ga_x$	I_2	920 → 590 … 800	Kos 2007
Ni:Ge	GaI_3	1000 → 800	Ger 1998
Ni_3Ge	GaI_3	1100 → 900	Ger 1998
Ni_5Ge_3	GaI_3	1000 → 800	Ger 1998
Ni_3Si	I_2	970 → 810 … 850	Son 2007
Ni_3Sn	I_2	900 → 700	Ger 1997
Ni_3Sn_2	I_2	900 → 700	Ger 1997
Ni:Ta	I_2	800 → 950	Ned 1997
Ni_8Ta	I_2	800 → 950	Ned 1997
Ni_7Ta_2	I_2	800 → 950	Ned 1997
Ni_3Ta	I_2	800 → 950	Ned 1997
Ni_2Ta	I_2	800 → 950	Ned 1997
NiTa	I_2	800 → 950	Ned 1997
$NiTa_2$	I_2	800 → 950	Ned 1997
$NiTa_9$	I_2	800 → 950	Ned 1997
Pd_xAl	I_2	600	Mer 1980
Pd_2Al	I_2	375 → 600	Sch 1975
	Cl_2, I_2, $TeCl_4$	1000 → 800	Dei 1998
PdAl	I_2	1050 → 850	Dei 1998
Pd_2Ga	$I_2 + GaI_3$	400 → 600	Dei 1998
	GaI_3	400 → 600	Kov 2008
PdGa	$I_2 + GaI_3$	400 → 600	Dei 1998
	$I_2 + AlI_3$	450 → 600	Dei 1998
Pd_3In	$I_2 + InI_3$	400 → 600	Dei 1998

Table 10.1 (continued)

Sink solid	Transport additive	Temperature / °C	Reference
Pd_2In	$I_2 + InI_3$	$400 \rightarrow 600$	Dei 1998
Pt/Si	I_2	800	Nic 1973
$ReSi_2$	Cl_2, I_2	$1050 \rightarrow 900$	Khr 1989
TaB_2	I_2	$1050 \rightarrow 900$	Arm 1978
$Ta{:}Co$	I_2	$800 \rightarrow 950$	Ned 1997
$Ta{:}Ge$	I_2	$800 \rightarrow 950$	Ned 1997
Ta_9Ge	I_2	$800 \rightarrow 950$	Ned 1997
Ta_5Ge_3	I_2	$800 \rightarrow 950$	Ned 1997
$TaGe$	I_2	$800 \rightarrow 950$	Ned 1997
$Ta{:}Ni$	I_2	$800 \rightarrow 950$	Ned 1997
Ta/Si	Br_2	1000	Nic 1971a
$TaSi_2$	Cl_2, Br_2, I_2	$1100 \rightarrow 950$	Nic 1966
$ThSb_x$	I_2	$T_1 \rightarrow T_2$	Hen 1977
TiB_2	Cl_2, Br_2	$1100 \rightarrow 950$	Nic 1966
	Br_2	1020	Nic 1973
	I_2	$1100 \rightarrow 950$	Nic 1966
	I_2, BI_3	$850 \rightarrow 950, 775 \rightarrow 925$	Feu 1979
	$BI_3, BBr_3, TiBr_4$	not specified	Ber 1981
	$TeCl_4, BBr_3$	$975 \rightarrow 845, 960 \rightarrow 830$	Feu 1979
	$Cl_2, Br_2, I_2, TeCl_4$	$1000 \rightarrow 900$	Mon 1987
TiC/Si	I_2	1100	Nic 1973
Ti/Ge	Cl_2, Br_2, I_2	750	Nic 1973
Ti_5Ge_3	I_2	$900 \rightarrow 700$	Wir 2000
$TiGe_2$	I_2	$900 \rightarrow 700$	Wir 2000
Ti/Si	I_2	1100	Nic 1973
$TiSi$	Cl_2	$1000 \rightarrow 900$	Hoa 1988
Ti_5Si_3	Cl_2, I_2	$1000 \rightarrow 900$	Hoa 1988
	Cl_2	$1000 \rightarrow 900$	Kra 1987
	I_2	$900 \rightarrow 1000$	Hoa 1988
	I_2	$1000 \rightarrow 900, 900 \rightarrow 1000$	Hoa 1988
	Cl_2	$1000 \rightarrow 900$	Kra 1987
$TiSi_2$	Cl_2	$1050 \rightarrow 850$	Nic 1966
	Cl_2	$1100 \ldots 900$ $\rightarrow 1000 \ldots 700$	Nic 1971a
	Cl_2	$1000 \rightarrow 900$	Kra 1987
	Cl_2, Br_2, I_2	$1000 \rightarrow 800$	Pes 1986
	Cl_2	$1000 \rightarrow 900$	Hoa 1988
	Cl_2, Br_2, I_2	1025	Kra 1991
	Br_2	$1050 \rightarrow 850$	Nic 1966
	Br_2	$1000 \rightarrow 800$	Nic 1971a
	I_2	$1050 \rightarrow 850$	Nic 1966
	I_2	$1000 \rightarrow 900, 900 \rightarrow 1000$	Hoa 1988

Table 10.1   (continued)

Sink solid	Transport additive	Temperature / °C	Reference
Ti_2Sn_3	I_2	500	Kle 2000
USb_x	I_2	$T_1 \rightarrow T_2$	Hen 1977
USb_2	I_2	$720 \rightarrow 860$	Hen 1979
U/Si	Br_2	1000	Nic 1971a
VB_2	Cl_2, Br_2, I_2	$1100 \rightarrow 950$	Nic 1966
V/Si	Br_2	1000	Nic 1971a
VSi_2	Cl_2	$1100 \rightarrow 950$	Nic 1966
	Cl_2	$1100 \ldots 900$ $\rightarrow 1000 \ldots 700$	Nic 1971a
	Cl_2	$1100 \rightarrow 1000$	Bar 1983
	Cl_2, I_2	1025	Kra 1991
	Br_2, I_2	$1100 \rightarrow 950$	Nic 1966
	Br_2	$1050 \rightarrow 800$	Nic 1971a
	Br_2	1000	Nic 1973
	$SiCl_4$	$1100 \rightarrow 1000$	Bar 1983
V_3Ga	I_2	$720 \rightarrow 600$	Das 1978
V_3Si_2	$Cl_2, SiCl_4$	$1100 \rightarrow 1000$	Bar 1983
V_5Si_3	$Cl_2, SiCl_4$	$1100 \rightarrow 1000$	Bar 1983
	Cl_2	1025	Kra 1991
W/Si	Br_2	1000	Nic 1971a
ZrB_2	Cl_2, Br_2, I_2	$1100 \rightarrow 950$	Nic 1966
ZrGeTe	$TeCl_4$	$900 \rightarrow 950$	Wan 1995
$ZrSb_2$	I_2	$500 \rightarrow 600$	Czu 2010
$ZrSi_{1-x}Ge_xTe$	$TeCl_4$	$900 \rightarrow 950$	Wan 1995
ZrSnTe	$TeCl_4$	$900 \rightarrow 950$	Wan 1995

Bibliography

Ack 1977   J. Ackermann, A. Wold, *J. Phys. Chem. Solids* **1977**, *38*, 1013.
Ala 1977   F. A. S. Al-Alamy, A. A. Balchin, *J. Cryst. Growth* **1977**, *39*, 275.
Ari 1971   T. Arizumi, M. Kakehi, R. Shimokawa, *J. Cryst. Growth* **1971**, *9*, 151.
Arm 1978   B. Armas, J. H. E. Jeffes, M. G. Hocking, *J. Cryst. Growth* **1978**, *44*, 609.
Arm 2007   M. Armbrüster, M. Schmidt, R. Cardoso-Gil, H. Borrmann, Y. Grin, *Z. Kristallogr. NCS* **2007**, *222*, 83.
Bar 1983   K. Bartsch, E. Wolf, *Z. Anorg. Allg. Chem.* **1983**, *501*, 27.
Beh 1997   G. Behr, J. Werner, G. Weise, A. Heinrich, A. Burkov, C. Gladun, *Phys. Stat. Sol.* **1997**, *169*, 549.
Beh 2001   G. Behr, L. Ivanenko, H. Vinzelberg, A. Heinrich, *Thin Solid Films* **2001**, *381*, 276.
Bei 1977   I. Beinglass, G. Dishon, A. Holzer, M. Schieber, *J. Cryst. Growth* **1977**, *42*, 166.

Ber 1979   C. Bernard, G. Constant, R. Feurer, *Proc. Electrochem. Soc.* **1979**, *79*, 368

Ber 1981   C. Bernard, G. Constant, R. Feurer, *J. Electrochem. Soc.* **1981**, *129*, 1377.

Bol 2003   R. Boldt, W. Reichelt, O. Bosholm, H. Oppermann, *Z. Anorg. Allg. Chem.* **2003**, *629*, 1839.

Bos 2000   O. Bosholm, H. Oppermann, S. Däbritz, *Z. Naturforsch.* **2000**, *55b*, 614.

Bos 2000a   O. Bosholm, H. Oppermann, S. Däbritz, *Z. Naturforsch.* **2000**, *55b*, 1199.

Bos 2000b   O. Bosholm, *Dissertation*, University of Dresden, **2000**.

Bos 2001   O. Bosholm, H. Oppermann, S. Däbritz, *Z. Naturforsch.* **2001**, *56b*, 329.

Czu 2010   A. Czulucki, *Dissertation*, University of Dresden, **2010**.

Das 1978   B. N. Das, J. D. Ayers, *J. Cryst. Growth* **1978**, *43*, 397.

Dei 1998   J. Deichsel, *Dissertation*, University of Hannover, **1998**.

Dru 2002   A. Druzhinin, E. N. Lavitska, I. I. Maryamova, H. W. Kunert, *Adv. Eng. Mater.* **2002**, *4*, 589.

Dru 2005   A. A. Druzhinin, I. P. Ostrovskii, Y. M. Khoverko, Y. V. Gij, *Functional Materials* **2005**, *12*, 738.

Fan 1972   A. K. L. Fan, G. H. Rosenthal, H. L. McKinzie, A. Wold, *J. Solid State Chem.* **1972**, *5*, 136.

Feu 1979   R. Feurer, G. Constant, *J. Less-Common Met.* **1979**, *67*, 107.

Ger 1995   S. Gerighausen, M. Binnewies, *Z. Anorg. Allg. Chem.* **1995**, *621*, 936.

Ger 1996   S. Gerighausen, E. Milke, M. Binnewies, *Z. Anorg. Allg. Chem.* **1996**, *622*, 1542.

Ger 1996a   S. Gerighausen, *Dissertation*, University of Hannover, **1996**.

Ger 1997   S. Gerighausen, R. Wartchow, M. Binnewies, *Z. Anorg. Allg. Chem.* **1997**, *623*, 1361.

Ger 1998   S. Gerighausen, R. Wartchow, M. Binnewies, *Z. Anorg. Allg. Chem.* **1998**, *624*, 1057.

Gru 1997   T. Grundmeier, W. Urland, *Z. Anorg. Allg. Chem.* **1997**, *623*, 1744.

Han 1967   J. J. Hanak, H. S. Berman, *Crystal Growth, Ed. H. S. Peiser (Pergamon, Oxford)* **1967**, *1*, 249.

Han 1976   J. J. Hanak, H. S. Berman, *International Conference on Crystal Growth*, Boston, **1966**, 249.

Hei 1996   A. Heinrich, A. Burkov, C. Gladun, G. Behr, K. Herz, J. Schumann, H. Powalla, *15th International Conference on Thermoelectrics* **1996**, 57.

Hen 1977   Z. Henkie, P. J. Markowski, *J. Cryst. Growth* **1977**, *41*, 303.

Hen 1979   Z. Henkie, A. Misiuk, *Krist. Techn.* **1979**, *14*, 539.

Hen 2001   Z. Henkie, A. Pietraszko, A. Wojakowski, L. Kępiński, T. Chichorek, *J. Alloys Compd.* **2001**, *317–318*, 52.

Hoa 1988   D. V. Hoanh, W. Bieger, G. Krabbes, *Z. Anorg. Allg. Chem.* **1988**, *560*, 128.

Hor 1971   R. Horyń, M. Dryś, *Krist. Techn.* **1971**, K 85.

Hua 1994   J. Huang, Y. Huang, T. Yang, *J. Cryst. Growth* **1994**, *135*, 224.

Hul 1968   F. Hulliger, *J. Less-Common Met.* **1968**, *16*, 113.

Jen 1966a   P. Jensen, A. Kjekshues, T. Skansen, *Acta Chem. Scand.* **1966**, *20*, 417.

Kan 2005   M. G. Kanatzidis, R. Pöttgen, W. Jeitschko, *Angew. Chem.* **2005**, *117*, 7156, *Angew. Chem. Int. Ed.* **2005**, *44*, 6995.

Khr 1989   M. Khristov, G. Gyurov, P. Peshev, *Cryst. Res. Technol.* **1989**, *24*, K 22.

Kje 1979   A. Kjekshus, P. G. Peterzéns, T. Rakke, A. F. Andresen, *Acta Chem. Scand.* **1979**, *33A*, 469.

Kle 2000   H. Kleinke, M. Waldeck, P. Gütlich, *Chem. Mater.* **2000**, *12*, 2219.

Koj 1975   T. Kojima, I. Nishida, *Jpn. J. Appl. Phys.* **1975**, *14*, 141.

Koj 1979   T. Kojima, I. Nishida, T. Sakata, *J. Cryst. Growth* **1979**, *47*, 589.

Kos 2007   A. V. Kosyakov, A. Yu. Zavrazhnov, A. V. Naumov, A. A. Nazarova, V. P. Zlomanov, *Inorg. Mater.* **2007**, *43*, 1199.

Kov 2008   K. Kovnir, M. Schmidt, C. Waurisch, M. Armbrüster, Y. Prots, Y. Grin, *Z. Kristallogr. – NCS* **2008**, *223, 7*.
Kra 1987   G. Krabbes, M. Ritschel, Do. V. Hoanh, *Acta Phys. Hung.* **1987**, *61*, 181.
Kra 1989   R. Krausze, M. Khristov, P. Peshev, G. Krabbes, *Z. Anorg. Allg. Chem.* **1989**, *579*, 231.
Kra 1990   R. Krausze, M. Khristov, P. Peshev, G. Krabbes, *Z. Anorg. Allg. Chem.* **1990**, *588*, 123.
Kra 1991   R. Krausze, G. Krabbes, M. Khristov, *Cryst. Res. Technol.* **1991**, *26*, 179.
Li 2003   Y. Li, L. Sun, L. Cao, J. Zhao, H. Wang, Y. Nan, Z. Gao, W. Wang, *Sci. China Ser.* **2003**, *G 46*, 47.
Loi 1972   R. Loitzl, C. Schüler, *German Patent* DE 2051404, **1972**.
Mas 1990   H. Massalski, Binary Alloy Phase Diagrams, 2nd Ed. ASN International, **1990**
Mer 1980   H. Merker, H. Schäfer, B. Krebs, *Z. Anorg. Allg. Chem.* **1980**, 662, 49.
Mon 1987   Y. Monteil, R. Feurer, G. Constant, *Z. Anorg. Allg. Chem.* **1987**, *545*, 209.
Mot 1975   S. Motojima, K Sugiyama, Y. Takahashi, *J. Cryst. Growth* **1975**, *30*, 233.
Mur 1995   K. Murase, K. Machida, G. Adachi, *J. Alloy. Compd.* **1995**, *217*, 218.
Ned 1996   R. Neddermann, M. Binnewies, *Z. Anorg. Allg. Chem.* **1996**, *622*, 17.
Ned 1996a   R. Neddermann, S. Gerighausen, M. Binnewies, *Z. Anorg. Allg. Chem.* **1996**, *622*, 21.
Ned 1997   R. Neddermann, *Dissertation*, University of Hannover, **1997**.
Ned 1998   R. Neddermann, M. Binnewies, *Z. Anorg. Allg. Chem.* **1998**, *624*, 733.
Nee 1971   H. Neels, W. Schmitz, H. Hottmann, R. Rössner, W. Topp, *Krist. Tech.* **1971**, *6*, 225.
Nic 1966   J. Nickl, M. Duck, J. Pieritz, *Angew. Chem.* **1966**, *17*, 822.
Nic 1971a   J. J. Nickl, J. D. Koukoussas, *J. Less-Common Met.* **1971**, *23*, 73.
Nic 1973   J. J. Nickl, J. D. Koukoussas, A. Mühlratzer, *J. Less-Common Met.* **1973**, *32*, 243.
Nie 1968   T. Niemyski, E. Kierzek-Pecold, *J. Cryst. Growth* **1968**, *3/4*, 162.
Nis 1980   M. Nishio, K. Tsuru, H. Ogawa, *Reports of the Faculty of Science and Engineering* **1980**, *8*, 59.
Nor 1990   M. L. Norton, N. Nevins, H. Tang, N. Chong, J. Scowyra, *Mater. Res. Bull.* **1990**, *25*, 257.
Okh 1978   Y. A. Okhrimenko, *Dielektr. Poluprovodr.* **1978**, *13*, 67.
Osa 2002   M. Osamura, *Report of researches, Nippon Institute of Technology* **2002**, *32*, 17.
Ouv 1972   J. Ouvrard, R. Wandji, B. Roques, *J. Cryst. Growth* **1972**, *13/14*, 406.
Pes 1986   P. Peshev, M. Khristov, *J. Less-Common Met.* **1986**, *117*, 361.
Pes 2000   P. Peshev, *J. Solid State Chem.* **2000**, *154*, 157.
Pla 2000   T. Plaggenborg, M. Binnewies, *Z. Anorg. Allg. Chem.* **2000**, *626*, 1478.
Ric 1967   M. Richardson, *Acta Chem. Scand.* **1967**, *21*, 2305.
Ros 1972   G. Rosenthal, R. Kershaw, A. Wold, *Mater. Res. Bull.* **1972**, *7*, 479.
Ros 1973   F. Rosenberger, M. C, DeLong, J. M Olson, *J. Cryst. Growth* **1973**, *19*, 317.
Ruc 1999   M. Ruck, *Z. Anorg. Allg. Chem.* **1999**, *625*, 2050.
Sau 1992   M. Sauer, A. Engel, H. Lueken, *J. Alloy. Compd.* **1992**, *183*, 281.
Sch 1962   H. Schäfer, *Chemische Transportreaktionen*, Verlag Chemie, Weinheim, **1962**.
Sch 1965   H. Schäfer, W. Fuhr, *J. Less-Common Met.* **1965**, *8*, 375.
Sch 1975   H. Schäfer, M. Trenkel, *Z. Anorg. Allg. Chem.* **1975**, *414*, 137.
Sch 1977   H. Schäfer, J. Nowitzki, *Z. Anorg. Allg. Chem.* **1977**, *435*, 49.
Schi 1976   M. Schieber, W. F. Schnepple, L. van den Berg, *J. Cryst. Growth* **1976**, *33*, 125.
Scho 1974   H. Scholz, *Acta Electron.* **1974**, *17*, 69.
Son 2007   Y. Song, S. Jin, *Appl. Phys. Lett* **2007**, *90*, 173122-1.

Szc 2007	J. R. Szczech, A. L. Schmitt, M. J. Bierman, S. Jin, *Chem. Mater.* **2007**, *19*, 3238.
Tej 1988	P. Tejedor, J. M. Olson, *J. Cryst. Growth* **1988**, *89*, 220.
Tej 1989	P. Tejedor, J. M. Olson, *J. Cryst. Growth* **1989**, *94*, 579.
Wan 1995	C. Wang, T. Hughbanks, *Inorg. Chem.* **1995**, *34*, 5524.
Wan 2001	C. H. Wannek, B. Harbrecht, *J. Solid State Chem.* **2001**, *159*, 113.
Wan 2002	C. H. Wannek, B. Harbrecht, *Z. Anorg. Allg. Chem.* **2002**, *628*, 1597.
Wan 2004	J. F. Wang, S. Y. Ji, K. Mimura, Y. Sato, S. H. Song, H. Yamane, M. Shimada, M. Isshiki, *Phys. Status Soidi.* **2004**, *201*, 2905.
Wan 2006	J. F. Wang, S. Saitou, S. Y. Ji, M. Isshiki, *J. Cryst. Growth* **2006**, *295*, 129.
Wan 2007	C. H. Wannek, B. Harbrecht, *Z. Anorg. Allg. Chem.* **2007**, *633*, 1397.
Web 1975	G. W. Webb, J. J. Engelhardt, *IEEE Trans. Magnetics* **1975**, *11*, 208.
Wil 2007	H. Wilhelm, M. Schmidt, R. Cardoso-Gil, U. Burkhardt, M. Hanfland, U. Schwarz, L. Akselrud, *Science and Technology of Advanced Materials* **2007**, *8*, 416.
Wir 2000	J. Wirringa, M. Binnewies, *Z. Allg. Anorg. Chem.* **2000**, *626*, 996.
Wir 2000a	J. Wirringa, R. Wartchow, M. Binnewies, *Z. Anorg. Allg. Chem.* **2000**, *626*, 1473.
Wir 2000b	J. Wirringa, M. Binnewies, *Z. Anorg. Allg. Chem.* **2000**, *626*, 1747.
Zha 1999	J. Zhao, Y. Li, R. Liu, X. Zhang, Z. Zhou, C. Wang, Y. Xu, W. Wang, *Chinese Phys. Lett.* **1999**, *16*, 208.

11 Gas Species and their Stability

It is of major importance for CVT reactions to know what happens during the reaction in the gas phase – more than in any other preparation method in solid-state chemistry. If one wants to use vapor transport reactions only in a prepara-tive way – without the purpose of understanding the course of the reaction in detail – it is often enough to detect qualitatively which solid can be transported with what kind of transport agent on an empirical basis. Knowledge about gase-ous, inorganic molecules and their stability is essential for a deeper understand-ing. A quantitative description of the transport action requires knowledge of the thermodynamic data of the condensed phases and gaseous molecules that are involved.

In this section, we will provide an overview of the different kinds of gaseous inorganic molecules that can occur during CVT reactions. In many cases, their thermodynamic data are known, can be estimated, or calculated by approxima-tion by means of quantum chemical methods so that thermodynamic model calcu-lations are possible. The following compilation makes no claim to be a complete overview of gaseous inorganic molecules. Those aspects are considered that are important for CVT. They are arranged according to substance groups: gaseous halogen compounds including oxide halogens, gaseous elements, gaseous hydro-gen compounds, and gaseous oxygen compounds.

11.1 Halogen Compounds

Metal halides and non-metal halides play a central role for CVT reactions. Halo-gens and many halogen compounds are effective and often-used transport agents. Halides built out of these are the most important transport-effective species. In the following, there will be an overview of halogens and halogen compounds as transport agents. After that, the role of halogen compounds as transport-effective gas species will be described.

Halogens and halogen compounds as transport agents The elemental halogens chlorine, bromine, and iodine are used frequently as transport agents. Fluorine, in contrast, is not suited for a number of reasons. On the one hand, the equilib-rium is extreme on the side of the reaction product in most cases. On the other hand, there are material problems because fluorine reacts with the container materials that are usually used for transport reactions.

Under transport conditions, chlorine, bromine, and iodine react with solids of different substance classes, e. g., with metals, intermetallic compounds, semi-met-als, metal oxides, metal sulfides, metal selenides, metal tellurides, metal nitrides, metal phosphides, metal arsenides, metal antimonides, metal silicides, metal germanides, some metal halogens, and more. In the process, gaseous metal halides, semi-metals halides, and the non-metal, respectively, are formed as a rule. In some cases, non-metal halides are formed. Thus during a reaction of metal phosphides with a halogen or a halogen compound, not only metal but also phosphoric halides can occur.

Due to the fact that halides show oxidizing characteristics, metal halides or oxide halides are often formed as transport-effective species in which the metal has a higher oxidation number than in the solid. Examples are the transport of chromium(III)-chloride with chlorine or tungsten(IV)-oxide with iodine, which are described in the following transport equations.

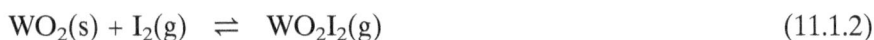

$$CrCl_3(s) + \frac{1}{2}Cl_2(g) \ \rightleftharpoons \ CrCl_4(g) \tag{11.1.1}$$

$$WO_2(s) + I_2(g) \ \rightleftharpoons \ WO_2I_2(g) \tag{11.1.2}$$

Hydrogen halides are versatile transport agents. The oxidation numbers of the metal in the solid and in the transport-effective gas species are generally equal because hydrogen halides do not have an oxidizing effect. Hydrogen halides are often used during the transport of oxides. Here the gaseous metal halide and water vapor are formed.

Halogen compounds, such as $TeCl_4$, PCl_5, $NbCl_5$, or $TaCl_5$ are also useful transport agents, especially for metal oxides. Reactions of the mentioned chlorides lead, on the one hand, to the formation of gaseous metal chloride or metal oxide chloride, on the other hand, oxygen is bound in form of volatile oxides (TeO_2, P_4O_6, P_4O_{10}) or volatile oxide chlorides ($TeOCl_2$, $POCl_3$, $NbOCl_3$, $TaOCl_3$). *Oppermann* was able to show that tellurium(IV)-chloride is a particular versatile transport agent (Opp 1975).

Halogen compounds as transport-effective species According to the basic works of *Schäfer*, gaseous metals or semi-metal halides, respectively, are formed as transport-effective species during the reaction of different solids with halogens or halogen compounds. In the following, some selected examples are shown (Sch 1962).

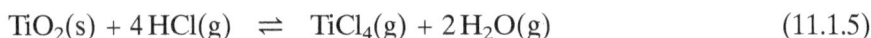

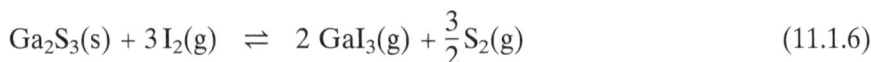

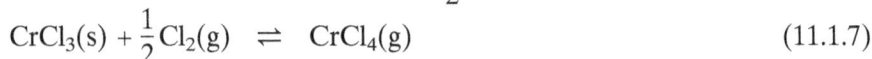

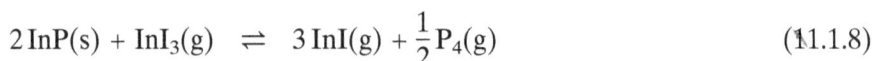

$$Co(s) + I_2(g) \ \rightleftharpoons \ CoI_2(g) \tag{11.1.3}$$

$$Si(s) + SiI_4(g) \ \rightleftharpoons \ 2\,SiI_2(g) \tag{11.1.4}$$

$$TiO_2(s) + 4\,HCl(g) \ \rightleftharpoons \ TiCl_4(g) + 2\,H_2O(g) \tag{11.1.5}$$

$$Ga_2S_3(s) + 3\,I_2(g) \ \rightleftharpoons \ 2\,GaI_3(g) + \frac{3}{2}S_2(g) \tag{11.1.6}$$

$$CrCl_3(s) + \frac{1}{2}Cl_2(g) \ \rightleftharpoons \ CrCl_4(g) \tag{11.1.7}$$

$$2\,InP(s) + InI_3(g) \ \rightleftharpoons \ 3\,InI(g) + \frac{1}{2}P_4(g) \tag{11.1.8}$$

The chlorides, bromides, and iodides of a metal usually show high volatility. Fluorides have higher boiling temperatures, and thus lower volatility, in most cases.

Table 11.1.1 Molecules in the vapor of the chlorides of main group elements.

1	2	13	14	15	16
LiCl, Li_2Cl_2	$BeCl_2$ Be_2Cl_4	BCl_3 BCl	CCl_4		
NaCl, Na_2Cl_2	$MgCl_2$	$AlCl_3$ Al_2Cl_6 AlCl	$SiCl_4$ $SiCl_2$	PCl_5 PCl_3	S_2Cl_2 SCl_2
KCl K_2Cl_2	$CaCl_2$	$GaCl_3$ Ga_2Cl_6 GaCl	$GeCl_4$ $GeCl_2$	$AsCl_3$	$SeCl_4$ $SeCl_2$
RbCl Rb_2Cl_2	$SrCl_2$	$InCl_3$ In_2Cl_6 InCl	$SnCl_4$ $SnCl_2$	$SbCl_3$	$TeCl_4$ $TeCl_2$
CsCl Cs_2Cl_2	$BaCl_2$	TlCl Tl_2Cl_2	$PbCl_2$	$BiCl_3$ BiCl	

Among the semi- and non-metal halides, in contrast, fluorides have particularly low boiling temperatures.

The thermodynamic stability of gaseous halides decreases from fluoride to iodide. The vast part of the halides evaporates undecomposed. The tendency to decomposition increases from fluoride to iodides. In Table 11.1.1, the most important gaseous main group chloride species are compiled.

A similar picture appears for bromides and iodides. However, the compounds become less stable and the tendency to form lower halides increases from fluorides to iodides. Thus phosphorus(V)-bromide decomposes to phosphorus(III)-bromide and bromine at room temperature; phosphorus(V)-iodide is unknown; sulfuric bromides are very unstable; and binary iodides of sulfur and selenium have not yet been detected.

In Table 11.1.2, the most important gaseous transition element chloride species are compiled. There is a similar picture for bromides and iodides. The behavior of the lanthanoid halides is similar to that of the lanthanum halides.

The vapor of metal halides can consist of monomeric, dimeric, and/or oligomeric molecules. The metal atoms in dimers and oligomers are generally linked to each other via a halogen bridge, sometimes there are metal/metal bonds (e. g., in Re_3X_9). The composition of the Al_2Cl_6 molecule is illustrated in Figure 11.1.1.

With M = Al, Ga, In, Fe, Sc, Y, Ln (Ln = lanthanoids), one can observe particularly large amounts of dimers M_2X_6. Trimers occur with copper(I)-halides and silver halides. The gas-phase composition of solid or liquid metal chloride was investigated in particular by *Schäfer* (Sch 1974a, Sch 1976). Rules were described to estimate the thermodynamic data of dimeric molecules. This can be useful if thermodynamic data of the considered compounds are unknown. One can assume that those rules that apply for metal chlorides also apply for bromides and iodides.

Metal halides of different metals A and B can react with each other in the gaseous state under the formation of **gas complexes**. Such gas complexes are also

Table 11.1.2 Molecules in the vapor of transitio metal chlorides.

3	4	5	6	7	8	9	10	11	12
$ScCl_3$ Sc_2Cl_6	$TiCl_4$ $TiCl_3$ Ti_2Cl_6 $TiCl_2$	VCl_4	$CrCl_4$ $CrCl_2$	$MnCl_2$ Mn_2Cl_4	$FeCl_3$ Fe_2Cl_6 $FeCl_2$ Fe_2Cl_4	$CoCl_2$ Co_2Cl_4	$NiCl_2$ Ni_2Cl_4	$CuCl,$ Cu_3Cl_3 Cu_4Cl_4	$ZnCl_2$ Zn_2Cl_4
YCl_3 Y_2Cl_6	$ZrCl_4,$	$NbCl_5,$ $NbCl_4$	$MoCl_5$ $MoCl_4$ $MoCl_3$	Tc_3Cl_9 $TcCl_4$	$RuCl_4$	$RhCl_3$	$PdCl_2$ Pd_6Cl_{12}	$AgCl,$ Ag_3Cl_3	$CdCl_2$
$LaCl_3$ La_2Cl_6	$HfCl_4$	$TaCl_5$ $TaCl_4$	WCl_6 WCl_4	Re_3Cl_9 $ReCl_5$	$OsCl_4$	$IrCl_3$	$PtCl_3$ $PtCl_2$ Pt_6Cl_{12}	$AuCl_3$ Au_2Cl_6 $AuCl$ Au_2Cl_2	$HgCl_2$ Hg_2Cl_2

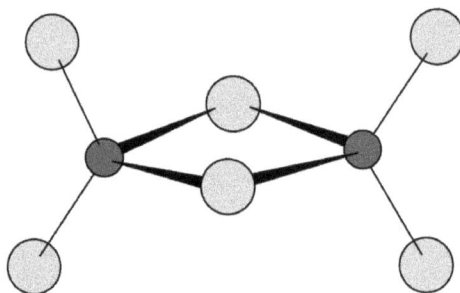

Figure 11.1.1 Constitution of Al_2Cl_6.

formed during heterogeneous reactions between solid and gaseous halides. For example, gaseous aluminum(III)-chloride reacts with a number of low volatility, solid metal chlorides under formation of gas complexes. This way their volatility is massively increased, often by some orders of magnitude. These reactions can be used preparatively as vapor transport reactions. The following reaction of cobalt chloride with aluminum(III)-chloride is an example of this (Bin 1977).

$$CoCl_2(s) + Al_2Cl_6(g) \rightleftharpoons CoAl_2Cl_8(g) \qquad (11.1.9)$$

This way, cobalt chloride can be transported with aluminum(III)-chloride far below the boiling temperature (e. g., 400 → 350 °C) (Sch 1974b). Today, we know numerous of these examples (Sch 1974b, Sch 1975, Sch 1976, Sch 1983). Tables 11.1.3 and 11.1.4 provide an overview of the detected chloride complexes (taken from (Sch 1976)). It is assumed that metal bromides and metal iodides form gas complexes in a similar way.

Thermodynamic data were determined for a considerable number of gas complexes (Bin 1974). *Schäfer* and *Binnewies* laid down empirical rules that can help estimating the thermodynamic data of gas complexes (Bin 1974, Sch 1976). A short description is found in section 12.2.2.

Table 11.1.3 Overview of bi-nuclear gas complexes of chlorides

Oxidation-numbers	Examples of gas complexes
I–I	$MM'Cl_2$ (M, M' = Li ... Cs, $NaCuCl_2$, $TlCuCl_2$, $MAgCl_2$ (M = Li ... Cs).
I–II	$NaBeCl_3$, $KBeCl_3$, $KMgCl_3$, $KCaCl_3$, $KSrCl_3$ $MCrCl_3$ (M = K, Rb, Cs) $MMnCl_3$ (M = Li ... Cs) $MFeCl_3$ (M = Li ... Cs) $MCoCl_3$ (M = Li ... Cs) $NaZnCl_3$, $KZnCl_3$ $MCdCl_3$ (M = Na ... Cs) $MSnCl_3$ (M = Na ... Cs) $MPbCl_3$ (M = Na ... Cs, Tl)
I–III	$MAlCl_4$ (M = Li ... .Cs), $CuAlCl_4$, $CuGaCl_4$, $BiAlCl_4$, $In^IIn^{III}Cl_4$, $MInCl_4$ (M = K, Cs) $TlInCl_4$, $CuInCl_4$, $TlTlCl_4$, $MFeCl_4$ (M = Na, K), $MScCl_4$ (M = Li ... Cs) $MYCl_4$ (M = Li, Na), $NaBiCl_4$, $MLnCl_3$ (M = Li ... Cs, Ln = lanthanoid, details see (Sch 1976))
I–IV	$KThCl_5$, $MThCl_5$ (M = Tl, Cu, Ag), $MUCl_5$ (M = Tl, Cu)
II–II	$CdPbCl_4$
II–III	$BeAlCl_5$, $BeFeCl_5$, $BeInCl_5$, $ZnInCl_5$, $SnInCl_5$, $HgAlCl_5$, $SnBiCl_5$
II–IV	$BeZrCl_6$, $BeUCl_6$, $PbThCl_6$
III–III	$AlGaCl_6$, $AlVCl_6$, $AlFeCl_6$, $AlSbCl_6$, $AlBiCl_6$, $GaInCl_6$, $AlInCl_6$, $FeAuCl_6$, $EuLuCl_6$
III–V	$AlUCl_8$, $InUCl_8$
IV–IV	$ThUCl_8$

Oxide halides Gaseous oxide halides are known for a few metals only, in particular some transition elements tend to form such compounds. *Ngai* and *Stafford* provide an overview (Nga 1971). Such molecules appear in particular in the cases of metals with high oxidation numbers: VOX_3, $NbOX_3$, $TaOX_3$ (X = F ... I), CrO_2X_2 (X = F ... Br), MoO_2X_2, WO_2X_2 (X = F ... I), $MoOX_4$, WOX_4 (X = F ... Br), $ReOCl_4$, OsO_2Cl_2, $OsOCl_4$, $RuOCl$, ReO_3X (X = F ... I). Gaseous oxide halides can play an important part as transport-effective species during the transport of oxides (Sch 1972). The following two examples are mentioned.

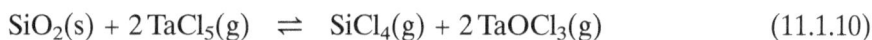

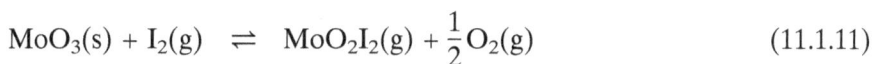

$$SiO_2(s) + 2\,TaCl_5(g) \;\rightleftharpoons\; SiCl_4(g) + 2\,TaOCl_3(g) \tag{11.1.10}$$

$$MoO_3(s) + I_2(g) \;\rightleftharpoons\; MoO_2I_2(g) + \frac{1}{2}O_2(g) \tag{11.1.11}$$

Some gaseous oxide halides of main group metals are known. Elements of group 13 form oxide halides, such as AlOCl, at very high temperatures around 2000 °C. A large number of oxide halides, such as Si_2OCl_6 or $Si_4O_4Cl_8$, are known

Table 11.1.4 Overview of multi-nuclear gas complexes of chlorides.

Oxidation-numbers	Examples of gas complexes
I–I	$LiCs_2Cl_3$, Li_2CuCl_3, $LiCu_2Cl_3$, $NaCu_2Cl_3$, $TlCu_2Cl_3$, $M_2Ag_2Cl_3$, MAg_2Cl_3 (M = Li … Cs)
I–II	$Na_2Zn_2Cl_6$, $Cs_2Cd_2Cl_5$
I–III	Cu_2AlCl_5, Cu_3AlCl_6, $CuAl_2Cl_7$, $Cu_2Al_2Cl_8$, Cu_2InCl_5, Cu_3InCl_6, $CuIn_2Cl_7$, Tl_2InCl_5
I–IV	Cu_2ThCl_6, $CuTh_2Cl_9$, Cu_2UCl_6, Tl_2ThCl_6, Tl_2UCl_6, TlU_2Cl_9
II–III	Be_2AlCl_7, $BeAl_2Cl_8$, $Be_2Al_2Cl_{10}$, $Be_3Al_2Cl_{12}$, Be_2InCl_7, $BeIn_2Cl_8$, Be_3InCl_9, $BeFe_2Cl_8$
	$Be_2Fe_2Cl_{10}$, $Be_3Fe_2Cl_{12}$, MAl_2Cl_8 (M = Mg, Ca, Mn, Co, Ni, Pb, Cr, Pd, Pt Cu)
	MFe_2Cl_8 (M = Mg, Ca, Sr, Ba, Mn, Co, Ni, Cd), $CoAl_4Cl_{14}$, VAl_3Cl_{11}
II–IV	Be_2UCl_8
III–III	$CrAl_3Cl_{12}$, $NdAl_3Cl_{12}$, $NdAl_4Cl_{15}$, VAl_2Cl_9, $FeAl_2Cl_9$,
III–IV	UAl_2Cl_{10}, UIn_2Cl_{10}

of silicon and germanium (Bin 2000, Lie 1997). The mentioned compounds, how-ever, do not play a noteworthy role during CVT reactions. In group 15, phosphorus forms several oxide chlorides and oxide bromides that are stable at high temperatures: POX_3, PO_2X, and POX (X = Cl, Br) (Bin 1990a). POF only exists at high temperatures (Ahl 1986). In the cases of arsenic and antimony AsOCl and SbOCl are stable at high temperatures; however, oxide halides of these elements in the oxidation stage V are not known (Bin 1990a). $TeOCl_2$ is the most important oxide halide of the main group elements. This gas species plays an important role during the transport of numerous oxides with tellurium(IV)-chloride (Opp 1975).

11.2 Elements in the Gaseous State

The importance of elemental halogens for CVTreactions has already been mentioned. Other elements also often occur in the gaseous state during transport reactions, especially gaseous non-metals. For example, gaseous elements of groups 15 and 16, respectively, form during the transport of pnictides or chalcogenides with halogens or halogen compounds in many cases. One example is the exothermic transport of zirconium arsenide, $ZrAs_2$, with iodine, which is described by the following equation:

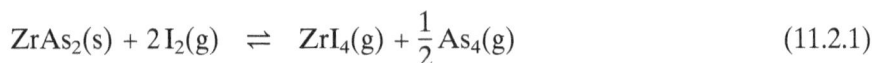

$$ZrAs_2(s) + 2\,I_2(g) \;\rightleftharpoons\; ZrI_4(g) + \frac{1}{2}As_4(g) \qquad (11.2.1)$$

In contrast to gaseous non-metals, metal vapors play a role in only a few reactions. This is particularly the case if the transport agent reacts with the non-metal instead of the metal atoms. The endothermic transport of zinc oxide with hydrogen, which can be described by the transport equation 11.2.2, exemplifies this:

$$ZnO(s) + H_2(g) \rightleftharpoons Zn(g) + H_2O(g) \qquad (11.2.2)$$

There are not many transport reactions of this kind, which is due to two reasons. Most of the transport agents used form more stable gaseous compounds with the metal atoms of the solid than with non-metal atoms. Additionally, even at high temperatures, the vapor pressure of most metals is too low to be exclusively gaseous. That is why the formation of an unsaturated metal vapor during transport reactions can only be expected if the boiling temperature of the metal is below approximately 1200 °C. This only applies to the following metals: Na (881 °C), K (763 °C), Rb (697 °C), Cs (657 °C), Mg (1093 °C), Zn (906 °C), Cd (766 °C), Hg (356 °C), Yb (1194 °C), and Te (989 °C).

In some cases, gaseous elements other than halogens can work as transport agents. Hence, oxygen can cause the transport of some platinum metals (Sch 1972). The transport is carried out by gaseous oxides of these metals, for example OsO_4. Sulfur can transport a series of transition metal sulfides (Sch 1968). Here gaseous polysulfides, such as TaS_3, are assumed to be the transport-effective species. There are similar observations for the CVT of some selenides. Sulfur is an effective transport agent for tellurium as well (Bin 1976). Compounds in which tellurium atoms were integrated in the different ring-shaped sulfur molecules were detected as transport-effective species. Phosphorus can transport gallium phosphide, GaP, and indium phosphide, InP, probably via GaP_5 and InP_5, respectively, as transport-effective species (Köp 2003). With the help of arsenic, the transport of gallium arsenide, GaAs, and indium arsenide, InAs, succeeds in a similar way (Köp 2004).

Composition of metal vapors With the help of mass spectrometric methods, gas phases of nearly all metals are well investigated, this was done by *Gingerich* in particular (Gin 1980). They predominantly consist of the atoms. The fraction of bi- or poly-atomic molecules in the saturated vapor is between 10^{-5} and 10 % (Gin 1980). Poly-atomic molecules, which contain different metal atoms, can appear in the vapor over melted alloys. However, their fraction is also clearly below that of the metal atoms. The thermodynamic data of poly-atomic molecules consisting of the same or different metal atoms are generally well known (Gin 1980). Nevertheless, these molecules do not play a role during CVT reactions.

Composition of non-metal vapors Under the condition of CVT reactions, the vapors of non-metals, apart from the noble gases, consist of very stable poly-atomic molecules, which appear in the gas phase in large fractions. In contrast to metals, atoms appear only subordinated. The fraction of different molecular species in the vapors of non-metals depends on the temperature and the pressure. Higher temperatures and lower pressures aid the formation of small molecules and atoms, respectively. Let us consider the example of unsaturated iodine vapor.

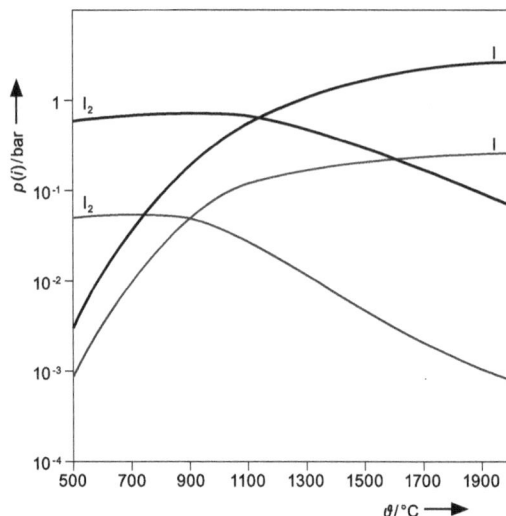

Figure 11.2.1 Partial pressure of I and I_2 at initial pressures (298 K) of 0.2 bar (black) and 0.02 bar (gray).

Table 11.2.1 Compilation of the gaseous species in non-metal vapors.

15	16	17
N_2	O_2	
P_4, P_2	S_2, S_3 ... S_7, S_8	Cl_2, Cl
As_4, As_2	Se_2, Se_3 ... Se_7, Se_8	Br_2, Br
Sb_4, Sb_2	Te_2, Te	I_2, I
Bi		

The influence of the temperature on the gas-phase composition for two different pressures is illustrated in Figure 11.2.1.

In both cases, the fraction of iodine atoms increases with rising temperatures. At an initial pressure of iodine of 0.2 bar, iodine atoms predominate above 1100 °C. At an initial pressure of 0.02 bar they do so already above 900 °C. Thus which gas species plays the predominate role depends on the respective reaction conditions. This must be checked in each individual case for deeper discussions. If elements of groups 15, 16, or 17 appear, one can assume that those species are formed that are compiled in Table 11.2.1. Deviations from this only appear under exceptional external conditions.

11.3 Hydrogen Compounds

A great number of gaseous hydrogen compounds of metals are described in the literature. In most cases, these are diatomic molecules of the composition MH (Gur 1974, quoted in Gin 1980). These molecules do not play any practical role for CVT reactions. However, the hydrogen compounds of non-metals could play an important role. Hydrogen halides, which are often used as transport agents, are particularly important; for example during the transport of metal oxides such as magnesium oxide.

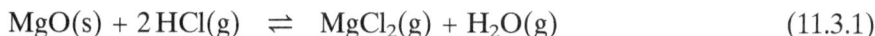

$$MgO(s) + 2\,HCl(g) \;\rightleftharpoons\; MgCl_2(g) + H_2O(g) \tag{11.3.1}$$

This example shows that water vapor, as the transport-effective species, is important during transport reactions. However, water can also function as transport agent. The transport of molybdenum(VI)-oxide and that of germanium with water are such examples.

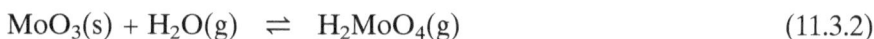

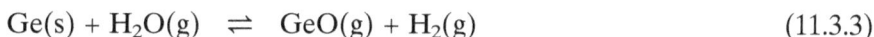

$$MoO_3(s) + H_2O(g) \;\rightleftharpoons\; H_2MoO_4(g) \tag{11.3.2}$$

$$Ge(s) + H_2O(g) \;\rightleftharpoons\; GeO(g) + H_2(g) \tag{11.3.3}$$

Water can also lead to the formation of transport-effective gaseous hydroxides:

$$Li_2O(s) + H_2O(g) \;\rightleftharpoons\; 2\,LiOH(g) \tag{11.3.4}$$

Water is of particular importance during the hydrothermal formation of numerous minerals in nature. The crystallization of rock crystal (α-quartz) under supercritical conditions is the most prominent example. This process, which is known from nature, is nowadays executed as an industrial procedure. It is assumed that under these special reaction conditions, silicon(IV)-oxide forms with water several gaseous silicic acids.

The transport reaction of tungsten with water and iodine is an important one in daily life. This reaction provides the basis of the operating mode of halogen lamps.

$$W(s) + 2\,H_2O(g) + 3\,I_2(g) \;\rightleftharpoons\; WO_2I_2(g) + 4\,HI(g) \tag{11.3.5}$$

Traces of water, often from the walls of the silicia tubes used during the transport, can be important for transport effects (Sch 1986).

Hydrogen sulfide and hydrogen selenide also appear during the transport of sulfides and selenides, respectively, with hydrogen halides. Hydrogen telluride is too unstable to develop under transport conditions. Ammonium chloride is particularly important. It decomposes to ammonia and hydrogen chloride during sublimation. Thus it is a hydrogen chloride source that is easy to handle and easy to dose. Ammonia decomposes to the elements at higher temperatures and thus creates a reducing atmosphere, which effects the equilibria involved in the transport in different ways.

11.4 Oxygen Compounds

The vast amount of metal oxides decomposes completely or partly while heating to high temperatures. *Gingerich* provides a compilation of gaseous metal oxides and their stability (Gin 1980). During decomposition reactions metals and oxygen can develop, or solid as well as gaseous oxides in lower oxidation states and oxygen. The thermal behavior of silicon(IV)-oxide is an example of this. A small fraction of approximately 1 % vaporizes congruently at temperatures above 1500 °C under the formation of $SiO_2(g)$; however, the decomposition to $SiO(g)$ and $O_2(g)$ dominates.

Some metal oxide vaporize congruently: CrO_3, MoO_3, WO_3, Re_2O_7, IrO_3, RuO_3, RuO_4, OsO_4, GeO, SnO, and PbO are examples of this. In the vapors of CrO_3, MoO_3, and WO_3 trimeric molecules appear, such as M_3O_9. SnO and PbO form dimers and trimers. All in all, gaseous metal oxides do not play a particular role during transport reactions. The following are mentioned as examples (Sch 1972, Sch 1960):

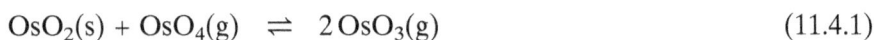

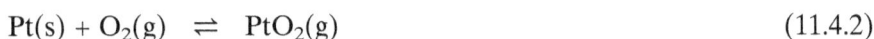

$$OsO_2(s) + OsO_4(g) \rightleftharpoons 2\,OsO_3(g) \tag{11.4.1}$$

$$Pt(s) + O_2(g) \rightleftharpoons PtO_2(g) \tag{11.4.2}$$

The role of non-metals is more important. Carbon monoxide can function as *transport agent* in different ways (Tam 1929, Tam 1931):

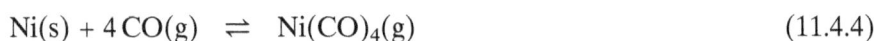

$$SnO_2(s) + CO(g) \rightleftharpoons SnO(g) + CO_2(g) \tag{11.4.3}$$

$$Ni(s) + 4\,CO(g) \rightleftharpoons Ni(CO)_4(g) \tag{11.4.4}$$

Often, non-metals occur as the *transport-effective species*, especially during the transport of oxides. Thus sulfuric vapor together with iodine can cause the transport of tin(IV)-oxide (Mat 1977).

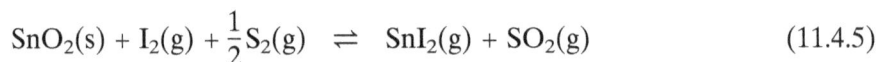

$$SnO_2(s) + I_2(g) + \frac{1}{2}S_2(g) \rightleftharpoons SnI_2(g) + SO_2(g) \tag{11.4.5}$$

During the CVT of metal oxides, tellurium(IV)-chloride plays a particular role, it reacts under formation of a metal chloride or a metal oxide chloride and binds oxygen in the form of $TeO_2(g)$ or $TeOCl_2(g)$ at the same time (Opp 1975).

$$TiO_2(s) + TeCl_4(g) \rightleftharpoons TiCl_4(g) + TeO_2(g) \tag{11.4.6}$$

$$MoO_3(s) + TeCl_4(g) \rightleftharpoons MoO_2Cl_2(g) + TeOCl_2(g) \tag{11.4.7}$$

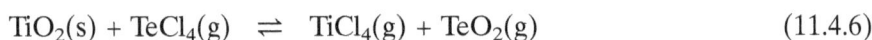

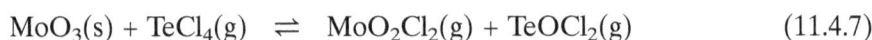

Further gaseous oxides that are important for the transport of oxide compounds are, among others: B_2O_3, SiO, P_4O_6, P_4O_{10}, As_4O_6, Sb_4O_6, SO_2, SeO_2, TeO, and TeO_2.

11.5 Other Substance Groups

During some transport reactions, gaseous sulfides, selenides, tellurides, or sulfide halides are of particular importance. Thus boron can be volatilized in the presence of sulfur or selenium. In this context, the molecules BS_2 (Bro 1973) and BSe_2 (Bin 1990b) have been detected. Aluminum also forms gaseous sulfides and selenides at high temperatures: Al_2S, AlS, Al_2S_2, Al_2Se, $AlSe$, and Al_2Se_2 (Mil 1974). Gaseous sulfides, selenides, and tellurides of group 14 are also known: SiS, $SiSe$, $SiTe$, GeS, $GeSe$, $GeTe$, SnS, $SnSe$, $SnTe$, PbS, $PbSe$, and $PbTe$ (Mil

1974). These molecules form dimers to a minor degree. Disulfides and diselenides, respectively, of these elements are less stable. The gaseous monosulfide, PS, of sulfur is described (Dre 1955).

Gaseous sulfide halides or selenide halides are known for only a few elements. For example, PSX_3 and PSX (X = F, Cl, Br) (Bin 1990a) for phosphorus are described. Niobium forms the gaseous sulfide halides and selenide halides $NbSCl_3$, $NbSBr_3$, $NbSeBr_3$, respectively (Sch 1966); for tantalum, $TaSCl_3$ and $TaSeBr_3$ are known (Sch 1966); the transport efficiency of MoSBr was mentioned (Kra 1981). Tungsten forms two volatile sulfide chlorides, WS_2Cl_2 (Nga 1971) and $WSCl_4$ (Bri 1970). The thermodynamic data of WS_2X_2 (X = Cl, Br, I) have been determined (Mil 2010). Gaseous $ReSCl_4$ is known (Rin1967). The transport efficiency of the hydrated halide oxide, $Bi(OH)_2X$ (X = Cl, Br, I), was reported (Opp 2000). Finally, $Pt(CO)_2Cl_2$ is a gaseous compound that becomes transport effective during the transport of platinum with carbon monoxide and chlorine (Sch 1971).

Bibliography

Ahl 1986	R. Ahlrichs, R. Becherer, M. Binnewies, H. Borrmann, M. Lakenbrink, S. Schunck, H. Schnöckel, *J. Am. Chem. Soc.* **1986**, *108*, 7905.
Bin 1974	M. Binnewies, H. Schäfer, *Z. Anorg. Allg. Chem.* **1974**, *407*, 327.
Bin 1976	M. Binnewies, *Z. Anorg. Allg. Chem.* **1976**, *422*, 43.
Bin 1977	M. Binnewies, *Z. Anorg. Allg. Chem.* **1977**, *435*, 156.
Bin 1990a	M. Binnewies, H. Schnöckel, *Chem. Rev.* **1990**, *90*, 32.
Bin 1990b	M. Binnewies, *Z. Anorg. Allg. Chem.* **1990**, *589*, 115.
Bin 2000	M. Binnewies, K. Jug, *Eur. J. Inorg. Chem.* **2000**, 1127.
Bri 1970	D. Britnell, G. W. A. Fowles, R. Mandycewsky, *Chem. Commun.* **1970**, 608.
Bro 1973	J. M. Brom, W. Weltner, *J. Mol. Spectrosc.* **1973**, *45*, 82.
Dre 1955	K. Dressler, *Helv. Phys. Acta.* **1955**, *28*, 563.
Gin 1980	K. A. Gingerich, E. Kaldis, Eds., *Current Topics in Material Science*, North-Holland Publishing Company, **1980**, *6*, 345.
Gur 1974	L. V. Gurvich, G. V. Karachevstev, V. N. Kondratyew, Y. A. Lebedev, V. A. Mendredev, V. K. Potatov, Y. S. Khodeev, *Bond Energies, Ionisation Potentials and Electron Affinities* Nauka, Moskau, **1974.**
Köp 2003	R. Köppe, J. Steiner, H. Schnöckel, *Z. Anorg. Allg. Chem.* **2003**, *62*, 2168.
Köp 2004	R. Köppe, H. Schnöckel, *Angew. Chem.* **2004**, *116*, 2222, *Angew. Chem., Int. Ed.* **2004**, *43*, 2170.
Kra 1981	G. Krabbes, H. Oppermann, *Z. Anorg. Allg. Chem.* **1981**, *481*, 13.
Lie 1997	S. Lieke und M. Binnewies, *Z. Anorg. Allg. Chem.* **1997**, *623*, 1705.
Mat 1977	K. Matsumoto, S. Kaneko, K. Katagi, *J. Cryst. Growth.* **1977**, *40*, 291.
Mil 1974	K. C. Mills, *Thermodynamic Data for Inorganic Sulfides, Selenides and Tellurides*, Butterworths, London, **1974**.
Mil 2010	E. Milke, R. Köppe, M. Binnewies, *Z. Anorg. Allg. Chem.* **2010**, 2010, 636, 1313.
Nga 1971	L. H. Ngai, F. E. Stafford in L. Eyring (Ed.) *Advances in High Temperature Chemistry*, Academic Press, New York **1971**, 313.
Opp 1975	H. Oppermann, *Kristall u. Techn.* **1975**, *10*, 485.

Opp 2000   H. Oppermann, M. Schmidt, H. Brückner, W. Schnelle, E. Gmelin, *Z. Anorg. Allg. Chem.* **2000**, *626* , 937.
Rin 1967   K. Rinke, M. Klein, H. Schäfer, *J. Less-Common Met.* **1967**, *12*, 497.
Sch 1960   H. Schäfer, A. Tebben, *Z. Anorg. Allg. Chem.* **1960**, *304,* 317.
Sch 1962   H. Schäfer, *Chemische Tranportreaktionen*, Verlag Chemie, Weinheim, **1962**.
Sch 1966   H. Schäfer, W. Beckmann, *Z. Anorg. Allg. Chem.* **1966**, *347*, 225.
Sch 1968   H. Schäfer, F. Wehmeier, M. Trenkel, *J. Less-Common Met.* **1968**, *16*, 290.
Sch 1971   H. Schäfer, U. Wiese, *J. Less-Common Met.* **1971**, *24*, 55.
Sch 1972   H. Schäfer, Nat. Bur. Standards, Spec. Publ. 364, *Solid State Chem. Proceed. 5ᵗʰ Materials Research Sympos.*, **1972**, 413.
Sch 1974a   H. Schäfer, M: Binnewies, *Z. Anorg. Allg. Chem.* **1974**, *410*, 251.
Sch 1974b   H. Schäfer, *Z. Anorg. Allg. Chem.* **1974**, *403*, 116.
Sch 1975   H. Schäfer, *J. Cryst. Growth*, **1975**, *31*, 31.
Sch 1976   H. Schäfer, *Angew. Chem.* **1976**, *88,* 775, *Angew. Chem., Int. Ed.* **1976**, *15*, 713.
Sch 1983   H. Schäfer, *Adv. Inorg. Radiochem.* **1983**, 26, 201.
Sch 1986   H. Schäfer, *Z. Anorg. Allg. Chem.* **1986**, *543*, 217.
Tam 1929   S. Tamaru, N. Ando, *Z. Anorg. Allg. Chem.* **1929**, *184*, 385,
Tam 1931   S. Tamaru, N. Ando, *Z. Anorg. Allg. Chem.* **1931**, *195*, 309.

12 Thermodynamic Data

Numerous condensed and especially gaseous substances can be involved in a chemical transport reaction. A deeper understanding of these reactions requires the knowledge of *which* substances are involved in the reactions. In order to describe quantitatively a transport reaction, it is necessary to know their thermodynamic data to be able to carry out model calculations of the transport reaction. This chapter deals with thermodynamic aspects of chemical transport reactions.

In order to describe the substance's thermodynamic stability, one needs to know the standard enthalpy of formation, the standard entropy, as well as the heat capacity as a function of the temperature, respectively, the Gibbs energy of formation. With the help of *Kirchhoff*'s law, the temperature-dependent numerical values of the standard enthalpy of formation, the standard entropy, and the Gibbs energy of formation can be calculated. The basic relations can be found in text books of physical chemistry and need not be repeated at this point.

12.1 Determination and Tabulation of Thermodynamic Data

The conventional methods of determining thermodynamic data are described in books by *Kubaschewski* and *Schmalzried*, which also offer bibliographies (Kub 1983, Schm 1978). While in *Kubaschewski*'s book, the application of thermodynamic data has priority, *Schmalzried* describes the theoretical background of solid-state thermodynamics in particular (Schm 1978). Conventional methods of determining thermodynamic data are, for example, the various calorimetric methods, methods of thermal analysis (DTA, DSC), electrochemical methods (EMF-measurements), and vapor pressure measurements with different methods (entrainment measurements, total pressure measurements, Knudsen cell measurements). These and other methods have certain advantages and disadvantages that are connected to their respective errors. Thus a substance's **standard enthalpy of formation**, which is recognized as "correct", can well be erroneous by 10 to 20 kJ · mol^{-1}. Standard enthalpies of formation, which have been confirmed by different authors using different methods in the ideal case, can be considered as particularly reliable. However, these reliable data are known for only a limited number of substances, especially those that play an important part in chemical technology. As a rule, one has to assume that there are deviations of the real value of 10 to 15 or even 20 kJ · mol^{-1} as far as the standard enthalpy of formation is concerned. In individual cases, the standard enthalpy of formation can be even more erroneous.

The standard entropies of many substances are quite reliable. If they are not determined by experiments, one can determine them by means of estimation or theoretical calculation. This also applies to the heat capacity.

The data that are required for the thermodynamic description of chemical reactions can be taken from tables. The most important are:

- Thermodynamic Data for Inorganic Sulfides, Selenides and Tellurides (Mil 1974)
- Termitscheskije Konstanti Veschtschestw (Glu 1999)
- Thermochemical Properties of Inorganic Substances (Kna 1991)
- Thermochemical Data of Pure Substances (Bar 1992)
- Nist-JANAF Thermochemical Tables (Cha 1998)
- Thermochemical Properties of Elements and Compounds (Bin 2002)

Additionally, one can look for further data in the original literature. Since ca. 1990 increasingly fewer works have been published that were aimed at determining thermodynamic data of inorganic compounds. The majority of these data are collected in newer tables (Bin 2002).

12.2 Estimation of Thermodynamic Data

If there is a lack of thermodynamic data for the modeling of a chemical transport reaction, these can also be estimated. There are methods that allow estimations with sufficient accuracy. In his book, *Kubaschewski* devotes an entire chapter to estimation methods (Kub 1983). *Schmalzried* also suggests methods for estimation of enthalpies (Schm 1978). Computer programs for modeling CVT experiments, which are available nowadays, allow estimations in a short time. A further limitation of the data is possible if the experimental observations are critically compared to the conducted thermodynamic calculations that are based on estimation.

12.2.1 Thermodynamic Data of Solids

The methods of estimation of the standard enthalpy of formation of a solid phase are generally not as reliable as the methods of estimating the standard entropy and heat capacity. There is no uniform method for estimating standard enthalpies of formation of random solids. Every method is restricted to a certain group of compounds. The estimation of standard enthalpies of formation on the basis of known data of compounds of a homologous series or chemically similar substances, respectively, appears to be an appropriate procedure. It is recommended that the estimation of the required value is carried out in several ways. In the following, some possibilities for estimating the standard enthalpies of formation are introduced.

Estimation of the standard enthalpy of formation of a binary solid compound
Let us consider the case that the standard enthalpy of formation of magnesium
selenide is to be estimated. To do so, the known data (in $kJ \cdot mol^{-1}$) of the
magnesium chalcogenides and the calcium chalcogenides can be compared:

MgO:	−601	CaO:	−635
MgS:	−346	CaS:	−473
MgSe:	required value	CaSe:	−368
MgTe:	−209	CaTe:	−272

The value for calcium selenide is between that of calcium sulfide and calcium
telluride. The numerical values show that the value of the standard enthalpy of
formation is fairly in the middle between that of calcium sulfide and calcium
telluride. A similar situation is expected for magnesium. A value of −270 to
−280 $kJ \cdot mol^{-1}$ would be expected. This corresponds approximately to the table
value of −293 $kJ \cdot mol^{-1}$. Frequently, graphic illustrations of enthalpies are used
to estimate unknown values by interpolation or extrapolation.

Estimation of the standard enthalpy of formation of a ternary solid compound
If one wants to estimate the enthalpy of formation of a ternary compound, its formation out of the binary base material is usually considered. This is clarified with the example of the formation of a ternary oxide.

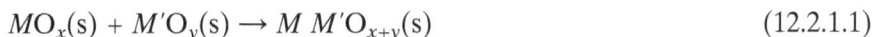

$$MO_x(s) + M'O_y(s) \rightarrow M\,M'O_{x+y}(s) \tag{12.2.1.1}$$

A reaction of this kind is principally exothermic. Typically the reaction enthalpy
is around −20 kJ per metal atom in the compound formed. This value will be
observed when the acid/base characteristics of the two reacting metal oxides are
similar. If, on the other hand, a basic oxide, such as CaO, reacts with an acid
oxide, such as SiO_2, a stronger exothermic reaction is observed. In this case, the
reaction enthalpy is −89 $kJ \cdot mol^{-1}$, thus approximately −45 kJ per constituent
cation in the calcium silicate $CaSiO_3$. *Schmidt* deals with this topic in detail
(Schm 2007). Other, often lower, guide values apply to other compound classes,
such as sulfides, selenides, or phosphides.

**Estimation of the standard enthalpy of formation of a ternary solid compound
with a complex anion** If a ternary solid compound contains a complex anion,
such as the phosphate, sulfate, or vanadate anion, the absolute value of the reaction enthalpy for its formation from the binary components is bigger than in the
previously mentioned case. This is clarified by the following example:

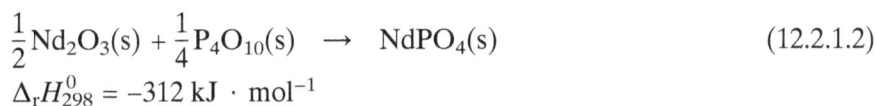

$$\frac{1}{2}Nd_2O_3(s) + \frac{1}{4}P_4O_{10}(s) \longrightarrow NdPO_4(s) \tag{12.2.1.2}$$
$$\Delta_r H^0_{298} = -312 \; kJ \cdot mol^{-1}$$

This corresponds to a value of −156 kJ per constituent cation (Nd, P).

The formation of $NdVO_4$ is another example:

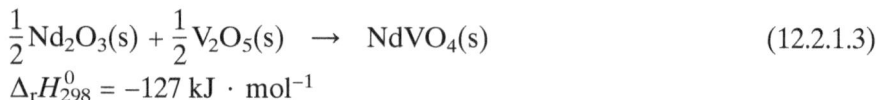

$$\frac{1}{2} Nd_2O_3(s) + \frac{1}{2} V_2O_5(s) \rightarrow NdVO_4(s) \tag{12.2.1.3}$$
$$\Delta_r H_{298}^0 = -127 \text{ kJ} \cdot \text{mol}^{-1}$$

One can see that the absolute value of the reaction enthalpy of 12.2.1.3 is clearly greater than for the formation of $CaSiO_3$, but a lot smaller than for the reaction with $NdPO_4$. The unusual big absolute numerical value for the formation of $NdPO_4$ is connected to the pronounced acidic character of P_4O_{10} (Kub 1983, Blu 2003, Schm 2007).

These examples provide indications for estimating unknown standard enthalpies of formation. General rules cannot be laid down. Every empirical rule refers to a certain substance group of chemically similar compounds.

Estimation of the heat capacity of a binary solid compound A substance's heat capacity can be estimated by means of the *Dulong–Petit* law. Accordingly, the heat capacity at 298 K per constituent atom is approximately 25 to 30 $J \cdot K^{-1} \cdot mol^{-1}$. Hence, for $C_{p,298}^0$, one expects a numerical value between 50 and 60 $J \cdot K^{-1} \cdot mol^{-1}$ (example: $C_{p,298}^0$ (FeS) = 50.5 $J \cdot K^{-1} \cdot mol^{-1}$) for an AB compound. When estimating, it is usual to consider that heat capacity increases with rising temperatures. The temperature dependency is usually described by the following polynomial:

$$C_{p,T}^0 = a + b \cdot T + c \cdot T^{-2} + d \cdot T^2 \tag{12.2.1.4}$$

With the help of the numerical values for a, b, c, and d, the temperature dependent enthalpy and entropy values can be calculated. To do so, equations 12.2.1.5 and 12.2.1.6 are used.

$$\begin{aligned} \Delta H_T^0 &= \Delta H_{298}^0 + \int_{298}^T C_p^0 \, dT \\ &= a[T - 298] + b[0.5 \cdot 10^{-3}(T^2 - 298^2)] + c[10^6(298^{-1} - T^{-1})] \\ &\quad + d\left[\frac{1}{3} 10^{-6}(T^3 - 298^3)\right] \end{aligned} \tag{12.2.1.5}$$

$$\begin{aligned} S_T^0 &= S_{298}^0 + \int_{298}^T C_p^0 \frac{dT}{T} \\ &= a \ln\frac{T}{298} + b \cdot 10^{-3}(T - 298) - \frac{1}{2}c \cdot 10^6 (T^{-2} - 298^{-2}) \\ &\quad + d[0.5 \cdot 10^{-6}(T^2 - 298^2)] \end{aligned} \tag{12.2.1.6}$$

Estimation of the heat capacity and the entropy of a ternary solid compound According to the *Dulong–Petit* law the heat capacity of a ternary or multi-component compound can principally be estimated. On the basis of experimentally determined values for binary starting compounds, however, one will get more exact results.

To do so, let us consider a typical solid-state reaction as in 12.2.1.7. It becomes apparent that the reaction entropy as well as the change of heat capacity of the system are near to zero. This is called the *Neumann–Kopp* law. On this basis, the standard entropy and the heat capacity, respectively, of a solid, can be estimated as the sum of the standard entropies and heat capacities, respectively, of their components.

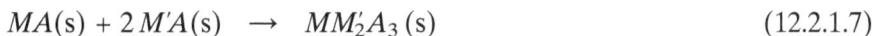

$$MA(s) + 2\,M'A(s) \quad \rightarrow \quad MM'_2A_3\,(s) \tag{12.2.1.7}$$

$$S^0_{298}(MM'_2A_3(s)) = S^0_{298}(MA(s)) + 2 \cdot S^0_{298}(M'A(s)) \tag{12.2.1.8}$$

$$C^0_{p,298}(MM'_2A_3(s)) = C^0_{p,298}(MA(s)) + 2 \cdot C^0_{p,298}(M'A(s)) \tag{12.2.1.9}$$

The deviation of these values for the entropy and heat capacity is usually less than approx. $\pm\, 8\ \text{J} \cdot \text{mol}^{-1} \cdot \text{K}^{-1}$ from the true values.

For approximations of equilibrium positions at high temperatures, one may use the values of enthalpy and entropy of all the reactants at 298 K. The consideration of molar heat capacities and their functions of temperature leads to similar results. However, for more exact calculations, the heat capacities have to be considered. Under no circumstances should one combine the thermodynamic data for 298 K of *one* reactant with that of *another* reactant at a *different* temperature. This leads to major errors in calculating the equilibrium position!

Estimation of thermodynamic data of mixed phases In the literature one can find only little information on the stability of solid mixed phases and their thermodynamic data, which are based on experiments. If such information is required, one is dependent on estimations. To do so, one uses models for ideal and real solid solutions. The model of the ideal solid solution is based on the assumption that there will not be any heat effects during the mixing procedure. The entropy of mixing is calculated by means of statistical thermodynamics. It always has a positive sign. The model of the real solution additionally contains a certain amount of the enthalpy of mixing. *Schmalzried* described details of these models (Schm 1978).

Consistency of a data set If there are several compounds of different compositions in a substance system, the consistency of the total data set plays an important role for thermodynamic considerations. Frequently there is no consistency due to a number of different experimental methods that lead to non-consistent values. When tabulated standard values are used, in most cases, the consistency check has already been carried out by the authors. A check on the consistency of the data used should precede thermodynamic calculations, especially if data from different references, such as measurements, estimations, tables and original literature, are put together.

The graphic illustration of the uniform standardized values of the standard enthalpy of formation, the enthalpies, and entropies, respectively, towards the composition (Figures 12.2.1.1 and 12.2.1.2) is particularly clear.

Figure 12.2.1.1 describes the situation of a binary system without a miscibility gap. The situation for the formation of compounds of the compositions A_3B, AB, and AB_3 are illustrated in Figure 12.2.1.2. Connecting individual data points of

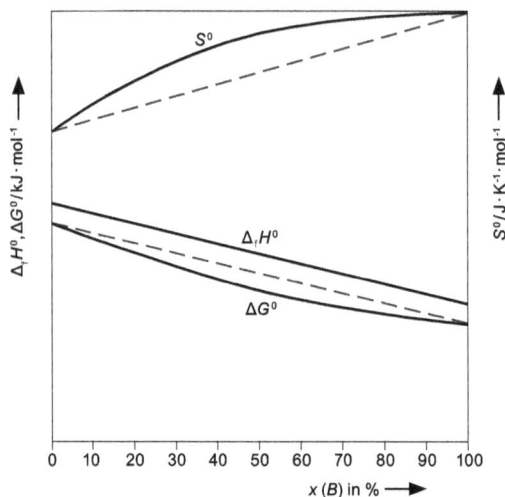

Figure 12.2.1.1 Schematic dependency of enthalpy, entropy, and Gibbs energy on the composition in a binary system without a miscibility gap.

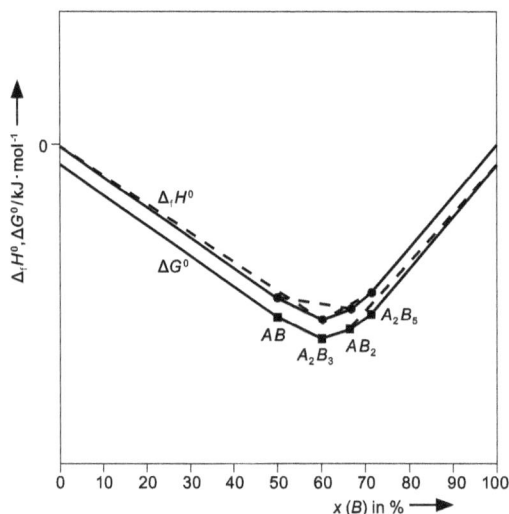

Figure 12.2.1.2 Schematic dependency of enthalpy, entropy, and Gibbs energy on the composition in a binary system with the formation of distinct compounds.

the Gibbs energy, the drawn angles have to be smaller than $180\ ^\circ C$ when the data are consistent. If there are larger angles, one of the compounds, which is placed on the intersection, would be formed endergonically from its co-existing adjacent phases. Thus it would not be thermodynamically stable. The same applies to the numerical checking of the formation reactions. If a compound is formed from its adjacent phases, the Gibbs energy must always be negative. This usually applies for the reaction enthalpy as well because the reaction entropy is close to zero.

$$A_3B(s) + AB_3(s) \longrightarrow 4\,AB(s)$$
$$\Delta_r G^0 < 0, \quad \Delta_r H^0 < 0 \tag{12.2.1.10}$$

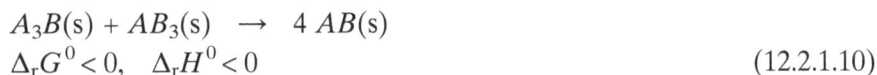

12.2.2 Thermodynamic Data of Gases

The enthalpies and entropies of formation of gaseous compounds can be estimated as well. While entropy values can be estimated with sufficient accuracy, caution is recommended when it comes to estimating enthalpies of formation. Frequently, simple rules do not apply and considerable errors must be expected.

Estimation of the standard enthalpy of formation of a binary gaseous compound The standard enthalpies of formation of gaseous molecules appear uniform within a homologous series in most cases. If one wants to estimate unknown enthalpy values by interpolation or extrapolation, one can use standard enthalpies of formation of analogous compounds as a comparison. However, it may be more helpful to compare not the standard enthalpies of formation but those from the obtainable enthalpies of atomization[1] $\Delta_{at}H_{298}^0$ (Opp 1975). The numerical value of the enthalpy of atomization is identical to the sum of the bond enthalpies $\Delta_b H_{298}^0$ of all the chemical bonds in the molecule considered. This way, one gets comparative data that are free of particular characteristics of the standard states of the elements of which the gaseous molecule considered consists. For example, the standard enthalpies of formation and the energies of atomization of chlorides, bromides, and iodides of aluminum, gallium, and indium show a uniform tendency. This makes it clear that an interpolation or an extrapolation, respectively, is admissible in this case.

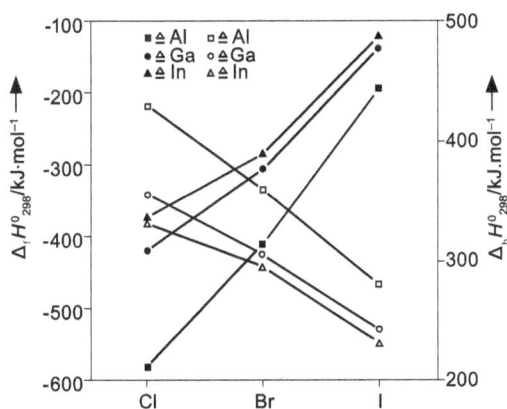

Figure 12.2.2.1 Standard enthalpies of formation and bond enthalpies of gaseous chlorides, bromides, and iodides of aluminum, gallium, and indium.

---

[1] The enthalpy of atomization is the enthalpy that is to apply in order to decompose a compound into the gaseous atoms from which it is constituted. It results from the standard enthalpy of formation of the compound considered and the standard enthalpies of formation of the respective atoms.

Table 12.2.2.1 Standard enthalpies of formation, enthalpies of atomization, and bond enthalpies of gaseous chlorides, bromides, and iodides of aluminum, gallium, and indium (Bin 2002).

Compound	$\Delta_f H_{298}^0/\text{kJ} \cdot \text{mol}^{-1}$	$\Delta_{at} H_{298}^0/\text{kJ} \cdot \text{mol}^{-1}$	$\Delta_b H_{298}^0/\text{kJ} \cdot \text{mol}^{-1}$
$AlCl_3$	−584.6	1278.2	426.1
$AlBr_3$	−410.5	1075.9	358.6
AlI_3	−193.3	843.4	281.1
$GaCl_3$	−422.9	1058.8	352.9
$GaBr_3$	−307.0	914.7	304.9
GaI_3	−137.6	730	243.3
$InCl_3$	−376.3	986.6	328.9
$InBr_3$	−285.2	885.3	295.1
InI_3	−120.5	687.5	232.5

Table 12.2.2.2 Standard enthalpies of formation, enthalpies of atomization, and bond enthalpies of the gaseous trichlorides of phosphorus, arsenic, antimony, and bismuth (Bin 2002).

Compound	$\Delta_f H_{298}^0/\text{kJ} \cdot \text{mol}^{-1}$	$\Delta_{at} H_{298}^0/\text{kJ} \cdot \text{mol}^{-1}$	$\Delta_b H_{298}^0/\text{kJ} \cdot \text{mol}^{-1}$
PCl_3	−288.7	986.5	328.8
$AsCl_3$	−261.5	927.2	309.1
$SbCl_3$	−313.1	942.5	314.2
$BiCl_3$	−265.3	838.8	279.6

Such an interpolation or extrapolation, respectively, can lead to values that are highly erroneous as can be seen in the comparison between the enthalpy of formation, the enthalpy of atomization, and the derived bond enthalpies $\Delta_b H_{298}^0$ of the trihalides of the elements of the group 15: PCl_3, $AsCl_3$, $SbCl_3$, and $BiCl_3$. The numerical values are compiled in Table 12.2.2.2.

One can see that the values of the standard enthalpies of formation and the enthalpies of atomization, respectively, of bonding do not show a uniform course. The bond enthalpy decreases from PCl_3 to $AsCl_3$, increases again marginally to $SbCl_3$, and then decreases again clearly to $BiCl_3$. One reason is that the elements phosphorus and arsenic have a non-metallic and semi-metallic character, respectively, while on the other hand antimony and bismuth have a metallic character. Therefore the bond character and thus the bond enthalpy and the stability of the chlorides change. If, for example, the enthalpy of formation of $SbCl_3$ was not known, one would get the wrong results by using interpolation. In these cases, one should determine the unknown value by means of quantum chemical methods. Consequently, a quantum chemical calculation of the standard enthalpies of formation of the mentioned molecules led to values that were near those that were experimentally determined (−295.0, −267.2, −314.6, −252.0) (Köp 2010).

Estimation of the enthalpies of formation of gas molecules with halogen bridges Metal halides can react in the gas phase in a number of ways. In this context, the most important reactions lead to the formation of metal/halogen compounds, in which two or more metal atoms are connected with each other via halogen bridges (see section 11.1). In the easiest case, dimeric metal halides are formed. As far as metal halides are concerned, these always exothermic dimerization reactions have been investigated in depth and the enthalpies of dimerization approximately determined (Sch 1974). In most cases, the numerical values of the enthalpy of dimerization are near those of the enthalpy of vaporization of the monomeric metal halide. One can find exceptions to this rule when some trichlorides, such as $GaCl_3$, $AlCl_3$, or $InCl_3$, are considered. Here, the dimeric molecules are particularly more stable, and the concentration of dimers is higher than in most other metal halide vapors. The particularly high stability of halogen bridges of the mentioned chlorides are used in different ways when aluminum chloride is used as a transport agent. The rules, which were worked out from the examples of many gaseous chlorides, should be transferable to the bromides and iodides. However, systematic experimental investigations have not been made.

If two different metal halides react with each other in the gas phase, so-called *gas complexes* are formed, in which two or more metal atoms are connected to each other by a halogen bridge. The reaction enthalpies of such reactions have been determined with the help of many examples. Empirical rules have been derived from these data. These show that during the reaction of a dimeric metal halide M_2Cl_{2n} with another $M_2'Cl_{2m}$ under the formation of the gas complex $MM'Cl_{m+n}$, the reaction enthalpy is near zero kJ · mol^{-1}. However, in certain cases, especially if one of the metal atoms is an alkali metal atom, the reactions are exothermic. The extensive data and the derived rules have been compiled by *Schäfer* (Sch 1976). The formation of gas complexes was investigated for chlorides in particular. Bromides and iodides offer a similar picture, so that the derived rules of stability could also be applied here.

Estimation of heat capacities of gaseous molecules The heat capacity and the entropy of a gas can be calculated by the methods of statistical thermodynamics from the molar mass, the geometry, and the oscillation frequencies of the respective molecule. Often, the geometry and the oscillation frequencies are not known by experimental investigations. They can either be estimated or calculated with good accuracy with the help of quantum chemical methods. The estimation of the heat capacity and the entropy of a gas become less important, and are replaced by quantum chemical calculations. The latter is recommended nowadays (see section 12.3). Here the classic rules of estimating the heat capacity $C_{p,298}^0$ will only briefly be summarized (Kub 1983):

- At all temperatures, the heat capacity $C_{p,298}^0$ of a mono-atomic gas is $5/2$ R = 20.8 J · mol^{-1} · K^{-1}.
- At room temperature, the heat capacity $C_{p,298}^0$ of a diatomic gas is approximately 30 J · mol^{-1} · K^{-1}. This value increases with temperature to a maximum of 38 J · mol^{-1} · K^{-1}. The heat capacity of diatomic molecules increases with the molar mass.

- The heat capacity of poly-atomic molecules is estimated with the help of known values of chemically similar molecules with similar molar masses.

Estimation of the entropies of gaseous compounds *Kubaschewski* provides empirical approximation formulae for estimating the entropies of gaseous molecules, which consist of *n* atoms, as a function of the molar mass *M*. These are summarized in Table 12.2.2.3.

Table 12.2.2.3 Empirical formula for estimating the standard entropies of *n*-atomic gaseous molecules.

n	$S^0_{298}/\mathrm{J} \cdot \mathrm{mol}^{-1} \cdot \mathrm{K}^{-1}$
1	$110.9 + 33 \cdot \lg M$
2	$101.3 + 68.2 \cdot \lg M$
3	$37.7 + 111.7 \cdot \lg M$
4	$-7.5 + 146.4 \cdot \lg M$
≥ 5	$-131.8 + 207.1 \cdot \lg M$

12.3 Quantum Chemical Calculation of Thermodynamic Data

During the past years, quantum chemical methods have been developed to efficient instruments that can be used to calculate thermodynamic quantities. *Ab initio* methods do not provide enthalpies of formation, but total energies at 0 K that result when a substance is built from the nuclei and electrons of the atom involved. These energies are indicated in hartree (1 hartree = 2625.5 kJ $\cdot$ mol^{-1}). They are always negative. The enthalpy of formation, on the other hand, does not refer to its formation from the atom's nuclei and electrons, but to the formation from the elements in the respective standard state. At first, the numerical values that are used in the sense of thermodynamics, are not comparable to those resulting from *ab initio* calculations. However, they become comparable if one considers the change of energy and enthalpy, respectively, during a chemical reaction and not the absolute values of distinct substances. The change of energy and enthalpy, respectively, during the formation of a crystalline solid out of other crystalline solids can be calculated in quantum chemical terms with certain accuracy without using experimental data. The change of energy and enthalpy, respectively, during the formation of a gaseous compound by a homogeneous gas-phase reaction can also be calculated. Reactions between condensed phases and gases, which form the basis of chemical transport reactions, cannot be dealt with (yet) by means of quantum chemical methods.

Ab initio calculations of (gaseous) molecules are highly accurate and there are program packages, which are generally available, to calculate certain characteristics of molecules. Such a program is TURBOMOLE (Ahl 1989). Molecule geometries and oscillation frequencies can be calculated. These allow the calculation of heat capacities and standard entropies with the help of statistical thermody-namics. It also allows the calculation of the reaction enthalpy of a homogeneous gas reaction. This again can be used to determine the unknown standard enthalpy of formation of a gaseous molecule. To do so, one considers a related reaction, in which the molecule is involved. The reaction's standard enthalpy of formation is to be calculated. The other reactants have to be chosen in a way that their enthalpies of formation are known with sufficient accuracy. Then the energy con-tents of all reactants are calculated, as well as the change of energy during the reaction, using TURBOMOLE. These values, at first calculated for 0 K, are re-lated to 298 K under the consideration of the heat capacities and converted into enthalpies. The coefficients of the polynomial that describes the temperature-dependent heat capacity (equation 12.2.1.4), are determined after the calculation of different values of $C_{p,T}^0$ in the desired temperature range with the help of a program, such as MAPLE (Mon 2005). With these coefficients, the temperature-dependent enthalpies (equation 12.2.1.5) and entropies (equation 12.2.1.6) can also be calculated. Let us have a look at this approach with the help of an exam-ple. While gaseous WO_2Cl_2 has been investigated and characterized in depth due to its importance in halogen lamps, the analogous sulfur compound WS_2Cl_2 has not be known until recently. It was observed mass spectrometrically as a product during the reaction of solid WS_2 with chlorine at 750 °C (Mil 2010) and could play an important role during the transport of WS_2 with chlorine. Quantum chemical calculations should help to calculate the enthalpy of formation, entropy, and heat capacity of the molecule. To do so, the following reaction is considered:

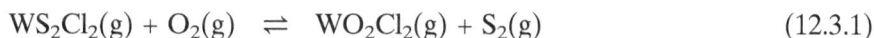

$$WS_2Cl_2(g) + O_2(g) \ \rightleftharpoons \ WO_2Cl_2(g) + S_2(g) \tag{12.3.1}$$

The calculation's result was a reaction enthalpy of $\Delta_r H_{298}^0 = -307.9$ kJ $\cdot$ mol^{-1}. With the help of the tabulated standard enthalpies of formation of the other reactants, a standard enthalpy of formation $\Delta_f H_{298}^0 (WS_2Cl_2(g)) = -235.0$ kJ $\cdot$ mol^{-1} is calculated. This value is in excellent agreement with the value of 230.8 kJ $\cdot$ mol^{-1} (Mil 2010) determined experimentally. It is assumed that the enthalpies of forma-tion can be calculated within an error range of ±15 kJ $\cdot$ mol^{-1} using this approach. One gets particularly exact values if a reaction is considered, as in this case, in which similar reactants on the educt and product side are involved. Heat capaci-ties and entropies can be calculated very exactly. The accuracy of the result often exceeds the values that were determined experimentally. Although the program is available for everybody, it is recommended that an expert with sufficient expe-rience in the area is consulted for the first calculations.

Bibliography

Ahl 1989	R. Ahlrichs, M. Bär, M. Häser, H. Horn, C. Kölmel, *Chem. Phys. Lett.* **1989**, *162*, 165.
Bar 1989	I. Barin, *Thermochemical Data of Pure Substances*, VCH, Weinheim, **1989**.
Bin 2002	M. Binnewies, E. Milke, *Thermochemical Data of Elements and Compounds*, Wiley-VCH, Weiheim, **2002**.
Blu 2003	M. Blum, K. Teske, R. Glaum, *Z. Anorg. Allg. Chem.* **2003**, *629*, 1709.
Cha 1998	M. W. Chase, *NIST-JANAF Thermochemical Tables*, ACS, **1992**.
Glu 1999	V. P. Glushko, V. A. Medvedev, L. V. Gurvich, *Thermal Constants of Substances*, Wiley, New York, **1999**.
Kna 1991	O. Knacke, O. Kubaschewski, K. Hesselmann, *Thermochemical Properties of Inorganic Substances*, Springer, Berlin, **1991**.
Köp 2010	R. Köppe, *upublished results*.
Kub 1983	O. Kubaschewski, C. B. Alcock, *Metallurgical Thermochemistry*, Pergamon Press, 5th Aufl. Oxford, **1983**.
Mil 1974	K. C. Mills, *Thermodynamic Data for Ionorganic Sulphides, Selenides and Tellurides*, Butterwords, London, **1974**.
Mil 2010	E. Milke, R. Köppe, M. Binnewies, *Z. Anorg. Allg. Chem.* **2010**, *636*, 1313.
Mon 2005	M. B. Monagan, K. O. Geddes, K. M. Heal, G. Labahn, S. M. Vorkoetter, J. McCarron, P. DeMarco, *Maple 10 Programming Guide*, Maplesoft, Waterloo ON, Canada, **2005**.
Opp 1975	H. Oppermann, M. Ritschel, *Krist. Tech.* **1975**, *10*, 485.
Sch 1974	H. Schäfer, M. Binnewies, *Z. Anorg. Allg. Chem.* **1974**, *410*, 251.
Sch 1976	H. Schäfer, *Angew. Chem.* **1976**, *88*, 775, *Angew. Chem., Int. Ed.* **1976**, *15*, 713.
Schm 1978	H. Schmalzried, A. Navrotsky, *Festkörperthermodynamik*, Akademie-Verlag, Berlin, **1978**.

13 Modeling of Chemical Vapor Transport Experiments: the Computer Programs TRAGMIN and CVTRANS

13.1 Purpose of Modeling of Chemical Vapor Transport Experiments

Beyond preparative applications, CVT reactions offer the acquisition of comparatively easily obtained important information on the thermodynamic characteristics of the solid at hand, and the gas phases involved, by relevant modeling. The modeling of CVT reactions can have different purposes:

- Exploratory calculations *before* the experiment can provide indications of favorable conditions concerning the temperature and the selection of transport agents.
- Modeling *after* the experiment, better after a series of experiments with a systematic variation of test conditions, provides information about the composition of the equilibrium gas phase, which is often not possible to gain by direct measurements.
- The critical comparison of observations with the results of the model calculations allow us to check or narrow down the thermodynamic data used[1].

All in all, thermochemical modeling leads to a detailed understanding of the chemical equilibria that occur in the transport system.

A possibly exact account of the observations of the transport system is aspired to by modeling. In doing so, one has to distinguish between those quantities that depend on the thermodynamic characteristics of the condensed phases and gas species involved, and those that are influenced or even determined by "non-thermodynamic" effects. The compounds of the equilibrium solid and the solution in the gas phase belong to the first group. Transport rates belong to the latter. Via the diffusion approach of *Schäfer* (see section 2.6), these are linked to the partial pressure differences between the source and the sink, and thus to the thermodynamic conditions in a system. However, they might be massively influenced by kinetic effects as well as by material flow by convection. According to the kind of experimental observations, one differentiates thermodynamic calculations and calculations that describe the speed of the material flow within the ampoule while modeling transport experiments.

---

[1] Examples are the narrowing down of the formation enthalpy of the solids VP (Gla 1989a), CrP, MnP (Gla 1989b), MNb_2O_6 (Roß 1992), CrOCl (Noc 1993), MSO_4, and $M_2(SO_4)_3$ (Dah 1992), as well as gaseous P_4O_6 (see section 6.2) (Gla 1999).

13.2 Equilibrium Calculations According to the G_{min} Method

As mentioned earlier in this book, the calculation of heterogeneous and homogeneous equilibria can take place through solving equilibrium systems that are based on the expression of mass action of given chemical reactions (K_p method) and fulfill certain (experimental) conditions (element balance, stoichiometry relations). The calculation of the equilibrium concentrations or pressures of a known solid is part of basic chemical training (Bin 1996) and is generally unproblematic. Basically the solids that appear under equilibrium conditions of a system, from a series of possibly condensed phases, can be calculated this way (if thermodynamic data are known). However, this exercise is more difficult in mathematical terms and requires clever programming. The questions of condensed phases that occur under certain conditions (pressure, temperature, element balances) in a system is of such great importance to many chemical problems that already more than 50 years ago *White, Johnson,* and *Dantzig* (Whi 1958) suggested another approach to calculate complex heterogeneous equilibria. The idea is based on the thought that the Gibbs energy of a balanced system is minimal. Thus the set of non-negative substance amounts is quested, which provides the lowest possible value for the entire Gibbs energy of a system and that at the same time fits the restriction of the balance of the available elements. An iterative method is used for the calculation of the equilibrium compositions. For the iterative method, start values, y^0, are estimated for the amounts of substance of the condensed phases and gas species that come into question. After a first iterative step, improved substance amounts x result compared to y^0, which are used for the calculation of improved start values on their part until the equilibrium composition is achieved. A new set of y-values is used for every iterative step.

The Gibbs energy G of a system is expressed by 13.2.1.

$$G = \sum_i n_i \cdot \mu_i \tag{13.2.1}$$

n_i stands for the amount of a substance, and μ_i, for its chemical potential, which is given by 13.2.2.

$$\mu = \mu_i^0 + R \cdot T \ln a_i \tag{13.2.2}$$

The activities a_i correspond to the partial pressures p_i for gas species that are treated as ideal (equation 13.2.3).

$$a_i = p_i = \left(\frac{n_i}{\sum n_i^g}\right) \sum p_i \tag{13.2.3}$$

In 13.2.3 $\sum n_i$ describes the total substance amount of all gas species, $\sum p_i$ is the total pressure. The condensed phases are to behave ideally, too; their activities are equal to one. A dimensionless number ($G/R \cdot T$) can be obtained with the above definitions (13.2.4).

$$G/R \cdot T = \sum_{i=1}^{m} n_i^g [(\mu^0/R \cdot T)_i^g + \ln \Sigma p_i + \ln (n_i^g/\Sigma n_i)]$$

$$+ \sum_{i=1}^{s} n_i^c (\mu^0/R \cdot T)_i^c \qquad (13.2.4)$$

The exponents g and c stand for the gas phase and the condensed phase. The number of gas species that are present under equilibrium conditions are termed m, those of the condensed phases s.

The quantity $(\mu^0/R \cdot T)$ is calculated for all gas species considered and the condensed phases by 13.2.5.

$$\mu^0/R \cdot T = (1/R) [(G^0 - H_{298}^0)/T] + \Delta_f H_{298}^0/R \cdot T \qquad (13.2.5)$$

Alternatively, numerical values for $(\mu^0/R \cdot T)$ can be obtained after 13.2.6.

$$\Delta(\mu^0/R \cdot T) = -\ln 10 \cdot \lg K \qquad (13.2.6)$$

The relation of the element balance can be written as in 13.2.7.

$$\sum_{i=1}^{m} z_{ij}^g \cdot n_i^g + \sum_{i=1}^{s} z_{ij}^c \cdot n_i^c = b_j \qquad (j = 1, 2, ..., l) \qquad (13.2.7)$$

In it, z_{ij} describes the number of atoms of the j-th element in the formula unit of the i-th substance. The total substance amount of the j-th elements is given by b_j and l is the number of elements.

The described G_{min} method for calculation of equilibria contains the search for the minimal Gibbs energy G respectively $(G/R \cdot T)$ of a system in consideration of the element balances (13.2.7). The method of the *Lagrange* multiplier is qualified to solve the problem. Thus, the following equations result.

$$(\mu^0/R \cdot T)_i^g + \ln \Sigma p_i + \ln (n_i^g/\Sigma n_i) - \sum_{j=1}^{l} \lambda_j \cdot z_{ij}^g$$
$$(i = 1, 2, ..., m) \qquad (13.2.8)$$

$$(\mu^0/R \cdot T)_i^c - \sum_{j=1}^{l} \lambda_j \cdot z_{ij}^c \qquad (i = 1, 2, ..., s) \qquad (13.2.9)$$

The λ_j are termed *Lagrange* multipliers.

The equations 13.2.7 and 13.2.8 can be developed to any point in a *Taylor* series $(y_1^g, y_2^g, ..., y_m^g; y_1^c, y_2^c, ..., y_s^c)$. In doing so, the terms of second or higher order can be neglected.

$$\sum_{i=1}^{m} z_{ij}^g \cdot y_i^g + \sum_{i=1}^{s} z_{ij}^c \cdot y_i^c - b_j + \sum_{i=1}^{m} z_{ij}^g (n_i^g - y_i^g) + \sum_{i=1}^{s} z_{ij}^c (n_i^c - y_i^c) = 0$$
$$(j = 1, 2, ..., l) \qquad (13.2.10)$$

$$(\mu^0/R \cdot T)_i^g + \ln p + \ln (y_i^g/Y) - \sum_{j=1}^{l} \lambda_j \cdot z_{ij}^g + (n_i^g/y_i^g) - (\Sigma n_i/Y) = 0$$

$$(j = 1, 2, ..., l) \tag{13.2.11}$$

with $Y = \sum_{i=1}^{m} y_i^g$

via 13.2.12 the numerical values of n_i^g are calculated:

$$n_i^g = -f_i + y_i^g \left[(\Sigma n_i/Y) + \sum_{j=1}^{l} \lambda_j \cdot z_{ij}^g \right] \quad (i = 1, 2, ..., m) \tag{13.2.12}$$

wherein signifies

$$f_i = y_i^g [(\mu^0/R \cdot T)_i^g + \ln \Sigma p_i + \ln (y_i^g/Y)] \quad (i = 1, 2, ..., m) \tag{13.2.13}$$

13.2.14 provides the summation of equation 13.2.12 via all i.

$$\sum_{j=1}^{l} \lambda_j \sum_{i=1}^{m} y_i^g \cdot z_{ij}^g = \sum_{i=1}^{m} f_i \tag{13.2.14}$$

The quantity C_j, which serves as a correction term in such cases in which the first assumed start values for the substance amounts do not meet the element balances, is defined as follows according to *Levine* (Lev 1962):

$$C_j = \sum_{i=1}^{m} z_{ij}^g \cdot y_i^g - b_j \quad (j = 1, 2, ..., l) \tag{13.2.15}$$

By substituting equations 13.2.11 and 13.2.14 in 13.2.9, one obtaines 13.2.16.

$$\sum_{k=1}^{l} \lambda_k \cdot r_{jk} + [(\Sigma n_i/Y) - 1] \sum_{i=1}^{m} z_{ij}^g \cdot y_i^g + \sum_{i=1}^{s} z_{ij}^c \cdot n_i^c = \sum_{i=1}^{m} z_{ij}^g \cdot f_i - C_j$$

$$(j = 1, 2, ..., l) \tag{13.2.16}$$

wherein signifies:

$$r_{jk} = r_{kj} = \sum_{i=1}^{m} (z_{ij}^g \cdot z_{ik}^g) y_i^g \quad (j, k = 1, 2, ..., l) \tag{13.2.17}$$

A system of $(l + s + 1)$ linear equations is formed from 13.2.16, 13.2.9, and 13.2.10. These consist of the $(l + s + 1)$ unknown quantities λ_j ($j = 1$, 2 ... 1), n_i^c ($i = 1, 2 ..., s$) and $\left[\left(\sum_{i=1}^{m} n_i^g/Y \right)/Y - 1 \right]$. In the following the latter expression is referred to as λ_{l+1} for reasons of clarity.

The system of equations can be solved according to *Gaussian* elimination. It should be noted that the solution delivers values for the substance amounts n_i^c, while substance amounts n_i^g are obtained by using the values λ_j ($j = 1, 2 \dots l + 1$) via equation 13.2.11. A singular matrix is obtained if a mix of two or more elements reacts off completely to a certain substance. The appearance of a singular matrix in the calculations can be avoided in two ways. On the one hand, the given element balance can be chosen in a way that it deviates slightly from the composition of the forming compound. On the other hand, one can include in the calculations traces of a new element, which is not defined to a stable compound.

Iteration procedure If all obtained substance amounts n_i are positive, they are used for the derivation of new start values of the next iteration cycle. If there are negative substance amounts, the difference between the start values y of the iteration and the values x, which are obtained in the iteration step, is reduced in order to get positive numerical values. The values for y_i' are set to zero for all substances with negative substance amounts. The equation 13.2.17 is valid for the y_i'. This way, y_i' can be calculated.

$$y_i' = y_i + \delta(x_i - y_i) \tag{13.2.18}$$

After this, the equation 13.2.18 is used in order to calculate better fitting, positive substance amounts. Therefore, a value $\delta = k \cdot \delta_{min}$ is used. It applies that $k < 1$ and δ_{min} be the smallest obtained value for δ. Usually, a numerical value near one is chosen for k, e. g., 0.99. The substance amounts y' are used for start values in the following iteration cycle. If the start values y of the substance amount fulfills the element balance, it also applies to the values of y' because the element balances are fulfilled by all values of δ. In order to avoid too many iteration steps, it became necessary to set a minimum level for the allowed start values of the substance amounts y. If the substance amount of one of the substances falls below this value, it is set to zero. This substance i is not considered in the following iteration steps because the numerical value x_i as well as the value for y_i' according to 13.2.12 and 13.2.18 assume the value zero. Because the substance amounts x^c are independent of the y^c values, the algorithms that minimize are usually formulated so that a condensed phase cannot determine the δ values before all other values x^g are positive. The quantity λ_{l+1}, which is a variable in the linear equation system that is to be solved, is used for checking if the Gibbs energy of the system has achieved the minimal value. If Σn_i approaches Y, so if the improved substance amounts, which are assumed as start values, become equal, the numerical value λ_{l+1} approaches zero. It has been proven that the numerical value $\lambda_{l+1} < 10^{-8}$ is a good choice to achieve a sufficient exactness for all substance amounts. If the minimum condition for G is not fulfilled, the calculated substance amounts are set in the equations 13.2.16, 13.2.14, and 13.2.9 and a new iteration cycle begins.

Treatment of condensed phases The above-described calculations are based on certain condensed phases whose presence was assumed under equilibrium conditions. Another set of condensed phases might provide a smaller value for the Gibbs energy of the system. That is why, it should be possible to remove or add, respectively, a condensed phase during the calculations until the correct composition is achieved. If this approach leads to a violation of the phase rule, the considered phases will be replaced by a new set of condensed phases. It is also possible to oppress individual condensed phases in the calculations. Thus, a condensed phase whose substance amount becomes more negative from iteration cycle to iteration cycle is removed from the calculations. The same happens if δ of a condensed phase approaches zero. The Gibbs energy of the system can be expressed by equation 13.2.19 in addition to 13.2.4.

$$G/R \cdot T = \sum_{j=1}^{l} \lambda_i \cdot b_j \qquad (13.2.19)$$

Equation 13.2.19 is obtained by substitution of 13.2.8 and 13.2.9 in 13.2.4. By considering 13.2.9 and 13.2.19, one can check whether a condensed phase, which has not been considered in the calculations, has to be included. This will be necessary if $(\mu^0/R \cdot T)_i^c$ becomes less than $\sum_{j=1}^{l} \lambda_j \cdot a_{ij}^c$. If no other condensed phase provides a smaller value for G, equilibrium is achieved. The number of possible combinations of the different condensed phases depends to a great extent on their total number and the number of elements. That is why the greater these numbers are, the more iteration steps are necessary to calculate the equilibrium composition. If a new set of condensed phases is considered in the calculation, the pressures of the gas species are recalculated in the following iteration steps as well. The $G_{\min}$ method as described above has been described by *Eriksson* (Eri 1971).

13.3 The Program TRAGMIN

The program TRAGMIN is designed for the thermodynamic calculation of heterogeneous equilibria where gas phases are involved. As a result, one determines the condensed phases, which are in equilibria, and their content of the substance amount balance as well as the composition of the equilibrium phase that belongs to it. On the basis of calculations of equilibria, the program allows a description of chemical vapor depositions (CVD: open system) and chemical vapor transport reactions (CVT: closed system).

The program is based on the solution procedure of calculating the equilibrium by minimizing the Gibbs energy of the system according to *Eriksson* (Eri 1971). The programming of the required calculation routine was realized by *Bieger, Selbmann, Sommer,* and *Krabbes* in the programming language Fortran (Som 1984). The modeling of the CVTs takes place in consideration of stationary relations that were formulated by *Krabbes, Oppermann, and Wolf* (Kra 1983) in the extended transport model, which describes the interrelation between source and sink chamber during CVT (section 2.4).

The specific values of the state variables temperature, volume, and pressure, as well as the substance amounts of the components (chemical elements) of the

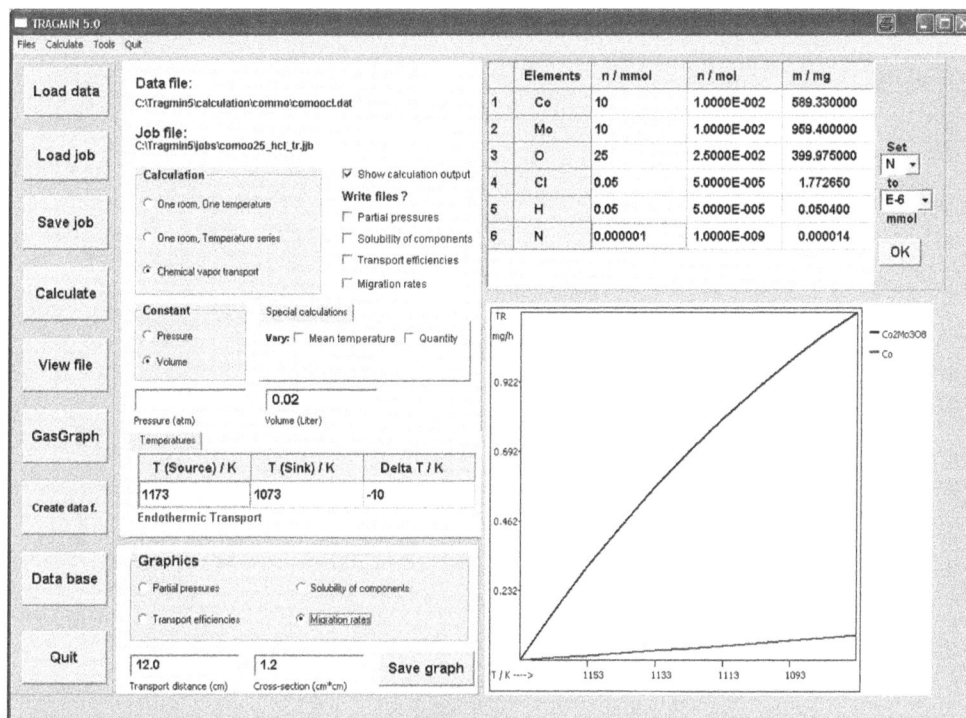

Figure 13.3.1 User interface of the TRAGMIN program showing sample calculation 2.

system are required as an input for TRAGMIN. In order to describe the equilibrium state, all conceivable condensed phases and gas species in the system with their thermodynamic data $\Delta_f H_T^0$, S_T^0 and C_p^0 are necessary set points. Compounds with defined composition as well as those with a homogeneity area can occur as condensed phases. At the moment, there are two different models that describe phases with homogeneity areas of the form AC_x and $A_{1-y}B_yC_x$.

When using the program, there is a principal distinction between calculation of the solid/gas-phase equilibria with a one-chamber model and the simulation of the gas-phase transport between zones that are spatially separated with different temperatures (two-chamber model). The modeling can be conducted assuming either a constant total pressure (open system) or a constant volume (sealed ampoule).

If a one-chamber model is used, calculations for a certain temperature or a series of temperatures (T_1 to T_2 with step width ΔT, sample calculation 1) are possible. Furthermore, series of calculations at constant temperature and variation of the content of a component or a gas species are possible.

The modeling of the gas-phase transport requires the input of the temperature of the source chamber and the variation of the sink temperature with a given step width (sample calculation 2). Furthermore, series with a gradual change of the medium temperature ($T_{sink} + T_{source})/2$ at constant step width, as well as calculations with constant temperature of the source and sink chamber and variation of the content of the component or a gas species are possible.

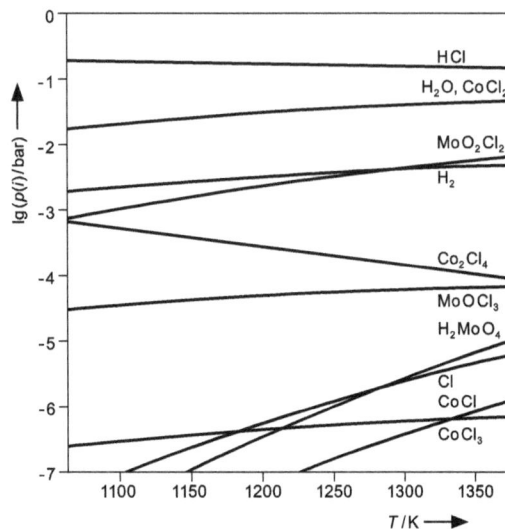

Figure 13.3.2 Composition of the equilibrium gas phase over a three-phase solid $Co_2Mo_3O_8/MoO_2/Co$ (transport agent HCl) according to *Steiner et al.* (Ste 2004).

The sub-program GASGRAPH is used for graphic illustrations. Here one gets information about the condensed phases that occur in the equilibrium, as well as the partial pressure of the gas species and gas-phase solubility of the components dependent on the temperature.

When modeling gas-phase transport, the compositions of the transported solids are indicated additionally and the transport efficiency of the gas species as well as the transport rates of the condensed compounds, which are dependent on the temperature, are graphically illustrated. The storage of the data is possible in ASCII format. Other sub-programs are for administration and collection of thermodynamic data, as well as for the automatic compiling of the input files that are required for TRAGMIN calculations.

Sample calculations in the Co/Mo/O/Cl/H system   In the mentioned system, for the calculation at $T > 1073 \, K$, the following important compounds were considered:

Condensed phases:	Co_3O_4, CoO, $CoCl_2$, Co, Mo, MoO_2, MoO_3, $CoMoO_4$, $Co_2Mo_3O_8$
Gaseous compounds:	CoCl, $CoCl_2$, Co_2Cl_4, $CoCl_3$, MoO_2Cl_2, $MoOCl_3$, $MoOCl_4$, Mo_3O_9, Mo_4O_{12}, Mo_5O_{15}, H_2MoO_4, O_2, Cl_2, Cl, HCl, H_2O, H_2, H.

Sample calculation 1   Solid–gas phase equilibrium (one-chamber model) at constant volume, calculation of a series between 1073 K and 1373 K with step width of 10 K. Further inputs: $V = 0.012 \, l$, $n(Co) = 0.01 \, mol$, $n(Mo) = 0.01 \, mol$, $n(O) = 0.025 \, mol$, $n(Cl) = 5 \cdot 10^{-5} \, mol$, $n(H) = 5 \cdot 10^{-5} \, mol$.

The substance amount, which is entered, corresponds to the composition of the three phase areas: $Co_2Mo_3O_8/MoO_2/Co$. Accordingly, when the model cal-

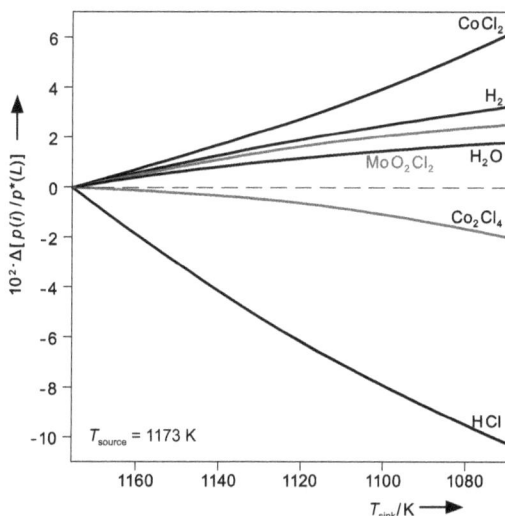

Figure 13.3.3 Transport efficiency of the individual gas species during the transport from a three-phase solid $Co_2Mo_3O_8/MoO_2/Co$ (transport agent HCl).

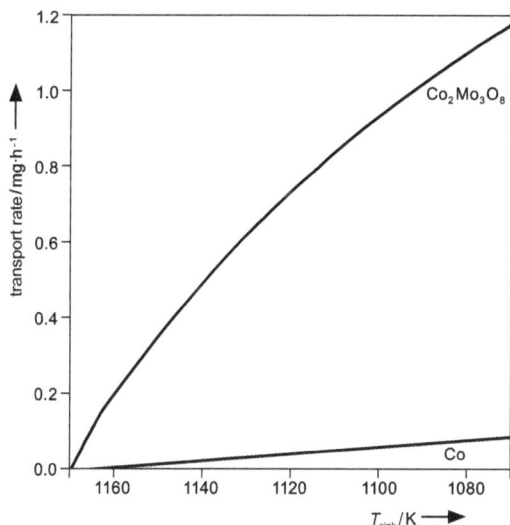

Figure 13.3.4 Calculated transport rate during the transport from a three-phase solid $Co_2Mo_3O_8/MoO_2/Co$ (transport agent HCl).

culations cover the entire investigated temperature range, these compounds are obtained as co-existing equilibrium phases in the source solid. Figure 13.3.2 shows the calculated composition of the equilibrium phase that belongs to it (gas species with partial pressure > 10^{-7} bar).

Sample calculation 2 Chemical vapor transport: transport agent HCl, calculation of a series with constant temperature T_{source} = 1173 K and variation of the sink temperature with a step width from -10 K to 1073 K; length of transport distance $s = 12$ cm; ampoule cross section $q = 1.2$ cm^2; further input analogous to calculation 1. According to the calculations, a simultaneous transport of $Co_2Mo_3O_8$ and elemental Co can be expected. The transport rate of $Co_2Mo_3O_8$ exceeds that of cobalt by 15 to 25 times (Figure 13.13.3). These results are in good agreement with the observations (Ste 2004).

TRAGMIN is available free of charge. The program is maintained and still further improved by U. *Steiner* (Dresden University of Applied Science).

Contact: Dr. Udo Steiner, Dresden University of Applied Science, Faculty of Mechanical Engineering/Process Engineering. e-mail: steiner@mw.htw-dresden.de; internet: www.tragmin.de.

13.4 The Program CVTRANS

Preliminary note. The computer program CVTRANS emerged from the programs EPC/EPLAM/EPDELT (Bec 1985). These again were based on the algorithms of *Eriksson*, *Noläng*, and *Richardson* (Eri 1971, Nol 1976, Ric 1977). Besides low operating comfort, this program package has shown some principal inadequacies in calculating simple heterogeneous equilibria and modeling of CVT experiments. Similar to all other programs that determine heterogeneous equilibria, the minimum of the Gibbs energy is determined iteratively (G_{min} method (Whi 1958, Eri 1971); see section 13.2). Calculations for systems in which the total pressure is limited by a saturation vapor pressure or a co-existing pressure, are hardly possible. Calculations appeared problematic in which the program has to determine the composition of a poly-phased equilibrium source solid out of a number of possible condensed phases. Furthermore, calculations that simulate the mass flow during a transport experiment dependent on time, so-called transport behavior (see section 2.5), could only be realized with laborious work, taking lots of time. However, these experiments, in particular if they are conducted with the help of the transport balance (see section 14.4), provide detailed information as far as deposition order, migration speed, as well as the composition of poly-phased equilibrium solids are concerned. That is why it seemed appropriate to develop a computer program that avoids the above-mentioned problems and offers operative comfort according to current developments.

General information on the program CVTRANS The program was originally developed for the operating system Windows 95 in the programming language Delphi 2.0 of Borland (Tra 1999); using it with updated operating systems (Windows XP, Vista, Windows 7) is, however, not a problem. The ideal screen resolution is $800 \cdot 600$ pixels. Every file and every value table with the results of series calculations, which are used by the program, are saved in ASCII-format. Thus they can be read by any text-processing or graphics program.

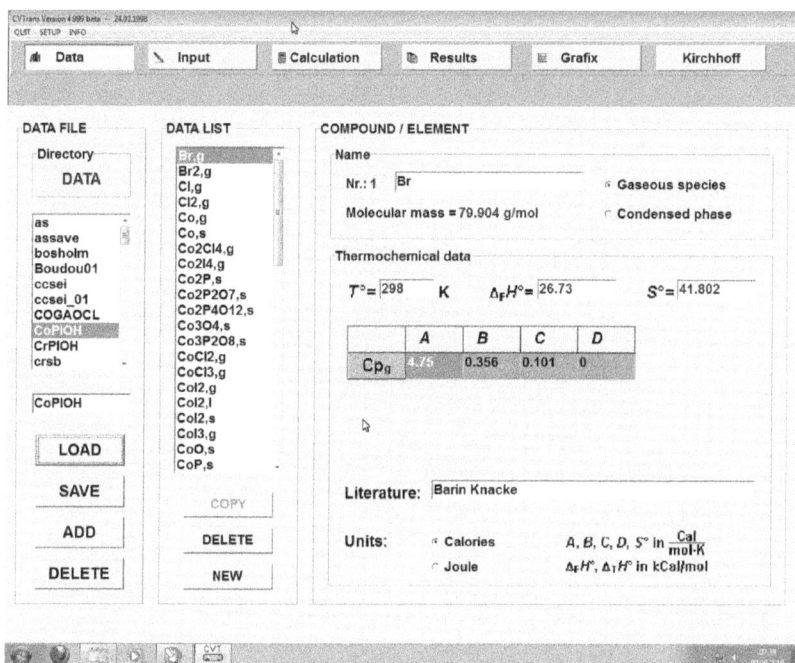

Figure 13.4.1 CVTRANS: "Data" window for the input of thermodynamic data.

According to the typical Windows screen, the program is arranged in different "windows" that are used during the application. The entry of the thermodynamic data window "Data" (Figure 13.4.1) is followed by the window "Input". Here two daughter windows open in which the user sets the compounds ("Compounds"; Figure 13.4.2) that are to be considered as well as other parameters ("Parameters"; Figure 13.4.3) of the calculation. The calculation starts in the window "Calculation" (Figure 13.4.4). "Results" allows the viewing and printing of the results via an internal editor (Figure 13.4.5) . The window "Grafix" allows a fast and clear graphic illustration of the results of series calculation (partial pressure, transport rates, composition of each solid as a function of the temperature or the amount of transport agent; Figure 13.4.6). Finally, the option "Kirchhoff" (Figure 13.4.7) offers a fast calculation of the thermodynamic data of a certain reaction.

Tips for performing of calculations Individual compounds are entered as empirical formulae in the usual chemical form in the window "Data" (see Figure 13.4.1). Brackets and fractional stoichiometry coefficients are not allowed. Thus, for example, Wustite at the lower phase limit, $Fe_{0.948}O$, is entered with the formula $Fe_{95}O_{100}$. The energy units joule and calorie are both supported. Temperatures can be entered either in °C or in K.

The compounds that shall be considered for the calculation of the equilibrium can be chosen in the window "Input – Compounds" (Figure 13.4.2). The compounds can be assigned start substance amounts of any number (in mg or mmol).

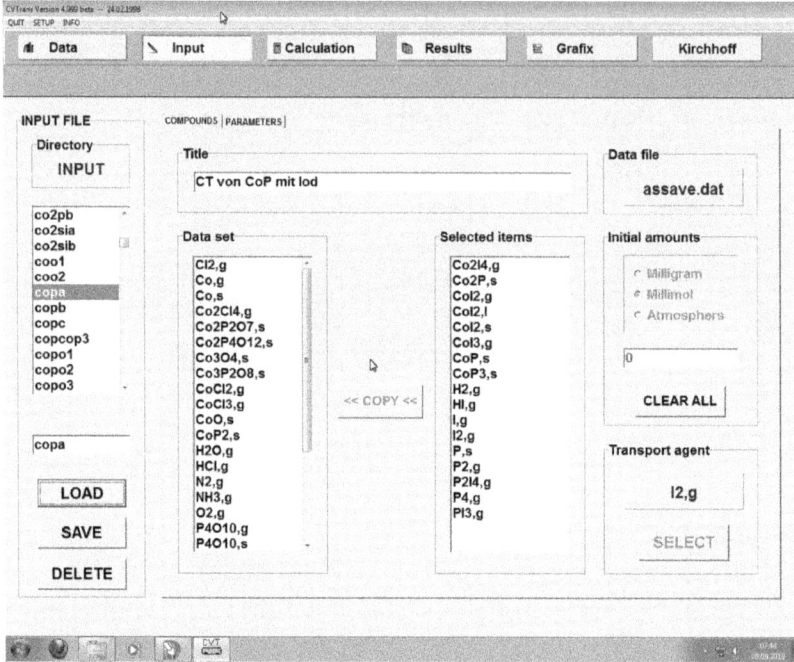

Figure 13.4.2 CVTRANS: "Input – Compounds" window for selecting the compounds and substance amounts for a calculation.

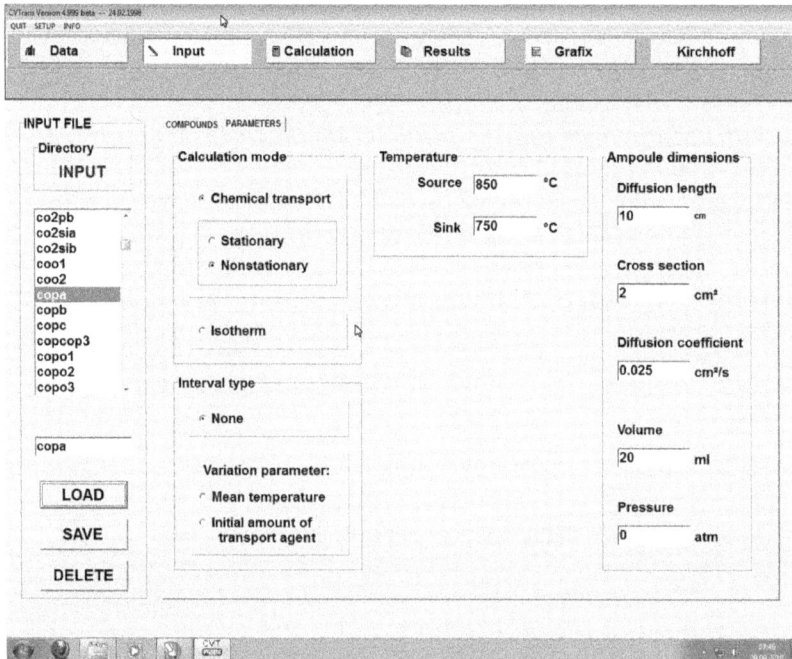

Figure 13.4.3 CVTRANS: "Input – Parameters" window for determining the experimental conditions.

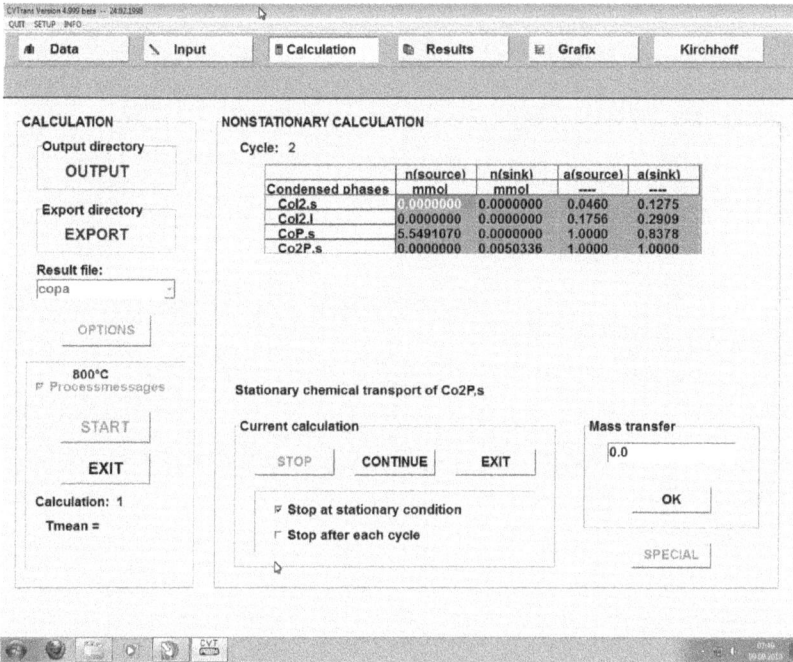

Figure 13.4.4 CVTRANS: "Calculation" window to start a calculation and choose a screen display during the calculation.

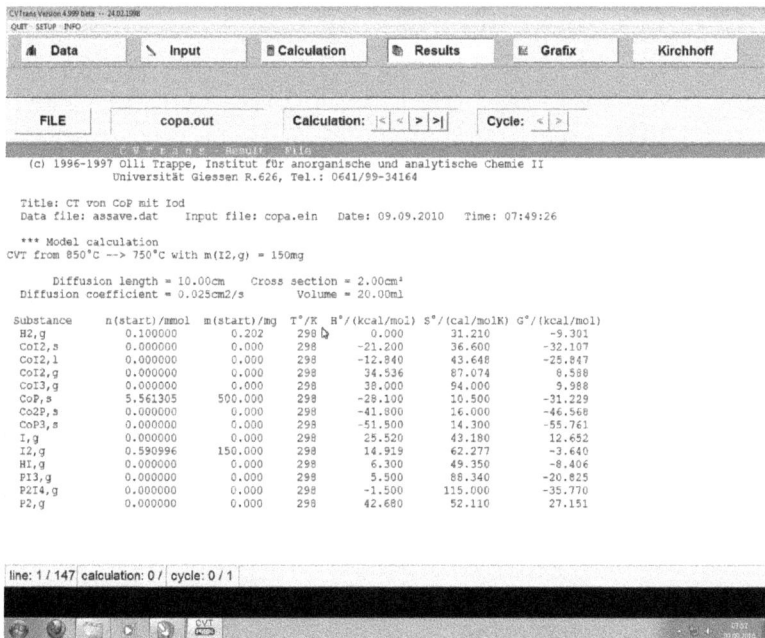

Figure 13.4.5 CVTRANS: "Results" window for the numerical output of results of the equilibrium calculations and the modeling of CVT experiments.

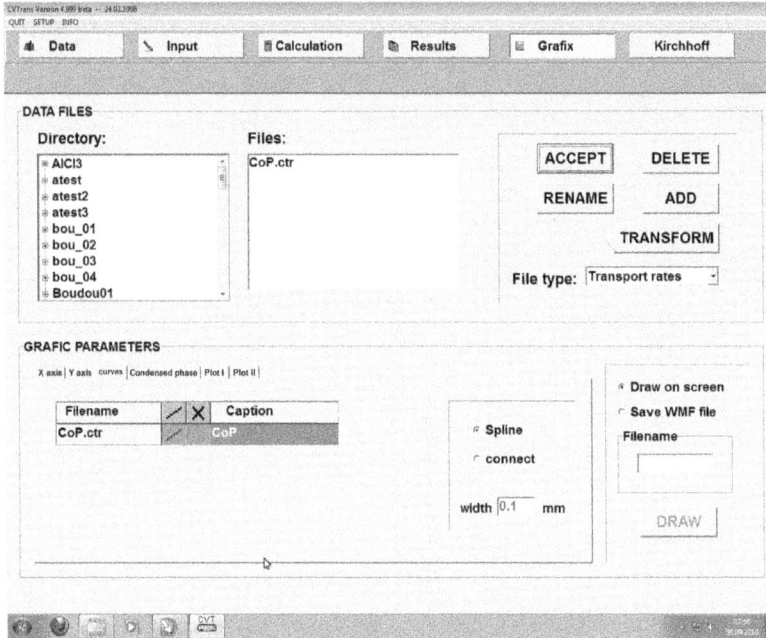

Figure 13.4.6 CVTRANS: "Graphix" window for graphic illustration of results of the equilibrium calculations and the modeling of CVT experiments.

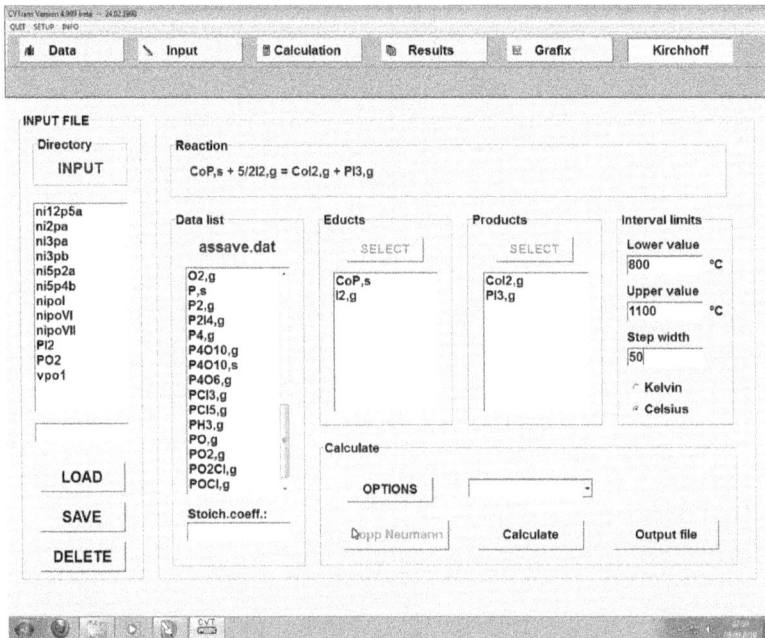

Figure 13.4.7 CVTRANS: "Kirchhoff" window for calculation of $\Delta_r G^0$, $\Delta_r H^0$, $\Delta_r S^0$, and K_p of the equilibrium reaction and display of results of interval calculations in ASCII format.

CVTRANS allows the calculation of heterogeneous equilibriums under different boundary conditions. Whether the equilibrium calculations to be carried out are under isothermal conditions (at constant pressure or constant volume) or transport experiments in the temperature gradient, can be set in the window "Input – Parameters" (Figure 13.4.2.3). Calculations for CVT can be carried out for simple, stationary systems with only one stationary state, as well as for complex, non-stationary experiments. The model calculations of non-stationary transport experiments are based on the co-operative transport model of *Schweizer* and *Gruehn* (Schw 1983a, Schw 1983b) (see section 2.5). In an easy way, this offers the possibility to simulate time-dependent substance flows in the ampoule. In the first cycle, one calculates as if the transport were stationary. If the composition of the sink solid differs from the source solid, a second cycle is started. The element balance of the equilibrium calculation of the source results from the substance amount, which was given at the beginning of the calculation, less the substance amount that was deposited in the sink (Figure 13.4.6). The calculation is finished when the solid is no longer on the source side, or its composition no longer changes. Here the criterion of stationarity can be checked by comparing the composition of the gas phase in the current calculation with the previous one. It is only a stationary transport of one single condensed phase if its substance amount is decreased in the source chamber and simultaneously increased in the sink chamber. If this applies to several condensed phases, it is a simultaneous transport of two or more condensed phases. Furthermore, the rare case can occur where the condensed phase, which dissolves at the source side, does not occur at all at the sink side (dismutation).

In order to calculate the transport rates according to *Schäfer*'s diffusion approach (see section 2.6, equation 2.6.11), the ampoule parameters (diffusion distance, cross section, volume) and the medium diffusion co-efficient of the gas species can be changed ("Input – Parameters"; Figure 13.4.2.3).

If CVTRANS has detected a stationary transport during the course of the calculations, the transport rate is automatically calculated.

The menu "Kirchoff" allows the calculation of the thermodynamic data for a given reaction equation. To do so, the data of the reactant are used, which are present in the data storage of CVTRANS. For the entered equation, the reaction data $(\Delta_r G^0, \Delta_r H^0, \Delta_r S^0, K_P)$ can be shown for a special temperature, or the data that were calculated for a temperature interval can be saved in an output file. Furthermore, it is also possible to create a value table with the data dependent on the temperature in an ASCII-file. Every value table that is created by CVTRANS can be graphically illustrated direct from the program.

CVTRANS is available for free for non-commercial users. Please enquire Prof. Dr. Robert Glaum, Institute of Inorganic Chemistry, University of Bonn. e-mail: rglaum@uni-bonn.de.

Bibliography

Bec 1985   J. Becker, *Computerprograms EPC, EPLAM, EPDELT und EPZEICH*, University of Gießen **1985**.

Bri 1946   S. R. Brinkley, *J. Chem. Phys.* **1946**, *14*, 563.

Dah 1992   T. Dahmen, R. Gruehn *Z. Anorg. Allg. Chem.* **1992**, *609*, 139.

Eri 1971   G. Eriksson, *Acta Chem. Scand.* **1971**, *25*, 2651.

Gla 1989a   R. Glaum, R. Gruehn *Z. Anorg. Allg. Chem.* **1989**, *568*, 73.

Gla 1989b   R. Glaum, R. Gruehn, *Z. Anorg. Allg. Chem.* **1989**, *573*, 24.

Kra 1983   G. Krabbes, H. Oppermann, E. Wolf, *J. Cryst. Growth* **1983**, *64*, 353.

Noc 1993   K. Nocker, R. Gruehn, *Z. Anorg. Allg. Chem.* **1993**, *619*, 699.

Nol 1976   B. I. Noläng, M. W. Richardson, *J. Crystal Growth* **1976**, *34*, 198.

Oth 1968   D. F. Othmer, H.-T. Chen, *Ind. Eng. Chem.* **1968**, *60*, 39.

Roß 1992   R. Roß, R. Gruehn, *Z. Anorg. Allg. Chem.* **1992**, *614*, 47.

Ric 1977   M. W. Richardson, B. I. Noläng, *J. Crystal Growth* **1977**, *42*, 90.

Sch 1962   H. Schäfer, *Chemische Transportreaktionen*, Verlag Chemie, Weinheim **1962**. H. Schäfer, *Chemical Transport Reactions*, Academic Press, New York, London, **1964**.

Schm 1995   A. Schmidt, R. Glaum, *Z. Anorg. Allg. Chem.* **1995**, *621*, 1693.

Scho 1989   H. Schornstein, R. Gruehn, *Z. Anorg. Allg. Chem.* **1989**, *579*, 173.

Scho 1990   H. Schornstein, R. Gruehn, *Z. Anorg. Allg. Chem.* **1990**, *587*, 129.

Schw 1983a   H.-J. Schweizer, *Dissertation*, University of Gießen, **1983**.

Schw 1983b   H.-J. Schweizer, R. Gruehn, *Angew. Chem.* **1983**, *95*, 80, *Angew. Chem. Int. Ed.* **1983**, *22*, 82.

Som 1984   K. H. Sommer, D. Selbmann, G. Krabbes, *scientific report, ZFW* Dresden, **1984**, Nr. 28.

Ste 2004   U. Steiner, S. Daminova, W. Reichelt, *Z. Anorg. Allg. Chem.* **2004**, *630*, 2541.

Tra 1994   O. Trappe, Diplomarbeit, University of Gießen, **1994**.

Tra 1999   O. Trappe, R. Glaum, R. Gruehn, *Computerprogram CVTRANS*, University of Gießen, **1999**.

Whi 1958   W. B. White, S. M. Johnson, G. B. Dantzig, *J. Chem. Phys.* **1958**, *28*, 751.

Zel 1968   F. J. Zeleznik, S. Gordon, *Ind. Eng. Chem.* **1968**, *60*, 27.

14 Working Techniques

This chapter provides information on the experimental realization of CVT experiments. Different research groups that deal with this subject use slightly different techniques. Each one offers advantages and disadvantages. The tips that are provided here must be seen as guidelines. What kind of technique is used depends on the aim of the experiment. If one wants to understand transport reactions precisely, particular procedures are necessary. For example, a transport ampoule can be baked out several days in vacuum before it is charged in order to remove traces of water. In such cases one would set high standards for the vacuum apparatus that is used. Also the purity of the chemicals used can play a role. Typically, for preparative or teaching purposes, these procedures are not necessary. Thus transport experiments can be conducted with different degrees of complexity.

14.1 Transport Ampoules and Transport Furnaces

In most cases, transport reactions are executed in closed ampoules of a suitable glass. The selection of the kind of glass is defined by the transport temperatures. Today borosilicate glass is frequently used, which is suitable up to 600 °C. Glass made from pure silicon dioxide and quartz, **silica glass**, is appropriate at higher temperatures and for more corrosive fillings. Silica tubes are stable at temperatures up to 1100 °C. It is important to note that water is released during the heating of silicia glass (water content up to 50 ppm). In order to avoid this, careful baking out of the ampoule in vacuum is recommended (see section 3.5).

In special cases different metal ampoules (Kal 1974) or containers made from ceramic material, such as Pythagoras mass or sintered corundum (Klo 1965), are used. However, sealing ampoules that are made from ceramic material is problematic. To avoid this, insets of sintered corundum or glassy carbon are integrated in a silicia ampoule instead (Figure 14.1.1). In particular, ceramic materials are used if the content of the ampoule and the SiO_2 of the silica glass ampoule react with each other. Alternatively the inner wall of the ampoule can be coated with carbon by pyrolysis of acetone (Ham 1993).

Vapor transport reactions take place in a temperature gradient. In order to set up the gradient in a controlled manner, tube furnaces with two, or sometimes three, independent heating coils are used. The transport furnace should be in a horizontal position in order to keep convection as part of the gas motion as small as possible. However, if the aim of the transport is the preparation of large amounts of substance by an endothermic transport, the furnace can be tilted so that the sink side is higher than the source side. This increases the transport rate. These experiments, however, cannot be described by the thermodynamic models that are based on gas motion by diffusion.

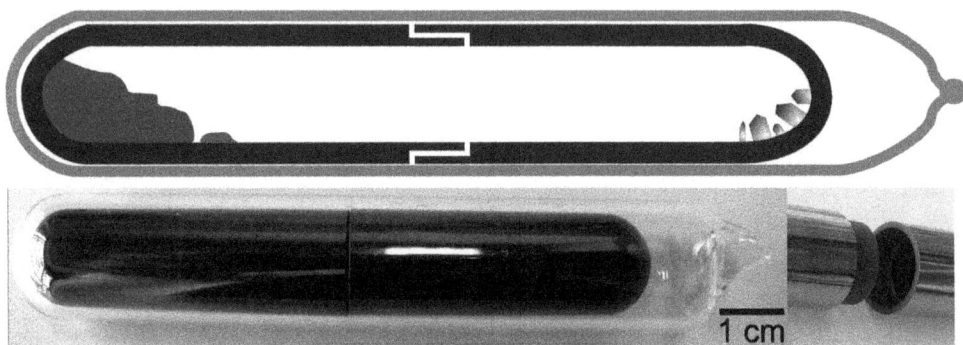

Figure 14.1.1 Transport ampoule with an inset of glassy carbon.

The so-called **short-distance transport**, which was described by *Krämer*, uses a high convective contribution to the gas motion (Krä 1974). The transport takes place in a vertical direction, over a distance of approximately 3 cm only. The ampoule cross section in such experiments is particularly large, 30 cm^2 (see Chapter 2). This way, CVT also succeeds in systems that otherwise have a low transport rate due to an unfavorable equilibrium position. This variation of the standard experimental set up is particularly effective for endothermic transport reactions. In cases of exothermic transport reactions, the convective part is omitted and the transport rate is exclusively determined by diffusion.

Experiments for the growth of larger crystals by means of CVT reactions are closely related to the short-distance transport. Here the transport ampoule is positioned vertically to provide a high convective mass flow. Furthermore, special ampoules for sorting out crystal seeds are used (Wil 1988).

A temperature controller and a thermosensor are required for each heating zone. Thermocouples *Type S* (Pt/Rh//Pt) are the sensors that are often used. At temperatures up to approximately 1000 °C, one could also use thermocouples of *Type K* (Ni/Cr//Ni). In practice, the thermocouples that measure temperatures T_1 and T_2 are part of the control circuit at the same time. There can be four thermocouples in a two-zone furnace; two in the actual control circuit and two to measure the temperatures of the source and sink chamber. The regular thermocouples are directly on the heating coil, the other two immediately next to the transport ampoule.

One has to make sure that the thermocouples measure the temperatures of the source and sink zone as accurately as possible. It is important that the tip of the thermocouple, the junction, is as close as possible to the source or the sink zone of the ampoule. Furthermore the other end of the thermocouple sticks out of the furnace so that the connection point between the actual **thermocouple** and the connection line to the temperature regulator is at room temperature; if not, measurement errors can occur. Figure 14.1.2 shows schematically the experimental set up for CVT in a two-zone furnace.

When the furnace construction is used for the first time, it is advisable to determine the horizontal temperature gradient of the furnace used with the help of a particularly long thermocouple at typical transport temperatures (e. g., 1000 → 900 °C). This way one gets the full picture of the temperature gradient and the zone of constant temperature in the furnace.

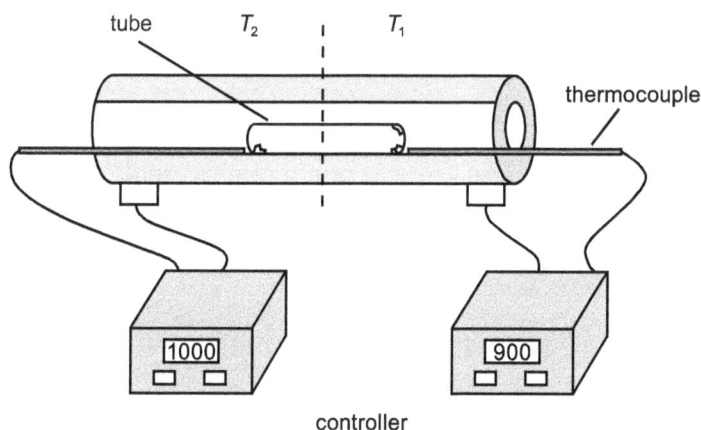

Figure 14.1.2 Experimental set up for chemical vapor transport in a two-zone furnace.

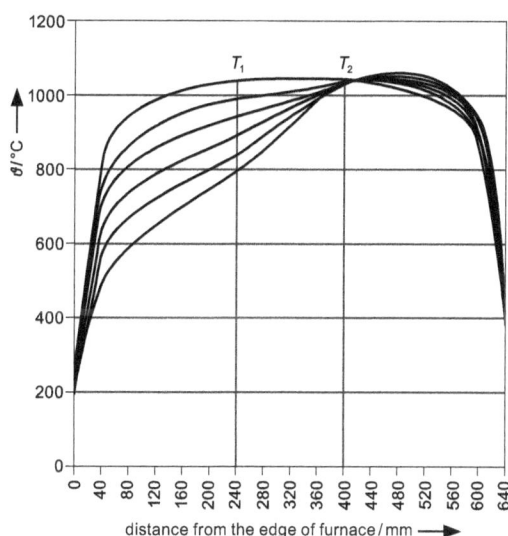

Figure 14.1.3 Typical temperature characteristic of a two-zone furnace (T_2 = constant; T_1 = variable).

14.2 Preparation of Transport Ampoules

The experimental procedures for preparing transport ampoules can be different. Above all, they are dependent on the physical and chemical properties of the transport agent. Three cases can be distinguished.

1. The transport agent is solid or liquid under normal conditions but has a considerable vapor pressure (example: iodine).
2. The transport agent is gaseous under normal conditions (examples: HCl, Cl_2).
3. The transport agent does not have a vapor pressure at room temperature, but has to be prepared in a pre-reaction (examples: $TeCl_4$, $AlCl_3$).

Figure 14.2.1 Transport ampoule with a ground joint.

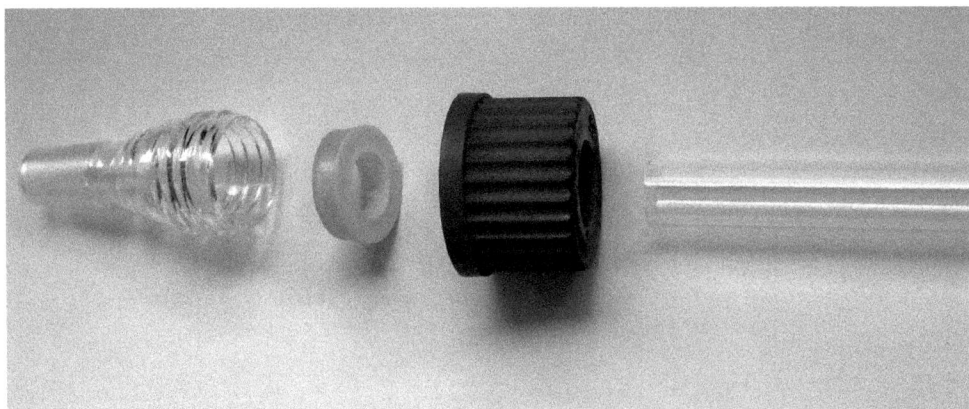

Figure 14.2.2 Transport ampoule with a quick-fit joint.

For 1 A tube closed on one side is prepared (diameter 10 to 20 mm). Approximately 15 to 20 cm from the closed end, the tube narrows so that the educts can be filled without any problems; the narrow part should be as narrow as possible in order to facilitate sealing but still wide enough to offer sufficient space for a long funnel. The open end of the tube should be arranged in such a way that a smooth and vacuum-tight junction to the vacuum line is possible. Figure 14.2.1 shows how a typical ampoule prepared in this way looks.

First, the prepared tubes are filled with approximately one gram of the solid that is to be transported. For this purpose one uses a funnel long enough that the outlet is near the ampoule bottom. This avoids small particles of the solid from sticking to the ampoule wall and then functioning as crystal seeds during deposition. This would lead to the unwanted formation of many small crystals. In the same way the transport agent can be added. Its amount is often selected so that

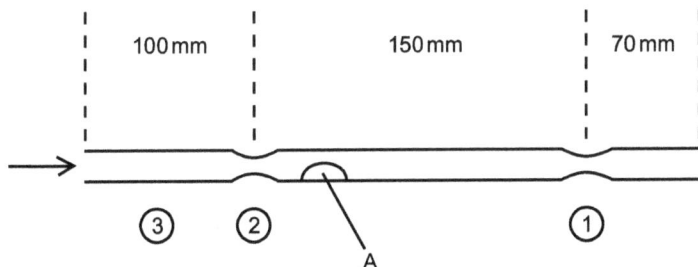

Figure 14.2.3 Transport ampoule to be filled with a gaseous transport agent.

the pressure (approximately expressed by the initial pressure of the transport agent) in the ampoule is 1 bar at the experiment temperature (gas law).

The transport ampoule and the vacuum line can be joined in different ways. Frequently, a ground-glass joint is chosen. However, certain glass-blowing skills are required. Alternatively, "quick-fit" joints have been established. This is less complex; however, it can lead to leaks in the joint when the ampoule is moved while sealing. For the purpose of teaching, this kind of joint is recommended because of its low cost.

The ampoule is attached to the vacuum line in a vertical arrangement; the valve stays shut. Usually the contents of the ampoule must be cooled with liquid nitrogen before evacuation in order to avoid vaporization or sublimation. For cooling, the bottom end of the vertically arranged ampoule is dipped 5 cm into the cooling medium, then after approximately 3 minutes the valve to the vacuum apparatus is opened; the reaction tube is evacuated and sealed. Cooling of the ampoule is necessary when the vapor pressure of one or more of the components of the filling is so high at room temperature that there is a danger of vaporization or sublimation. If iodine is used as transport agent, cooling is obligatory.

Safety advice: It is absolutely essential to avoid the condensation of liquid oxygen or moisture within the ampoule prior to sealing. A leak between the interior of the ampoule and the atmosphere, particularly when the contents have already been cooled with liquid nitrogen but not yet evacuated, must be prevented! If the ampoule is sealed in this state, a strong explosion will be the result after the removal of the cooling agent due to very high pressure in the ampoule. For safety reasons, liquid nitrogen that is used for cooling should be kept in a Dewar vessel made of metal.

For 2 The transport agent is gaseous under normal conditions (examples: HCl, Cl_2). In such a case, different working techniques are possible. Using a technique, that was used in earlier times especially, a special glass tube is prepared. This is schematically illustrated in Figure 14.2.3.

Solid A is inserted in the tube at the marked spot. Then, for some time, a stream of the transport agent is passed through the tube. This way the air is displaced. Subsequently, it is sealed; first at spot (1), then at (2). It is important to check that the transport agent can escape through a pressure valve. Otherwise, through heating of the tube, internal pressure develops, which makes sealing impossible. The advantage of this technique is its easy realization. However, there are disadvantages as well.

- Unnecessary amounts of the transport agent are needed. Possible precautions concerning disposal have to be taken.
- The initial pressure of the transport agent is always 1 bar at room temperature; other initial pressures cannot be set.

Another way to introduce a certain amount of gaseous transport agent into the ampoule is via condensation. Often chlorine is used as the transport agent. If one wants to avoid the handling of a steel cylinder filled with liquid chlorine, the element can be evolved easily by thermal decomposition of platinum(II)-chloride at 525 °C. One can also use anhydrous copper(II)-chloride, which can be obtained from the dihydrate by heating in a drying cabinet at 140 °C. These chlorides decompose according to the following reactions:

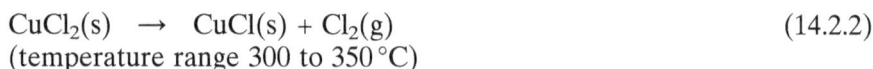

$$PtCl_2(s) \rightarrow Pt(s) + Cl_2(g) \tag{14.2.1}$$
(above 525 °C)

$$CuCl_2(s) \rightarrow CuCl(s) + Cl_2(g) \tag{14.2.2}$$
(temperature range 300 to 350 °C)

If $CuCl_2$ is used as the chlorine source, one has to make sure that the temperature does not rise above 350 °C because the vapor pressure of the CuCl formed is already considerable then. The required amount of $PtCl_2$ or $CuCl_2$, respectively, depends on the setting of the initial pressure, the temperature, and the volume of the ampoule. It can be calculated with the help of the ideal gas law. In order to set up a pressure of 1 bar at room temperature in a volume of one milliliter, about 0.01 g $PtCl_2$ is needed.

$$p \cdot V = n \cdot R \cdot T = \frac{m}{M} R \cdot T$$

$$m = \frac{p \cdot V \cdot M}{R \cdot T} \tag{14.2.3}$$

$$m = \frac{1 \text{ bar} \cdot 10^{-3} \, 1 \cdot 266 \text{ g} \cdot \text{mol}^{-1}}{0.08314 \, 1 \cdot \text{bar} \cdot K^{-1} \cdot \text{mol}^{-1} \cdot 298 \text{ K}} = 1.073 \cdot 10^{-2} \text{ g}$$

An apparatus that is suited to fill a transport ampoule with chlorine is schematically illustrated in Figure 14.2.4.

The transport ampoule is filled with the source solid as described above. The previously calculated amount of $PtCl_2$ or $CuCl_2$, respectively, is filled in the right side of the apparatus. Then the apparatus is built together and evacuated. Subsequently, $PtCl_2$ or $CuCl_2$, respectively, is heated carefully with the burner until the decomposition is finished. As a next step, the bottom end of the transport ampoule is cooled with liquid nitrogen and the ampoule is sealed. Usually the metallic platinum that is formed during decomposition does not interfere with the

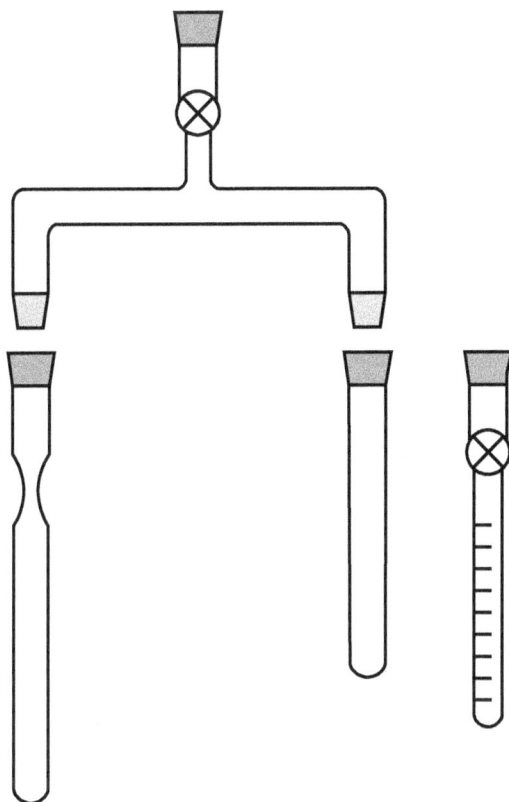

Figure 14.2.4   Apparatus to fill a transport ampoule with a defined amount of chlorine.

transport experiment. Alternatively one can introduce small chlorine-filled silica capillaries into the transport tube. An apparatus for particularly ambitious work, which allows filling the transport ampoules with different gases of high purity, is described in the literature (Ros 1979).

In order to fill a transport ampoule with a defined amount of bromine, the following working technique has proven successful. A graduated pipette of 1 ml volume is melted off at the bottom and a Teflon plug is set at the upper end. The prepared pipette is then filled with dry bromine. If the tube for the decomposition of $PtCl_2$ or $CuCl_2$, respectively, which is illustrated in Figure 14.2.4, is replaced by this pipette, a defined amount of bromine can easily be re-condensed in the transport ampoule. $PtBr_2$ can serve as a bromine source as well. It decomposes to the elements above 475 °C.

The corresponding ammonium halides are usually used as the source for the hydrogen halides, as these are particularly easy to dose. One has to keep in mind, however, that the ammonia that is formed first, decomposes into nitrogen and hydrogen at temperatures above 700 °C and a reducing atmosphere is generated. To avoid this, the hydrogen halides have to be inserted with a special working technique.

Conveniently, hydrogen chloride is formed by the reaction of ammonium chloride and concentrated sulfuric acid. Hydrogen bromide forms during the conversion of ammonium bromide with concentrated phosphoric acid. Sulfuric acid is not suited due to its oxidizing character. Hydrogen iodide is thermally unstable; HI decomposes to a large extent in H_2 and I_2 at transport temperatures. In practically all cases, ammonium iodide is used as the hydrogen iodide source because the additional proportion of hydrogen that is formed by the decomposition can be tolerated. Alternatively, hydrogen halides can be taken from small, commercially available lecture bottles.

A certain amount of water can easily be provided by heating $BaCl_2 \cdot 2\,H_2O$ to 150 °C. Oxygen forms during heating of gold(III)-oxide to approximately 200 °C.

For 3 It is particularly easy to feed the transport ampoule if the transport agent has practically no vapor pressure at room temperature, for example aluminum(III)-chloride and tellurium(IV)-chloride. However, these additives, which are frequently used as transport agents, are very moisture sensitive. In order to insert them into transport ampoules, different working techniques are used.

a) Small, thin-walled silica glass capillaries are filled with the necessary amount of the transport additive under suitable protective measures (glove box). These are sealed under vacuum. The capillaries are scarified with a glass knife and inserted in the transport ampoule. After sealing the ampoule, it is shaken until the capillary breaks. Alternatively, the transport additive can be inserted directly into the transport ampoule in the glove box. It is recommended that these transport additives are synthesized, aluminum(III)-chloride in particular, because commercial products often do not meet the requirements (see section 7.2). By shaking the ampoule, the contents are moved to the sink side as well. Any unwanted crystallization seeds are removed by putting the ampoule in an ultrasonic bath in vertical position for a few minutes.

b) An ampoule is used as shown in Figure 14.2.3. At 3, a combustion boat containing tellurium or aluminum, respectively, is inserted. The metal reacts with a suitable gas (Cl_2, HCl) and the reaction product is sublimed in the part of the ampoule between 1 and 2. One has to make sure that the flow velocity of the gas is fitting. After complete conversion (a few minutes), one seals at 1 and the reaction tube is connected to the vacuum apparatus. It is important at this stage to exclude the access of air and, in particular, moisture under all circumstances. Subsequently, the tube is evacuated and sealed at 2. The vacuum pump is then run by a gas ballast for some minutes in order to remove the aggressive Cl_2 (or HCl). Precautions with regard to the pump are not necessary due to the small amounts of substance.

14.3 The Transport Experiment

The prepared transport ampoule is placed in the furnace so that the middle of the furnace and the middle of the ampoule are at the same position. Before the actual transport temperatures are set, the ampoule is exposed to a reversed temperature gradient for approx. one day. This way, the ampoule walls on the

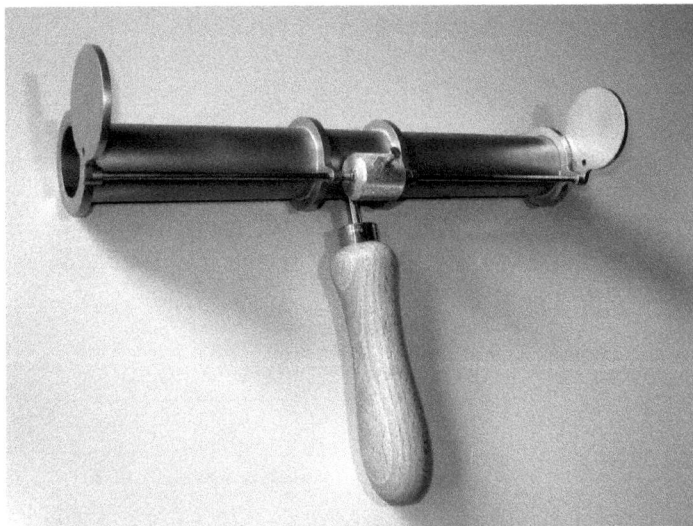

Figure 14.3.1   An "ampoule catcher".

sink side are freed of small crystallization seeds. This approach is called **back transport** or **transport in a reverse temperature gradient**. The ampoule is taken out carefully after finishing the experiment. In order to obtain crystals without being contaminated by the condensed gas phase, one has to make sure that the gas phase condenses on the source side. Different approaches have been established to achieve this.

a) The transport ampoule is moved off the furnace on its source side so far that the ampoule sticks out approximately 5 cm. Within a few minutes the gas phase condenses there and the ampoule can be taken out.

> **Safety advice:** Mainly there is danger of explosion. That is why the end of the ampoule that sticks out must be secured with safety glass as a splinter shield. The ampoule should be handled with suitable protective gear until it is opened.

b) The transport ampoule is shifted to the source side by a so-called "ampoule catcher" (Figure 14.3.1). The "ampoule catcher" is closed and the source side is dipped into cold water.

Opening of the ampoule can be done in different ways. Either the ampoule is scarified at a suitable spot, wrapped with a firm cloth, and broken open, or an approximate 1 mm slot is cut in the ampoule with a suitable tool. After securing with a cloth the ampoule is then opened by levering with a screwdriver. Finally, the sink side is rinsed out with a suitable solvent.

14.4 The Transport Balance

The transport balance is a measuring device that allows the recording of the time dependence of transport experiments. In the process, the changes of the tracking force of the balance is recorded and graphically represented during the entire transport experiment by means of a computer. This way, the transport action can be followed online. The schematic set up of the measuring arrangement is illustrated in Figure 14.4.1. *Plies* described the set up and function principle of the transport balance for the first time (Pli 1989).

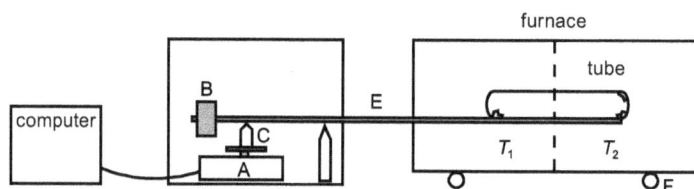

Figure 14.4.1 Schematic set up of the transport balance (A: scale; B: counterweight; C: plug; D: edge; E: lever; F: furnace).

Using a transport balance, the ampoule is positioned on a lever of two corundum bars. The other end of the lever presses on an electronic scale by a counterweight. In doing so, the lever must have a possibly low tracking force. The balance is connected to a computer via a data output. During the transport process, the ampoule lies on the lever that reaches into the furnaces in a levitating manner. During this experiment, a shift of the mass m_1 along the transport distance Δs within the transport ampoule leads to a change of the tracking force m_2 of the scale. The quantity of the detected weight change is dependent on the length of the lever l (= edge/plug distance). The following relation results:

$$m_2 = \frac{m_1 \cdot \Delta s}{l} \qquad (14.4.1)$$

The quotient $\Delta s/l$ should be about one. This way, transport rates can be recorded that are greater than or equal to $1 \text{ mg} \cdot \text{h}^{-1}$.

Experimental

1. The transport ampoule is put on the lever. The tracking mass of the lever on the scale is chosen so that the scale is about 1 g. The edge/plug distance is measured.
2. The furnace is moved above the lever with the ampoule. It is checked that the lever is suspended in air.
3. The furnace is heated; first a reverse temperature gradient is applied.
4. The computer program for recording the data is started. As a rule a measurement is taken every 1 to 3 minutes.
5. Transport in a reverse temperature gradient is carried out until changes in weight are no longer registered.

6. The furnace is set to the medium transport temperature on both sides. If it is reached, the temperature gradient is set. When the transport temperatures are reached, the zero point (tare) is set at the transport balance.
7. The data are shown online on the screen during the transport experiment.
8. After the ampoule has cooled down, the transport distance is measured.
9. The measured values are corrected by means of equation 14.4.1 and the transported mass is graphically illustrated against the transport duration. The transport rates (mg $\cdot$ h^{-1}) can be calculated from the gradient of the curve.

The following points have to be heeded during experiments with the transport balance:

- Avoid shattering, air draught, or solar radiation.
- Temperature fluctuations should be as small as possible, e. g., by sealing the furnace outlet with an insulating material and/or putting up protective shields (lever side).

14.5 High Temperature Vapor Transport: Transport under Plasma Conditions

Two further variations of the CVT method are "high-temperature vapor growth" (HTVG) (Kal 1974a, Kal 1975) and the CVT under plasma conditions (Vep 1971). Both techniques are considerably laborious.

Kaldis described the CVT at high temperatures (around 2000 °C) for different compounds, such as SiO_2, Al_2O_3, EuO, Eu_2SiO_4, Eu_3SiO_5, LaS, SmS, EuS, EuSe, EuTe, YbSe, YbTe, GdP, and HoP (Kal 1968, Kal 1971, Kal 1972, Kal 1981).

Besides CVT, *Kaldis* understands by HTVG in general, all methods for crystallization via the gas phase at high temperatures, including sublimation, decomposition, sublimation, and auto transport, respectively. Single-crystal growth of the compounds LaS, NdSe, NdTe, SmSe, SmTe, GdS, GdSe, GdTe, HoTe, ErTe, and, in particular, of rare earth nitrides EuN, GdN, HoN, DyN, as well as TbN, are successful examples (Kor 1972a). The experiments were executed exclusively as endothermic transports in vertically arranged, welded metal ampoules made of molybdenum or tungsten under high vacuum. The heating was achieved by means of high-frequency technology.

The CVT under the influence of non-isothermal, low-pressure plasmas has been investigated and described by *Veprek* in particular (Vep 1971, Vep 1976, Vep 1980, Vep 1988). The transport can take place from the plasma to neutral gas, from neutral gas to plasma, and from plasma to plasma. Compared to isothermal plasma, for example, when the neutral gas and the electrons have very high temperatures, the temperature of the neutral gas and the temperature of the electrons of non-isothermal plasma are two relatively independent, variable parameters. The plasma involved in CVT is called "cold plasma".

Plasmas feature a considerably higher internal energy than a neutral gas in the thermodynamic equilibrium at the same temperature, because a plasma consists of charged and neutral particles. The internal energy of a non-isothermal low-pressure plasma basically corresponds to the sum of the excitation energy of all the particles. Thus the kinetic energy of gaseous hydrogen at 25 °C is approx. $8 \cdot kJ \cdot mol^{-1}$; that of a hydrogen plasma at 25 °C and a degree of dissociation of 80 % is approx. 340 kJ $\cdot$ mol^{-1} (Vep 1971).

The effect of plasmas on CVT can be assigned to two categories. One category contains the kinetic aspect; the other category contains the thermodynamic aspect. The first category covers all reactions that should take place according to thermodynamic consideration but which are kinetically inhibited, for example, the deposition of carbon via the *Boudouard* reaction. In these cases, the plasma has a catalyzing effect. The second category covers cases in which the plasma state influences the position of the thermodynamic equilibrium in a way that CVT becomes possible. Examples are the transport of aluminum nitride, titanium nitride, and zirconium nitride, respectively, around 1000 °C in a nitrogen/chlorine low-pressure plasma or the transport of carbon in hydrogen, oxygen, or nitrogen plasma. In this context we want to refer particularly to the deposition of diamond layers from a hydrogen/methane low-pressure plasma (98 to 99 % H_2; 1 to 2 % CH_4; 100 to 200 mbar total pressure) at 800 to 900 °C (Reg 2001, Ale 2003). The deposition of polycrystalline silicon by CVT in hydrogen plasma at 60 °C is another example in which kinetic as well as thermodynamic aspects are important.

Bibliography

Ale 2003	V. D. Aleksandrov, I. V. Sel'skaya, *Inorg. Mater.* **2003**, *39*, 455.
Ham 1993	A. Hammerschmidt, *Dissertation*, University of Münster, **1993**.
Kal 1968	E. Kaldis, *J. Cryst. Growth* **1968**, *3/4*, 146.
Kal 1971	E. Kaldis, *J. Cryst. Growth* **1971**, *9*, 281.
Kal 1972	E. Kaldis, *J. Cryst. Growth* **1972**, *17*, 3.
Kal 1974	E. Kaldis, C. H. L. Goodman, Eds., *Crystal Growth, Theory and Techniques*, Plenum Press, New York, **1974**.
Kal 1974a	E. Kaldis, *J. Cryst. Growth* **1974**, *24/25*, 53.
Kal 1981	E. Kaldis, W. Peteler, *J. Cryst. Growth* **1981**, *52*, 125.
Klo 1965	H. Klotz, *Vakuum-Technik* **1965**, *3*, 63.
Kor 1972a	J. Kordis, K. A. Gingerich, R. J. Seyse, E. Kaldis, R. Bischof, *J. Cryst. Growth* **1972**, *17*, 53.
Krä 1974	V. Krämer, R. Nitsche, M. Schumacher *J. Cryst. Growth* **1974**, *24/25*, 179.
Reg 2001	L. Regel, W. Wilcox, *Acta Astronaut.* **2001**, *48*, 129.
Ros 1979	F. Rosenberger, J. M. Olson, M. C. Delong, *J. Cryst. Growth* **1979**, *47*, 321.
Schö 1980	E. Schönherr, *Crystals, Growth, Properties, Applications*, 2, Springer, Berlin, **1980**.
Vep 1971	S. Veprek, C. Brendel, H. Schäfer, *J. Cryst. Growth* **1971**, *9*, 266.
Vep 1976	S. Veprek, *Pure & Appl. Chem.* **1976**, *48*, 163.
Vep 1980	S. Veprek, *Chimia* **1980**, *12*, 489.
Vep 1988	S. Veprek, *J. Less-Common Met.* **1988**, *137*, 367.
Wil 1988	K.-Th. Wilke, J. Bohm, *Kristallzüchtung*, Harri Deutsch, Frankfurt **1988**.

15 Selected Experiments for Practical Work on Chemical Vapor Transport Reactions

In the previous chapters, essential principles, models, and numerous examples of CVT reactions have been discussed and described. In the following, we will illustrate some selected experiments that are to intensify the theoretical understanding of CVT reactions in a comprehensible way, and which are easily feasible in the laboratory. The experiments are suitable in particular for practical training (or lectures) for advanced students. The given examples shall convey the physical and chemical background as well as provide general information about chemical transport reactions as a synthesis method.

15.1 Transport of WO$_2$ with HgX$_2$ (X = Cl, Br, I)

During preparation for this experiment, one should think about the following questions. With the help of the clear example of the transport of WO$_2$ with mercury halides, the following questions can be tackled with your own calculations and experiments.

- What are CVT reactions and which transport agent is suitable?
- What is the basic precondition for CVT reactions?
- What is the average transport temperature that is suitable for the system under investigation?
- How can the transport direction be determined?
- How is the transport rate determined?

What are chemical vapor transport reactions and which transport agent is suitable?

Chemical vapor transport reactions can be used for purification of substances, for synthesis of crystalline compounds, and for doping of materials. Chemical vapor transports are characterized by the reversible reaction of a solid or liquid source material with a transport agent in a heterogeneous gas-phase reaction under the formation of gaseous species only. These are transferred along a temperature gradient to the sink. There the back-reaction under formation of the initial solid or liquid substance takes place.

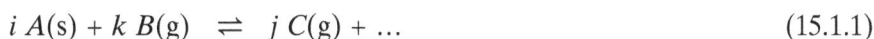

$$i\,A(s) + k\,B(g) \;\rightleftharpoons\; j\,C(g) + \dots \tag{15.1.1}$$

At first sight, the reaction is similar to sublimation. But the substance $A(s)$ does not have its own, transport-effective partial pressure $p(A)$ in the relevant tempe-

Figure 15.1.1 Crystal of WO$_2$.

rature range. The dissolution in the gas phase is bound to the presence of a transport agent.

The solid WO$_2$ does not have its own measurable vapor pressure, which would be suitable to transfer the compound to the gas phase in the sense of a sublimation. The phase rather decomposes at 1000 K with an oxygen partial pressure of 10^{-20} bar towards metallic tungsten. Here the addition of a transport agent is necessary:

As a general rule, the halogens chlorine, bromine, and iodine, or halogen compounds, such as the hydrogen halides HX (X = Cl, Br, I), are suitable as transport agents. For the transport of WO$_2$ the gas species WO$_2X_2$ will be effective for vapor transport.

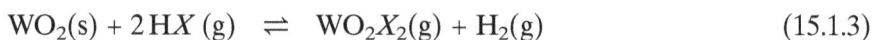

$$WO_2(s) + X_2(g) \ \rightleftharpoons \ WO_2X_2(g) \tag{15.1.2}$$

$$WO_2(s) + 2\,HX\,(g) \ \rightleftharpoons \ WO_2X_2(g) + H_2(g) \tag{15.1.3}$$

Also, adding mercury halides, which are gaseous at elevated temperatures, is potentially suitable to transport both components of the solid phase – tungsten as well as oxygen – via the gas phase.

$$WO_2(s) + HgX_2(g) \rightleftharpoons WO_2X_2(g) + Hg(g) \qquad (15.1.4)$$

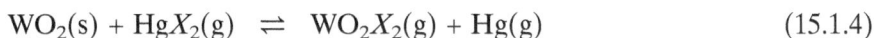

At temperatures above 300 °C the mercury halides evaporate completely. Afterwards the gaseous species WO_2X_2 is formed in addition to gaseous mercury.

What is the basic precondition for chemical vapor transport reactions?

The basic precondition for CVT reactions is a balanced equilibrium. For reactions that are described by *one* independent reaction equation, transport reactions can be expected for equilibrium constants K_p in the range from 10^{-4} up to 10^4, respectively, Gibbs energies $\Delta_r G^0$ of approximately –100 to 100 kJ $\cdot$ mol^{-1}. The partial pressure gradient Δp as a driving force for material transport between the dissolution and deposition site is achieved by a temperature gradient.

A highly exergonic reaction $\Delta_r G^0 < -100$ kJ $\cdot$ mol^{-1} ($K_p > 10^4$) shows a large dissolution of the solid into the gas phase. That sounds great. But the back-reaction under deposition of the solid phase is impossible in thermodynamic terms. That means that on the source side the compound to be transported is almost completely transferred into the gas phase without deposition on the sink side.

During a highly endergonic reaction $\Delta_r G^0 > 100$ kJ $\cdot$ mol^{-1} ($K_p < 10^{-4}$) hardly any solid is transferred into the gas phase, thus transport cannot take place.

With the help of thermodynamic data of the substances involved in the reaction (see Table 15.1.1), the values of the Gibbs energy and the equilibrium constants, respectively, of possible transport reactions can be calculated.

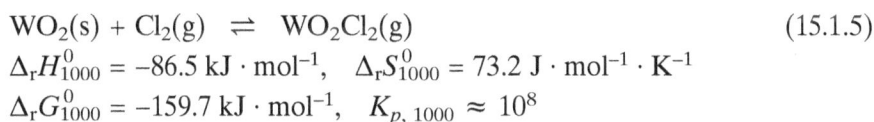

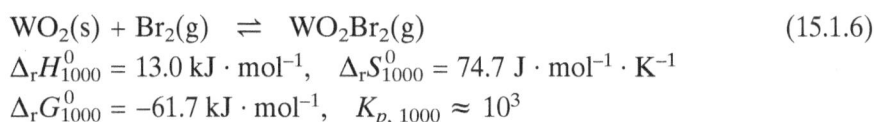

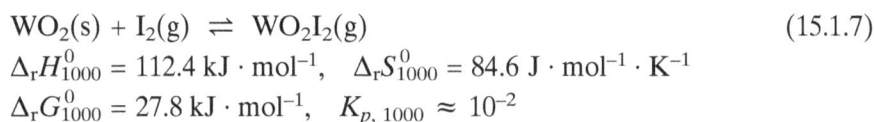

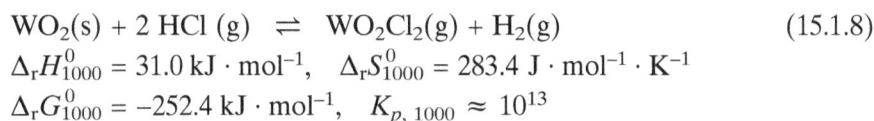

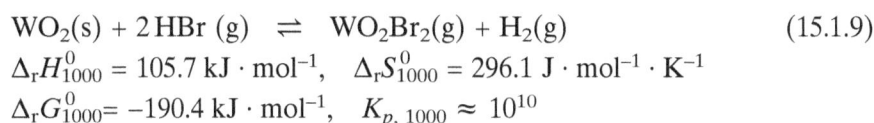

$$WO_2(s) + Cl_2(g) \rightleftharpoons WO_2Cl_2(g) \qquad (15.1.5)$$
$$\Delta_r H^0_{1000} = -86.5 \text{ kJ} \cdot \text{mol}^{-1}, \quad \Delta_r S^0_{1000} = 73.2 \text{ J} \cdot \text{mol}^{-1} \cdot \text{K}^{-1}$$
$$\Delta_r G^0_{1000} = -159.7 \text{ kJ} \cdot \text{mol}^{-1}, \quad K_{p,\,1000} \approx 10^8$$

$$WO_2(s) + Br_2(g) \rightleftharpoons WO_2Br_2(g) \qquad (15.1.6)$$
$$\Delta_r H^0_{1000} = 13.0 \text{ kJ} \cdot \text{mol}^{-1}, \quad \Delta_r S^0_{1000} = 74.7 \text{ J} \cdot \text{mol}^{-1} \cdot \text{K}^{-1}$$
$$\Delta_r G^0_{1000} = -61.7 \text{ kJ} \cdot \text{mol}^{-1}, \quad K_{p,\,1000} \approx 10^3$$

$$WO_2(s) + I_2(g) \rightleftharpoons WO_2I_2(g) \qquad (15.1.7)$$
$$\Delta_r H^0_{1000} = 112.4 \text{ kJ} \cdot \text{mol}^{-1}, \quad \Delta_r S^0_{1000} = 84.6 \text{ J} \cdot \text{mol}^{-1} \cdot \text{K}^{-1}$$
$$\Delta_r G^0_{1000} = 27.8 \text{ kJ} \cdot \text{mol}^{-1}, \quad K_{p,\,1000} \approx 10^{-2}$$

$$WO_2(s) + 2\,HCl\,(g) \rightleftharpoons WO_2Cl_2(g) + H_2(g) \qquad (15.1.8)$$
$$\Delta_r H^0_{1000} = 31.0 \text{ kJ} \cdot \text{mol}^{-1}, \quad \Delta_r S^0_{1000} = 283.4 \text{ J} \cdot \text{mol}^{-1} \cdot \text{K}^{-1}$$
$$\Delta_r G^0_{1000} = -252.4 \text{ kJ} \cdot \text{mol}^{-1}, \quad K_{p,\,1000} \approx 10^{13}$$

$$WO_2(s) + 2\,HBr\,(g) \rightleftharpoons WO_2Br_2(g) + H_2(g) \qquad (15.1.9)$$
$$\Delta_r H^0_{1000} = 105.7 \text{ kJ} \cdot \text{mol}^{-1}, \quad \Delta_r S^0_{1000} = 296.1 \text{ J} \cdot \text{mol}^{-1} \cdot \text{K}^{-1}$$
$$\Delta_r G^0_{1000} = -190.4 \text{ kJ} \cdot \text{mol}^{-1}, \quad K_{p,\,1000} \approx 10^{10}$$

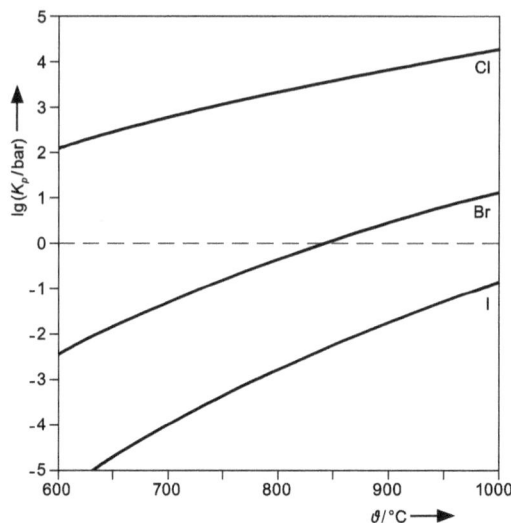

Figure 15.1.2 Equilibrium constants of transport reactions of $WO_2(s)$ with $HgX_2(g)$ (X = Cl, Br, I).

$$WO_2(s) + 2\,HI\,(g) \;\rightleftharpoons\; WO_2I_2(g) + H_2(g) \tag{15.1.10}$$
$$\Delta_r H_{1000}^0 = 174.6\;\text{kJ}\cdot\text{mol}^{-1}, \quad \Delta_r S_{1000}^0 = 314.3\;\text{J}\cdot\text{mol}^{-1}\cdot\text{K}^{-1}$$
$$\Delta_r G_{1000}^0 = -139.7\;\text{kJ}\cdot\text{mol}^{-1}, \quad K_{p,\,1000} \approx 10^7$$

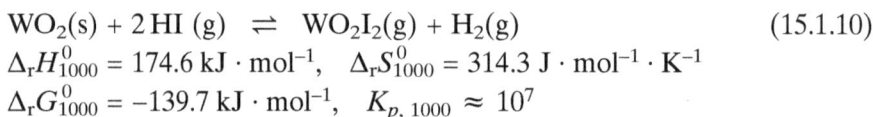

The calculations' results give a realistic outlook on the prospective results of transport experiments. Using halogens, transport with iodine seems to be promising (see section 5.2.6). In the case of bromine, transport seems at least possible; whereas chlorine causes an extreme equilibrium under the formation of $WO_2Cl_2(g)$ – so that transport should not be possible.

With the hydrogen halides equilibria are far on the side of the reaction products. This is due to a clearly higher gain of entropy during the reaction. Although one can observe gradations in the equilibrium position for transports with HI and HBr compared to HCl, transports are mainly not expected.

The transport of WO_2 with mercury halides seems possible for all three transport agents HgX_2 (X = Cl, Br, I). The equilibrium constants are within the limits of $10^{-4} < K_p < 10^4$. Thus these systems are ideal to foster the understanding of a systematic approach and particularly to extend the understanding of CVTs. In the following, we shall focus on the transport of WO_2 with mercury halides.

Using mercury bromide for the transport, the equilibrium position is least extreme – in this case, the best transport results can be expected, see Figure 15.1.2. Using $HgCl_2(g)$ for the transport, the equilibrium is shifted to the right; that means solid WO_2 is transferred well into the gas phase. However, the deposition of the solid on the sink side is only possible to a limited degree. Even when the temperature is decreased, the equilibrium is still on the product side.

The equilibrium constant for transport with $HgI_2(g)$ indicates that the solid is hardly dissolved – the equilibrium is shifted to the left. Thus these are adverse conditions for transport.

$$WO_2(s) + HgCl_2(g) \rightleftharpoons WO_2Cl_2(g) + Hg(g) \qquad (15.1.11)$$
$$\Delta_rH^0_{1000} = 115.5 \text{ kJ} \cdot \text{mol}^{-1}, \quad \Delta_rS^0_{1000} = 171.9 \text{ J} \cdot \text{mol}^{-1} \cdot \text{K}^{-1}$$
$$\Delta_rG^0_{1000} = -56.4 \text{ kJ} \cdot \text{mol}^{-1}, \quad K_{p,\,1000} \approx 10^3 \text{ bar}$$

$$WO_2(s) + HgBr_2(g) \rightleftharpoons WO_2Br_2(g) + Hg(g) \qquad (15.1.12)$$
$$\Delta_rH^0_{1000} = 190.5 \text{ kJ} \cdot \text{mol}^{-1}, \quad \Delta_rS^0_{1000} = 170.5 \text{ J} \cdot \text{mol}^{-1} \cdot \text{K}^{-1}$$
$$\Delta_rG^0_{1000} = 20.0 \text{ kJ} \cdot \text{mol}^{-1}, \quad K_{p,\,1000} \approx 10^{-1} \text{ bar}$$

$$WO_2(s) + HgI_2(g) \rightleftharpoons WO_2I_2(g) + Hg(g) \qquad (15.1.13)$$
$$\Delta_rH^0_{1000} = 249.6 \text{ kJ} \cdot \text{mol}^{-1}, \quad \Delta_rS^0_{1000} = 179.2 \text{ J} \cdot \text{mol}^{-1} \cdot \text{K}^{-1}$$
$$\Delta_rG^0_{1000} = 70.4 \text{ kJ} \cdot \text{mol}^{-1}, \quad K_{p,\,1000} \approx 10^{-4} \text{ bar}$$

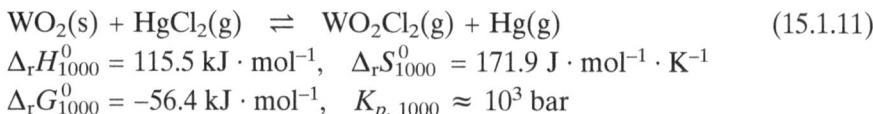

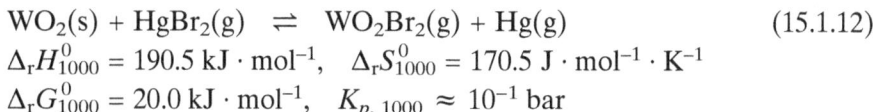

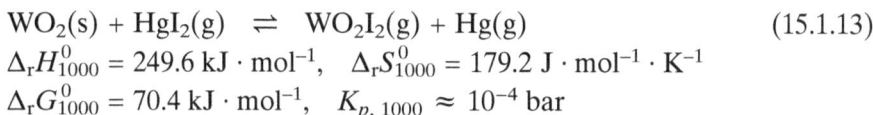

What is the average transport temperature that is suitable for the system under investigation?

The optimum, average temperature $[(T_2+T_1)/2]$ for CVT reactions, results from the requirement of $\Delta_rG^0 \approx 0$. If the thermodynamic data of the reaction are known, which can easily be obtained from the values of the involved species according to Hess's law, the optimum average temperature can be calculated from the quotient of the heat of reaction and the entropy. The better the data, the more realistic are the results. With the help of standard data given for 298 K, the first estimation of the optimum transport temperature can be made. The results of this calculation are not to be met to the exact degree. One rather finds a range of ±100 K that is suitable for the transport.

$$\Delta_rG^0 = \Delta_rH^0_T - T \cdot \Delta_rS^0_T \qquad (15.1.14)$$
$$0 = \Delta_rH^0_T - T \cdot \Delta_rS^0_T$$

$$T_{\text{opt.}} = \frac{\Delta_rH^0_T}{\Delta_rS^0_T} \qquad (15.1.15)$$

Through differences in the temperatures of the source and sink side, the equilibrium is brought towards the gaseous products when dissolving and shifted towards the solid when depositing.

Calculations of the equilibrium constants were first made for an average temperature of 1000 K. If the temperatures vary, one will get the typical courses of the curve (see Figure 15.1.2). If the temperature is decreased, the equilibrium position in the transport system with HgCl$_2$ becomes less extreme. In contrast, the equilibrium position for the transport with HgI$_2$ becomes more favorable when the temperature is increased above 1000 K.

The optimum, average temperature resulting from the quotient of the reaction enthalpy and entropy for the transport with HgCl$_2$ are at 700 K and 400 °C, respectively; with HgBr$_2$ at 1100 K and 800 °C, respectively; and with HgI$_2$ 1400 K and 1100 °C, respectively. In this case, the calculation of the temperature on the basis of the standard values at 298 K as well as of the derived values for

1000 K lead to the same results, which means that an estimation is possible with simple calculations.

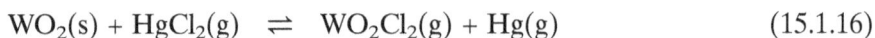

$$WO_2(s) + HgCl_2(g) \rightleftharpoons WO_2Cl_2(g) + Hg(g) \tag{15.1.16}$$
$$\Delta_r H^0_{1000} = 122.5 \text{ kJ} \cdot \text{mol}^{-1}, \quad \Delta_r S^0_{1000} = 183.6 \text{ J} \cdot \text{mol}^{-1} \cdot \text{K}^{-1}$$
$$T_{opt} = 125\,500 \text{ J} \cdot \text{mol}^{-1}/183.6 \text{ J} \cdot \text{mol}^{-1} \cdot \text{K}^{-1}$$
$T_{opt} \approx 700$ K and 400 °C, respectively

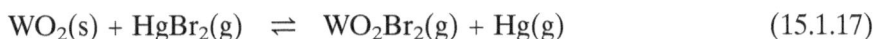

$$WO_2(s) + HgBr_2(g) \rightleftharpoons WO_2Br_2(g) + Hg(g) \tag{15.1.17}$$
$$\Delta_r H^0_{1000} = 188.1 \text{ kJ} \cdot \text{mol}^{-1}, \quad \Delta_r S^0_{1000} = 170.0 \text{ J} \cdot \text{mol}^{-1} \cdot \text{K}^{-1}$$
$T_{opt} \approx 1100$ K and 800 °C, respectively

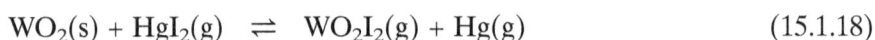

$$WO_2(s) + HgI_2(g) \rightleftharpoons WO_2I_2(g) + Hg(g) \tag{15.1.18}$$
$$\Delta_r H^0_{1000} = 238.4 \text{ kJ} \cdot \text{mol}^{-1}, \quad \Delta_r S^0_{1000} = 165.4 \text{ J} \cdot \text{mol}^{-1} \cdot \text{K}^{-1}$$
$T_{opt} \approx 1400$ K and 1100 °C, respectively

How can the transport direction be determined?

If a transport operation can be described in good approximation by *one* reaction, the direction of the transport results from the heat balance of the heterogeneous equilibrium according to the van't Hoff equation and the Clausius–Clapeyron relation, respectively:

$$\frac{d \ln K_p}{d \frac{1}{T}} = \frac{-\Delta_r H^0_T}{R} \tag{15.1.19}$$

In a reaction with negative reaction enthalpy (an exothermic dissolving reaction), the equilibrium constant K_p increases with decreasing temperatures – thus dissolution takes place at low, and deposition at high, temperatures. To put it another way: the transport is directed to the hotter zone ($T_1 \rightarrow T_2$).

$$\Delta_r H^0_T < 0; \quad d \ln K_p \sim d\frac{1}{T} \tag{15.1.20}$$

In a reaction with a positive heat of reaction (an endothermic dissolving reaction), K_p increases with increasing temperatures – so that dissolution takes place at higher, and deposition at lower, temperatures. Now, the transport proceeds to the cooler zone ($T_2 \rightarrow T_1$).

$$\Delta_r H^0_T > 0; \quad d \ln K_p \sim dT \tag{15.1.21}$$

Since the transport direction results only from the heat of reaction the conclusion for all three investigated transport systems of WO_2 is clear: the reaction enthalpy is positive in each case – a transport to the less hot zone results. The total amount of the heat of reaction does not affect whether a transport is carried out. If the reaction enthalpy is close to zero one has to check the accuracy of the data used as they can contain errors of 10 to 20 kJ $\cdot$ mol^{-1}.

How is the transport rate determined?

Substance transport via gas motion between the dissolution and deposition site takes place by diffusion or convection. If the ampoule lies horizontally and if the total pressure is between 10^{-3} bar and 3 bar (Opp 1987, Sch 1962), substance transport is mainly caused by diffusion. In most cases, diffusion is the rate-determining step, as it is much slower than the heterogeneous reactions between the solid and the transport agent. At pressures above 3 bar, convection becomes dominant.

For transport reactions that can be described by a *single* reaction (15.1.1), one can describe the transport rate by *Schäfer*'s equation (1.5.4) given that the CVT is solely determined by diffusion (Sch 1962, Sch 1973b). High values of Δp result in a high transport rate. A large cross section increases the transport rate positively as does a short transport distance (diffusion path). According to the transport equation a high average temperature is formally advantageous for the transport rate; however, the influence of the temperature on the equilibrium constant, and thus Δp, is more essential.

Finally, in selecting the transport agent, the temperature, and the temperature gradient, respectively, one should consider the aim of the transport. A high transport rate is undoubtedly advantageous for the synthesis of a compound or its purification. If crystals are to be grown, keep in mind the crystal quality, and therefore rather choose a lower transport rate.

The calculation of transport rates for WO$_2$ by *Schornstein* and *Gruehn* (Sch 1988, Sch 1989) at first shows a clear dominance of transports with HgBr$_2$ in the average temperature range: the expected transport rates are ten times higher than for transports with HgCl$_2$ and HgI$_2$. We have already explained why. Due to the

Table 15.1.1 Thermodynamic data of gas species involved in the transport reaction between WO$_2$(s) and the mercury halides HgX$_2$(g) (Kna 1991).

	$\Delta_f H_{298}^0$ / kJ · mol^{-1}	$\Delta_f H_{1000}^0$ / kJ · mol^{-1}	S_{298}^0 / J · mol^{-1} · K^{-1}	S_{1000}^0 / J · mol^{-1} · K^{-1}
WO$_2$(s)	−589.7	−540.5	50.5	132.9
Cl$_2$(g)	0	25.5	223.1	266.7
Br$_2$(g)	30.9	57.0	245.4	290.3
I$_2$(g)	62.2	88.6	260.2	305.5
H$_2$(g)	0	20.7	130.7	166.3
HCl(g)	−92.3	−71.3	186.9	222.8
HBr(g)	−36.4	−15.0	198.7	235.2
HI(g)	26.5	47.1	206.6	242.1
Hg(g)	61.4	76.0	175.0	200.2
HgCl$_2$(g)	−143.3	−100.5	294.8	368.2
HgBr$_2$(g)	−87.8	−44.5	320.2	394.7
HgI$_2$(g)	−16.2	27.4	336.2	411.1
WO$_2$Cl$_2$(g)	−671.5	−601.5	353.9	472.8
WO$_2$Br$_2$(g)	−550.8	−470.5	365.7	497.9
WO$_2$I$_2$(g)	−428.9	−339.5	377.1	523.0

balanced equilibrium, high differences of partial pressures occur between the source and the sink. This way, the driving force for diffusion of the gas species is high and thus for substance transport as well.

For transport with $HgCl_2$ the transport rate decreases with increasing temperature. As we have already seen, equilibrium 15.1.16, which is far to the right side, is responsible for it. Only if the temperature decreases, can the equilibrium move to the left. The resulting, higher differences of partial pressures between the dissolution and deposition side cause increasing transport rates at low temperatures.

Using mercury iodide as transport agent, the equilibrium is on the side of the source material at low temperatures. By increasing the temperature, the equilibrium is shifted to the side of the reaction products, and the transport rate increases (Figure 15.1.2).

Experiments on the transport of WO_2 with HgX_2 (X = Cl, Br, I)

Chemicals required	Equipment required
WO_2 approx. 1g	silica tubes
W approx. 0.05 g	long powder funnel
$HgCl_2$ 0.08 g	two-zone furnace
$HgBr_2$ 0.1 g	protective gloves, safety glasses,
HgI_2 0.13 g	"ampoule catcher"

Experimental Approximately 1 g WO_2 with the additive W and the exactly weighed amount of the transport agent are loaded into a silica ampoule (approx. 100 ... 150 mm length and 16 mm diameter), which was previously baked out under vacuum at 900 °C. By adding elemental tungsten a defined oxygen partial pressure is achieved for the transport system by the equilibrium 15.1.22.

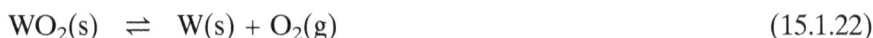

$$WO_2(s) \ \rightleftharpoons \ W(s) + O_2(g) \tag{15.1.22}$$

This way, a potential surplus of oxygen from the residual gas or remaining moisture in the ampoule walls is avoided, and the formation of the more oxygen-rich phase $W_{18}O_{49}$ is prevented; $W_{18}O_{49}$ would be transported besides WO_2 as blue-purple needles and thus contaminate the transported WO_2. After evacuating ($p < 10^{-3}$ mbar), the transport ampoule is sealed under vacuum. The ampoule is exposed to a reversed temperature gradient for approximately 12 hours. This way crystallization nuclei, which adhere to the wall of the sink side, are transported to the source. During the actual transport reaction, there will be fewer but therefore larger crystals. This approach is called **transport in a reverse temperature gradient**. Afterwards the transport ampoule is exposed to a defined temperature gradient in a two-zone furnace. After the transport time, the ampoule is taken out of the furnace and quenched with cold water. In order to get clean crystals, it is recommendable to quench the source side first and thus condense the gas phase with the transport agent there. Only then, should the sink side with the crystals be cooled down.

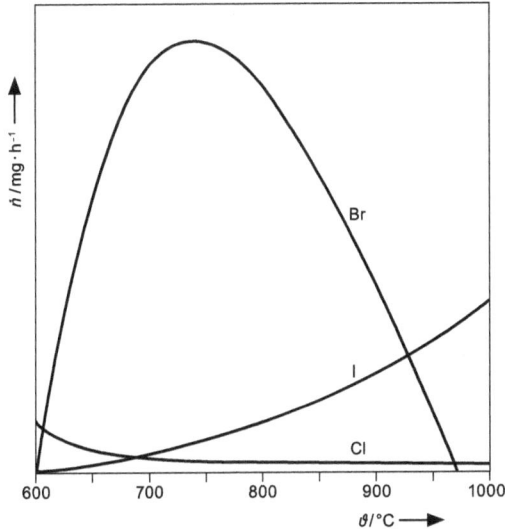

Figure 15.1.3 Progression of transport rates during the transport of WO$_2$ with HgX$_2$ (X = Cl, Br, I) according to *Schornstein* and *Gruehn* (Sch 1988, Sch 1989).

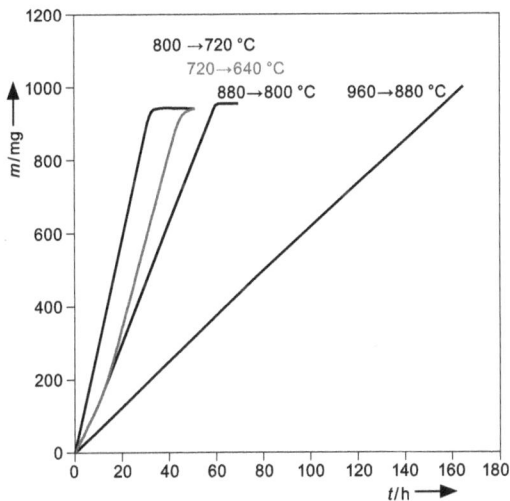

Figure 15.1.4 Mass transport during the chemical vapor transport of WO$_2$ with HgBr$_2$, observed by a transport balance.

The ampoules can be taken out of the furnace after one or two days. The decision about the duration of the experiments depends on the transport system and the temperature. Transports with HgBr$_2$ should take 24 to 36 hours at the maximum, transports with HgCl$_2$ and HgI$_2$ can last longer than 48 hours. If the transport rate is to be determined, one has to make sure that the source solid is not completely transported to the sink. The transporting solid must be carefully dissolved out of the tip of the ampoule. If the obtained crystals are weighed out, one will have an average transport rate for the duration of the transport. In doing

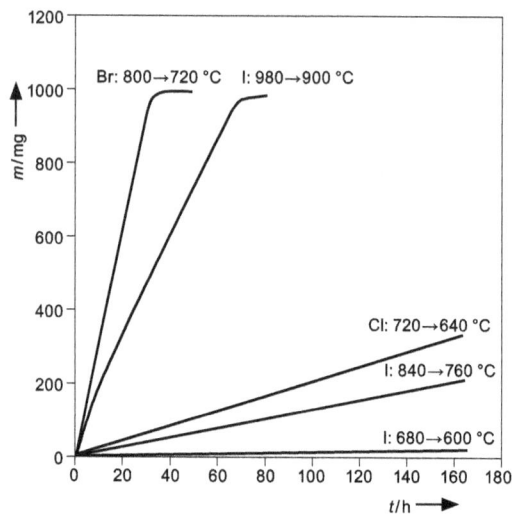

Figure 15.1.5 Mass transport during the chemical vapor transport of WO$_2$ with HgX$_2$ (X = Cl, Br, I), observed by a transport balance.

Figure 15.1.6 Typical crystal morphology of the formation of rose-shaped, intergrown crystal agglomerations during the transport of WO$_2$ with HgBr$_2$.

so, compare the transport rates at different temperature ranges of the optimum average transport reaction temperature, which you have calculated.

Figure 15.1.7 Typical crystal morphology of single crystallites with an edge length of up to 1 mm formed during the transport of WO$_2$ with HgI$_2$.

Observations Corresponding to the simple estimation of the transport behavior of WO$_2$ with mercury halides, one gets the best results with the addition of HgBr$_2$. Chemical vapor transport by mercury bromide is possible in a wide temperature range. Transport rates above 30 mg · h^{-1} are achievable. Temperatures at the source side of around 800 °C and at the sink side of 720 °C prove optimum. This result confirms the estimation of the optimum transport temperature. Due to the shift of the equilibrium, the transport rate decreases at both rising temperatures (880 → 800 °C and 960 → 880 °C, respectively; Figure 15.1.3) and falling temperatures (720 → 640 °C; Figure 15.1.3).

Transport reactions with HgCl$_2$ and HgI$_2$ clearly show lower transport rates. Experiments with mercury iodide must be carried out at higher temperatures according to the estimation. Temperatures up to 1000 °C are practicable; above this the silica ampoule will be heavily damaged by re-crystallization. Using an average transport temperature of 940 °C, transport rates of up to 15 mg · h^{-1} (Figure 15.1.4) can be achieved. The transport rate decreases drastically with falling temperatures. With an average temperature of 640 °C the rate is even lower than 1 mg · h^{-1}. Transport experiments with HgCl$_2$ show the worst results as far as the transport rate is concerned: according to the calculation, lower temperatures are principally more favorable; however, in the range from 500 to 700 °C the transport rates are only in the range of 1 mg · h^{-1} (Figure 15.1.5). Transport almost grinds to a halt at higher temperatures.

One can come to a completely different evaluation if the quality of the crystals instead of the transport rate is given prominence. Relatively high transport rates cause uncontrolled nucleation and crystal growth. As a consequence, one gets highly intergrown and rose-shaped crystal agglomerations for transports with $HgBr_2$. Frequently, *one* compact solid of these intergrown crystallites of WO_2 is found in the sink (Figure 15.1.6).

Using average temperatures of approx. 800 °C in the transport system with HgI_2, one gets isolated, rod-shaped crystals of up to 1 mm edge length (Figure 15.1.7). The preparation of single-crystals for crystal structure analysis is possible from these experiments, even though not every crystal is suitable. The low transport rate of 1 to 2 mg · h^{-1} (Figure 15.1.5) favors undisturbed nucleation and crystal growth in this case. In the process, smaller highly defective crystallites are dissolved in favor of other individuals.

Conclusion Chemical vapor transport reactions are predictable. Simple estimations concerning the feasibility and the course of transport reactions are already possible with a basic understanding of the method and its thermodynamic background. It is worth the effort in every case – one avoids unnecessary experiments using "trial and error" procedure. As the book at hand proves, conclusions by analogy between similar transport systems are helpful as a first guide; however, the factual favorable parameter for a transport should be estimated carefully in advance.

15.2 Transport of $Zn_{1-x}Mn_xO$ Mixed-crystals

Zinc oxide crystallizes in the wurtzite structure. In this structure, zinc atoms can be substituted by other metal atoms in the oxidation state II. The ZnO/MO phase diagrams describe the mutual solubility as a function of temperature. Although MnO crystallizes in a rock salt structure, considerable amounts of MnO dissolve in ZnO. The phase diagram of the system is illustrated in Figure 15.2.1.

Mixed-crystals, in which the zinc atoms are partly substituted by manganese atoms, appear particularly impressive. They have an intense wine-red color and form particularly distinct needle-shaped crystals during the transport. Furthermore, due to certain physical characteristics, $Zn_{1-x}Mn_xO$ is of particular research interest, especially in solid-state physics (keyword: "spintronics").

Zinc oxide can be transported with different transport agents (see section 2.1; Table 5.1). Mixed-crystals, in which a certain amount of the zinc atoms are substituted by other atoms of oxidation state II, can be obtained with the help of chemical transport reactions, too. The transport of solid solutions of $Zn_{1-x}Mn_xO$ is a particularly attractive experiment for practical work, because well formed, intensely colored crystals of a material with interesting physical properties develop.

Zinc oxide is commercially available in sufficiently pure form. Virtually every zinc oxide material is suitable for this experiment. Particular precautions are not necessary.

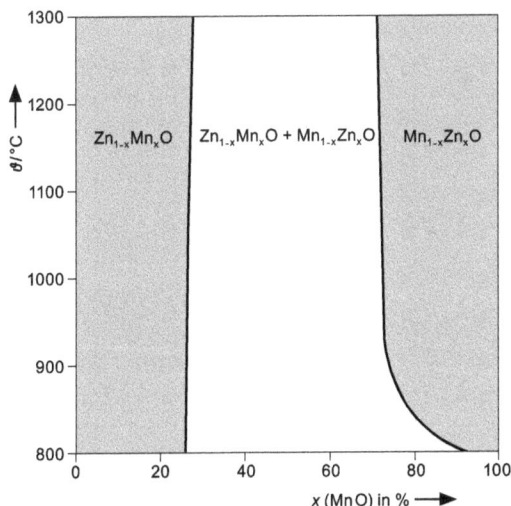

Figure 15.2.1 Phase diagram of the ZnO/MnO system according to *White* and *McIlfvried* (Whi 1965).

In binary manganese oxides, the manganese atoms can take on oxidation numbers from II to VII. In order to make sure that the manganese atoms of oxidation state II can be integrated into the zinc oxide lattice, one synthesizes MnO by decomposition of commercially available manganese(II)-carbonate in hydrogen flow as a first step. Manganese(II)-oxide formed this way has an olive-green color.

Choosing the transport agent, one must not select oxidizing substances as these could cause oxidation of the manganese(II)-ions. Because of this, the elemental halogens chlorine and bromine are ruled out. Hydrogen chloride is preferable and ammonium chloride is even better suited as transport additive. Ammonium chloride decomposes into ammonia and hydrogen chloride. At temperatures above 600 °C ammonia decomposes into the elements: hydrogen arises which creates a reducing atmosphere and prevents the oxidation of manganese(II)-ions. Ammonium chloride is also easy to handle and to dose.

Experimental The transport of mixed-crystals is described in the literature (Loc 1999). Commercially available zinc oxide is carefully dried in a drying cabinet because the fine powder usually contains considerable amounts of adsorbed water. Manganese(II)-oxide is prepared by the decomposition of manganese(II)-carbonate in a hydrogen flow at 550 °C. A horizontally arranged tube furnace that heats a glass or silica-glass tube that is furnished with ground joints at both ends is used for this. The required amount of manganese carbonate is put in a combustion boat and placed in the center of the furnace. Hydrogen is dried with sulfuric acid (gas wash bottle!) and passed through the tube. After the hydrogen/oxygen explosion test has proved negative, the furnace is heated to a temperature of 550 °C, the temperature is held for 5 hours and the furnace is then cooled down in a hydrogen flow. The olive-green manganese(II)-oxide is removed and kept in a desiccator because it readily oxidizes at room temperature. The prepared transport ampoule is fed with a ZnO/MnO mixture. Ammonium chloride is added so that the initial pressure, calculated for $NH_4Cl(g)$, is about 0.5 bar at a medium transport temperature of 825 °C. Ammonium chloride decomposes at the transport temperature according to 15.2.1.

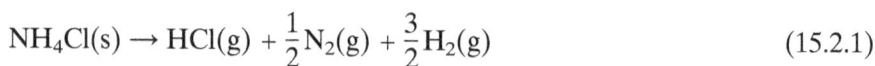

$$NH_4Cl(s) \rightarrow HCl(g) + \frac{1}{2}N_2(g) + \frac{3}{2}H_2(g) \qquad (15.2.1)$$

The actual initial pressure is three times higher than that calculated for $NH_4Cl(g)$.

After application of an inverse temperature gradient for one day, the transport ampoule that was previously heated out and loaded as described is exposed to a temperature gradient from 900 °C to 750 °C for five days. After cooling down, the ampoule is opened and the product in the form of wine-red needles of up to 1 cm length is recovered.

This experiment is ideal as a group experiment in which the proportion of manganese varies between 0.5 % and approx. 15 %. The color intensifies with increasing quantity of manganese and can be seen clearly with the naked eye. The intense red color is traced back to the fact that the integration of manganese(II)-ions in the lattice leads to a large reduction of the band gap. This results in absorption of light in the visible range (Saa 2009a, Saa 2009b).

Further investigations

X-ray characterization of the product

⇒ $Zn_{1-x}Mn_xO$ ($0 \leq x \leq 0.25$) mixed-crystals adopt the wurtzite structure.

Grinding the product

⇒ Powder brightens, depending on the content of manganese, the powder has a light-red or orange color (particle-size effect).

Analytical investigations

The product is analyzed by a suitable method.
⇒ The content of manganese is determined and from it the composition of the $Zn_{1-x}Mn_xO$ mixed-crystals, which is then compared to the phase diagram (Figure 15.2.1).

Magnetic measurements

The magnetic behavior of products of different compositions is investigated.
⇒ At low manganese content the magnetic behavior of Mn^{2+} ions corresponds to the high spin $3d^5$ electronic configuration. If the content of manganese is increased, co-operative antiferromagnetic interaction between the Mn^{2+} ions gets stronger; thus, the magnetic susceptibility decreases.

Thermodynamic discussion

The transport behavior of $Zn_{1-x}Mn_xO$ is described by model calculations. The transport behavior is very similar to that of pure ZnO. Still, an interesting point for closer consideration would be the influence of the mixed-crystals formation on the transport behavior of the system.

Chemicals required		Equipment required
ZnO	approx. 1 g	silica tubes
$MnCO_3$	approx. 0.5 g	long powder funnel
NH_4Cl	approx. 0.1 g	two-zone furnace
hydrogen		*protective gloves, safety glasses,*
		"ampoule catcher"

15.3 Transport of Rhenium(VI)-oxide

In the rhenium/oxygen system there exist the compounds rhenium(IV)-oxide (ReO_2), rhenium(VI)-oxide (ReO_3), and rhenium(VII)-oxide (Re_2O_7). Due to its extraordinarily high transport rate and the favorable temperature range, the CVT of rhenium(VI)-oxide is ideally suited for experiments in lectures or demonstration experiments. This is because these experiments can be conducted in a two-zone furnace at rather low temperatures in a short time (Figure 15.3.1). By adding mercury(II)-halides HgX_2 (X = Cl, Br, I), rhenium(VI)-oxide can be transported with transport rates in the range from 20 to 25 mg · h^{-1}. Generally, less well developed crystals are deposited in the process. At low transport rates, rhenium(VI)-oxide is obtained as well developed red, metallic, shiny crystals with an edge-length up to several millimeters. Their crystal form can be rod-shaped, prismatic, or cubic.

Figure 15.3.1 Two-zone furnace with viewing window (HTM Reetz GmbH).

Table 15.3.1 Thermodynamic data of the species involved in the transport reaction of $ReO_3(s)$ with mercury halides $HgX_2(g)$ according to *Oppermann* and *Feller et al.* (Opp 1985, Fel 1998).

	$\Delta_f H^0_{298}/$ kJ $\cdot$ mol^{-1}	$\Delta_f H^0_{800}/$ kJ $\cdot$ mol^{-1}	$S^0_{298}/$ J $\cdot$ mol^{-1} $\cdot$ K^{-1}	$S^0_{800}/$ J $\cdot$ mol^{-1} $\cdot$ K^{-1}
$ReO_3(s)$	−589.1	−542.5	69.3	158.7
$Hg(g)$	61.4	71.8	175.0	195.5
$HgCl_2(g)$	−143.3	−112.8	294.8	354.4
$HgBr_2(g)$	−87.8	−56.9	320.2	380.8
$HgI_2(g)$	−16.2	14.9	336.2	397.2
$ReO_3Cl(g)$	−536.4	−490.2	315.9	406.8
$ReO_3Br(g)$	−497.1	−450.8	327.6	418.5
$ReO_3I(g)$	−443.9	−391.4	334.7	438.0

Transport experiments with mercury(II)-chloride in the temperature range from 500 to 400 °C proved particularly suitable for demonstration purposes. Due to the favorable equilibrium position, compared to other mercury halides, one can apply the lowest transport temperature and achieve high transport rates at the same time. Furthermore, the low deposition temperature is necessary because above 400 °C, the decomposition of rhenium(VI)-oxide to rhenium(IV)-oxide and gaseous Re_2O_7 begins. Using mercury(II)-chloride, crystal growth is observed more easily because the gas phase is not as intensely colored as when the iodide was used.

The transport of rhenium(VI)-oxide can be described in good approximation with the help of the following transport equations:

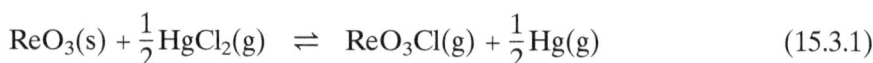

$$ReO_3(s) + \frac{1}{2}HgCl_2(g) \ \rightleftharpoons \ ReO_3Cl(g) + \frac{1}{2}Hg(g) \qquad (15.3.1)$$

$\Delta_r H^0_{800} = 144.6$ kJ $\cdot$ mol^{-1}, $\quad \Delta_r S^0_{800} = 168.6$ J $\cdot$ mol^{-1} $\cdot$ K^{-1}

$\Delta_r G^0_{800} = 9.7$ kJ $\cdot$ mol^{-1}. $\quad K_p = 2.3 \cdot 10^{-1}$ bar, $\quad T_{opt} \approx 860$ K

$$ReO_3(s) + \frac{1}{2}HgBr_2(g) \ \rightleftharpoons \ ReO_3Br(g) + \frac{1}{2}Hg(g) \qquad (15.3.2)$$

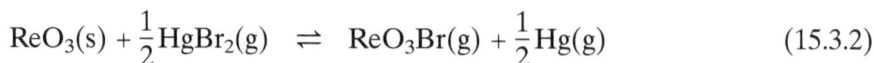

$\Delta_r H^0_{800} = 156.0$ kJ $\cdot$ mol^{-1}, $\quad \Delta_r S^0_{800} = 167.2$ J $\cdot$ mol^{-1} $\cdot$ K^{-1}

$\Delta_r G^0_{800} = 22.3$ kJ $\cdot$ mol^{-1}, $\quad K_p = 3.5 \cdot 10^{-2}$ bar, $\quad T_{opt} \approx 930$ K

$$ReO_3(s) + \frac{1}{2}HgI_2(g) \ \rightleftharpoons \ ReO_3I(g) + \frac{1}{2}Hg(g) \qquad (15.3.3)$$

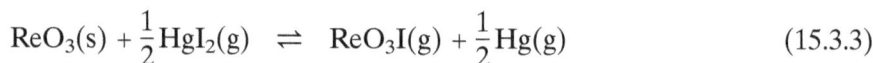

$\Delta_r H^0_{800} = 179.6$ kJ $\cdot$ mol^{-1}, $\quad \Delta_r S^0_{800} = 178.4$ J $\cdot$ mol^{-1} $\cdot$ K^{-1}

$\Delta_r G^0_{800} = 36.8$ kJ $\cdot$ mol^{-1}, $\quad K_p = 4 \cdot 10^{-3}$ bar, $\quad T_{opt} \approx 1010$ K.

The endothermic character of these transport reactions agrees with the observed transport direction. The comparison of the Gibbs energy and of the equilibrium constants, respectively, with different transport agents, shows that the equilibrium position for transport with mercury(II)-chloride is most favorable at 800 K.

Experimental The transport of rhenium(VI)-oxide with mercury halides as well as its thermodynamic characteristics are described in the literature (Opp 1985, Fel 1998). The required chemicals, rhenium(VI)-oxid and mercury(II)-chloride, are commercially available. It is advisable to dry the rhenium(VI)-oxide in a vacuum at room temperature before using it. Mercury(II)-chloride is slightly hygroscopic, which is why it is a good idea to keep the substance either in a desiccator or a glove box. However, handling of all substances in air is no problem.

The prepared transport ampoule is filled with approximately 1 g ReO$_3$ and with HgCl$_2$ (1 mg $\cdot$ cm^{-3}). The charged and vacuum-sealed transport ampoule is subjected to transport in an inverse temperature gradient 400 to 500 °C for one day in order to minimize the number of crystal nuclei. After this time, the demonstration experiment can be conducted by reversing the temperature gradient. Alternatively, after transport in an inverse temperature gradient one can quench the ampoule beginning at the source side. The regular transport can be started later; the formation of the gas phase is clearly visible when the ampoule is heated up slowly.

Chemicals required	Equipment required
ReO$_3$ approx. 1 g HgCl$_2$ approx. 0.02 g	silica ampoule long powder funnel two-zone furnace *protective gloves, safety glasses* *"ampoule catcher"*

15.4 Transport of Nickel

The CVT of nickel has been much investigated and documented in both closed systems and flowing systems. This is because of two reasons. First, the purification of nickel according to the *Mond–Langer* process is a CVT reactions of technical interest. Second, the migration of nickel in a temperature gradient with carbon monoxide is one of the model systems that was used for the development of the theory of CVT reactions. This is valid in particular for the development of a diffusion model for closed reaction systems (see Chapter 2).

The exothermic chemical transport of nickel with carbon monoxide can be described with the help of the following transport equation:

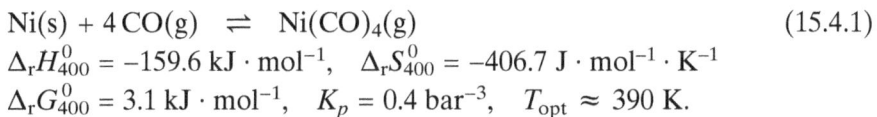

$$\text{Ni(s)} + 4\,\text{CO(g)} \ \rightleftharpoons \ \text{Ni(CO)}_4\text{(g)} \tag{15.4.1}$$

$$\Delta_r H^0_{400} = -159.6\ \text{kJ} \cdot \text{mol}^{-1}, \quad \Delta_r S^0_{400} = -406.7\ \text{J} \cdot \text{mol}^{-1} \cdot \text{K}^{-1}$$

$$\Delta_r G^0_{400} = 3.1\ \text{kJ} \cdot \text{mol}^{-1}, \quad K_p = 0.4\ \text{bar}^{-3}, \quad T_{\text{opt}} \approx 390\ \text{K}.$$

The thermodynamic parameters heat of reaction and Gibbs energy, which are characteristic for the transport reaction, determine the transport direction from T_1 to T_2 as well as the very favorable equilibrium constant. The calculated optimum transport temperature of 390 K (117 °C) is in good agreement with the

Table 15.4.1 Thermodynamic data of the species involved in the transport reactions of nickel with carbon monoxide.

	$\Delta_f H^0_{298}/$ kJ $\cdot$ mol^{-1}	$\Delta_f H^0_{400}/$ kJ $\cdot$ mol^{-1}	$S^0_{298}/$ J $\cdot$ mol^{-1} $\cdot$ K^{-1}	$S^0_{400}/$ J $\cdot$ mol^{-1} $\cdot$ K^{-1}
Ni(s)	0	2.8	29.9	37.9
CO(g)	–110.5	–107.6	197.7	206.3
Ni(CO)$_4$	–602.9	–587.1	410.6	456.2

medium temperature of the experimentally proven transport gradient. In flow systems, the dissolution of the source material occurs at temperatures between 50 and 80 °C (T_1), and deposition between 190 and 200 °C (T_2). In closed systems (ampoule) a temperature gradient of 80 to 200 °C is ideal. The pressure of the transport agent carbon monoxide is approximately 1 bar. During transport a thin nickel layer is obtained at the sink side.

Experimental The transport of nickel with carbon monoxide can be started from compact material (foil, wire, beads) as well as from powder. In any case, the reaction between nickel and carbon monoxide to form volatile tetracarbonyl-nickel, which acts as transport-effective gas species, is kinetically inhibited at 80 °C. The kinetic inhibition is more dominant if compact material is used instead of powder. The dissolution of the source material is almost uninhibited by kinetics when the nickel powder has been prepared immediately prior to its use by reduction in a flow of hydrogen. The reduction of NiC$_2$O$_4$ $\cdot$ 2 H$_2$O in a hydrogen flow at 500 °C is another way of obtaining very reactive nickel powder. When massive material is used, the kinetic inhibition can be reduced by adding small amounts of sulfur (approx. 5 mg) because sulfur catalyzes the conversion of nickel with carbon monoxide.

A prepared transport ampoule is filled with 1 g nickel and evacuated. After that, the ampoule is filled with carbon monoxide ($p \approx 800$ mbar) and eventually sealed. A transport ampoule charged in this way is exposed to a temperature gradient in a two-zone furnace. When nickel is used, the back transport (transport in an inverse temperature gradient) from 200 (source material) to 80 °C takes place first for, one or two days. After this time, the actual transport reaction can be executed by reversing the temperature gradient. When massive material is used, back transport is not necessary. Using the usual transport ampoules (length of diffusion path 100 to 150 mm; diameter 12 to 15 mm), transport rates from 5 to 10 mg $\cdot$ h^{-1} can be expected.

Chemicals required	*Equipment required*
Nickel (1 g foil, wire or powder)	ampoule
NiC$_2$O$_4$ $\cdot$ 2 H$_2$O	long powder funnel
carbon monoxide	two-zone furnace
possibly hydrogen	*protective gloves, safety glasses*
possibly sulfur (approx. 5 mg)	*"ampoule catcher"*

Safety advice: Carbon monoxide is an extremely toxic gas, which may be used only with the appropriate sensor technique and safety measures (fume hood). Also, the tetracarbonylnickel produced is a volatile and toxic substance.

15.5 Transport of Monophosphides MP (M = Ti to Co)

The monophosphides of the 3d metals (M = Ti to Ni) crystallize in the nickel arsenide structure type (TiP (Scho 1954), VP (Fje 1986)) and in the derived MnP-structure type (CrP (Sel 1972a), MnP (Fje 1984), FeP, and CoP (Run 1962)), respectively. This transition from one structure type to the other is a textbook example of structural change, which can be traced back to electronic reasons and is discussed in detail in the literature (Tre 1986, Sil 1986, Bur 1995). Crystals of the monophosphides are suitable for further measurements of their magnetic and electric characteristics.

Monophosphides are accessible in well crystallized form by CVT experiments using iodine as the transport agent (Figure 15.5.1). It is interesting that the transport conditions vary significantly along the series from Sc to Co, despite the close chemical and crystal similarity of the compounds.

General experimental procedure Synthesis and CVT of the phosphides are carried out as a "one-pot synthesis" in closed, evacuated silica ampoules with the approximate dimensions (l = 12 cm; d = 1.5 cm). The phosphides are synthesized from the elements. An amount of 500 to 1000 mg of source material is reasonable for the above-mentioned dimensions. The reaction between metal and phosphorus (establishment of equilibrium prior to transport) takes several hours. In order to avoid too high an internal pressure in the ampoules at the beginning of an experiment, a pre-reaction initiated by heating up using a Bunsen burner is recommended. (Attention! Danger of explosion! Estimate the maximum internal pressure!) After that the ampoules are exposed to a reversed temperature gradi-

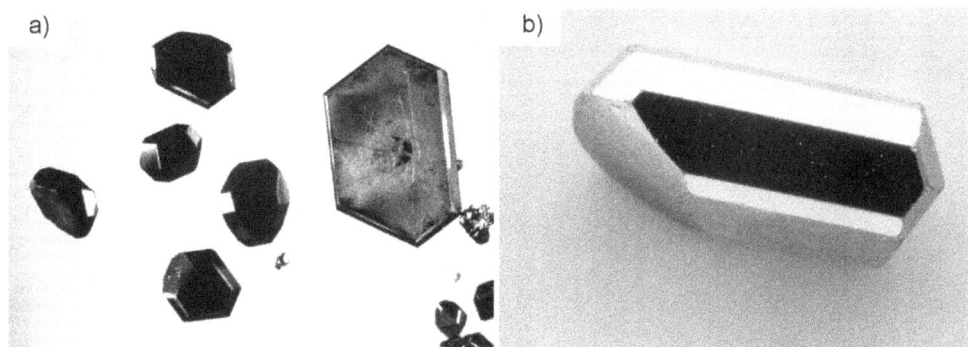

Figure 15.5.1 Crystals of VP (a) and CrP (b) from transport experiments (Gla 1990).

Table 15.5.1 Temperature gradient and iodine concentration for the CVT of monophosphides MP (M = Ti to Co).

Phosphide	Temperature/°C	$\beta(Iodine)/mg \cdot cm^{-3}$	Comments
TiP	$800 \rightarrow 900$	2	Unproblematic (Gla 1990)
VP	$800 \rightarrow 900$	7	Transport is susceptible to moisture and oxygen and might be suppressed by these (Gla 1989a)
CrP	$1000 \rightarrow 900$	2	Unproblematic; higher amounts of iodine and lower temperatures lead to the depositon of $CrI_2(l)$ (Gla 1989b)
MnP	$900 \rightarrow 1050$	0.25	Iodine reacts almost quantitatively to $MnI_2(l)$; transport rates are low (Gla 1989b)
FeP	$800 \rightarrow 550$	2	Unproblematic; the unusual temperature gradient provides indeed the best results (Sel 1972b, Gla 1990)
CoP	$1000 \rightarrow 900$	2	Unproblematic; high iodine sample and low temperatures lead to the deposition of $CoI_2(l)$ (Schm 1995)
NiP	$960 \rightarrow 860$	2.5	Problematic due to the narrow thermal stability range of NiP; often, only the formation of NiP_2 and Ni_5P_4 is observed

ent for one day (clearing transport). In the process, the establishment of an equilibrium within the solid and between the solid and the gas phase occurs. Transport toward the empty end of the ampoule cannot take place. During this process the end of the ampoules is cleared of crystallization nuclei. Thus usually larger crystals are formed during the transport experiments. It has proven useful to place the ampoules in an asymmetric way in the temperature gradient during the transport experiments (2/3 of the ampoule in the range of the source temperature; 1/3 at the deposition temperature). This way one gets a narrow deposition zone.

Iodine is always used as the transport additive. However, one has to consider that it is not always identical with the transport agent in the case of the transport of phosphides (see section 9.1).

After the experiment, the ampoule is taken out of the furnace with the source side first ("ampoule catcher"). It is quenched with water for the purpose of condensation of the gas phase at this side, too. After opening the ampoule, the crystals are recovered. They are weighed out after washing with diluted $NaOH/H_2O_2$, water, and acetone. The same is done with the remaining, non-transported solid in the source.

The deposited crystals can grow to edge lengths up to several millimeters. The transport rates of the phosphides are in the range of several mg · h^{-1}.

Thermodynamic considerations and transport equilibrium The experimental observations during the CVT of monophosphides can be retraced in thermodynamic model calculations using the computer programs TRAGMIN or CVTRANS, described in Chapter 13. The required thermodynamic data are compiled in Table 15.5.2. Small amounts of hydrogen (0.01 mmol) that are released by the reaction of the remaining moisture in the ampoules and the phosphides should be taken into account in the model calculations.

In line with the model calculations, the transport determining equilibria 15.5.1 to 15.5.6 can be formulated (see section 9.1).

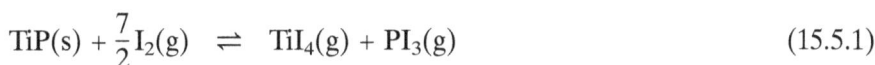

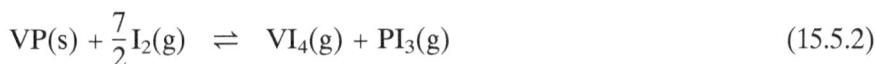

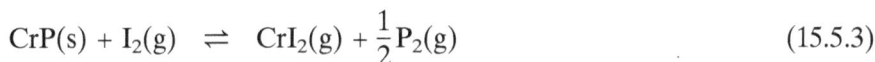

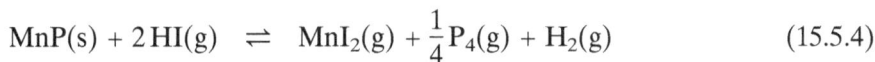

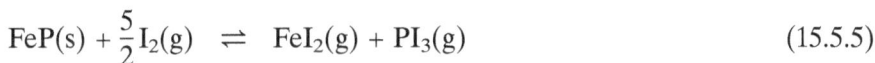

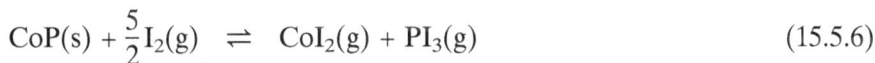

$$\text{TiP(s)} + \frac{7}{2}\text{I}_2(\text{g}) \;\rightleftharpoons\; \text{TiI}_4(\text{g}) + \text{PI}_3(\text{g}) \tag{15.5.1}$$

$$\text{VP(s)} + \frac{7}{2}\text{I}_2(\text{g}) \;\rightleftharpoons\; \text{VI}_4(\text{g}) + \text{PI}_3(\text{g}) \tag{15.5.2}$$

$$\text{CrP(s)} + \text{I}_2(\text{g}) \;\rightleftharpoons\; \text{CrI}_2(\text{g}) + \frac{1}{2}\text{P}_2(\text{g}) \tag{15.5.3}$$

$$\text{MnP(s)} + 2\,\text{HI(g)} \;\rightleftharpoons\; \text{MnI}_2(\text{g}) + \frac{1}{4}\text{P}_4(\text{g}) + \text{H}_2(\text{g}) \tag{15.5.4}$$

$$\text{FeP(s)} + \frac{5}{2}\text{I}_2(\text{g}) \;\rightleftharpoons\; \text{FeI}_2(\text{g}) + \text{PI}_3(\text{g}) \tag{15.5.5}$$

$$\text{CoP(s)} + \frac{5}{2}\text{I}_2(\text{g}) \;\rightleftharpoons\; \text{CoI}_2(\text{g}) + \text{PI}_3(\text{g}) \tag{15.5.6}$$

Suggestions for further experiments and model calculations The above-described transport experiments with the thermodynamic considerations can be expanded in different directions.

- "Aluminothermic" phosphide synthesis. The synthesis of phosphides is started from phosphorus and the metal oxide, which is reduced *in situ* by aluminum. The separation of the phosphide from Al_2O_3 can be achieved by chemical transport.
- Adding $CrPO_4$ during the transport and simultaneous transport of $Cr_2P_2O_7$/ CrP (see section 6.2).
- Synthesis of mixed-crystals (e. g., FeP, $Fe_{1-x}Co_xP$, CoP) by chemical transport experiments.
- Carry out calculations for the transport system that you worked on and change the experimental conditions in the calculations. How would the transport behavior change?
- Compare the results of model calculations with and without considering hydrogen.

Table 15.5.2 Thermodynamic data for modeling the transport behavior of monophosphides MP (M = Ti to Co).

Com-pound	$\Delta_f H^0_{298}/$ kJ $\cdot$ mol^{-1}	$\Delta_f H^0_{298}/$ J $\cdot$ mol^{-1} $\cdot$ K^{-1}	a[a]	b	c	Reference
H$_2$(g)	0	130.6	27.28	3.26	0.50	(Bar 1973)
HI(g)	26.4	206.5	26.32	5.94	0.92	(Bar 1973)
I(g)	105.6	180.7	20.39	0.28	0.28	(Bar 1973)
I$_2$(g)	62.4	260.6	37.40	0.57	0.62	(Bar 1973)
P$_2$(g)[b]	178.6	218.0	36.30	0.80	−4.16	(Bar 1973)
P$_4$(g)	128.7	279.9	81.34	0.68	−13.44	(Bar 1973)
P$_2$I$_4$(g)	−6.3	481.2	126.36	5.56	–	(Nol 1978)
PI$_3$(g)	23.0	369.6	82.72	0.53	−3.97	(Nol 1978)
TiI$_2$(s)	−269.4	123.2	84.05	7.29	0.004	(Nol 1978)
TiI$_2$(g)	−57.7	329.3	62.32	0.004	−1.56	(Nol 1978)
TiI$_3$(g)	−150.2	379.9	82.87	0.14	−5.25	(Nol 1978)
TiI$_4$(g)	−287.0	433.0	108.01	0.04	−3.36	(Nol 1978)
TiP(s)	−282.8	50.2	44.98	0.04	–	(Gla 1990)
VI$_2$(s)	−262.0	146.4	84.06	7.28	0.004	(Gla 1989a)
VI$_2$(g)	−23.3	320.1	62.32	0.02	−1.56	(Gla 1989a)
VI$_3$(s)	−280.3	211.5	113.32	8.90	0.73	(Gla 1989a)
VI$_4$(g)	−135.6	461.9	108.01	0.04	−3.36	(Gla 1989a)
VP(s)	−255.2	50.2	44.98	10.46	–	(Gla 1989a)
CrI$_2$(s)	−158.2	154.4	89.12	2.94	−12.35	(Gla 1989b)
CrI$_2$(l)[c]	−59.7	293.3	108.80	–	–	(Gla 1989b)
CrI$_2$(g)	107.2	353.4	60.67	2.44	–	(Gla 1989b)
Cr$_2$I$_4$(g)	10.1	573.6	129.70	4.90	–	(Gla 1989b)
CrI$_3$(g)	48.8	408.1	68.83	3.26	−16.74	(Gla 1989b)
CrI$_4$(g)	8.3	467.3	108.03	0.04	−3.36	(Gla 1989b)
CrP(s)	−106.7	46.0	49.87	6.78	–	(Gla 1989b)
MnI$_2$(l)[e]	−173.0	288.4	108.78	–	–	(Gla 1989b)
MnI$_2$(g)	−54.4	336.1	59.88	3.26	0.12	(Gla 1989b)
Mn$_2$I$_4$(g)	−271.8	538.1	128.13	6.52	0.24	(Gla 1989b)
MnP(s)	−96.2	52.3	44.98	10.46	–	(Gla 1989b)
FeI$_2$(s)	−104.6	167.4	82.94	2.46	–	(Nol 1978)
FeI$_2$(l)	−71.1	195.4	112.97	–	–	(Nol 1978)
FeI$_2$(g)	87.9	349.4	60.53	2.17	−0.21	(Nol 1978)
Fe$_2$I$_4$(g)	8.4	543.5	130.09	3.76	0.23	(Nol 1978)
FeI$_3$(g)	43.7	412.5	79.24	2.51	–	(Nol 1978)
Fe$_2$I$_6$(g)	−57.4	675.8	175.23	5.02	–	(Nol 1978)
FeP(s)	−105.4	50.2	49.87	6.79	–	(Nol 1978)
CoI$_2$(s)	−88.7	153.1	66.11	32.22	–	(Schm 1995)
CoI$_2$(l)[d]	26.2	315.1	105.81	−14.21	−0.62	(Schm 1995)
CoI$_2$(g)[d]	191.2	442.2	59.88	3.26	0.12	(Schm 1995)
Co$_2$I$_4$(g)[d]	187.1	733.3	130.09	3.76	0.23	(Schm 1995)
CoI$_3$(g)	159.0	393.3	79.24	2.51	–	(Schm 1995)
CoP(s)	−117.6	43.9	46.02	8.79	−1.51	(Schm 1995)

[a] $C_P = a + b \cdot 10^{-3} \cdot T + c \cdot 10^5 \cdot T^{-2}$.
[b] The enthalpy of formation of all compounds containing phosphorus correspond to $\Delta_f H^0_{298}$ (P$_{red}$) = 0 kJ $\cdot$ mol^{-1}.
[c] The values correspond to T = 1066 K.
[d] The values correspond to T = 1050 K.
[e] The values correspond to T = 911 K.

Chemicals required	Equipment required
respective metal or metal oxide (approx. 1 g foil, wire or powder)	silica ampoule
	long powder funnel
phosphorus (red)	two-zone furnace
where appropriate, aluminum (sheet metal or foil, not powder)	*protective gloves, safety glasses*
	"ampoule catcher"
iodine as transport additive	
diluted lye and acetone for washing the products	

15.6 Numeric Calculation of a Co-existence Decomposition Pressure

Numerous metals are known to form several compounds with different compositions in a binary systems M/A. Hence, iron forms the oxides "FeO" (wustite), "Fe_3O_4" (magnetite), and "Fe_2O_3" (hematite). If we want to synthesize one of these phases with the help of a CVT reaction, the oxygen partial pressure in the system determines which one is formed on the sink side: The higher the oxygen pressure, the more oxygen rich is the deposited solid. Where there is no other oxygen source than the solid used, $p(O_2)$ will be set by its co-existence decomposition pressure. At exactly this oxygen partial pressure there are two solids, in their composition adjacent solid phases, e. g., Fe_2O_3 next to Fe_3O_4. If the oxygen partial pressure is higher than the co-existence decomposition pressure, only one phase is stable, in this case Fe_2O_3. If the oxygen partial pressure is lower, only Fe_3O_4 is formed. These considerations do not only apply for oxides, in adapted form they are generally valid for systems that contain one volatile component.

In particular, many solid phases appear in the vanadium/oxygen system. There CVT reactions have proven to be *the* preparation process to synthesize certain compounds free of phase impurities. This is discussed in detail in Chapter 5. Now we want to calculate exemplary the co-existence decomposition pressure of V_2O_3. During the decomposition of V_2O_3, the solid phase VO and gaseous oxygen are formed. For simplicity, the equation for the decomposition reaction is set up where the stoichiometric coefficient for O_2 is unity:

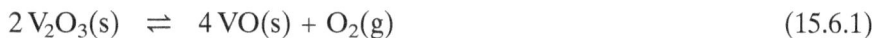

$$2\,V_2O_3(s) \;\rightleftharpoons\; 4\,VO(s) + O_2(g) \tag{15.6.1}$$

$$K_p = p(O_2) \tag{15.6.2}$$

The equilibrium constant K can be calculated from the thermodynamic data of 15.6.1.

$$\ln K = -\frac{\Delta_r H^0}{R \cdot T} + \frac{\Delta_r S^0}{R} \tag{15.6.3}$$

The calculated, dimensionless equilibrium constant K corresponds to the following mass action term:

$$K = \frac{a^4(\text{VO}) \cdot a(\text{O}_2)}{a^2(\text{V}_2\text{O}_3)} \tag{15.6.4}$$

Because by definition the activities of pure, condensed phases equal one, the expression can be simplified:

$$K = a(\text{O}_2) \tag{15.6.5}$$

The activity of a gas can be linked to its partial pressure by the following expression in good approximation:

$$a(i) = \frac{p(i)}{p^0} \tag{15.6.6}$$

The standard pressure p^0 is 1 bar. The tabulated thermodynamic data usually correspond to this pressure. Thus the following relation applies:

$$a(i) = \frac{p(i)}{\text{bar}} \tag{15.6.7}$$

Hence the oxygen partial pressure results, in bar:

$$K = \frac{p(\text{O}_2)}{\text{bar}} \quad \Rightarrow \quad p(\text{O}_2) = K \cdot \text{bar} \tag{15.6.8}$$

The thermodynamic data of all the reaction participants are needed for numerical calculation. This way phase barograms can be set up by calculation of the corresponding pressure/temperature pairs. The phase barograms allow a derivation of suitable transport conditions (see Chapter 5). This cannot only be applied to oxide systems but can generally be transferred to the set up of phase barograms of different classes of substance. In all cases, a decomposition of a solid phase in the co-existing adjacent phase takes place under the formation of a gas phase.

	$\Delta_f H_{298}^0 /$ kJ $\cdot$ mol^{-1}	$\Delta_B H_{2000}^0 /$ kJ $\cdot$ mol^{-1}	$S_{298}^0 /$ J $\cdot$ mol$^{-1} \cdot$ K^{-1}	$S_{2000}^0 /$ J $\cdot$ mol$^{-1} \cdot$ K^{-1}
V$_2$O$_3$(s)	−1218.8	−976.4	98.1	353.3
VO(s)	−431.8	−325.5	39.0	149.6
O$_2$(g)	0	59.2	205.1	268.7

The following values result from these data:

$$\Delta_r H_{2000}^0 = 710.0 \text{ kJ} \cdot \text{mol}^{-1} \qquad \Delta_r S_{2000}^0 = 160.5 \text{ J} \cdot \text{mol}^{-1} \cdot \text{K}^{-1}$$

$$\ln K = -\frac{710\,000 \text{ J} \cdot \text{mol}^{-1}}{8.3145 \text{ J} \cdot \text{mol}^{-1} \cdot \text{K}^{-1} \cdot 2000 \text{ K}} + \frac{160.5 \text{ J} \cdot \text{mol}^{-1} \cdot \text{K}^{-1}}{8.3145 \text{ J} \cdot \text{mol}^{-1} \cdot \text{K}^{-1}}$$

$$\ln K = -23.392$$

$$K = p(O_2)/\text{bar}$$

$$p(O_2) = 6.9 \cdot 10^{-11} \text{ bar}$$

Bibliography

Bab 1977 A. V. Babushkin, L. A. Klinkova, E. D. Skrebkova, *Izv. Akad. Nauk SSSR Neorg. Mater.* **1977**, *13*, 2114.

Bar 1973 I. Barin, O. Knacke, *Thermochemical Properties of Inorganic Substances*, Springer, Berlin, **1973**.

Ben 1969 L. Ben-Dor, L. E. Conroy, *Isr. J. Chem.* **1969**, *7*, 713.

Ben 1974 L. Ben-Dor, Y. Shimony, *Mater. Res. Bull.* **1974**, *9*, 837.

Bur 1995 J. K. Burdett, *Chemical Bonding in Solids*, Oxford University Press, New York, **1995**.

Det 1969 J. H. Dettingmeijer, J. Tillack, H. Schäfer, *Z. Anorg. Allg. Chem.* **1969**, *369*, 161.

Fel 1998 J. Feller, H. Oppermann, M. Binnewies, E. Milke, *Z. Naturforsch.* **1998**, *53 b*, 184.

Fje 1984 H. Fjellvag, A. Kjekshus, *Acta Chem. Scand.* **1984**, *38A*, 563.

Fje 1986 H. Fjellvag, A. Kjekshus, *Monatsh. Chem.* **1986**, *117*, 773.

Gla 1989a R. Glaum, R. Gruehn, *Z. Anorg. Allg. Chem.* **1989**, *568*, 73.

Gla 1989b R. Glaum, R. Gruehn, *Z. Anorg. Allg. Chem.* **1989**, *573*, 24.

Gla 1990 R. Glaum, *Dissertation*, University of Gießen, **1990**.

Kna 1991 O. Knacke, O. Kubaschewski, K. Hesselmann; *Thermochemical Properties Of Inorganic Substances*, 2nd Ed., Springer, **1991**.

Loc 1999 S. Locmelis, M. Binnewies, *Z. Anorg. Allg. Chem.* **1999**, *625*, 1573.

Mon 1890 L. Mond, C. Langer, F. Quincke, *J. Chem. Soc.*, **1890**, 749.

Nol 1978 B. I. Noläng, M. W. Richardson, *Free Energy Data for Chemical Substances*, university Uppsala, Schweden, **1978**.

Opp 1985 H. Oppermann, *Z. Anorg. Allg. Chem.* **1985**, *523*, 135.

Opp 1987 H. Oppermann, *Freiberger Forschungshefte*, VEB Deutscher Verlag für Grundstoffindustrie **1987**, *A 767*, 97.

Rog 1969 D. B. Rogers, R. D. Shannon, A. W. Sleight, J. L. Gillson, *Inorg. Chem.* **1969**, *8*, 841.

Run 1962 S. Rundqvist, *Acta Chem. Scand.* **1962**, *16*, 287.

Saa 2009a H. Saal, M. Binnewies, M. Schrader, A. Börger, K.-D. Becker, V. A. Tikhomirov, K. Jug, *Chem. Eur. J.* **2009**, *15*, 6408.

Saa 2009b H. Saal, T. Bredow, M. Binnewies, *Phys. Chem. Chem. Phys.* **2009**, *11*, 3201.

Sch 1962 H. Schäfer, *Chemische Transportreaktionen*, Verlag Chemie GmbH, Weinheim, **1962**.

Sch 1973a H. Schäfer, T. Grofe, M. Trenkel, *J. Solid State Chem.* **1973**, *8*, 14.

Sch 1973b H. Schäfer, *Z. Anorg. Allg. Chem.* **1973**, *400*, 242.

Sch 1982 H. Schäfer, *Z. Anorg. Allg. Chem.*, **1982**, *493*, 17.

Schm 1995 A. Schmidt, R. Glaum, *Z. Anorg. Allg. Chem.* **1995**, *621*, 1693.

Scho 1988 H. Schornstein, R. Gruehn, *Z. Anorg. Allg. Chem.* **1988**, *561*, 103.

Scho 1989 H. Schornstein, R. Gruehn, *Z. Anorg. Allg. Chem.* **1989**, *579*, 173.

Scho 1954	N. Schoenberg, *Acta Chem. Scand.* **1954**, *8*, 226.
Sel 1972a	K. Selte, A. Kjekshus, A. F. Andresen, *Acta Chem. Scand.* **1972**, *26*, 4188.
Sel 1972b	K. Selte, A. Kjekshus, *Acta Chem. Scand.* **1972**, *26*, 1276.
Sil 1986	J. Silvestre, W. Tremel, R. Hoffmann, *J. Less-Common. Met.* **1986**, *108*, 5174.
Tre 1986	W. Tremel, R. Hoffmann, J. Silvestre, *J. Amer. Chem. Soc.* **1986**, *108*, 5174.
Ull 1979	*Ullmanns Encyclopädie der technischen Chemie*, Bd. 17., Verlag Chemie Weinheim **1979**, S. 259.
Whi 1965	W. B. White, K. W. McIlfvried, *Trans. Brit. Ceram. Soc.* **1965**, *64*, 523.

16 Appendix

16.1 Important Thermodynamic Equations

$$\Delta_r G^0 = \Delta_r H^0 - T \cdot \Delta_r S^0$$

$$\Delta_r G^0 = -R \cdot T \cdot \ln K$$

$$\ln K = -\frac{\Delta_r H^0}{R \cdot T} + \frac{\Delta_r S^0}{R}$$

$$C_{p,T}^0 = a + b \cdot T + c \cdot T^{-2} + d \cdot T^2$$

$$\Delta H_T^0 = \Delta H_{298}^0 + \int_{298}^{T} C_p^0 \, \mathrm{d}T$$
$$= a[T - 298] + b[0.5 \cdot 10^{-3} (T^2 - 298^2)] + c[10^6 (298^{-1} - T^{-1})]$$
$$+ d\left[\frac{1}{3} 10^{-6} (T^3 - 298^3)\right]$$

$$S_T^0 = S_{298}^0 + \int_{298}^{T} C_p^0 \frac{\mathrm{d}T}{T}$$
$$= a \ln \frac{T}{298} + b \cdot 10^{-3} (T - 298) - \frac{1}{2} c \cdot 10^6 (T^{-2} - 298^{-2})$$
$$+ d[0.5 \cdot 10^{-6} (T^2 - 298^2)]$$

16.2 Selected Physical Units, Constants, and Conversions

Table 16.2.1 Basic Units of the International System of Units (SI)

Physical variable	Symbol	SI unit	Symbol
length	l	Meter	m
mass	m	Kilogram	kg
time	t	Second	s
electric current	I	Ampere	A
thermodynamic temperature	T	Kelvin	K
amount of substance	n	Mole	mol
luminus intensity	Iv	Candela	cd

Table 16.2.2 Selected Derived Units

Physical variable	Symbol	Unit	Symbol	Relation to SI unit
frequency	ν	hertz	Hz	$1\ \text{Hz} = 1\ \text{s}^{-1}$
energy, work, heat	E, W, Q	joule	J	$1\ \text{J} = 1\ \text{N} \cdot \text{m} = 1\ \text{W} \cdot \text{s}$ $= 1\ \text{kg} \cdot \text{m}^2 \cdot \text{s}^{-2}$
force	F	newton	N	$1\ \text{N} = 1\ \text{J} \cdot \text{m}^{-1}$ $= 1\ \text{kg} \cdot \text{m} \cdot \text{s}^{-2}$
pressure	p	pascal	Pa	$1\ \text{Pa} = 1\ \text{N} \cdot \text{m}^{-2}$ $= 1\ \text{kg} \cdot \text{m}^{-1} \cdot \text{s}^{-2}$
pressure	p	bar	Bar	$1\ \text{bar} = 10^5\ \text{Pa}$ $= 10^5 \cdot \text{kg} \cdot \text{m}^{-1} \cdot \text{s}^{-2}$
power	P	watt	W	$1\ \text{W} = 1\ \text{A} \cdot \text{V} = 1\ \text{J} \cdot \text{s}^{-1}$ $= 1\ \text{kg} \cdot \text{m}^2 \cdot \text{s}^{-3}$
electrical charge	Q	coulomb	C	$1\ \text{C} = 1\ \text{A} \cdot \text{s} = 1\ \text{J} \cdot \text{V}^{-1}$
electrical voltage	U	volt	V	$1\ \text{V} = 1\ \text{W} \cdot \text{A}^{-1} = 1\ \text{J} \cdot \text{C}^{-1}$ $= 1\ \text{kg} \cdot \text{m}^2 \cdot \text{A}^{-1} \cdot \text{s}^{-3}$
electrical resistivity	R	ohm	Ω	$1\ \Omega = 1\ \text{V} \cdot \text{A}^{-1} = 1\ \text{S}^{-1}$ $= 1\ \text{kg} \cdot \text{m}^2 \cdot \text{A}^{-2} \cdot \text{s}^{-3}$
electrical conductivity	G	siemens	S	$1\ \text{S} = \Omega^{-1} = 1\ \text{A} \cdot \text{V}^{-1}$ $= 1\ \text{A}^2 \cdot \text{s}^3 \cdot \text{kg}^{-1} \cdot \text{m}^{-2}$
electrical capacity	C	farad	F	$1\ \text{F} = 1\ \text{C} \cdot \text{V}^{-1} = 1\ \text{J} \cdot \text{V}^{-2}$ $= 1\ \text{A}^2 \cdot \text{s}^4 \cdot \text{k}\ \text{g}^{-1} \cdot \text{m}^{-2}$
magnetic flux	Φ	weber	Wb	$1\ \text{Wb} = 1\ \text{V} \cdot \text{s}$ $= 1\ \text{kg} \cdot \text{m}^2 \cdot \text{A}^{-1} \cdot \text{s}^{-2}$
magnetic flux density	B	tesla	T	$1\ \text{T} = 1\ \text{Wb} \cdot \text{m}^{-2}$ $= 1\ \text{V} \cdot \text{s} \cdot \text{m}^{-2}$ $= 1\ \text{kg} \cdot \text{A}^{-1} \cdot \text{s}^{-2}$
inductance	L	henry	H	$1\ \text{H} = 1\ \text{Wb} \cdot \text{A}^{-1}$ $= 1\ \text{V} \cdot \text{s} \cdot \text{A}^{-1}$ $= 1\ \text{kg} \cdot \text{m}^{-2} \cdot \text{A}^{-2} \cdot \text{s}^{-2}$

Table 16.2.3 Fundamental Constants

Term	Symbol	Value
atomic mass constant	m_u	$1.6605402 \cdot 10^{-27}$ kg = 1 u
mass of a proton	$m(\mathrm{p})$	$1.6726231 \cdot 10^{-27}$ kg
mass of an electron	$m(\mathrm{e})$	$9.1093897 \cdot 10^{-31}$ kg
elementary charge	e	$1.60217733 \cdot 10^{-19}$ C
Bohr's radius	a_0	$5.29177249 \cdot 10^{-11}$ m
Boltzmann's constant	k	$1.380658 \cdot 10^{-23}$ J $\cdot$ K^{-1}
Avogadro's constant	N_A	$6.02214 \cdot 10^{23}$ mol^{-1}
Faraday's constant	$F\ (= N_\mathrm{A} \cdot e)$	$9.6485309 \cdot 10^{4}$ C $\cdot$ mol^{-1}
gas constant	$R\ (= N_\mathrm{A} \cdot k)$	8.31451 J $\cdot$ K^{-1} $\cdot$ mol^{-1}
		$= 0.0831451$ l $\cdot$ bar $\cdot$ K^{-1} $\cdot$ mol^{-1}
speed of light (in a vacuum)	c	$2.99792458 \cdot 10^{8}$ m $\cdot$ s^{-1}
dielectric constant in a vacuum	ε_0	$8.85418782 \cdot 10^{-12}$ F $\cdot$ m^{-1}
Planck's constant	h	$6.6260755 \cdot 10^{-34}$ J $\cdot$ s
Bohr's magneton	μ_B	$9.2740154 \cdot 10^{-24}$ J $\cdot$ T^{-1}

Table 16.2.4 Conversions of Some Other Units

Term	Relation	
length	1 Å $= 10^{2}$ pm $= 10^{-10}$ m	(Å: ångström)
energy	1 cal $= 4.184$ J	
	1 eV $= 1.6022 \cdot 10^{-19}$ J $\ \hat{=}\ 96.4852$ kJ $\cdot$ mol^{-1}	
	1 cm$^{-1} = 1.9865 \cdot 10^{-23}$ J $\ \hat{=}\ 1.1963 \cdot 10^{-2}$ kJ $\cdot$ mol^{-1}	
dipole moment	1 D $= 3.336 \cdot 10^{-30}$ C $\cdot$ m	(D: debye)
pressure	1 bar $= 10^{5}$ Pa	
	1 atm $= 760$ torr $= 1.01325 \cdot 10^{5}$ Pa ($= 1013$ hPa)	
	1 Torr $= 133.32$ Pa	

16.3 Abreviations

a	Activity
a_i	activity of substance i
a, b	oxidation number
A,B,C	substances, respective species
a, b, c, d	polynomial coefficients for description of the temperature-dependent heat capacity
B^0	initial concentration
b_j	total amount of substance of the element j
c	Concentration
$C_{p,T}^0$	heat capacity at constant pressure
CSVT	close spaced vapor transport
CVD	chemical vapor deposition
CVT	chemical vapor transport
D	diffusion coefficient
$\overline{D}$	mean diffusion coefficient
D^0	binary diffusion coefficient
δ	phase range
δ	correction factor for the calculation of the starting values for the following iteration cycle
ΔG_T^0	Gibbs energy of formation at T
$\Delta_r G_T^0$	Gibbs energy of reaction at T
$\Delta_b H_T^0$	bond enthalpy at T
$\Delta_f H_T^0$	standard heat of formation at T
$\Delta_r H_T^0$	standard heat of reaction at T
$\Delta\lambda$	solubility difference
Δp	partial pressure difference
$\Delta_r S_T^0$	reaction entropy at T
ΔT	temperature difference
η	Viscosity
ε	stationary principle
F	number of the degrees of freedom in system
F	flux factor
f'	flow rate
g	phase symbol "gaseous"
Gr	Grashof number
HTVG	high temperature vapor growth
i, k, j	stochiometric coefficients
J	Flux
k	number of components in a system
K	equilibrium constant
K_p	equilibrium constant
k_B	Boltzmann's constant
L	Solvent
l	phase symbol "liquid"

l	diffusion length
l	total number of elements in a system
$l(\mathrm{w})$	section of length
Ln	Lanthanoid
λ	solubility in the gas phase
λ_j	Lagrange's multiplier
M	molar mass
m	stoichiometric coefficient
m	mass
M	metal
M'	metal
MS	mass spectrometry
μ_i	chemical potential of substance i
n	stoichiometric coefficient
n	molar number
n^*	balance of molar numbers
$\dot{n}(A)$	transport rate of A
n_i	molar number of i
ν	stoichiometric coefficient
p	pressure
p^0	standard pressure
p_i	partial pressure of species i
p^*	balance pressure
P	number of phases in a system
PVD	physical vapour deposition
Q	non-metal (except halogen)
q	diffusion cross section
QBK	sources condensed phase(s); from the German "Quellenboden-körper"
R	gas constant
r	radius
RE	rare earth metal
$\mathrm{r_u}$	number of independent equilibria in a system
s	phase symbol "solid"
s	number of gaseous species in a system
s	diffusion length
SBK	sink's condensed phase(s); from the German "Senkenboden-körper"
Sc	Schmidt number
S_T^0	standard entropy at T
Σp	total pressure
$\Sigma n_i^{\mathrm{g}}(X)$	sum of molar numbers of all gaseous species
σ	particle diameter
t	time
T	thermodynamic temperature (K)
T^0	standard temperature
T_1	lower temperature of the gradient

T_2	higher temperature of the gradient
T_{opt}	optimum transport temperature
T	mean temperature
ϑ	temperature (°C)
ϑ_m	melting temperature (°C)
V	volume
$w(i)$	transport efficiency of species i
W	flow rate
x	molar fraction
x	stoichiometric coefficient
X	halogen
x	molar number after the first iteration cycle (in G_{min} procedure)
y	stoichiometric coefficient
y	initial value of the molar number for further iteration cycles
y^0	initial value of the molar number for the first iteration cycle
$z_{ij}^g, z_{ij}^c \ (a_{ij}^g, a_{ij}^c)$	number of atoms of element j in a formula unit

Index

Plate Section

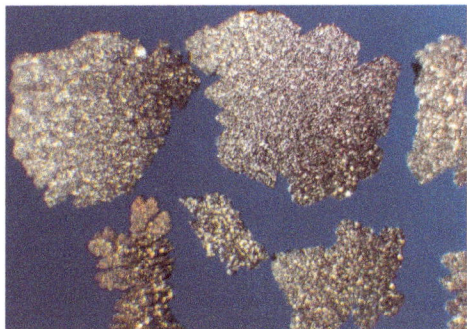

MnO_{1+x}

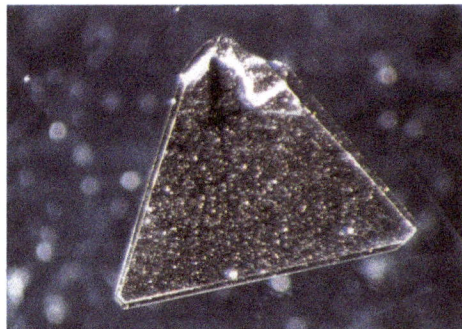

V_2O_3

$V_{1-x}Nb_xO_2$

V_6O_{13}

1.0 mm

V_2O_5

Ta_2O_5

Mo_4O_{11}

γ-Mo_4O_{11}

MoO_2

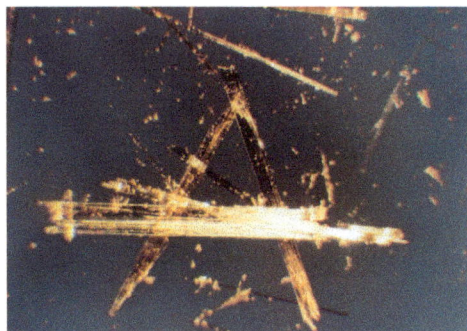

MoO_3

$W_{18}O_{49}$

WPO_5

ReO$_2$

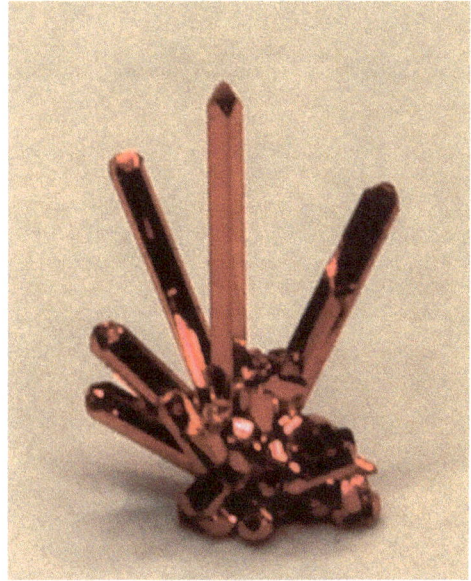

ReO$_3$

Fe$_3$O$_4$

Ru$_{1-x}$Sn$_x$O$_2$

ZnCr$_2$O$_4$

ZnO$_{1-x}$

VOSO$_4$

Fe$_2$(SO$_4$)$_3$

N-CoSO$_4$

ZnSO$_4$

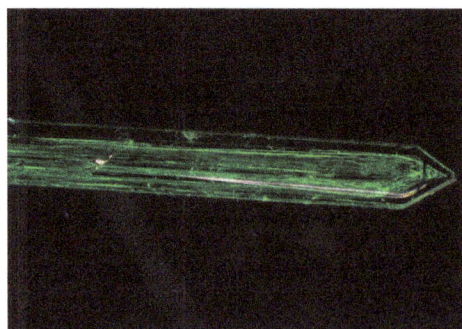

CuSe$_2$O$_5$

CuTe$_2$O$_5$

BPO_4

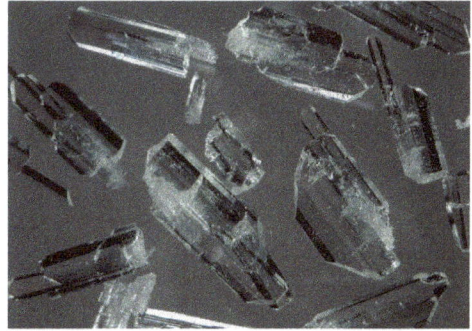

$CePO_4$

$NdPO_4$

$Nd_{1-x}Pr_xPO_4$

$Nd_{1-x}Sm_xPO_4$

$NdAsO_4$

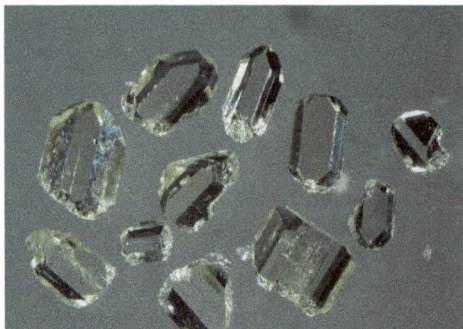

$Sm_{1-x}Gd_xPO_4$

$GdPO_4$

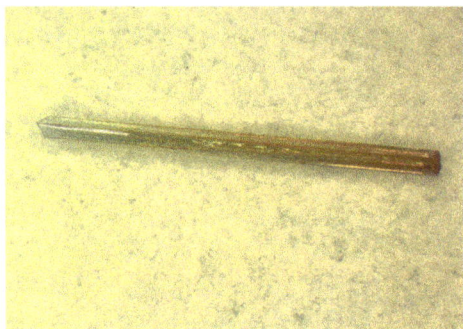

$EuSbO_4$

$NdVO_4$

$RhVO_4$

CuV_2O_6

$TiPO_4$

$Ti(PO_3)_3$ *c*-type

β-VPO_5 partially reduced

$Cr_2P_2O_7$

β-$CrPO_4$

$FeTi_4(PO_4)_6$

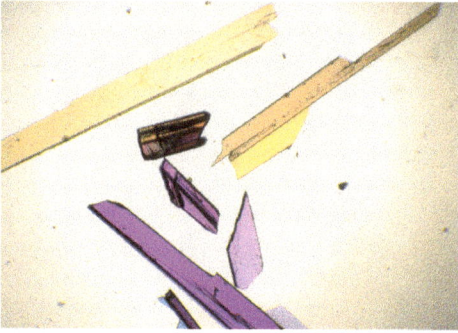

$Co_2P_2O_7$

$Co_2Si(P_2O_7)_2$

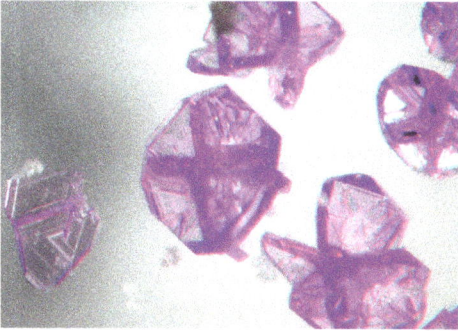

$CoTi_4(PO_4)_6$

$\delta\text{-}Ni_2P_2O_7$

$NiTi_2O_2(PO_4)_2$

$Cu_2P_2O_7$

$Fe_{0.97}Mn_{0.03}S$

$Fe_{0.97}Mn_{0.03}S$

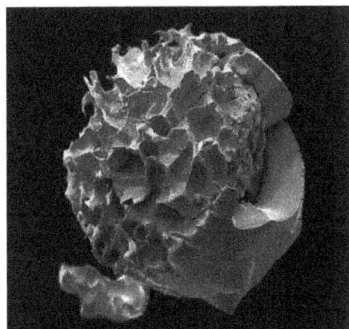

Ag_2Se

PtS

ZnS

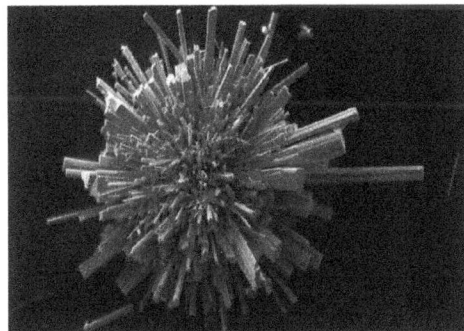

$Pd_{0.62}Pt_{0.38}S$

CrP

VP

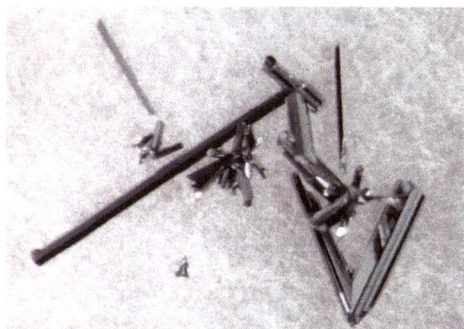

WP$_2$

CoP$_2$

CoP$_3$

HfAs$_2$

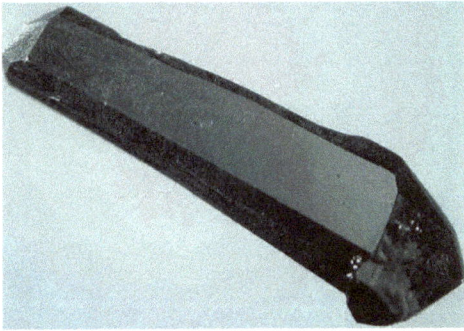

CdGeAs$_2$

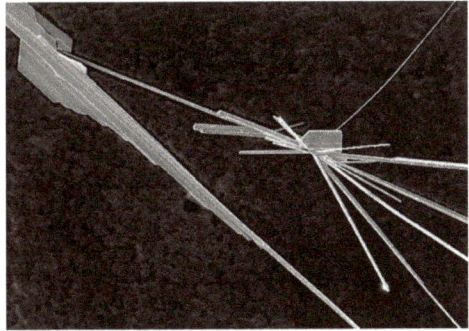

ZrSb$_2$

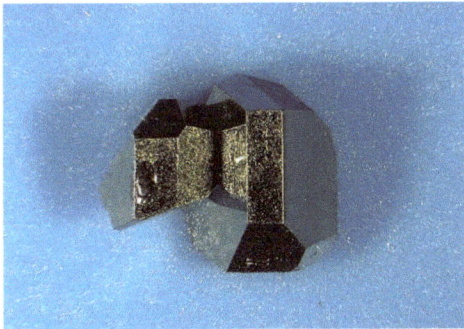

FeSb$_2$

Pd$_2$Ga

CrGe

FeGe

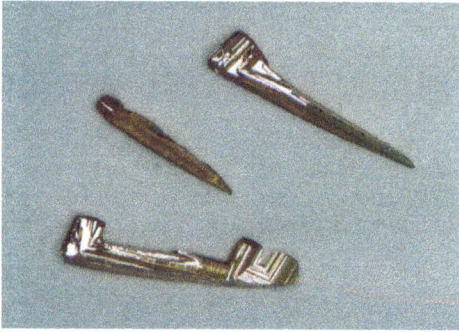

$Bi_7O_9I_3$

$Bi_{24}O_{31}Cl_{10}$

BiOCl

BiOI

BiSeI

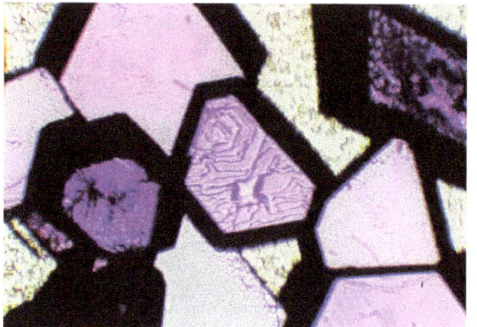

$CoCl_2$